Springer Series in Materials Science

Volume 242

More information about this series at http://www.springer.com/series/856

The Springer Series in Materials Science covers the complete spectrum of materials physics, including fundamental principles, physical properties, materials theory and design. Recognizing the increasing importance of materials science in future device technologies, the book titles in this series reflect the state-of-the-art in understanding and controlling the structure and properties of all important classes of materials.

Markus G.R. Sause

In Situ Monitoring of Fiber-Reinforced Composites

Theory, Basic Concepts, Methods, and Applications

Springer

Markus G.R. Sause
Institute of Physics and Institute of Materials
Resource Management
University of Augsburg
Augsburg, Germany

ISSN 0933-033X ISSN 2196-2812 (electronic)
Springer Series in Materials Science
ISBN 978-3-319-30953-8 ISBN 978-3-319-30954-5 (eBook)
DOI 10.1007/978-3-319-30954-5

Library of Congress Control Number: 2016940044

Printed on acid-free paper

This Springer imprint is published by Springer Nature
The registered company is Springer International Publishing AG Switzerland

To Kerstin
and my parents Brigitte and Dieter

Preface

The design of materials with essentially new properties by modification of existing materials is an ongoing journey in human history. Today, more than one hundred thousand engineering materials are known and are used to fabricate the goods used in our daily life and to enable some of the most challenging technological applications. Still there is a strong desire to increase the performance of some materials or to substitute existing materials to overcome technological limitations or to reduce the ecological footprint. In materials science, one powerful trend always has been to compose materials by incorporation of spurious dopants in an otherwise pure material. This rather simple concept gave rise to the large amount of metallic alloys, ceramic compounds, and polymer blends we know today. One specific type of microscopic dopants to incorporate in a material is reinforcement fibers. These are often found to substantially increase the mechanical performance of the so-obtained composite material. Hence, this concept of fiber reinforcement has led to unique materials, exhibiting some of the most extraordinary properties available for today's engineering applications. While those fiber-reinforced composites have now been in use for a long time in aerospace applications, other industry branches are still at the edge to use these materials for their products. Some of the main challenges of these materials are the production process itself, but also the theoretical description of the material. Still the failure behavior of this class of materials is subject to numerous investigations, and theoretical predictions of their ultimate limits lack in accuracy. One reason for this lack of theoretical description is the complexity of material structure found in fiber-reinforced materials. Apart from the existence of two (or more) material ingredients with different properties, a technical composite exhibits a hierarchical structure covering at least four orders of magnitude in length scale. The material itself typically exhibits nonlinear mechanical behavior and nonlinear temperature response as well as pronounced anisotropy of all material properties.

This naturally causes a strong need for adequate test methods that are able to provide the relevant materials properties on every length scale required for the description of the material. In the past decades, many new test procedures have been established to specifically address the need of characterization of composite materials. With the advent of modern theories to describe the complex failure mechanisms occurring in composite materials, even more properties are required to be measured nowadays. While this has already led to several new test concepts, one particular field of research, which has seen a strong interest within the last years, is the in situ characterization of composite materials. As a major benefit compared to other test concepts, the in situ approach does not just provide a single value after the measurement, but instead is able to record the full evolution of damage in the material until the occurrence of final failure. Consequently, all this information can be used to understand the evolution of failure in composite materials.

The aim of this book is to provide an overview of some of the emerging in situ test concepts and method combinations to demonstrate their applicability to assist in research of composite materials. It is thought to provide a first insight into the operational principles of each method and to present the different ways of interpretation of the obtained results. It further provides an overview of some practical applications ready to be used in quality control and discusses the applicability of the individual methods as well as their strengths and weaknesses.

The book is organized in seven chapters. Chapter 1 provides an introduction to the relevance of in situ testing for fiber-reinforced composites. Chapter 2 briefly reviews the present state of failure theories available to predict failure of fiber-reinforced polymers, since a majority of the experimental methods need a thorough understanding of the failure mechanisms occurring in these materials in order to interpret their sensitivity for detection. Chapters 3–6 introduce different experimental methods used for in situ characterization. In all cases, a focus is given on the operation principles and recent developments. Reference to standard literature is made as much as possible to facilitate the reading. Chapter 3 deals with the applications of digital image correlation (DIC), and Chap. 4 emphasizes the developments in acoustic emission (AE) analysis. Chapter 5 has its focus on electromagnetic emission (EME) analysis, while Chap. 6 discusses possibilities of in situ computed tomography (CT). In Chap. 7, a comparison of the capabilities of the methods in application to in situ testing of composite materials is made, and suitable method combinations are discussed.

As supplementary data, Appendix A provides a comprehensive list of common experimental parameters influencing comparability of acoustic emission measurements. Appendix B lists all of the materials properties used for the modeling work throughout the chapters. Appendix C provides an overview of acoustic emission parameter definitions, and common definitions and abbreviations are found in Appendix D.

Augsburg, Germany Markus G.R. Sause

Acknowledgments

First of all, I would like to express my gratitude to Prof. Dr. Siegfried Horn, Prof. Dr. Ferdinand Haider, and Prof. Dr. Marvin Hamstad for guiding me in the process of Habilitation and for providing every level of understanding and support I needed during this process.

My gratitude is also to Prof. Kanji Ono, Prof. Stepan Lomov, and Prof. Ian Sinclair to review this work as part of my Habilitation thesis review. My gratitude is also to Prof. Dr. Marvin Hamstad and Prof. Dipl.-Ing. Dr. techn. Roland Hinterhölzl for proofreading and reviewing parts of this book and thus providing me the opportunity to improve the content.

I would also like to thank all of the students, who have contributed to this book by providing experimental or modeling data or by intensive discussion of their interpretation. The work would not have been possible without the valuable contribution of numerous Ph.D. students. I would like to thank Stefan Richler for micromechanical test data, Sebastian Gade for performing EME measurements and modeling work, Sinan Kalafat for acoustic emission measurements and in situ computed tomography measurements, Thomas Guglhör for computed tomography imaging, Andreas Monden and Michael Greisel for digital image correlation data, and Ursula Weiss for accompanying modeling work.

Further, I am indebted to all other students who have contributed within their master's or bachelor's thesis. I would like to thank Marina Plöckl, Marcus Bornschlegl, and Wolfgang Skopalik for acoustic emission measurements, Eva Laukmanis for micromechanical test data, Florian Staab for electromagnetic emission data, Nora Schorer for a majority of the digital image correlation data, Uli Buchner and Tobias Krones for computing tomography investigations, and Philipp Potstada and Benjamin Alaca for electromagnetic emission measurements.

Further thanks also go to the technical staff for their ongoing support in testing issues and for taking care of the students' needs. I would like to thank Stefan Schmitt for the good laboratory organization, Sabine Bessel for sensor calibration, and Katja Ziebe-Eiffler for digital image correlation and mechanical testing data as well as for specimen fabrication.

Next, I would like to thank all academic colleagues for their collaboration in the various fields covered by this book. Substantial parts of my present work were inspired by the work of Dr.-Ing. habil. Jürgen Bohse, Dr. Andreas Brunner, Prof. Dr.-Ing. habil. Dipl.-Geophys. Christian Große, Prof. Emmanuel Ramasso, Prof. Vincent Placet, Dr.-Ing. Tobias Müller, Prof. Dr.-Ing. Alexander Horoschenkoff, and Prof. Antonios Kontsos. The contributions of various modeling aspects by my colleague Dr.-Ing. Andreea-Manuela Zelenyak are highly appreciated. Many thanks are also to those professionals from industry for their constant feedback regarding applicability of the method in industrial environments. In particular, I would like to thank Birte Höck, Joachim Scharringhausen, and Dr. Gunter Geiss from MT Aerospace AG, Dr. Matthias Goldammer and Dr. Detlef Rieger from Siemens AG, and Dr. Rui de Oliveira from BASF AG. Special recognition is also noted for all persons allowing me to coauthor their scientific publications, namely Dr. Wolfgang Müller, Dr. Judith Moosburger-Will, Dr. Günter Obermeier, Torben Priess, Michael Greisel, and Florian Henne. My deep gratitude is to Dr. Lidewei Vergeynst, Dr. Franziska Baensch, and Prof. Dr. Peter Niemz for the opportunity to get a deep insight into in situ methods application to wood materials.

I would also like to thank all of those I have forgotten to mention and all of those who have used their valuable time to discuss any scientific aspects of this work.

Contents

1 Introduction . . . 1
References . . . 4

2 Failure of Fiber-Reinforced Composites . . . 5
2.1 Classification of Failure Mechanisms . . . 6
2.1.1 Microscale . . . 7
2.1.2 Mesoscale . . . 13
2.1.3 Macroscale . . . 16
2.2 Failure Theories for Fiber-Reinforced Composites . . . 18
2.2.1 Quasi-Static Failure . . . 19
2.2.2 Quasi-Static Failure Including Growth of Damage, Damage Mechanics, and Degradation . . . 21
2.2.3 Long-Term Behavior . . . 28
2.2.4 High Velocity . . . 31
2.3 Challenges in Mechanical Testing of Fiber-Reinforced Materials . . . 32
2.3.1 Detection of First Failure Onsets . . . 34
2.3.2 Tracking Failure Evolution . . . 35
2.3.3 Ductile Matrix Materials . . . 35
2.4 What Can In Situ Methods Contribute to Mechanical Testing? . . . 36
2.4.1 In Situ Microscopy . . . 36
2.4.2 Digital Image Correlation . . . 37
2.4.3 X-Ray Methods . . . 39
2.4.4 Thermography . . . 40
2.4.5 Shearography . . . 42
2.4.6 Ultrasonic Measurements . . . 42
2.4.7 Acoustic Emission . . . 43
2.4.8 Electromagnetic Emission . . . 44
References . . . 45

3 Digital Image Correlation . . . 57
3.1 Principle of Operation . . . 58
3.2 System Accuracy . . . 63
3.2.1 Error Sources . . . 64
3.2.2 Resolution . . . 67
3.3 Strain Concentration . . . 69
3.3.1 Measurement of Strain Concentration Due to Internal Defects . . . 71
3.3.2 FEM Modeling of Strain Concentration Due to Internal Defects . . . 80
3.3.3 Detectability of Defects Using DIC . . . 97
3.4 Application to Composites . . . 111
3.4.1 DIC as Optical Extensometer . . . 111
3.4.2 Detection of Failure Onsets . . . 118
3.4.3 Detectability of Failure Mechanisms . . . 124
References . . . 126

4 Acoustic Emission . . . 131
4.1 Principle of Operation . . . 131
4.2 Source Mechanics . . . 133
4.2.1 AE Rise-Times and Plate Waves . . . 135
4.2.2 AE Source Model Implementation for Fiber Reinforced Materials . . . 138
4.2.3 Case Studies for Acoustic Emission Sources in Fiber Reinforced Materials . . . 147
4.2.4 Detectability of Failure Mechanisms . . . 187
4.3 Wave Propagation . . . 192
4.3.1 Attenuation . . . 202
4.3.2 Influence of Geometry . . . 206
4.3.3 External and Internal Obstacles . . . 214
4.4 Detection of Acoustic Emission Signals . . . 226
4.4.1 Comparison of Sensor Types . . . 230
4.4.2 Sensor Modeling . . . 233
4.4.3 Waveguides . . . 239
4.4.4 Other Factors Affecting Sensor Sensitivity . . . 245
4.5 Signal Classification . . . 250
4.5.1 Recommended Practices Before Starting Signal Classification . . . 252
4.5.2 Pattern Recognition Method to Detect Natural Clusters . . . 256
4.5.3 Uncertainty of Classification . . . 260
4.5.4 Factors of Influence . . . 264
4.6 Source Localization . . . 282
4.6.1 Determination of Signal Onset . . . 285
4.6.2 Classical Localization Methods . . . 287

4.6.3 Source Localization Methods Based on Neural Networks 291
4.7 Application to Composites 305
4.7.1 AE Source Identification Using FEM Results 305
4.7.2 Detection of Failure Onset 309
4.7.3 Comparison to Failure Criteria Predictions 317
4.7.4 Tracking Failure Evolution 327
References 349

5 Electromagnetic Emission 361
5.1 Principle of Operation 361
5.2 Source Mechanism 363
5.2.1 Modeling of Electromagnetic Emission Sources 371
5.2.2 Test Sources for Electromagnetic Emission 389
5.3 Signal Propagation 396
5.4 Detection of Electromagnetic Emission Signals 400
5.4.1 EME Detector Concepts 401
5.4.2 Electromagnetic Shielding 413
5.5 Application to Composites 427
5.5.1 Measurement of EME Due to Crack Formation 427
5.5.2 Measurement of Fracture Surface Orientation 445
5.5.3 Detectability of Failure Mechanisms 450
References 453

6 Computed Tomography 457
6.1 Principle of Operation 458
6.2 Detail Visibility 461
6.2.1 Object Resolution 462
6.2.2 Artifacts 464
6.2.3 Detectability of Defects in Fiber Reinforced Materials 475
6.3 Volumetric Inspection of Materials 481
6.3.1 Concepts for Ex Situ Loading 481
6.3.2 Concepts for In Situ Loading 482
6.4 Digital Volume Correlation 493
6.5 Application to Composites 498
6.5.1 Ex Situ Testing 502
6.5.2 In Situ Testing 515
References 527

7 Combination of Methods 533
7.1 Comparison of In Situ Methods 534
7.1.1 In Situ Capabilities 536
7.1.2 Detectability of Failure Mechanisms 538
7.1.3 Detection Sensitivity 540
7.1.4 Extension of Methods to Large Scale and Field Testing 543

7.2 Established Method Combinations . . . 548
7.2.1 Imaging Methods and Acoustic Emission . . . 548
7.2.2 Digital Image Correlation and Acoustic Emission . . . 555
7.2.3 Thermography and Acoustic Emission . . . 558
7.2.4 Computed Tomography and Acoustic Emission . . . 563
7.2.5 Computed Tomography and Digital Image Correlation . . . 590
7.2.6 Electromagnetic Emission and Acoustic Emission . . . 590
7.2.7 Acousto-Ultrasonics . . . 596
7.3 Future Developments . . . 601
7.3.1 Digital Image Correlation . . . 602
7.3.2 Acoustic Emission . . . 602
7.3.3 Electromagnetic Emission . . . 604
7.3.4 In Situ Computed Tomography . . . 605
7.3.5 Method Combinations . . . 606
References . . . 606

Appendix A: Acoustic Emission—Parameters of Influence . . . 611

Appendix B: Material Properties . . . 617

Appendix C: Acoustic Emission Signal Parameters . . . 621

Appendix D: Definitions and Abbreviations . . . 625

Index . . . 629

Financial Support

Financial support for the research described in this book was provided by a number of sources. I gratefully acknowledge the financial support by the Federal Ministry of Education and Research (BMBF) for funding the joint research projects MAI zfp and MAI plast within the leading edge cluster MAI Carbon, the joint research project FORCiMA funded by the Bavarian Research Foundation, the German Research Foundation (DFG) for funding the research project HO 955/8-1 and HO 955/8-2, the Free state of Bavaria for funding of the project ComBo within the program "BayernFIT," and the Free state of Bavaria for funding the project CFK-Integral within the program "Neue Materialien für Bayern."

About the Author

Markus Sause studied Physics at the University of Augsburg and earned his doctoral degree in 2010 in Experimental Physics at the same institution. He received the "Erich-Krautz-Preis" in 2010 for his outstanding contribution to the interpretation of acoustic emission of fiber-reinforced materials. In 2015, he was awarded his Habilitation in Experimental Physics. He is lecturer at the University of Augsburg and head of the Materials Testing Research Group at the Institute of Physics and at the Institute of Materials Resource Management. Since 2014, he has been a member of the EWGAE executive committee and active in several other committees dedicated to the testing and analysis of fiber-reinforced materials. His research interests span the mechanics of fiber-reinforced composites, their destructive and nondestructive testing, as well as numerical methods to interpret the materials behavior. A special focus is given to bridge the gap between destructive testing approaches and nondestructive inspection to perform in situ analysis of materials failure.

Chapter 1
Introduction

Failure of engineering materials is a phenomenon, which has been present since the early days of mankind and has been investigated on a modern scientific basis starting at the advent of the industrial age. Only in recent decades it has become convenient to build structures based on previous numerical calculations to predict failure on the material level rather than based on empirical knowledge. The pioneering work of those researchers with a focus on engineering materials like metals, concrete, ceramics, and polymers has enabled us to allow construction of technically safe structures out of these materials. In recent time it has become a significant trend in material science to fabricate composite materials in order to combine benefits of different material classes. Such composite materials are applied for a variety of reasons. Some material combinations aim to improve electrical conductivity by inclusion of particles, others try to improve the mechanical behavior by adding reinforcing elements. The increasing number of interfaces inherent to composite materials generally induces a more complex failure behavior than encountered in pure materials.

Fiber reinforced polymers (FRPs) are a class of materials that show an extraordinary strength-to-weight ratio and stiffness-to-weight ratio. However, the current limited predictability of material failure typically has resulted in high safety margins for the allowable design limits in construction of FRP structures. Therefore, in safety relevant structures, the full potential of this material class is seldom used. Many attempts have been made to transfer concepts of fracture mechanics and failure prediction from pure materials to such heterogeneous and very often anisotropic materials. Since structural failure in FRP is a complex evolution of various microscopic failure mechanisms, understanding of their evolution is key to understand global failure and to reach reliable predictions. Also, the structural integrity is affected differently by the occurrence of different failure mechanisms. Thus, it is a requirement to distinguish between different types of failure mechanisms that occur within the structure under load.

One comprehensive attempt to assess the status of today's failure theories for FRPs is known as world-wide failure exercise. The organizers of this round-robin

M.G.R. Sause, *In Situ Monitoring of Fiber-Reinforced Composites*,
Springer Series in Materials Science 242, DOI 10.1007/978-3-319-30954-5_1

test invited leading researchers world-wide to predict failure of selected laminate configurations for a set of load cases. This comparative assessment of theoretical failure prediction has brought together researchers working on the same field but using fairly different approaches. Due to the complexity of failure found in fiber reinforced materials, it was one of the essential results of the "World-Wide Failure Exercise I" [1–4] and "World-Wide Failure Exercise II" [5–7], that for some load cases theoretical predictions of the different participants were differing by a large amount relative to each other and relative to the experimental data. Since in the world-wide failure exercises the benchmark was the experimentally obtained values, it was also argued [8, 9] whether the experimental data is reliable enough to be used for comparison.

While the required parameters for failure prediction in metals, ceramics, or polymers are in many cases straight forward to measure, the testing of fiber reinforced composites remains a challenge by itself. Moreover, due to the anisotropy the number of required material parameters is substantially larger in fiber reinforced materials than for isotropic engineering materials.

There is an ongoing discussion regarding the testing methods involved to obtain material properties of FRPs. Present standards to obtain material properties for fiber reinforced composites focus on the signatures of load–displacement curves or stress–strain curves to deduce angle dependent moduli or strength values. However, for some cases such global stress–strain curves have only limited relevance to the determination of the true (or required) material properties. Since microscopic failure precedes the ultimate failure of the test specimen on the macroscopic scale, it is doubtful if maximum values of stress–strain curves are the correct way to derive relevant material properties. The heterogeneity and anisotropy on the mesoscopic scale can often induce extremely different stress and strain states than expected from the macroscopic loading condition. In particular, the individual layers of a laminate are subject to constraining effects of the neighboring layers and thus exhibit stress states distinctly different to the global average. After initiation of first failure on the microscopic scale, a complex interplay of failure mechanisms initiates that involves mechanisms on length-scales spanning from the atomistic scale to the dimensions of the test specimen. The loading conditions, the fiber architecture, fiber volume fraction, and the fracture behavior of the matrix material all contribute to the ability of the material to survive upon further loading until finally ultimate failure occurs on the macroscopic scale. Due to the statistically driven initiation and evolution of microscopic failure, FRPs have the reputation to typically exhibit a large scatter of material properties. The significance of this statistical distribution of initiation sites for failure is known as a volume effect and has been intensively investigated [10].

The present work does not aim to improve the mechanical testing part or the failure theories themselves, but to provide the necessary experimental methods to allow monitoring of the progression of failure in FRPs. In this sense, failure refers to the occurrence of microscopic or macroscopic networks of cracks in the material. If failure initiation can be detected in an appropriate way, imaging methods can be used to capture the accumulated damage state present in the material at that stage.

Such imaging can then be incorporated in a modeling environment and be used to allow a detailed forward prediction of the failure behavior or it can be used to deduce more relevant material properties as an input for failure theories.

The work presented herein is thus motivated by two possibilities of composite analysis enabled by the presence of validated experimental methods for:

1. **Detection of failure initiation**
   This deals with the applicability of experimental methods to detect the first onset of a particular failure mechanism in a fiber reinforced composite. If such failure mechanisms can be detected with certain reliability, the resulting local or global stress and strain levels can be related to this failure onset to derive more representative material properties.
2. **Tracking failure evolution**
   After failure initiation, the proposed experimental methods should allow the continuous monitoring of failure evolution in FRPs. This allows for continuous comparison of the experimental damage states relative to theoretical predictions. Moreover, this allows deducing subsequent stages of damage progression in order to improve the understanding of failure in general.

These two possibilities are exemplified in application to detect crack initiation and crack propagation in Fig. 1.1. To investigate initiation and growth of cracks in FRPs, a variety of experimental methods are applied. Imaging methods are frequently used to monitor the damage progress after loading and unloading of test specimens. This allows visualization of the damage progress in discrete steps, but the unloading of the specimens causes existent cracks to close and makes them

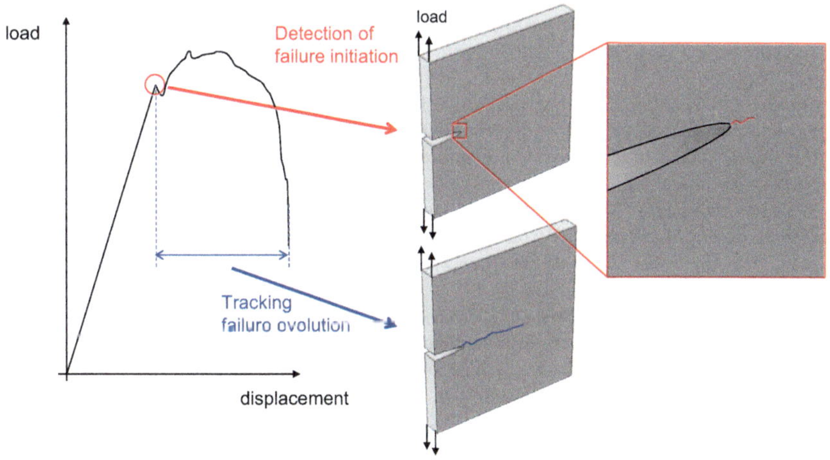

Fig. 1.1 Detection of failure initiation and tracking of failure evolution illustrated for the occurrence of crack formation on the microscopic scale and subsequent propagation on the macroscopic scale

harder to spot. To overcome this problem, one possibility is the application of in situ microscopy. This yields continuous observation, but the imaging process is either restricted to surface observations using optical microscopy or electron microscopy and is limited to the specimen size or resolution as in computed tomography (CT). As an alternative for surface and near surface, digital image correlation (DIC) can add valuable information to spot damaged areas by sudden changes in the strain field. In the same way as the ear complements the eye, the detection of acoustic emission (AE) can act as complementary method to imaging methods in order to detect failure initiation and to track growth of damage in the full volume of the specimen. Also the generation of electromagnetic emission (EME) during crack initiation and crack growth is a nondestructive method useful for online monitoring of failure in FRPs covering the full volume of the specimen.

References

1. Soden, P.D., Hinton, M.J., Kaddour, A.S.: Biaxial test results for strength and deformation of a range of E-glass and carbon fibre reinforced composite laminates: failure exercise benchmark data. Compos. Sci. Technol. **62**, 1489–1514 (2002)
2. Hinton, M.J., Kaddour, A.S., Soden, P.D.: Evaluation of failure prediction in composite laminates: background to "part B" of the exercise. Compos. Sci. Technol. **62**, 1481–1488 (2002)
3. Hinton, M.J., Soden, P.D.: Predicting failure in composite laminates: the background to the exercise. Compos. Sci. Technol. **58**, 1001–1010 (1998)
4. Soden, P.D., Hinton, M.J., Kaddour, A.S.: Lamina properties, lay-up configurations and loading conditions for a range of fibre-reinforced composite laminates. Compos. Sci. Technol. **58**, 1011–1022 (1998)
5. Kaddour, A.S., Hinton, M.J.: Maturity of 3D failure criteria for fibre-reinforced composites: comparison between theories and experiments: part B of WWFE-II. J. Compos. Mater. **47**, 925–966 (2013)
6. Hinton, M., Kaddour, A.: The background to part B of the Second World-Wide Failure Exercise: evaluation of theories for predicting failure in polymer composite laminates under three-dimensional states of stress. J. Compos. Mater. **47**, 643–652 (2013)
7. Hinton, M., Kaddour, A.: Triaxial test results for fibre reinforced composites: the Second World-Wide Failure Exercise benchmark data. J. Compos. Mater. **47**, 925–966 (2013)
8. Christensen, R.M.: A physically based cumulative damage formalism. In: Major Accomplishments in Composite Materials and Sandwich Structures: An Anthology of ONR Sponsored Research, pp. 51–65 (2009)
9. Cuntze, R.: Comparison between experimental and theoretical results using Cuntze's "failure mode concept" model for composites under triaxial loadings-part B of the Second World-Wide Failure Exercise. J. Compos. Mater. **47**, 893–924 (2013)
10. Wisnom, M.R.: Size effects in the testing of fibre-composite materials. Compos. Sci. Technol. **59**, 1937–1957 (1999)

Chapter 2
Failure of Fiber-Reinforced Composites

Generally, we start to refer to failure of a material, when the material no longer fulfills its intended use. Therefore, the physical phenomenon associated with the failure is subject to the anticipated use as well. A scratch in the hood of an automobile is an annoying thing but may not necessarily stop the hood from operating in its intentional use. Still the scratch may cause initiation of corrosion, since the protective coating may have been partially removed. In contrast to the automobile hood, a scratch in a painting directly affects the product quality. The painting may not look as beautiful anymore and therefore may fail in its intended use. In the context of this book, the term "failure" always refers to the structural failure of the material due to mechanical load. The primary source for this failure are cracks forming within a material causing it to reduce its effective cross section or introducing instability causing buckling, kinking, or likewise failure modes.

The mechanisms of crack formation and propagation within materials is a physical phenomenon which is still not fully understood due to the complex interplay between the local microstructure of the material and its impact on the macroscopic values of stress fields. In fracture mechanics, the calculation of material specific quantities like critical strain energy release rates is based on macroscopically observable quantities like forces and displacements. Microscopically, the strain energy required for crack propagation is converted to surface energy, heat, elastic energy release, and plastic deformation. Still one of the unsolved problems in fracture mechanics is the description of tensile extensions ahead of the crack tip and the dynamic displacement field surrounding the crack tip [1]. This is owed to the fact that a direct observation of crack nucleation and propagation is usually not possible, for example, by optical means, due to the crack propagation velocities in solid materials. Also the influence of dynamical instabilities is not fully understood [2].

Despite of such missing scientific details, high-quality phenomenological approaches exist to describe failure of typical engineering materials due to crack formation. The fracture mechanics concepts of Griffith and Irwin [3, 4] form the basis for the description of crack initiation and propagation in metals, polymers and

M.G.R. Sause, *In Situ Monitoring of Fiber-Reinforced Composites*,
Springer Series in Materials Science 242, DOI 10.1007/978-3-319-30954-5_2

ceramics. However, in fiber-reinforced materials, the hierarchical and heterogeneous microstructure superimposes the isotropic and homogenous properties of the matrix materials. Additional effects arise due to crack branching, crack deflection, and mixed mode conditions. Therefore, some simple assumptions inherent to the classical descriptions are no longer fulfilled. Hence, a broad variety of theories is currently applied for the description of failure in fiber-reinforced polymers.

This chapter starts first with an overview on failure mechanisms in composites and introduces their hierarchical classification. The second part of this chapter briefly reviews failure theories currently applied in the field of fiber-reinforced composites. The third part of this chapter then has its focus on the mechanical methods used to provide material properties necessary as input data for these failure theories. In particular, test methods still controversially discussed in the community are critically reviewed. The last section then finally provides a first overview on "secondary" methods as being used to improve the "primary" mechanical methods to obtain valid material data.

2.1 Classification of Failure Mechanisms

During mechanical loading of fiber-reinforced plastics various microscopic damage mechanisms occur. The accumulated damage finally results in failure of the composite. As seen in Fig. 2.1a, on the microscale cracks can evolve within the matrix material, breakage of the fibers can occur, or cracks develop at the interface between both constituents. From the fracture mechanics point of view, the situation grows more complex, since each of these fundamental mechanisms can be induced by various types of macroscopic loading conditions. On the mesoscopic scale seen in Fig. 2.1b, as a function of external loading, these mechanisms can interact, grow, and can form even more complex mechanisms. Finally their growth develops into a critical state, which then causes ultimate failure of the composite structure due to the macroscopic loading condition as seen on the macroscopic scale of Fig. 2.1c. In the following, an introduction is given for the different failure mechanisms

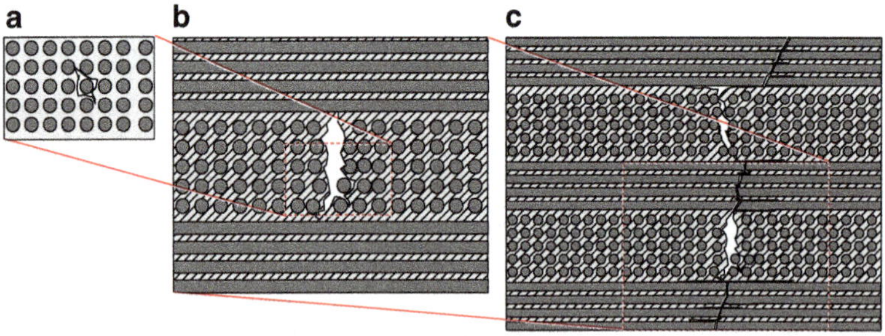

Fig. 2.1 Schematic illustration of failure evolution in a cross-ply used to define microscale (**a**), mesoscale (**b**), and macroscale (**c**)

according to a hierarchical classification with respect to the length scale of occurrence. Parts of the following discussion are based on the considerations found in standard literature [5–7] as well as the author's previous work [8].

2.1.1 Microscale

Fundamentally, failure initiates at the atomistic level. As consequence of excessive mechanical forces, bonds are separated and cracks can initiate at weak spots or grow further. There are valuable attempts found in literature to conduct ab initio predictions of failure based on Monte–Carlo simulations [9, 10]. However, most of these approaches are still limited in the amount of complexity being captured by the algorithms. In particular, the consideration of an explicit microstructure is usually beyond the capabilities of such approaches, that is, solid media may be assumed to be homogeneous.

In the context of fiber-reinforced composites, the microscale is determined by the typical sizes of the constituents. In this case, the diameter of the fibers and their characteristic length are used to define this scale. Therefore, the microscale is in the range of micrometers, which is considerably larger than the atomistic scales. Since this allows distinguishing different failure mechanisms beyond failure of the individual constituents, this is the first relevant scale to explicitly discuss for fiber-reinforced materials.

Stress components causing the corresponding failure in unidirectional composites are noted using subset directions indicating parallel ($\|$) and transverse ($\perp$) orientation relative to the fiber axis. On this scale, the distribution of the reinforcement fibers within the matrix material is of particular relevance. The fraction of fiber filaments embedded in the matrix material is usually referred to as fiber volume fraction. Other than on the macroscale, the fiber volume fraction in a real composite material may change substantially as a function of position. Especially for fiber architectures such as woven fabrics or for materials with comparatively low fiber volume fractions, the extreme cases of regions of pure matrix and dense fiber volume fractions may occur (cf. Fig. 2.2). In order to quantify these effects, automated image processing of cross sections [11] as well as volumetric analysis of fiber distributions by computed tomography have emerged as suitable methods [12].

One of the key hypotheses in fracture mechanics is the fact that anomalies are always present in real materials. In the context of fiber-reinforced materials, there are several characteristic defects located at the microscale, which can be understood as such anomalies. In Figs. 2.3, 2.4, and 2.5 such typical defects found in real fiber-reinforced composites are shown.

As a residue of the thermal processing, polymer matrix materials often show porosity at various scales. These pores can result from gases of the chemical curing process or from missing consolidation pressure during the solidification. The type of porosity smaller in extent than fiber diameters is often referred to as

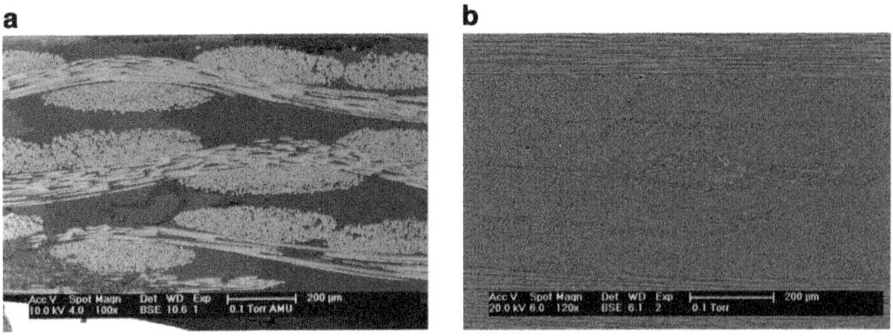

Fig. 2.2 Examples of fiber distribution on the microscale in glass/epoxy-woven fabrics (**a**) and unidirectional carbon/epoxy (**b**)

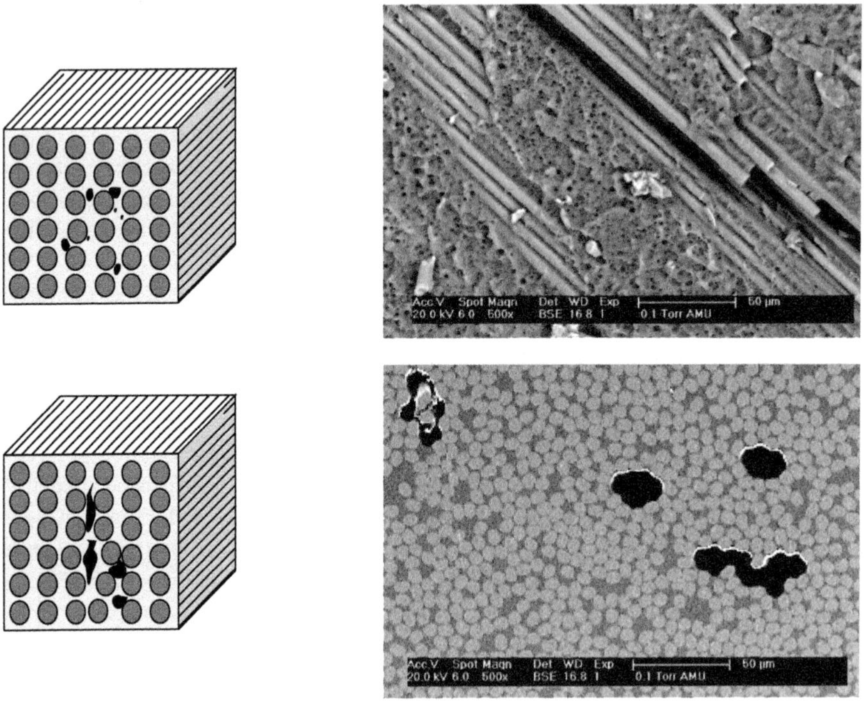

Fig. 2.3 Schematic illustration of microporosity and mesoporosity (*left*) and according scanning electron microscopy images of carbon/epoxy material (*right*)

microporosity. At this scale, many small pores reside within the matrix material, sometimes resulting in foam-like microstructure as seen in Fig. 2.3.

At the next length scale (size of fiber diameters), the pores often stick to the fiber filaments or fiber bundles and therefore cause regions, where the fibers are not supported by the matrix material. This can cause preliminary buckling if the

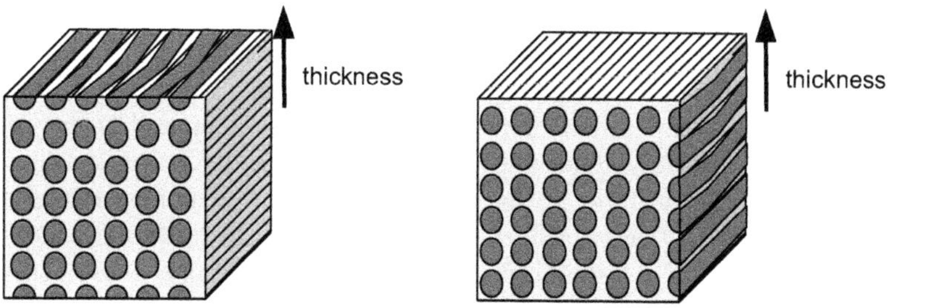

Fig. 2.4 Schematic illustration of in-plane (*left*) and out-of-plane (*right*) undulations

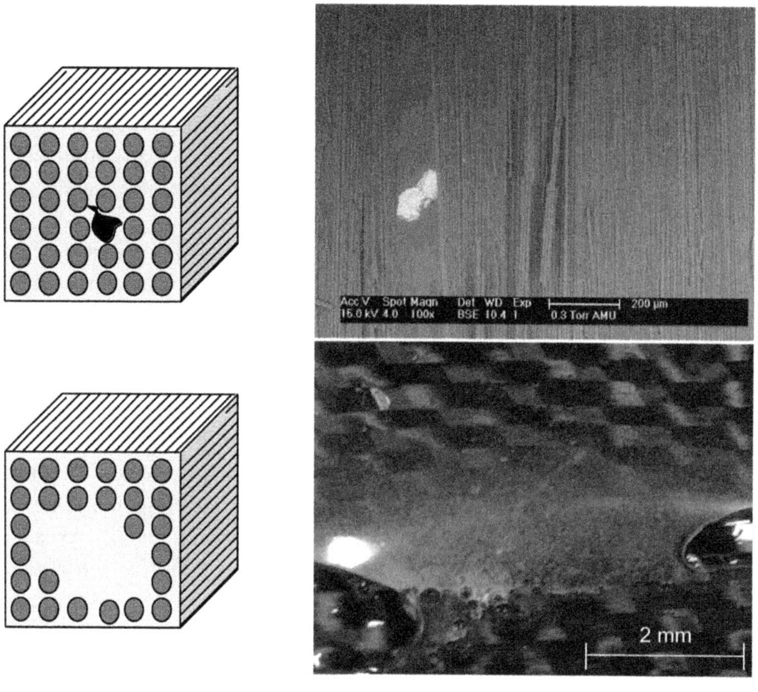

Fig. 2.5 Schematic illustration of particle inclusion, resin rich areas and illustrative microscopy images of a carbon/epoxy material

composite is subject to compressive load or shear load. In deviation to the terminology of length scale used herein, such types of pores are already referred to as mesoporosity. This is to reflect the length scale of microporosity found below.

Another type of defect which is always present to some extent within fiber-reinforced materials is undulation (waviness) of the fibers. As seen in Fig. 2.4, one can distinguish between in-plane and out-of-plane undulations relative to the thickness direction of the laminate. Due to the waviness of the fiber filaments or fiber bundles, the local stress exposure can dramatically change. For a tensile load, the neighboring fiber filaments may thus be subject to higher stress levels, while in compressive load the waviness may induce early fiber buckling.

Other types of anomalies frequently encountered in fiber-reinforced composites are shown in Fig. 2.5. Inclusions of dust particles, resin-rich areas, binder yarns or fiber knots all share the common impact of disturbing the local fiber distribution. This is either realized in form of fiber undulations or in affecting the local fiber volume fraction. Both effects considerably change the local stress exposure of fiber and matrix material and therefore can cause preliminary failure. However, some of these defects can also cause benign failure behavior. One of the main improvements in terms of damage tolerance is the ability of the fibers to cause crack deflection of interfiber cracks. Here, every crack deflection or crack branching is advantageous, since the additional energy necessary to elongate the crack propagation path contributes directly to the macroscopic fracture toughness.

But even fiber-reinforced materials free of defects have intrinsic weak spots. Due to production process at elevated temperatures, most fiber-reinforced polymers are subject to a residual thermal stress between fibers and matrix. Here, the matrix material between two adjacent fibers is exposed to a geometrically induced stress concentration and therefore acts as weakest link when loading the material transverse to the fiber axis [5, 13–15].

Perhaps, the most critical type of failure induced within fiber-reinforced polymers is fiber failure. As the reinforcement part of the composite, this inherently degrades the structures ability to bear increased load. For uniaxial tensile loading parallel to the fiber axis, fiber breakage is observed if the local strength of the fiber filament is exceeded. However, such single fiber breakage does not necessarily result in ultimate failure of the composite. On the microscopic scale fibers show a broad distribution in their ultimate strength due to variations of the fiber diameter and internal flaws [5, 16]. This can result in early failure of numerous fibers, while the remaining fibers still act as load-bearing elements. Due to this strength distribution of the fibers, weak fibers are the first to fail. As a consequence of the fracture of a single fiber, the stress concentration at the position of rupture initiates debonding between fiber and matrix as well as fracture of the surrounding matrix material. This type of failure is schematically shown in Fig. 2.6 and is denoted fiber breakage in the following. For comparison, a respective scanning electron microscopy image of fiber breakage caused by tensile loading is shown as well.

For compressive loading, the failure behavior of fibers is characteristically different. Due to the compressive load, the reinforcement fibers typically buckle, once the matrix material cannot support their orientation any longer. After that buckling point, band kinking occurs, which finally results in crushing of the fibers at distinct positions [17] as schematically shown in Fig. 2.7. For comparison, a corresponding scanning electron microscopy image of compressive fiber failure is shown in Fig. 2.7. As indicated in the schematic illustration, this type of failure naturally results in significant fiber fragmentation, when observed at the surface level.

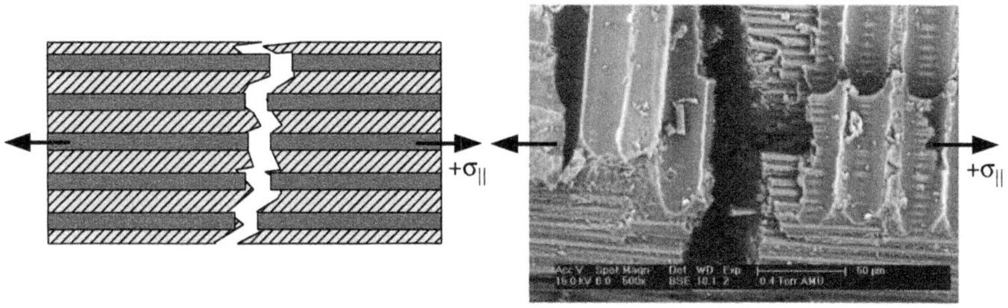

Fig. 2.6 Schematic illustration of fiber breakage due to tensile loading and corresponding electron microscopy image of a carbon/epoxy material in surface view (based on [8])

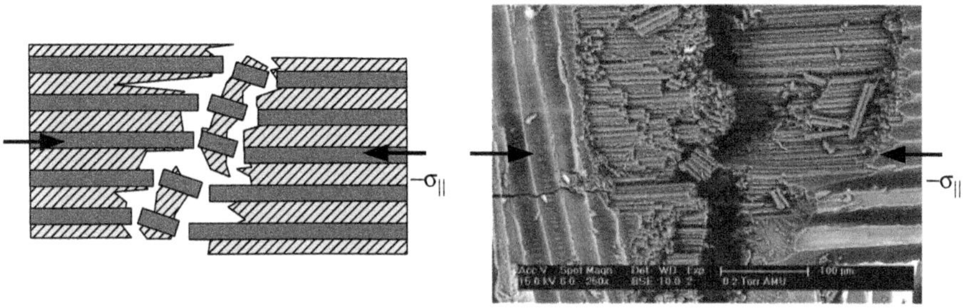

Fig. 2.7 Schematic illustration of fiber crushing due to compressive loading and corresponding electron microscopy image of a carbon/epoxy material in surface view (based on [8])

The type of failure observed most often in fiber-reinforced structures is interfiber fracture. As schematically shown in Fig. 2.8, this failure of the matrix material can originate from various loading conditions. In all cases, cracks initiate or propagate within the matrix or at the interface between fiber and matrix, if the respective loading condition of normal stresses or shear stresses exceeds the local strength of the matrix or interface, respectively.

In principle, the process of interfiber fracture can be further decomposed into different microscopic failure mechanisms. While fracture of the matrix material itself is always involved to some extent, the failure of the fiber–matrix interface is also likely to occur. At the microscopic scale, one can distinguish between the process of fiber–matrix debonding and the fiber–matrix pull-out.

As debonding the partial or full loss of contact between fiber and matrix due to external load is understood, which can occur in various load configurations (cf. Fig. 2.9).

As second failure process at the interface between fiber and matrix, the fiber–matrix pull-out is often discussed in literature [18–23]. As seen in Fig. 2.10, this involves a fiber–matrix debonding process as initial damage state, but then is followed by a sliding of the fiber filament relative to the matrix. The filament is not necessarily destroyed, but often spans the evolving crack in the matrix material. However, this failure mechanism is only likely to occur for tensile loading in

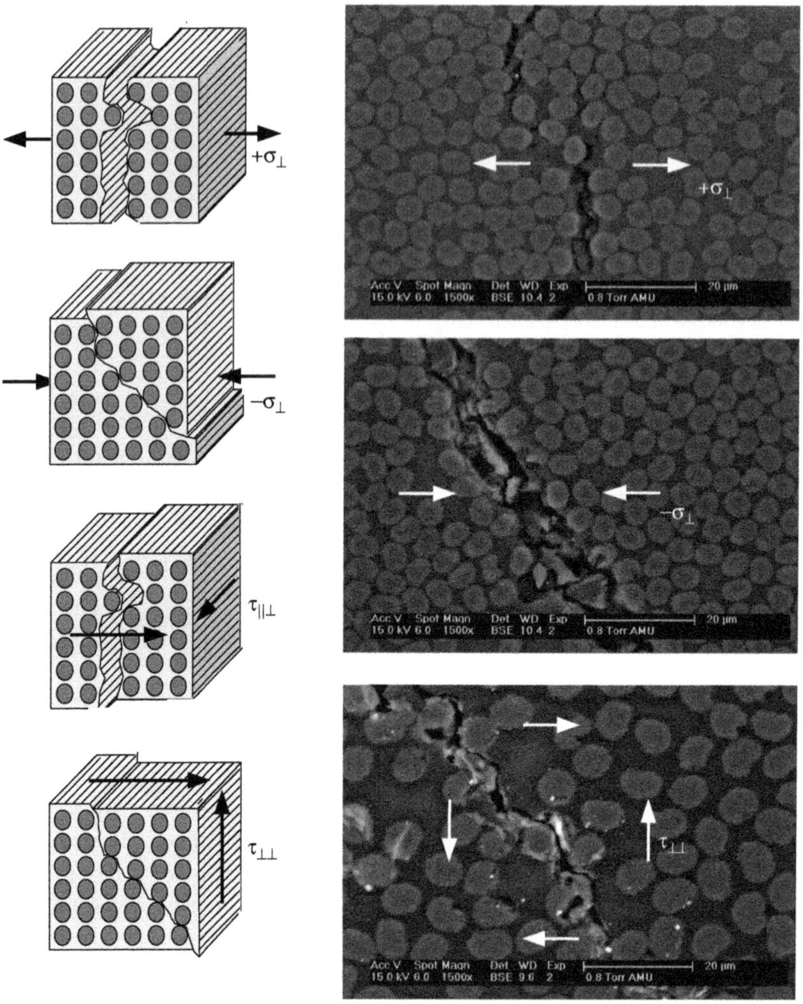

Fig. 2.8 Schematic illustration of interfiber fracture (matrix cracking) and corresponding electron microscopy cross-sectional images of a carbon/epoxy material for typical load cases (based on [8])

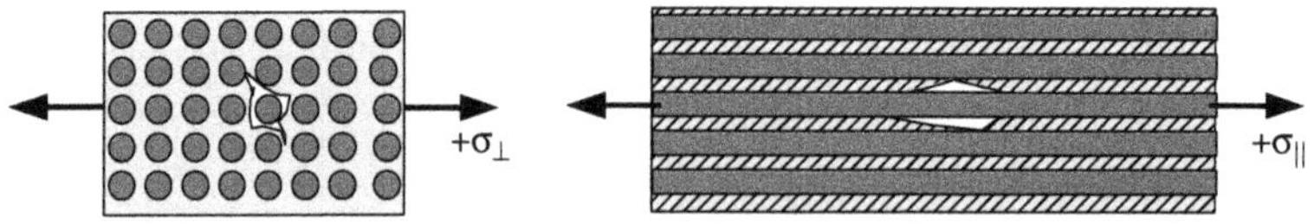

Fig. 2.9 Schematic illustration of fiber–matrix debonding due to transverse stress and due to fiber parallel stress

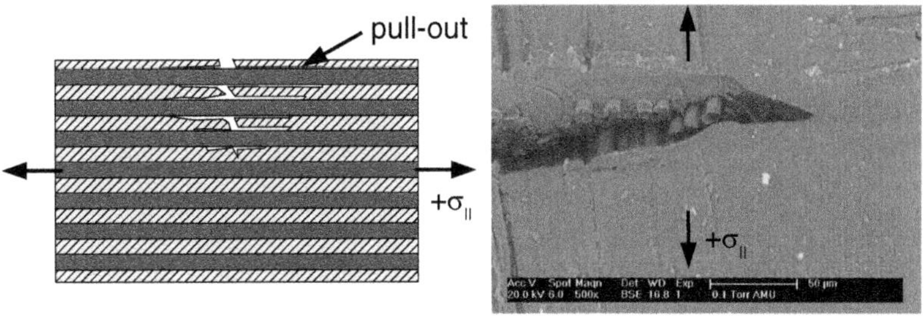

Fig. 2.10 Schematic illustration of fiber–matrix pull-out due to fiber parallel stress with respective scanning electron microscopy image of a carbon/epoxy material

parallel to the fiber axis and for fiber–matrix combinations with relatively low bonding strength. If the adhesive bonding between fiber and matrix is too large, no debonding can initiate due to shear forces at the interface and fiber failure is observed instead.

2.1.2 Mesoscale

Beyond the length scale defined by the composite constituents, the mesoscale is found. In fiber-reinforced materials, the relevance of this scale typically arises due to the textile architecture chosen to build laminates. In technical applications, laminates with only one fiber orientation are usually not relevant since they exhibit very poor mechanical properties transverse to the fiber orientation. Therefore, laminates are frequently fabricated as stacks of individual plies consisting of unidirectional layers in various orientations, woven fabrics, other textiles, or mixtures thereof. In the following, the term mesoscale is used to define the length scale of interaction between these different plies forming the laminate; thus, it is located in the range of several hundred micrometers.

For loading in the out-of-plane direction of the laminate, intralaminar delamination (splitting) and interlaminar delamination (inter-ply) can occur. The relevant material property determining inter-ply failure is the fiber–matrix adhesion. As seen in the microscopy images in Fig. 2.12, the magnitude of the bonding strength between fiber and matrix dominates the laminates interlaminar fracture resistance. As a function of increased bonding strength, the fracture surfaces show a pronounced hackle structure [7], which is caused by an increased amount of crack deflection and crack branching and therefore requires more energy for the same amount of macroscopically relevant crack propagation length (cf. Fig. 2.12).

For the interlaminar failure, the weak link is the individual ply itself. As for the inter-ply delamination, the relevant material property is the fiber–matrix adhesion. If the individual ply is exposed to high out-of-plane stress, it is also likely that failure will occur within a layer. As seen in Fig. 2.13, there is no predescribed fracture plane, so interlaminar delamination typically is associated with substantial

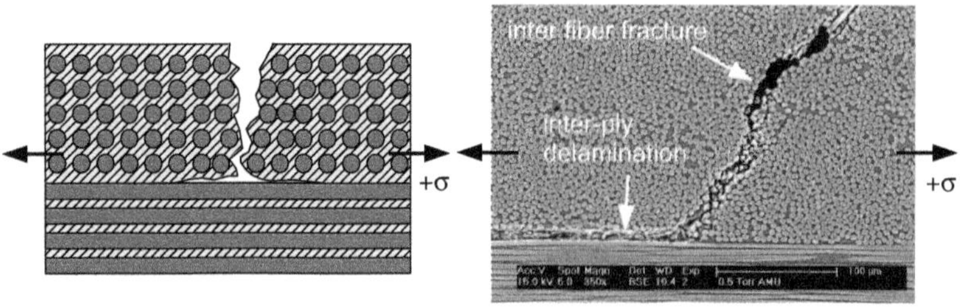

Fig. 2.11 Schematic illustration of inter-ply delamination with corresponding electron microscopy images of a carbon/epoxy material in cross-sectional view (based on [8])

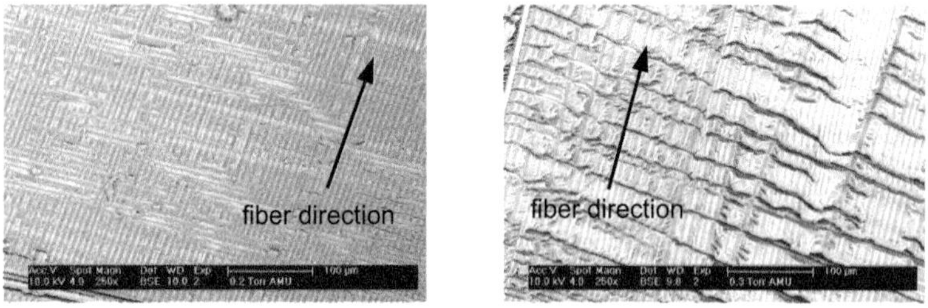

Fig. 2.12 Influence of fiber–matrix adhesive strength on delamination behavior in scanning electron microscopy images of a carbon/epoxy material in top view

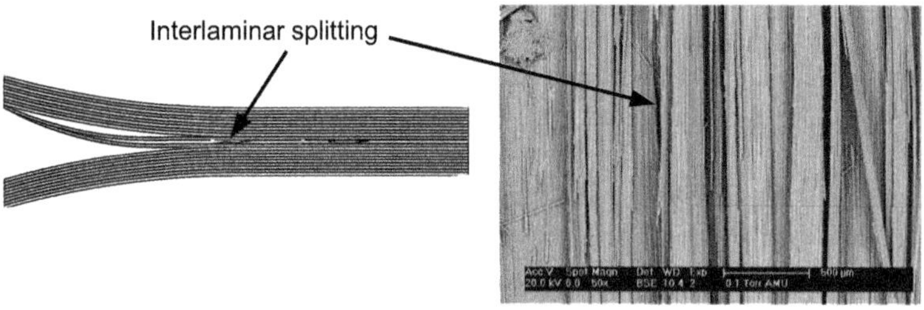

Fig. 2.13 Schematic illustration of interlaminar delamination (splitting) in unidirectional fiber-reinforced layers and corresponding fracture surface of a carbon/epoxy material in scanning electron microscopy image

splitting of the fiber layers and rough fracture surfaces as seen from the scanning electron microscopy image.

If such laminates are subject to uniaxial stresses, the individual plies are exposed to fairly different stress levels. As long as there is no damage at the interface between the plies, the assumption of strain coupling is fulfilled [5]. Since the elastic properties and strength values of the individual plies vary significantly as function of fiber orientation, characteristically different stress states are reached.

As seen in Fig. 2.14, tensile loading of a laminate often causes early interfiber cracks in the off-axis oriented plies. Dependent on the respective external load

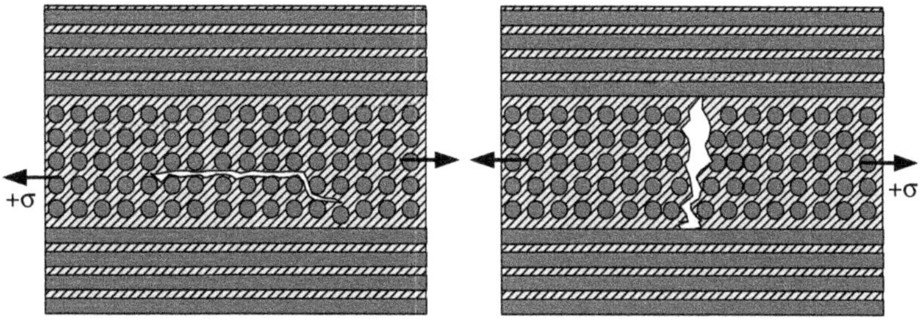

Fig. 2.14 Schematic illustration of interfiber cracks in off-axis layers

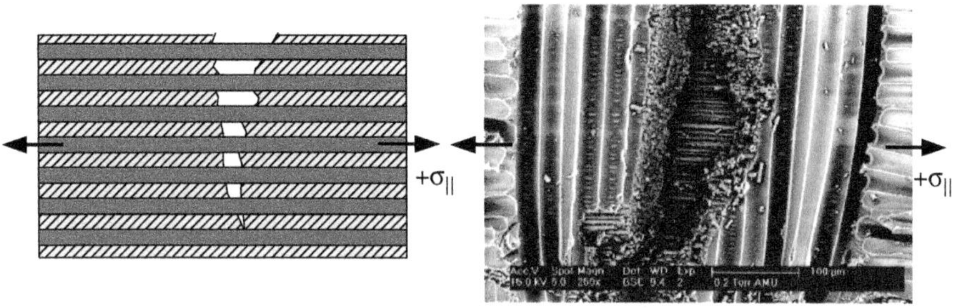

Fig. 2.15 Schematic illustration of fiber bridging together with corresponding electron microscopy images of a carbon/epoxy material in top view (based on [8])

configuration and the stacking sequence, these interfiber cracks can occur either in parallel or transverse to the load axis. Adjacent plies of different fiber orientation typically act as crack stopper, since the presence of fibers oriented with the load axis avoids failure of the matrix material due to local strain coupling. However, such interfiber cracks can induce further damage as shown in Fig. 2.11. The existence of such cracks typically causes strong stress concentration at the crack tip. This is a typical site for initiation of inter-ply delamination propagating along the interface between two plies.

Another process of mesoscopic failure, linked to the strength of the fiber–matrix interface is the phenomenon of fiber bridging. If the fiber–matrix bonding is sufficiently low, fiber pull-out can occur if the laminate is subject to tensile loading. If the fiber filaments are strong enough, they can survive the pull-out to a certain extend and span a mesoscopic crack propagating transverse to the fiber axis as schematically shown in Fig. 2.15. In this case the crack progress is somewhat constrained, since the intact fibers still bear the applied load. This failure mechanism is also visible in the scanning electron microscopy image in Fig. 2.15.

Even at the mesoscale porosity can exist as a residue of the production process. As seen from Fig. 2.16, these pores are located between adjacent plies and can have a substantial impact on the mechanical properties of the laminate. Since the pores do not carry any load, the laminate is subject to high local stress concentration and may fail due to the onset of delamination, due to buckling, or due to the increased stress level in the reduced cross section.

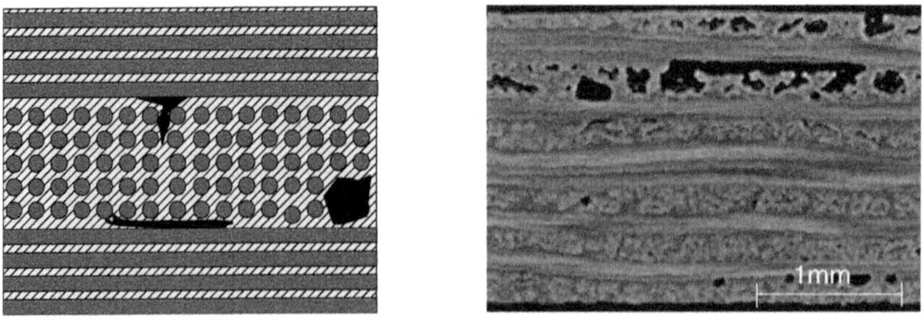

Fig. 2.16 Schematic illustration of macroporosity including corresponding computed tomography image of a carbon/epoxy material in cross-sectional view

2.1.3 *Macroscale*

Due to these large number of failure mechanisms on the microscale and mesoscale, the definition of macroscopic composite failure usually is subject to the type of application and the type of loading. The scale associated with this term ranges from the thickness of the laminate to the full dimension of the structure. Therefore, the length scale is within the range of few millimeters to some meters.

Although interfiber fracture does not usually cause ultimate failure of the composite on the macroscopic scale and may even be acceptable in some applications, it can still alter the mechanical properties, for example, the stiffness of the structure. In addition, such failure can change the surface properties of the composite or can raise secondary damage from aggressive media penetrating the structure or an increased rate of moisture absorption. Also, microscopic fiber breakage does not necessarily result in ultimate failure. In addition, the damage progression induced by the stacking sequence often causes the global structure to compensate the failure of one full layer.

An example of a macroscopic fracture surface is shown in Fig. 2.17 as originating from T-Pull testing. As seen from this example, all of these macroscopic fracture surfaces are composed of different microscopic failure mechanisms, whereas the final type of fracture is often dominated by the processes on the mesoscale.

Another example to demonstrate the link between microscale, mesoscale, and macroscale is the damage pattern observed in a woven fabric structure. As seen from the computed tomography image in Fig. 2.18, the macroscopically visible crack ranging through the laminate thickness develops many branches on the mesoscale causing vast inter-ply delamination and microbuckling of individual layers as seen by the cross-sectional images. On the length scale below the individual plies (the microscale), all sorts of interfiber cracks and fiber breakage can be observed.

One particularly relevant macroscopic failure mode for fiber-reinforced composites is the impact damage. As seen schematically in Fig. 2.19, the damage state

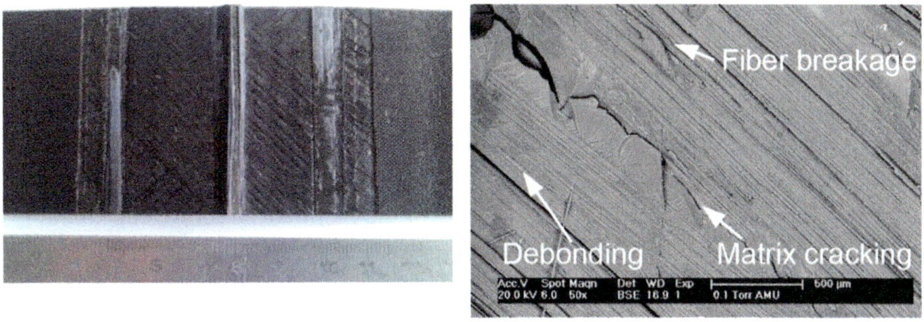

Fig. 2.17 Typical example of a fracture surfaces. Photography of full fracture surface and detail of fracture surface as electron microscopy images of a carbon/epoxy material in top view

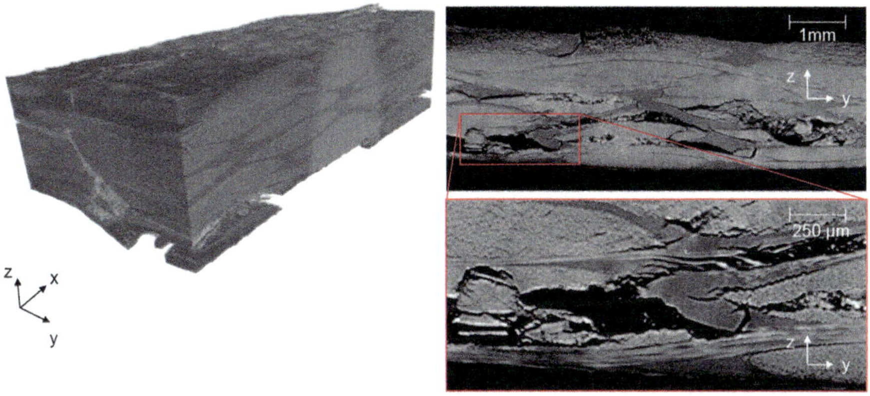

Fig. 2.18 3D image of damaged woven fabric and virtual 2D cross sections of the volume showing details of the damage state

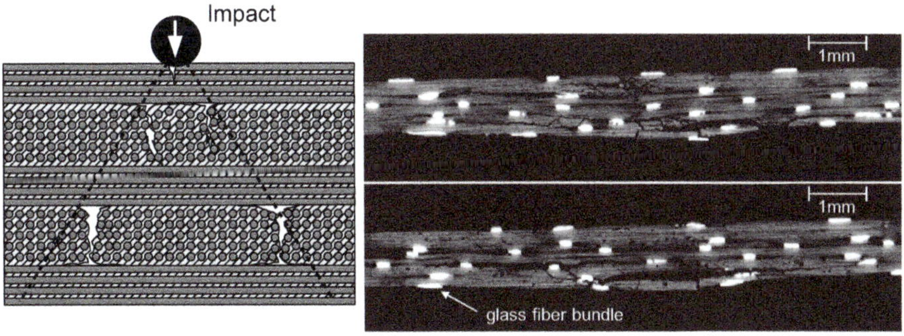

Fig. 2.19 Schematic drawing of impact damage and respective computed tomography scan of impact zone (carbon/epoxy with embedded glass fiber bundles)

reached within a multiaxial laminate due to a low velocity impact is rather severe. On the surface level, the impactor sometimes causes only small damage in the form of local matrix damage or breakage of some surface plies. Therefore, such low velocity impacts are often barely visible from the outside. However, on the inside of the laminate, the large bending stresses during impact cause considerable interlaminar shear stresses resulting in initiation and growth of interfiber failure and delamination. This weakens the stiffness of the laminate structure considerably and therefore comprises a major damage type that is well within the macroscopic range.

2.2 Failure Theories for Fiber-Reinforced Composites

Within the last decades many research groups developed criteria to predict failure in fiber-reinforced composites. As discussed in Sect. 2.1, there are various microscopic failure mechanisms causing macroscopic failure. Moreover, their occurrence and evolution is distinctly different for the type of macroscopic loading condition. Hence, numerous theories are found in literature aiming to describe the complex initiation and growth of damage in composites. In order to provide an overview of those different theories, a first categorization can be made with respect to their range of application. One can distinguish between theories focusing on description of:

- Quasi-static failure.
- Quasi-static failure including growth of damage, damage mechanics, and degradation.
- Long-term behavior (creep, stress rupture, and fatigue).
- High-velocity (high strain rates).

In the following, a brief review on the most important developments in these areas is given. Focus rests on recent developments and the usage of experimentally obtained values as input parameters for the theories.

2.2.1 Quasi-Static Failure

For the range of quasi-static loading conditions, many failure criteria have been developed to predict the first onset of failure (first ply failure) as well as the respective ultimate failure (last ply failure). In 1986, Reifsnider and Nahas presented and compared already 30 different failure criteria for that purpose [24]. Since that time many refinements have been proposed and essentially new criteria have been developed.

In principle, stress and strain failure criteria are equivalent descriptions since the two quantities are directly linked by the material's constitutive law. But as pointed out by Puck, strain criteria are more difficult in their mathematical description, since the superposition by transverse strains causes an inherent complexity of the theoretical formulation [25]. Moreover, to obtain the respective input data a uniaxial strain state is required to measure valid fracture strains. This is not feasible in many experimental configurations.

In general, a distinction can be made between global criteria and individual criteria differentiating between failure mechanisms. The global criteria apply classical continuum mechanics theory to describe fiber-reinforced composites as homogenous and anisotropic solid. In contrast, the differentiating criteria consider the microstructure of the material and take into account the failure of fibers or matrix material.

In 1971, Tsai and Wu published a generalized failure theory for anisotropic solids known as the Tsai–Wu criteria [26]. Three years later, Wu demonstrated that all global criteria published until that point could be directly derived from the Tsai–Wu criteria [27]. Later Hashin [28] discussed limitations of those theories in terms of the physical principles and pointed out the lack of consistency found for the Tsai–Wu criteria under some stress states. Essentially this initiated the development of further criteria which explicitly distinguish between fiber failure and interfiber failure.

The type of failure criteria based on this failure hypothesis are called differentiating criteria. They provide separate failure conditions for fiber failure and for interfiber failure. In most failure criteria formulations simple maximum stress criteria are used to describe fiber failure. But for interfiber failure it quickly turned out that more sophisticated failure criteria are required to describe the fracture behavior under different loading conditions. First attempts have been made by Hashin [28], who adapted the fracture criterion of Mohr [29] formulated for failure of brittle materials. The basic rule in Mohr's concept states that fracture is caused by the stress in the fracture plane. While this concept is mathematically straightforward to formulate in isotropic materials, it is much more difficult to formulate for fiber-reinforced materials. Here, the detection of the correct fracture plane can only be achieved by iterative algorithms in the general case. Therefore, formulations are usually given for the simplified case as introduced by the presence of transversely isotropic symmetry in unidirectional fiber-reinforced materials. This yields the formulation of the failure criterion as quadratic–additive interaction criteria based on four of the five stress invariants given by the transversal isotropy. An alternative formulation for the interfiber failure criterion is given by Cuntze [30], who uses five stress invariants. Based on this formulation, as function of external loading, three independent interfiber failure modes for composites can occur as shown in Fig. 2.20. However, the transition range between the different failure modes is challenging, so probabilistic methods are applied in the formulation by Cuntze.

As alternative approach, Mohr's fracture criterion uses a solid physical basis, so Puck started to continue Hashin's work to determine the correct fracture plane throughout a series of publications [25, 31–36]. He established a parabolic interaction criterion for interfiber failure based on Mohr's fracture criterion and three

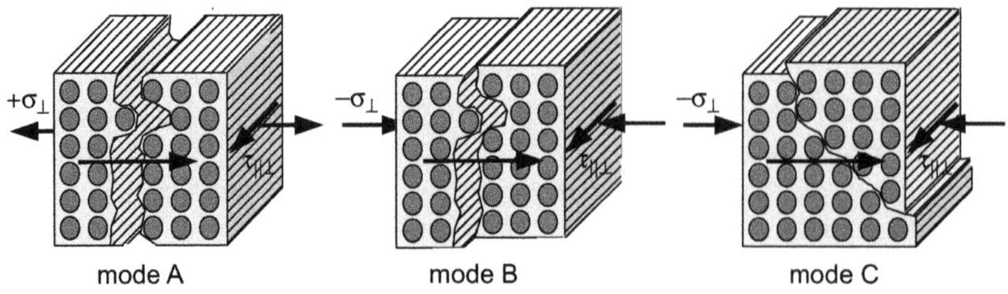

Fig. 2.20 Fracture modes in plane stress conditions according to Puck

different interfiber failure modes for plane stress conditions as shown in Fig. 2.20 [25].

Fracture mode A occurs due to a tensile stress component $\sigma_\perp$ superimposed by a shear stress component $\tau_{\parallel\perp}$. The resulting crack growth in a unidirectional layer is of a mixed crack opening mode (mode I) and shear mode (mode II) caused by the superimposed stress conditions. Fracture mode B is caused by a compressive stress component $-\sigma_\perp$ superimposed by a shear stress component $\tau_{\parallel\perp}$. Here, no crack opening mode occurs, so crack growth is mostly due to the shear mode (mode II). Finally, fracture mode C is caused by a compressive stress component $-\sigma_\perp$ superimposed by a shear stress component $\tau_{\parallel\perp}$. In contrast to mode B, $-\sigma_\perp$ is the dominating stress component. As consequence crack growth occurs angled to the load axis of the compressive stress component as seen in Fig. 2.20. The fracture angle is a result of the additional shear stress component acting in the inclined fracture plane.

However, in addition to the four basic material strengths, the definition of the slopes linking the tensile and compressive stress states requires the measurement of the so-called inclination parameters.

One noteworthy effort to compare the predictive capabilities of different failure criteria was carried out by Hinton, Soden, and Kaddour throughout a series of exercises. These became known as "World-Wide Failure Exercise I" [37–40] and "World-Wide Failure Exercise II" [41–43] due to the multitude of participants from countries all over the world. The procedure chosen by the organizers comprises two stages. During part A of the exercise, the participants conduct blind predictions given a set of material properties and loading conditions. At the beginning of part B, the experimental data is revealed to the participants and they are allowed to adjust their predictions accordingly. Afterwards, comparison is made between the prediction results of the different theories applied by the participants.

Within the "World-Wide failure exercise I" the focus was on the response of continuous fiber laminates subject to in-plane biaxial loading situations, in the absence of stress concentrations. As a short summary of the World-Wide failure exercise I, the failure criteria using differentiating criteria performed reasonably well as compared to the global failure criteria. In particular, the failure criteria given by Puck [35, 44], Zinoviev [45, 46], and Cuntze [47] were found to yield highest predictive capabilities [48].

One of the major gaps of the theories present at that time was the lack of predictive capabilities under 3D (i.e., triaxial) states of stress. Such stress states are typically found in thick composite components after impact or as a result of stress concentrators. Therefore, in the "World-Wide Failure exercise II" 3D states of stress including triaxial failure and associated theories were also considered.

As the outcome of this second exercise was more diverse than for the first exercise, an overall recommendation for a particular theory seems premature. According to [41], the overall impression of the 3D failure theories provided by Pinho [49, 50], Carrere [51, 52], Puck [53, 54], and Cuntze [30] were ranked best. However, many discussions have accompanied the "World-Wide Failure Exercise II" regarding the reliability of the experimental data provided. In particular, for those test cases dealing with 3D stress states many involved researchers commented on the experimental issues associated with the measurements [30, 41, 49, 54–57]. Beyond those failure theories compared in the "World-Wide Failure exercise" other groups also presented derivatives of existing theories or proposed essentially new theories [58, 59]. Such new approaches are particularly relevant for the case of textile composites. In these materials, the undulations of the fiber yarns and the pronounced mesostructure cause further challenges to prediction of damage progression and ultimate failure.

2.2.2 *Quasi-Static Failure Including Growth of Damage, Damage Mechanics, and Degradation*

As mentioned in Sect. 2.1, there are various microscopic failure mechanisms, which finally cause ultimate failure of a macroscopic structure. The consideration of their contribution to degradation of macroscopic properties and to the initiation of subsequent damage is a challenge, which goes beyond the prediction of global stress or strain limits. Several failure theories presented in Sect. 2.1.1 already take into account the degradation of laminate properties due to occurrence of damage inside the plies.

However, the description of the macroscopic failure in terms of accumulation of microscopic damage should theoretically be possible. More specifically, the microscopic scale is the correct length scale to account for the microstructure introduced in fiber-reinforced composites [5, 60–64]. However, the derivation of accurate microscopic mechanical properties is experimentally challenging. In particular, the measurement of strength values for small microscopic volumes is very difficult and predictions based on microscopic measurements extrapolating to the macroscopic scale are also influenced by the statistical distribution of flaws [65]. Moreover, using a micromechanical approach in practice typically comes with high computational intensity. Therefore, analytical methods using homogenized properties, defect-free unidirectional plies, orthotropic symmetry, and macroscopic stress–strain relations are typically preferred. In this context, the concept known as classical laminate theory is usually applied.

This concept uses a layer-wise failure of plies and applies respective degradation models to account for damage accumulated in a layer. For this approach, the question arises, if macroscopically measured strength values of unidirectional layers are applicable to the plies inside a laminate. As reported by various authors [66–70], crack propagation inside such internal layers depends on the thickness of the ply as well as on the stiffness of the neighboring layers. For sufficiently thin layers, strain coupling to the adjacent layers can inhibit the opening of existing cracks. According to Parvizi [68] this effect is most pronounced for thin layers, but vanishes for ply thicknesses above 0.4 mm.

To describe the propagation of existing cracks inside a solid usually concepts of fracture mechanics are applied. These approaches could in principle also be used to describe the damage progression in fiber-reinforced composites. The formulations of the stress-intensity concept, the critical strain energy release or the J-Integral are well-established concepts for pure materials [71–75]. Since flaws are also present in fiber-reinforced materials, a direct transfer of the fracture mechanics concepts seems straightforward. However, the computational intensity and the lack of consistent micromechanical parameters have prevented the application of micromechanical fracture mechanics for a long time. In recent years, some approaches used fracture mechanics concept as hybrid approaches [50, 58] or in their pure form [76] to describe damage progression within laminates. Similar to the strength based failure criteria, there is still room for improvement of the experimental methods to determine microscopic fracture toughness values for various load conditions [15, 77, 78].

As a consequence of the results of the "World-Wide Failure exercise II" more focus is then given to the prediction of damage. Within the "World-Wide Failure exercise III," the prediction of matrix crack initiation, delamination initiation due to transverse cracks, deformation, and final fracture is investigated. While this exercise is still going on, first results of the blind predictions indicate a clear diversity of predictions regarding the effects of ply thickness, the stacking sequence and interaction of existing cracks with neighboring layers, in particular to induce delamination [79]. However, a larger tendency is seen to use finite element modeling [80–86] and micromechanical approaches [80, 81, 85, 87–89] to obtain failure predictions. A specific challenge for the existing theories is the forward prediction of crack density and crack saturation of interfiber cracks. In fact, many theories require the crack density of some cases as input parameter, rather than allowing a physically based forward prediction. Therefore, experimental methods are required which allow suitable tracking of this crack evolution as function of loading for the reference cases, but as well to benchmark failure theory predictions for arbitrary laminate configurations and load conditions.

Similarly, many efforts have been commenced in the last years in order to improve the prediction of damage initiation and damage progression for textile composites [90–95]. For these materials, even the prediction of first failure onset is comparatively complex because of the explicit microstructure and mesostructure of textile composites. The textile architecture of the fiber reinforcement (e.g., woven fabrics, noncrimp fabrics, and braids) cause a complex multiaxial residual stress state after curing which needs to be accounted for. Since this involves calculations on various length scales, classical homogenization approaches can effectively be combined with finite element based routines to account for such multiscale problems [90, 93–96].

In the following, a brief review is made for three particular failure modes typically addressed in current failure theories for laminates composed of unidirectional plies. The failure criteria given by Puck, Schürmann, and Deuschle [5, 53, 54, 97] is taken as an example[1] to discuss current trends in the development of failure theories and to discuss the relation to recent experimental findings on damage evolution in fiber-reinforced composites.

For the consideration of fiber failure due to tensile or compressive loading σ_1, usually simple maximum stress criteria are applied. In recent years, 3D failure criteria evolved, which partially consider the influence of the stress components transverse to the load axis. The impact of the off-axis components σ_2 and σ_3 on fiber failure has already been noted by Edge [98] and Hart-Smith [99], but was neglected in the original work by Puck [25]. One example of an improved theoretical formulation was proposed by Puck, Schürmann, and Deuschle [5, 53, 54, 97]. The simple maximum stress criterion using the strength of the unidirectional layer under tensile load $R_{\parallel}^{+}$ and under compressive load $R_{\parallel}^{-}$ is given as:

$$\frac{|\sigma_1|}{R_{\parallel}^{+}} = 1 \quad \text{for} \quad \sigma_1 \geq 0 \tag{2.1a}$$

$$\frac{|\sigma_1|}{R_{\parallel}^{-}} = 1 \quad \text{for} \quad \sigma_1 < 0 \tag{2.1b}$$

The stress exposure in a single fiber filament is given by the relation applying the empirical magnification factor m (e.g., 1.3 for GFRP and 1.1 for CFRP) to respect the influence of the off-axis components σ_2 and σ_3.

$$\sigma_{f\parallel} = E_{f\parallel} \cdot \varepsilon_1 + \nu_{f\perp\parallel} \cdot m \cdot (\sigma_2 + \sigma_3) \tag{2.2}$$

Due to strain coupling, the laminate strain equals the filament strain, therefore:

$$\varepsilon_1 = \frac{\sigma_1}{E_{\parallel}} - \frac{\nu_{\perp\parallel}}{E_{\parallel}}\sigma_2 - \frac{\nu_{\perp\parallel}}{E_{\parallel}}\sigma_3 \tag{2.3}$$

Combining (2.2) and (2.3), this yields the refined form of the fiber failure criterion:

[1] There is still much discussion regarding the general applicability of such failure criteria, so the reader should always be aware of the potential drawbacks of a particular theory.

$$\left|\frac{\sigma_1}{R_{\parallel}^{\pm}}\right| - \left(\nu_{\perp\parallel} - \nu_{f\perp\parallel}\frac{E_{\parallel}}{E_{f\parallel}} \cdot m\right) \cdot \left|\frac{\sigma_2 + \sigma_3}{R_{\parallel}^{\pm}}\right| = 1 \qquad (2.4)$$

However, in the meaning of strength-based failure criteria, fiber failure is usually understood as failure of fiber bundles or failure of full layers. In the context of first failure analysis by secondary methods, it is worth to have a closer look on the failure evolution causing macroscopic failure due to loading of a unidirectional material parallel to the fiber axis. It is well known that fiber filament strength is subject to a broad strength distribution. Therefore, even in a defect-free laminate it is likely that failure of some fiber filaments precedes failure of the full layer. As demonstrated in recent work by Scott et al. [100, 101] and Bunsell et al. [102, 103], there is indeed an evolution of fiber filament failure preceding the ultimate failure of a layer. According to their findings, failure of a single filament is likely to occur after a certain local stress exceeds the strength of a fiber filament, that is, the failure position is of statistical nature. The surrounding filaments are able to compensate the breakage of the individual, so one fiber filament failure is not of large relevance to the durability of the structure. The situation changes if other fiber filaments fail in close vicinity. This increases the stress exposure of the surrounding filaments.

As quantified by in situ synchrotron tomography, a critical health state of fiber-reinforced materials is reached if the number N of localized fiber failures reaches a critical value. Such a set of N localized fiber failures was entitled N-plets by Scott et al. [101]. A scheme of the evolution of single fiber filament failures starting at a 1-plet and growing to a 3-plet and 6-plet configuration is shown in a cross-sectional view in Fig. 2.21. However, in reality the cumulative N-plet count refers to a 3D local volume rather than just a local 2D cross section of the material. Remarkably enough N was found to typically stay < 20 before ultimate failure of the structure occurs.

As discussed in Sect. 2.2.1, the failure of fiber-reinforced composites due to interfiber fracture is much more challenging to capture in a failure theory. Based on the results of the World-Wide Failure exercise I and II Puck's and Cuntze's failure,

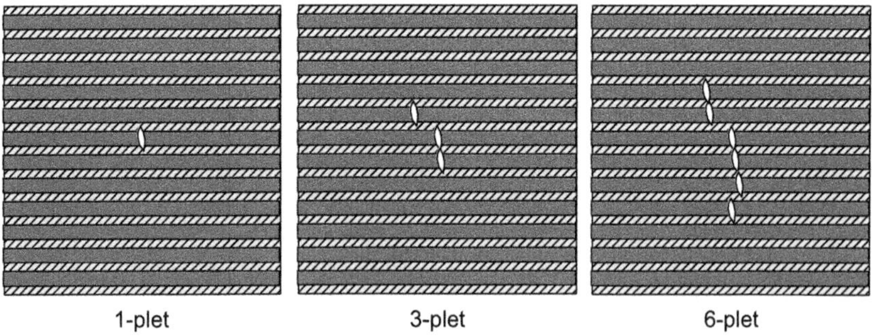

Fig. 2.21 Example of local accumulation of fiber failures starting at 1-plet (**a**) and increasing to 3-plet (**b**), and 6-plet (**c**) configuration

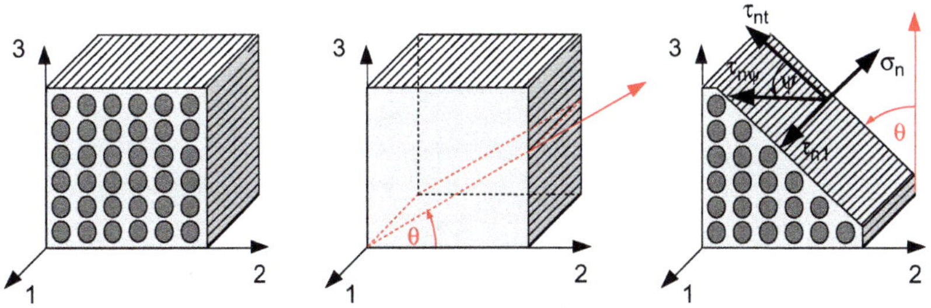

Fig. 2.22 Definition of fracture plane at angle θ and definition of stress components acting at fracture plane

criteria were ranked among the best available [41, 48, 104]. In the following, the 3D failure criterion as given in [5, 53, 54, 97] is used as an example.

As shown in Fig. 2.22 and discussed in Sect. 2.2.1, every cross-sectional plane perpendicular to the fiber axis is a potential fracture plane. Therefore, only three stress components (and their superposition) are causing interfiber failure. These are the components σ_n, τ_{nt} and τ_{n1} with reference to the normal (n), tangential (t), and fiber parallel (1) axis of the fracture plane as shown in Fig. 2.22.

For such an arbitrary fracture plane orthogonal to σ_n, the superposition of the shear stress components τ_{nt} and τ_{n1} can be replaced by an effective shear stress component $\tau_{n\psi}$ using:

$$\frac{\tau_{n1}}{\tau_{nt}} = \tan\psi \tag{2.5}$$

$$\tau_{n\psi} = \sqrt{\tau_{n1}^2 + \tau_{nt}^2} \tag{2.6}$$

Therefore, interfiber failure is governed by the stress σ_n and $\tau_{n\psi}$ acting on the fracture plane. The failure criterion can thus be formulated using the respective unidirectional material strength values $R_\perp^+$, $R_{\perp\perp}^A$, and $R_{\perp\parallel}$ and the inclination parameters $p_{\perp\psi}^+$ and $p_{\perp\psi}^-$ (cf. Fig. 2.23) defined as:

$$\left.\frac{d\tau_{n\psi}}{d\sigma_n}\right|_{\sigma_n=0} = -p_{\perp\psi}^+ \tag{2.7a}$$

$$\left.\frac{d\tau_{n\psi}}{d\sigma_n}\right|_{\sigma_n=0} = -p_{\perp\psi}^- \tag{2.7b}$$

In the tensile load regime ($\sigma_n > 0$), one can define:

$$\left(\frac{\tau_{n\psi}}{R_{\perp\psi}^A}\right)^2 + 2\frac{p_{\perp\psi}^+\sigma_n}{R_{\perp\psi}^A} + \left(1 - 2\frac{p_{\perp\psi}^+ R_\perp^+}{R_{\perp\psi}^A}\right)\cdot\frac{\sigma_n^2}{\left(R_\perp^+\right)^2} = 1 \tag{2.8}$$

For compressive load ($\sigma_n < 0$)

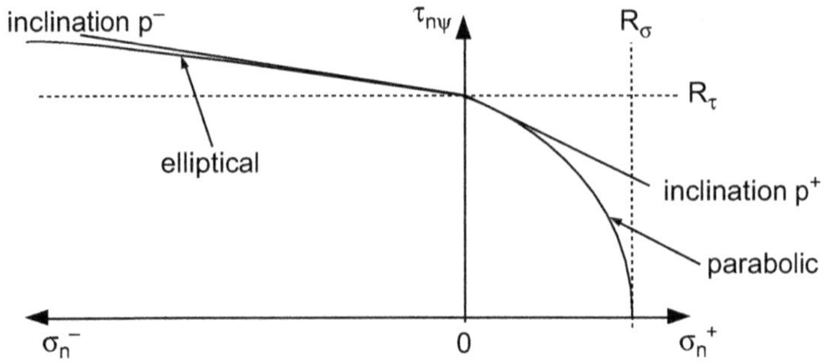

Fig. 2.23 Definition of the inclination parameters

$$\left(\frac{\tau_{n\psi}}{R^A_{\perp\psi}}\right)^2 + 2\frac{p^-_{\perp\psi}\sigma_n}{R^A_{\perp\psi}} = 1 \tag{2.9}$$

And for pure shear ($\sigma_n = 0$)

$$\left(\frac{\tau_{n\psi}}{R^A_{\perp\psi}}\right)^2 = \left(\frac{\tau_{nt}}{R^A_{\perp\perp}}\right)^2 + \left(\frac{\tau_{n1}}{R^A_{\perp\parallel}}\right)^2 \tag{2.10}$$

The combination of (2.8), (2.9), and (2.10) then finally yields Puck's formulation of the interfiber failure criterion for an arbitrary fracture plane.

For $\sigma_n \geq 0$:

$$\sqrt{\left[\left(\frac{1}{R^+_\perp} - \frac{p^+_{\perp\psi}}{R^A_{\perp\psi}}\right)\cdot\sigma_n\right]^2 + \left(\frac{\tau_{nt}}{R^A_{\perp\perp}}\right)^2 + \left(\frac{\tau_{n1}}{R_{\perp\parallel}}\right)^2} + \frac{p^+_{\perp\psi}}{R^A_{\perp\psi}}\cdot\sigma_n = 1 \tag{2.11a}$$

For $\sigma_n < 0$:

$$\sqrt{\left(\frac{\tau_{nt}}{R^A_{\perp\perp}}\right)^2 + \left(\frac{\tau_{n1}}{R_{\perp\parallel}}\right)^2 + \left(\frac{p^-_{\perp\psi}}{R^A_{\perp\psi}}\cdot\sigma_n\right)^2} + \frac{p^-_{\perp\psi}}{R^A_{\perp\psi}}\cdot\sigma_n = 1 \tag{2.11b}$$

Up to this point only failure initiation is captured by the failure criterion. In order to predict damage progression the theory has to take into account the growth of such cracks, their impact on triggering of new damage and their impact on macroscopic stress and strain states. In the sense of Puck's failure theory, fiber failure comprises the ultimate limit of the laminate. Therefore, description of damage progression spans the range between the first occurrence of interfiber failure and final occurrence of fiber failure.

There are two ways to represent interfiber cracks in failure theories. Either those are represented as explicit description in a micromechanical or multiscale approach

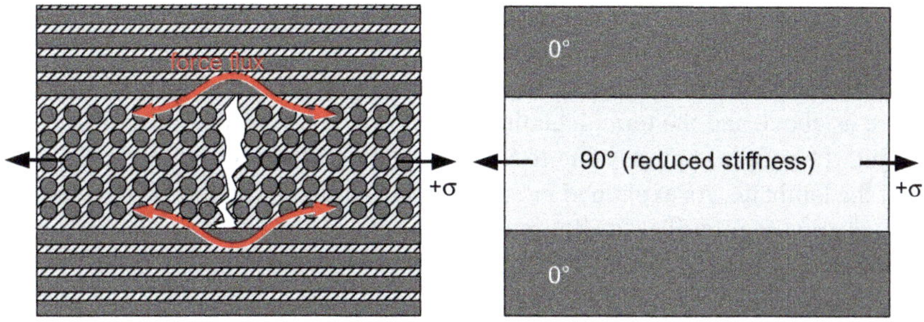

Fig. 2.24 Representation of interfiber cracks in discrete and smeared form

or the cracks are taken into account as smeared representation in a continuum damage mechanics approach. A comparison of both concepts is shown in Fig. 2.24.

The approach taken by Deuschle and Puck [53, 54, 97] is the smeared representation allowing a layer-wise degradation of properties. At the occurrence of interfiber failure spanning the full thickness of internal layers, this usually does not disintegrate the whole laminate. Instead a sequential increase of crack density occurs. Accordingly, the laminate properties (i.e., stiffness) degrade. For the example shown in Fig. 2.24, the occurrence of interfiber cracks increases the amount of load to be carried by the neighboring layers. Therefore, such a damaged laminate will deform more under the same load as the same laminate configuration without any damage. Hence, the degradation of the laminate properties due to such interfiber cracks is represented by a reduction of stiffness using the degradation parameter $\tilde{\eta}$ in the interval [1; 0]

$$\text{reduced stiffness} = \tilde{\eta} \cdot \text{original stiffness} \tag{2.12}$$

For degradation the parameter $\tilde{\eta}$ goes down from 1 (no damage) to 0 if saturated crack density is reached. Particular rules for more complex degradation rules for the 3D case are discussed by Deuschle [97]. Such smeared crack representations facilitate fast computation of the degradation process, but obviously introduce assumptions regarding their impact on the local interaction with neighboring layers.

Recently, Trappe et al. applied X-ray refraction techniques to measure the occurrence of interfiber cracks in glass-fiber-reinforced composites [105–108]. According to their findings, the inhomogeneous microstructure of fiber-reinforced composites triggers a statistical distribution of the occurrence of interfiber cracks. Therefore, occurrence of such cracks cannot be predicted by layer-wise failure analysis. As demonstrated by Scott et al., additional aspects come into play if the existence of voids in the laminate is considered [109]. Using high-resolution synchrotron tomography, the interfiber cracks were shown to interact to a substantial degree with such defects, since they tend to intersect with voids and grow directed along such weak spots. Those aspects are hardly reflected by layer-wise degradation analysis, but should be considered if the void content is of significant contribution.

It is noted that there is some ambiguity between the usage of the term interfiber fracture as above and the term delamination. However, delamination is different to other interfiber failure modes due to its orientation relative to the thickness direction of the laminate. As explained in Sect. 2.1, the laminate can fail by separation of two layers due to interfiber crack growth in parallel to the in-plane direction. Such delamination is mostly caused by through-thickness tension components or shear stress components. Usually, the initiation and growth of delamination is a problem well formulated by fracture mechanics approaches using mode I, mode II, mode III, or superimposed fracture modes. In experiments, it is frequently observed that delamination often initiates at the interface between adjacent plies at the position of the crack tip of an interfiber crack. This is due to the stress concentration at the crack tip, which has been accounted for in various theories (e.g., [69]).

2.2.3 Long-Term Behavior

Much progress has been made to develop failure criteria applicable for quasi-static loading conditions. However, for investigation of the long-term behavior of composite structures fewer investigations have been conducted so far. In the following, a brief review is made comprising the particular challenges of failure prediction in the fields of creep, stress rupture, and fatigue, and the associated experimental approaches to assist in the understanding of failure for these load cases. Following the classification used by [110], strain rates below $10^{-6}\ \mathrm{s}^{-1}$ are considered to be creep loading and strain rates at or below $10^{-3}\ \mathrm{s}^{-1}$ are considered to represent quasi-static deformations. As an extension beyond the quasi-static range, strain rates above $10^{2}\ \mathrm{s}^{-1}$ are classified as high strain rates, strain rates above $10^{4}\ \mathrm{s}^{-1}$ are called very high strain rates, and strain rates above $10^{6}\ \mathrm{s}^{-1}$ are referred to as ultra-high strain rates.

2.2.3.1 Creep and Stress Rupture

For fiber-reinforced polymer materials, creep loading is naturally associated with severe drawbacks based on the general experience of pure polymers. However, creep and relaxation effects are not always considered to be of negative impact for the composite. As discussed by Schürmann [5], the stress relaxation within the matrix material can cause significant reduction of residual thermal stress and a redistribution of stress into the fiber reinforcements. Still, the structure is subject to permanent deformation, which might not be acceptable in some applications. For pure polymers, the time–temperature superposition principle is usually applied to allow extrapolation of long-term stability based on short-term measurements. Thermosets are the dominating type of polymer used as matrix material, but there is a growing interest in thermoplastic matrix materials. Although thermoplastic matrix materials are not essentially new in applications and despite of the higher creep rates faced in thermoplastic polymers, less investigations have been conducted. A comparison between creep rates of a fiber-reinforced epoxy resin and fiber-reinforced Polyetheretherketone has been performed by [111]. As further demonstrated for glass-fiber-reinforced Polyamide [112], carbon-fiber-reinforced polyimide [113], and glass-fiber-reinforced polycarbonate [114], the application of the Arrhenius type time–temperature superposition principle is applicable in these cases. In particular, if influences of physical and thermal ageing are taken into account explicitly in analytical equations, stable predictions of creep strength as function of crystallinity and fiber volume fraction seem feasible [112].

For the prediction of stress rupture, there is no fundamental difference to the considerations presented in Sects. 2.2.1 and 2.2.2. In principle, the same strength-based formulations or fracture mechanics approaches can be adopted. However, one has to explicitly consider the viscoelastic nature of the polymer matrix material in the case of creep loading and the evolution of damage in the static load case. Although this is straightforward in theory, the according mathematical formulation is not trivial in most cases. Only few attempts have been made to explicitly account for viscoelastic material behavior in present failure theories. Guedes reviewed different formulations using continuum damage mechanics and cumulative damage formulations [115, 116]. For the case of stress-rupture behavior, comparison was made between maximum work stress criterion (Reiner–Weissenberg criterion [117] and modifications thereof), maximum strain criterion, and linear cumulative damage formulation of Miner [118] as well as Christensen's cumulative damage formalism [119]. For the purpose of lifetime prediction under constant stress, it was demonstrated that the different formulations discussed produce relatively similar predictions [115]. However, experimental evidence indicates the limited predictability of stress rupture in real composite structures such as pressure vessels [120]. Failure of such safety relevant structures are very critical and the predictive capability of existent theories not yet sufficient to decrease safety margins. It is well known that glass fibers, carbon fibers, and Kevlar fibers all suffer from long-term static loads. Even for the best performant carbon fibers, the maximum allowable pressure needs to be reduced to less than 60 % of the static burst pressure when designed for more than 100,000 h operation and an intended probability of stress-rupture failure less than 0.001 [121]. Therefore, accelerated testing using the time–temperature superposition discussed by [122–124] needs further validation to assess its relevance for stress-rupture predictions.

From the viewpoint of secondary methods in conjunction with creep and stress-rupture experiments, the biggest improvement might be further shortening of the test duration. At present, the overall duration of a test campaign to obtain creep properties for only one laminate configuration can still take more than 10,000 h to extrapolate the stability for two decades [115]. Therefore, in situ methods being able to detect the onset of secondary or tertiary stages of creep as well as the ability to differentiate between different mechanisms during stress-rupture testing are much appreciated.

2.2.3.2 Fatigue

Another load scenario fiber-reinforced materials are facing in application is fatigue loading. From the testing point of view, the term fatigue is usually associated with the test procedure first proposed by Wöhler [125]. Here, a specimen is cyclically loaded with constant amplitude levels at a constant frequency. In fatigue testing, the load configurations are usually determined by the R-ratio, which is the minimum load level divided by the maximum load level of the specimen. One can distinguish between:

- Tensile–tensile fatigue (upper and lower amplitude ≥ 0, R-Ratio ≥ 0).
- Tensile–compression fatigue (upper amplitude > 0, lower amplitude < 0, R-Ratio ≤ 0).
- Compression–compression fatigue (upper and lower amplitude ≤ 0, R-Ratio ≥ 0).

Other specific load cases are known as unidirectional loading (upper amplitude > 0, lower amplitude $= 0$, R-Ratio $= 0$) and fully reversed loading (magnitude of upper amplitude equal to magnitude of lower amplitude, R-Ratio $= -1$).

In all cases, the testing is continued until a significant stiffness reduction or ultimate failure occurs. The test campaign is carried out for different amplitude levels (S) between the static amplitude limit and sufficiently small amplitude values. The number of cycles to failure (N) is plotted versus the amplitude level. This S–N curve is used to establish the long-term limit of materials, that is, to define the amplitude value below no failure of the material will occur for an infinite number of cycles.

In isotropic materials, the fatigue life is often estimated from the original version of the linear cumulative damage law (LCD), referred to as Miner's rule [118] or improvements thereof [126, 127]. But even for isotropic materials, it has been questioned, if simple LCD approaches will yield realistic lifetime predictions in fatigue loading [126].

Some fiber-reinforced composites, such as CFRP, are associated with superior lifetime in fatigue. However, this is not essentially true as far as first damage initiation and damage progression is concerned [128]. One of the key problems with composites is the sequence of amplitudes faced in the loading–unloading cycles [115, 129–131]. Since the real "load spectrum" differs from application to application, it is hard to standardize. But for fiber-reinforced composites, the load spectrum has significant impact on the number of load cycles to failure. In a fiber-reinforced material, damage can initiate at comparatively low load levels and continuously grow until final failure is reached. The type of damage growth may change as function of the applied load and may also interact with other damage existent in the material (e.g., small matrix cracks initiating large growth of delamination). Therefore, preliminary initiation of damage at higher load levels can extend further in subsequent cycles with low load levels. This can cause growth of damage at load levels, which usually would not result in damage initiation or growth by themselves as been evident from a series of bucket truck failures [132].

Therefore, different failure theories have been proposed to explicitly account for the load spectrum effect. A good review on the major developments is given by Degrieck [128] and a more recent comparison of selected failure theory predictions is presented by Guedes [115]. A basic distinction in terms of failure criteria can be made according to Sendeckyj between macroscopic strength criteria, criteria based on residual strength, those based on residual stiffness, and failure criteria taking into account an explicit formulation of an actual failure mechanism [133].

One class of models comprises the so-called fatigue life models, which are essentially based on the S–N curve behavior as discussed above. These models usually neglect the interaction of different failure mechanisms such as interfiber failure or fiber breakage. Examples for these failure models are given in [122, 134–141]. As a result, these types of models yield the fatigue life of the composite studied based on the extrapolation of the measured S–N curves. The second class of models take into account the degradation of composite properties based on phenomenological descriptions. Some of these degradation models are formulated for the residual stiffness [129, 142–148], others are using the residual strength based on deterministic descriptions or based on statistical strength distributions [149–156]. The last type of models uses an explicit formulation of the damage progression correlating one or more damage variables to the damage extent. This quantitatively accounts for the damage progression in a physically sound way. Distinction can be made between models able to predict the damage growth [157–162] and models relating the damage growth to property degradation [163–172].

A common challenge of the models provided until today is the extrapolation of data acquired in model specimens to real structures. Usually, input data for failure theories is collected using a specific stacking sequence and uniaxial loading in one test configuration. The corresponding theories may be able to describe these situations quite accurately, but for the general situation of multiaxial loading conditions in stacking sequences changing as function of position in a real structure, their predictive capabilities are still questionable.

For the experimental approaches, comprehensive reviews in application to composites have been provided in [133, 173]. One of the key challenges yet to overcome is the ability of test methods to accompany long-term testing with nondestructive assessment of failure evolution. At least for the models using explicit formulations of damage growth such input parameters are essential to further advance the reliability of the model and to improve their predictive capabilities for generalized load situations. Another factor is similar to the main challenge in creep experiments, which is the shortening of the test duration. Fatigue testing with low amplitudes can take several weeks. In order to extrapolate the lifetime such testing is still required and since composites lack from the applicability of the Miner's rule, there is currently no suitable approach to circumvent realistic load spectra. Alternative termination criteria other than residual stiffness values or ultimate failure are highly appreciated. In particular, for testing of structures accompanying in situ measurements are suitable tools to improve the cost effectiveness of the test campaigns.

2.2.4 High Velocity

The increasing application of fiber-reinforced materials in crash-relevant structures requires corresponding experimental methods to characterize these materials under high strain rates. The failure modes found for high strain rates cannot be expected to be the same as for the strain rates used in quasi-static conditions. During fast damage accumulation, no relaxation effects and, hence, stress redistribution can occur. Moreover, elastic properties and failure behavior of the matrix material can change significantly due to transition from ductile to brittle failure modes.

The failure modes considered in high strain rate experiments are typically of ultimate nature, that is, the experiments result in immediate failure of the laminate rather than partial or sequential damage accumulation. Therefore, description of material failure does not consider evolution of damage to the extent as the criteria used for creep, fatigue, or quasi-static load situations. As a function of strain rates one can distinguish between experimental configurations considered being in the creep domain $< 10^{-6}\ s^{-1}$, the quasi-static deformation range $< 10^{-3}\ s^{-1}$, and those reached by servohydraulic test rigs $< 10^{0}\ s^{-1}$. Above this limit specialized equipment, such as Kolsky bars (also known as Split–Hopkinson bars) are used to reach strain rates $> 10^{6}\ s^{-1}$ [174]. For even higher strain rates, impactors driven by gravity or projectiles shot by gas cannons are used to test materials subject to high-velocity impact. A good review on the recent developments of the experimental techniques in high-velocity testing is given by Ramesh [110] and a critical review of the Kolsky bars is given by Gama [175].

From the view point of failure theories, recently Kawai compared the predictive capabilities of classical failure theories of Tsai–Wu, Tsai–Hill, and Hashin–Rotem with respect to the off-axis strength values of unidirectional laminates under high strain rate tension and compression [176]. A detailed discussion of the failure modes and fracture angles was given and some key modifications to the failure criteria were proposed.

For fiber-reinforced composites various experimental efforts have been made to characterize material properties under tensile, compressive, and shear loading [176–183]. Until recent times, the focus of the experimental work was to develop suitable test rigs, assuring the anticipated load situation at the constant strain rates. Secondary methods evolved with the usage of high-speed cameras to directly observe the fracture of fiber-reinforced materials [184]. Improvements have been made by [181–183, 185] introducing digital image correlation systems to measure the full strain field during the test. This validated fundamental assumptions regarding the principle of operation of Kolsky bars. Beyond the point of image acquisition and processing, secondary methods are not used extensively. Beyond a certain strain rate, acoustic methods are generally problematic, since the load is introduced by an acoustic wave itself. Secondary methods like acoustic emission, guided wave testing, vibrometry, or ultrasonic tomography are hence not of big relevance for the experiments used in the field of materials testing at strain rates up to $10^{6}\ s^{-1}$. However, applications exist, where impacts themselves are monitored by acoustic emission sensors to localize their position as well as to deduce the impact energy [186–189].

2.3 Challenges in Mechanical Testing of Fiber-Reinforced Materials

In general, mechanical testing of fiber-reinforced materials is used to obtain material properties that are used for quality control or for designing composite structures. While quality control often solely requires comparative measurements carried out under reproducible conditions, the design and engineering requires the measurement of true material properties. Therefore, material testing has to assure that the obtained data is in accordance with the boundary conditions and assumptions of the respective theory. For the linear elastic range of the material, the major requirement is the generation of homogeneous stress–strain fields within the test specimen. Moreover, suitable test specimen geometries with defined superimposed loading conditions are still challenging. For the measurement of the material properties after first onset of damage inside the material several challenges arise. The first challenge is the detection of first damage itself. Subsequent to this load level, the second challenge has its focus on the usage of appropriate specimen geometries to assure the intended type of failure. Many times, the specimens fail in an unexpected fashion due to failure modes different from the intended mode.

Almost all of these challenges stem from the hierarchical structure of fiber-reinforced materials. Compared to "pure" engineering materials as metals, polymers, or ceramics, fiber-reinforced materials intrinsically have at least two additional characteristic length scales relevant for failure progression. The first is the length scale of their fiber reinforcements and the second is the length scale of the plies composed into a laminate. For textile composites, an additional length scale defined by the characteristic dimensions of the elementary cell of the textile reinforcement also comes into play.

Therefore, during testing one is always subject to two fundamental questions:

- Which is the appropriate scale to obtain material parameters?
- What are valid "material" parameters?

In contrast to many classical engineering materials, which are mostly homogenous in their properties on several length scales, the fiber reinforcements in composites induce heterogeneity, inhomogeneity, or both. The relevance of the fiber distribution is linked to the observation scale. While on the macroscopic scale, one can hardly resolve the individual fiber filaments, their distribution and orientation is quite obvious on the length scale of micrometers. Therefore, the measurable properties can be completely different for measurements on the microscopic scale. As consequence, one has to provide different material properties for failure theories driven by microscopic volume elements or failure theories based on a macroscopic homogenized continuum representation.

Also, the fiber reinforcements induce a certain degree of anisotropy in the material properties. Compared to isotropic materials this inevitably causes a larger amount of required testing methods. Therefore, specific test geometries are required, which allow for testing under load conditions typically not applied for classical engineering materials.

While many of these aspects are sufficiently taken into account for some test procedures, for other cases there still is an active discussion in the community regarding the reliability and comparability of different test procedures. As discussed by Bleier [190], Cuntze [30], Pinho [49, 84], Deuschle [97], and Basan [191], there is a strong need to provide true material properties as input for modern failure criteria of fiber-reinforced composites. Juhasz points out that failure prediction generally should be possible without prior knowledge of the exact load configuration and according experimental test results of a specific structure [59].

Therefore, in order to verify different failure criteria, there is a strong need for testing methods assuring the validity of superimposed loading conditions and for methods allowing tracking the evolution of failure until the point of rupture.

In the following, three typical challenges for presently used test standards are presented. A particular focus is given to the applicability of secondary methods to improve the understanding of the test results and their relevance to the work presented herein.

2.3.1 Detection of First Failure Onsets

As discussed in Sect. 2.2.1 various failure theories aim to predict the first occurrence of a particular failure mode within fiber-reinforced materials. But for many laminate configurations, there is no possibility to quantify the onset of interfiber failure in the individual plies from load–displacement curves. Moreover, the detection of the onset of fiber failure is also not feasible using stress–strain curve information unless this is considered to be identical to ultimate failure. Apart from the failure theories, there are several other standardized methods, which rely on the information obtained from the first onset of damage.

One test setup, which requires the determination of the initiation of interlaminar delamination, is the short-beam-shear test according to the standards ISO 14130, ASTM D 2344, DIN EN 2377, and DIN EN 2563. But sometimes the occurrence of delamination onset is not identical to the maximum load or the first significant load drop as demonstrated in Chap. 3. However, the latter is a requirement to obtain the apparent interlaminar shear strength. Despite of the usage of this method mostly in quality control, this already indicates the ambiguous level of confidence associated with the proposed data reduction schemes.

Another experimental setup, which monitors crack propagation, is the configuration to deduce mode I interlaminar fracture toughness according to ISO 15024, ASTM D 5528, prEN 6033, or AITM 1.0005. Besides the valuable data reduction routines to deduce the fracture toughness included with the ASTM D 5528 standard several methods are noted to quantify the level of the initial fracture toughness. This is owed to the fact that none of the onset criteria included within the standard is applicable for all types of materials. Therefore, secondary methods can be valuable tools to assist in the interpretation of such methods and to obtain true material properties.

Similarly the standards ASTM D7905, prEN 6034, and AITM 1.0006 used to obtain mode II fracture toughness by the End-Notched-Flexure (ENF) configuration attempt to derive the onset of interlaminar crack growth using signatures of force-displacement curves. Here, the type of crack insert, the measurement of the crack length, and the load fixture itself can have significant impact on the fracture toughness value [192]. But the critical fracture toughness value is also affected by the load value picked for the first onset of delamination. Here, again the load maximum is not necessarily the correct value to choose for the onset as demonstrated in Chap. 4. Another method aiming to deduce valid mode II fracture toughness values is the Transverse Crack Tension (TCT) specimen. Although the TCT tests are currently not covered by any standard, the key challenge associated with this configuration stems again from the validity of the detected onset of delamination.

2.3.2 *Tracking Failure Evolution*

From a general point of view, there are two good reasons to detect the failure evolution in a test specimen made from fiber-reinforced material. The first reason is to compare the failure evolution predicted by a failure theory to an actual measurement. The second reason is the interpretation of data from stress–strain curves after onset of initial failure.

The comparison of predicted and measured failure evolution is most interesting to validate attempts of degradation analysis or to detect subsequent onsets of particular failure mechanisms in angle-ply laminates. This type of comparison makes even more sense for a more complex structure than a test specimen. Here, simple laminate degradation analysis faces more challenges. However, this requires validated experimental methods capable of recording the evolution of failure. A comparison of the individual capabilities and specific combinations of such in situ methods are made in Chap. 7.

In order to deduce meaningful data from stress–strain curves, one has to assure that the underlying assumptions are fulfilled. One prominent example is tensile testing of unidirectional specimens with load axis parallel to the fiber axis. In such configurations, first failure in form of interfiber fracture occurs already at 70 % of the ultimate load [25]. In addition, fiber alignment is far from perfect and therefore fiber failure initiates at the edge of the specimen due to off-axis loading or stress concentration effects. Hence, the measured ultimate strength used as first-ply failure strength in this load configuration is a value obtained in a material state, which is typically far away from the undamaged state. This is in contrast to the usual assumptions made in failure theories. But the impact of these preliminary failures on the achievable ultimate strength is certainly dependent on the individual material type and production process. In such cases, methods capable to track the evolution of damage are suitable to assist the interpretation of stress–strain curves.

2.3.3 *Ductile Matrix Materials*

Traditionally, fiber-reinforced polymers are fabricated from brittle matrix systems, such as epoxy resins. Consequently, many standards used for testing of fiber-reinforced polymers implicitly assume brittle failure behavior of the matrix system. Therefore, some of the assumptions made for the data reduction routines cannot be directly transferred to matrix systems with ductile failure behavior, for example, due to nonlinearities in the stress–strain behavior of these materials.

In recent years, an increasing interest in thermoplastic matrix materials has been evident. However, the testing of fiber-reinforced thermoplastic polymers is not a straightforward procedure using established standards. Nevertheless, some of the existing standards are still applicable if the data reduction routine is accompanied by additional information using in situ methods.

For all test methods, the validity of specimen geometries is subject to discussion. In many thermoplastic materials, the interlaminar properties are substantially higher than for typical epoxy resins. Therefore, many test specimens prepared to fail via interlaminar or transverse stresses may fail in unintended failure modes. Suitable adjustment of the specimen geometries or changes to the concept of load introduction can be used to avoid such failure. Assistance by secondary methods monitoring the occurrence of failure outside the intended position is of great practical usage for such reinvention of testing procedures.

For fracture mechanics applications, a particular challenge is the suitable detection of first delamination onsets, since the increased ductility of the matrix material tends to reduce the stress concentration at the crack tip. At the same time, the validity of boundary conditions given by linear elastic fracture mechanics is subject to discussion. Secondary methods can be used to provide better insight into the progression of damage and therefore allow better interpretation of the respective crack propagation energies.

2.4 What Can In Situ Methods Contribute to Mechanical Testing?

Based on the difficulties to obtain reliable material properties from stress–strain or load–displacement curves several attempts have been made to improve the level of reliability. In general, the idea is to aid the interpretation of the material behavior by a secondary method independent in its measurement principle from the macroscopic stress–strain response. The relevance and suitability of the secondary experimental methods stems from the intended purpose and is briefly reviewed for several methods in the following.

2.4.1 In Situ Microscopy

The application of in situ microscopy during testing has become a standard in some test configurations for fiber-reinforced materials. Namely, testing under mode I condition facilitates usage of a travelling microscope to measure the accurate crack propagation length. But also other test methods can benefit from accompanied in situ microscopy as seen in Fig. 2.25, for a simple unidirectional test specimen with fiber orientation parallel to the load axis. The simultaneous recording of images can easily aid to interpret load drops preceding the maximum load. In this test configuration, such behavior is often found for preliminary fiber failure induced at the edges.

However, the major drawback of such imaging techniques is the lack of sensitivity associated with their field of view. Naturally, this kind of imaging is limited to detection of failure at the surface level unless the internal failure is so severe that their occurrence causes corresponding damage at the surface. Also, a proper trade-off between a high resolution and a large field of view has to be made beforehand.

2.4.2 Digital Image Correlation

Digital image correlation also counts as an in situ imaging technique, but is discussed independently, since it comprises various new aspects to be used for detection of failure in fiber-reinforced materials. In the context of materials testing, the method is used to deduce the spatially resolved strain field using a series of

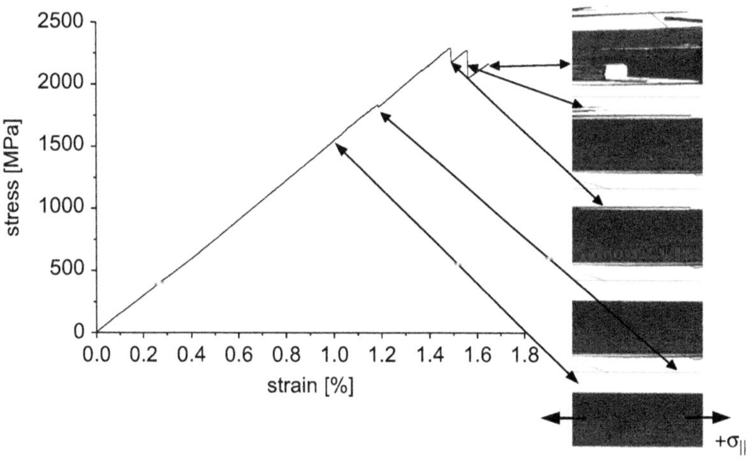

Fig. 2.25 Stress–strain curve from tensile testing of unidirectional fiber-reinforced polymer. Visual observation of the test region at distinct load steps. *Arrows* indicate position of corresponding failure at each load level

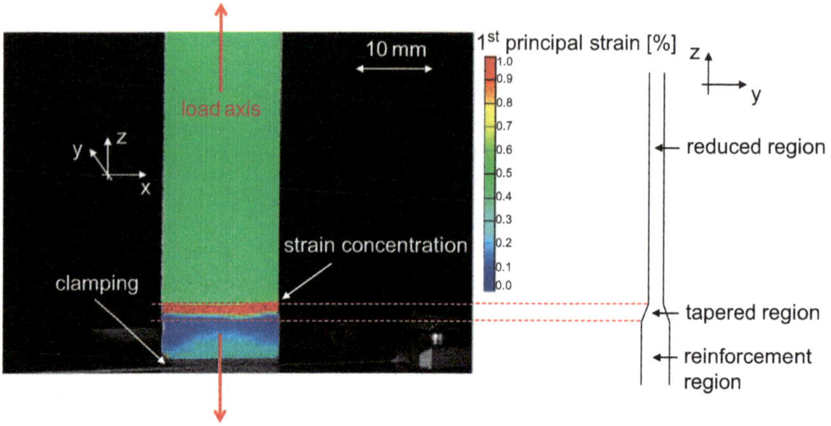

Fig. 2.26 DIC image of inhomogeneous strain field at wedge between reinforcement region and reduced region of tensile specimen

sequential images with speckle-type patterns (see also Chap. 3 for a detailed description). In terms of the challenges associated with testing of fiber-reinforced materials, there is one major improvement by application of DIC. This is the possibility to perform full-field measurements of strain values to deduce material properties [95, 193, 194]. As seen in the example in Fig. 2.26, the DIC method is able to compute the strain values not only between two particular points, but within all parts of the specimen covered by a speckle pattern. This effectively allows selecting only a particular area to obtain strain values for data reduction. Therefore, regions with inhomogeneous strain fields (e.g., at the reinforcements) can be excluded and valid strain values (e.g., within the reduced section) can be obtained. At the same time, such full-field measurements allow to optimize specimen geometries to assure the anticipated strain state is reached in every part of the test volume.

In recent years, DIC using high-resolution camera systems became available. This provides even further possibilities for the analysis of composite materials. In addition to the possibility to perform full-field measurements, one can observe disturbances and anomalies of the strain field during such experiments. Intuitively one would associate such signatures with other anomalies such as the occurrence of failure. As seen in Fig. 2.27, the presence of these signatures in strain field measurements encourages the investigation of the limits of the DIC method in this aspect as presented in Chap. 3.

2.4.3 X-Ray Methods

Complementary to the conventional in situ imaging methods, several research groups demonstrated the use of X-ray imaging to obtain information on the damage progress in fiber-reinforced materials. The main advantage of X-Rays compared to

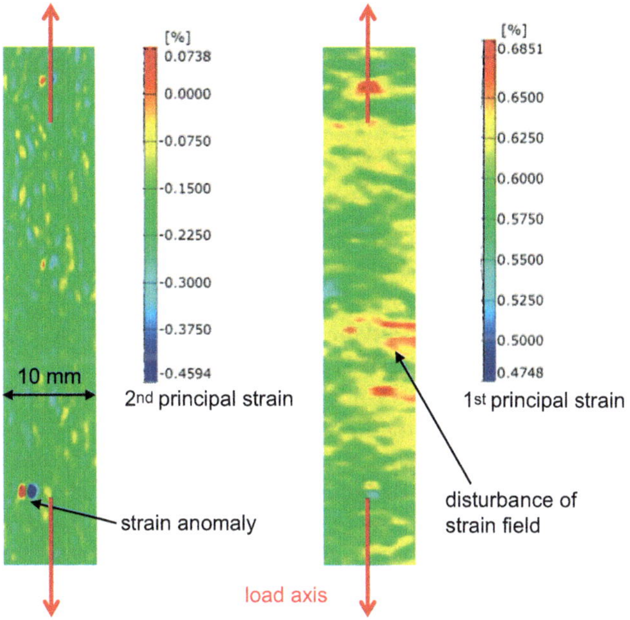

Fig. 2.27 Anomalies and disturbances of the strain field as readily visible by DIC systems

light is their ability to penetrate the specimen and therefore to obtain volumetric information. The possibility to use computed tomography for visualization of internal damage states has become a standard method already.

One application of X-rays considers the possibility to detect the onset and growth of interfiber failure during mechanical loading [106, 195]. Using X-ray refraction,[2] these groups were able to detect the onset and growth of internal defects such as fiber–matrix debonding due to changes in the refraction index of the specimen.

Other groups have been using synchrotron radiation in combination with in situ loading stages to carry out volumetric imaging of miniature specimens under mechanical load [101, 109, 196, 197]. Such microscopic imaging of the specimen intuitively allows to track the initiation of damage in the interior and to deduce the interaction between different failure mechanisms at increasing load levels. As seen in Fig. 2.28 in an exemplary image using a commercial X-ray computed tomography device, the level of detail reached in small specimens is sufficiently high to visualize details of the fracture mode and even to spot single fiber filaments. This gave rise to numerous investigations to track the occurrence and accumulation of failure in composite materials under mechanical load. Other groups have adopted

[2] X-Ray refraction is similar to refraction of the visible light and is not identical to the established techniques of X ray diffraction (see [195]).

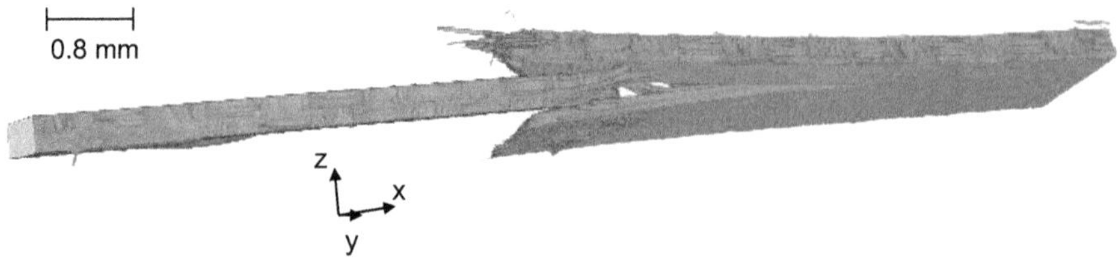

Fig. 2.28 3D image of small carbon/epoxy composite specimen with details of the fracture mode

this analysis routine and investigated the failure of fiber-reinforced materials under various load conditions [94, 101, 103, 109, 197, 198].

In particular, it is technically feasible to combine every mechanical test configuration with a computed tomography device in situ. The term "in situ" in this context typically refers to load-hold cycles with intermediate scanning. That way damage is first introduced and existing damage states are kept under load. For imaging stress relaxation effects have to be considered in general and creep effects of the matrix polymer have to be taken into account in particular. Moreover, there is currently no feasible technical solution to scale the method to allow specimen dimensions larger than the few millimeters of cross section currently used. But the possibilities of this approach to assist in interpretation of failure of fiber-reinforced materials are so encouraging that a particular focus on this topic is given in Chap. 6.

2.4.4 Thermography

Another type of in situ measurement that is applied in combination with mechanical testing of fiber-reinforced materials is thermography imaging. As the name indicates, this also allows for continuous imaging of the specimen under load, but the origin of the visual information is fairly different to the previously presented techniques. In contrast to the imaging methods discussed before, the thermography cameras record the spatial generation of heat via detection of the generated infrared radiation. The generation of heat can be attributed to two source mechanisms. The first one is the heat release during fracture of a material. For strong failure events, this dissipative energy contribution is readily detectable by commercial equipment as seen in Fig. 2.29. Here, tensile testing of an open-hole specimen was monitored by thermography imaging. The heat release during macroscopic fracture saturates the measurement range of the equipment. Before ultimate failure, several heat signatures could be spotted in the area surrounding the hole, which allows monitoring evolution of damage to a certain extent. The second type of heat source is friction inside the material or between the support and the test specimen. While the generation of heat due to the latter is a general problem especially during fatigue testing of composites [199–201], the friction inside the material is key to assist in

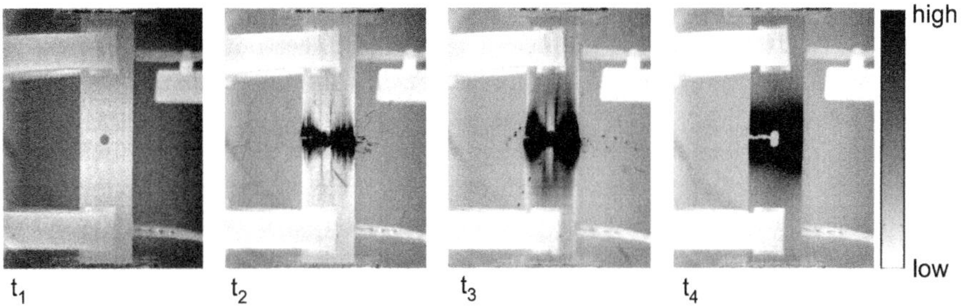

Fig. 2.29 Thermography images of four subsequent times ($t_1 < t_2 < t_3 < t_4$) during fracture of an open-hole tensile test

interpretation of material failure. During periodic cyclic loading experiments, existing crack surfaces can rub against each other and therefore generate a noticeable amount of heat. Using thermography imaging, this allows spotting damaged areas and to quantify their evolution to a certain extent (Fig. 2.29).

However, there are two problems associated with both types of heat sources. As mentioned above, the transient heat pulse due to material fracture is only detectable for larger failure events. In addition, the heat transfer from the damaged spot to the surface causes strong dispersion of the spatial information as function of subsurface position and time. This effect is also relevant for the generation of heat by cyclic friction of crack surfaces, but is superimposed additionally by the accumulation of heat in the specimen itself. Therefore no attempts have been made to directly correlate the occurrence of individual failure types in fiber-reinforced materials to signatures in thermography images. However, it is worth noting that the resolution of commercial infrared cameras is well suited to measure temperature rises of few milli Kelvin with sufficient spatial resolution.

2.4.5 *Shearography*

Shearography is another optical method, which is distinct to the above mentioned methods in the way it is used for in situ monitoring. The specimen under investigation is illuminated using an expanded laser beam and images are taken using a CCD camera via an optical shearing element [202]. The shearing element allows a coherent superposition of two laterally displaced images of the surface of the specimen in the image plane. The lateral displacement introduced by the shearing element is called the shear of the images. The superposition of the two images is called the shearogram, which is an interferogram of an object wave with the sheared object wave as a reference wave. To allow for defect detection, two shearograms are recorded for different loading conditions of the sample. The type of loading should induce some deformation of the specimen, which is visible at the surface. In practice, thermal, acoustical, or mechanical loading are applied in a static or dynamic way [203–208]. Due to internal referencing shearography is inherently insensitive to the environmental vibrations and therefore applicable to testing in field environments.

The absolute difference of the shearograms recorded at different loading states results in an interference fringe pattern, which is directly correlated to the local slope of the specimen deformation. Hence, anomalies in the strain field can be detected and interpreted in a similar fashion as discussed in Sect. 2.4.2 for digital image correlation. This has been actively pursued for defect detection in composite materials in the past years but has been mostly directed towards field inspection and to detect mostly macroscopic defect sizes [206–209].

2.4.6 Ultrasonic Measurements

For the field of ultrasonic measurements, various approaches are reported in literature to aid the interpretation of occurrence of damage in fiber-reinforced materials. Well known are the inspection methods using active pulse-echo or through-transmission configurations to visualize damage inside test specimens. However, the imaging mode of these configurations (i.e., B-scans or C-scans) are not straightforward to be used in situ with mechanically loaded specimens. But due to the close relation between the elastic properties of a material and the ultrasound velocity and attenuation, guided wave ultrasound pulses can be used to track the damage state in a fiber-reinforced material. The sensitivity of guided acoustic waves to such internal defects is very high as long as the wavelength is sufficiently small to interact with the defect. Such in situ monitoring using two or more sensors is always limited to the occurrence of damage within the full propagation path between the sensors. New attempts apply ultrasonic tomography and use sensor arrays to localize damaged areas by beam forming techniques [210–213]. This allows a substantial improvement of the spatial resolution to detect the occurrence of failure in a material, but the temporal resolution is still limited to the acquisition rate of the active pulses.

Another method suitable to detect the occurrence of failure in situ is the application of (nonlinear) vibrometry. In this configuration an active pulser is used to generate cyclic acoustic waves in the specimen under load. One can image the defect positions by observation of the acoustic wave field using a scanning laser vibrometer. The occurrence of failure mechanisms of a characteristic size will cause a noticeable change in the resonances visible as acoustic wave field. For a particular position this can be quantified in terms of amplitude changes and frequency or phase shifts [214–216].

2.4.7 *Acoustic Emission*

Many approaches are reported in literature to monitor acoustic emission during testing of fiber-reinforced materials [93, 217–234]. In general, all mechanisms described in Sect. 2.1 generate acoustic emission during their formation. The detection of the respective acoustic emission waves by several sensors allows the inverse calculation of the source position based on the arrival time of the signals at the different sensors. At the same time the detection and accumulation of signals allows an in situ monitoring of the damage progress in fiber-reinforced materials. In this context, it is worth noting that the sensitivity of the acoustic emission method is extraordinary high, which easily allows detection of single fiber filament failure as demonstrated in Chap. 4. However, the level of detail seen by acoustic emission is sometimes hard to relate to the level of description by failure theories or the level of observation of other experimental methods. For instance, it is arguable whether the onset of fiber failure as within a failure theory is identical to the first detection of fiber failure by acoustic emission. Also, the propagation of the acoustic emission signals in a test specimen has significant impact on the interpretability of the signal. Therefore, only a thorough understanding of the acoustic emission source, the signal propagation, and the detection process can lead to a reliable interpretation of the measurement data. Chapter 4 presents recent developments in this field and is used to provide an overview of test configurations in conjunction with acoustic emission acquisition.

Beyond the application of acoustic emission to obtain properties of fiber-reinforced materials, a large amount of work has also been conducted to use the method in situ for inspection of structures [132, 235–238] or to conduct failure prediction. Most noteworthy are the approaches to predict burst pressures in fiber-reinforced materials [231, 239, 240]. In this context, the usage of acoustic emission analysis to support interpretation of material failure has already been covered by several internationally recognized standards such as ISO, EN, ASTM, DIN, or other national or proprietary standards.

2.4.8 *Electromagnetic Emission*

Similarly to acoustic emission, the origin of electromagnetic emission is the active occurrence of damage in the material. As demonstrated in Chap. 5, typical failure types in fiber-reinforced composites also cause detectable electromagnetic emission.

However, the details of the source mechanism of electromagnetic emission due to material failure are still under controversial discussion in literature. Also, today no commercial equipment is available to readily detect electromagnetic emission signals. Therefore, electromagnetic emission itself is currently understood as a method under development and is not yet covered by any technical standards.

However, the method is very promising for failure analysis of fiber-reinforced materials and might emerge as a future in situ testing method.

Similar to acoustic emission, the electromagnetic emission benefits from the high discrimination capability, that is, it only senses active material failure. There are also indications of even higher discriminative capabilities than for acoustic emission, since it might be speculated if friction does provide electromagnetic emission signals. Since friction sources form a typical, but unwanted acoustic emission source, this discrimination would help to improve the reliability of signal interpretation. Also, the electromagnetic signals show negligible influence of the propagation medium and sensor systems are available that show relatively flat frequency response. Source orientations can be detected using sensor arrays and can be used to quantify the resulting orientations of interfiber crack planes. As a technical drawback, the present systems signal-to-noise ratio is one to two orders of magnitude lower than for commercially available acoustic emission measurement systems. Moreover, the signal amplitude shows a strong decay as function of source sensor distance. Despite of these drawbacks, Chap. 5 demonstrates first attempts dedicated to use electromagnetic emission analysis to monitor failure in fiber-reinforced materials.

References

1. Yamada, I., Masuda, K., Mizutani, H.: Electromagnetic and acoustic emission associated with rock fracture. Phys. Earth Planet. Inter. **57**, 157–168 (1989)
2. Frid, V., Rabinovitch, A., Bahat, D.: Fracture induced electromagnetic radiation. J. Phys. D Appl. Phys. **36**, 1620–1628 (2003)
3. Griffith, A.A.: The phenomena of rupture and flow in solids. Philos. Trans. R. Soc. Lond. Ser. A **221**, 163–198 (1921)
4. Irwin, G.: Analysis of stresses and strains near the end of a crack traversing a plate. J. Appl. Mech. **24**, 361–364 (1957)
5. Schürmann, H.: Konstruieren mit Faser-Kunststoff-Verbunden. Springer, Berlin (2005)
6. Matthews, F.L., Rawlings, R.D.: Composite Materials: Engineering and Science. Woodhead, Cambridge (1999)
7. Greenhalgh, E.: Failure Analysis and Fractography of Polymer Composites. Woodhead, Oxford (2009)
8. Sause, M.: Identification of Failure Mechanisms in Hybrid Materials Utilizing Pattern Recognition Techniques Applied to Acoustic Emission Signals. mbv-Verlag, Berlin (2010)
9. Kermode, J.R., Albaret, T., Sherman, D., Bernstein, N., Gumbsch, P., Payne, M.C., Csányi, G., De Vita, A.: Low-speed fracture instabilities in a brittle crystal. Nature **455**, 1224–1227 (2008)
10. He, F., Tan, C.M., Zhang, S., Cheng, S.: Monte Carlo simulation of fatigue crack initiation at elevated temperature. In: 13th Conference on Fracture, pp. 1–10. Beijing, China (2013)
11. Guglhoer, T., Manger, F., Sause, M.G.R.: Quantification of local fiber distribution for optimization of tape laying techniques. In: ECCM16—16th European Conference on Composite Materials, pp. 1–9. Sevilla, Spain (2014)
12. Salaberger, D., Arikan, M., Paier, T., Kastner, J.: Characterization of damage mechanisms in glass fibre reinforced polymers using x-ray computed tomography. In: 11th European Conference on Non-Destructive Testing (ECNDT 2014), pp. 1–9. Prague, Czech Republic (2014)

13. Hobbiebrunken, T., Hojo, M., Fiedler, B., Tanaka, M., Ochiai, S., Schulte, K.: Thermomechanical analysis of micromechanical formation of residual stresses and initial matrix failure in CFRP. JSME Int. J. Ser. A **47**, 349–356 (2004)
14. Hobbiebrunken, T., Fiedler, B., Hojo, M., Ochiai, S., Schulte, K.: Microscopic yielding of CF/epoxy composites and the effect on the formation of thermal residual stresses. Compos. Sci. Technol. **65**, 1626–1635 (2005)
15. Greisel, M., Jäger, J., Moosburger-Will, J., Sause, M.G.R., Mueller, W.M., Horn, S.: Influence of residual thermal stress in carbon fiber-reinforced thermoplastic composites on interfacial fracture toughness evaluated by cyclic single-fiber push-out tests. Compos. Part A Appl. Sci. Manuf. **66**, 117–127 (2014)
16. Tanaka, T., Nakayama, H., Sakaida, A., Horikawa, N.: Estimation of tensile strength distribution for carbon fiber with diameter variation along fiber. Mater. Sci. Res. Int. **5**, 90–97 (1999)
17. Jumahat, A., Soutis, C., Jones, F., Hodzic, A.: Fracture mechanisms and failure analysis of carbon fibre/toughened epoxy composites subjected to compressive loading. Compos. Struct. **92**, 295–305 (2010)
18. Alhashmi, H., Kumar, S.: Micro-mechanical modeling of fiber pull-out stresses in an axi-symmetric composite system. In: ECCM16—16th European Conference on Composite Materials, pp. 1–9. Sevilla, Spain (2014)
19. Yang, L., Thomason, J.L.: Interface strength in glass fibre-polypropylene measured using the fibre pull-out and microdebond methods. In: ICCM 17—17th International Conference on Composite Materials, pp. 1–9, Edinburgh (2009)
20. Lin, Y., Scheuring, T., Friedrich, K.: Matrix morphology and fibre pull-out strength of T700/PPS and T700/PET thermoplastic composite. J. Mater. **30**, 4761–4769 (1995)
21. Pisanova, E., Zhandarov, S., Mäder, E.: How can adhesion be determined from micromechanical tests? Compos. Part A Appl. Sci. Manuf. **32**, 425–434 (2001)
22. Novak, J., Pearce, C.J., Grassl, P., Yang, L., Thomason, J.: Analysis of the microbond test using nonlinear fracture mechanics. In: ICCM 17—17th International Conference on Composite Materials, pp. 1—13. Edinburgh (2009)
23. Fuentes, C.A., Brughmans, G., Verpoest, I., van Vuure, A.W., Tran, L.Q.N., Dupont-Gillain, C.: Effect of roughness on the interface in natural fibre composites: physical adhesion and mechanical interlocking. In: ECCM16—16th European Conference on Composite Materials. Sevilla, Spain (2014)
24. Reifsnider, K., Sendeckyj, G., Wang, S., Johnson, W., Stinchcomb, W., Pagano, N., Nahas, M.: Survey of failure and post-failure theories of laminated fiber-renforced composites. J. Compos. Technol. Res. **8**, 138 (1986)
25. Puck, A.: Festigkeitsanalyse von Faser-Matrix-Laminaten Modelle für die Praxis. Carl Hanser Verlag, Munich, Germany (1996)
26. Tsai, S.W., Wu, E.M.: A general theory of strength for anisotropic materials. J. Compos. Mater. **5**, 58–80 (1971)
27. Wu, E.: Phenomenological anisotropic failure criteria. Mech. Compos. Mater. **2**, 353–431 (1974)
28. Hashin, Z.: Failure criteria for unidirectional fiber composites. J. Appl. Mech. **47**, 329 (1980)
29. Mohr, O.: Welche Umstände bedingen die Elastizitätsgrenze und den Bruch eines materials? Z. Ver. Dtsch. Ing. **44**, 1524–1530 (1900)
30. Cuntze, R.: Comparison between experimental and theoretical results using Cuntze's "failure mode concept" model for composites under triaxial loadings—part B of the second world-wide failure exercise. J. Compos. Mater. **47**, 893–924 (2013)
31. Puck, A.: Festigkeitsberechnung an Glasfaser/Kunststoff-Laminaten bei zusammengesetzter Beanspruchung. Kunststoffe **59**, 780–787 (1969)
32. Puck, A.: GFK-Drehrohrfedern sollen höchstbeanspruchte Stahlfedern ersetzen. Kunststoffe **80**, 1380–1384 (1990)
33. Puck, A.: Ein Bruchkriterium gibt die Richtung an. Kunststoffe **82**, 607–610 (1992)

34. Puck, A., Kopp, J., Knops, M.: Guidelines for the determination of the parameters in Puck's action plane strength criterion. Compos. Sci. Technol. **62**, 371–378 (2002)
35. Puck, A., Schürmann, H.: Failure analysis of FRP laminates by means of physically based phenomenological models. Compos. Sci. Technol. **62**, 1633–1662 (2002)
36. Puck, A., Mannigel, M.: Physically based non-linear stress–strain relations for the inter-fibre fracture analysis of FRP laminates. Compos. Sci. Technol. **67**, 1955–1964 (2007)
37. Soden, P.D., Hinton, M.J., Kaddour, A.S.: Biaxial test results for strength and deformation of a range of E-glass and carbon fibre reinforced composite laminates: failure exercise benchmark data. Compos. Sci. Technol. **62**, 1489–1514 (2002)
38. Hinton, M.J., Kaddour, A.S., Soden, P.D.: Evaluation of failure prediction in composite laminates: background to "part B" of the exercise. Compos. Sci. Technol. **62**, 1481–1488 (2002)
39. Hinton, M.J., Soden, P.D.: Predicting failure in composite laminates: the background to the exercise. Compos. Sci. Technol. **58**, 1001–1010 (1998)
40. Soden, P.D., Hinton, M.J., Kaddour, A.S.: Lamina properties, lay-up configurations and loading conditions for a range of fibre-reinforced composite laminates. Compos. Sci. Technol. **58**, 1011–1022 (1998)
41. Kaddour, A.S., Hinton, M.J.: Maturity of 3D failure criteria for fibre-reinforced composites: comparison between theories and experiments: part B of WWFE-II. J. Compos. Mater. **47**, 925–966 (2013)
42. Hinton, M., Kaddour, A.: The background to part B of the second world-wide failure exercise: evaluation of theories for predicting failure in polymer composite laminates under three-dimensional states of stress. J. Compos. Mater. **47**, 643–652 (2013)
43. Hinton, M., Kaddour, A.: Triaxial test results for fibre reinforced composites: the second world-wide failure exercise benchmark data. J. Compos. Mater. **47**, 925–966 (2013)
44. Puck, A.: Failure analysis of FRP laminates by means of physically based phenomenological models. Compos. Sci. Technol. **58**, 1045–1067 (1998)
45. Zinoviev, P.A., Lebedeva, O.V., Tairova, L.P.: A coupled analysis of experimental and theoretical results on the deformation and failure of composite laminates under a state of plane stress. Compos. Sci. Technol. **62**, 1711–1723 (2002)
46. Zinoviev, P.: The strength of multilayered composites under a plane-stress state. Compos. Sci. Technol. **58**, 1209–1223 (1998)
47. Cuntze, R.G., Freund, A.: The predictive capability of failure mode concept-based strength criteria for multidirectional laminates. Compos. Sci. Technol. **64**, 343–377 (2004)
48. Soden, P.D., Kaddour, A.S., Hinton, M.J.: Recommendations for designers and researchers resulting from the world-wide failure exercise. Compos. Sci. Technol. **64**, 589–604 (2004)
49. Pinho, S., Vyas, G., Robinson, P.: Material and structural response of polymer-matrix fibre-reinforced composites: part B. J. Compos. Mater. **47**, 679–696 (2013)
50. Pinho, S., Darvizeh, R., Robinson, P., Schuecker, C., Camanho, P.: Material and structural response of polymer-matrix fibre-reinforced composites. J. Compos. Mater. **46**, 2313–2341 (2012)
51. Carrere, N., Laurin, F., Maire, J.-F.: Micromechanical-based hybrid mesoscopic 3D approach for non-linear progressive failure analysis of composite structures. J. Compos. Mater. **46**, 2389–2415 (2012)
52. Carrere, N., Laurin, F., Maire, J.-F.: Micromechanical-based hybrid mesoscopic three-dimensional approach for non-linear progressive failure analysis of composite structures—part B: comparison with experimental data. J. Compos. Mater. **47**, 743–762 (2012)
53. Matthias Deuschle, H., Kroplin, B.-H.: Finite element implementation of Puck's failure theory for fibre-reinforced composites under three-dimensional stress. J. Compos. Mater. **46**, 2485–2513 (2012)
54. Deuschle, H.M., Puck, A.: Application of the Puck failure theory for fibre-reinforced composites under three-dimensional stress: comparison with experimental results. J. Compos. Mater. **47**, 827–846 (2013)

55. Kaddour, A., Hinton, M., Smith, P., Li, S.: The background to the third world-wide failure exercise. J. Compos. Mater. **47**, 2417–2426 (2013)
56. Huang, Y., Jin, C., Ha, S.K.: Strength prediction of triaxially loaded composites using a progressive damage model based on micromechanics of failure. J. Compos. Mater. **47**, 777–792 (2013)
57. Zhang, D., Xu, L., Ye, J.: Prediction of failure envelopes and stress–strain curves of fiber composite laminates under triaxial loads: comparison with experimental results. J. Compos. Mater. **47**, 763–776 (2013)
58. Tessmer, J.: Theoretische und algorithmische Beitraege zur Berechnung von Faserverbundschalen. Ph.D. Thesis, University of Hannover (2000)
59. Juhasz, T.: Ein neues physikalisch basiertes Versagenskriterium für schwach 3D-verstärkte Faserverbundlaminate. Ph.D. Thesis, University Carolo-Wilhelmina Braunschweig (2003)
60. Axelrad, D.: Micromechanics of Solids. Elsevier Scientific, Amsterdam (1978)
61. Huang, Z.M., Ramakrishna, S.: Micromechanical modeling approaches for the stiffness and strength of knitted fabric composites: a review and comparative study. Compos. Part A Appl. Sci. Manuf. **31**, 479–501 (2000)
62. Kuksenko, V.S., Tamuzs, V.P.: Fracture Micromechanics of Polymer Materials. Springer, Dordrecht (1981)
63. Stellbrink, K.: Micromechanics of Composites: Composite Properties of Fibre and Matrix Constituents. Carl Hanser Verlag, Munich (1996)
64. Voyiadjis, G.Z., Deliktas, B.: A coupled anisotropic damage model for the inelastic response of composite materials. Comput. Meth. Appl. Mech. Eng. **183**, 159–199 (2000)
65. Wisnom, M.R.: Size effects in the testing of fibre-composite materials. Compos. Sci. Technol. **59**, 1937–1957 (1999)
66. Flaggs, D.L., Kural, M.H.: Experimental determination of the in-situ transverse lamina strength in graphite/epoxy laminates. J. Compos. Mater. **16**, 103–116 (1982)
67. Ogin, S.L., Smith, P.A.: Fast fracture and fatigue growth of transverse ply cracks in composite laminates. Scr. Metall. **19**, 779–784 (1985)
68. Parvizi, A., Garrett, K.W., Bailey, J.E.: Constrained cracking in glass fibre-reinforced epoxy cross-ply laminates. J. Mater. Sci. **13**, 195–201 (1978)
69. Nairn, J., Hu, S.: The initiation and growth of delaminations induced by matrix microcracks in laminated composites. Int. J. Fract. **57**, 1–24 (1992)
70. Nairn, J.A., Hu, S.: The formation and effect of outer-ply microcracks in cross-ply laminates: a variational approach. Eng. Fract. Mech. **41**, 203–221 (1992)
71. Anderson, T.L.: Fracture Mechanics: Fundamentals and Applications, 3rd edn. Taylor & Francis Group, Boca Raton (2005)
72. Groos, D.: Bruchmechanik. Springer, Berlin (2007)
73. Rice, J.R.: A path independent integral and the approximate analysis of strain concentration by notches and cracks. J. Appl. Mech. **35**, 379 (1968)
74. Rice, J.R.: Elastic fracture mechanics concepts for interfacial cracks. J. Appl. Mech. **55**, 98 (1988)
75. Dowling, N.E.: Mechanical Behavior of Materials. Prentice Hall, Upper Saddle River (1999)
76. Doudican, B.M., Zand, B., Amaya, P., Butalia, T.S., Wolfe, W.E., Schoeppner, G.A.: Strain energy based failure criterion: comparison of numerical predictions and experimental observations for symmetric composite laminates subjected to triaxial loading. J. Compos. Mater. **47**, 847–866 (2013)
77. Mueller, W.M., Moosburger-Will, J., Sause, M.G.R., Horn, S.: Microscopic analysis of single-fiber push-out tests on ceramic matrix composites performed with Berkovich and flat-end indenter and evaluation of interfacial fracture toughness. J. Eur. Ceram. Soc. **33**, 441–451 (2013)
78. Jäger, J., Sause, M.G.R., Burkert, F., Moosburger-Will, J., Greisel, M., Horn, S.: Influence of plastic deformation on single-fiber push-out tests of carbon fiber reinforced epoxy resin. Compos. Part A Appl. Sci. Manuf. **71**, 157–167 (2015)

79. Kaddour, A., Hinton, M., Smith, P., Li, S.: A comparison between the predictive capability of matrix cracking, damage and failure criteria for fibre reinforced composite laminates: part A of the third world-wide failure exercise. J. Compos. Mater. **47**, 2749–2779 (2013)
80. Chamis, C.C., Abdi, F., Garg, M., Minnetyan, L., Baid, H., Huang, D., Housner, J., Talagani, F.: Micromechanics-based progressive failure analysis prediction for WWFE-III composite coupon test cases. J. Compos. Mater. **47**, 2695–2712 (2013)
81. Laurin, F., Carrere, N., Huchette, C., Maire, J.-F.: A multiscale hybrid approach for damage and final failure predictions of composite structures. J. Compos. Mater. **47**, 2713–2747 (2013)
82. Daghia, F., Ladeveze, P.: Identification and validation of an enhanced mesomodel for laminated composites within the WWFE-III. J. Compos. Mater. **47**, 2675–2693 (2013)
83. Flatscher, T., Schuecker, C., Pettermann, H.: A constitutive ply model for stiffness degradation and plastic strain accumulation: its application to the Third World Wide Failure Exercise (part A). J. Compos. Mater. **47**, 2575–2593 (2013)
84. Pinho, S., Vyas, G., Robinson, P.: Response and damage propagation of polymer-matrix fibre-reinforced composites: predictions for WWFE-III part A. J. Compos. Mater. **47**, 2595–2612 (2013)
85. Sapozhnikov, S.B., Cheremnykh, S.I.: The strength of fibre reinforced polymer under a complex loading. J. Compos. Mater. **47**, 2525–2552 (2013)
86. Forghani, A., Zobeiry, N., Poursartip, A., Vaziri, R.: A structural modelling framework for prediction of damage development and failure of composite laminates. J. Compos. Mater. **47**, 2553–2573 (2013)
87. Singh, C.V., Talreja, R.: A synergistic damage mechanics approach to mechanical response of composite laminates with ply cracks. J. Compos. Mater. **47**, 2475–2501 (2012)
88. McCartney, L.: Derivations of energy-based modelling for ply cracking in general symmetric laminates. J. Compos. Mater. **47**, 2641–2673 (2013)
89. McCartney, L.: Energy methods for modelling damage in laminates. J. Compos. Mater. **47**, 2613–2640 (2012)
90. Gorbatikh, L., Ivanov, D., Lomov, S., Verpoest, I.: On modelling of damage evolution in textile composites on meso-level via property degradation approach. Compos. Part A Appl. Sci. Manuf. **38**, 2433–2442 (2007)
91. Lomov, S., Ivanov, D., Verpoest, I., Zako, M., Kurashiki, T., Nakai, H., Hirosawa, S.: Meso-FE modelling of textile composites: road map, data flow and algorithms. Compos. Sci. Technol. **67**, 1870–1891 (2007)
92. Whitcomb, J.D., Chapman, C.D., Tang, X.: Derivation of boundary conditions for micromechanics analyses of plain and satin weave composites. J. Compos. Mater. **34**, 724–747 (2000)
93. Mikhaluk, D.S., Truong, T.C., Borovkov, A.I., Lomov, S.V., Verpoest, I.: Experimental observations and finite element modelling of damage initiation and evolution in carbon/epoxy non-crimp fabric composites. Eng. Fract. Mech. **75**, 2751–2766 (2008)
94. Ivanov, D.S., Baudry, F., Van Den Broucke, B., Lomov, S.V., Xie, H., Verpoest, I.: Failure analysis of triaxial braided composite. Compos. Sci. Technol. **69**, 1372–1380 (2009)
95. Lomov, S.V., Ivanov, D.S., Verpoest, I., Zako, M., Kurashiki, T., Nakai, H., Molimard, J., Vautrin, A.: Full-field strain measurements for validation of meso-FE analysis of textile composites. Compos. Part A Appl. Sci. Manuf. **39**, 1218–1231 (2008)
96. Ivanov, S.G., Gorbatikh, L., Lomov, S.V.: Interlaminar fracture behaviour of textile composites with thermoplastic matrices. In: ECCM16—16th European Conference on Composite Materials, pp. 1–8. Sevilla, Spain (2014)
97. Deuschle, H.M.: 3D failure analysis of UD fibre reinforced composites: Puck's theory within FEA. Ph.D. Thesis, University of Stuttgart (2010)
98. Edge, E.C.: Does transverse and shear loading affect the compression strength of unidirectional CFC? A reply to Dr Hart-Smith. Composites **25**, 159–161 (1994)
99. Hart-Smith, L.: Fibrous composite failure criteria-fact and fantasy. In: 7th International Conference of Composite Structures. Paisley, (1993)

100. Scott, A.E., Mavrogordato, M., Wright, P., Sinclair, I., Spearing, S.M.: In-situ fibre fracture measurement in carbon-epoxy laminates using high resolution computed tomography. Compos. Sci. Technol. **71**, 1471–1477 (2011)
101. Scott, A.E., Sinclair, I., Spearing, S.M., Mavrogordato, M., Bunsell, A.R., Thionnet, A.: Comparison of the accumulation of fibre breaks occurring in a unidirectional carbon/epoxy composite identified in a multi-scale micro-mechanical model with that of experimental observations using high resolution computed tomography. In: Matériaux 2010, pp. 1–9. Nantes, France (2010)
102. Fuwa, M., Harris, B., Bunsell, A.R.: Acoustic emission during cyclic loading of carbon-fibre-reinforced plastics. J. Phys. D Appl. Phys. **8**, 1460–1471 (2001)
103. Scott, A.E., Sinclair, I., Spearing, S.M., Thionnet, A., Bunsell, A.R.: Damage accumulation in a carbon/epoxy composite: comparison between a multiscale model and computed tomography experimental results. Compos. Part A Appl. Sci. Manuf. **43**, 1514–1522 (2012)
104. Kaddour, A.S., Hinton, M.J., Soden, P.D.: A comparison of the predictive capabilities of current failure theories for composite laminates: additional contributions. Compos. Sci. Technol. **64**, 449–476 (2004)
105. Ortwein, H.-P., Bohse, J., Trappe, V.: Untersuchung der Mikrorissbildung in Faserkunststoffverbunden mittels Röntgen- refraktions- und Schallemissionsmessung. In: 18. Kolloquium Schallemission, pp. 1–7. Wetzlar, Germany (2011)
106. Trappe, V., Günzel, S., Hickmann, S.: Non-destructive evaluation of micro cracking in short fibre reinforced thermoplastics with X-ray-refraction. In: ICCM 17—17th International Conference on Composite Material. Edinburgh (2009)
107. Trappe, V., Hickmann, S., Sturm, H.: Bestimmung des Zwischenfaserbruchversagens in textilfaserverstärkten Glasfaserkunststoff mittels Refraktionstopographie. Mater. Test. **50**, 615–622 (2008)
108. Metzkes, K., Trappe, V.: Damage Evolution in short fibre reinforced polyamide caused by biaxial loading. In: ECCM16—16th European Conference on Composite Materials, pp. 1–8. Sevilla, Spain (2014)
109. Scott, A.E., Sinclair, I., Spearing, S.M., Mavrogordato, M.N., Hepples, W.: Influence of voids on damage mechanisms in carbon/epoxy composites determined via high resolution computed tomography. Compos. Sci. Technol. **90**, 147–153 (2014)
110. Ramesh, K.T.: High strain rate and impact experiments. In: Springer Handbook of Experimental Solid Mechanics, pp. 1–31. Springer, Berlin (2008)
111. Horoschenkoff, A.: Beitrag zur Charakterisierung des nichtlinear thermoviskoelastischen Kriechverhaltens von Faserverbundwerkstoffen. Ph.D. Thesis, Technical University of Munich (1995)
112. Sakai, T., Somiya, S.: Analysis of creep behavior in thermoplastics based on visco-elastic theory. Mech. Time-Dependent Mater. **15**, 293–308 (2011)
113. Somiya, S.: Creep behavior of a carbon-fiber-reinforced thermoplastic polyimide resin. J. Thermoplast. Compos. Mater. **7**, 91–99 (1994)
114. Sakai, T., Somiya, S.: Analysis of creep behavior in thermoplastic based on visco-elastic theory. In: Proceedings of the SEM Annual Conference. Albuquerque, New Mexico (2009)
115. Guedes, R.M.: Creep and fatigue lifetime prediction of polymer matrix composites based on simple cumulative damage laws. Compos. Part A Appl. Sci. Manuf. **39**, 1716–1725 (2008)
116. Guedes, R.M.: Relationship between lifetime under creep and constant stress rate for polymer-matrix composites. Compos. Sci. Technol. **69**, 1200–1205 (2009)
117. Reiner, M., Weissenberg, K.: A thermodynamic theory of the strength of the materials. Rheol. Leafl. **10**, 12–20 (1939)
118. Miner, M.: Cumulative damage in fatigue. J. Appl. Mech. **12**, 159–164 (1945)
119. Christensen, R.: A physically based cumulative damage formalism. Int. J. Fatigue **30**, 595–602 (2008)

120. Saulsberry, R.L., Greene, N.J., Hernandez, L.: Stress rupture nondestructive evaluation of composite overwrapped pressure vessels. In: Lulla, K. (ed.) 2011 Biennial Research and Technology Development Report, pp. 287–289. (2011)
121. Chang, J.B.: Implementation guidelines for ANSI/AIAA S-081: space systems composite overwrapped pressure vessels. In: Aerospace Report No. TR-2003(8504)-1, AD A413531, p. 83. (2003)
122. Miyano, Y., Nakada, M., Sekine, N.: Accelerated testing for long-term durability of FRP laminates for marine use. J. Compos. Mater. **39**, 5–20 (2005)
123. Miyano, Y., Nakada, M., Nishigaki, K.: Prediction of long-term fatigue life of quasi-isotropic CFRP laminates for aircraft use. Int. J. Fatigue **28**, 1217–1225 (2006)
124. Nakada, M., Okuya, T., Miyano, Y.: Statistical prediction of tensile creep failure time for unidirectional CFRP. Adv. Compos. Mater. **23**, 451–460 (2014)
125. Wöhler, A.: Über die Festigkeits-Versuche mit Eisen und Stahl. Z. Bauwesen **1–3**, 74–106 (1870)
126. Haibach, E.: Betriebsfestigkeit. Springer, Berlin (2006)
127. Palmgren, A.: Die Lebensdauer von Kugellagern. Z. Ver. Dtsch. Ing. **68**, 339–341 (1924)
128. Degrieck, J., Van Paepegem, W.: Fatigue damage modeling of fibre-reinforced composite materials: review. Appl. Mech. Rev. **54**, 279 (2001)
129. Hwang, W., Han, K.S.: Cumulative damage models and multi-stress fatigue life prediction. J. Compos. Mater. **20**, 125–153 (1986)
130. Barnard, P.M., Butler, R.J., Curtis, P.T.: Composite Structures 3. Springer, Dordrecht (1985)
131. Farrow, I.R.: Damage accumulation and degradation of composite laminates under aircraft service loading: assessment and prediction. Ph.D. Thesis, Cranfield Institute of Technology (1989)
132. Bingham, A.H., Ek, C.W.: Acoustic emission testing of aerial devices and associated equipment used in the utility industries, Ausgabe 1139. ASTM International (1992)
133. Sendeckyj, G.P.: Life prediction for resin-matrix composite materials. In: Reifsnider, K.L. (ed.) Fatigue of Composite Materials. Elsevier B.V., Amsterdam (1990)
134. Hashin, Z., Rotem, A.: A fatigue failure criterion for fiber reinforced materials. J. Compos. Mater. **7**, 448–464 (1973)
135. Ellyin, F., El-Kadi, H.: A fatigue failure criterion for fiber reinforced composite laminae. Compos. Struct. **15**, 61–74 (1990)
136. Reifsnider, K., Gao, Z.: A micromechanics model for composites under fatigue loading. Int. J. Fatigue **13**, 149–156 (1991)
137. Wu, C.M.L.: Thermal and mechanical fatigue analysis of CFRP laminates. Compos. Struct. **25**, 339–344 (1993)
138. Jen, M.: Strength and life in thermoplastic composite laminates under static and fatigue loads. Part II: formulation. Int. J. Fatigue **20**, 617–629 (1998)
139. Jen, M.: Strength and life in thermoplastic composite laminates under static and fatigue loads. Part I: experimental. Int. J. Fatigue **20**, 605–615 (1998)
140. Philippidis, T.P., Vassilopoulos, A.P.: Fatigue strength prediction under multiaxial stress. J. Compos. Mater. **33**, 1578–1599 (1999)
141. Castillo, E., Fernández-Canteli, A., Hadi, A.S.: On fitting a fatigue model to data. Int. J. Fatigue **21**, 97–106 (1999)
142. Hwang, W., Han, K.S.: Fatigue of composites–fatigue modulus concept and life prediction. J. Compos. Mater. **20**, 154–165 (1986)
143. Sidoroff, F., Subagio, B.: Fatigue damage modelling of composite materials from bending tests. In: Sixth International Conference on Composite Materials (ICCM-VI) & Second European Conference on Composite Materials (ECCM-II), vol. 4, pp. 4.32–4.39. London (1987)
144. Van Paepegem, W., Degrieck, J.: Experimental set-up for and numerical modelling of bending fatigue experiments on plain woven glass/epoxy composites. Compos. Struct. **51**, 1–8 (2001)

145. Yang, J.N., Lee, L.J., Sheu, D.Y.: Modulus reduction and fatigue damage of matrix dominated composite laminates. Compos. Struct. **21**, 91–100 (1992)
146. Lee, L.J., Fu, K.E., Yang, J.N.: Prediction of fatigue damage and life for composite laminates under service loading spectra. Compos. Sci. Technol. **56**, 635–648 (1996)
147. Brøndsted, P., Lilholt, H., Andersen, S.I.: Fatigue damage prediction by measurements of the stiffness degradation in polymer matrix composites. In: 8th International Spring Meeting of ICFC. Paris, France (1997)
148. Whitworth, H.A.: A stiffness degradation model for composite laminates under fatigue loading. Compos. Struct. **40**, 95–101 (1997)
149. Chou, P., Croman, R.: Degradation and sudden-death models of fatigue of graphite/epoxy composites. In: Composite Materials: Testing and Design (5th Conf) ASTM STP, (1979)
150. Halpin, C., Jerina, K., Johnson, T.: Characterization of composites for the purpose of reliability evaluation. In: Analysis of the Test Methods for High Modulus Fibers and Composites: A Symposium Presented at a Meeting of Committee D-30 on High Modulus Fibers and Their Composites. (1973)
151. Yang, J., Jones, D.: Load sequence effects on the fatigue of unnotched composite materials. In: Fatigue of Fibrous Composite Materials, pp. 213–232. ASTM STP, (1981)
152. Daniel, I.M., Charewicz, A.: Fatigue damage mechanisms and residual properties of graphite/epoxy laminates. Eng. Fract. Mech. **25**, 793–808 (1986)
153. Rotem, A.: Fatigue and residual strength of composite laminates. Eng. Fract. Mech. **25**, 819–827 (1986)
154. Schaff, J.R., Davidson, B.D.: Life prediction methodology for composite structures. Part II—spectrum fatigue. J. Compos. Mater. **31**, 158–181 (1997)
155. Caprino, G.: Predicting fatigue life of composite laminates subjected to tension–tension fatigue. J. Compos. Mater. **34**, 1334–1355 (2000)
156. Yao, W.X., Himmel, N.: A new cumulative fatigue damage model for fibre-reinforced plastics. Compos. Sci. Technol. **60**, 59–64 (2000)
157. Biner, S.B., Yuhas, V.C.: Growth of short fatigue cracks at notches in woven fiber glass reinforced polymeric composites. J. Eng. Mater. Technol. **111**, 363 (1989)
158. Bergmann, H.W., Prinz, R.: Fatigue life estimation of graphite/epoxy laminates under consideration of delamination growth. Int. J. Numer. Meth. Eng. **27**, 323–341 (1989)
159. Dahlen, C., Springer, G.S.: Delamination growth in composites under cyclic loads. J. Compos. Mater. **28**, 732–781 (1994)
160. Bartley-Cho, J., Gyu Lim, S., Hahn, H.T., Shyprykevich, P.: Damage accumulation in quasi-isotropic graphite/epoxy laminates under constant-amplitude fatigue and block loading. Compos. Sci. Technol. **58**, 1535–1547 (1998)
161. Bucinell, R.B.: Development of a stochastic free edge delamination model for laminated composite materials subjected to constant amplitude fatigue loading. J. Compos. Mater. **32**, 1138–1156 (1998)
162. Schön, J.: A model of fatigue delamination in composites. Compos. Sci. Technol. **60**, 553–558 (2000)
163. Highsmith, A., Reifsnider, K.: Stiffness-reduction mechanisms in composite laminates. In: Damage in Composite Materials, pp. 103–117. ASTM International, Philadelphia, (1982)
164. Reifsnider, K.L.: The critical element model: a modeling philosophy. Eng. Fract. Mech. **25**, 739–749 (1986)
165. Subramanian, S., Reifsnider, K.L., Stinchcomb, W.W.: A cumulative damage model to predict the fatigue life of composite laminates including the effect of a fibre-matrix interphase. Int. J. Fatigue **17**, 343–351 (1995)
166. Talreja, R.: Stiffness properties of composite laminates with matrix cracking and interior delamination. Eng. Fract. Mech. **25**, 751–762 (1986)
167. Ogin, S.L., Smith, P.A., Beaumont, P.W.R.: Matrix cracking and stiffness reduction during the fatigue of a (0/90)s GFRP laminate. Compos. Sci. Technol. **22**, 23–31 (1985)

168. Spearing, S.M., Beaumont, P.W.R., Ashby, M.F.: Fatigue damage mechanics of composite materials. II: a damage growth model. Compos. Sci. Technol. **44**, 169–177 (1992)
169. Spearing, S.M., Beaumont, P.W.R.: Fatigue damage mechanics of composite materials. I: experimental measurement of damage and post-fatigue properties. Compos. Sci. Technol. **44**, 159–168 (1992)
170. Ladevèze, P.: A damage computational method for composite structures. Comput. Struct. **44**, 79–87 (1992)
171. Shokrieh, M.M., Lessard, L.B.: Progressive fatigue damage modeling of composite materials. Part I: modeling. J. Compos. Mater. **34**, 1056–1080 (2000)
172. Shokrieh, M.M., Lessard, L.B.: Progressive fatigue damage modeling of composite materials. Part II: material characterization and model verification. J. Compos. Mater. **34**, 1081–1116 (2000)
173. Reifsnider, K.: Fatigue of Composite Materials. Elsevier, Amsterdam (1990)
174. Field, J.E., Walley, S.M., Proud, W.G., Goldrein, H.T., Siviour, C.R.: Review of experimental techniques for high rate deformation and shock studies. Int. J. Impact Eng. **30**, 725–775 (2004)
175. Gama, B.A., Lopatnikov, S.L., Gillespie, J.W.: Hopkinson bar experimental technique: a critical review. Appl. Mech. Rev. **57**, 223 (2004)
176. Kawai, M., Saito, S.: Off-axis strength differential effects in unidirectional carbon/epoxy laminates at different strain rates and predictions of associated failure envelopes. Compos. Part A Appl. Sci. Manuf. **40**, 1632–1649 (2009)
177. Vogler, T.J., Kyriakides, S.: Inelastic behavior of an AS4/PEEK composite under combined transverse compression and shear. Part I: experiments. Int. J. Plast. **15**, 783–806 (1999)
178. Hsiao, H.M., Daniel, I.M., Cordes, R.D.: Strain rate effects on the transverse compressive and shear behavior of unidirectional composites. J. Compos. Mater. **33**, 1620–1642 (1999)
179. Hosur, M.V., Alexander, J., Vaidya, U.K., Jeelani, S.: High strain rate compression response of carbon/epoxy laminate composites. Compos. Struct. **52**, 405–417 (2001)
180. Tsai, J.-L., Sun, C.T.: Strain rate effect on in-plane shear strength of unidirectional polymeric composites. Compos. Sci. Technol. **65**, 1941–1947 (2005)
181. Koerber, H., Xavier, J., Camanho, P.P.: High strain rate characterisation of unidirectional carbon-epoxy IM7-8552 in transverse compression and in-plane shear using digital image correlation. Mech. Mater. **42**, 1004–1019 (2010)
182. Körber, H., Camanho, P.P.: Characterisation of unidirectional carbon-epoxy IM7-8552 in longitudinal compression under high strain rates. In: DYMAT 2009—9th International Conferences on the Mechanical and Physical Behaviour of Materials Under Dynamic Loading, pp. 185–191. EDP Sciences, Les Ulis, France (2009)
183. Koerber, H., Vogler, M., Kuhn, P., Camanho, P.P.: Experimental characterisation and modelling of non-linear stress–strain behaviour and strain rate effects for unidirectional carbon-epoxy. In: ECCM16—16th European Conference on Composite Materials, pp. 1–8. Sevilla, Spain (2014)
184. Lee, D., Tippur, H., Bogert, P.: Quasi-static and dynamic fracture of graphite/epoxy composites: an optical study of loading-rate effects. Compos. Part B Eng. **41**, 462–474 (2010)
185. Gilat, A., Schmidt, T.E., Walker, A.L.: Full field strain measurement in compression and tensile split hopkinson bar experiments. Exp. Mech. **49**, 291–302 (2008)
186. Ciampa, F., Meo, M., Barbieri, E.: Impact localization in composite structures of arbitrary cross section. Struct. Health Monit. **11**(6), 643–655 (2012)
187. Ciampa, F., Meo, M.: Impact detection in anisotropic materials using a time reversal approach. Struct. Health Monit. **11**, 43–49 (2011)
188. Gorman, M.R., Ziola, S.M.: Hypervelocity impact (HVI). WLE Small-Scale Fiberglass Panel Flat Target C-2, vol. 4. (2007)
189. Gorman, M.R., Ziola, S.M.: Hypervelocity impact (HVI). WLE High Fidelity Specimen Fg (RCC)-2, vol. 6. (2007)

190. Bleier, A.: Prüfverfahren zur Bestimmung exakter Werkstoffkennwerte einer unidirektionalen Schicht unter besonderer Berücksichtigung physikalischer Nichtlinearitäten. Ph.D. Thesis, Technical University Darmstadt (2011)
191. Basan, R.: Untersuchung der intralaminaren Schubeigenschaften von Faserverbundwerkstoffen mit Epoxidharzmatrix unter Berücksichtigung nichtlinearer Effekte. Ph.D. Thesis, Technical University Berlin (2011)
192. Blackman, B.R.K., Brunner, A.J., Williams, J.G.: Mode II fracture testing of composites: a new look at an old problem. Eng. Fract. Mech. **73**, 2443–2455 (2006)
193. Lomov, S.V., Boisse, P., Deluycker, E., Morestin, F., Vanclooster, K., Vandepitte, D., Verpoest, I., Willems, A.: Full-field strain measurements in textile deformability studies. Compos. Part A Appl. Sci. Manuf. **39**, 1232–1244 (2008)
194. Pierron, F.: Application of full-field measurement techniques to composite materials and structures. Compos. Part A Appl. Sci. Manuf. **39**, 1193 (2008)
195. Müller, B.R., Hentschel, M.P., Harbich, K.-W., Lange, A., Schors, J.: X-ray refraction topography for non-destructive evaluation of advanced materials. In: 15th World Conference on NDT, Roma, Italy (2000)
196. Baensch, F., Zauner, M., Sanabria, S.J., Sause, M.G.R., Pinzer, B.R., Brunner, A.J., Stampanoni, M., Niemz, P.: Damage evolution in wood: synchrotron radiation micro-computed tomography (SRµCT) as a complementary tool for interpreting acoustic emission (AE) behavior. Holzforschung **69** (2015)
197. Scott, A.E., Hepples, W., Kalantzis, N., Wright, P., Mavrogordato, M.N., Sinclair, I., Spearing, S.M.: High resolution damage detection of loaded carbon/epoxy laminates using synchrotron radiation computed tomography. In: ICCM-18 18th International Conference on Composite Materials, pp. 1–6. ICC Jeju, Korea (2011)
198. Patel, A.J., Sottos, N.R., Wetzel, E.D., White, S.R.: Autonomic healing of low-velocity impact damage in fiber-reinforced composites. Compos. Part A Appl. Sci. Manuf. **41**, 360–368 (2010)
199. Bagheri, Z.S., Tavakkoli Avval, P., Bougherara, H., Aziz, M.S.R., Schemitsch, E.H., Zdero, R.: Biomechanical analysis of a new carbon fiber/flax/epoxy bone fracture plate shows less stress shielding compared to a standard clinical metal plate. J. Biomech. Eng. **136**, 091002 (2014)
200. Abello, L.S., Marco, Y., Le Saux, V., Robert, G., Charrier, P.: Fast prediction of the fatigue behavior of short fiber reinforced thermoplastics from heat build-up measurements. Procedia Eng. **66**, 737–745 (2013)
201. Gornet, L., Wesphal, O., Burtin, C., Bailleul, J.-L., Rozycki, P., Stainier, L.: Rapid determination of the high cycle fatigue limit curve of carbon fiber epoxy matrix composite laminates by thermography methodology: tests and finite element simulations. Procedia Eng. **66**, 697–704 (2013)
202. Francis, D., Tatam, R.P., Groves, R.M.: Shearography technology and applications: a review. Meas. Sci. Technol. **21**, 102001 (2010)
203. Hung, Y.Y., Liang, C.Y.: Image-shearing camera for direct measurement of surface strains. Appl. Opt. **18**, 1046–51 (1979)
204. Hung, Y.Y.: Shearography for non-destructive evaluation of composite structures. Opt. Lasers Eng. **24**, 161–182 (1996)
205. Hung, Y.Y.: Shearography: a new optical method for strain measurement and nondestructive testing. Opt. Eng. **21**, 213391 (1982)
206. Wang, W.-C., Su, C.-W., Liu, P.-W.: Full-field non-destructive analysis of composite plates. Compos. Part A Appl. Sci. Manuf. **39**, 1302–1310 (2008)
207. Gryzagoridis, J., Findeis, D.: Benchmarking shearographic NDT for composites. Insight non-destructive test. Cond. Monit. **50**, 249–252 (2008)
208. De Angelis, G., Meo, M., Almond, D.P., Pickering, S.G., Angioni, S.L.: A new technique to detect defect size and depth in composite structures using digital shearography and unconstrained optimization. NDT E Int. **45**, 91–96 (2012)

209. Pezzoni, R., Krupka, R.: Laser-shearography for non-destructive testing of large-area composite helicopter structures. In: 15th World Conference on NDT, Roma, Italy (2000)
210. Leonard, K.R., Malyarenko, E.V., Hinders, M.K.: Ultrasonic Lamb wave tomography. Inverse Probl. **18**, 1795–1808 (2002)
211. Hay, T.R., Royer, R.L., Gao, H., Zhao, X., Rose, J.L.: A comparison of embedded sensor Lamb wave ultrasonic tomography approaches for material loss detection. Smart Mater. Struct. **15**, 946–951 (2006)
212. Jansen, D.P., Hutchins, D.A., Mottram, J.T.: Lamb wave tomography of advanced composite laminates containing damage. Ultrasonics **32**, 83–90 (1994)
213. Prasad, S.M., Balasubramaniam, K., Krishnamurthy, C.: V: Structural health monitoring of composite structures using Lamb wave tomography. Smart Mater. Struct. **13**, N73–N79 (2004)
214. Solodov, I., Wackerl, J., Pfleiderer, K., Busse, G.: Nonlinear self-modulation and subharmonic acoustic spectroscopy for damage detection and location. Appl. Phys. Lett. **84**, 5386 (2004)
215. Solodov, I., Busse, G.: Nonlinear air-coupled emission: the signature to reveal and image microdamage in solid materials. Appl. Phys. Lett. **91**, 251910 (2007)
216. Krohn, N., Stoessel, R., Busse, G.: Acoustic non-linearity for defect selective imaging. Ultrasonics **40**, 633–637 (2002)
217. Pappas, Y.Z., Kostopoulos, V.: Toughness characterization and acoustic emission monitoring of a 2-D carbon/carbon composite. Eng. Fract. Mech. **68**, 1557–1573 (2001)
218. Maire, E., Carmona, V., Courbon, J., Ludwig, W.: Fast X-ray tomography and acoustic emission study of damage in metals during continuous tensile tests. Acta Mater. **55**, 6806–6815 (2007)
219. Bohse, J., Kroh, G.: Micromechanics and acoustic emission analysis of the failure process of thermoplastic composites. J. Mater. Sci. **27**, 298–306 (1992)
220. Haselbach, W., Lauke, B.: Acoustic emission of debonding between fibre and matrix to evaluate local adhesion. Compos. Sci. Technol. **63**, 2155–2162 (2003)
221. Bohse, J.: Damage analysis of polymer matrix composites by acoustic emission testing. In: EWGAE 2004—26th European Conference on Acoustic Emission Testing, pp. 339–348. Berlin, Germany (2004)
222. Cuadra, J., Vanniamparambil, P.A., Hazeli, K., Bartoli, I., Kontsos, A.: Damage quantification in polymer composites using a hybrid NDT approach. Compos. Sci. Technol. **83**, 11–21 (2013)
223. Unnthorsson, R., Runarsson, T., Jonsson, M.: Monitoring the evolution of individual AE sources in cyclically loaded FRP composites. J. Acoust. Emiss. **25**, 253–259 (2007)
224. Surgeon, M., Vanswijgenhoven, E., Wevers, M., Van Der Biest, O.: Acoustic emission during tensile testing of SiC-fibre-reinforced BMAS glass-ceramic composites. Compos. Part A Appl. Sci. Manuf. **28**, 473–480 (1997)
225. Ramirez-Jimenez, C.R., Papadakis, N., Reynolds, N., Gan, T.H., Purnell, P., Pharaoh, M.: Identification of failure modes in glass/polypropylene composites by means of the primary frequency content of the acoustic emission event. Compos. Sci. Technol. **64**, 1819–1827 (2004)
226. Narisawa, I., Oba, H.: An evaluation of acoustic emission from fibre-reinforced composites. J. Mater. Sci. **19**, 1777–1786 (1984)
227. Huguet, S., Godin, N., Gaertner, R., Salmon, L., Villard, D.: Use of acoustic emission to identify damage modes in glass fibre reinforced polyester. Compos. Sci. Technol. **62**, 1433–1444 (2002)
228. Anastassopoulos, A.A., Philippidis, T.P.: Clustering methodology for the evaluation of acoustic emission from composites. J. Acoust. Emiss. **13**, 11–21 (1995)
229. Giordano, M., Calabro, A., Esposito, C., D'Amore, A., Nicolais, L.: An acoustic-emission characterization of the failure modes in polymer-composite materials. Compos. Sci. Technol. **58**, 1923–1928 (1998)

230. Scholey, J.J., Wilcox, P.D., Wisnom, M.R., Friswell, M.I.: Quantitative experimental measurements of matrix cracking and delamination using acoustic emission. Compos. Part A Appl. Sci. Manuf. **41**, 612–623 (2010)
231. Waller, J.M., Nichols, C.T., Wentzel, D.J., Saulsberry, R.L., Thompson, D.O., Chimenti, D. E.: Use of modal acoustic emission to monitor damage progression in carbon fiber/epoxy composites. In: AIP Conference Proceedings, pp. 919–926. San Diego, (2011)
232. Plöckl, M., Sause, M.G.R., Scharringhausen, J., Horn, S.: Failure analysis of nol-ring specimens by acoustic emission. In: 30th European Conference on Acoustic Emission, pp. 1–12. Granada, Spain (2012)
233. Sause, M.G.R., Müller, T., Horoschenkoff, A., Horn, S.: Quantification of failure mechanisms in mode-I loading of fiber reinforced plastics utilizing acoustic emission analysis. Compos. Sci. Technol. **72**, 167–174 (2012)
234. Ono, K., Gallego, A.: Research and applications of AE on advanced composites. J Acoust. Emiss. **30**, 180–229 (2012)
235. Anastassopoulos, A., Tsimogiannis, A., Kouroussis, D.: Unsupervised classification of acoustic emission sources from aerial man lift devices. In: 15th World Conference on NDT, Roma, Italy (2000)
236. Anastassopoulos, A.A., Kouroussis, D.A., Nikolaidis, V.N., Proust, A., Dutton, A.G., Blanch, M.J., Jones, L.E., Vionis, P., Lekou, D.J., van Delft, D.R. V, Joosse, P.A., Philippidis, T.P., Kossivas, T., Fernando, G.: Structural integrity evaluation of wind turbine blades using pattern recognition analysis on acoustic emission data. In: Proceedings of the 25th European Conference on Acoustic Emission Testing. Prague, Czech Republic (2002)
237. Grosse, C.U., Ohtsu, M.: Acoustic Emission Testing. Springer, Berlin (2008)
238. Weihnacht, B., Schulze, E., Frankenstein, B.: Acoustic emission analysis in the dynamic fatigue testing of fiber composite components. In: 31st Conference of the European Working Group on Acoustic Emission, pp. 1–8. Dresden, Germany (2014)
239. Abraham, A.R.A., Johnson, K.L., Nichols, C.T., Saulsberry, R.L., Waller, J.M.: Use of statistical analysis of acoustic emission data on carbon-epoxy COPV materials-of-construction for enhanced felicity ratio onset determination—JSC-CN-26080. (2011)
240. Downs, K.S., Hamstad, M.A.: Acoustic emission from depressurization to detect/evaluate significance of impact damage to graphite/epoxy pressure vessels. J. Compos. Mater. **32**, 258–307 (1998)

Chapter 3
Digital Image Correlation

In order to detect the occurrence of failure in a material, visual observation is a very intuitive approach. Although global failure is many times easier to observe visually, more dedicated analysis is required to detect small changes indicating the occurrence of first failure. One specific type of visual observation follows the idea to track and measure changes between subsequently acquired images of an object. The concept used for that purpose is based on digital image acquisition of a moving and deforming object and is known as digital image correlation (DIC). The background of this technique stems from the basic principles of optics, perspective imaging, and digital image analysis.

The aim of this chapter is not to cover all aspects of the topic of DIC, but to introduce the key concepts and point out the relevant principles for measurement of strain fields of solid materials. For the interested reader, a comprehensive review on the topic is given by Sutton, Orteu, and Schreier [1] and a concise review on the application of 2D-DIC is also found in [2].

One of the first attempts to use image correlation as a measurement technique to register features in images of different views was made by Hobrough in the 1950s. A few years later in 1961 he also developed a system to perform first quantitative positional measurements extracted from an image correlation process. With the availability of digitized images as provided by CMOS or CCD cameras in the 1960s and 1970s until the present days the branch of vision-based algorithms with a focus on DIC techniques has become a growing field of research. The development of the algorithms of the DIC method is driven by the accompanying applications in material science. Since the first experimental application in materials science by Peters and Ranson in 1982 [3], many experimental studies in the field of fiber reinforced composites were accompanied by DIC as a measurement technique [4–12]. In recent decades, suitable approaches have been developed to measure deformations with high magnification (e.g., inside a scanning electron microscope) and high-speed camera systems have been used to allow digital image correlation during high strain rate testing [13, 14].

M.G.R. Sause, *In Situ Monitoring of Fiber-Reinforced Composites*,
Springer Series in Materials Science 242, DOI 10.1007/978-3-319-30954-5_3

This chapter starts with a description of the mathematical and physical principles used by 2D-DIC and 3D-DIC. Focus is given on the parameters that limit detectability for state-of-the-art equipment and comparison is made between theoretical predictions and measurements under laboratory conditions. The second part of the chapter has its focus on the theoretical description of strain concentration due to internal flaws, especially for those types found in fiber reinforced composites. Measurements with a variety of defect types are compared to calculations using finite element modeling. Detectability of failure mechanisms in fiber reinforced composites is discussed as function of failure type, failure size, and depth position below surface. The last section elaborates some typical applications of DIC measurements in conjunction with testing of fiber reinforced composites.

3.1 Principle of Operation

The idea of DIC techniques is to yield a quantitative measure of motion, rotation, and deformation occurring between two subsequently acquired images of an object as seen in Fig. 3.1.

In general, it is not possible to find the correspondence of one pixel in a subsequent image, since the gray value of the single pixel can be found at many other pixel positions in the subsequent image too. Therefore, there is no unique correspondence of one single pixel position in subsequently acquired images. Instead, a small neighborhood around the pixel of interest is considered to form a more characteristic pattern to detect in the subsequent images (see Fig. 3.1). This small neighborhood is usually referred to as subset, window, or facet. However, if repeating structures such as grids of points, lines, or otherwise regular patterns are used, there is still no unique correspondence between the images. This matching problem is known as the correspondence problem. One effective way to circumvent this lack of unique correspondence is to use a non-periodic, stochastic pattern as the speckle pattern shown in Fig. 3.1. In order to allow for comparable spatial

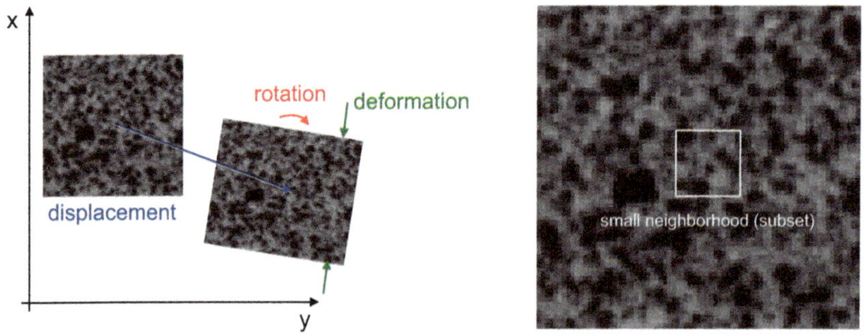

Fig. 3.1 Tracking deformation states between images and speckle pattern including small neighborhoods (subset) used for image matching process

resolution within the image a narrow size distribution of the image features (e.g., black dots on white background) is required.

For the matching process between two images, various algorithms are applied in practice [1, 15]. In the following, the Lucas-Kanade tracker algorithm [16] is used as an example to demonstrate the basic principles of the digital image correlation process. Using the small neighborhoods around the pixel of interest, a subset of the full image is defined. The intensity distribution of the reference subset is denoted as $\widetilde{F}(\boldsymbol{r})$ and the intensity distribution of the subset after displacement by $\widetilde{G}(\boldsymbol{r})$. Now, the goal is to find the 2D-displacement vector $\boldsymbol{u}_{\text{opt}}$ between these two subsets, which can be done by minimization of the sum of squares deviation:

$$\boldsymbol{u}_{\text{opt}} = \min \sum \left| \widetilde{G}(\boldsymbol{r} + \boldsymbol{u}) - \widetilde{F}(\boldsymbol{r}) \right|^2 \tag{3.1}$$

Such optimization routines are carried out using a cost function χ written as first-order Taylor series

$$\chi^2(u_x + \Delta_x,\ u_y + \Delta_y) = \sum \left| \widetilde{G}(\boldsymbol{r} + \boldsymbol{u}) - \frac{\partial \widetilde{G}}{\partial x}\Delta_x - \frac{\partial \widetilde{G}}{\partial y}\Delta_y - \widetilde{F}(\boldsymbol{r}) \right|^2 \tag{3.2}$$

This starts with the present estimates u_x and u_y for the initial 2D-displacement vector and the incremental motion updates Δ_x and Δ_y in the current iteration. Taking partials of (3.2) with respect to Δ_x and Δ_y and setting them to zero results in a linear equation system being solved incrementally

$$\begin{pmatrix} \Delta_x \\ \Delta_y \end{pmatrix} = \begin{pmatrix} \sum \left(\frac{\partial \widetilde{G}}{\partial x} \right)^2 & \sum \frac{\partial \widetilde{G}}{\partial x} \frac{\partial \widetilde{G}}{\partial y} \\ \sum \frac{\partial \widetilde{G}}{\partial x} \frac{\partial \widetilde{G}}{\partial y} & \sum \left(\frac{\partial \widetilde{G}}{\partial y} \right)^2 \end{pmatrix}^{-1} \begin{pmatrix} \sum \frac{\partial G}{\partial x} \left(\widetilde{F} - \widetilde{G} \right) \\ \sum \frac{\partial G}{\partial y} \left(\widetilde{F} - \widetilde{G} \right) \end{pmatrix} \tag{3.3}$$

This optimization procedure iteratively refines the estimate for $\boldsymbol{u}_{\text{opt}}$ in p iterations using $\boldsymbol{u}^{p+1} = \boldsymbol{u}^p + \boldsymbol{\Delta}$ until convergence is achieved. This algorithm can suitably detect arbitrarily large 2D-displacements as long as the initial estimate is within the convergence radius of the method. However, in many applications the image matching process needs to account for high distortions, such as additional rotations of the image. This is taken into account using shape functions $\xi(\boldsymbol{r},\boldsymbol{p})$ that transform pixel coordinates in the reference subset into the coordinates of the deformed subset. This requires modification of the cost function of (3.2) in the following way

$$\chi^2(\boldsymbol{p}) = \sum (\tilde{G}\left(\xi(\boldsymbol{r},\boldsymbol{p})\right) - \tilde{F}(\boldsymbol{r}))^2 \tag{3.4}$$

Optimization is carried out for the parameter vector $\boldsymbol{p}$ of the shape function. Taking derivatives of (3.4) with respect to all parameters of $\boldsymbol{p}$ for the affine shape functions, the parameter update can be written as

$$\boldsymbol{\Delta p} = \boldsymbol{H}^{-1}\boldsymbol{q} \tag{3.5}$$

Using the symmetric Hessian matrix

$$H = \begin{pmatrix} \sum \tilde{G}_x^2 & \sum \tilde{G}_x\tilde{G}_y & \sum \tilde{G}_x^2 x & \sum \tilde{G}_x^2 y & \sum \tilde{G}_x\tilde{G}_y x & \sum \tilde{G}_x\tilde{G}_y y \\ & \sum \tilde{G}_y^2 & \sum \tilde{G}_x\tilde{G}_y x & \sum \tilde{G}_x\tilde{G}_y y & \sum \tilde{G}_y^2 x & \sum \tilde{G}_y^2 y \\ & & \sum \tilde{G}_x^2 x^2 & \sum \tilde{G}_x^2 xy & \sum \tilde{G}_x\tilde{G}_y x^2 & \sum \tilde{G}_x\tilde{G}_y xy \\ & & & \sum \tilde{G}_x^2 y^2 & \sum \tilde{G}_x\tilde{G}_y xy & \sum \tilde{G}_x\tilde{G}_y y^2 \\ & & & & \sum \tilde{G}_y^2 x^2 & \sum \tilde{G}_y^2 xy \\ & & & & & \sum \tilde{G}_y^2 y^2 \end{pmatrix} \tag{3.6}$$

and

$$\boldsymbol{q} = \begin{pmatrix} \sum \tilde{G}_x\left(\tilde{F} - \tilde{G}\right) \\ \sum \tilde{G}_y\left(\tilde{F} - \tilde{G}\right) \\ \sum \tilde{G}_x x\left(\tilde{F} - \tilde{G}\right) \\ \sum \tilde{G}_x y\left(\tilde{F} - \tilde{G}\right) \\ \sum \tilde{G}_y x\left(\tilde{F} - \tilde{G}\right) \\ \sum \tilde{G}_y y\left(\tilde{F} - \tilde{G}\right) \end{pmatrix} \tag{3.7}$$

denoting partial differentiations by subscripts, e.g., $\tilde{G}_x = \frac{\partial \tilde{G}}{\partial x}$.

The coordinates (x_i, y_j) and (x_i^*, y_j^*) are related by the deformation that occurs between the two images. If the motion is perpendicular to the optical axis of the camera, then the relation between (x_i, y_j) and (x_i^*, y_j^*) can be approximated by a shape function resembling a 2D affine transformation:

$$x^* = x + u + \frac{\partial u}{\partial x}\Delta x + \frac{\partial u}{\partial y}\Delta y \tag{3.8a}$$

$$y^* = y + v + \frac{\partial v}{\partial x}\Delta x + \frac{\partial v}{\partial y}\Delta y \tag{3.8b}$$

Here u and v are translations of the center of the sub-image in the x and y direction, respectively. The distances from the center of the sub-image to the point (x, y) are denoted by Δx and Δy, and $\frac{\partial u}{\partial x}, \frac{\partial u}{\partial y}, \frac{\partial v}{\partial x}, \frac{\partial v}{\partial y} \cdots$ are the displacement gradients. The latter are straightforward to be interpreted as components of the local strain tensor. If evaluated relative to the global x- and y-direction the displacement gradients can readily be converted to the axial and transverse strain components with respect to the axial load axis of a test specimen. Another usual conversion of the local strain tensor is to calculate a principal component analysis to obtain the first, second, and third[1] principal strain values. The principal strain values have been found to be particularly useful to detect strain anomalies (see Sect. 3.3), whereas for homogeneous strain fields the first principal strain value simply coincides with the axial strain value (if this is the dominant deformation direction) and the second principal strain coincides with the transverse strain value, respectively.

Additional considerations are required if offsets and scaling of the lighting conditions occur during acquisition of the images. In practical applications these effects will always be present to some extent and may falsify the matching process due to systematic shifts or scales of the recorded gray values. As has been shown in [1] such effects can be compensated using a modified gray value distribution $\widetilde{G}' = \widetilde{a}\widetilde{G} + \widetilde{b}$ introducing an offset $\widetilde{b}$ and a scale parameter $\widetilde{a}$. Consequently, the cost function formulated in (3.4) is modified to yield

$$\chi^2 = \sum \left(\widetilde{a}\widetilde{G}_i + \widetilde{b} - \widetilde{F}_i\right)^2 \tag{3.9}$$

Therefore the iterative minimization procedure has to account for two additional parameters $\widetilde{a}$ and $\widetilde{b}$. Optimal estimators of these two parameters can be derived from two equations

$$\frac{\partial \chi^2}{\partial \widetilde{a}} = 2\sum \left(\widetilde{a}\widetilde{G}_i + \widetilde{b} - \widetilde{F}_i\right)\widetilde{G}_i \tag{3.10a}$$

$$\frac{\partial \chi^2}{\partial \widetilde{a}} = 0 \rightarrow \widetilde{a}_{\text{opt}} = \frac{\sum (\widetilde{F}_i - \widetilde{b})\widetilde{G}_i}{\sum \widetilde{G}_i^2}$$

$$\frac{\partial \chi^2}{\partial \widetilde{b}} = 2\sum \left(\widetilde{a}\widetilde{G}_i + \widetilde{b} - \widetilde{F}_i\right) \tag{3.10b}$$

$$\frac{\partial \chi^2}{\partial \widetilde{b}} - 0 \rightarrow \widetilde{b}_{\text{opt}} = \frac{\sum \widetilde{F}_i - \widetilde{a}\widetilde{G}_i}{\sum 1}$$

Using $\left\langle \widetilde{F}_i \right\rangle = \widetilde{F}_i - \left\langle \widetilde{F} \right\rangle$ and $\left\langle \widetilde{G}_i \right\rangle = \widetilde{G}_i - \left\langle \widetilde{G} \right\rangle$ one can write the optimal parameter estimates as

[1] The third principal strain component is only applicable for 3D-DIC measurements.

$$\tilde{a}_{\text{opt}} = \frac{\sum \left\langle \tilde{F}_i \right\rangle \left\langle \tilde{G}_i \right\rangle}{\sum \left\langle G_i \right\rangle^2} \tag{3.11a}$$

$$\tilde{b}_{\text{opt}} = \left\langle \tilde{F} \right\rangle - \left\langle \tilde{F}\tilde{G} \right\rangle \frac{\sum \left\langle \tilde{F}_i \right\rangle \left\langle \tilde{G}_i \right\rangle}{\sum \left\langle \tilde{G}_i \right\rangle^2} \tag{3.11b}$$

which allows for writing the cost functional solely in terms of the gray value distributions $\tilde{F}$ and $\tilde{G}$

$$\chi^2 = \sum \left(\left(\frac{\sum \left\langle \tilde{F}_i \right\rangle \left\langle \tilde{G}_i \right\rangle}{\sum \left\langle G_i \right\rangle^2} \tilde{G}_i - \left\langle \tilde{G} \right\rangle \frac{\sum \left\langle \tilde{F}_i \right\rangle \left\langle \tilde{G}_i \right\rangle}{\sum \left\langle \tilde{G}_i \right\rangle^2} \right) - \tilde{F}_i + \left\langle \tilde{F} \right\rangle \right)^2 \tag{3.12}$$

Until this point it was assumed that the object shows only negligible out-of-plane motion between subsequent images. Using a single camera system noticeable out-of-plane motion of the object with respect to the camera will introduce virtual displacement gradients. These will corrupt the in-plane measurements and may make it difficult or impossible to separate the true deformations and "pseudo image deformations" introduced by the out-of-plane motion. As shown in [17] a relative motion away from the image plane decreases image magnification and introduces a negative normal strain. In [17] it was demonstrated that measurements by a regular 2D-DIC system result in a relationship of the normal strain ε to the out-of-plane motion w in a linear fashion resulting in a proportionality of $\varepsilon = 1524 \cdot 10^{-6} \cdot w$.

In the typical 2D-DIC configuration only one camera is used to acquire images of the object. Such a single camera performs a perspective transformation of a 3D object point to a 2D image point. Therefore, information of the third dimension is irreversibly reduced during this operation. As shown in Fig. 3.2 this causes two points $\overline{\boldsymbol{O}}_1$ and $\overline{\boldsymbol{O}}_2$ being projected to the same image point $\overline{\boldsymbol{P}}_1 = \overline{\boldsymbol{P}}_2$. In contrast, the usage of two camera systems with different viewing angles as shown in Fig. 3.2 allows for recovery of the 3D position of the object point.

Using such a stereo vision system the measurement error due to out-of-plane motion can almost be compensated. Experiments for a 3D-DIC system demonstrated the measured strain error to be linear with the out-of-plane motion by a factor of only $1.31 \cdot 10^{-6}$ [17]. This is four orders of magnitude lower than for the recordings using a 2D-DIC system.

However, for such stereo matching processes, further modifications to (3.10a) and (3.10b) are required, since the nonlinearity of perspective projection does not allow usage of affine transformations even if the object is a plane (for details see [1]).

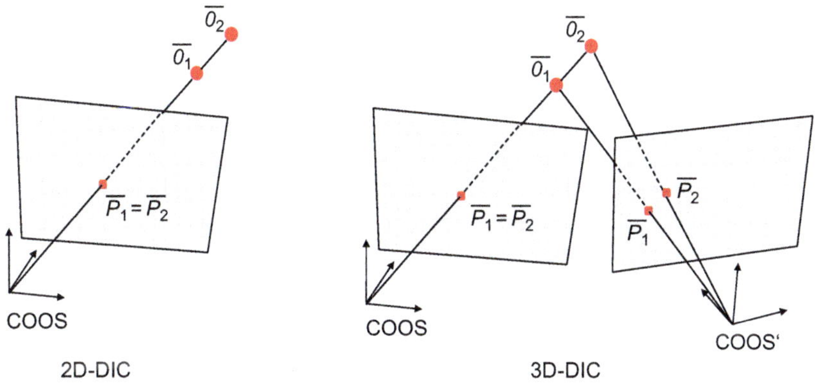

Fig. 3.2 Camera configuration as usually used in 2D-DIC (**a**) and stereo vision process to measure 3D object positions used in 3D-DIC (**b**)

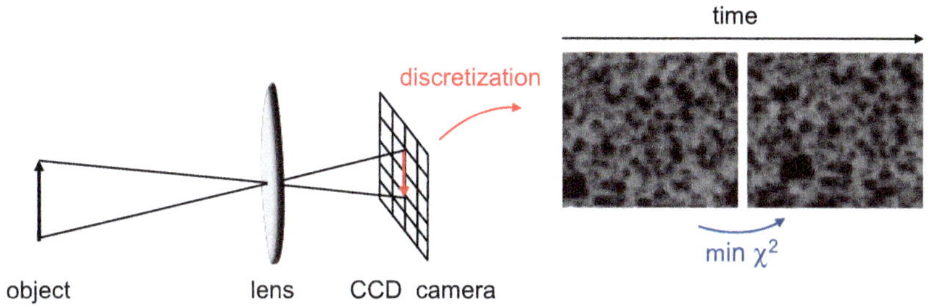

Fig. 3.3 Acquisition chain of DIC comprising steps of optical imaging using lens system, digital image acquisition, and image correlation

3.2 System Accuracy

For the purpose of materials testing, the extraction of quantitative displacement and strain values out of the acquired images is of largest practical interest. Therefore concise knowledge of the measurement accuracy is required. In this context one can distinguish between several factors, which influence the finally obtained measurement values. As seen in Fig. 3.3, one part deals with the measurement errors introduced by the optics of the measurement system. Lens distortions such as spherical aberration, coma distortion, astigmatism, curvature of field, and linear distortion are introduced by the system optics used in the image acquisition process. However, these effects can be sufficiently compensated using accurate calibration procedures to map the object coordinates and the image coordinates [1]. Descriptions of these effects are extensively covered in standard literature on optics [18, 19]. In the following, the discussion is thus focused on the measurement errors arising due to the digital image acquisition, due to the image correlation process and the computations in the stereo-vision process.

3.2.1 Error Sources

The DIC method is known to measure displacements with sub-pixel accuracy and lateral surface strains in the range $<10^{-3}$. There are several sources influencing the system resolution, which can be due to intrinsic noise of the acquired images, statistical and systematical errors introduced by the system calibration, sub-pixel effects resulting from the camera resolution [1, 20–22] or due to intrinsic uncertainties of the image correlation algorithm.

As first category, the correlation errors describe uncertainties of the correlation of corresponding subset positions in subsequent images. These correlation errors can be further divided in systematic and random errors. A second category arises for the usage of 3D-DIC methods. Here, the 3D-coordinates reconstruction method introduces further error sources for the finally computed values. Although the displacement and strain values are the quantities of interest for the purpose of materials testing, the primary measurement value in DIC is the pixel (px). The link between these pixel shifts and the computed displacement and strain values will thus be established in Sect. 3.2.2.

3.2.1.1 Systematic Errors

Systematical errors are caused by sub-pixel effects, which occur because of the discretization of the speckle pattern by the CCD camera pixels. In order to obtain sub-pixel accuracy of the image correlation, the cost function χ^2 has to be evaluated at non-integer locations of the pixel array. Therefore, gray values have to be interpolated between the sample points (pixels). As interpolation is a method of approximating a value between two samples, the use of gray value interpolation in the matching algorithm can be expected to introduce errors. Another factor likely to introduce errors in the sub-pixel image correlation is the highly nonlinear distortions of the subsets.

As has been investigated in detail in [1, 21, 23] there are three main factors considered as sources for systematic errors:

- Image noise level
- Choice of subset size
- Choice of speckle pattern size

The typical procedure to assess the influence of these parameters is the evaluation of synthetic or experimental speckle patterns. These speckle patterns are moved in one direction by a known value resulting in a sub-pixel displacement. Therefore, the reference displacement is known and the mismatch of the values calculated by the DIC method can be compared relative to that reference. A qualitative result of such an investigation is shown in Fig. 3.4. As reported previously [1, 21, 23, 24], there is a typical sinusoidal fluctuation of the matching error, which is common to most of the DIC algorithms. For most of the DIC

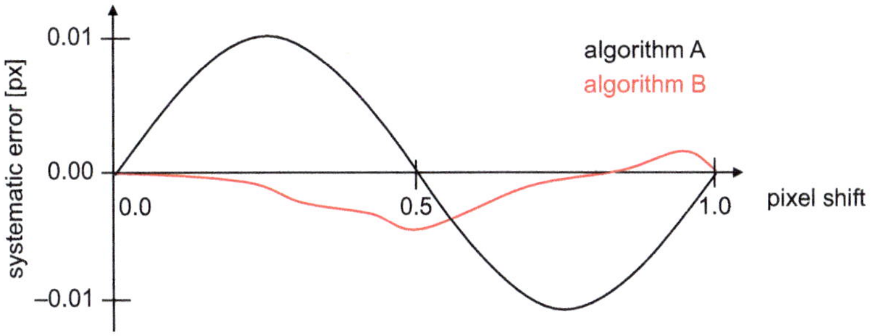

Fig. 3.4 Schema representation of the systematic matching error as function of the DIC matching algorithm. Absolute values of pixel mismatch and detailed behavior depend on the respective algorithm (see [21, 23] for details)

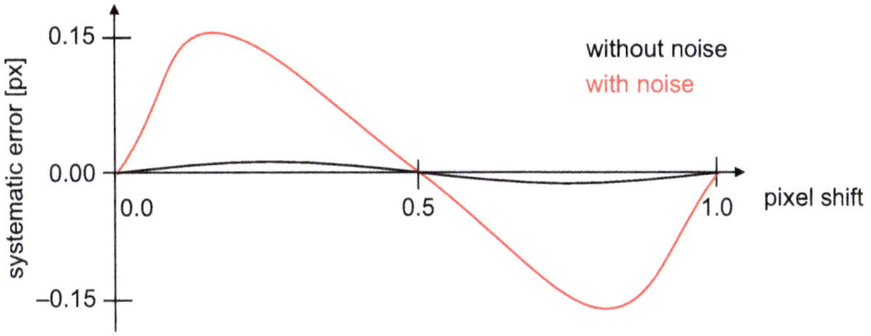

Fig. 3.5 Schema representation of the systematic matching error as function of noise present in the images. Absolute values of pixel mismatch and detailed behavior depend on the respective algorithm (see [21, 23] for details)

algorithms investigated the matching error approaches zero at the integer positions, i.e., at 0 and 1 px displacement. Some of the algorithms also approach zero matching error at 0.5 px but others tend to fluctuate without noticeable structure. Therefore, Fig. 3.4 is only understood as qualitative representation, whereas the quantitative values of the matching error depend on the respective DIC algorithm and on the subset shape functions used therein.

For random noise added to the image, the overall performance of the matching process starts to decrease. As has been shown by [21, 23], there are some DIC algorithms, which are almost unaffected by the presence of random noise in the images, while other algorithms show almost linear dependence on random noise superimposed to a displaced speckle pattern (cf. Fig. 3.5). For the typical noise values as expected in real measurements a broad range of matching errors was found. Whilst the best algorithm was found to be constantly below 0.005 px mismatch, some of the algorithms result in maximum values of up to 0.25 px mismatch.

Similar effects have been found for the subset sizes. Most of the algorithms yield almost no characteristics in the mismatch error as function of different subset sizes (e.g., spanning a typical range from 6 to 32 px). However, some algorithms yield pronounced dependence on the subset size, potentially causing a magnification of the mismatch error by a factor of four for a decreasing subset size.

In general, the same effect could be expected for the speckle pattern size. Since the subset is the discretized version of the real speckle pattern, both size choices can have a similar impact. Therefore it is more convenient to investigate the effect of the speckle pattern normalized by the subset size to assess the effects of this discretization. The findings of [21, 23] can be compared to those derived from the expression of the systematic error developed by [22]. In their expression, the systematic error is calculated taking into account the gray value gradients of the initial and translated (or deformed) images as well as the noise level. Hence, there is a common tendency of most of the algorithms to increase their sensitivity to the presence of noise, when coarse speckle patterns are used.

3.2.1.2 Random Errors

Random matching errors occur because of the limited number of pixels and corresponding gray values in each subset, and the sub-pixel matching process. The corresponding statistical error decreases with the square root of the number of subset pixels used for the matching process. Additional random error sources are statistical noise of the gray values, different illumination settings for the two cameras, image contrast, and size of the speckle pattern on the specimen surface. Such random matching errors can often be compensated by smoothing operations or by averaging of the acquired images.

The random matching error is evaluated using the same procedures as the systematic matching error. The main difference is the evaluation of the error quantity. While the systematic matching error is understood as global displacement error of the image correlation process, the random matching error is understood as standard deviation of all of the individual subset results.

As demonstrated by [21, 23], the mean random matching error generally decreases with the subset size and increases with the noise level as schematically seen in Fig. 3.6. Roux and Hild [25] have proposed an analytical expression of the random matching error showing that it scales linearly with image noise and pixel size and with the inverse of the mean square gradient and subset size.

However, based on the different implementations of DIC algorithms, some of the algorithms show a mean random matching error independent of the speckle size. As for the systematic matching error, the random matching error is a function of the interpolation process which generates individual gray value gradients. Therefore, each algorithm tends to show unique sensitivity on these factors, causing intrinsic complexity to the prediction of the random matching error.

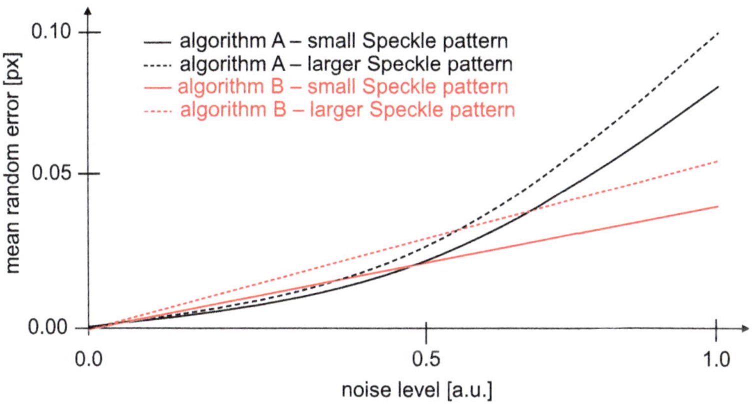

Fig. 3.6 Schema representation of the mean random matching error as function of the speckle pattern size and the noise level. Absolute values of pixel mismatch and detailed behavior depend on the respective algorithm (see [21, 23] for details)

3.2.1.3 Stereo Vision

As indicated in Fig. 3.2, the 3D-DIC approach uses the stereo vision concept to derive the 3D position of an object. The errors introduced by this matching procedure add to the errors discussed for the image matching process. Here, the uncertainties of the calibration parameters translate into errors, when reconstructing the 3D coordinates from correlated subsets of the two cameras.

Calibration errors appear in a systematical manner as a function of the subset positions in the camera images, causing local distortions of the reconstructed 3D space. In order to get an understanding of the impact on the evaluated data, the distortion effects can be described by the following approximation:

If $\overline{\boldsymbol{P}} = (x, y, z)$ is the real position of a reconstructed point, the measured coordinate $\overline{\boldsymbol{P}_{\text{exp}}}$ will differ from $\overline{\boldsymbol{P}}$ by a deviation vector $\Delta\overline{\boldsymbol{P}}$. In the vicinity of a point $\overline{\boldsymbol{P}}_0 = (x_0, y_0, z_0)$ the deviation vector can be approximated using the distortion matrix $\boldsymbol{K}$:

$$\Delta\overline{\boldsymbol{P}}\left(\overline{\boldsymbol{P}}_0 + \boldsymbol{r}\right) = \Delta\overline{\boldsymbol{P}}\left(\overline{\boldsymbol{P}}_0\right) + \boldsymbol{K} \cdot \boldsymbol{r} \tag{3.13}$$

The evaluation of the components of the distortion matrix $\boldsymbol{K}$ and the known displacement $\boldsymbol{r}$ thus allows the calculation of the deviation vector $\Delta\overline{\boldsymbol{P}}$ associated with the error in reconstruction of the 3D position.

3.2.2 *Resolution*

For the final interpretation of measured displacement and strain values, the resolution of the imaging system has to be considered. Based on the principles of optics, there is a straightforward derivation of the pixel resolution. As seen for the simple

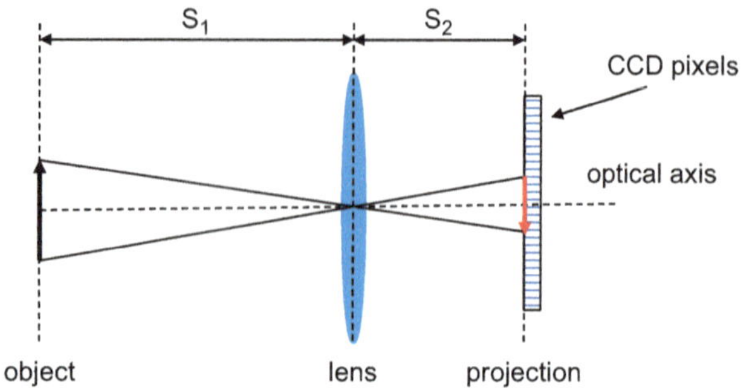

Fig. 3.7 Principle of pinhole camera establishing relation between size of object and single pixel size on camera system

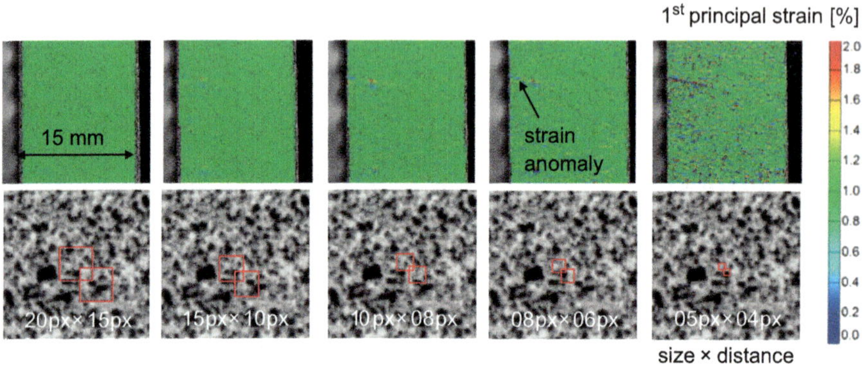

Fig. 3.8 Example of strain field calculated for identical images with 0.028 mm/px resolution (*upper row*) and different subset window sizes in magnified section of speckle pattern (*lower row*)

example of a pinhole camera in Fig. 3.7, the measured object is projected to the detector using a lens system. Accordingly, the system resolution is given by the size of the object, divided by the number of pixels in the used array. This assumes, the projected image is within the depth of field of the camera system (i.e., is projected without blurring) and the whole configuration is calibrated to compensate for the effect of lens distortions.

This theoretical system resolution is thus given as pixels per unit length and is very dependent on the selected configuration. Choosing lens systems with different magnification factors and adjusting the distance to the object can be used to optimize this value. Therefore, the measured displacement value is directly obtained from the change in pixels times the resolution. Similarly, the measured strain values are usually derived from the change in pixels relative to the local reference length. Therefore, the latter sometimes is more tolerable to the occurrence of distortions, since the displacement and the reference length are affected by such distortions in the same way.

Although the DIC systems provide sub-pixel accuracy as discussed in Sect. 3.2.1, this does not necessarily allow interpretation of strain values at this length scale. Beyond the single pixel resolution, the limiting factor is the choice of subset sizes. As seen in Fig. 3.8 with 0.028 mm/px, a certain trade-off between the increasing spatial resolution and the increasing matching errors has to be achieved. Here the same image data was processed using different subset window sizes. Beyond the effects of interaction between the subset size and the speckle pattern size, a clear trend is observed. For larger subset sizes no anomalies are visible in the strain field. With decreasing subset size, such strain anomalies appear and tend to grow sharper and more detailed, which is caused by the increased spatial resolution of the strain field as provided by the increased amount of subset windows. However, for very low subset window sizes (such as 5 px) the increasing matching errors start to dominate the appearance of the strain field. In addition to the increased error of the absolute values this also prevents the detection of such strain anomalies and poses further challenges to overcome when detecting the occurrence of failure in fiber reinforced materials.

As a rough estimate of current systems capabilities, for a typical commercial 3D-DIC system the respective accuracy as function of different lens systems was evaluated by [26]. Taking into account the error sources introduced in Sect. 3.2.1 it was observed that displacement errors are present on the order of less than 0.02 px, translating into strain errors being limited to 0.05×10^{-3}, when using a lens with 50 mm focal length, or 0.2×10^{-3} strain error for a 17 mm lens for the selected field of view. If the displacement between images is small (less than 50 px), the respective errors were found to scale linearly. For a wide angle lens system with a focal length of 4.8 mm, distortions introduced by the camera could not be sufficiently compensated by the calibration routine.

3.3 Strain Concentration

Based on the history of solid mechanics, the presence of geometric discontinuities is usually associated with the concept of stress concentration. As seen in Fig. 3.9, the presence of a hole causes a change of stress flux in the residual cross section when subject to a tensile load. This is a well-known concept in structural mechanics and an established routine for design of structural components [27, 28]. The same principle holds valid for the presence of internal defects such as cracks. These cracks cause a stress concentration at the crack tip and may propagate after exceeding a critical stress value. For typical engineering materials like metals, polymers, or ceramics the critical size of these cracks and the critical stress values can directly be obtained using the concepts of fracture mechanics [29–31]. Due to the complexity of the microstructure and the shape of the cracks, similar approaches for fiber reinforced composites are still subject to recent research efforts [27, 32–34].

The maximum stress near a crack occurs in the area of lowest radius of curvature. For the simple case of an elliptical crack of length $2a$ and width $2b$ (cf. Fig. 3.9) for an applied external stress σ, the stress at the ends of the major axes of the crack is given by:

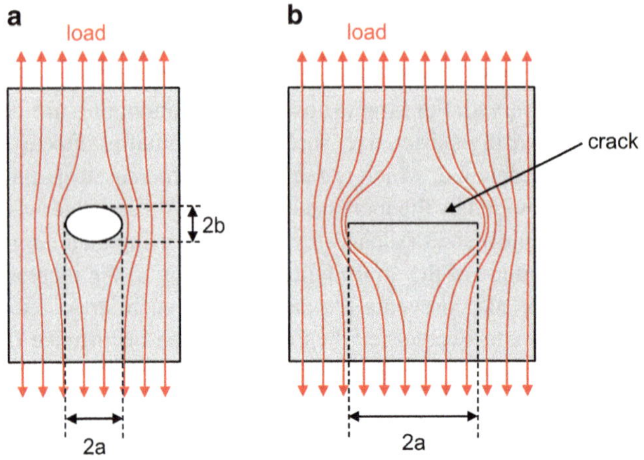

Fig. 3.9 Schema of stress flux around an elliptical hole (**a**) and stress flux around a crack (**b**)

$$\sigma_{\max} = \sigma\left(1 + 2\frac{a}{b}\right) = \sigma\left(1 + 2\sqrt{\frac{a}{r_c}}\right) \tag{3.14}$$

This uses r_c as radius of curvature of the crack tip. The so-called stress concentration factor is the ratio of the highest stress ($\sigma_{\max}$) to a reference stress (σ) of the gross cross section. As the radius of curvature approaches zero, the maximum stress approaches infinity. It is worth noting that the stress concentration factor is only a function of the shape of a crack but not of the absolute crack length.

Based on the stress–strain relation implied by Hooke's law, there is an expected relation between the stress concentration and the strain concentration. Although linear elastic material behavior does not hold valid for the region close the crack tip, there is still a distinct proportionality between the values of stress concentration and strain concentration as discussed in detail for the case of fiber reinforced materials by [27, 32–34].

Various experimental methods have been applied to derive stress or strain concentration, such as photoelastic stress analysis, shearography, strain gages, or digital image correlation. On the theoretical side, many analytical formulations have been provided for design of structures, most of them going back to [28]. However, for the complexity faced in composite materials in terms of anisotropy, boundary conditions, and heterogeneity, numerical methods based on finite element modeling have been recently proposed [33, 34].

Combining these simple considerations on strain concentration and the availability of high resolution cameras for full-field strain measurements, it is feasible to evaluate the capabilities of DIC systems to detect strain concentrations due to typical failure types in fiber reinforced composites. This is motivated by recent literature indicating strain concentration due to the microstructure of the material [35] or the formation of internal cracks [36].

However, for the existence of strain concentrations at a certain load level, there are several options how these indications may relate to the formation of (micro-) cracks on the interior of the material. All kinds of defects, such as voids, particle inclusions, or fiber disorientations, are well known to act as crack initiation sites. The presence of these defects will cause a local disturbance of the strain field due to the presence of geometric discontinuities (voids) or due to the locally different stiffness properties (fiber orientation, inclusions).

In all cases, after crack initiation there is a geometrical discontinuity located in the composite, which is given by the presence of the crack itself. Accordingly, the strain field is affected by the presence of this discontinuity. The effect of the latter will substantially depend on the size of the crack and the type of stiffness reduction (e.g., fiber failure vs. inter-fiber failure).

In order to estimate the visibility of such defects in full-field strain measurements, Sects. 3.3.1 and 3.3.2 present results of recent studies used to assess the applicability of presently available high resolution DIC systems [37, 38]. For these first studies, focus is given to the typical specimen dimensions used in materials testing, spanning a range of several centimeters in size.

3.3.1 Measurement of Strain Concentration Due to Internal Defects

For the experimental part of the study tensile specimens are prepared, which are either free of defects (reference) or comprise certain model defects in various depth positions below the surface. The respective specimen dimensions are shown in Fig. 3.17 including the position of the speckle pattern. As material system a Sigrafil CE1250-230-39 prepreg system was used, which was cured according to the material supplier's specifications. All experiments are carried out by calibrated 3D-DIC measurements using an ARAMIS 12 M system applying the acquisition parameters reported in Tables 3.1 and 3.2. A photographic representation of a typical experimental configuration is shown in Fig. 3.10. The tensile specimens are subject to a tensile load using a universal testing machine with a cross-head velocity of 1 mm/min.

To assess the limits of detectability for specific failure mechanisms in composites it is necessary to prepare defects of a known size at a certain depth of the laminate. Since this is impractical to be done by using mechanical experiments (i.e., the size and position of a defect is arbitrary in such a case), artificial defects are prepared to represent the existence of real defects. To this end, defects are embedded in the specimens during the fabrication process.

To fabricate an artificial fiber breakage one ply of the stacking sequence is cut for the full thickness for a certain length. As seen in the computed tomography image in Fig. 3.11 after curing of the laminate, the fiber cuts are clearly visible and accurate dimensional measurements are possible to obtain the real depth position

Table 3.1 Acquisition settings of DIC system used for tensile tests to detect strain concentration effects

Measurement setup	Value
Camera	Toshiba CMOS camera CSC12M25BMP19-01B with 4096 × 3072 pixels
Lenses	Titanar, focal length 5.6/100 mm
Lighting	KSP 0495-0001A LED 20 W/24 V white 30°
Filter	Polarization-filter Schneider-Kreuznach
Field of view	(55.9 × 41.7) mm^2
Scale factor	0.0136 mm/px
Subset size	21 × 21 pixel
Acquisition rate	2 Hz
Software	ARAMIS
Configuration	3D

Table 3.2 Acquisition settings of DIC system for tensile tests on unidirectional fiber reinforced composites

Measurement setup	Value
Camera	Toshiba CMOS camera CSC12M25BMP19-01B with 4096 × 3072 pixels
Lenses	Titanar, focal length 2.8/50 mm
Lighting	KSP 0495-0001A LED 20 W/24 V white 30°
Filter	Polarization-filter Schneider-Kreuznach
Field of view	(113.4 × 86.7) mm^2
Scale factor	0.028 mm/px
Subset size	19 × 19 pixel
Acquisition rate	2 Hz
Software	ARAMIS
Configuration	3D

and size of the artificial defect. In the example the nominal size of the fiber cuts was chosen to be 10 mm in length and one ply thickness of 0.22 mm. However, the cut is not always perfectly oriented inside the laminate, i.e., it develops an angle relative to the z-axis as seen from the right image in Fig. 3.11.

Also, the crack front is not exactly linear, but tends to deviate slightly from an ideal line as seen from the xy-cross section in the left image (this shows a cross section inside the laminate, i.e., the top-surface seen in the left image is not the outer surface of the laminate).

To fabricate an artificial delamination, inserts of thin Ethylene tetrafluoroethylene (ETFE) sheets are used. In the present case two ETFE sheets of 10 μm thickness are molded together at the edges using a hot wire processing technique. This yields ETFE balloons of well-defined cross-sectional dimensions filled with air. These balloons are embedded in the laminate during fabrication and are easy to detect in computed tomography images of the cured laminate (cf. Fig. 3.12). In the left image the yz-cross section reveals the width of the delamination (nominally 10 mm) and the wavy orientation of the embedded balloon

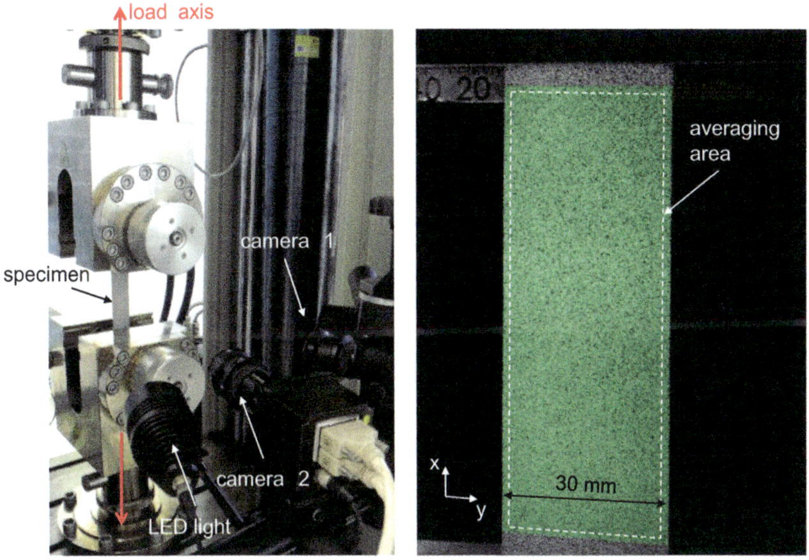

Fig. 3.10 Typical experimental setup and specimen dimensions including speckle pattern position and area marked by *dashed line* as used to obtain mean strain values

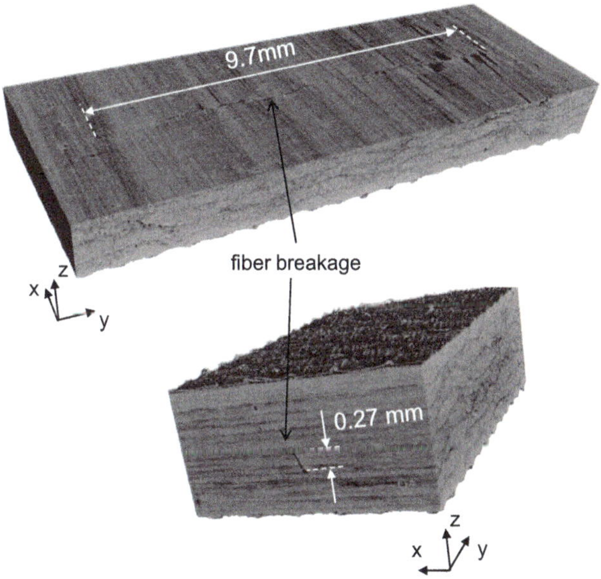

Fig. 3.11 3D-slices of computed tomography images of artificially prepared fiber breakage in a unidirectional laminate

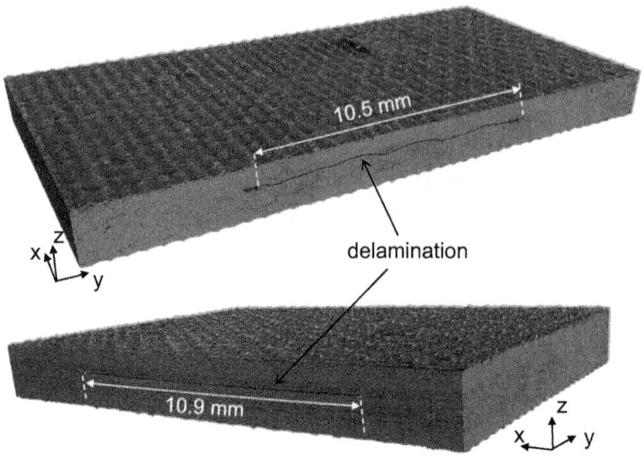

Fig. 3.12 3D-slices of computed tomography images of artificially prepared inter-ply delamination in a unidirectional laminate

along the *z*-direction. This is a consequence of the material contraction during laminate curing and thus cannot be avoided using this fabrication process. However, the orientation in parallel to the fiber axis is nearly a straight line as seen from the *xz*-cross section in the lower right figure. Here the nominal length of the delamination was also chosen to be 10 mm, yielding a cross section of the prepared inter-ply delamination of 100 mm^2.

Comparatively challenging is the fabrication of artificial defects resembling inter-fiber cracks. Using 50 μm thin Polyimide (PI) stripes it is possible to embed an out-of-plane discontinuity as seen in Fig. 3.13. As seen from the *xy*-cross section in the left image, the nominal length of this inter-fiber crack was chosen to be 10 mm, which can be implemented rather accurately. However, the angle of this discontinuity in the *yz*-plane is hard to control during laminate fabrication. For the example shown in Fig. 3.13, the resulting angle was evaluated to be approximately 61° as seen from the *yz*-cross section in the right image.

It was found to be particularly challenging to fabricate different angles in a well-defined orientation. In a real composite, the occurrence of inter-fiber cracks at other angles than 90° is readily expected from failure theory and experimental observations. Accordingly, the orientation will also cause distinct changes with respect to detectability of such inter-fiber cracks as demonstrated in Sect. 3.3.2.

Based on these three representative examples of artificially prepared defects, a systematic study was conducted. In all cases investigated, the artificially prepared defects are meant to represent an already existing crack surface. Therefore, the values discussed in the following are understood as strain concentration effects due to the existence of these defects, not due to the formation of these defects. The implication of this assumption will be discussed in Sect. 3.3.3.

For each failure type artificial defects of different size and depth position were prepared. The systematic aspects of this study will be discussed in Sect. 3.3.2. As a

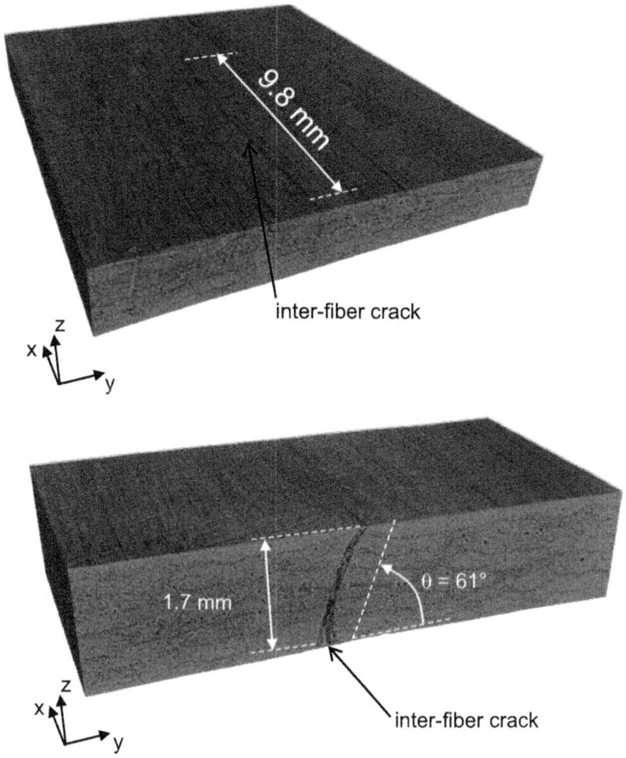

Fig. 3.13 3D-slices of computed tomography images of artificially prepared inter-fiber crack at 61° angle in a unidirectional laminate

representative example of the resultant experimental measurements, the case of fiber breakage embedded in a unidirectional laminate will be presented in the following.

As seen from the false-color representation of Fig. 3.14, the presence of an artificial fiber breakage within the laminate causes a readily visible strain concentration at its position. All examples shown in Fig. 3.14 are shown for a mean (global) strain value of 0.42 % as indicated by the false-color histogram of the x-strain (in parallel to load axis). Each measurement represents a different length of fiber breakage starting at nominally 10 mm (Fig. 3.14a) and decreasing to 7.5 mm (Fig. 3.14b), 5.0 mm (Fig. 3.14c), and 2.5 mm (Fig. 3.14d). All fiber breakages were located at the second ply (i.e., 0.22 mm below the surface). For each measurement the strain values above and below the crack position exceed the surrounding (global) strain value by 0.10–0.27 %. As seen in the figures, there is a distinct relationship between the size of the crack and the height of the strain value exceeding the background value. As expected from (3.14), smaller defect dimensions cause smaller strain concentration.

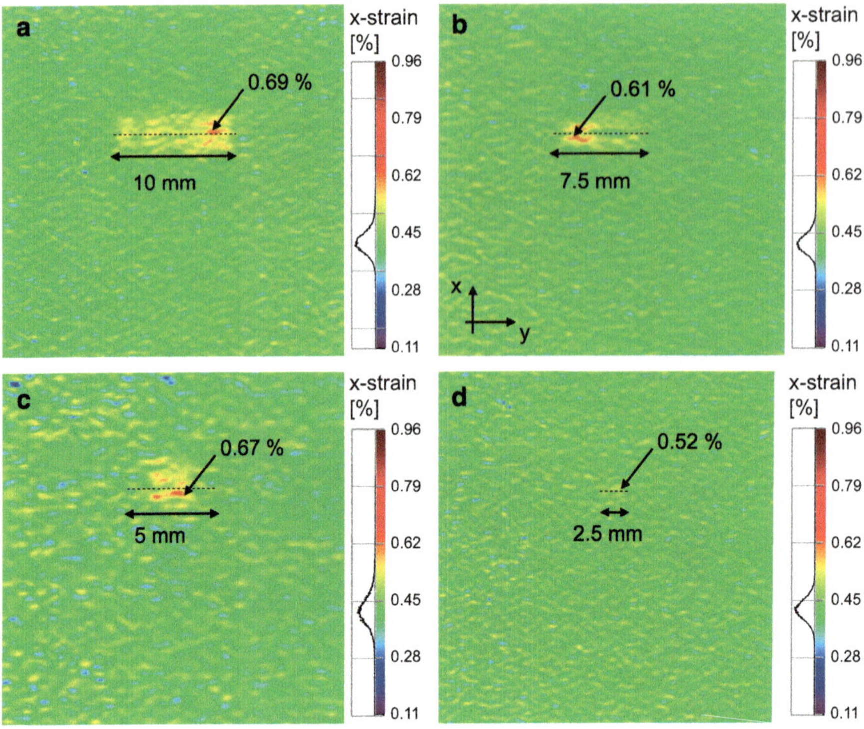

Fig. 3.14 Typical measurement result for artificial fiber breakage in a unidirectional laminate located at 0.22 mm below surface (2nd ply) at 0.42 % mean strain and changing spatial size ranging from 10 mm (**a**) to 7.5 mm (**b**), 5.0 mm (**c**), and 2.5 mm (**d**)

Keeping the spatial dimensions constant at 10 mm size, the depth position of the fiber breakage is now systematically varied. As seen from the measured strain fields of Fig. 3.15 at a constant mean strain value of 0.42 %, this causes a significant drop in the measured strain concentration when moving from near surface positions to deeper within the laminate. For convenience, the depth position of the artificial fiber breakage is noted in terms of the cut ply assuming an individual ply thickness of 0.22 mm. Thus the 2nd ply is located 0.22 mm below the surface, the fiber breakage itself spanning the full extent of the second ply, i.e., ranging to 0.44 mm. As indicated in Fig. 3.11, these values constitute only nominal values, but are good enough to discuss the overall trend. Evaluating the images in Fig. 3.15 according to (3.15), the strain concentration value at the defect position was found to exceed the mean value by 0.27 % for the 2nd ply, by 0.23 % for the 3rd ply, and by 0.08 % for the 4th ply position depending on the exact position used for the strain quantification. The reason for this decay will be discussed using the respective modeling results in Sect. 3.3.2. With respect to the fluctuation of strain values in the surrounding area, the question arises if the measured strain concentration is significant relative to the background noise. This requires a discussion of the background

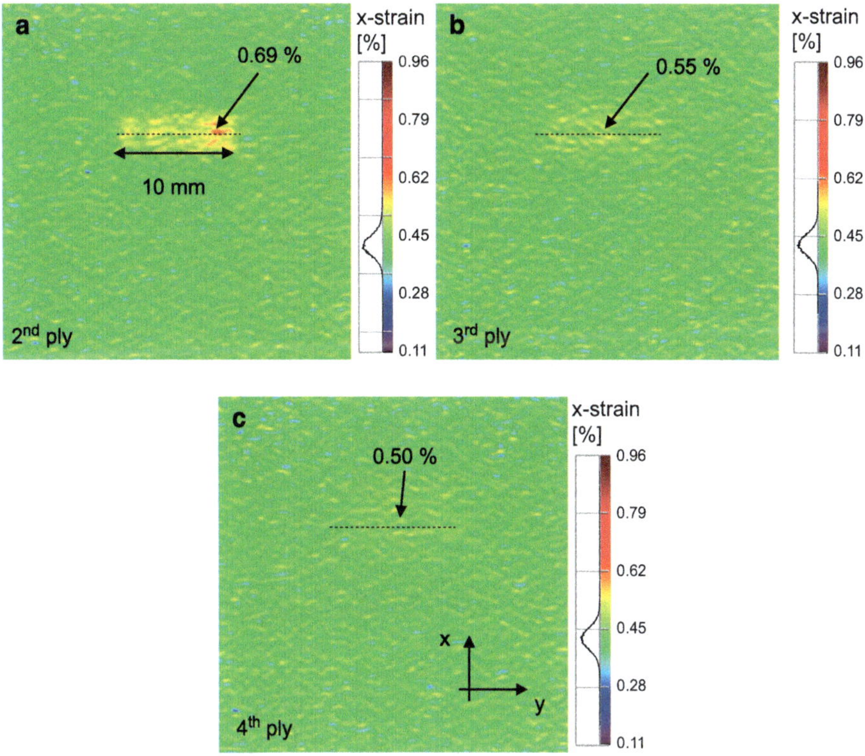

Fig. 3.15 Typical measurement result for artificial fiber breakage in a unidirectional laminate at 0.42 % mean strain and 10 mm size located at different depth positions below surface ranging from 2nd ply (**a**) to 3rd ply (**b**) and 4th ply (**c**)

noise level and a refinement of the strategy to quantify strain concentration. To this end, the discussion of the background noise floor is found in Sect. 3.3.3. In order to quantify a representative strain concentration value for each defect, the experimentally used data reduction strategy is introduced next.

The experimental measurements were carried out using static tensile tests with an acquisition rate of 2 Hz yielding a large data base for a single type of defect. For a representative subset of the images as function of the applied stress values, two measurements are made each. First, the average tensile stress within the cross-hatched area of Fig. 3.17 is evaluated by calculating the arithmetic average of all subset values excluding the region of the embedded defect. This is denoted as strain average $\langle\varepsilon\rangle_{avg}$ in Fig. 3.16a and is plotted as function of the applied stress value.

Second, at the positions of the embedded artificial defect the strain concentration value $\langle\varepsilon\rangle_{conc}$ is evaluated at distinct points. As seen from the linear regression line in Fig. 3.16a, the strain concentration values show almost linear behavior as function of the applied stress value, but exceed the strain average at all times. For the given example of a fiber breakage of Fig. 3.14a, this is indicated by the region of

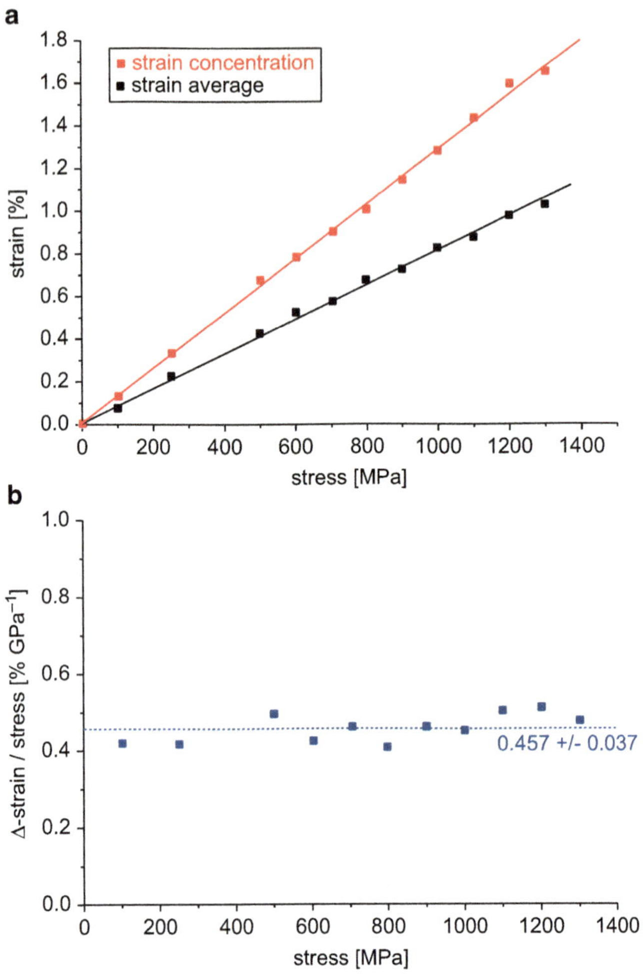

Fig. 3.16 Evaluation of strain average and strain concentration of artificial fiber breakage of 10 mm size located at 2nd ply below surface as function of stress value (**a**) and respective quantification of the strain exceedance normalized by the applied stress as function of stress (**b**)

increased compliance in the vicinity of the cut fibers. The origin of this local change of compliance is straight forward to understand by the partial absence of fibers contributing to the stiffness of the laminate.

While the example of Fig. 3.16a is rather obvious in terms of absolute values, the same finding was valid for each defect investigated. All artificial defects investigated exhibit similar behavior, being either systematically greater or lower than the strain average, depending on the type of artificial defect and the strain direction used for evaluation (see Sect. 3.3.2).

As measure of strain concentration relative to the surrounding area $\Delta\varepsilon$ the following equation is evaluated for each stress value:

$$\Delta\varepsilon = \left| \langle\varepsilon\rangle_{\text{conc}} - \langle\varepsilon\rangle_{\text{avg}} \right| \tag{3.15}$$

using the averaging procedure

$$\langle\varepsilon\rangle_{\text{conc}} = \frac{1}{m}\sum_{j=1}^{m} \varepsilon_j \tag{3.15a}$$

$$\langle\varepsilon\rangle_{\text{avg}} = \frac{1}{n}\sum_{k=1}^{n} \varepsilon_k \tag{3.15b}$$

where m is the total number of subsets evaluated in the global strain field and n is the number of evaluation points of strain concentration in the region of the defect. Taking the absolute values in (3.15) allows to interpret $\Delta\varepsilon$ as positive scalar deviation relative to $\langle\varepsilon\rangle_{\text{avg}}$. This definition was chosen since the detectability of a defect does not distinguish between positive or negative exceedance of the strain average.

For the values of the example in Fig. 3.16a, the value of $\Delta\varepsilon$ is observed to increase with the applied stress value, thus indications of such embedded defects are easier to observe at higher stress values. By dividing the value of $\Delta\varepsilon$ by the applied stress σ the normalized value $\Delta\varepsilon/\sigma$ shown in Fig. 3.16b is obtained. Remarkably, this value is almost independent of the applied stress, which is attributed to a linear elastic relationship of both $\langle\varepsilon\rangle_{\text{conc}}$ and $\langle\varepsilon\rangle_{\text{avg}}$ to the applied stress. Taking the arithmetic average of these values and their according standard deviation this allows to obtain a characteristic value for this type of artificial defect at its depth position and evaluates for this example as (0.457 ± 0.037) [% GPa^{-1}]. The specific units of $\Delta\varepsilon/\sigma$ are chosen to be percentage per Gigapascal to provide numeric values close to one, which are easier to discuss in the following. However, the use of GPa^{-1} should not be misinterpreted as compliance value of the laminate, since the used strain value does not reflect the global or local compliance value, but the difference between both.

To reach a statistically more significant representation for one single defect, the same evaluation is performed at n different points of strain concentration, i.e., Fig. 3.16 shows only one of these evaluations. Thus, the n locations for each defect evaluation are chosen at spots of highest strain concentration, since the extreme values will determine visibility of the strain signature. In contrast, averaging of many points in the area of the embedded defects (i.e., $n > 50$) was not found to yield appropriate values, since the extreme values of strain concentration will be substantially reduced in the averaging procedure. Thus $n = 5$ positions were typically used for the evaluation.

Based on this type of measurements, the following section is used to validate a modeling strategy to predict the strain concentration effects in a laminate due to the existence of typical defects before starting the discussion of system detectability limits in Sect. 3.3.3.

3.3.2 FEM Modeling of Strain Concentration Due to Internal Defects

In order to begin a systematic study of the strain concentration caused by internal defects in fiber reinforced materials, a finite element modeling approach is developed and validated using DIC measurements on reference specimens, open-hole tensile specimens, and specimens with embedded artificial defects.

For the finite element method modeling the "structural mechanics module" of the commercial software package Comsol Multiphysics is used. A full 3D representation of the experimentally used geometries is implemented and discretized by a mesh using linear order shape functions and a maximum edge length of 0.5 mm with substantial mesh refinement down to 10 μm when approaching the crack tips. As material model, the data of the experimentally used Sigrafil CE 1250-230-39 material given in Table B.1 in Appendix B is used in the following. The individual plies of the test specimen are chosen as homogenous unidirectional material with an orientation in accordance with the experimental configuration. The reinforcement sections are modeled as stacking sequence of $[+45/-45/0]_{sym}$ using the Sigrafil CE 1250-230-39 material data. For modeling of the strain concentration, a stationary tensile load is applied as force at the top and bottom surfaces of the reinforcements on one end of the specimen as seen in Fig. 3.17a. The opposite side of the specimen uses fixed displacement conditions at the surface of the reinforcements corresponding to the experimental configuration. All geometric dimensions of the tensile specimens are matched to the experimental values, with their nominal dimensions shown in Fig. 3.17.

To validate the finite element model in its global stress–strain prediction, a first comparison of calculation results is made relative to experimental data for a case study without incorporated defects. Figure 3.18 shows a comparison between experimental results of a unidirectional tensile specimen with load axis in parallel to the fiber axis and the respective modeling result. For the experimental results, the strain values are obtained from the 3D-DIC measurements using the acquisition parameters noted in Table 3.1. The experimental strain values are obtained as mean value from area inside the dashed line as marked in Fig. 3.10 on the surface of the specimen. Comparison is made to the simulated mean strain value as calculated on the cross-hatched surface in the area indicated in Fig. 3.17. As seen from Fig. 3.18, there is very good agreement of the experimental and simulated axial and transverse stress–strain curves. Based on this validation, the linear-elastic material model is

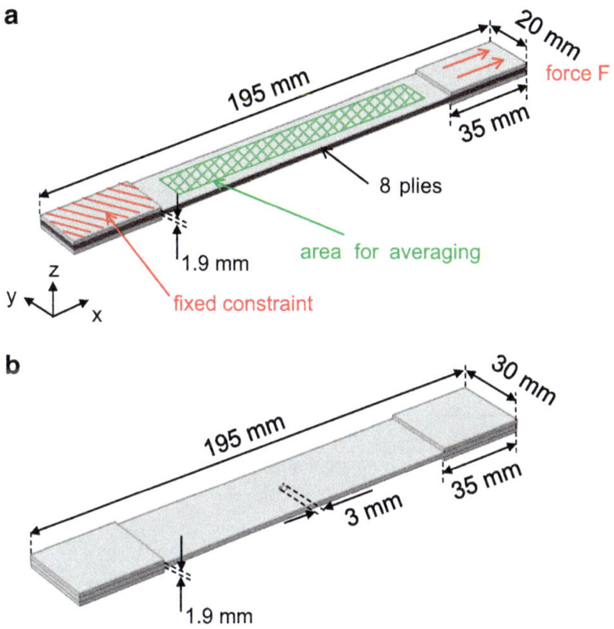

Fig. 3.17 Model setup for validation of material properties and cross-hatched area for averaging of strain values (**a**) and model setup for open-hole tensile test including dimensions (**b**)

assumed to be valid for the investigated stress limit of 800 MPa, an axial strain limit of 0.65 %, and a transverse strain limit of −0.21 %.

3.3.2.1 Full-Field Comparison

To further validate the accuracy of the FEM model, a typical well-known macroscopic discontinuity is introduced to cause a strain concentration at a preferred position. This is done in the form of an open-hole tensile test using a drilled hole of nominally 3 mm diameter as seen in Fig. 3.17. Figure 3.19 shows the axial and transverse strain fields of the open-hole tensile test at 346 MPa. Comparison is made in Fig. 3.19 for the strain concentration in x- and y-direction in the vicinity of the hole as obtained from 3D-DIC measurements (b,d) and as computed by finite element modeling (a,c). As seen from the figures, the absolute values of strains of measurement and computation agree well to each other. Some differences arise due to the noise being present in the experimental measurement causing fluctuations of the strain values and the spatial resolution of the measurement causing less resolved peak structures when compared to the computation result. Using this full-field validation it may thus be further concluded that the chosen modeling parameters, such as discretization settings, are sufficient to compute similar strain concentration values as experimentally measured.

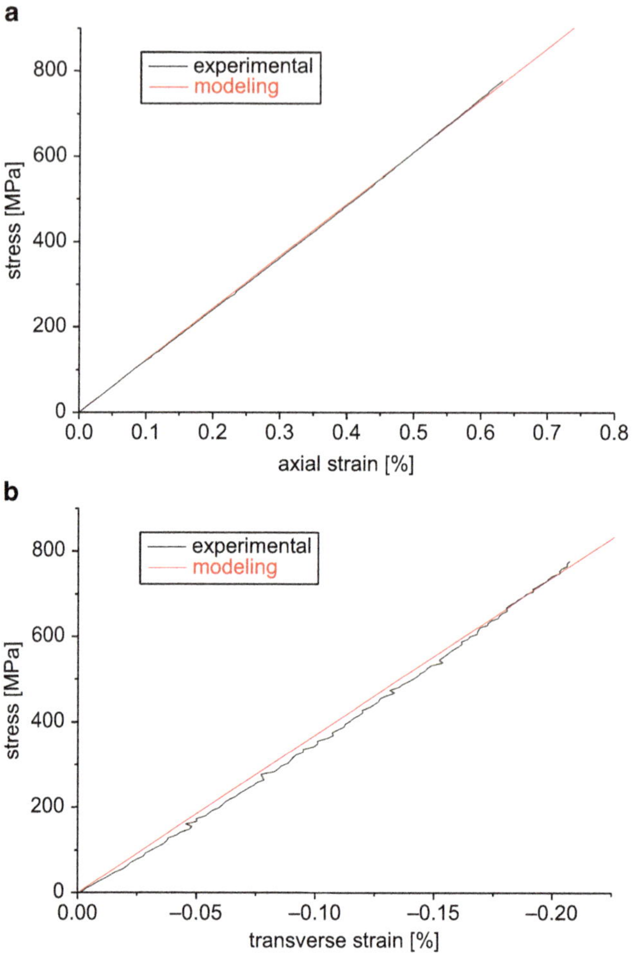

Fig. 3.18 Comparison of axial (**a**) and transverse (**b**) strain curves as obtained by 3D-DIC and by finite element modeling

Due to the good agreement of the computed strain values for the open-hole discontinuity, it is assumed that further finite element based predictions of strain concentrations due to internal defects are valid within a certain level of accuracy. The validated material model is thus used to calculate strain concentration effects due to embedded artificial defects in fiber reinforced materials. To this end, one representative example of each defect is shown in Figs. 3.20, 3.21, and 3.22 each.

As first example, Fig. 3.20 presents a comparison of an experimentally measured strain concentration of a fiber breakage of 7.5 mm size located at the 2nd ply below the surface at 0.42 % mean strain and the corresponding modeling result. Due to the identical strain scales of both figures (measured in parallel to the load axis), a direct discussion of the observed differences is possible. To model the cut fibers the approach makes use of a “thin elastic layer” oriented in the *yz*-plane. The stiffness of the “thin elastic layer” is set to zero, resembling fully separated surfaces, i.e., a fully developed crack at this position. Due to the presence of the resin after specimen fabrication, this assumption may only be fulfilled after exceeding a

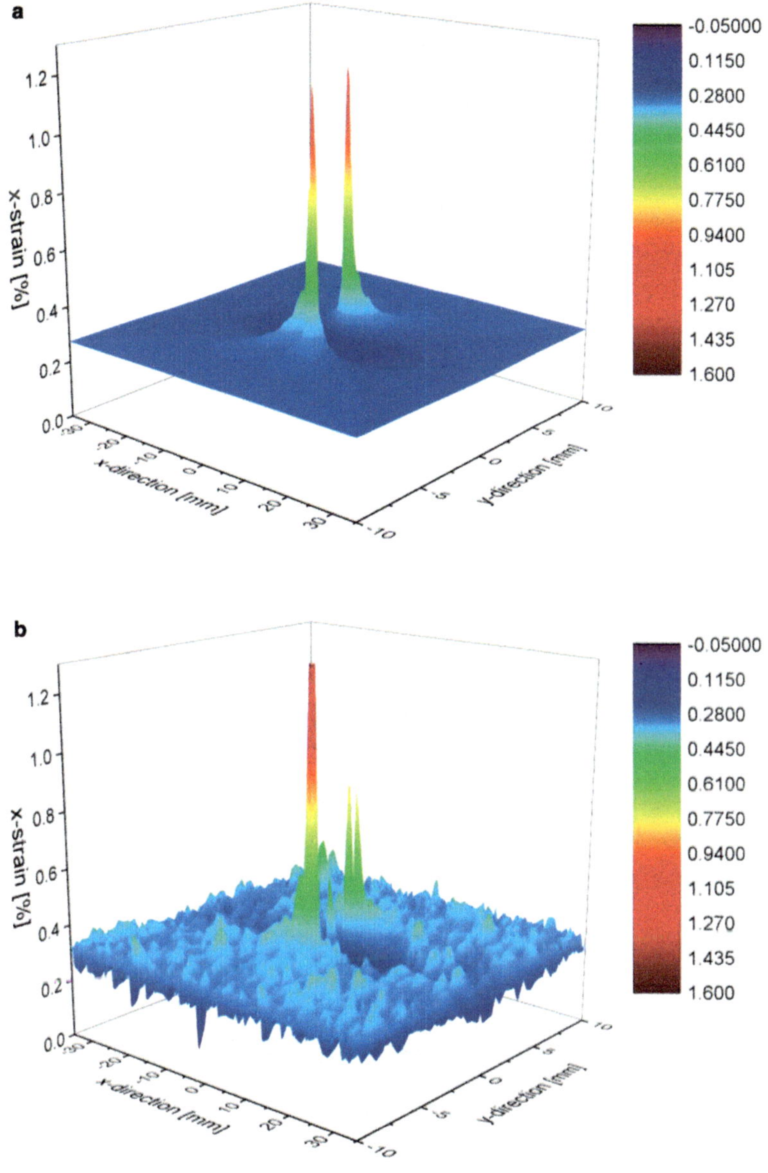

Fig. 3.19 Comparison of the strain-concentration at 346 MPa of open-hole tensile test in *x*-strain field and *y*-strain field as computed by finite element modeling (**a**,**c**) and as obtained from 3D-DIC measurements (**b**,**d**)

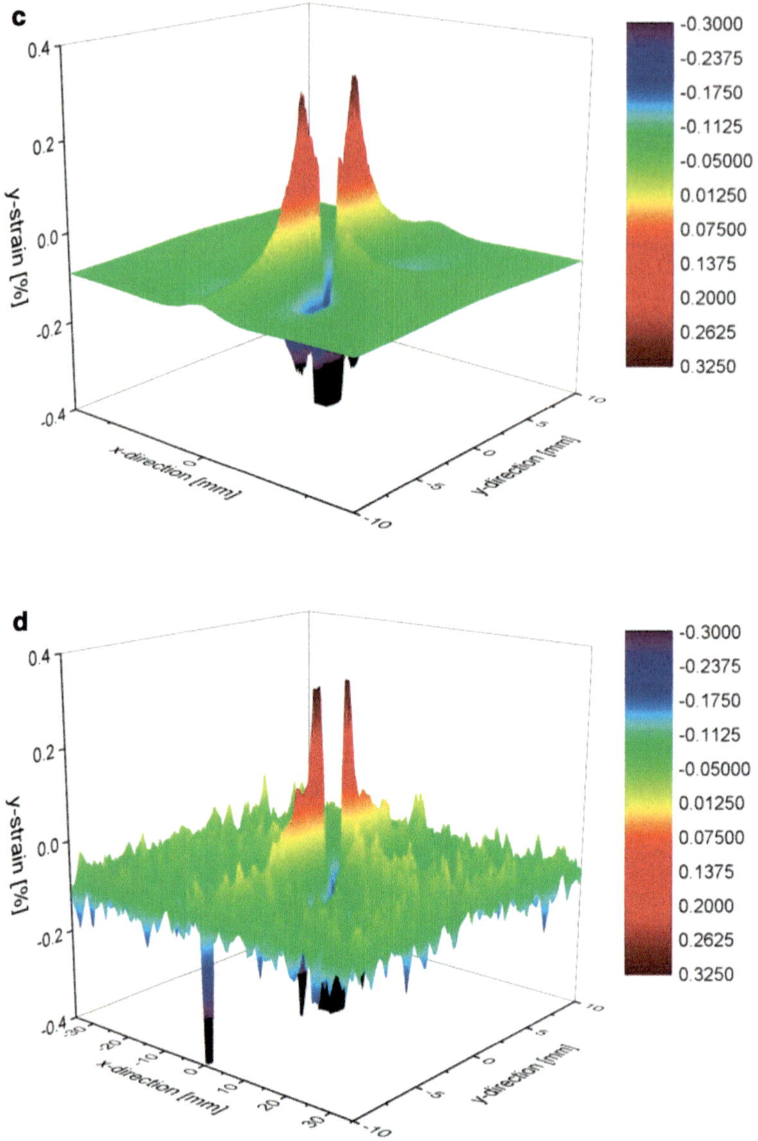

Fig. 3.19 (continued)

certain minimum stress. This is required to crack the resin at this position and to open a gap representative for broken fibers. At the applied 569 MPa stress this is likely the case. The modeling result shows a primary strain concentration above and below the position of the modeled crack. Highest values occur at the center of the crack, i.e., at 3.75 mm distance to each edge of the crack. In contrast, the example of Fig. 3.20a yields highest strain concentration values at the left edge of the crack. This can likely be due to imperfections during the preparation, since the cut fibers

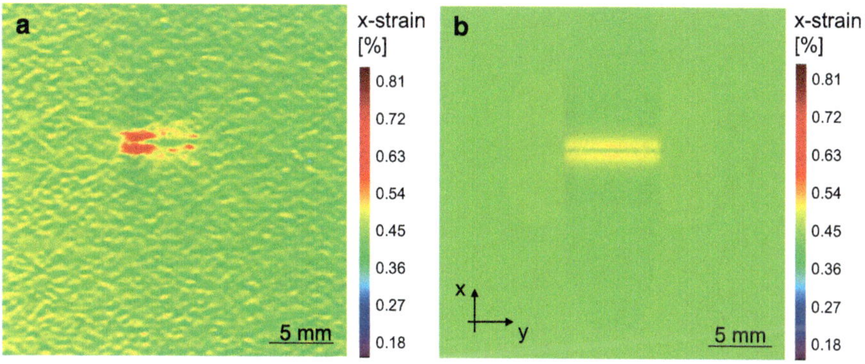

Fig. 3.20 Comparison of the strain-concentration at 569 MPa for an artificial fiber breakage of 7.5 mm size in a unidirectional laminate located at the 2nd ply below surface as measured (**a**) and as computed by finite element modeling (**b**) (based on [37])

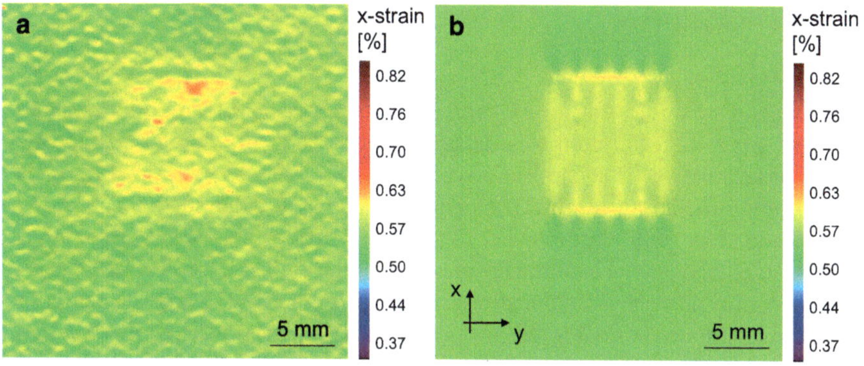

Fig. 3.21 Comparison of the strain-concentration at 695 MPa of an artificial delamination of 10 mm × 10 mm size in a unidirectional laminate located at the interface between 1st and 2nd ply below surface as measured (**a**) and as computed by finite element modeling (**b**) (based on [37])

tend to deviate from an ideal orientation along the x-axis as revealed by the 3D images of Fig. 3.11. The orientation within the xz-plane may develop an angle, which is not identical at each position along the y-axis. These effects may cause distinctly different strain concentration values, since their absolute orientation exhibit some intrinsic scatter. Moreover, secondary damage might be present at the measurement position or might develop during specimen loading. For the case shown in Fig. 3.20a, there is a certain likelihood of delamination formation during laminate curing as consequence of the cut fibers. Furthermore, such delamination may also initiate at low load levels, as e.g. used as measurement principle in transverse-crack tension specimens (see Sect. 4.7 or [39]). Neglecting these effects, the overall match of strain concentration predicted by the finite element model was found to be in good agreement to the experimental measurement.

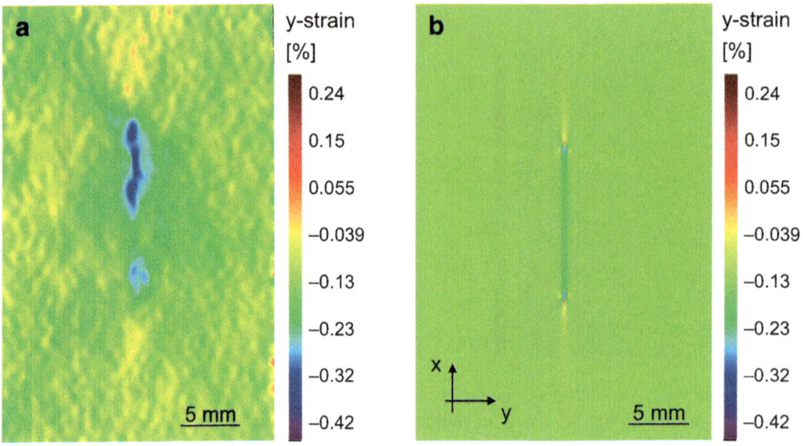

Fig. 3.22 Comparison of the strain-concentration at 535 MPa of an artificial inter-fiber crack of 10 mm size in a unidirectional laminate as measured (**a**) and as computed by finite element modeling (**b**) (based on [37])

As second example, Fig. 3.21 presents a comparison of an experimentally measured strain concentration of a delamination of 10 mm × 10 mm size located at the interface between 1st and 2nd ply below surface stressed by 695 MPa and the corresponding modeling result. To model such inter-ply delamination, the defect was embedded as "thin elastic layer" of same dimensions in x- and y-direction as the experimental defect. To mimic the presence of the embedded ETFE balloon, the "thin elastic layer" was given the stiffness properties of ETFE along the z-axis for compressive stresses normal to the surface and zero stiffness in x- and y-direction. Use of a perfectly flat rectangular representation was found to underestimate the measured strain concentration at least by a factor of two. However, the true shape of the artificially embedded defect exhibits certain waviness along the z-axis as seen from the computed tomography image of Fig. 3.12. Using a geometry with similar wavelength and z-amplitude as for the experimental defect, the computation result of Fig. 3.20b was obtained. Given the identical strain value range there is good agreement between experiment and model. Primary differences between both figures are attributed to the non-periodic waviness of the experimentally prepared delamination and the difference in edge representation of the embedded ETFE balloon. The latter exhibits a pronounced topology in the experiment, but was merely accounted for as straight line in the model. This may likely cause the higher strain concentration values seen in the model result. However, for the overall quantification of strain concentration, this will be negligible as demonstrated in the following.

As third exemplary representation a comparison is made between an experimentally prepared artificial inter-fiber crack and the corresponding modeling result. Figure 3.22 shows the comparison for a stress value of 535 MPa and 10 mm size of the inter-fiber crack. As described above, this type of defect is prepared using an embedded PI stripe reaching from the front to the back of the laminate, i.e., spanning the full z-extent of the laminate. Accordingly, the model uses an embedded cube of 50 µm length along the y-axis, 10 mm length along the x-axis, and 1.8 mm along the z-axis (full laminate thickness). The domain is given the elastic properties of PI as noted in Table B.1 in Appendix B. Using this representation, the image of Fig. 3.22b is computed, yielding good agreement to the experimental result of Fig. 3.22a.

Corresponding to the experiment, the higher $\Delta\varepsilon$ values are observed for the y-axis providing a better visible signature. Thus in the regions of the embedded inter-fiber crack larger Poisson contraction seems to occur, translating into lower absolute y-strain values. The main differences between experimental and modeling results are the waviness and position of the signature. The waviness may readily be understood from the computed tomography images of Fig. 3.13 and is due to the same preparation artifacts as previously described for fiber breakage. In the experiment, the strain concentration seems to occur preferentially at the positions close to the edge. This is also seen in the model result, yet the zone of highest strain concentration was found to be smaller and the transition more continuous. Given the geometric imperfections of the prepared inter-fiber crack it is thus likely that this overrides the continuous transition of strain values as expected from the model predictions.

3.3.2.2 Signatures of Artificial Defects

To investigate a broad range of possible configurations in unidirectional laminates, variations of defect type, depth position, and size as shown in Fig. 3.23 are studied. The positions of discontinuities represented by the embedded defects are marked as colored lines in Fig. 3.23. All modeling results are compared to measurement values of experimentally prepared artificial internal defects following the data reduction strategy explained previously. For extraction of quantitative values from the modeling results a different strategy was applied, since the data reduction does not benefit from a statistical approach, given the model uses a numerically convergent approach. Thus the extreme values of strain concentration may directly be extracted from the model results at sufficient accuracy without further averaging.

To demonstrate the procedure, a typical result from a modeled fiber breakage located at the 4th ply below surface is shown in Fig. 3.24. For each model, five different dimensions have been tested. As indicated by the strain concentration in close vicinity to the crack position, there is a clear visibility of the discontinuity. To quantitatively interpret the modeling results, the following procedure was applied. First, a cross-section line is drawn at the surface level of the specimen as indicated in Fig. 3.24. The resulting evaluation of the x-strain and y-strain as function of position along this line is shown in Fig. 3.25a, b, respectively.

Due to the presence of the reinforcements at the specimen ends and the applied boundary conditions (cf. Fig. 3.17), there is a characteristic background deformation for the x-strain and the y-strain. The presence of the fiber breaks causes some

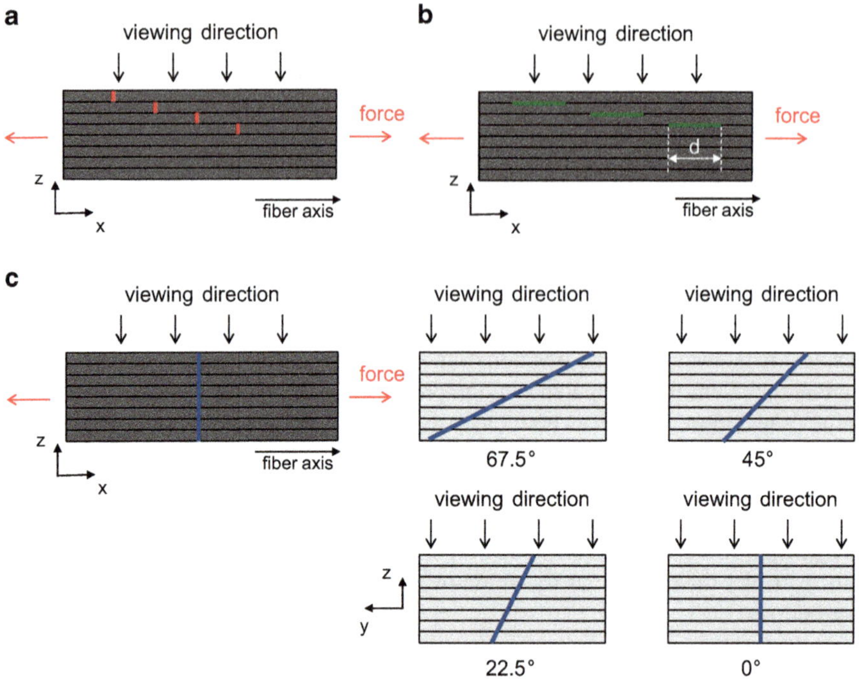

Fig. 3.23 Depth configuration and orientation of defects to study strain concentration in unidirectional materials for fiber breakage (**a**), inter-ply delamination (**b**) and inter-fiber failure in *xz*- and *yz*-cross section (**c**)

readily visible indications, which are seen as spikes in Fig. 3.25a and as dips in Fig. 3.25b. The strain concentration was evaluated as maximum height of the peak relative to the strain average following (3.15). The cross-section line was chosen according to the position of maximum strain concentration. The strain average was obtained from the mean values of the cross-hatched area of the reference specimen free of defects at the same stress level.

As first type of defects, the resultant strain concentration for fiber breakage is evaluated. Figure 3.26 shows the $\Delta\varepsilon/\sigma$ evaluated from the strain in x-direction for the modeled cases of fiber breakage as solid rectangles. In addition, experimental results are added to the same graph by open circles including their measurement error as discussed above. The colors indicate the depth position below surface, with the 1st ply being the surface ply. As seen from Fig. 3.26, the presence of fiber breakage at the surface ply causes strain concentration of approximately equal height. As long as the spatial resolution of the DIC system is sufficient to detect the fiber breakage, this should still be valid even for smaller sizes than the 1.25 mm investigated. For the model results there is even a small trend towards higher $\Delta\varepsilon$ for smaller defect sizes. For instance, in Fig. 3.25a there is a zone of strain concentration at each edge of the modeled fiber break. When decreasing the size of the defect these two individual regions of strain concentration merge into one larger signature.

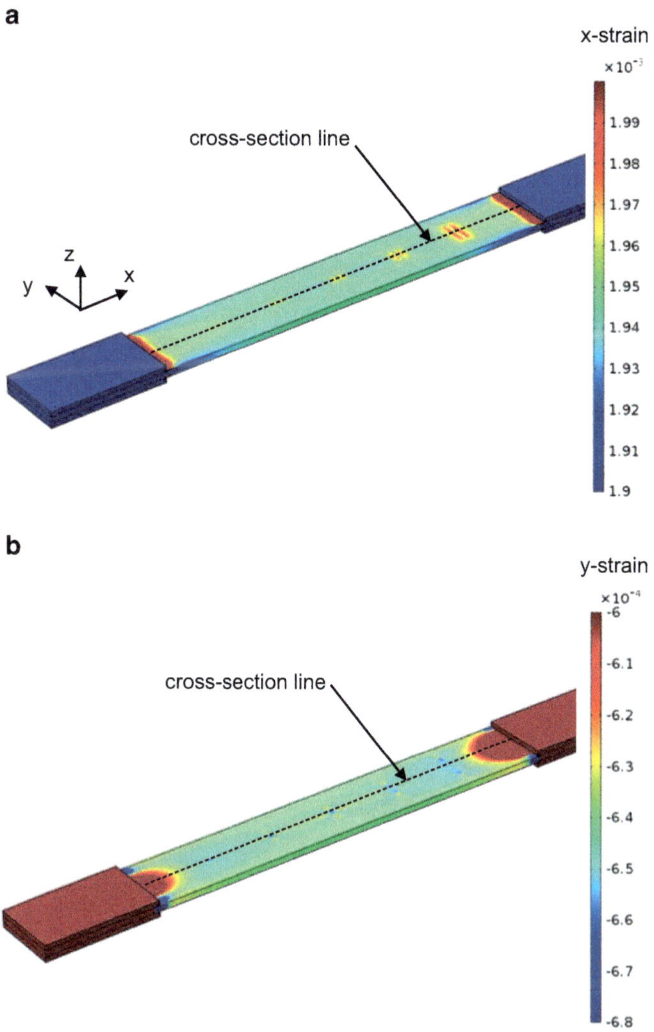

Fig. 3.24 x-component of strain field (**a**) and y-component of strain field (**b**) with presence of fiber breakage of varying defect size buried at 4th ply below surface of laminate

This causes a noticeable increase for the detectable $\Delta\varepsilon$ in this case (Fig. 3.30a shows a respective example for the case of inter-ply delamination).

The situation changes for fiber breakages covered by one or more intact plies, i.e., the fiber breakage is located within the volume of the material and does not reach the surface. For all depth positions studied herein, there is a general trend from large to small lateral dimensions. In all cases, the strain concentration decreases with decreasing defect dimensions. According to (3.14) the behavior as function of defect size should obey a square root law, which is found to be applicable in the present case and is shown as solid line fit to the modeled data in

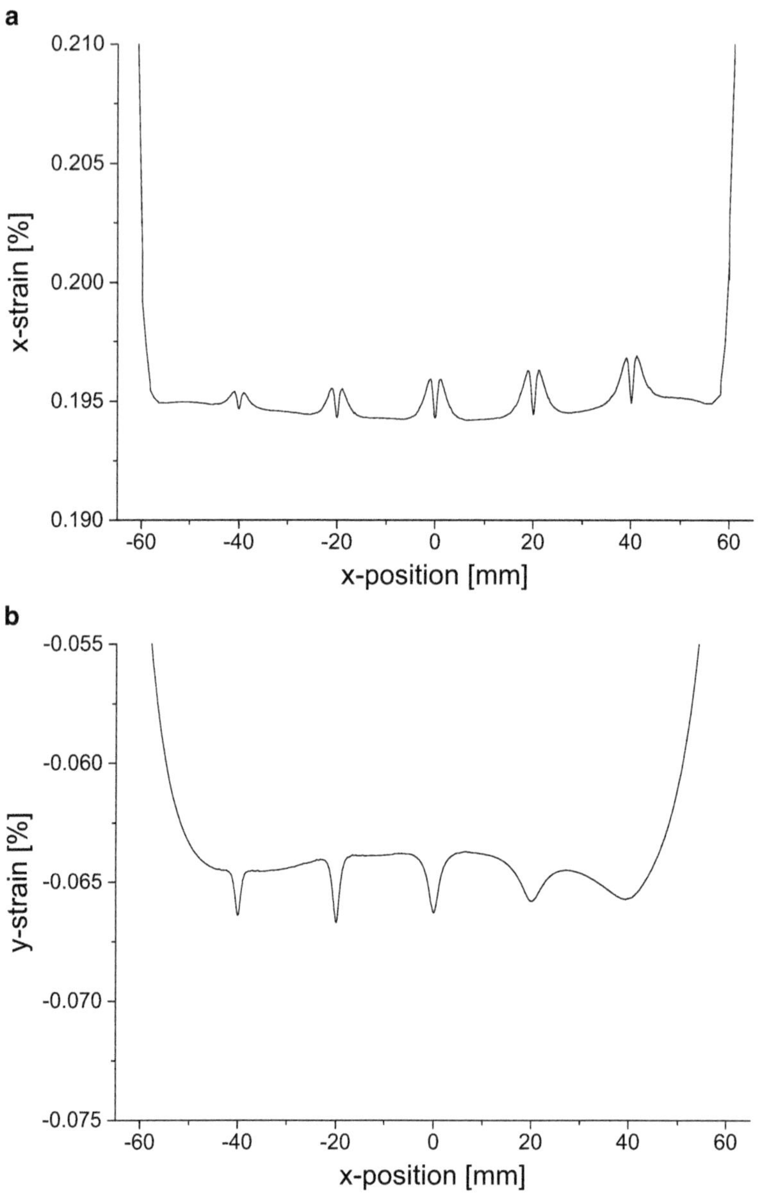

Fig. 3.25 Evaluation of *x*-strain (**a**) and *y*-strain (**b**) along cross-section line of Fig. 3.24

Fig. 3.26. The modeled data is supported well by the experimental data given the measurement errors of the experimental data. Due to the implication discussed in the model validation above, it is likely that not all experimentally prepared defects conform with the nominal values used in Fig. 3.26. Especially for the smaller

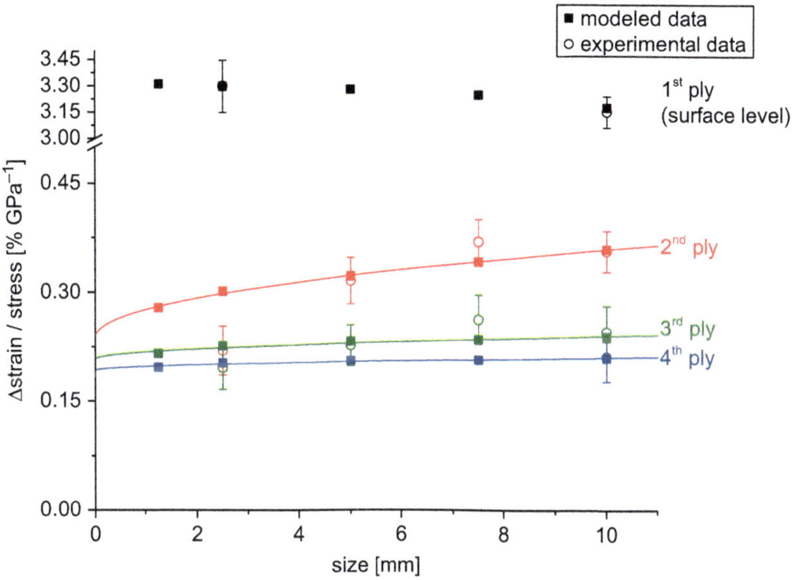

Fig. 3.26 $\Delta\varepsilon/\sigma$ due to fiber breakage of varying defect size buried at several depths below surface (based on [37])

defects, the preparation may easily cause a false orientation of the intended fracture plane. Especially for the 2.5 mm fiber cut located in the 2nd ply such effects may be the reason for the substantial deviation to the predicted trend. Furthermore, the same size defect located in the 3rd ply agrees well with the model prediction and is found at a similar level of strain concentration as the 2nd ply defect. The latter contradicts the expected behavior as function of defect depth as will be discussed in the following. Thus it may be speculated that the measured strain concentration value of 2.5 mm at the 2nd ply might underestimate the real value for such defect size.

For a constant defect size there is a distinct behavior expected as function of defect depth. For defects buried deeper inside the material a respective decrease of detectable strain concentration is expected at the surface level. As observed from the trend in Fig. 3.26 for each defect size, this observation is found to be applicable. To better quantify this decrease with defect depth, Fig. 3.27 shows the x-stress component and the x-strain component in the xz-cross section of the model of a fiber break with 10 mm extent in y-direction. The cross section is chosen at the center of the crack, i.e., with 5 mm distance to each edge along the y-direction. As seen from these cross sections, stress field and strain field exhibit very similar signatures, which becomes even more apparent when computing a measure[2] of the local

[2] Due to the anisotropy of the material and the Poisson's contraction effect, a specific discussion of the local stiffness response by means of the components $\partial\boldsymbol{\sigma}/\partial\boldsymbol{\varepsilon}$ essentially yields the same result.

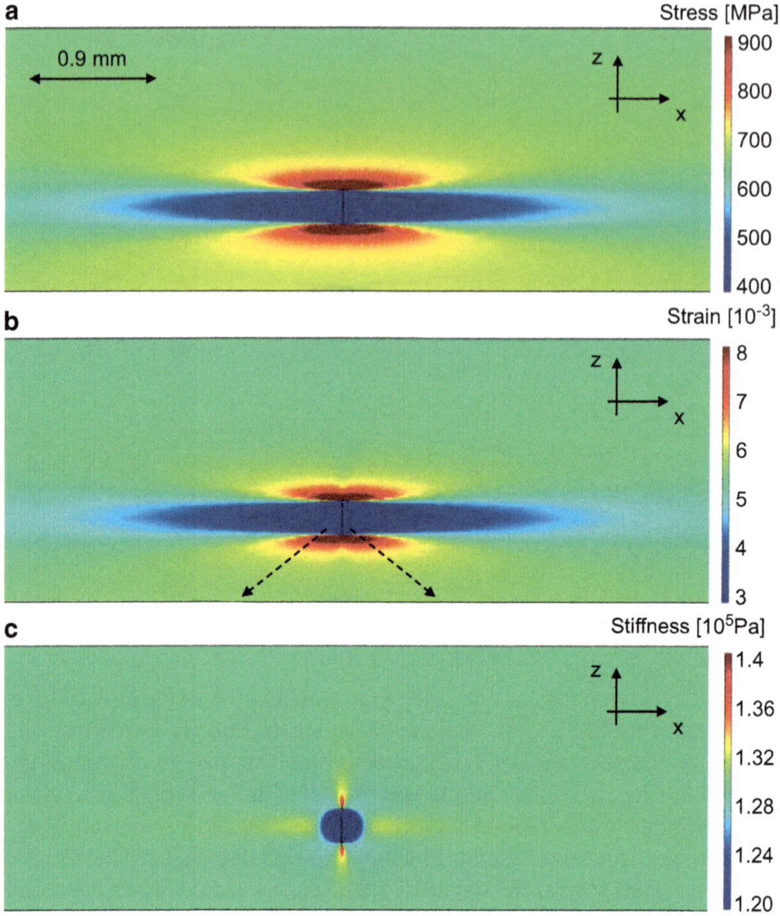

Fig. 3.27 Zone of strain field (**a**), stress field (**b**), and local stiffness (**c**) around modeled fiber breakage of 10 mm size in y-direction plotted in xz-cross section at stress value of 694 MPa

stiffness response by evaluating $\sigma_x/\Delta\varepsilon_x$ as seen in Fig. 3.27c. Except for the expected locations close to the edges of the crack tip the stiffness response is almost constant indicating the direct proportionality of stress and strain concentration. In particular, for the direction of highest strain concentration (indicated by dashed lines in Fig. 3.27b) the stress–strain response is almost linear. Thus, the use of a linear elastic fracture mechanics description seems applicable to discuss the behavior of strain concentration. For an isotropic material the stress field around such a mode I crack would decrease in x-direction σ_x as function of z-distance z to the crack tip following

$$\sigma_x = K_I \cdot \frac{1}{\sqrt{2\pi z}} \tag{3.16}$$

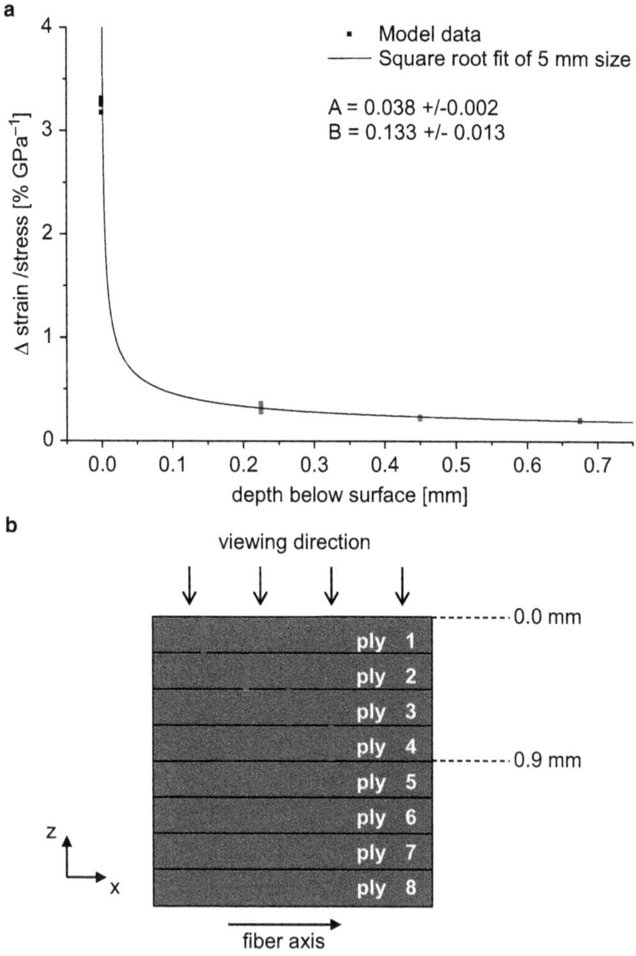

Fig. 3.28 Change in $\Delta\varepsilon/\sigma$ due to fiber breakage as function of defect depth with colors corresponding to depth position (**a**) and respective scheme to quantify depth position of defect (**b**)

For the given case of an anisotropic material a similar relationship, may be assumed which will give rise to a distinctly different stress concentration factor K_I. However, this is expected to be a constant material property, which does not inhibit the following discussion. As function of depth position z, the strain concentration can thus be estimated to be proportional to the stress concentration. Consequently, a decay of strain concentration $\Delta\varepsilon$ as function of $z^{-1/2}$ is derived as first order approximation from analytical theories.

Figure 3.28 presents the same data as Fig. 3.26, now plotted as function of defect depth below surface. As measure of defect depth, the edge of the crack closest to the surface is selected as indicated in the scheme of Fig. 3.27b (first ply is at $z = 0$ mm).

As seen from the square root fit to the modeled data in Fig. 3.28a, the decrease of $\Delta\varepsilon/\sigma$ with defect depth is well described by the relationship $\Delta\varepsilon/\sigma = \widetilde{A} + \widetilde{B}z^{-1/2}$ using the fit parameters $\widetilde{A}$ and $\widetilde{B}$. For better visibility, Fig. 3.28a shows only the fit of the 5 mm size fiber cut, since the remaining fits essentially provide the same trend, but almost superimpose in Fig. 3.28a. Thus the consideration regarding the general trend described by (3.14) seems readily applicable for the case of fiber cuts studied herein.

As second type of defects, the occurrence of inter-ply delamination was studied. Figure 3.29 shows the strain concentration $\Delta\varepsilon/\sigma$ evaluated from the strain in x-direction for the modeled cases of inter-ply delamination as solid rectangles. Experimental results are added to the same graph by open circles including their measurement error. The colors indicate the depth position of the interface studied.

Although, inter-ply delamination in unidirectional materials may occur at arbitrary depth positions, the discretization steps taken for the analysis are chosen in accordance with the interface positions of the plies used to build the laminate. This is due to the experimental preparation of the artificial defects, since the ETFE balloons may only be prepared between the individual plies. Thus the colors in Fig. 3.29 indicate the interface location, i.e., ply 1-2 indicates the position between 1st and 2nd ply. All defects are prepared inside the volume of the material, so there is no defect, which extends to the surface. As consequence, all $\Delta\varepsilon$ values obey the same trend as function of defect size following the square root law as given in (3.14). However, the fit values used to obtain the lines of Fig. 3.29 return distinctly different values than for fiber breakage. This is due to the geometric difference

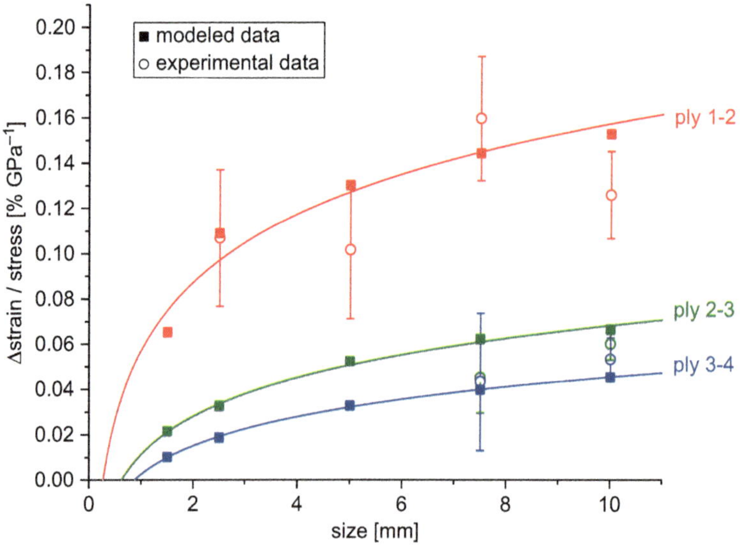

Fig. 3.29 $\Delta\varepsilon/\sigma$ due to inter-ply delamination of varying defect size buried at several depths below surface as strain concentration per GPa external stress (based on [37])

between the modeled defects. Although the inter-ply delamination does produce a noticeable strain concentration, the modeled waviness and its size along the x-axis cause a distinctly different signature. As seen from the evaluation of the x-strain in Fig. 3.30a for five different sizes of inter-ply delamination, the signature basically

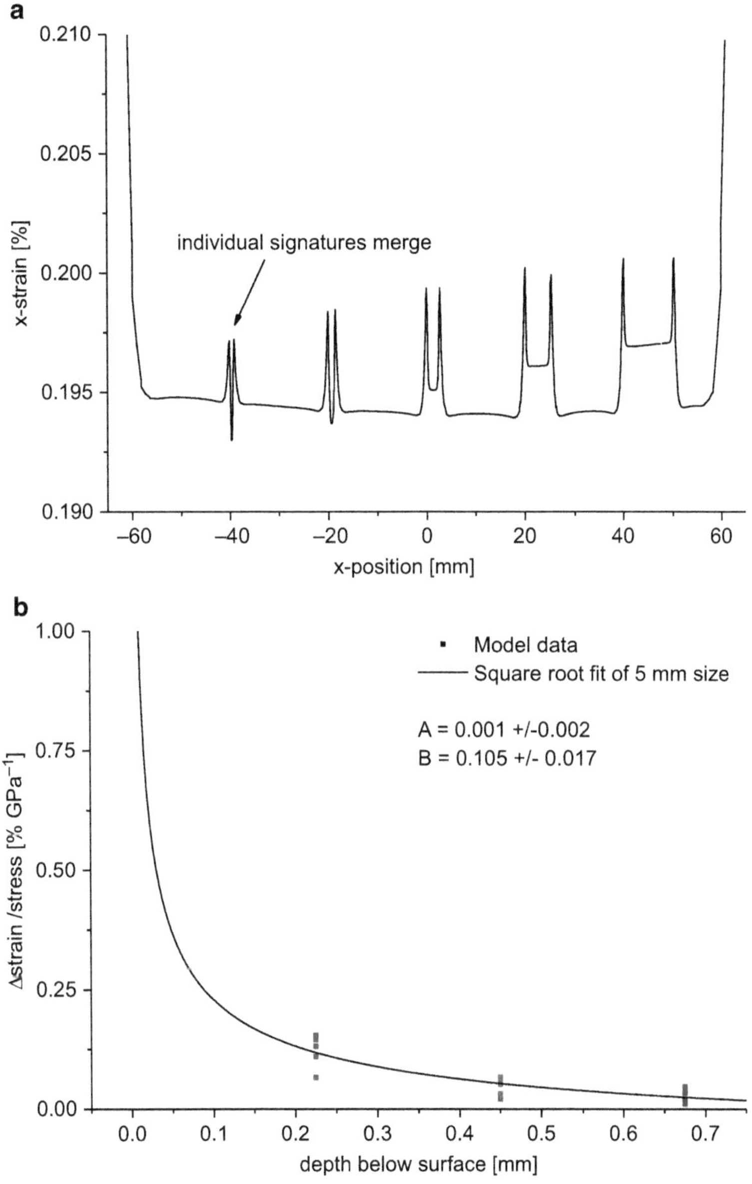

Fig. 3.30 Evaluation of x-strain along cross-section line according to Fig. 3.24 for depth position 1-2 (**a**) and evaluation of $\Delta\varepsilon/\sigma$ as function of depth position (**b**)

consists of two sharp strain peaks located at the edge of the delamination and a plateau of increased strain values in between. Since the waviness is implemented along the y-axis, there is no strain value modulation seen in Fig. 3.30a. For smaller defect sizes the two peaks start to approach each other, merging into one signature below 1.25 mm delamination size and thus possibly deviate from the square root dependency of (3.14).

Plotting the data of Fig. 3.29 as function of defect depth below surface returns the graphical representation shown in Fig. 3.30b. As defect depth position, the mean depth with respect to the sinusoidal waviness amplitude was chosen which directly corresponds to the position of the interface between the individual plies. For better visibility only the fit for the 5 mm size delamination is shown in Fig. 3.30b as solid line. Similar to the case study of fiber cuts above the trend of $\Delta\varepsilon/\sigma$ as function of defect depth is well captured by the relationship $\Delta\varepsilon/\sigma = \widetilde{A} + \widetilde{B}z^{-1/2}$. The resulting fit parameters $\widetilde{A}$ and $\widetilde{B}$ are characteristically different to the fiber cuts, which is dominated by the absolute difference of $\Delta\varepsilon/\sigma$. Nevertheless, the overall trend confirms the applicability of the abovementioned considerations regarding the decay of $\Delta\varepsilon/\sigma$ with $z^{-1/2}$.

As further defect type, the occurrence of inter-fiber failure has been investigated. Since inter-fiber failure in the case of a unidirectional material is assumed to span the full thickness of the laminate, the depth position is not varied for this case, i.e., the experimentally prepared and modeled defects always reach to the surface for angles $<90°$. For the specific case of an inclination angle of 90° the defect is oriented within the xy-plane with its fracture plane as seen in Fig. 3.23. This would correspond to a delamination at the midplane of the laminate. The evaluation of $\Delta\varepsilon/\sigma$ as function of defect size and inclination angle is shown in Fig. 3.31 for the cases of inclination angles of 0°, 22.5°, 45°, and 67.5° following the angle definition given in the scheme of Fig. 3.23. For the angle 90° experimental data and modeled data is added to the figure obtained from a respective delamination case. Evidently, the observed behavior in Fig. 3.31 is distinctly different to the findings for fiber cuts and delamination which requires some further discussion.

Based on the previous discussions, the value of $\Delta\varepsilon/\sigma$ is influenced by the size of the defect and the depth position. For the strain concentration as function of defect size, Fig. 3.31 reveals opposing trends above and below the inclination of 45°. The behavior observed at 67.5° and 90° clearly corresponds to the trend observed for delamination obeying a square root law as function of size. At an inclination angle of 45° almost no change as function of defect size is found. Above 45° the behavior cannot be described by a square root dependency and $\Delta\varepsilon/\sigma$ is instead found to increase with decreasing defect size.

This is due to the definition of $\Delta\varepsilon$ in (3.15) using the absolute value of strain concentration. For the cases above 45°, the highest $\Delta\varepsilon$ are obtained for the y-strain values $\langle\varepsilon\rangle_{\text{conc}}$ higher than $\langle\varepsilon\rangle_{\text{avg}}$ in good agreement with the observations for delamination. For inclination angles smaller than 45°, the highest $\Delta\varepsilon$ signatures occur for $\langle\varepsilon\rangle_{\text{conc}}$ smaller than $\langle\varepsilon\rangle_{\text{avg}}$ as indicated in Fig. 3.32b. Thus, a turnover of

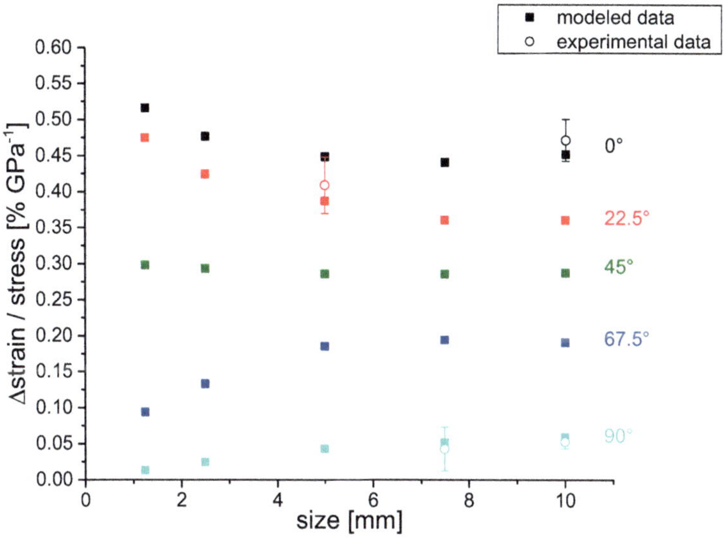

Fig. 3.31 $\Delta\varepsilon/\sigma$ due to inter-fiber cracks as function of varying defect size and inclination angles (based on [37])

the behavior may be expected at an angle of 45° resulting in the almost constant values as function of defect size.

To investigate this effect more precisely, the data of Fig. 3.31 is plotted as function of inclination angle in Fig. 3.32. The strain signatures are best described by a cosine fit $\Delta\varepsilon/\sigma = \widetilde{A} + \widetilde{B}\cos\left(\varphi - \widetilde{C}\right)$ of the 5 mm defect size using fit parameters $\widetilde{A}, \widetilde{B}$, and $\widetilde{C}$ which is shown as solid line. This fit is found to describe the observed dependency very well. Thus, the decay of strain concentration as function of angle may readily be interpreted as sinusoidal change between the two extreme cases at 0° orientation and 90° orientation. The change in dimension of the embedded crack is then translated into different amplitudes B.

3.3.3 *Detectability of Defects Using DIC*

Based on the results of Sects. 3.3.1 and 3.3.2, the purpose of this section is the discussion of the proposed measurement routine for the application to evaluate composite failure. Beyond the quantification of the strain concentration presented in the previous section, this requires further considerations to be made. First, the influence of measurement parameters and experimental issues intrinsic to the measurement principle are discussed. Second, the predictive capabilities of the presented modeling approach and the applicability to real defects are reviewed.

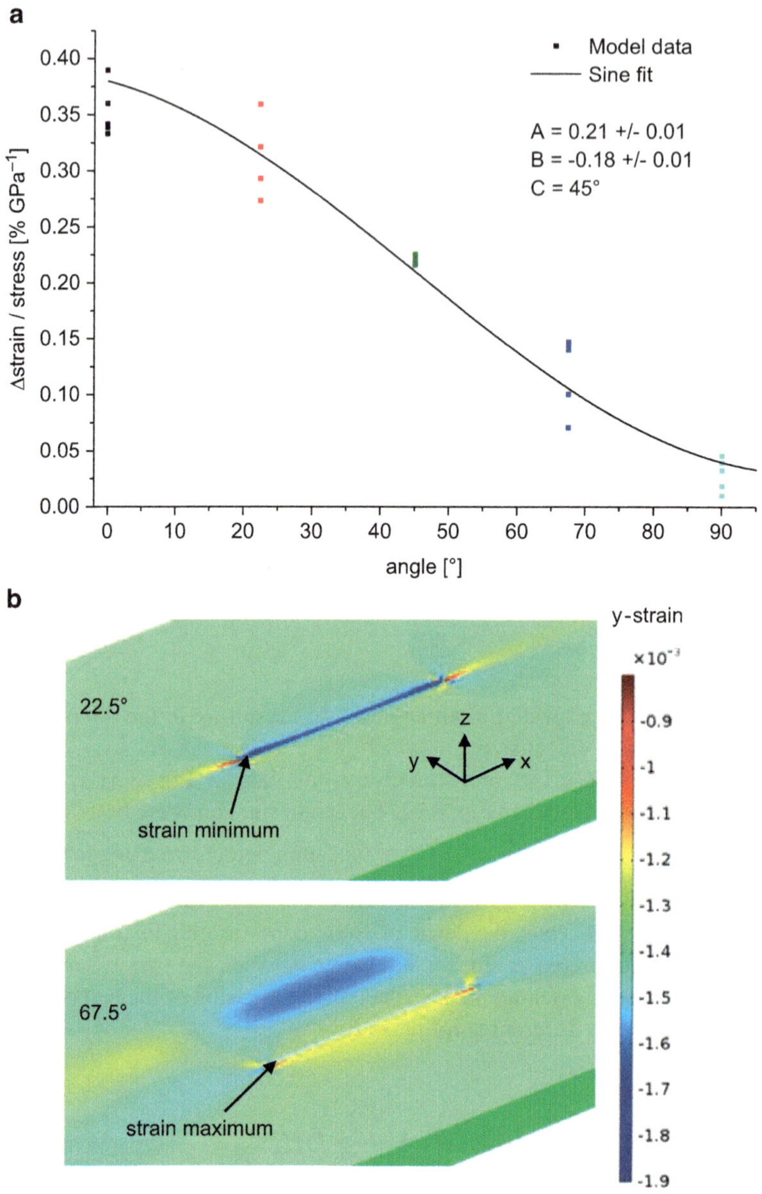

Fig. 3.32 Evaluation of $\Delta\varepsilon/\sigma$ as function of inclination angle (**a**) and regions for evaluation above and below 45° (**b**)

Finally, the system limits of presently used 3D DIC systems are established and compared to the experimental findings of Sect. 3.3.2 to assess the detectability limit of the DIC method to spot buried defects.

3.3.3.1 Experimental Parameters

For the experimental measurement of full-field strain distributions, several aspects need to be considered when discussing signatures attributed to strain concentrations. For the exemplary measurement in Fig. 3.33a at a mean strain of 0.44 % there is an obvious gradient along the y-direction in the x-strain values. The origin of this gradient can be manifold, starting with specimen imperfections, and bad bonding of reinforcement tabs or wrong specimen alignment. In all cases, the consequence can be a systematic variation of the resultant strain level as function of position. This is not only bad for measurement of material properties, but also poses further challenges for the interpretation of strain concentration. In this context, the obtained strain reference $\langle \varepsilon \rangle_{avg}$ may need to be interpreted differently. To demonstrate this effect, Fig. 3.33b shows a model result of the same 7.5 mm fiber breakage cases in the 2nd and 3rd ply below surface with a superimposed gradient in the tensile force. This change in loading conditions causes a visible strain gradient similar to the experimental data. Nevertheless, the strain concentration due to the embedded defects is still clearly visible for both cases. However, for application of (3.15) the question of the correct interpretation of $\langle \varepsilon \rangle_{avg}$ arises. Since the load gradient is balanced, the same strain average $\langle \varepsilon \rangle_{avg}$ is calculated as for the case of Fig. 3.26. Thus, for specimens with unbalanced strain gradients, the correct average strain should be taken from the local neighborhood of the defect, since the strain concentration relative to this reference will define the detectability.

Consequently, the occurrence of such strain gradients does not affect the general possibilities to spot defects as strain concentration in full-field strain measurements. Moreover, almost the same quantitative information may be extracted, if the strain reference is taken from the local neighborhood for cases of inhomogeneous (global)

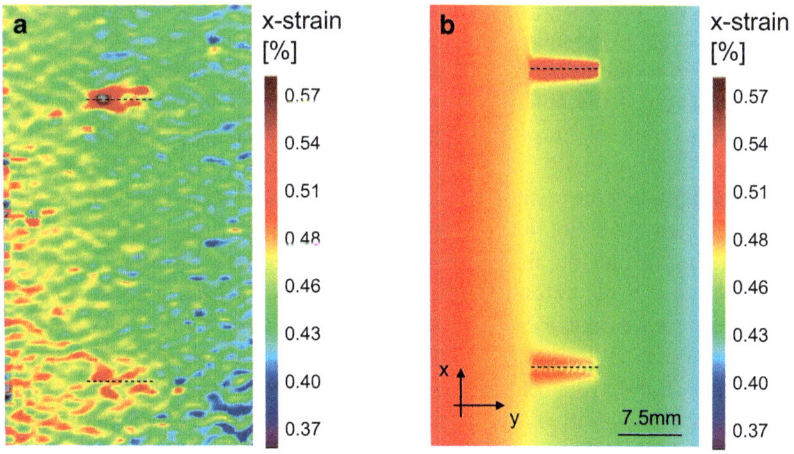

Fig. 3.33 Experimental data of embedded fiber breaks exhibiting gradient in x-strain (**a**) and corresponding model with embedded defects (**b**)

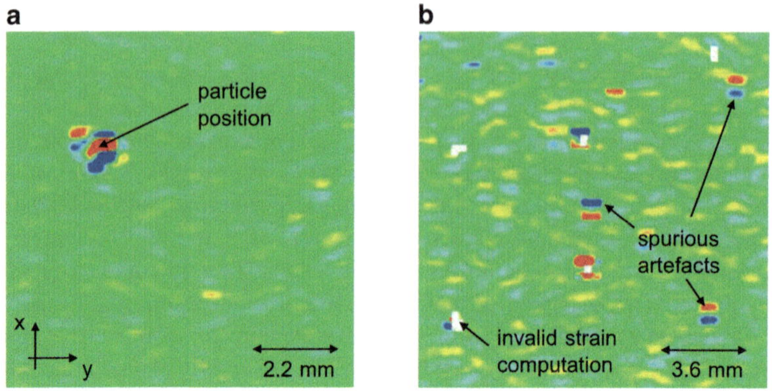

Fig. 3.34 Dust particle artifact (**a**) and artifacts due to incorrect subset tracking (**b**)

strain fields. The latter is especially relevant for real structures, because these will hardly exhibit equal strain values at all locations on the measured surface.

Another important effect to consider for measurement of strain concentrations arises from the image correlation algorithm and the speckle pattern used for the measurement. As seen in the images in Fig. 3.34, there are characteristic signatures, which might readily be interpreted as local strain concentration.

There are various effects, which might cause a false computation of a local strain value leading to characteristic artifacts:

1. Dust particles on lens
2. Incorrect subset tracking
3. Corrupt speckle patterns
4. Wrong choice of subset size relative to speckle pattern

For the first effect, the principle of DIC explained in Sect. 3.2 may readily be applied to understand the impact. If a dust particle sits on the camera objective or at any other position included within the optical path of the system it may readily cause a distortion of the optical beam path. This directly translates into a wrong computation of the real strain value at the position of the dust particle within the field of view. Consequently, the position of this artifact will not change during the measurement, i.e., the position of the strain signature will be constant during the measurement and will not move with the specimen during loading. Thus, the position of the artifact within the image will be constant for the case of small deformations. For larger deformations, its position may still start to displace, due to the wrong reference image taken with the dust particle in place. As additional source to generate similar artifacts in the strain field, the occurrence of dead pixels on the camera chip may be named. In such cases, one or more pixels on the camera chip are not operating with equivalent sensitivity compared to the remaining pixels or do not operate at all. For some systems it is possible to eliminate such dead pixels by calibration routines, but for systems without such elimination the recorded images have systematic deficiencies at the locations of dead pixels. The consequence will be an erroneous computation of the strain values for those subsets involving the dead pixels. Thus, the position of the artifact within the image will stay at a constant position and will not move with the loaded specimen.

Another class of artifacts, which may readily be interpreted as strain concentration, are those due to incorrect subset tracking. As seen in the example of Fig. 3.34b, this may cause two characteristic effects. If the subsets are not recovered by both cameras in a 3D-DIC configuration, some software packages indicate this by returning a distinctly different color in the image (white colored subsets in the case shown in Fig. 3.34b). This marks subsets with obviously invalid strain values. For high resolution measurements, spurious artifacts tend to appear as well. These are found as characteristic pair of high and low strain values in the direct vicinity of one subset. Their occurrence is sometimes not permanent and may vary in each step of image correlation. This is most likely due to imperfections of the speckle pattern causing incorrect subset tracking from image to image and, consequently, unrealistically high strain values.

Even more computation artifacts may occur, when a corrupt speckle pattern is used. The definition of "corrupt" covers several aspects, all of them resulting in an erroneous measurement of strain values. First of all, the speckle pattern may show unintended variation of the particle sizes. For sprayed patterns this may typically occur due to formation of larger drops or irregular particle distribution (cf. Fig. 3.35a). Moreover, for speckle patterns which do not permanently stick to the specimen the pattern might get harmed during handling of the specimen prior to testing or during mounting of the specimen in a test machine (cf. Fig. 3.35b). For all these cases, the speckle pattern no longer fulfills the quality required for the image correlation process and the derived displacement vectors will suffer in accuracy to a substantial extent. Consequently, the strain values calculated for such areas will not provide sufficient precision to evaluate the small strain concentrations expected for most of the defects studied in Sect. 3.3.2.

Another source for artifacts is the choice of subset size relative to the applied speckle pattern particle distribution size. For the purpose of detection of failure

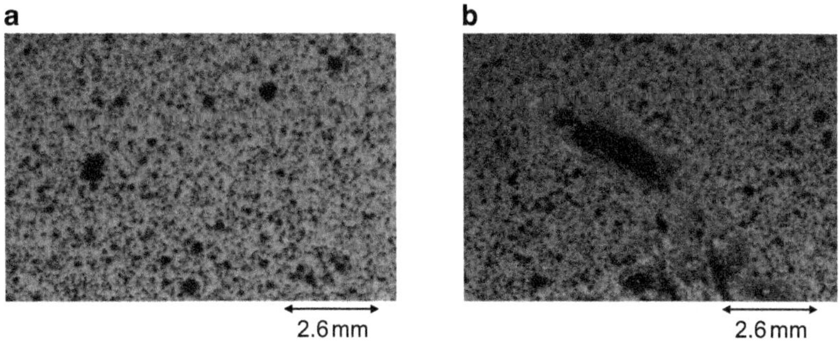

Fig. 3.35 Corrupt speckle pattern showing irregular particle distribution (**a**) and partially destroyed speckle pattern (**b**)

types as presented in Sect. 3.3.2 a sufficiently high spatial resolution is required to spot smaller defect sizes. This relationship will be discussed in Sect. 3.3.3.3 in more detail. However, a smaller subset size will give rise to several effects. Very small subset sizes tend to generate unreliable strain values as seen for the subset size cases of 10, 20, and 30 px seen in Fig. 3.36. Even for a perfect speckle pattern, this is

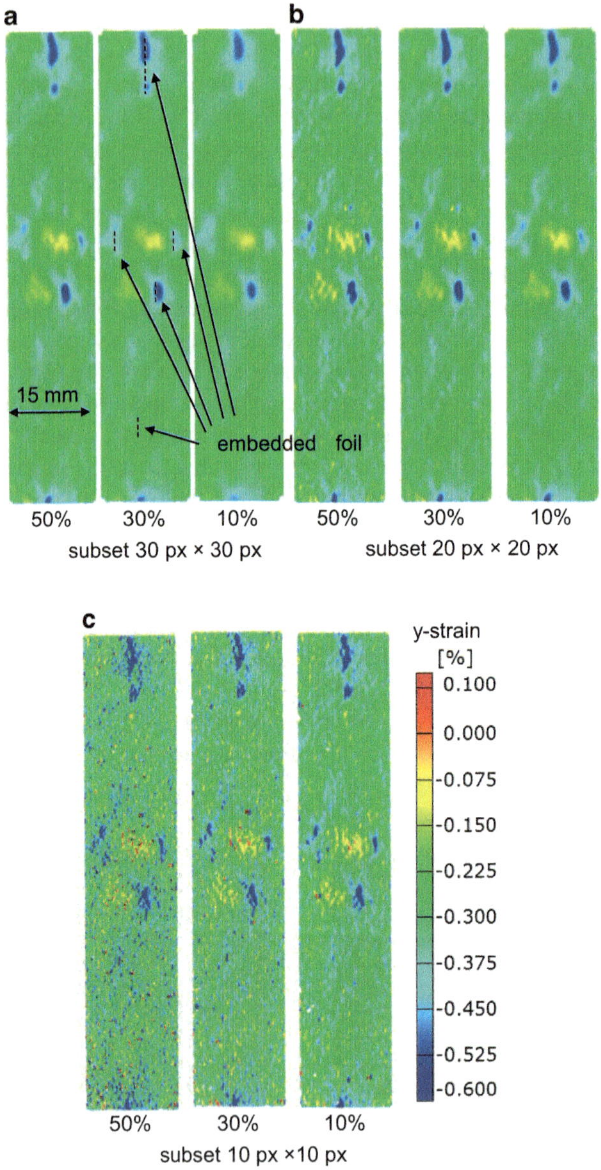

Fig. 3.36 Influence of subset size and subset intersection on strain noise and visibility of strain signatures for five embedded inter-fiber cracks of various dimensions

owed to the fact that the image correlation process is based on a correlation algorithm, which generally benefits from a larger number of data points to correlate (cf. Sect. 3.2). Thus, subset values lower than 20 px need to be used with care when interpreting the computed values. Furthermore, the stochastic pattern used for the image correlation procedure needs to match the chosen subset size in order to avoid erroneous computations. Clearly, if the speckle pattern exhibits additional inhomogeneity as discussed above, the local deviations will also give rise to artifacts, which will appear more often for smaller subset sizes, even if the subset size is well adopted to the global average.

3.3.3.2 Applicability of the Modeling Approach to Real Defects

As pointed out in Sect. 3.3.2, the artificial defects are prepared by means of embedded foils or cut fibers. This naturally opens the discussion, whether or not these defects are representative for real failure mechanisms as observed in composites.

For the artificial fiber failure, the filaments are cut during the fabrication of the laminate and thus a resin rich area is expected in between the position of the filaments (cf. Fig. 3.11). This is an obvious difference to a real fiber failure, where the space between the filament ends would be filled by a cavity. The model uses an almost realistic assumption of broken filaments, since the thin elastic layer is fully decoupled. As indicated in Sect. 3.3.2 the strain concentration measured for the fiber cuts agrees reasonably well to the model predictions. Thus the estimated values of $\Delta\varepsilon/\sigma$ are expected to be also applicable for real fiber failure cases. However, the general drawback for this prepared defect arises from the lateral sizes investigated. For real laminates, the expected defect sizes for fiber failure prior to ultimate failure are mostly in the order of few tens of micrometers. This is due to the fact that clustering of only few fibers in immediate vicinity will induce global failure of the respective ply [40, 41]. Since these defect sizes are difficult to prepare by the experimental means of manual cutting, it is difficult to evaluate smaller defect sizes experimentally. However, even for smaller sizes of fiber failure than investigated experimentally the model predicts almost comparable values of $\Delta\varepsilon/\sigma$. Given a reasonable dimension of fiber failure of a rectangular area of 20 μm × 20 μm (i.e., approximately 5–9 fiber filaments, dependent on filament diameter and fiber volume fraction) a corresponding model was evaluated to provide a strain concentration of 0.06 % at 695 MPa. This is close to the system accuracy as will be discussed below and, therefore, may be at the edge of detectability.

For the case of delamination, the investigated defect sizes are quite realistic. The main discrepancy to a real delamination originates from the use of an ETFE foil as embedded balloon. Whilst this approach seems to provide a reasonable approximation at first, it does cause a substantial bias for the local strain concentration. Figure 3.37 compares a computation result of the x-strain evaluated for a 10 mm size delamination including the embedded ETFE foil as used for the validation in

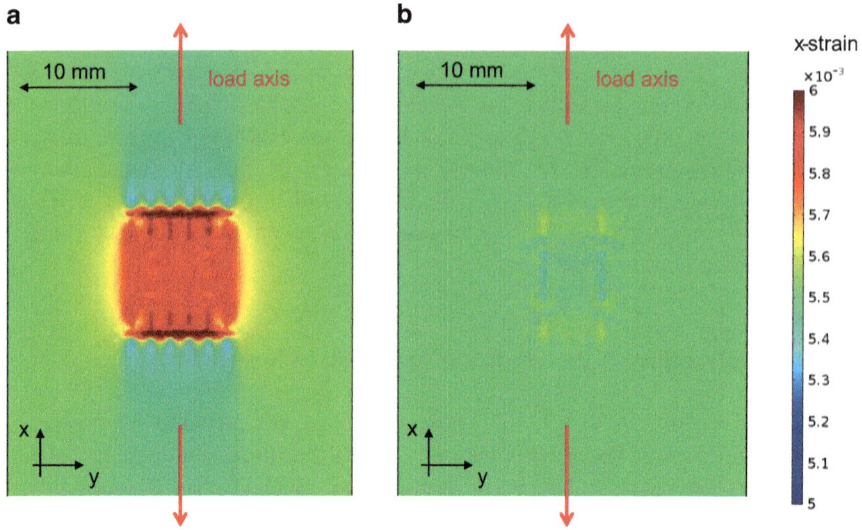

Fig. 3.37 Modeling result for inter-ply delamination at interface between 1st and 2nd ply with embedded ETFE (**a**) and without (**b**)

Sect. 3.3.2 and without such a foil as expected for a real delamination. Given the same color range of the x-strain in Fig. 3.37, the difference is quite obvious. Clearly, the ETFE foil acts like an intermediate layer of reduced stiffness, causing high strain concentration in the region of the embedded balloon. In comparison, the layer modeled in Fig. 3.37b uses the same waviness as in Fig. 3.37a, but the out-of-plane compressive stiffness of the layer is chosen to be compatible with the remaining composite as expected for a real delamination. As consequence, the calculated strain concentration drops from $\Delta\varepsilon = 0.108\ \%$ to $\Delta\varepsilon = 0.007\,\%$. Clearly, this will cause a substantial reduction of the possibilities to spot the occurrence of delamination failure in real laminates regardless of their size.

Similar to the previous case, the dimensions of the artificial inter-fiber failure defects studied in Sect. 3.3.2 are in a typical range as expected for real composites. Again the main discrepancy to the real failure mode arises from the inclusion of a thin PI foil to mimic the presence of an internally separated material. Although the tensile and shear load transfer by this thin foil is negligible, it still provides an artificial layer of low stiffness under compressive load. This local change in compliance is responsible for the generation of the high transverse strain values evaluated as strain concentration effect in Sect. 3.3.2. However, if there is no such layer in a real specimen, the appearance of the inter-fiber crack will likely change. Thus, in a tensile load scenario the model of an inter-fiber crack with identical stiffness of the thin elastic layer as for the surrounding composite did not even provide a strain signature exceeding the numerical noise level. However, since the perfect orientation of such inter-fiber failure is quite unlikely, the result of a computation with a minimal rotation of 2° around the z-axis is shown in

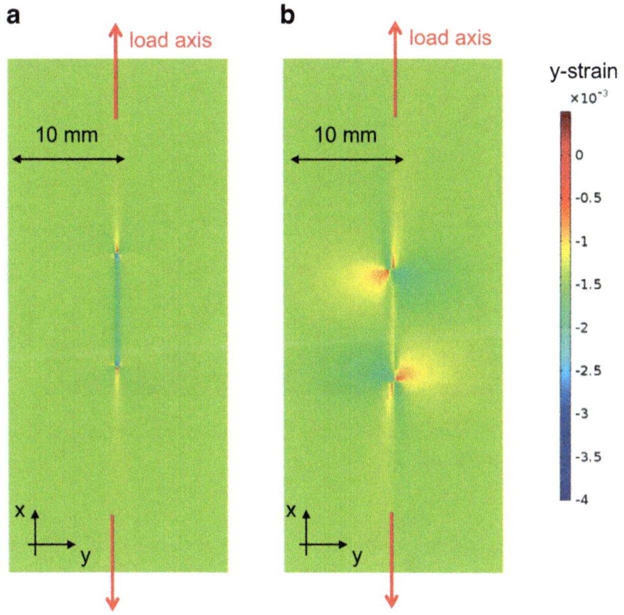

Fig. 3.38 Modeling result for inter-fiber failure with 0° inclination angle and embedded PI (**a**) and without PI, but additional rotation of 2° around *z*-axis (**b**)

Fig. 3.38b instead. As seen due to the same *y*-strain color range in Fig. 3.38, this immediately yields a strain concentration signature of equivalent intensity as for the embedded PI foil. Given the typical deviations of the inter-fiber crack orientations in a real laminate it may thus be assumed that this type of failure is readily visible by means of 3D DIC observations. For the macroscopically visible crack shown in Fig. 3.52 in Sect. 3.4, this finding may also be directly concluded from visual observations during the measurement.

However, more interesting applications arise for those inter-fiber cracks not reaching the surface since this inhibits a direct visual monitoring of their occurrence. As seen in the measurements of the *x*-strain field of a multi-axial laminate in Fig. 3.39 characteristic strain concentrations aligned with the ply directions (45° and 90°) indicate the occurrence of inter-fiber failure in these layers. Thus it may be concluded from modeling and experiment that the observation of inter-fiber failure should generally be possible given the system accuracy is sufficient as will be discussed in the next section.

3.3.3.3 Limitations Due to System Accuracy

So far, some first limitations of the experimental measurement technologies were presented and the visibility of defects was mostly discussed in terms of modeling results. However, the modeling results come with zero background noise and

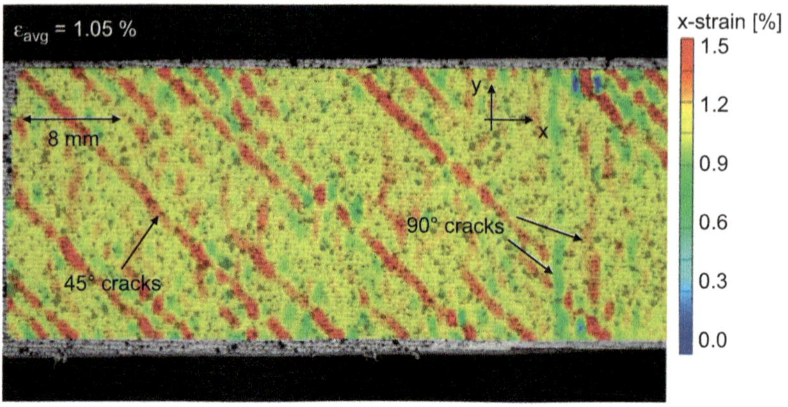

Fig. 3.39 Experimental tensile tests showing clear signatures of inter-fiber cracks of 45° and 90° orientation to load-axis (x-direction)

neglects aperture effects which arise due to the experimentally used subset sizes. Thus, the final step to judge on the detectability of specific failure mechanisms consists in the assessment of limitations of the measurement system. To this end, there are two essential influences to account for. First, the detectability will be discussed in terms of the detectable strain concentration $\Delta\varepsilon$ relative to the background noise. Second, the aperture effect given by the subset size will be evaluated to establish the limit of spatial resolution of the system.

With reference to system accuracy, the measurement equipment needs to detect very small local increases of strain concentration $\Delta\varepsilon$ relative to surrounding areas. The absolute limit of detectability may be defined in terms of the values $\langle\varepsilon\rangle_{\text{avg}} \pm \Delta\langle\varepsilon\rangle_{\text{avg}}$ and $\langle\varepsilon\rangle_{\text{conc}} \pm \Delta\langle\varepsilon\rangle_{\text{conc}}$ taking into account the measurement uncertainty of the global average strain value $\Delta\langle\varepsilon\rangle_{\text{avg}}$ and the local strain value $\Delta\langle\varepsilon\rangle_{\text{conc}}$.

Detectability of a signature is then defined as

$$\langle\varepsilon\rangle_{\text{conc}} - \left(\langle\varepsilon\rangle_{\text{avg}} + \Delta\langle\varepsilon\rangle_{\text{conc}} + \Delta\langle\varepsilon\rangle_{\text{avg}}\right) > 0 \tag{3.17}$$

The average value of $\langle\varepsilon\rangle_{\text{conc}}$ thus needs to be larger than the value of $\langle\varepsilon\rangle_{\text{avg}}$ plus the uncertainties of both strain measurements $\Delta\langle\varepsilon\rangle_{\text{avg}}$ and $\Delta\langle\varepsilon\rangle_{\text{conc}}$. This definition assumes the validity of (3.15), i.e., $\langle\varepsilon\rangle_{\text{conc}} > \langle\varepsilon\rangle_{\text{avg}}$ needs to be generally applicable.

The overall strain concentration $\Delta\varepsilon$ will determine the absolute detectability of a failure mechanism. If the strain concentration values get small, the detectability will be mostly limited by the uncertainty $\Delta\langle\varepsilon\rangle_{\text{avg}}$ (cf. Fig. 3.40).

However, given the investigation in Sect. 3.3.2 it is evident that the strain concentration may be approximated using a linear elastic material behavior. Consequently, the absolute value of $\Delta\varepsilon$ will scale linearly with the applied stress. Hence, defects which fall below the detection limit at low load levels may readily

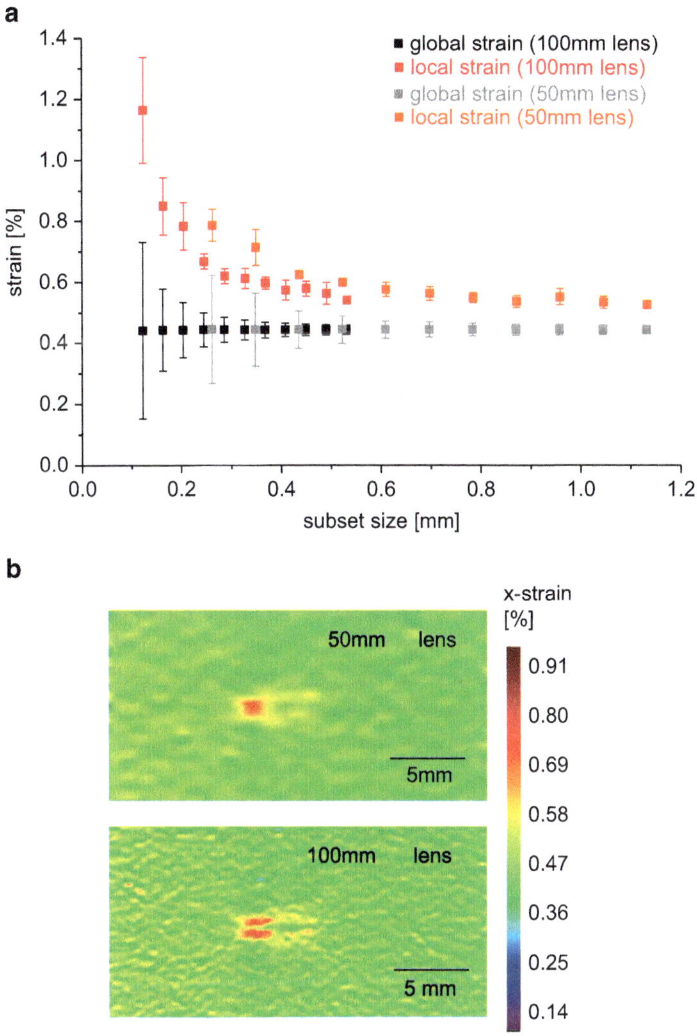

Fig. 3.40 Comparison of evaluated strain values as function of subset size for two lenses (**a**) and corresponding full-field measurement showing the strain concentration close to the measured fiber cut (**b**) (in part based on [37])

be visible at higher load levels, since the values of $\Delta\langle\varepsilon\rangle_{\mathrm{avg}}$ and $\Delta\langle\varepsilon\rangle_{\mathrm{conc}}$ were not found to increase with the applied stress.

For the experimental investigation of Sect. 3.3.2, the settings according to Table 3.1 allow an estimate of the absolute strain limit based on Fig. 3.40. Given the subset size of 21 px and the resolution of 0.0136 mm/px the minimum value of $\Delta\varepsilon$ to be detected may be estimated based on the uncertainty of $\Delta\langle\varepsilon\rangle_{\mathrm{avg}} = 0.029\,\%$ and $\Delta\langle\varepsilon\rangle_{\mathrm{conc}} = 0.023\,\%$ and therefore evaluates as $\Delta\varepsilon = 0.052\,\%$.

The spatial resolution of the measured strain field is determined by the camera resolution and the choice of subset parameters. The camera resolution first of all depends on the pixel density of the camera sensor and the measurement distance to the object. Due to the fact that modern cameras already provide 5–50 Megapixels on their camera chips, an increase of pixel density by one order of magnitude is not readily expected. Except for some specific fields of applications like astronomy, there is no strong market requirement driving a development towards larger pixel densities.

For a particular system a direct benefit towards higher spatial resolution may thus be obtained by an increase of the magnification factor. Given the general laws of optics as introduced in Sect. 3.1 an increased magnification at constant detector size comes with the drawback of decreased field of view. So for experimental work using 3D DIC, the following discussion will be limited to object sizes of reasonable extent, i.e., in the range of typical laboratory size specimens.

To demonstrate the effect of the magnification factor on the system accuracy, Fig. 3.40a provides a comparison of measurements of a fiber cut monitored by the same DIC system using a 50 mm lens and a 100 mm lens. The respective magnification factors correspond to 0.0291 and 0.0136 mm/px, respectively. For this study, the size of the chosen subsets d_{subset} was systematically varied keeping the intersection of subsets at approximately 50 %. To allow direct comparison of both lenses, subset size values are reported as metric values instead of pixels. Whilst the global strain average $\langle\varepsilon\rangle_{\text{avg}}$ is almost unaffected by the changed lens for all subset sizes, a distinct change in the values of local strain $\langle\varepsilon\rangle_{\text{conc}}$ is observed for subset sizes of lengths <0.6 mm. This is caused by the increased spatial resolution provided by the 100 mm lens.

However, the findings shown in Fig. 3.40 indicate that the spatial resolution of the strain field is also the result of the size of the chosen subsets and thus will add to the ability to detect smaller defects. As discussed previously, the intuitive assumption is that smaller subset sizes will provide better spatial resolution and, therefore, detectability. However, a decrease of subset size comes with the price of increased artifacts and less stable strain values. As consequence, an optimum trade-off between the accuracy of the strain value and the spatial resolution is required.

This relationship is already visible in Fig. 3.40a and is indicated by the decreasing uncertainty $\Delta\langle\varepsilon\rangle_{\text{avg}}$ as function of subset size. Thus it is not advisable to go below a certain subset size in pixels. For the 50 mm lens a subset length of 0.6 mm and for the 100 mm lens a subset size of 0.25 mm may thus be used as a proper choice.

Since some systems average the local strain value by usage of neighboring subsets, an additional aperture effect needs to be considered for the strain evaluation. As seen in the schema in Fig. 3.41a, the strain value at a certain subset position is evaluated from the average of a local neighborhood (a 3×3 neighborhood with 50 % intersection in this example). Therefore, assuming a typical Gaussian weight function, the spatial resolution of the system may be estimated from the half-width

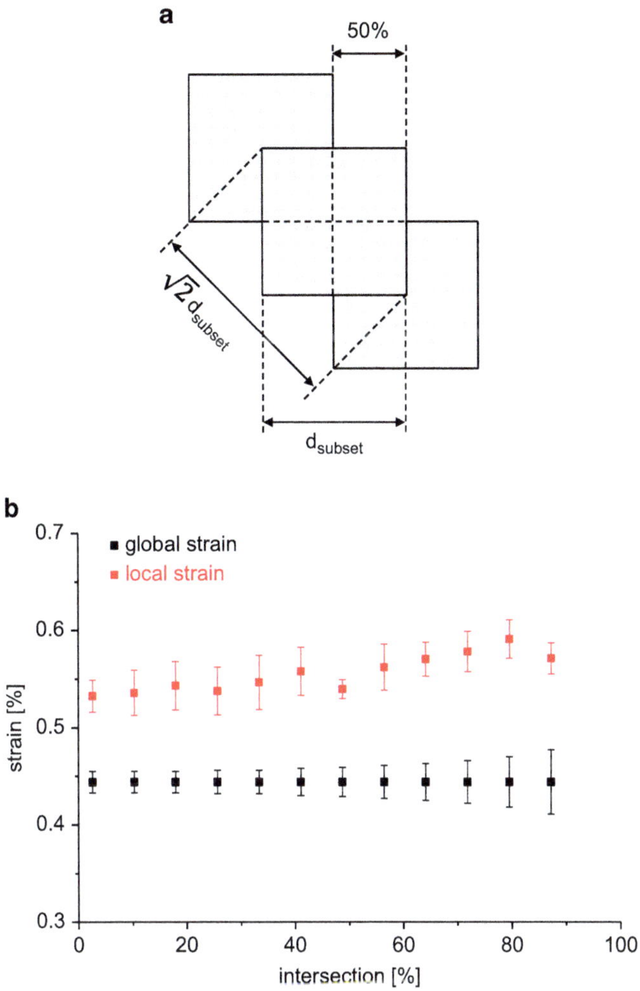

Fig. 3.41 Schema to evaluate spatial relation of subsets (**a**) and evaluated strain values as function of subset intersection for example of Fig. 3.40 (**b**) (in part based on [37])

at half-height. The evaluation of the strain value at the position of the embedded fiber cut shown in Fig. 3.41b clearly reveals the importance of this contribution to the visibility of the defect. As function of subset intersection the global strain value $\langle\varepsilon\rangle_{avg}$ is not affected much, but the local strain value $\langle\varepsilon\rangle_{conc}$ undergoes an almost linear increase from 0 to 95 % subset intersection. This is due to the highly localized strain concentration observed for this case and thus provides an improvement in detectability by a factor of two. Therefore an increase in subset intersection constitutes the better approach to improve spatial resolution as compared to a decrease of subset size. For the limit of intersection approaching 100 % the size of the single subset starts to dominate the measurement again. Consequently, the subset size will constitute the absolute limit in terms of spatial resolution, i.e., the minimum size d_{conc} of a strain concentration, which can be detected may be approximated for a quadratic subset size d_{subset} as:

$$d_{\mathrm{conc}} > \sqrt{2} d_{\mathrm{subset}} \tag{3.18}$$

However, the computation of full-field strain values with small subset sizes and high intersection values approaches the limits of presently available computation capacities. As more realistic estimate, one may thus instead consider an intersection of approximately 50 % leading to a modification of (3.18) by accounting for a 3×3 neighborhood according to the scheme in Fig. 3.41a:

$$d_{\mathrm{conc}} > 2\sqrt{2} d_{\mathrm{subset}} \tag{3.19}$$

For the type of high-resolution system ARAMIS 12 M used in the study in Sect. 3.3.2 the subset size of 21 pixels corresponds to a physical length of 0.29 mm, which turns into an expected lower limit for defect visibility of 0.81 mm.

Combining the limitations with respect to detectability given by (3.17) and (3.19), some final conclusions can be obtained.

Due to the spatial resolution of presently available systems, it will be difficult to detect strain concentrations as indications in the range of one millimeter or below. This does not necessarily imply that smaller defects may not be visible, since the size of the strain concentrations does typically exceed the physical size of the defect by at least a factor of two (cf. Figs. 3.20, 3.21, and 3.22). Therefore many failure types occurring in fiber reinforced composites may be sufficiently resolved.

A more critical parameter for the visibility of a buried defect the visible strain concentration value at the surface of the material was evaluated. The present approach using leading edge image correlation algorithms is at the limit of detectability when considering the small strain concentrations expected for some of the smaller size failure mechanisms. Especially for the case of delamination, a low expectation for detectability is given by the low values of strain concentration. For such cases, the use of shearography as described, i.e., in ASTM E 2581 was found to be much more sensitive [42–45]. However, large size defects will cause noticeable strain concentration values and can therefore be used as valuable tool to monitor the occurrence of failure in composites as will be presented in Sect. 3.4. Another natural extension of the presented approach considers the use of digital volume correlation to evaluate strain fields not at the surface of a composite material, but on the inside as will be demonstrated in Sect. 6.4.

3.4 Application to Composites

As discussed in Sect. 3.3 the possibility to detect the occurrence of particular defects in composite materials is a question of the spatial resolution of the measured strain field as well as the depth position of occurrence. In this section, some traditional fields of application for DIC during testing of composites and some new developments are presented to apply DIC for evaluation of material properties of composite materials.

3.4.1 DIC as Optical Extensometer

One of the traditional fields of application for DIC is the replacement of strain gages or mechanical extensometers in mechanical testing. Several drawbacks associated with the use of strain gages are well known and are found in standard literature (i.e., [46]):

- There is always a mismatch of thermal expansion coefficients at least in one direction. Since the outermost layer usually has anisotropic thermal properties, changes of temperature during testing may have substantial impact on the measurement accuracy.
- The high strain levels sometimes reached in fiber reinforced materials may cause preliminary failure of the strain gage due to overstraining.
- Another factor is the possible interaction between periodic structures of the composite (e.g., woven fabric) and the grid length, causing error-prone measurements.
- The long-term stability during creep or fatigue experiments are well-known challenges for strain gages as well.
- For extensometers, knife edges may act as stress concentrators causing preliminary failure at the mounting positions.

On the positive side, strain gages and extensometers benefit from a long tradition in mechanical testing, which is accompanied by well-developed technical equipment and skilled personnel.

Some of the possibilities of optical measurement techniques were already presented in Chap. 2.4 demonstrating the selection of appropriate regions to derive mechanical properties. In this context it is worth noting that the accuracy of 3D-DIC methods has nowadays reached the same level as strain gages [1, 47–50]. Moreover, the evaluation of the full strain field allows distinguishing between global strains (corresponding to strain gage values) and local strains (as caused by the presence of strain concentrators). Measurement of the full set of material properties as required by the advanced failure criteria presented in Sect. 2.3 requires a comprehensive amount of material testing. In the following, some examples are presented how DIC techniques can be used to obtain material properties of composites.

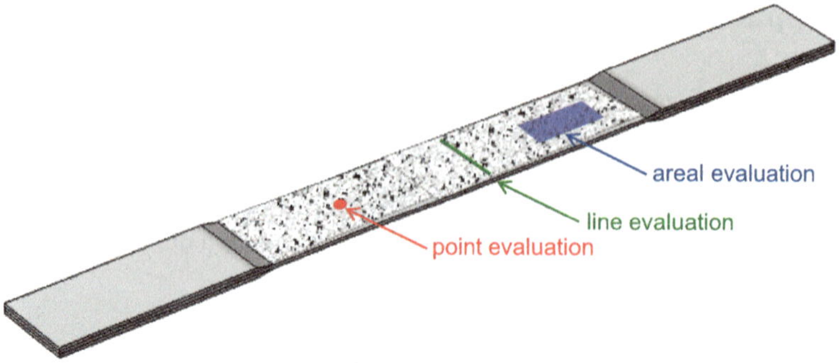

Fig. 3.42 Drawing of typical tensile test specimen and scheme of possible strain component evaluations using DIC

Since the measurement by DIC results in full-field information of the strain components of the material under test, there is a variety of possibilities for offline analysis. As schematically shown in Fig. 3.42 the computed strain components can be evaluated, e.g., at specific points or averaged along lines or areas of interest. Using areal or line evaluations strain components may also be visualized based on their position to yield results as seen in Fig. 3.25 or Fig. 3.19.

3.4.1.1 Tensile Testing of Unidirectional Fiber Reinforced Polymers

As a basic example for characterization of composites, the tensile test specimen and the measurement result of a unidirectional fiber reinforced polymer tested with fiber orientation in parallel to the load axis is shown in Fig. 3.43. For that purpose standard specimen types as required by DIN EN ISO 527-5, DIN EN 2561, or similar standards are used. The speckle pattern is applied to the surface of the test specimen as shown in Fig. 3.43. To produce the speckle pattern, a white base coat with subsequent application of a black spray pattern has proven to be useful for materials characterization of composites. The opposite approach using a white pattern directly applied on the naturally black color of carbon fiber reinforced composites does not provide equivalent results due to the opaqueness of the outer surface. Thus a well-defined background such as given by the white base coating is the best choice for a good signal-to-noise ratio of the measurement. In application to composites, the quality of the surface finish can be expected to have an impact on the measurement signal. Due to the rough residues of peel-plies as compared to a smooth finish from tooling surfaces different surface roughness values are found in practice. Comparative measurements of surfaces with peel-ply residues and surfaces subject to subsequent machining did not reveal substantial differences in the measured absolute strain values. However, slight differences were found in the width of the strain distribution (i.e., the scatter around the absolute value) within the same area used for averaging.

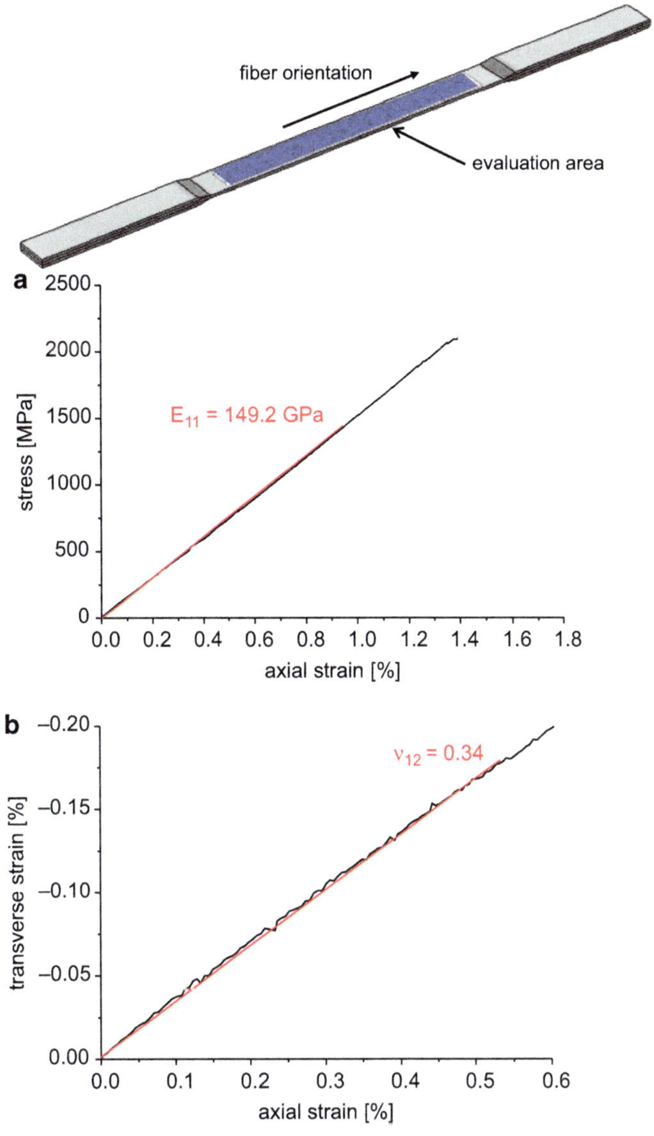

Fig. 3.43 Drawing of specimen used for tensile testing and exemplary stress–strain curve to evaluate modulus (**a**) and exemplary diagram of axial and transverse strain (**b**) to evaluate Poisson's ratio υ_{12}

As exemplary evaluation, strain values in parallel to the fiber axis and perpendicular to the fiber axis are averaged in the marked area and are used to evaluate the modulus and the Poisson's ratio υ_{12} as shown in Fig. 3.43.

For tensile testing perpendicular to the fiber axis 3D-DIC can be used to measure the transverse modulus and the Poisson's ratio υ_{21} or the Poisson's ratio υ_{23}. For the

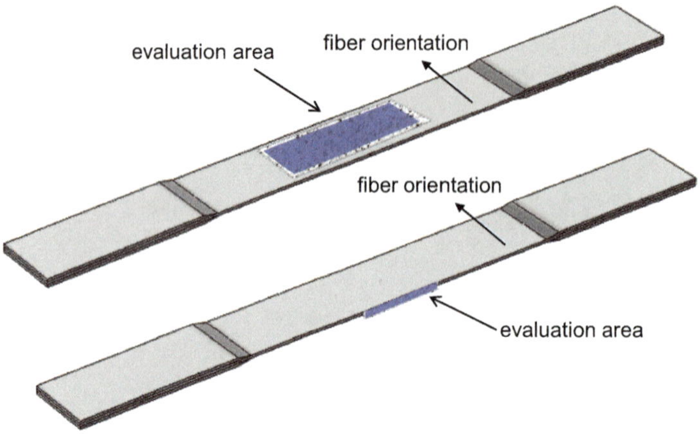

Fig. 3.44 Position of speckle pattern used to measure values for transverse modulus and Poisson's ratio υ_{21} (*top*) and the Poisson's ratio υ_{23} (*bottom*) including evaluation area

measurement, standard specimens in accordance with DIN-EN-ISO 527-5, DIN EN 2597, or similar standards are used. As seen for the specimen configuration in Fig. 3.44, a speckle pattern can be applied to the flat side of the laminate to measure the Poisson's ratio υ_{21}. Applying a speckle pattern to the thin side of the laminate allows to measure the Poisson's ratio υ_{23} (cf. Fig. 3.44 bottom). This measurement is particularly relevant to establish the 3D elasticity tensor of a unidirectional fiber reinforced polymer. For measurement of strain fields on the thin side of the laminate a sufficiently high resolution of the subset regions across the 2 mm thickness direction is required to avoid influences of the free edges. Therefore the averaging area chosen slightly offset to the side surfaces, so a border of approximately the size of the plate thickness is neglected for the averaging procedure.

An exemplary configuration for the 3D-DIC setup used for such measurements is given in Table 3.2. To derive the global strain field, the given resolution is comparatively high. As shown in [1] lower resolutions will still yield comparable results to conventional extensometers.

3.4.1.2 Compressive Testing of Unidirectional Fiber Reinforced Polymers

Similar in concept to tensile testing, evaluation of the strain field in compression testing as function of the fiber orientation and side surface allows measuring the Poisson's ratios υ_{12}, υ_{13}, υ_{21}, and υ_{23}. It is worth noting that the Poisson ratio's υ_{13} and υ_{23} are usually not measured in experimental setups according to ASTM D 6641, DIN EN 2850, DIN EN ISO 14126, or similar standards. As seen in Fig. 3.45, the specimen types correspond to the respective standards and the speckle pattern is

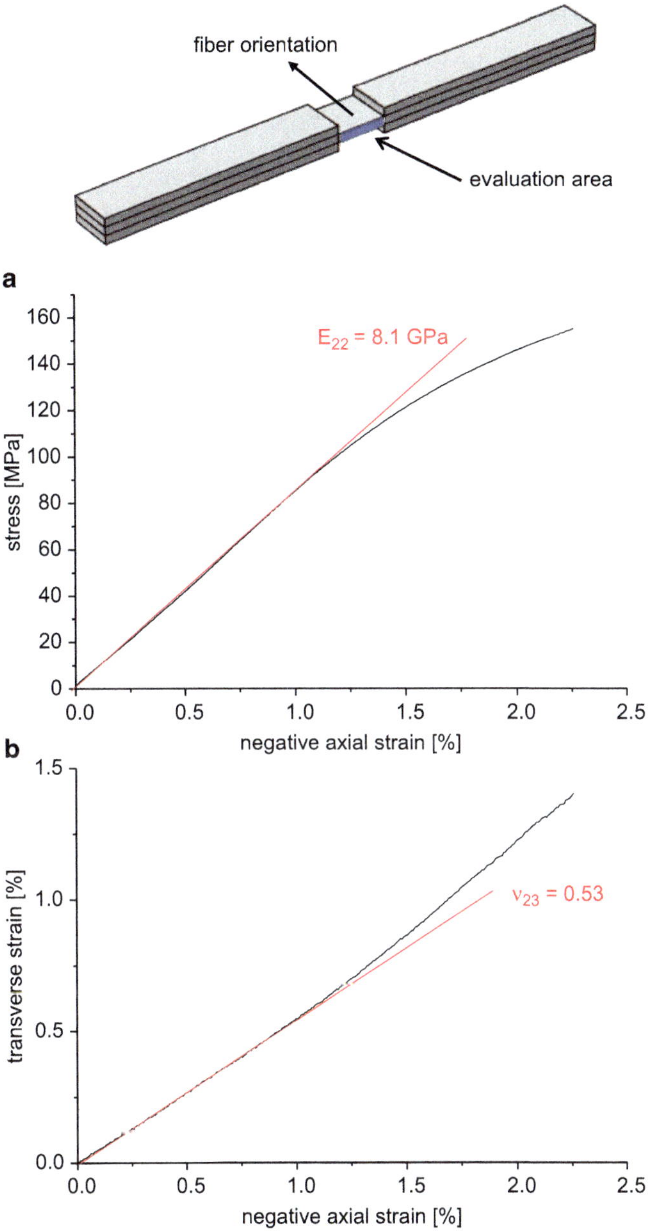

Fig. 3.45 Drawing of specimen used for compressive testing and exemplary stress–strain curve to evaluate transverse modulus (**a**) and exemplary diagram of axial and transverse strain to measure Poisson's ratio υ_{23} (**b**)

applied to the respective surfaces of the specimen as described before. An exemplary stress–strain curve to obtain the transverse compressive modulus and a plot of axial and transverse strain to obtain the Poisson's ratio υ_{23} is shown at the bottom of

Table 3.3 Acquisition settings of DIC system for compression tests of unidirectional fiber reinforced composites

Measurement setup	Value
Camera	Toshiba CMOS camera CSC12M25BMP19-01B with 4096 × 3072 pixels
Lenses	Titanar, focal length 2.8/50 mm
Lighting	KSP 0495-0001A LED 20 W/24 V white 30°
Filter	Polarization-filter Schneider-Kreuznach
Field of view	(103.4 × 77.4) mm^2
Scale factor	0.025 mm/px
Subset size	19 × 19 pixel
Acquisition rate	2 Hz
Software	ARAMIS
Configuration	3D

Fig. 3.45. In this test configuration it is highly recommended to use 3D-DIC measurements instead of 2D-DIC measurements. The reason for this recommendation is the strong tendency of fiber reinforced polymers to show unintended buckling modes during compressive loading, which might cause substantial out-of-plane motion with respect to the focal plane of the camera system. This could cause potentially high errors in 2D-DIC measurements. At the same time, some of the established standards involve some monitoring of the specimen buckling. ASTM D 6641 requires measurement of the deformation values using two strain gages mounted on each of the two specimen sides. If more than 10 % deviation between the deformations of the two strain gages is observed, the test results are rejected. Similar measurements can be performed using 3D-DIC and can be used to assure, the specimen has failed under acceptable compressive strain states. Typical DIC configurations confirm with those given in Table 3.3. In comparison to the tensile test the main difference is the smaller area used for extraction of the strain field. Therefore, a fine speckle pattern and higher resolution in px/mm is required for these measurements.

3.4.1.3 V-Notched Rail Shear Testing of Unidirectional Fiber Reinforced Polymers

Another reasonable application for full strain field measurements is to assure specific strain conditions during testing. Especially for testing under shear loading there is a long history in development of test rigs suitable for producing constant shear stress as recently reviewed by [51, 52]. One of the key requirements for unidirectional fiber reinforced polymers is the homogeneity of the produced shear stress field. While this can be achieved with various rail shear test fixtures, the usage of bolts to fix the specimen often induced preliminary failure. The test concept favored by ASTM D 7078 uses a combination of a rail-shear type movement and Iosipescu-type notch geometry to induce a zone of theoretically pure shear stress in

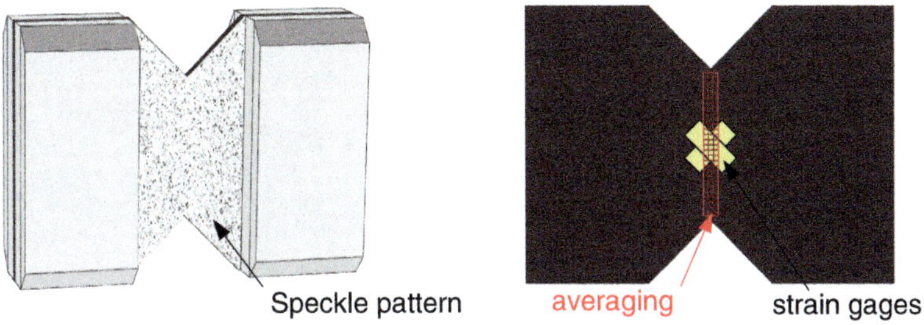

Fig. 3.46 Speckle pattern position in V-Notched rail shear configuration (*left*) and strain gage positions with averaging area for DIC measurements marked in *red* (*right*)

Table 3.4 Acquisition settings of DIC system for V-notched rail shear tests of unidirectional fiber reinforced composites

Measurement setup	Value
Camera	Toshiba CMOS camera CSC12M25BMP19-01B with 4096x3072 pixels
Lenses	Titanar, focal length 2.8/50 mm
Lighting	KSP 0495-0001A LED 20 W/24 V white 30°
Filter	Polarization-filter Schneider-Kreuznach
Field of view	$(81.9 \times 61.4)\ \text{mm}^2$
Scale factor	0.02 mm/px
Subset size	19×17 pixel square
Acquisition rate	2 Hz
Software	ARAMIS
Configuration	3D

the region between the notches. However, the standard recommends measurement of the shear strain value in the center of the specimen using two strain gages mounted in ±45° orientation relative to the vertical direction. The resulting measurement of the strain values is likely error-prone if the shear stress zone is not as sharp as theoretically expected. Figure 3.47 shows a comparison between two specimens of the same unidirectional fiber reinforced polymer with fiber orientation in parallel to the movement of the fixture (vertical direction in Fig. 3.46) measured using the DIC setup given in Table 3.4.

Whereas one specimen shows a rather undefined shear strain field (specimen A), the other specimen exhibits a sharp concentration of the shear strain field as expected from theory (specimen B). Failure in specimen A occurs preliminary due to crack initiation offset to the notch position, which is likely due to defects present at that position. If the shear strain is evaluated at the marked position, the resulting stress–strain curves shown in Fig. 3.47 are also falsified by this shear strain field. Therefore, direct DIC observation of the strain field can be used as additional factor to judge on the validity of the measurement.

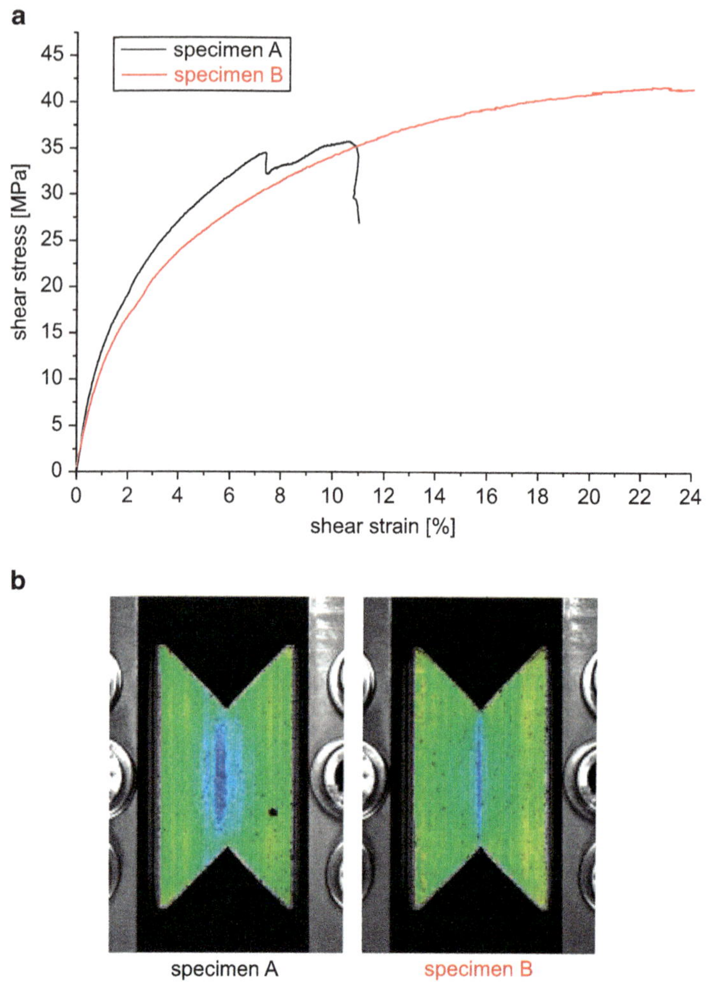

Fig. 3.47 Measured stress–strain curve for V-notched rail shear specimens including shear strain field for two different specimens at a shear strain of 4 %

3.4.2 Detection of Failure Onsets

In the context of fracture mechanics, the detection of first failure onsets is usually of huge relevance to obtain fracture toughness values for composite materials. Especially for brittle matrix materials such as epoxy, the first onset of crack growth is often used to deduce the critical fracture toughness value as relevant material property. For more ductile matrix materials such as some thermoplastic polymers, some of the established experimental setups in fracture mechanics may fail in their concept or in their interpretation. In the following some examples of test standards are presented to demonstrate these challenges. Therefore, using the detection of the full strain field during loading, the central aim in the next applications is not the quantitative measurement of the strain field or the replacement of strain gages, but the correct interpretation of the strain concentration shortly before and after crack growth. This is somewhat different to approaches to quantitatively measure the strain field around crack tips to directly calculate the J-integral values during crack growth [1].

3.4.2.1 Short-Beam Shear Tests

One example of configurations being used frequently in composites testing is the short-beam shear test. Although this test configuration is not meant to measure a true material property relevant to fracture mechanics, it still is a well-suited method being used in quality assurance and screening of materials. In the experiment the load is applied at the center of the specimen utilizing a load nose midways between the supports (cf. Fig. 3.48).

The apparent interlaminar shear strength value τ_{ILSS} is related to the onset of delamination under shear forces parallel to the layers of the laminate and is calculated by

$$\tau_{\text{ILSS}} = \frac{3F_{\text{onset}}}{4bh} \tag{3.20}$$

using the load at the moment of delamination onset F_{onset}, the width of the specimen b, and the thickness of the specimen h.

During the test, samples are loaded with a constant displacement rate of the loading nose of 1 mm/min and at the same time the speckle pattern on the side surface of the specimens was monitored by 3D-DIC techniques as seen in Fig. 3.49 using the parameters given in Table 3.5.

As shown for the image in Fig. 3.49, sudden increase of the shear angle in positive and negative direction indicates interlaminar failure at the center plane. The first signature in the measured load–displacement curve coincides with the delamination onset. With delamination onset (and growth), the load drops to a lower value. As the bending stiffness of the specimen is reduced, the slope of the force–displacement curve decreases after delamination onset. Usually further load drops can be observed when loading the specimen further. Although this observation does seem trivial, the usual procedure recommended in standards ISO 14130,

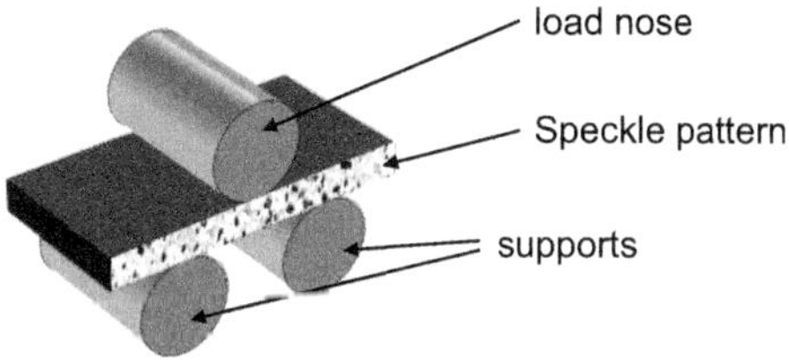

Fig. 3.48 Schema of short beam shear test configuration including speckle pattern position

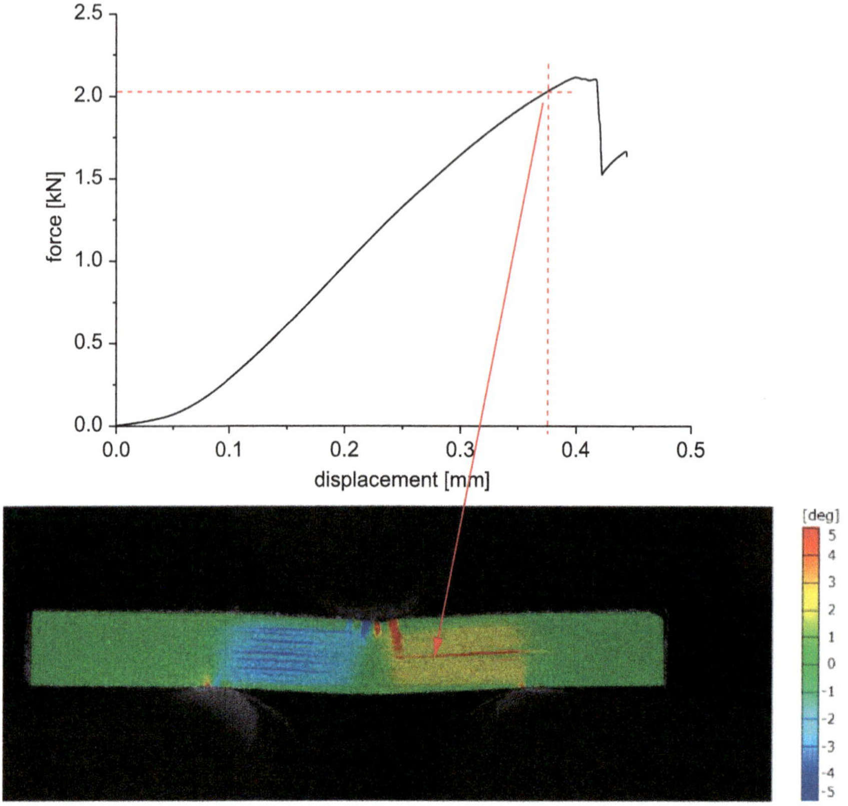

Fig. 3.49 Force–displacement curve including DIC image at onset of interlaminar crack growth with false-color indicating shear angle. First onset is marked in *red* in force–displacement curve

Table 3.5 Acquisition settings of DIC system for short beam shear tests

Measurement setup	Value
Camera	Toshiba CMOS camera CSC12M25BMP19-01B with 4096 × 3072 pixels
Lenses	Titanar, focal length 2.8/50 mm
Lighting	KSP 0495-0001A LED 20 W/24 V white 30°
Filter	Polarization-filter Schneider-Kreuznach
Subset size	11 × 8 pixel square
Acquisition rate	1 Hz
Software	ARAMIS
Configuration	2D

ASTM D 2344, DIN EN 2377, and DIN EN 2563 is the usage of the force maximum of the curve as value of F_{onset}. This systematically overestimates the apparent interlaminar shear strength for the present example by 4.9 %. Also, ductile matrix materials often show substantial deviations to the sequence of damage expected by

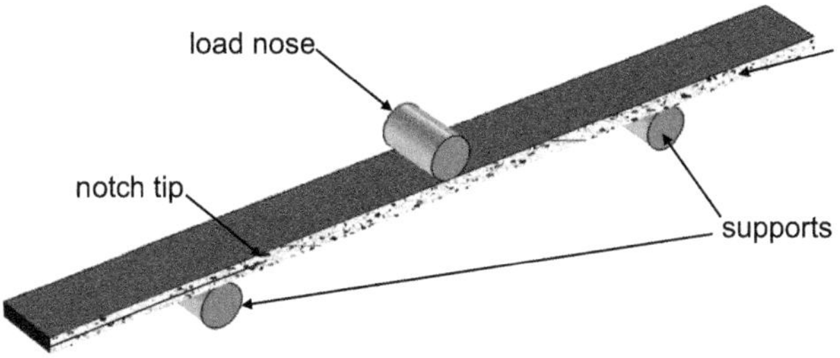

Fig. 3.50 Schema of ENF test configuration including speckle pattern position

the test standards and therefore cannot be interpreted based solely on load–displacement curves. For such cases, the DIC measurements are a suitable alternative to securely assign signatures in the load–displacement profile to the occurrence of interlaminar failure [38].

3.4.2.2 End-Notched Flexure Tests

Another typical test configuration used in fracture mechanics of fiber reinforced composites is the end-notched flexure (ENF) test as seen in Fig. 3.50. The aim of the test is to initiate an interlaminar crack under a mode II load situation. Although there is an ongoing discussion if this test configuration does produce a pure mode II load situation [53, 54], it is used here as another representative example for measurement of material properties utilizing DIC techniques.

Conceptually similar to the interpretation of the short-beam shear test, the ENF test uses the force value F_{onset} at the onset of interlaminar crack growth to derive the critical mode II fracture toughness according to

$$G_{IIc} = \frac{3m_c {F_{\text{onset}}}^2 {a_0}^2}{2b} \tag{3.21}$$

This uses the initial crack length a_0, the width of the specimen b, and a compliance calibration factor m_c for the specimen evaluated prior to testing. Therefore, the only critical parameter to obtain during the final test procedure is the value of F_{onset}.

To demonstrate the usage of DIC in this context, ENF tests were carried out using a layup consisting of 16 layers of Sigrafil CE1250-230-39 unidirectional epoxy-based prepreg with a pre-crack prepared at the interface at the midplane. Samples of 170 mm length, 25 mm width, and 3.6 mm thickness were cut from the cured plate. Simultaneously, images were recorded for DIC following the parameters reported in Table 3.6.

As seen for the example in Fig. 3.51, the force–displacement curve is straightforward to interpret in accordance with existing standards. Indicated by the two subsequent DIC images at the time of maximum force, the first load drop is clearly associated with the onset of interlaminar crack growth. However, for some cases it

Table 3.6 Acquisition settings of DIC system for ENF tests

Measurement setup	Value
Camera	Toshiba CMOS camera CSC12M25BMP19-01B with 4096 × 3072 pixels
Lenses	Titanar, focal length 2.8/50 mm
Lighting	KSP 0495-0001A LED 20 W/24 V white 30°
Filter	Polarization-filter Schneider-Kreuznach
Subset size	11 × 11 pixel
Acquisition rate	1 Hz
Software	ARAMIS
Configuration	2D

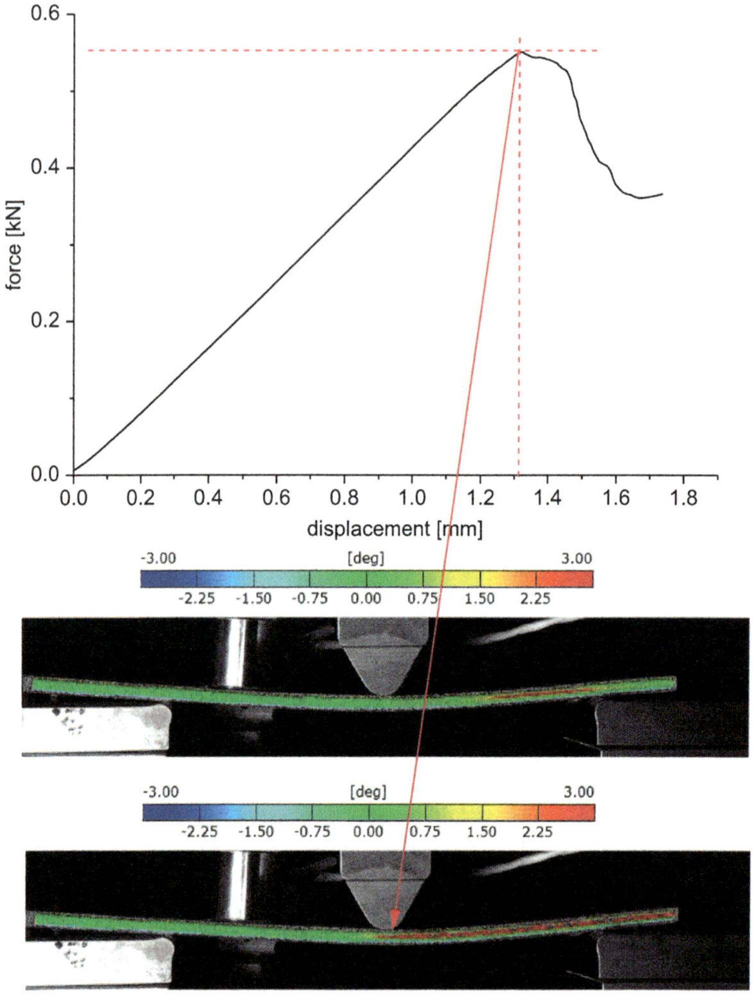

Fig. 3.51 Force–displacement curve including two DIC images before and after onset of interlaminar crack growth with false-color indicating shear angle. Onset of crack growth is marked in *red* in force–displacement curve

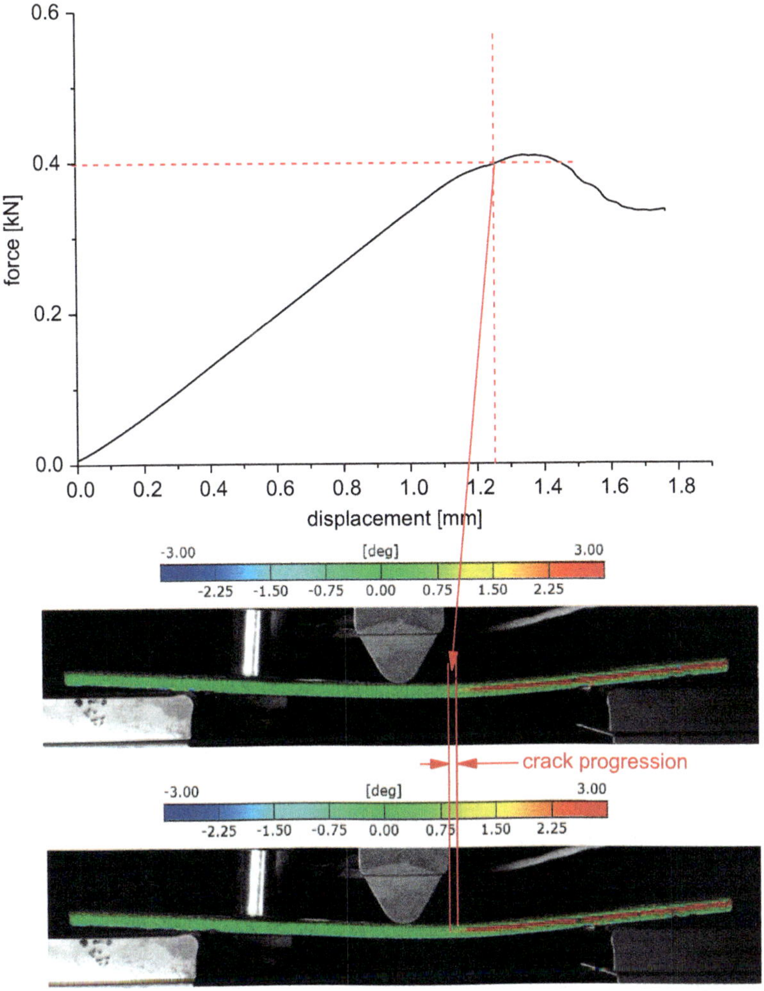

Fig. 3.52 Force–displacement curve including two DIC images before and after onset of interlaminar crack growth with false-color indicating shear angle. Onset of crack growth is marked in *red* in force–displacement curve

is harder to determine the correct onset of crack growth as shown in the second example of a unidirectional laminate in Fig. 3.52. Here the change of slope in the load–displacement curve indicates a change in the material, but is hard to evaluate in quantitative terms. The DIC images clearly reveal the onset of crack growth at the position marked in the load–displacement curve. For this example the onset of interlaminar crack growth was found to precede the force maximum. Therefore the evaluation according to common test standards would overestimate the onset load by 6.3 %, while the combination with DIC allows for evaluation of a more relevant and conservative load level.

3.4.3 Detectability of Failure Mechanisms

In Sect. 3.3 it was demonstrated that certain types of defects in fiber reinforced materials can be found by DIC. However, these investigations were limited to artificial defects as strain concentrators. In the following, some results are presented for tensile loading of carbon fiber reinforced materials monitored by 3D-DIC with parameters as given in Table 3.2. Tensile tests on unidirectional fiber reinforced polymers were carried out in accordance with DIN EN 2561. As shown in the sequence of images in Fig. 3.53, it is easy to spot macroscopic crack growth in the DIC images due to the local decrease of the 2nd principal strain. In this tensile test parallel to the fiber axis, such crack growth is not intended prior to ultimate failure, since such splicing significantly affects the stress distribution in the cross section of the test specimen. Hence ultimate failure below the real material strength is quite likely. In this context, DIC imaging can act as indicator to spot such failure modes long before (i.e., at $\varepsilon_{z,avg} = 0.30\,\%$) the ultimate rupture strain of 1.30 % for the present example. In comparison to DIC, the detection of such crack growth by conventional optical methods suffers from the weak perceptibility of inter-fiber cracks in carbon fiber reinforced materials. However, the detection of this type of crack is trivial by DIC, since the inter-fiber cracks typically span the whole

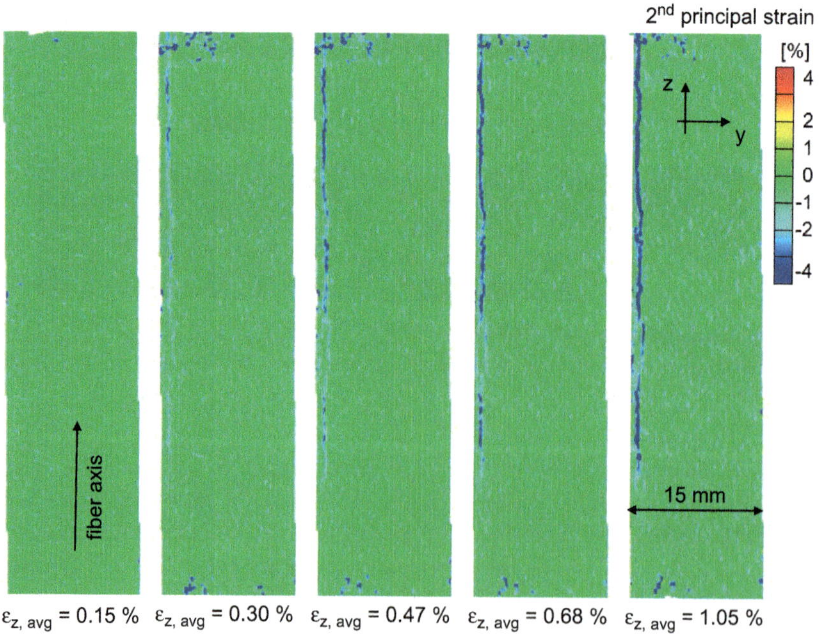

Fig. 3.53 Strain field of 2nd principal strain at various load levels indicated by the strain averaged over the full area seen $\varepsilon_{z,avg}$. Growth of inter-fiber crack parallel to fiber axis is clearly seen by presence of areas with strain values < 0

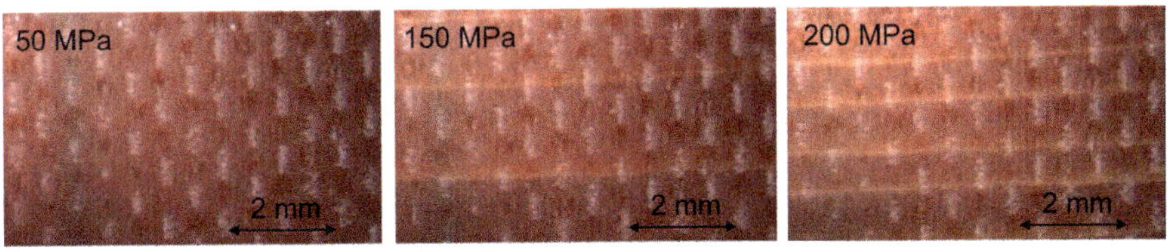

Fig. 3.54 Optical microscopy results for $[0/90_2]_{sym}$ stacking sequence of GFRP. Crack growth in the 90° layers at three different load levels is seen well due to the change in reflectivity of the crack surface

Table 3.7 Acquisition settings of DIC system for tensile tests on $[0/90]_{sym}$ fiber reinforced composites

Measurement setup	Value
Camera	Toshiba CMOS camera CSC12M25BMP19-01B with 4096 × 3072 pixels
Lenses	Titanar, focal length 2.8/50 mm
Lighting	KSP 0495-0001A LED 20 W/24 V white 30°
Filter	Polarization-filter Schneider-Kreuznach
Field of view	(107.4 × 82.1) mm^2
Scale factor	0.026 mm/px
Subset size	30 × 20 pixel square
Acquisition rate	1 Hz
Software	ARAMIS
Configuration	3D

thickness of the unidirectional laminate in this load situation. Therefore the crack is also at the surface level, which is a situation easy to detect by DIC as discussed in Sect. 3.3.

As discussed in Sect. 3.3, the detectability of failure types below the first ply or even deeper below the surface of a laminate is much more challenging than the previous example. For glass fiber reinforced materials the optically transparent fibers and matrix systems allow a direct observation of inter-fiber cracks in the plies 90° to the load axis. This is shown in exemplary microscopy investigations in Fig. 3.54. With increasing load, an increasing number of inter-fiber cracks can be observed due to the change in optical reflectivity of the crack surface. For carbon fiber reinforced materials, such direct observation is not possible. Therefore DIC can be used as suitable alternative.

To this end tensile tests of an $[0/90]_{sym}$ stacking of an epoxy prepreg system were monitored using the DIC parameters given in Table 3.7. As shown in the sequence of images in Fig. 3.55, the formation of cracks in the 90° plies is clearly observed by the strain concentrations transverse to the load axis (vertical direction in Fig. 3.55). In this case, the visibility originates from the increase in z-strain values at the surface level, i.e., in the 0° layer. Due to the presence of the inter-fiber crack below this surface layer an increased stress exposure occurs in the 0° layers,

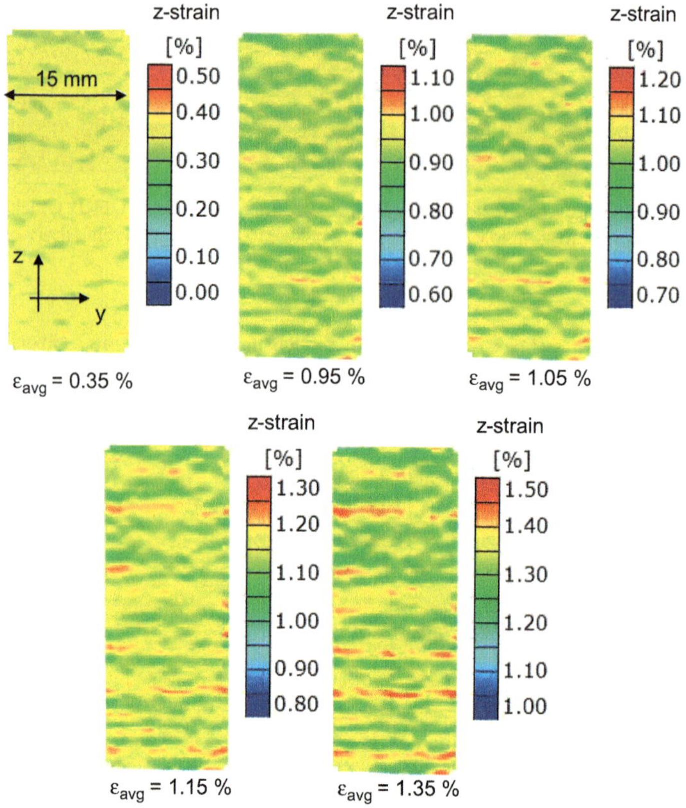

Fig. 3.55 Strain field of strain in z-axis at various load levels indicated by the strain averaged over the full area seen ε_{avg}. Growth of inter-fiber cracks in the 90° layers is clearly seen by the presence of areas of increased strain values

which translates into a local increase of strain. However, the appearance of these cracks is smoother than for optical imaging. This is due to the resolution of the subsets used in DIC and the smearing of the strain field throughout the surface ply. Therefore detectability of such cracks is limited to the outermost plies as discussed in Sect. 3.3. Moreover, the onset of inter-ply delamination at the crack tip further reduces the strain concentration and therefore may reduce the visibility of the existing crack.

References

1. Sutton, M.A., Orteu, J.J., Schreier, H.: Image Correlation for Shape, Motion and Deformation Measurements: Basic Concepts, Theory and Applications. Springer, Berlin (2009)
2. Pan, B., Qian, K., Xie, H., Asundi, A.: Two-dimensional digital image correlation for in-plane displacement and strain measurement: a review. Meas. Sci. Technol. **20**, 062001 (2009)

3. Peters, W.H., Ranson, W.F.: Digital imaging techniques in experimental stress analysis. Opt. Eng. **21**, 213427 (1982)
4. Fouinneteau, M.R.C., Pickett, A.K.: Shear mechanism modelling of heavy tow braided composites using a meso-mechanical damage model. Compos. Part A Appl. Sci. Manuf. **38**, 2294–2306 (2007)
5. Fouinneteau, M.R.C., Pickett, A.K.: Failure characterisation of heavy tow braided composites using digital image correlation (DIC). In: Applied Mechanics and Materials, pp. 399–406, (2006). doi:10.4028/www.scientific.net/AMM.5-6.399
6. Geers, M.G.D., de Borst, R., Peijs, T.: Mixed numerical-experimental identification of non-local characteristics of random-fibre-reinforced composites. Compos. Sci. Technol. **59**, 1569–1578 (1999)
7. Machida, K., Sato, M., Ogihara, S.: Application of stress analysis by digital image correlation and intelligent hybrid method to unidirectional CFRP laminate. In: Key Engineering Materials, pp. 1121–1124, (2007). doi:10.4028/www.scientific.net/KEM.345-346.1121
8. Nicoletto, G., Anzelotti, G., Riva, E., Roncella, R.: Mesomechanic strain fields in woven composites: experiments vs. FEM modeling. In: Proceedings of 3rd Workshop on Optical Measurement Techniques for Structures and Systems (OPTIMESS'2007). Leuven, Belgium (2007)
9. McGowan, D.M., Ambur, D.R., Hanna, T.G., McNeill, S.R.: Evaluating the compressive response of notched composite panels using full-field displacements. J. Aircr. **38**, 122–129 (2001)
10. Périé, J.-N., Calloch, S., Cluzel, C., Hild, F.: Analysis of a multiaxial test on a C/C composite by using digital image correlation and a damage model. Exp. Mech. **42**, 318–328 (2002)
11. Rae, P.J., Palmer, S.J.P., Goldrein, H.T., Lewis, A.L., Field, J.E.: White-light digital image cross-correlation (DICC) analysis of the deformation of composite materials with random microstructure. Opt. Lasers Eng. **41**, 635–648 (2004)
12. Ruggy, K.L., Cox, B.N.: Deformation mechanisms of dry textile preforms under mixed compressive and shear loading. J. Reinf. Plast. Compos. **23**, 1425–1442 (2004)
13. Koerber, H., Xavier, J., Camanho, P.P.: High strain rate characterisation of unidirectional carbon-epoxy IM7-8552 in transverse compression and in-plane shear using digital image correlation. Mech. Mater. **42**, 1004–1019 (2010)
14. Koerber, H., Vogler, M., Kuhn, P., Camanho, P.P.: Experimental characterisation and modelling of non-linear stress-strain behaviour and strain rate effects for unidirectional carbon-epoxy. In: ECCM16—16th European Conference on Composite Materials, pp. 1–8. Sevilla, Spain (2014)
15. Xavier, J., Sousa, A.M.R., Morais, J.J.L., Filipe, V.M.J., Vaz, M.: Measuring displacement fields by cross-correlation and a differential technique: experimental validation. Opt. Eng. **51**, 043602 (2012)
16. Lucas, B.D., Kanade, T.: An iterative image registration technique with an application to stereo vision. In: Proceedings of the 1981 DARPA Image Understanding Workshop, pp. 121–130, (1981)
17. Sutton, M.A., Yan, J.H., Tiwari, V., Schreier, H.W., Orteu, J.J.: The effect of out-of-plane motion on 2D and 3D digital image correlation measurements. Opt. Lasers Eng. **46**, 746–757 (2008)
18. Fowles, G.R.: Introduction to Modern Optics. Dover, New York (2012)
19. Pedrotti, F.L., Pedrotti, L.S.: Introduction to Optics, 2nd edn. Prentice Hall, Upper Saddle River (1993)
20. Schreier, H.W.: Systematic errors in digital image correlation caused by intensity interpolation. Opt. Eng. **39**, 2915 (2000)
21. Dupré, J.C., Bornert, M., Robert, L., Wattrisse, B.: Digital image correlation: displacement accuracy estimation. In: ICEM 14—14th International Conference on Experimental Mechanics. Poitiers, France (2010)

22. Wang, Y.Q., Sutton, M.A., Bruck, H.A., Schreier, H.W.: Quantitative error assessment in pattern matching: effects of intensity pattern noise, interpolation, strain and image contrast on motion measurements. Strain **45**, 160–178 (2009)
23. Amiot, F., Bornert, M., Doumalin, P., Dupré, J.-C.C., Fazzini, M., Orteu, J.-J.J., Poilâne, C., Robert, L., Rotinat, R., Toussaint, E., Wattrisse, B., Wienin, J.S.: Assessment of digital image correlation measurement accuracy in the ultimate error regime: main results of a collaborative benchmark. Strain **49**, 483–496 (2013)
24. Tong, W.: An evaluation of digital image correlation criteria for strain mapping applications. Strain **41**, 167–175 (2005)
25. Roux, S., Hild, F.: Stress intensity factor measurements from digital image correlation: post-processing and integrated approaches. Int. J. Fract. **140**, 141–157 (2006)
26. Becker, T., Splitthof, K., Siebert, T., Kletting, P.: Technical Note—T-Q-400-Accuracy-3DCORR-003-EN, (2006)
27. Yeh, H.-Y., Chern, C.: The Yeh-Stratton criterion for stress concentrations in fiber-reinforced composite materials. J. Compos. Mater. **32**, 141–157 (1998)
28. Pilkey, W.D., Pilkey, D.F.: Peterson's Stress Concentration Factors. Wiley, Hoboken (2008)
29. Griffith, A.A.: The phenomena of rupture and flow in solids. Philos. Trans. R. Soc. Lond. Ser. A **221**, 163–198 (1921)
30. Irwin, G.: Analysis of stresses and strains near the end of a crack traversing a plate. J. Appl. Mech. **24**, 361–364 (1957)
31. Tada, H., Paris, P.C., Irwin, G.R.: The Stress Analysis of Cracks Handbook, 3rd edn. ASME, New York, (2000)
32. O'Higgins, R.M., McCarthy, M.A., McCarthy, C.T.: Comparison of open hole tension characteristics of high strength glass and carbon fibre-reinforced composite materials. Compos. Sci. Technol. **68**, 2770–2778 (2008)
33. Kushch, V.I., Sevostianov, I., Mishnaevsky, L.: Stress concentration and effective stiffness of aligned fiber reinforced composite with anisotropic constituents. Int. J. Solids Struct. **45**, 5103–5117 (2008)
34. Chang, K.-Y., Llu, S., Chang, F.-K.: Damage tolerance of laminated composites containing an open hole and subjected to tensile loadings. J. Compos. Mater. **25**, 274–301 (1991)
35. Nicoletto, G., Anzelotti, G., Riva, E.: Mesoscopic strain fields in woven composites: experiments vs. finite element modeling. Opt. Lasers Eng. **47**, 352–359 (2009)
36. Godara, A., Raabe, D., Bergmann, I., Putz, R., Müller, U.: Influence of additives on the global mechanical behavior and the microscopic strain localization in wood reinforced polypropylene composites during tensile deformation investigated using digital image correlation. Compos. Sci. Technol. **69**, 139–146 (2009)
37. Schorer, N., Sause, M.G.R.: Identification of failure mechanisms in CFRP laminates using 3D digital image correlation. In: 20th International Conference on Composite Materials, pp. 1–8. Copenhagen, Denmark (2015)
38. Monden, A., Sause, M.G.R., Hartwig, A., Hammerl, C., Karl, H., Horn, S.: Evaluation of surface modified CFRP-metal hybrid laminates. In: Euro Hybrid Materials and Structures 2014, pp. 1–8. Stade, Germany (2014)
39. Priess, T., Sause, M.G., Fischer, D., Middendorf, P.: Detection of delamination onset in laser-cut carbon fiber transverse crack tension specimens using acoustic emission. J. Compos. Mater. **49**, 2639–2647 (2015)
40. Scott, A.E., Sinclair, I., Spearing, S.M., Mavrogordato, M., Bunsell, A.R., Thionnet, A.: Comparison of the accumulation of fibre breaks occurring in a unidirectional carbon/epoxy composite identified in a multi-scale micro-mechanical model with that of experimental observations using high resolution computed tomography. In: Matériaux 2010, pp. 1–9. Nantes, France (2010)
41. Scott, A.E., Mavrogordato, M., Wright, P., Sinclair, I., Spearing, S.M.: In-situ fibre fracture measurement in carbon-epoxy laminates using high resolution computed tomography. Compos. Sci. Technol. **71**, 1471–1477 (2011)

42. De Angelis, G., Meo, M., Almond, D.P., Pickering, S.G., Angioni, S.L.: A new technique to detect defect size and depth in composite structures using digital shearography and unconstrained optimization. NDT E Int. **45**, 91–96 (2012)
43. Pezzoni, R., Krupka, R.: Laser-shearography for non-destructive testing of large-area composite helicopter structures. In: 15th World Conference on NDT. Roma, Italy (2000)
44. Lomov, S.V., Ivanov, D.S., Verpoest, I., Zako, M., Kurashiki, T., Nakai, H., Molimard, J., Vautrin, A.: Full-field strain measurements for validation of meso-FE analysis of textile composites. Compos. Part A Appl. Sci. Manuf. **39**, 1218–1231 (2008)
45. Hung, Y.Y.: Shearography for non-destructive evaluation of composite structures. Opt. Lasers Eng. **24**, 161–182 (1996)
46. Hoffmann, K.: An Introduction to Measurements Using Strain Gages. Hottinger Baldwin Messtechnik GmbH, Darmstadt, Germany (1987)
47. Robert, L., Nazaret, F., Cutard, T., Orteu, J.-J.: Use of 3-D digital image correlation to characterize the mechanical behavior of a fiber reinforced refractory castable. Exp. Mech. **47**, 761–773 (2007)
48. Grédiac, M., Pierron, F., Avril, S., Toussaint, E.: The virtual fields method for extracting constitutive parameters from full-field measurements: a review. Strain **42**, 233–253 (2008)
49. Yoneyama, S., Morimoto, Y., Takashi, M.: Automatic evaluation of mixed-mode stress intensity factors utilizing digital image correlation. Strain **42**, 21–29 (2006)
50. Hild, F., Roux, S.: Digital image correlation: from displacement measurement to identification of elastic properties—a review. Strain **42**, 69–80 (2006)
51. Basan, R.: Untersuchung der intralaminaren Schubeigenschaften von Faserverbundwerkstoffen mit Epoxidharzmatrix unter Berücksichtigung nichtlinearer Effekte. Ph.D. Thesis, Technical University Berlin (2011)
52. Trappe, V., Basan, R., Grasse, F.: Stiffness and fracture of shear loaded laminates with unidirectional and biaxial fibre orientation investigated with a picture frame test. In: 16th European Conference on Composite Materials, pp. 22–26. Seville (2014)
53. Blackman, B.R.K., Brunner, A.J., Williams, J.G.: Mode II fracture testing of composites: a new look at an old problem. Eng. Fract. Mech. **73**, 2443–2455 (2006)
54. Brunner, A.J., Blackman, B.R.K., Davies, P.: A status report on delamination resistance testing of polymer-matrix composites. Eng. Fract. Mech. **75**, 2779–2794 (2008)

Chapter 4
Acoustic Emission

Acoustic emission (AE) analysis is about the detection and interpretation of ultrasonic waves caused by rapid internal displacements. In the context used herein, the formation and propagation of cracks in fiber reinforced materials are understood as the most relevant acoustic emission source. This acoustic emission source is generally described as spatial crack surface oscillation [1, 2]. During propagation of the emitted acoustic wave, the characteristics of the signal (e.g., frequency content) suffer from attenuation, dispersion, and propagation in guiding media. In addition, the characteristics of the signals detected at the surface of the solid are further altered by the detection process using piezoelectric sensors.

This chapter starts with a short introduction to the principle of operation followed by a sequential review on the acoustic emission source, aspects of wave propagation in guided media, and the signal detection process. Subsequently, signal classification techniques and source localization algorithms are discussed with a particular focus on recent developments. The chapter closes with some applications of acoustic emission as in situ technique to monitor failure of fiber reinforced composites.

4.1 Principle of Operation

AE signals originate from sudden microscopic displacements inside solid media. One major source for AE in fiber reinforced materials is the initiation and growth of damage. Microscopically this relates to the generation and propagation of cracks inside the matrix material, along the interface between matrix and fiber or the rupture of the fiber filaments. Another relevant AE source found in fiber reinforced polymers is friction of existent crack walls, which is frequently encountered in unloading and reloading of damaged composites. In any case, in situ monitoring by AE sensors on the structure under loading can identify active areas in the composite and can identify their time of occurrence. In the past, the detection of AE signals and the according interpretation has led to a variety of approaches and analysis

M.G.R. Sause, *In Situ Monitoring of Fiber-Reinforced Composites*,
Springer Series in Materials Science 242, DOI 10.1007/978-3-319-30954-5_4

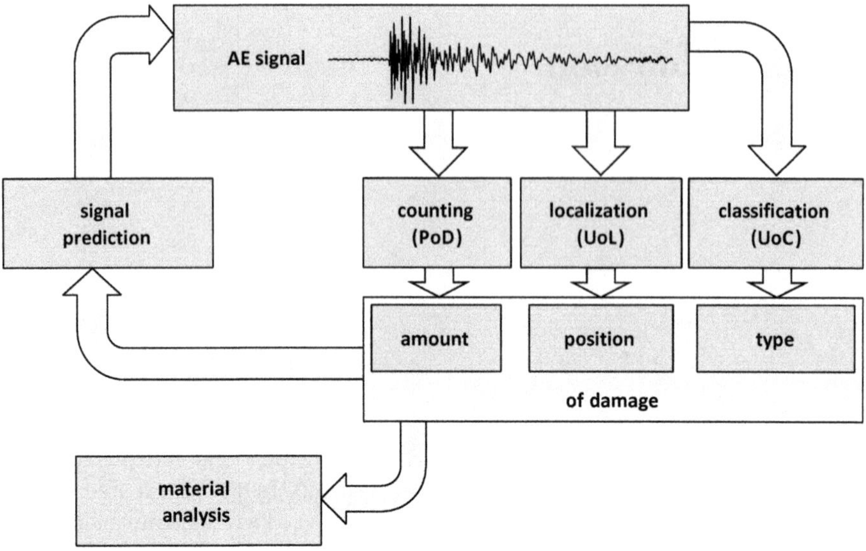

Fig. 4.1 Analysis methods to use with AE signals and AE signal prediction as forward approach

methods. One possibility to categorize the potential of AE analysis techniques is shown in Fig. 4.1. Starting with the detected AE signals, a first distinction can be made between counting the AE activity, localizing the AE source position, and identifying classes of AE signals. If carried out successfully, this yields the amount of damage, the position of the damage zone, and the type of damage. Thus, the combination of these three types of information can yield valuable information for the interpretation of failure of fiber reinforced composites. The application of AE to this intent is not essentially new and has seen substantial development in the past decades. A good review on the present status in this field can be found in, e.g., [3] for the work until 2000 or in [4] for the ongoing research since 2000.

For the analysis options illustrated in Fig. 4.1, counting the number of signals is considered to be a standard technique in AE analysis since this is already inherent to the detection of the AE signals. However, the counting of the signals is associated with an uncertainty, given by the probability of detection (PoD).

The localization of source positions is already well established but comes with additional challenges, when acoustically anisotropic and inhomogeneous media like fiber reinforced polymers are considered [5]. Established localization techniques can yield useful source positions in laboratory size fiber reinforced polymer specimens. However, their application in realistic structures including hybrid joints or multi-material combinations is still challenging; therefore, every calculated source location is subject to an uncertainty of location (UoL).

The identification of a particular type of damage in fiber reinforced polymers has seen several substantial advancements in the last two decades. For discussion of this analysis method, it is suitable to distinguish between signal classification and source identification. The first task is the grouping of signals based on their similarity relative to each other. This is often achieved by analysis of signal features, i.e., parameters calculated from the detected signals. The methods to group AE signals range from very simple approaches comprising discrete feature values to automated or semiautomated pattern recognition strategies. However, every classification result is subject to uncertainty of the assignment of the signals to a particular group. This is named uncertainty of classification (UoC).

After grouping of the signals, the second task is the assignment of one group of signals to a particular source mechanism. This is typically done by phenomenological observations [6], by comparative measurement of test specimens with known types of AE sources [7, 8] or subsequent microscopy [9]. Recent advances also allow the forward prediction of the emitted AE signal of a specific source type by analytical methods [10] or finite element modeling (FEM) [11–13].

As shown in Fig. 4.1, the possibility to predict the AE signal of a known type of failure is key to establish traceable acoustic emission analysis techniques and to add a sound physically based theory for generation of acoustic emission signals.

4.2 Source Mechanics

The formation and propagation of cracks in solid media are a field of research that has been active for decades. Still, the theoretical description of the physics at the crack tip and the crack dynamics are active fields of research [14, 15]. The generation of acoustic emission signals due to the crack motion is a closely related topic. Despite of the broad range of technical applications, only little work has been performed recently to advance the understanding of the physical processes involved in the generation of acoustic emission.

In order to interpret the detected acoustic emission signals in terms of their relevance to material failure, it is required to have concise knowledge of the underlying source mechanics. In the past, various valuable attempts have been made to provide a theoretical description of acoustic emission sources.

In general, the excitation process of an elastic wave as a result of crack surface movement is given by the elastic wave equation as long as the deformation remains elastic, i.e., for small displacements [16]. In the case of crack growth, the stress level at the newly formed surface area drops from the initial value σ_{frac} to zero. As a result the crack surface is deflected within a characteristic time t_e and starts to oscillate around its new energetic steady state [17]. The crack kinematic behavior can then be described by a crack motion vector $\boldsymbol{b}$ (Burgers vector) and a vector normal to the crack surface $\boldsymbol{d}$. As demonstrated by [18], the characteristics of crack deflection are fully described by the moment tensor concept, which couples the crack kinematic behavior with the elastic properties:

$$\widetilde{\boldsymbol{M}} = \int_V \boldsymbol{C} \cdot \boldsymbol{b} \cdot \boldsymbol{d} \, \mathrm{d}V \tag{4.1}$$

The resulting excitation of a small elastic wave released in the surrounding medium is then derived from the theory of elastic waves [19].

It is important to note that the duration for the initial deflection t_e is strictly linked to the elastic properties of the cracking material and the local loading conditions. This implies that the excited elastic wave is a characteristic fingerprint for the failure process and thus a valuable source of information that is accessible macroscopically by AE measurements. In general the duration for the initial deflection decreases with the brittleness of the material [20] and thus determines the oscillation frequency. Since the initial deflection time t_e typically is in the order of magnitude between 10^{-8} and 10^{-4} s, the frequencies of the excited elastic waves at the crack tip are located in the ultrasonic range between 10 kHz and 100 MHz.

While the correlation of t_e to the elastic properties may seem obvious, the influence of the local loading conditions on the excited elastic wave is more complex. Because crack formation can result from different loading conditions, the oscillation of the crack surface is strongly influenced in intensity, direction of radiation, and damping behavior by the type of failure.

To this end, M. Ohtsu and K. Ono have examined the correlation between the response function on the surface of linear elastic solids and dislocations arising from acoustic emission sources located within [21, 22]. According to the generalized theory of acoustic emission, the elastic wave amplitude observed at position $\boldsymbol{r}$ due to a microscopic displacement $\boldsymbol{d}(t)$ located at $\boldsymbol{r}_{\text{source}}$ in an isotropic medium can be written as

$$\boldsymbol{u}(\boldsymbol{r}, t) = \Gamma(\boldsymbol{r}, \boldsymbol{r}_{\text{source}}, t) \otimes \boldsymbol{C} \cdot \Delta V \cdot \boldsymbol{d}(\boldsymbol{r}_{\text{source}}, t) \tag{4.2}$$

Here $\otimes$ denotes the convolution integral between the Green function $\Gamma(\boldsymbol{r}, \boldsymbol{r}_{\text{source}}, t)$ and elasticity tensor $\boldsymbol{C}$ times crack volume ΔV. The crack kinematics are taken into account by the source-time function using a magnitude of displacement $\boldsymbol{d}$ and a rise-time t_e. Various step-function descriptions exist, which are used to define the three-dimensional spatial displacement of the crack surface during crack formation [2, 20, 22–24]. In particular the rise-time of the initial crack surface displacement is an essential parameter to model the crack surface motion [25].

However, there are no reports in literature of successful measurements of rise-times of real acoustic emission sources, e.g., due to cracking. Instead, the rise-time is typically estimated based on the elastic properties of the bulk material. Despite of that drawback, this type of source modeling has been successfully applied to many cases and the basic concept has been used within the generalized theory of acoustic emission by Ono and Ohtsu [21, 22], the work of Scruby and Wadley [26–28], and numerous other analytical descriptions [2, 10, 20, 29, 30].

In recent years it has become convenient to use numerical methods to model acoustic emission sources. In this field, Prosser, Hamstad, and Gary applied FEM to simulate acoustic emission sources based on body forces acting as point source in a solid [31, 32]. Hora and Cervena investigated the difference between nodal sources, line sources, and cylindrical sources to build geometrically more representative acoustic emission sources [33]. At the same time, a finite element approach was proposed utilizing an acoustic emission source model taking into account the geometry of a crack and the inhomogeneous elastic properties in the vicinity of the acoustic emission source [24].

In the following several investigation results as achieved by means of FEM are used to demonstrate some of the key principles in acoustic emission source mechanics. The main advantage of FEM in this context is that precise control of the source characteristics (such as rise-time and source orientation), source location, and propagation distance and direction is feasible. Another closely related advantage is the possibility to compute the surface displacement directly, whereas in the experiment, this evaluation suffers from the characteristics of the AE sensor.

4.2.1 AE Rise-Times and Plate Waves

As the first step towards a better understanding of the interaction between source types and the detected acoustic emission signals, the relationship between the source and the guided wave formation shall be investigated. To this end, there is a limited set of factors to consider, whose influence shall be discussed first:

- Source magnitude
- Source rise-time
- Source-sensor distance (x source position)
- Source-sensor angle (x and y source position)
- Source depth position (z source position)

To address these different factors, a simple case of an aluminum plate with buried dipole sources is considered. This kind of investigation follows the established procedures of previous work conducted by Hamstad et al. [25, 31, 34]. The model configuration uses a 2D-axisymmetric representation of a plate with 4.7 mm thickness and material properties of AlMg3 as given in Table B.1 in Appendix B. As source, a dipole of 150 μm axis length using a cosine-bell source function with varying rise-time t_e and strength of 1 N was modeled. The out-of-plane displacement is calculated for a distance of 240 mm using mesh settings with maximum edge length of 0.5 mm and a time step of 0.1 μs. All these settings were found to yield convergent results based on the coherence level of >98 % in the frequency range up to 1 MHz, when compared to a solution with higher mesh resolution and lower time step. Also the computations using these settings were found to be in very good agreement to the respective calculations in [25, 31, 34], therefore extending these previous findings.

The influence of the source magnitude is covered by the generalized theory of acoustic emission and is addressed in terms of FEM work in [24, 32, 35]. Since this is not of large relevance for the discussion as long as linear elastic computations are used, the reader is referred to the respective original work, but also to Sect. 4.2.2 for more recent work in this field. For the present configuration, the source magnitude was kept constant. The source rise-time dominates the bandwidth of the source and therefore is explicitly accounted for in the present investigation following the previous work of [25]. Effects of changes of the source-sensor distance and the source-sensor angle are also excluded from the present investigation by focusing on a 2D problem, evaluating the signals at constant distance. However, the respective effects due to different wave propagation paths are extensively discussed in Sect. 4.3 for various configurations as faced in composite materials. As pointed out by Downs et al., the source depth position causes a systematic shift of the intensity contribution of the A_0 wave mode as soon as the source position moves away from the center plane of the plate [31, 32]. Therefore, even for the same type of source, there is a systematic change of the detected mode ratio.

This influence of the depth position is emphasized in Fig. 4.2 in two Choi-Williams distributions (CWDs) as calculated for the respective relative depth

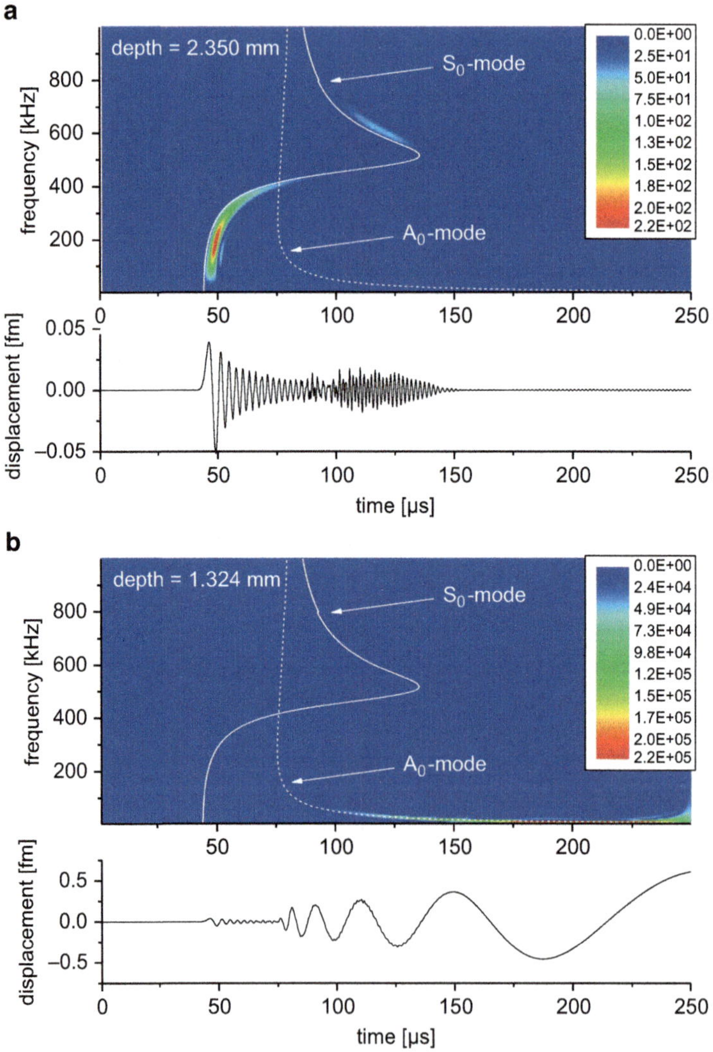

Fig. 4.2 Choi-Williams distribution of two signals at 240 mm distance due to an in-plane dipole source positioned at 2.350 mm (**a**) and 1.324 mm (**b**) below the surface

positions. The superimposed dispersion curves clearly allow identifying the individual modal contributions in the present configuration, which are the fundamental symmetric mode (S_0) and antisymmetric mode (A_0), often referred to as extensional and flexural mode.

Another important factor of influence for the modal composition of an acoustic emission signal was found to be the rise-time. As visualized in Fig. 4.87 in Sect. 4.4, the bandwidth of the source determines if a particular wave mode can be excited or not. If the bandwidth of the source cuts off at too low frequency, there is no possibility to excite guided wave modes propagating preferentially above that cutoff frequency. The bandwidth of an acoustic emission source is dominantly influenced by its rise-time, a factor which was also investigated previously using FEM [25]. For the present case of buried in-plane dipoles, the influence of the rise-time was studied and is shown in exemplary CWDs for an in-plane source at the same depth position in Fig. 4.3a, b. As reported previously, the change in rise-time substantially changes the excited wave modes [25]. For the shortest rise-time, the bandwidth is sufficiently high, so higher-order modes are excited. But with decreasing bandwidth, the S_0 mode cannot be excited beyond certain intensity and the signal is now dominated by the A_0 mode.

From the viewpoint of source identification procedures, one interesting question is the relative influence of these two factors. Therefore, signals within a range of depth position and a range of rise-times were calculated and signals evaluated at a distance of 240 mm. For each of the signals, the CWD was calculated, and the maximum intensity of the coefficients for each of the wave modes was determined to yield a measure of intensity. Subsequently, the matrix of rise-times and relative depth positions was interpolated to yield the 3D representation of the coefficient intensity as shown in Fig. 4.4 for the S_0 mode and in Fig. 4.5 for the A_0 mode.

As seen from the intensity map of the S_0 mode in Fig. 4.4, there is some influence of the relative depth position, which is expected from the according shift from the center plane to the surface of the plate. However, the influence of the rise-time is apparently much more relevant, since it is almost impossible to generate a reasonable intensity of the S_0 mode above a rise-time of 2.5 μs.

For the A_0 mode, the behavior is strictly different. Here the relative depth position shows a pronounced effect as expected by the shift of source position from the center plane to the surface of the plate. Within the selected range of rise-times, there is no substantial influence of the rise-time for a constant depth position. This relates to the dominant propagation of the A_0 mode at low frequencies. This frequency range is hard to cut off since it is always found in the bandwidth of sources operating with steplike source functions.

Although these results in combination with previous publications [25, 31, 32, 34] are only reported for an isotropic material, the key principles are still applicable to anisotropic materials as fiber reinforced laminates. However, there are numerous additional effects, which have to be considered during wave propagation, which are discussed in more detail in Sect. 4.3.

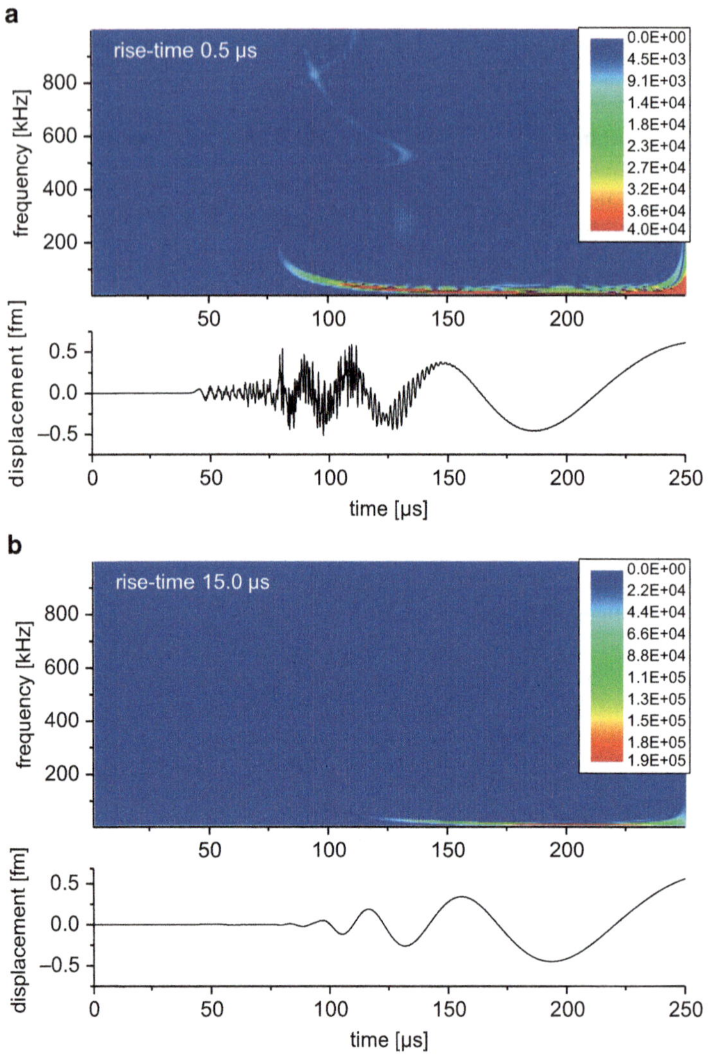

Fig. 4.3 Choi-Williams distribution of two signals at 240 mm distance due to an in-plane dipole source positioned at 1.324 mm below the surface using a rise-time of 0.5 µs (**a**) and 15.0 µs (**b**)

4.2.2 AE Source Model Implementation for Fiber Reinforced Materials

Despite the high value of dipole-source representation to deduce some basic conclusions on source operation, such description still lacks in the level of detail, when compared to real acoustic emission sources. Therefore different approaches

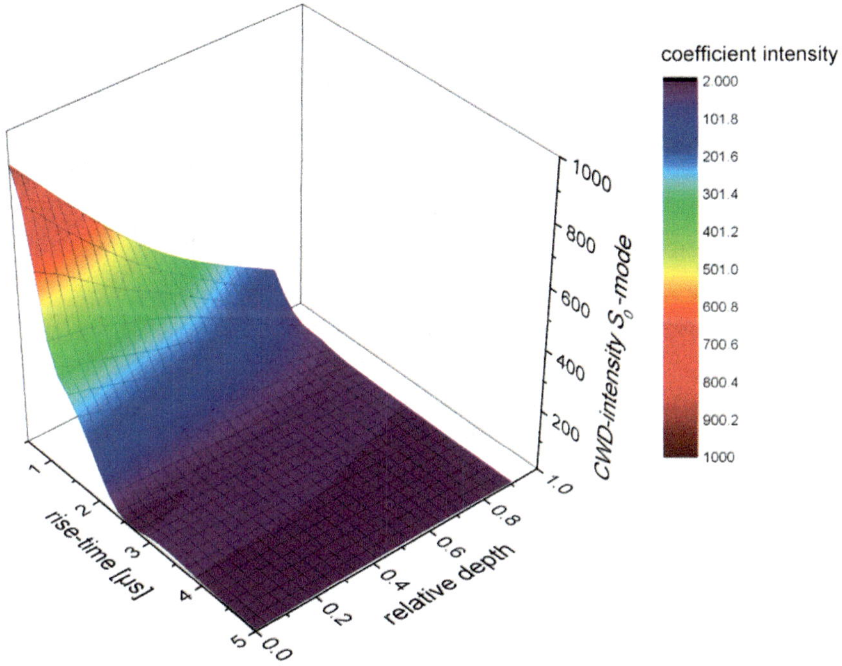

Fig. 4.4 Intensity of CWD coefficient of the S_0 mode as function source depth and rise-time

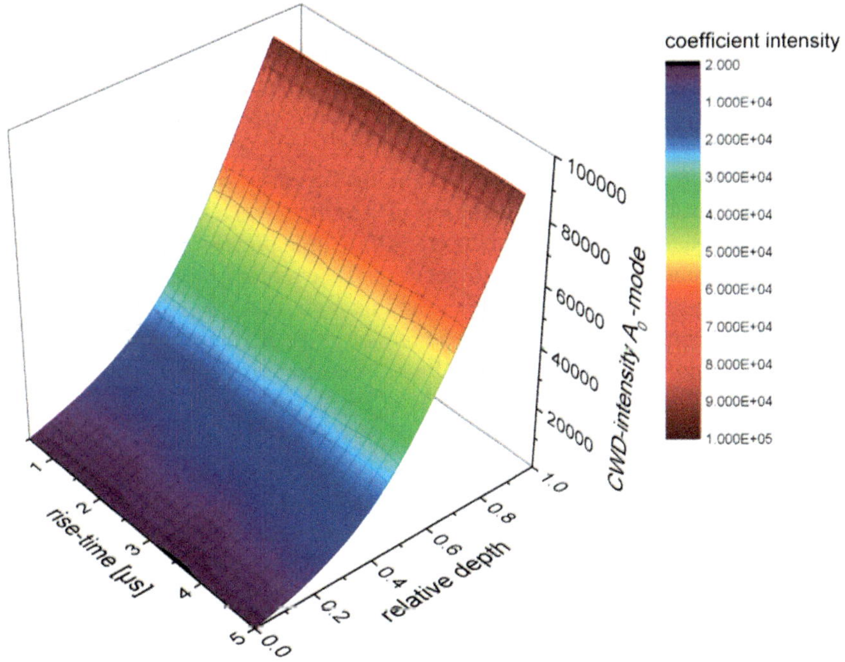

Fig. 4.5 Intensity of CWD coefficient of the A_0 mode as function source depth and rise-time

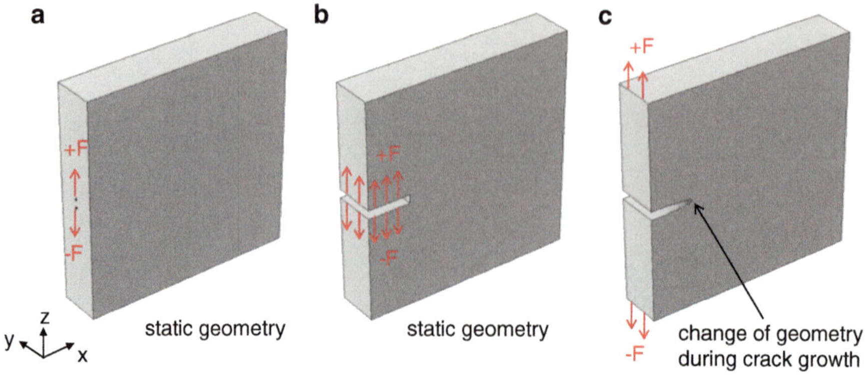

Fig. 4.6 Different types of acoustic emission source model descriptions used in literature employing point sources (**a**) or extended sources (**b**) in conjunction with analytic source functions. New source model description presented herein using dynamic changes of the source geometry based on fracture mechanics (**c**) (reprinted from [13])

have been published in the past in order to improve the accuracy of acoustic emission source models. Based on previous investigations, one can categorize the different modeling strategies to describe acoustic emission sources of crack propagation as shown in Fig. 4.6. The first type of source models considers point-like sources explicitly defining the source dynamics utilizing analytical source functions (cf. Fig. 4.6a). As the second type, we can name those attempts that have been made to incorporate more accurate source geometries, while the modeled crack dynamics are still based on analytical source functions (cf. Fig. 4.6b). The third type uses accurate artificial source geometries and does not need an analytical source function to generate acoustic emission. Instead, this type of source model is capable to generate the crack dynamics based on experimentally accessible parameters and fracture mechanics laws.

Currently most source models proposed in the literature are of type one or type two, thus requiring an explicit definition of a source function. Accordingly, no details of the dynamics arising from the crack formation process and the subsequent crack surface motion are taken into account. To overcome these limitations, a source model description of type three would be preferred.

From a mathematical modeling and simulation point of view, there are two main challenges in providing a numerically based acoustic emission source model of the third type. The first challenge consists of the different scales involved in the problem (crack length of the order of microns versus signal wavelength of the order of millimeters to centimeters) and the proper scale bridging. Owing to the vastly different observations scales, a full multi-scale approach is thus necessary. The second challenge stems from the calculation of temporal and spatial evolution of the surfaces of the crack. This is a level of detail that is typically not accounted for in modeling approaches used to describe crack formation by means of cohesive zone elements, extended finite element methods, or similar implementations.

Throughout the publications [13, 36, 37], a new source model description was developed and validated. Its application to fiber reinforced materials is presented in the following. This source modeling approach consists of three sequential modeling steps as schematically presented in Fig. 4.7. The first step is derived from classical structural mechanics. Suitable displacement boundary conditions are defined for the geometry considered to restrict some of the displacement components on one end. The other end is loaded by a force high enough to initiate fracture at the crack plane considered. If the respective force value is unknown, the implementation of a fracture criterion (e.g., fracture toughness, max. stress, etc.) to deduce the onset load for crack initiation is a straightforward procedure using a stationary solver sequence with incremental loading. If the external force is known from experiments, the measured force value can directly be used for the stationary solver. For the example shown in Fig. 4.7, the presence of the notch causes stress concentration at the tip of the notch, which will cause crack initiation at this position.

In the second step, the initial conditions for the displacement $\boldsymbol{u}$ and stress state $\boldsymbol{\sigma}$ are chosen to be identical to the static values $\boldsymbol{u}_{\text{static}}$ and $\boldsymbol{\sigma}_{\text{static}}$ as calculated in the previous step. Boundary conditions for restricted displacement components and external loads are kept identical to the previous step. In contrast to the previous step, a transient calculation of the displacement field is now performed. In addition, boundary conditions at the crack plane are selected to allow for crack opening according to a fracture mechanics law. The duration of this transient calculation t_{frac} is chosen to be sufficient until crack propagation has come to a rest. As seen from Fig. 4.7, the presence of the static displacement field causes crack propagation with an accompanying excitation of an acoustic emission wave. This spatial movement is seen best in the velocity field, since the static displacement dominates the displacement scale and therefore inhibits to observe the very small displacements caused by the acoustic emission wave.

For the third[1] step, the initial conditions ($t = 0$) for the displacement, velocity, and stress states of the last time value of the previous step (t_{frac}) are used. Boundary conditions for restricted displacement components and external loads are kept identical to the first and second step. The boundary conditions applied at the crack plane are chosen to allow for independent movement of the new crack surfaces without allowing penetration of each other. This transient calculation is continued for a sufficient duration t_{end} to allow for signal propagation in the test geometry as shown in Fig. 4.7.

The present implementation of crack growth requires the definition of a fracture plane, similar to conventional cohesive zone modeling. In the example given in Fig. 4.7, the fracture plane is the horizontal xy-plane located at the z-position of the notch tip. The extension of the fracture plane in x-direction is chosen sufficiently large that the crack will not grow beyond the end of that plane.

[1] The third step is only required due to the technical implementation chosen. For the selected modeling environment, different solver settings were required for the second and the third computation step making it necessary to split this part of the computation in two steps.

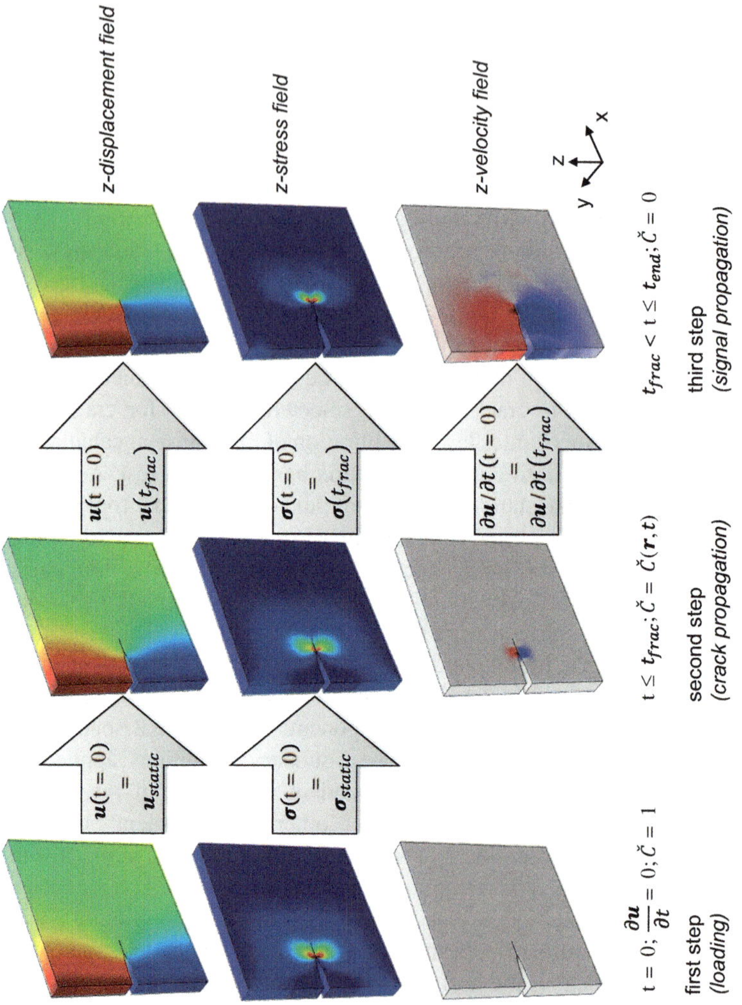

Fig. 4.7 Schematic of source model description using three subsequent modeling steps (reprinted from [13])

The stiffness vector $\widetilde{\boldsymbol{k}}$ of this fracture plane is written in terms of the boundary coordinate system (t_1, t_2, n), the normal modulus E_n of the fracture plane, and the in-plane shear modulus $G_{t_{12}}$ and the in-plane Poisson's ratio $\nu_{t_{12}}$ as

$$\widetilde{\boldsymbol{k}} = \begin{pmatrix} k_{t1} \\ k_{t2} \\ k_n \end{pmatrix} \tag{4.3}$$

$$k_{t1} = k_{t2} = \frac{G_{t_{12}}}{h} \tag{4.4}$$

$$k_n = \frac{E_n(1 - \nu_{t_{12}})}{h(1 + \nu_{t_{12}})(1 - 2\nu_{t_{12}})} \tag{4.5}$$

The parameter $\boldsymbol{h}$ is an effective thickness associated with the internal surface. The thickness value $\boldsymbol{h}$ is chosen sufficiently small, so that the value of $\widetilde{\boldsymbol{k}}$ has negligible influence on the overall compliance of the model.

To model crack growth, the stiffness vector is multiplied by a degradation function $\check{C}(\boldsymbol{r})$ evaluated as a function of the position on the fracture surface $\boldsymbol{r}$.

4.2.2.1 Source Models for Microscale

In [13, 36] the von Mises equivalent stress σ_v was chosen to model failure in isotropic materials. For a general stress state, σ_v is written in terms of the normal stresses σ_i and shear stresses τ_{ij} as

$$\sigma_v = \sqrt{\sigma_x^2 + \sigma_y^2 + \sigma_z^2 - \sigma_x\sigma_y - \sigma_x\sigma_z - \sigma_y\sigma_z + 3\left(\tau_{xy}^2 + \tau_{xz}^2 + \tau_{yz}^2\right)} \tag{4.6}$$

Degradation of the stiffness vector $\widetilde{\boldsymbol{k}}$ thus occurs if σ_v exceeds the materials tensile strength R.

For technical reasons, it is required to track the maximum value $\sigma_{\max}$ of σ_v throughout the computation. Therefore, the current implementation evaluates whether the fracture condition is fulfilled in the present time step i or if it was fulfilled in any previous time step. Therefore the degradation function is written in terms of the maximum value of either $\sigma_{\max}$ or σ_v:

$$(\boldsymbol{r}) = \begin{cases} 1 & \text{if} \quad \max(\sigma_{\max}(\boldsymbol{r}),\ \sigma_v(\boldsymbol{r})) < R \\ 0 & \text{if} \quad \max(\sigma_{\max}(\boldsymbol{r}),\ \sigma_v(\boldsymbol{r})) \geq R \end{cases} \tag{4.7}$$

While this procedure was found to yield valid results in application to failure of homogeneous and isotropic media, such as pure matrix polymer or carbon fiber filaments [13], further modifications are required to apply this approach to AE sources in fiber reinforced materials.

First, the microstructural complexity of the material has to be accounted for. Since some of the failure mechanisms discussed in Sect. 2.1 start on the scale of several micrometers and specimens are in the length scale of several centimeters or meters, a suitable multi-scale approach needs to be developed to bridge those scales. Second, the time scales of fracture for fiber materials (less than nanoseconds) and the typical time scales for signal propagation (many microseconds) are different by several orders of magnitude. Therefore, suitable adoption of the time resolution is required during the progress of the simulation.

To achieve the transition between the inhomogeneous material properties on the microscopic scale and the homogeneous material properties on the macroscopic scale, the usage of a perfectly matched layer region has been proposed in [12, 24]. Based on the wavelengths faced in the simulation and the fast divergence of the waves (i.e., fast increase of wavelength) when moving away from the source, the resolution of the microstructure can be limited to some micrometers. For a fiber reinforced material, this comprises the typical representative volume element (RVE) approach as seen in Fig. 4.8. In previous investigations values between 30 and 75 μm were found to be sufficient to explicitly model the microstructure and the crack process in this RVE. With reference to (4.8) and the scheme in Fig. 4.9, the radius of this RVE with pure material properties $\boldsymbol{C}_0$ (i.e., fiber and matrix) is designated r_{pure}. Adjacent to this inner zone, a perfectly matched layer (PML) sphere with radius r_{pml} is modeled. Here the microstructure is still geometrically resolved, but the material properties $\boldsymbol{C}'$ are gradually changed using intermediate properties as function of radial position r (with $r = 0$ at the center of the RVE). The material properties in the PML domains $\boldsymbol{C}'$ for $r_{\mathrm{pure}} \leq r \leq r_{\mathrm{pml}}$ are defined as

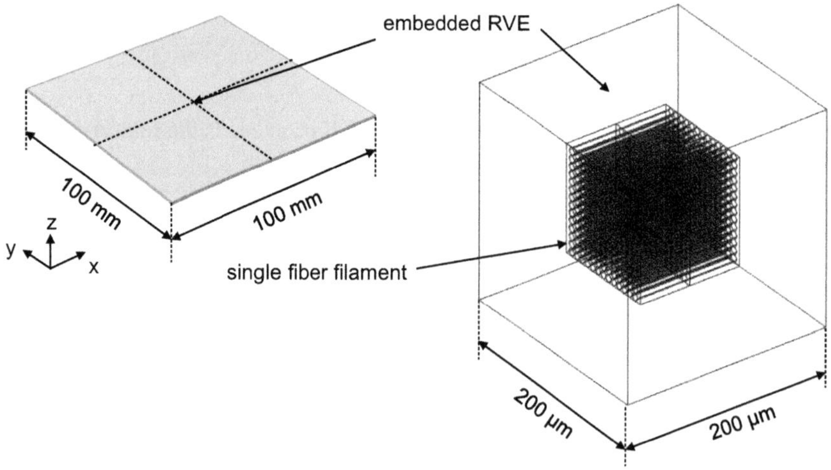

Fig. 4.8 Geometry of RVE embedded in macroscopic geometry for wave propagation. The RVE includes a PML domain allowing smooth transition of properties between the microstructural part and the homogenized layers. Calculations within the embedded RVE and the macroscopic geometry are carried out simultaneously

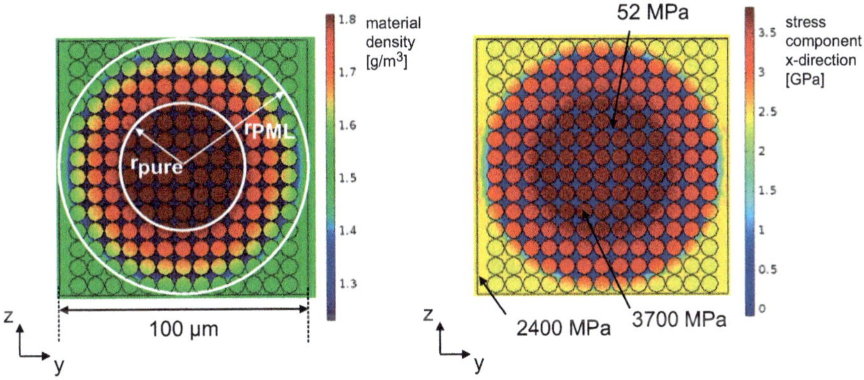

Fig. 4.9 Visualization of the transition of material properties between r_{pure} and r_{PML} using material density (*left*) and according tensile stress states for macroscopically applied tensile load in x-direction (*right*)

$$C' = C_0 + \left(\left(\frac{r}{r_{\text{pure}}} - 1 \right) \cdot (C_1 - C_0) \right) \tag{4.8}$$

The required size of this RVE zone is governed by the fiber architecture, the fiber volume fraction, the size of the modeled crack growth, and the wavelengths faced for the respective failure mechanism. For the inner dimensions of the RVE stated above, dimensions of r_{pml} in the range between 50 and 100 μm were found to yield convergent solutions for the cases presented in the following. On the macroscopic scale (outside the RVE and PML region), the individual layers of the composite are modeled as homogenized anisotropic continuum C_1. A visualization of the transition of material properties between r_{pure} and r_{PML} is shown in Fig. 4.9 using the material density in the yz-cross-section of the RVE as an example. The operation of the RVE in terms of the stress distribution is visualized in Fig. 4.9 as well. For a macroscopically applied tensile stress of 2400 MPa in x-direction, the stresses within the RVE are based on the individual stiffness of fiber and matrix and thus evaluate as 3700 MPa within the modeled fiber filaments and 52 MPa within the matrix.

Before fracture occurs, materials deform with substantial contributions of plastic deformation. While this contribution is almost negligible for most of the reinforcement fibers (i.e., carbon, glass, Kevlar), the polymeric matrix materials typically undergo plastic deformation to a certain extent. Although the AE release (i.e., generation of elastic waves) is dominated by the elastic energy release, the influence of plastic deformation should still be considered in the material models within the RVE. However, the extent of the RVE needs to be enlarged to fully include the size of the plastic deformation zone inside r_{pure}, since it is not practicable to rewrite (4.8) in a trivial way to consider the transition of plastic material models between the microscale and the macroscale.

For the change in time resolution, the wavelengths or frequencies faced in the source modeling procedure are the relevant quantity to discuss. Since breakage of a single filament of a carbon fiber is in the order of one nanosecond, this requires a sufficiently small time step (e.g., in the order of 10^{-11} s). While this time step is still straightforward to use during crack growth, this high time resolution is impracticable for calculation of signal propagation with durations of several 100 μs. With reference to Fig. 4.7, this requires a modification of the approach in the third step of the modeling dealing with signal propagation. Based on publications on AE source modeling [12, 24], one way is an iterative increase in the time step until the final resolution is reached. For modeling of fiber breakage with a duration of crack growth of $t_{\text{crack}} = 1\,\text{ns}$, this could be done in a sequence of resolutions doubling the time step every 100 time increments starting at 0.01 ns and ending at a 10 ns time step.

4.2.2.2 Source Models for Meso- and Macroscale

While the RVE approach is well suited to describe the propagation of cracks within the size scale of several fiber diameters, the approach becomes numerically impractical beyond a certain length scale. This is due to the fact that the RVE itself comprises most of the computational intensity, and thus it is not reasonable to extend much beyond the dimensions shown in Fig. 4.8. Also, the failure on the mesoscopic or macroscopic length scale in fiber reinforced materials is expected to be governed more by the properties of the homogenized anisotropic solid rather than the individual fiber and matrix materials contributing to the fracture surface. To this end another approach was used in the following to describe the occurrence of failure on this length scale.

As introduced in Sect. 2.2.1, Puck's failure criteria for inter-fiber failure and fiber failure are frequently applied to describe the occurrence of failure in fiber reinforced materials. Considering now a homogeneous anisotropic solid in the following, (2.11a) and (2.11b) are implemented within the "structural mechanics module" of the software package COMSOL Multiphysics and are used to describe layer-wise failure in unidirectional fiber reinforced materials on the mesoscopic and macroscopic scale. For the stiffness vector $\widetilde{\boldsymbol{k}}$ of the fracture plane, the values of normal modulus E_{n}, the shear modulus $G_{t_{12}}$, and the Poisson's ratio $\nu_{t_{12}}$ are approximated based on the orientation of the fracture plane with respect to the orientation of the modeled layer. Although this yields only a first-order approximation of the values, a proper choice of h was found to be much more relevant to yield negligible compliance of the closed crack surface.

For the mesoscale and macroscale, the degradation function is written in terms of the PUCK criteria as follows:

$$(\boldsymbol{r}) = \begin{cases} 1 & \text{if} \quad \max(\text{PUCK}_{\max}(\boldsymbol{r}),\ \text{PUCK}(\boldsymbol{r})) < 1 \\ 0 & \text{if} \quad \max(\text{PUCK}_{\max}(\boldsymbol{r}),\ \text{PUCK}(\boldsymbol{r})) \geq 1 \end{cases} \tag{4.9}$$

Similar to (4.7) this makes use of the maximum value $\text{PUCK}_{\max}$ of Puck's failure criteria at a given position to compare the present value with the results of all previous solutions.

4.2.3 *Case Studies for Acoustic Emission Sources in Fiber Reinforced Materials*

One advantage of the newly proposed type three approach compared to other formulations for acoustic emission source models is the experimental availability of the input parameters. In the proposed model, crack growth and acoustic emission are solely defined by the macroscopic loading condition and the failure criterion used. In particular, no explicit source function comprising internal forces or rise-times is necessary to initiate an acoustic emission signal. In this context, the validity of the new source modeling approach has been verified against experimental results from micromechanical experiments and was found to be in good agreement [13, 37]. This validated modeling approach is used in the following to demonstrate the excitation process of acoustic emission signals in several load cases as encountered in fiber reinforced materials.

To systematically address the different AE source mechanisms found in fiber reinforced composites, a convenient approach is to begin with a unidirectional fiber reinforced material and then extend the approach to more complex configurations. For all modeling work presented in this context in the following, the material properties of T700/PPS as reported in Table B.1 in Appendix B were used. The elastoplastic material properties for the PPS matrix were derived from tensile tests of the pure PPS material.

In order to avoid any influence of the thickness, or the geometry of the propagation medium, a simple plate as shown in Fig. 4.10 was used as macroscopic geometry for all investigations. The 1 mm thickness of the plate only allows propagation of the S_0 and A_0 Lamb wave modes. As stacking sequence, unidirectional plates with fiber axis in parallel to the x-axis and a [0/90/90/0] cross-ply stacking were used. For evaluation of the AE signals, the out-of-plane displacement as detected in 0°, 45°, and 90° orientation relative to the x-axis is used. The observation positions were chosen at 25 mm distance to the source position located at the center of the plate (see Fig. 4.16). All calculations were performed for 100 μs

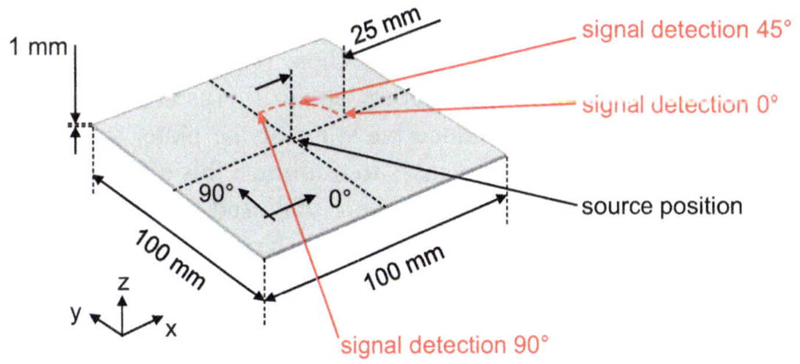

Fig. 4.10 Model configuration used to investigate AE signals of typical failure mechanisms in fiber reinforced materials including detection positions of the out-of-plane displacement signals

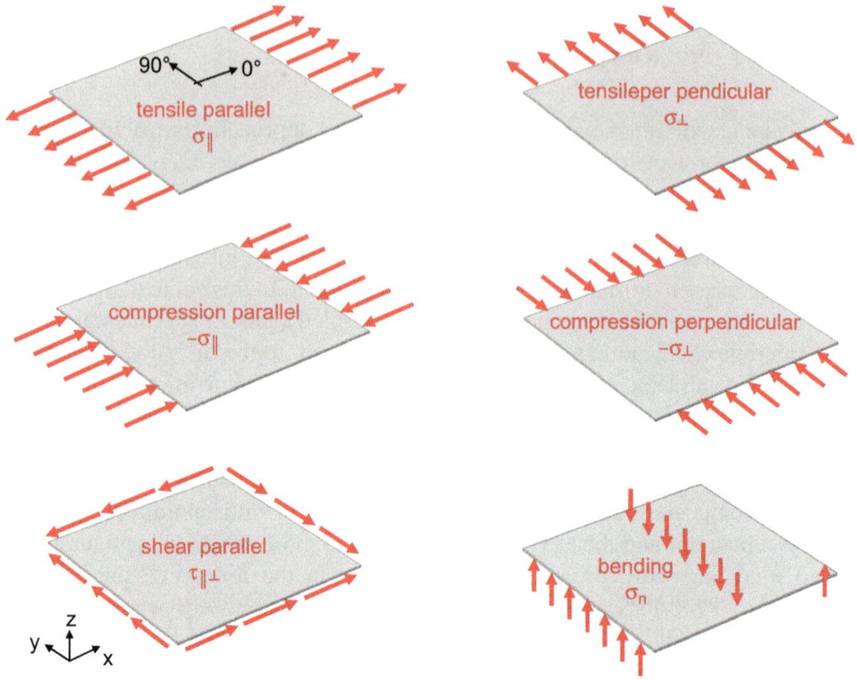

Fig. 4.11 Scheme of macroscopic load cases studied to induce AE sources in fiber reinforced materials

duration of signal propagation. For convenience, the displacement offset due to the applied macroscopic load is subtracted from the simulated out-of-plane displacements in the following to plot AE signals starting at zero amplitude.

In the following, various failure mechanisms as typical in fiber reinforced materials are modeled. In order to induce these different failure types, various load conditions are required. The chosen macroscopic load situations are summarized in Fig. 4.11. Based on the considerations of failure theories as discussed in Sects. 2.1 and 2.2, there are specific failure mechanisms expected for each of the load cases shown in Fig. 4.11. An overview of these failure mechanisms is found in Table 4.1 including a sketch of their micromechanical representation.

Since some of the load cases induce buckling of the plate, suitable boundary conditions were chosen for all of the cases to minimize this effect and to yield the intended stress components on the macroscopic scale and on the microscopic scale within the RVE. The macroscopic load levels of each case were gradually increased until the failure condition within the RVE or at the designated fracture plane was fulfilled. In addition, the depth position of the microscopic failure mechanisms was varied to study its influence in analogy to Sect. 4.2.1. Details of the respective source configurations are provided in the following subsections.

Table 4.1 Summary of load cases and expected failure mechanisms including micromechanical representation

Load case	Failure mechanism	Micromechanical representation
Tensile parallel $\sigma_{\parallel}$	Fiber breakage	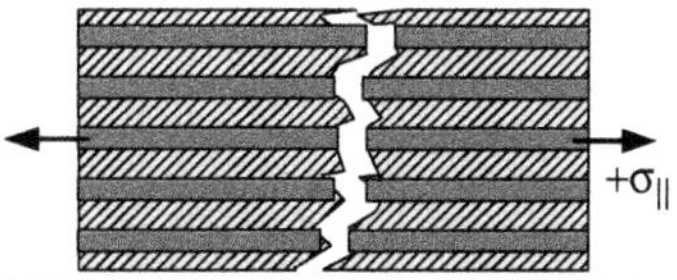
	Fiber bridging	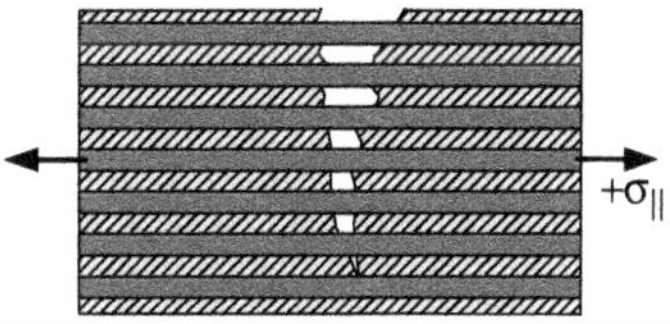
Tensile perpen-dicular $\sigma_{\perp}$	Inter-fiber failure	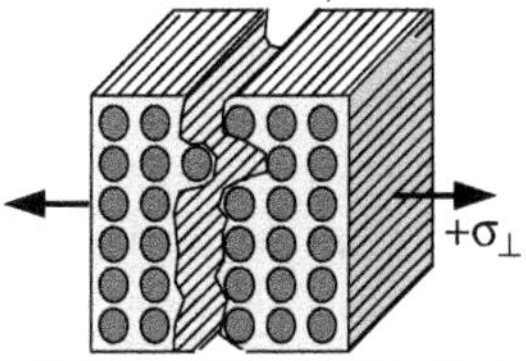
Compression parallel $-\sigma_{\parallel}$	Fiber breakage	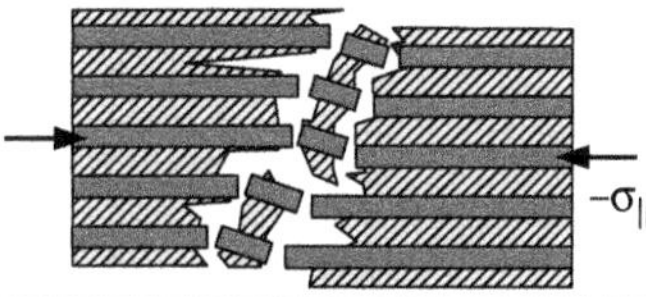
Compression perpendicular $-\sigma_{\perp}$	Inter-fiber failure	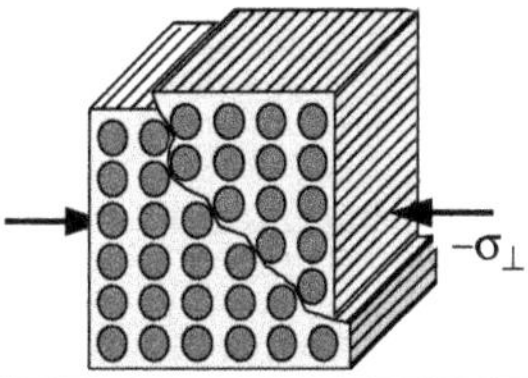
Shear parallel $\tau_{\parallel\perp}$	Inter-fiber failure	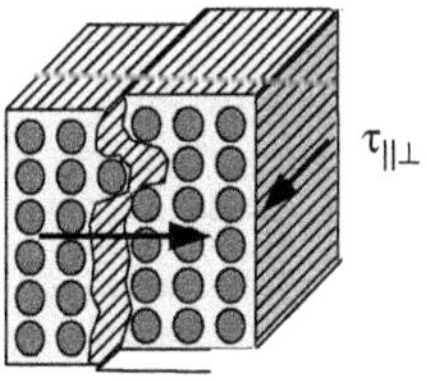

(continued)

Table 4.1 (continued)

Load case	Failure mechanism	Micromechanical representation
	Inter-ply delamination	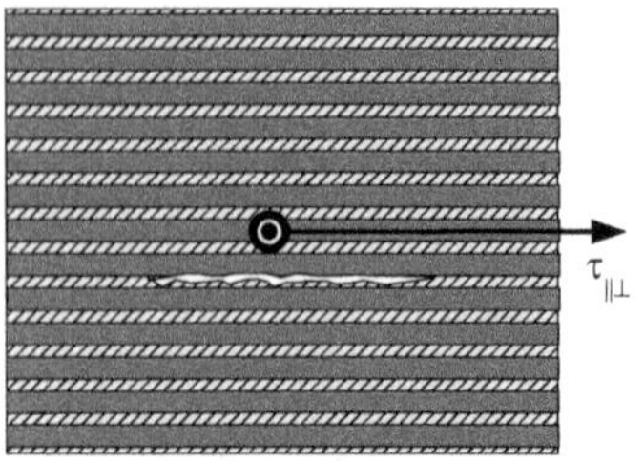
Bending σ_n	Inter-fiber failure in 90° plies	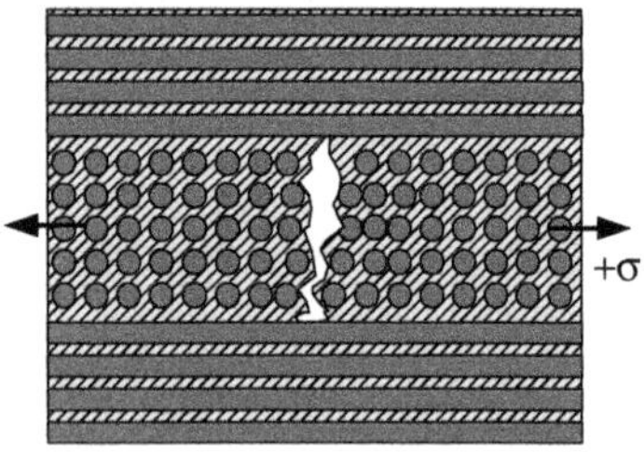
	Inter-ply delamination at interface between 90° and 0° ply	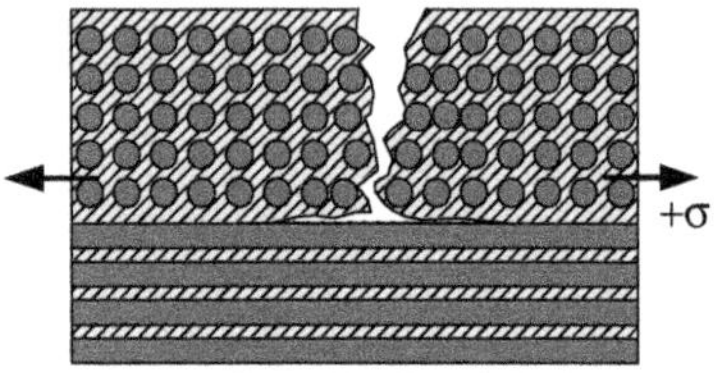

4.2.3.1 Inter-fiber Failure

As one of the fundamental failure mechanisms in fiber reinforced composites, inter-fiber failure is understood as formation of cracks in the matrix region between the fibers and along the interface between fiber and matrix (cf. Sect. 2.1). For a unidirectional fiber reinforced material, a broad distribution of crack sizes is expected [38]. As shown in Fig. 2.1, such inter-fiber failure may occur on the length scale of few fiber diameters, but may also span the full thickness of a ply and may propagate several centimeters in one step. Therefore several AE source models need to be investigated in terms of the dimensions of the inter-fiber cracks. In the following, the AE source models were implemented as described in the previous Sect. 4.2.2 employing a thin elastic layer in combination with Puck's inter-fiber failure criterion. Before starting the modeling work, an analysis using Puck's failure criterion is made to which of the above load situations are able to induce inter-fiber failure.

As the first load case, the tensile parallel $(\sigma_{\|})$ situation is studied. As discussed by Puck [39], inter-fiber failure in such a load situation is usually not covered by most of the failure theories. Still it is evident from experimental observations that such load situations may sometimes cause inter-fiber failure in the form of splitting of the fiber layers. This is likely due to the generation of shear stresses at imperfections of the material or other breakdowns of the perfect tensile parallel load case. However, for the AE source modeling applying Puck's failure criterion, it was found impossible to generate inter-fiber failure at any orientation of the fracture plane before the occurrence of fiber failure. Thus it is not feasible to calculate AE sources due to inter-fiber failure following Puck's failure criterion in the tensile parallel $(\sigma_{\|})$ load case and similar in the compressive parallel $(-\sigma_{\|})$ load case. Consequently, only results of the remaining four load cases will be discussed in the following.

Tensile Perpendicular $\sigma_{\perp}$

As the first example, tensile loading perpendicular to the fiber axis $(\sigma_{\perp})$ is presented and used as an example to demonstrate some key aspects of acoustic emission source operation. For a unidirectional composite material, this load case typically results in ultimate failure due to macroscopic inter-fiber failure. However, the ultimate (macroscopic) separation of the material is not the focus of this study. Herein inter-fiber failure is assumed to stop after a certain propagation distance to generate an AE signal for comparison with other failure mechanisms. Hence, the crack growth was confined to a distinct size. The geometry of the crack and the according load configuration is shown in Fig. 4.12. To initiate the crack at the center position of the plate, an artificial flaw of 10 μm length spanning the full thickness direction was included to act as a precrack.

In the stationary model, the static load $F_{\perp}$ was increased to 1000 N before Puck's inter-fiber failure criterion indicated crack growth at the position of the edge of the

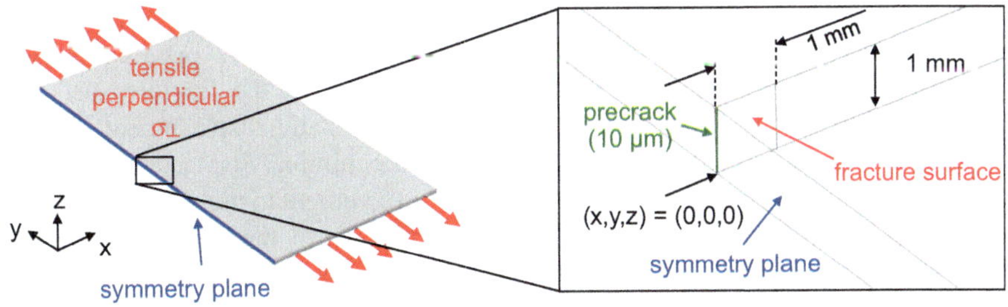

Fig. 4.12 Model configuration including position of symmetry plane and details of crack region

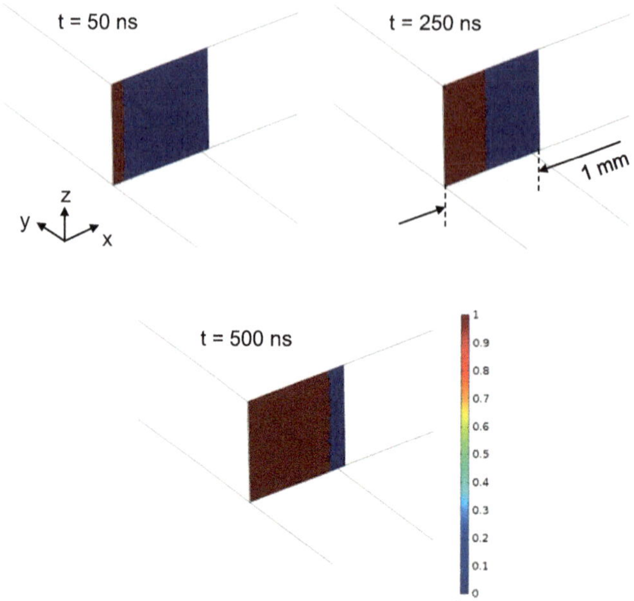

Fig. 4.13 Extension of fracture surface during crack growth at $t = 50$ ns, $t = 250$ ns, and $t = 500$ ns after initiation of crack propagation

precrack yielding $\sigma_{\perp} = 20\,\text{MPa}$. Due to the presence of the static load, a static displacement field exists within the specimen. As pointed out in the previous section, these static load and displacement field are transferred to a second computation step. This initiates the transient degradation of the stiffness vector of the thin elastic layer based on the evaluation of Puck's failure criterion. A sequence of images during the duration of crack propagation is shown in Fig. 4.13. Here the value of the degradation function is plotted for $t = 50$ ns, $t = 250$ ns, and $t = 500$ ns after initiation of crack propagation. The false-color range of the degradation function $\check{C}(\boldsymbol{r})$ indicates the extension of the crack by red areas, while blue colors resemble areas, where the material is still in contact.

Due to the orientation of the local stress components, the crack walls are subject to a dominant mode I type load situation. This causes a crack opening during propagation of the crack, and the resulting movement of the fracture surfaces constitutes the actual source mechanism. Since the static displacement values dominate the displacement scale, it is hard to spot the minimal displacement values superimposed to this background. Thus it is more convenient to use diagrams of the acoustic velocity magnitude to visualize the generation of the acoustic emission wave around the crack. As seen from Fig. 4.14, the near field of this crack type consists of a radiation perpendicular to the crack surfaces, which extends along the x-axis with the progression of the crack.

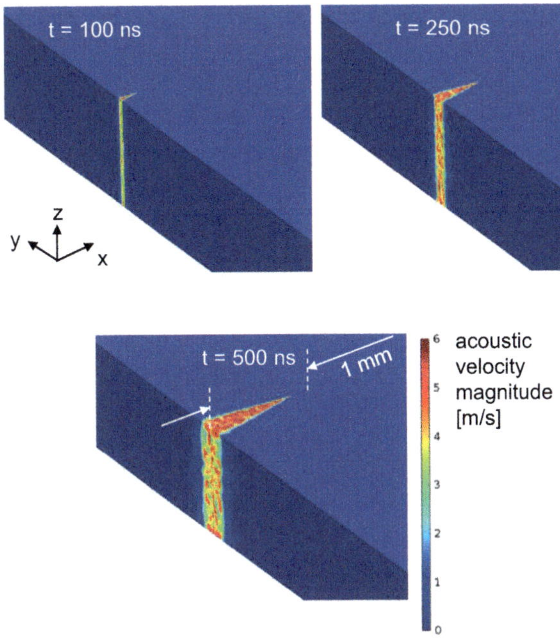

Fig. 4.14 Plot of acoustic velocity magnitude during crack growth at $t = 100$ ns, $t = 250$ ns, and $t = 500$ ns after initiation of crack propagation

In order to obtain the source function of the acoustic emission source, the y-displacement at the position $(x, y, z) = (0, 0, 0.5)$ mm is evaluated. This position is at the center of the growing crack and thus the position with the maximum crack opening. As seen in Fig. 4.15, during the duration of crack propagation, the y-displacement increases continuously. However, at the moment of crack arrest, the y-displacement does not stop or cease, but increases further until a maximum value is reached. Subsequently, the crack surface starts to vibrate and finally settles at a new equilibrium position.

Based on the theory of Green and Zerna [40], the expected position of the crack surface in a static crack opening situation of a crack length a may be estimated for an isotropic material with modulus E and Poisson's ratio ν for a given stress σ as

$$u_{\text{theory}} = \frac{4 \cdot (1 - \nu)}{\pi \cdot E} \cdot \sigma \cdot a \tag{4.10}$$

Following the intent of (4.10) for a mode I type load, the calculated value using the material stiffness $E_{22}, \nu_{21}, \sigma_{\perp} = 20\,\text{MPa}$, and $a = 2\,\text{mm}$ is marked as dashed line in Fig. 4.15b. As can be seen from the comparison, in the initial part, these values are systematically lower than the present model predictions. This originates from the underlying assumptions in the theory. In [40] static crack opening is assumed, i.e., the length of the crack is already present as flaw in the material and merely opens due to an external force. If the same assumptions are made in the present model,

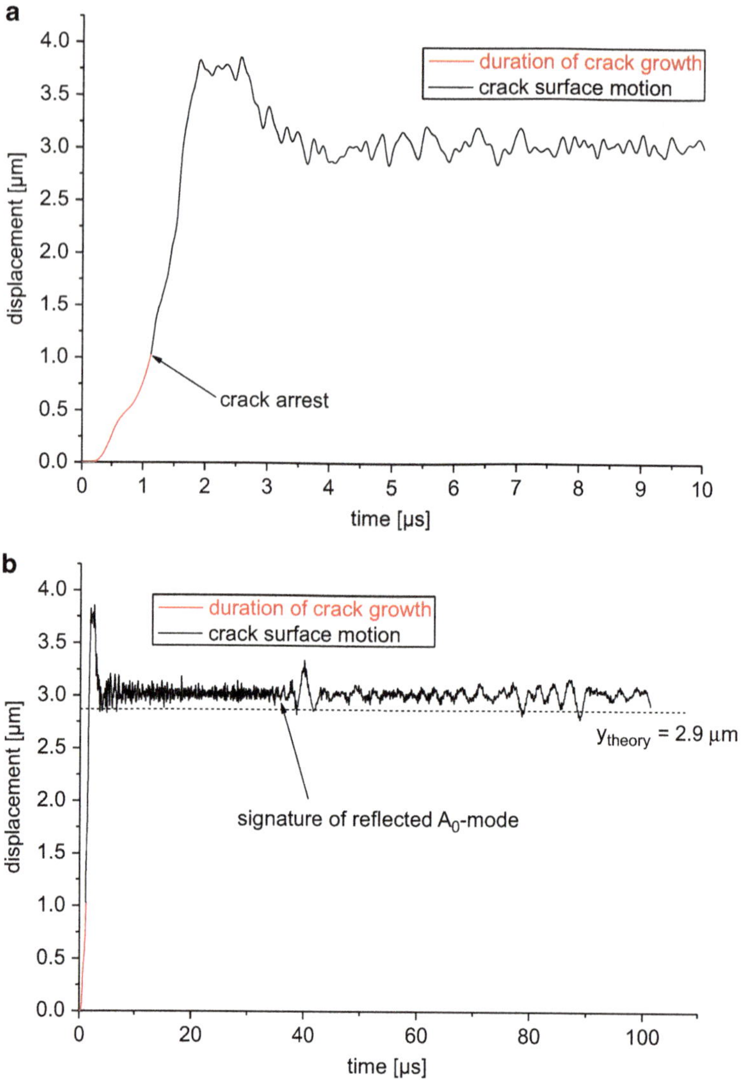

Fig. 4.15 Evaluation of source displacement in *y*-direction during crack growth (**a**) and for full duration of computation (**b**)

the achieved deformation state is in very good agreement to the values predicted by [40]. However, the dynamic crack propagation seems to generate initial acoustic emission amplitudes being larger than predicted by the analytical theory.

At the moment of crack arrest at $t = 1.1$ µs, the model starts with the third computation step to calculate the spreading of the acoustic emission signal in the plate. In the far field, the displacement and velocity components generated by the crack source are converted into plate waves and start to propagate toward the detection positions. As seen from the acoustic velocity fields in Fig. 4.16 at six

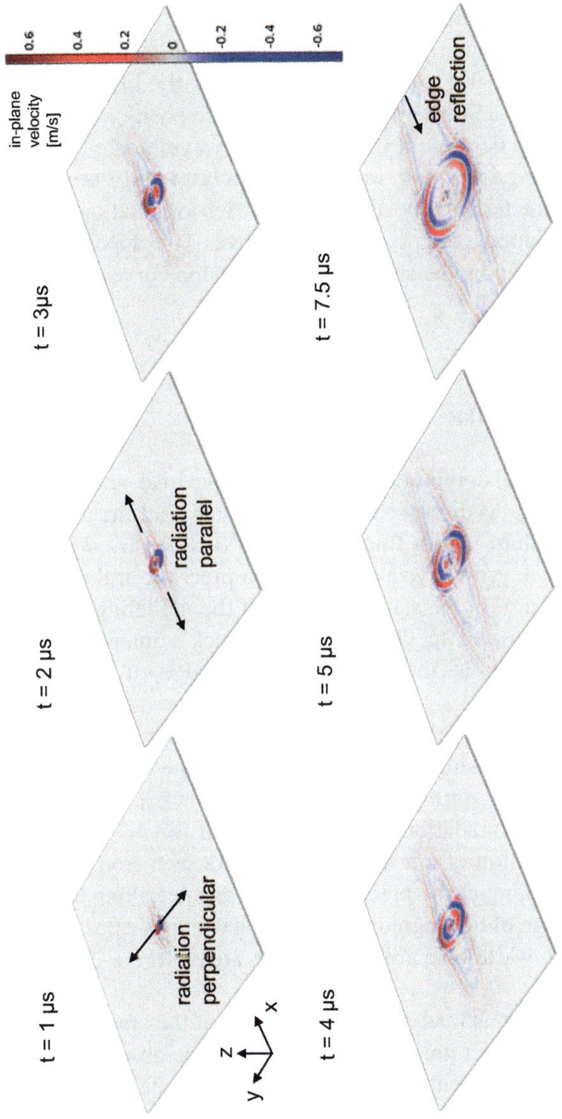

Fig. 4.16 Propagation of acoustic waves in the far field around the crack at distinct time steps after initiation of crack propagation

distinct time steps, there are two major contributions to the wave field. The first part is the radiation of the wave in orientation perpendicular to the fracture surface. This constitutes the dominant contribution to the wave field. As the second part, the propagation of the crack causes a wave in parallel to the direction of crack growth. This less-intense contribution spreads out faster due to the acoustic anisotropy of the unidirectional fiber reinforced plate.

Due to the different radiation of the source in each direction, the resulting signals obtained at 0°, 45°, and 90° start to appear significantly different. A comparison of the signals detected at 25 mm distance to the source position is shown in Fig. 4.17. Based on the angular dependency of the guided wave modes, the initial arrival of the S_0 mode is observed first in 0° orientation and last for 90° orientation. The mode intensities are characteristically different for each orientation, which is a consequence of the radiation pattern discussed above. The appearance of the modes themselves is governed by the shape of the dispersion curves as shown in Fig. 4.59 in Sect. 4.3.

Compression Perpendicular $-\sigma_\perp$

As the next type of load configuration, a compressive load perpendicular to the fiber axis $-\sigma_\perp$ was studied. As discussed in Sect. 2.1, such load situations are expected to cause inter-fiber failure with a fracture surface inclined by $\theta = 53°$ as shown in Fig. 4.18. Here crack growth was initiated by a precrack at the center of the crack with a width of 10 μm. In the stationary step of the modeling procedure, the static load $-F_\perp$ was increased to 9000 N before Puck's inter-fiber failure criterion indicated crack growth at the position of the edge of the precrack yielding $-\sigma_\perp = 90\,\text{MPa}$.

Due to the presence of the compression load, a static displacement field exists within the specimen causing crack propagation at the designated fracture plane. A sequence of images during the duration of crack propagation is shown in Fig. 4.19. Here the degradation function is shown for $t = 50$ ns, $t = 250$ ns, and $t = 500$ ns after initiation of crack propagation. As seen from Fig. 4.19, the crack starts to grow beginning at the precrack and propagates along the x-axis direction. The false-color range of the degradation function $\check{C}(\boldsymbol{r})$ indicates the extension of the crack by red areas, while blue colors resemble areas, where the material is still in contact.

Similar to the previous load case, the near field of the crack is shown as velocity magnitude at three distinct time steps in Fig. 4.20. The observed crack propagation causes a motion perpendicular to the formation of the crack surface which is inclined by 53°. In comparison to the load case $\sigma_\perp$, the intensity of the surface motion is not balanced along the y-axis at each edge of the crack surface. Instead, there is an asymmetric intensity observed radiating in the positive y-axis direction at the top surface of the plate and in the negative y-axis direction on the bottom of the plate (cf. Fig. 4.20).

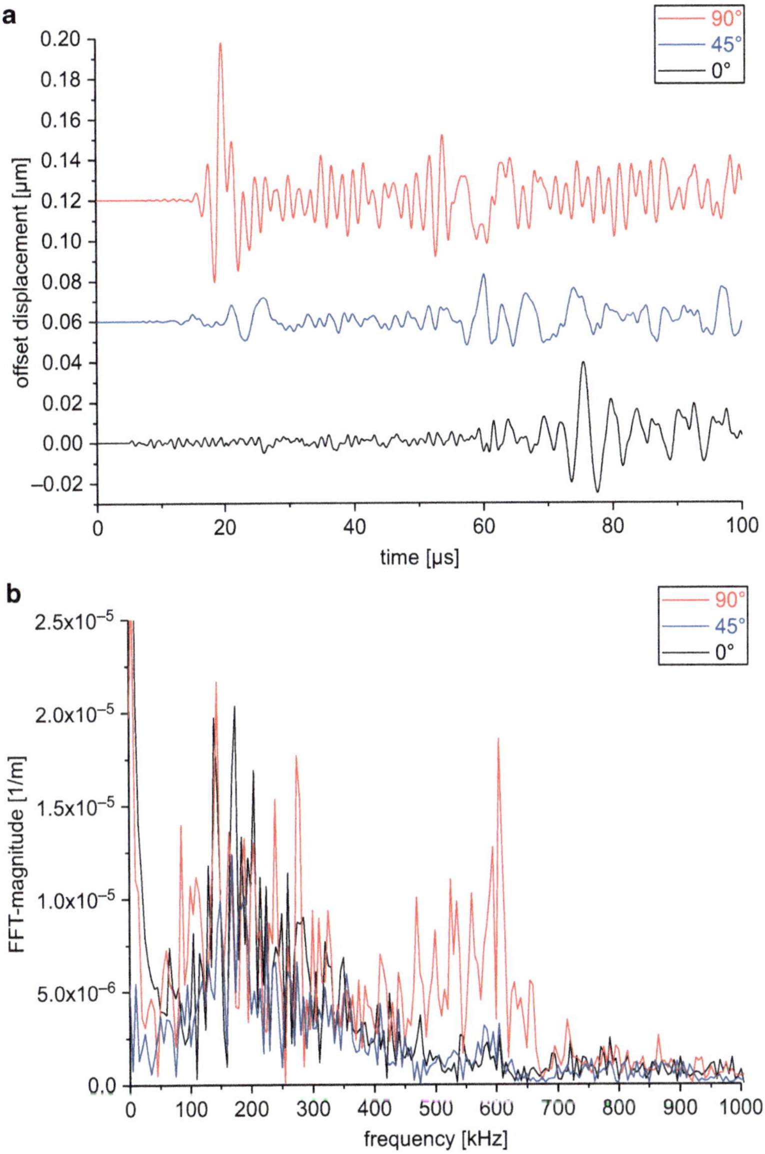

Fig. 4.17 Acoustic emission signals of inter-fiber failure due to $\sigma_{\perp}$ detected in unidirectional plate at 0°, 45°, and 90° relative to fiber orientation at 25 mm distance to source position in time domain (**a**) and frequency domain (**b**)

The acoustic emission signals detected in 25 mm distance to the source position are shown in Fig. 4.21 for three different angles relative to the source position. Again, the combination of the radiation pattern of the source and the anisotropy of

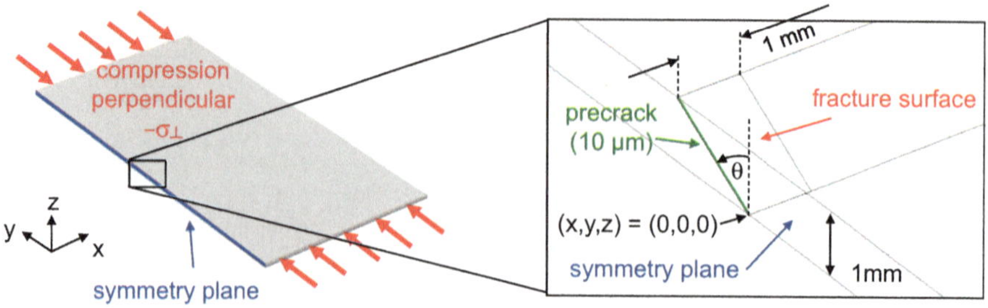

Fig. 4.18 Model configuration including position of symmetry plane and details of crack region

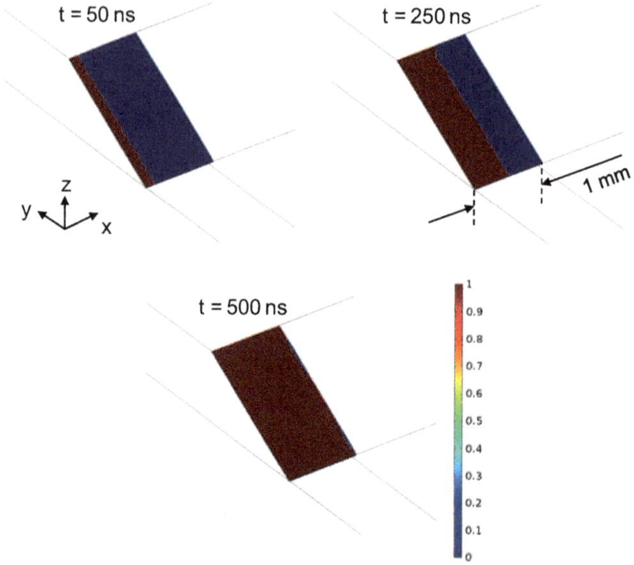

Fig. 4.19 Extension of fracture surface during crack growth at $t = 50$ ns, $t = 250$ ns, and $t = 500$ ns after initiation of crack propagation

the propagation medium causes signals, which look distinctly different although they emanate from the same source. This finding is not unexpected and was also experimentally verified (cf. Sect. 4.5.4.3). In direct comparison to the $\sigma_\perp$ load case, the signals detected at each of the angles show reasonable similarity.

Shear Parallel $\tau_{\parallel\perp}$

As further load case relevant for inter-fiber failure, shear loading parallel $\tau_{\parallel\perp}$ to the fiber axis is investigated. This shear stress situation indicated by the load arrows in Fig. 4.22 is expected to yield a fracture surface orientation as for the $\sigma_\perp$ case.

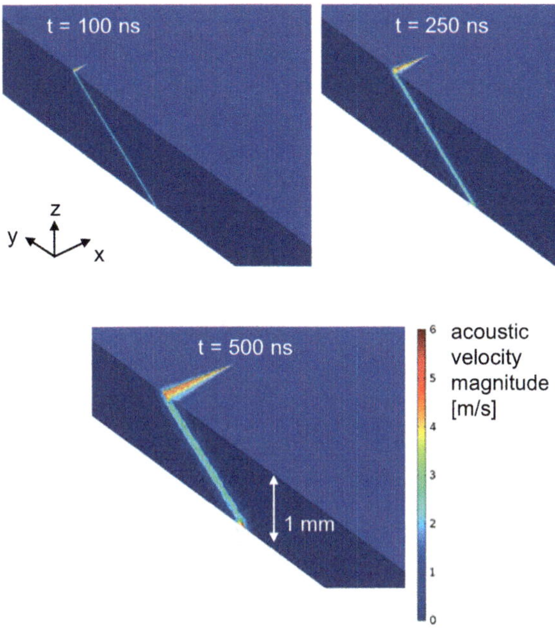

Fig. 4.20 Plot of acoustic velocity magnitude during crack growth at $t = 100$ ns, $t = 250$ ns, and $t = 500$ ns after initiation of crack propagation

As shown in the schematically drawing in Fig. 4.22, a precrack of 10 μm length was chosen to facilitate crack initiation. To allow for better comparison with the other inter-fiber failure cases, the extension of the fracture surface was chosen to be 1 mm along the x-axis. A macroscopic load of $F_{\parallel\perp} = 8000\,\text{N}$ results in a shear stress value of $\tau_{\parallel\perp} = 80\,\text{MPa}$ at the position of the fracture surface.

A visualization of crack propagation according to Puck's failure criterion is shown in three distinct time steps in Fig. 4.23. Similar to the previous load cases, the crack front advances steadily along the direction of the x-axis due to the presence of the macroscopic stress field. While the speed of crack propagation along the x-axis is almost unaltered compared to the previous cases, slight differences arise in the shape of the crack front. This is attributed to the different load components faced in this shear load case.

The acoustic velocity magnitude is shown as false-color plot in Fig. 4.24 at three distinct time steps after crack initiation. While the first impression of the crack front resembles some close similarity to the near field emanating from the crack growth under $\sigma_\perp$, there are some substantial differences. In the $\tau_{\parallel\perp}$ load case, the extent of the motion along the y-axis does reach as far as for the $\sigma_\perp$ case, and some noticeable additional motions with higher frequencies are observed. The latter are seen best as shades ahead of the intense part of the near-field zone spreading out in x and y-direction.

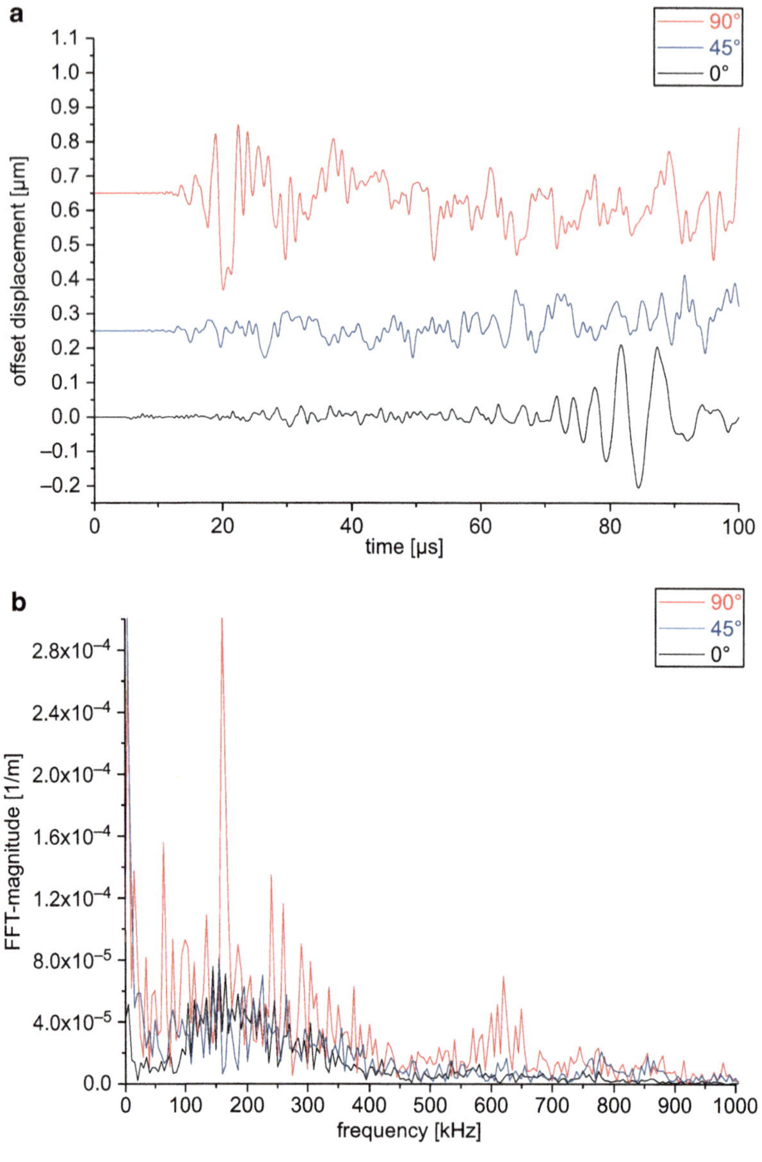

Fig. 4.21 Acoustic emission signals of inter-fiber failure due to $-\sigma_{\perp}$ detected in unidirectional plate at 0°, 45°, and 90° relative to fiber orientation at 25 mm distance to source position in time domain (**a**) and frequency domain (**b**)

The change in source operation as compared to the $\sigma_{\perp}$ load case also translates into distinct differences in the detected acoustic emission signals. In Fig. 4.25 a comparison of the signals detected at three different angles relative to the source position is shown. In contrast to the signals of Figs. 4.17 and 4.21, these acoustic

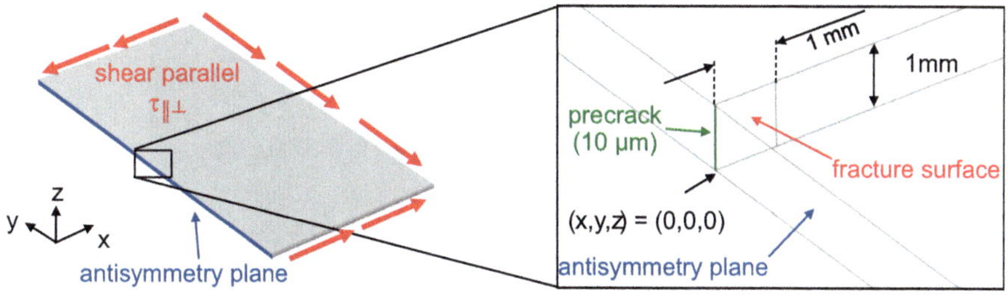

Fig. 4.22 Model configuration including position of antisymmetry plane and details of crack region

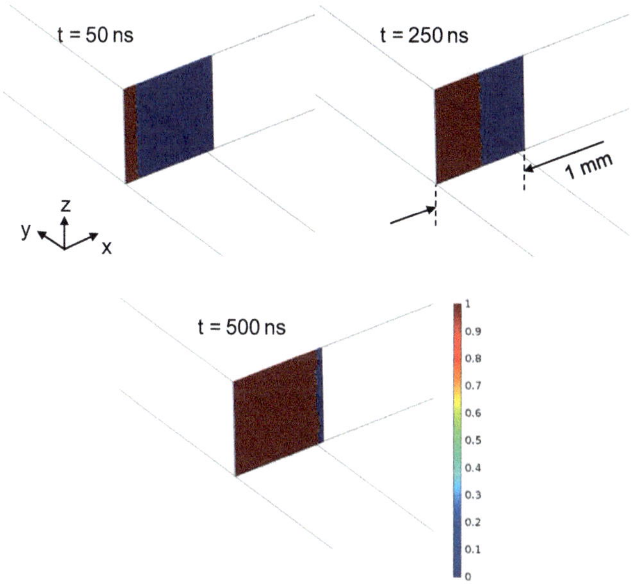

Fig. 4.23 Extension of fracture surface during crack growth at $t = 50$ ns, $t = 250$ ns, and $t = 500$ ns after initiation of crack propagation

emission signals share the common aspect of lack of lower frequency components. This is due to the confined possibilities of crack surface movement. Since the macroscopic $\tau_{\parallel\perp}$ load does not cause a mode I type crack opening movement, but rather a mode II type sliding crack propagation, the dominant vibration of the crack surface is a movement in the xz-plane.

Bending σ_n

As the last type of load scenario for the occurrence of inter-fiber failure, a bending load σ_n is investigated. Here, a different laminate type is selected to induce

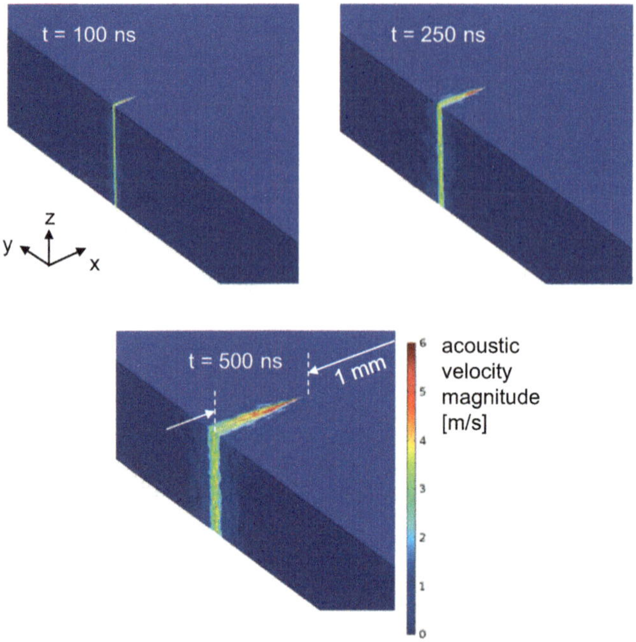

Fig. 4.24 Plot of acoustic velocity magnitude during crack growth at $t = 100$ ns, $t = 250$ ns, and $t = 500$ ns after initiation of crack propagation

inter-fiber failure in a $[0/90]_{sym}$ stacking sequence. The thickness was chosen to be 1 mm in accordance with the previously studied unidirectional plate. A schematically view of the load configuration with details of the crack model is shown in Fig. 4.26. The inter-fiber failure fracture plane is defined directly below the position of applied load and to occur within the full thickness of the 90° layer. For better comparability with the previous cases, the size of the crack along the y-axis direction is artificially constrained to yield a length of 500 μm.

To initiate crack growth in the designated fracture plane, a precrack of 10 μm length in the z-axis direction was modeled at the top edge of the 90° layer. The required external load was evaluated to be $F_n = 500\,\mathrm{N}$ so that Puck's failure criterion indicated crack initiation at the position of the precrack. The crack propagation is visualized for three distinct time steps in Fig. 4.27 using the false-color representation of the degradation function. In this load case, the crack initiates at the precrack position and starts to propagate with an almost plane crack front along the z-axis direction until it reaches the position of the bottom 0° layer.

The near field of the growing crack is shown in Fig. 4.28 for three distinct time steps as plot of the acoustic velocity magnitude. Since the crack propagation direction is along the z-axis direction, the view of the acoustic velocity magnitude within the xz-plane needs some further explanation. Because the crack starts at the top of the 90° layers, this is also the start position of the crack surface motion. With increasing time the motion of the crack walls turn on further displacements along

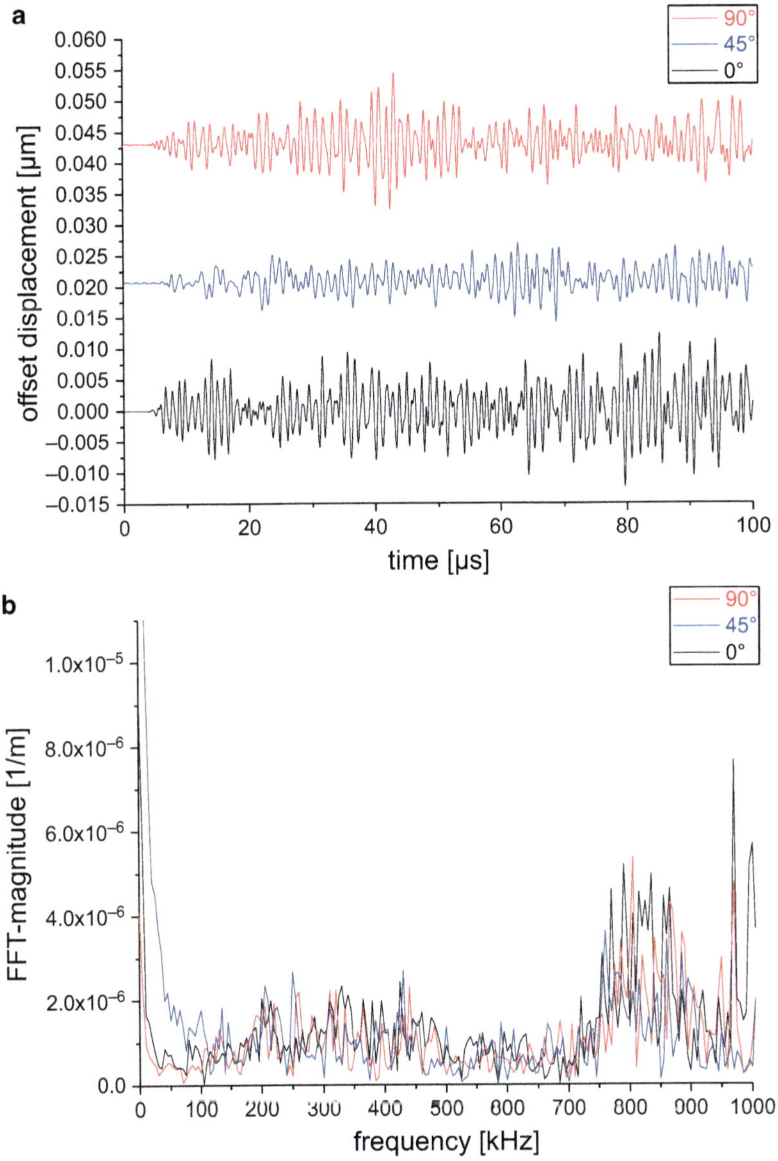

Fig. 4.25 Acoustic emission signals of inter-fiber failure due to $\tau_{\parallel\perp}$ detected in unidirectional plate at 0°, 45°, and 90° relative to fiber orientation at 25 mm distance to source position in time domain (**a**) and frequency domain (**b**)

the crack propagation direction causing a stripe-like appearance of the near-field zone. After the crack has reached the bottom of the 90° layers, the motion solely spreads out from this position. The effect of the different fiber orientations in the 0° and 90° layers also adds to the formation of the near field. Since wave propagation along the *x*-axis is faster in the top and bottom 0° layer, the motion in this part

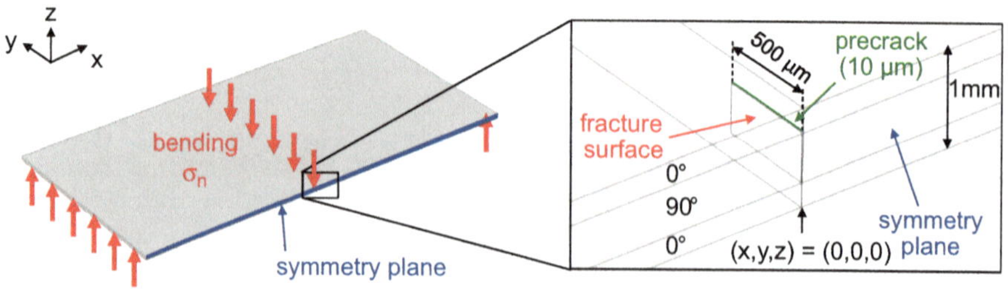

Fig. 4.26 Model configuration including position of symmetry plane and details of crack region

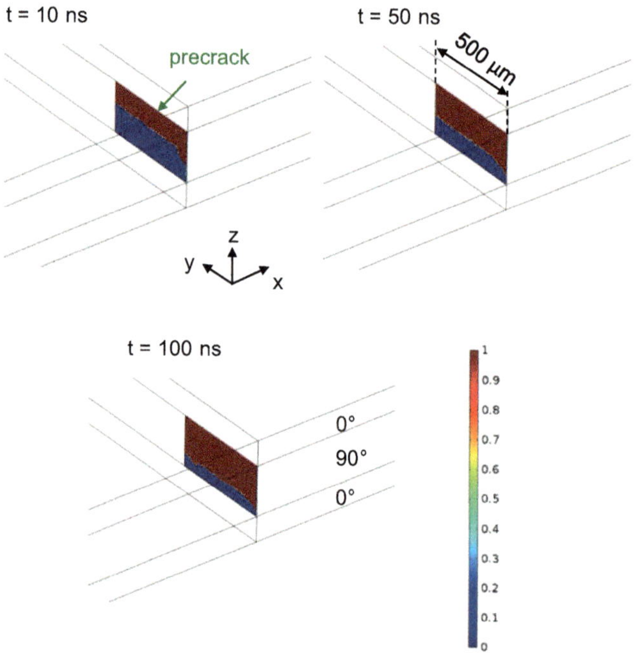

Fig. 4.27 Extension of fracture surface during crack growth at $t = 10$ ns, $t = 50$ ns, and $t = 100$ ns after initiation of crack propagation

of the plate propagates faster than in the 90° layer before all motions finally turn into a common guided wave mode.

As expected from the different stacking sequence, but also due to the different source type, the acoustic emission signals appear to be different to the previous cases. As shown for three distinct angles, the acoustic emission signals detected in 25 mm distance to the source position show a pronounced high-frequency vibration superimposed by an out-of-plane movement with comparatively low frequency.

The high-frequency part originates from similar confined crack motions as discussed for the $\tau_{\|\perp}$ case. In the bending case, the upper half of the fracture

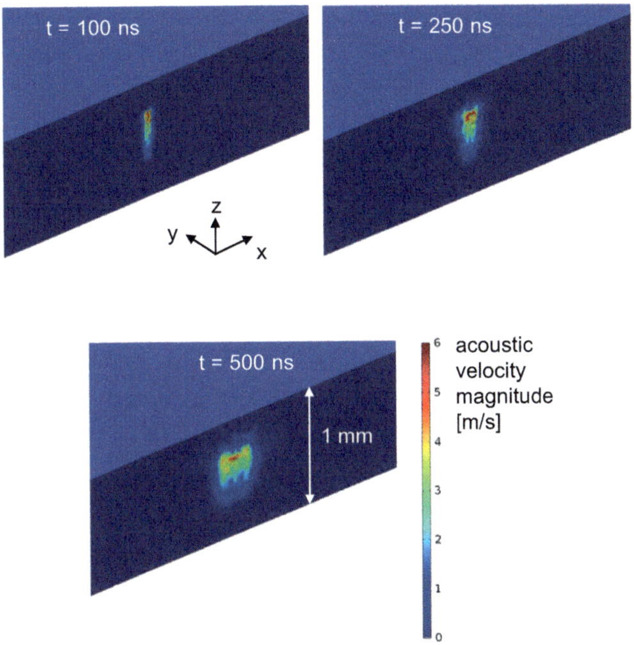

Fig. 4.28 Plot of acoustic velocity magnitude during crack growth at $t = 100$ ns, $t = 250$ ns, and $t = 500$ ns after initiation of crack propagation

plane is subject to a dominant compressive normal load, while the lower part is subject to a tensile normal load. In combination with the maximum shear load at the neutral axis of the plate, this leads to superimposed constraints with respect to the movement of the fracture surface. These constraints inhibit a free vibration of the full fracture surface (as, i.e., observed in [13]) and cause the vibration seen in Fig. 4.29. The low-frequency contribution however has a distinctly different origin. Due to the formation of the crack, the plate loses its stiffness, which results in a macroscopic settlement at the center of the plate. This settlement causes a plate motion, which is directed along the 0° and 90° axis orientation and hence visible in the contributions of the respective signals.

4.2.3.2 Delamination

As the second category of failure mechanisms studied, the focus is now on the occurrence of inter-ply delamination. Many failure theories do not distinguish between inter-ply delamination and inter-fiber failure, since crack growth is occurring between the fibers in both cases. However, due to the different orientation of the fracture surface relative to the thickness direction of the plate, this failure type is acoustically different to the cases studied in Sect. 4.2.3.1.

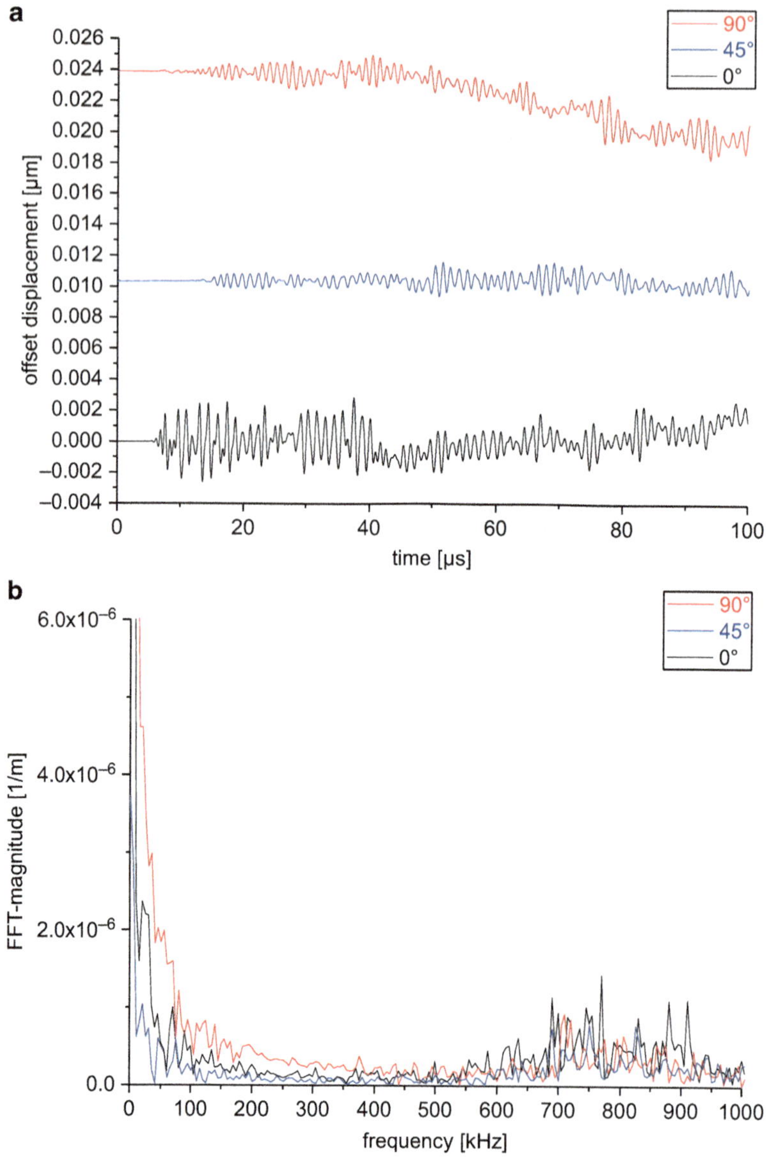

Fig. 4.29 Acoustic emission signals of inter-fiber failure due to σ_n detected in $[0/90]_{\text{sym}}$ plate at 0°, 45°, and 90° relative to fiber orientation at 25 mm distance to source position in time domain (**a**) and frequency domain (**b**)

Using Puck's failure criterion to evaluate the occurrence of delamination (i.e., inter-fiber failure with $\theta = 90°$), the same considerations as in Sect. 4.2.3.1 apply. For the load cases $\sigma_{\|}$ and $-\sigma_{\|}$, there is no inter-fiber failure observed before the occurrence of fiber failure. Similarly, the fracture plane orientation for the load case

$\sigma_{\perp}$ in a unidirectional plate is 0°, and for the load case $-\sigma_{\perp}$, the fracture plane orientation is calculated as $\theta = 53°$. Consequently, none of those load cases is discussed in the following. Only the $\tau_{\|\perp}$ load case was found to yield a delamination type failure in a unidirectional material. In addition, the important scenario of inter-ply delamination in a multiaxial laminate is investigated. To this end, an analysis using a $[0/90]_{\mathrm{sym}}$ laminate subject to bending load σ_n is presented.

Shear Parallel $\tau_{\|\perp}$

As shown in Fig. 4.30, the fracture surface is oriented parallel to the top and bottom surface ($\theta = 90°$) of the laminate plate. The width of inter-ply delamination was chosen as 1 mm, and crack growth was confined to a maximum length of 1 mm allowing better comparison to the inter-fiber failure cases studied previously. To allow crack initiation at the designated position, a precrack of 10 μm width was modeled at the edge to the *yz*-plane.

At an applied force of $F_{\|\perp} = 6300\,\mathrm{N}$, Puck's failure criterion indicates crack growth at the precrack position. A visual representation of crack growth found for this load configuration is shown in Fig. 4.31. At three distinct time steps, the value of the degradation function is plotted as false-color values. Unlike the inter-fiber failure cases, the crack propagation in the present case seems to be substantially slower.

For this load case, the direction of crack propagation is along the *x*-axis. The type of crack motion seen in the acoustic velocity magnitude images in Fig. 4.32 is mostly due to the perpendicular motion of the crack surface. As indicated from the multiple wave peaks in Fig. 4.32, the crack motion is subject to a sequence of crack arrest and crack initiations. This turns into a high-frequency contribution as seen in the acoustic emission signals in Fig. 4.33 and a characteristic resonance like frequency. However, due to Poisson's contraction, the crack wall motion along the *z*-axis is energetically limited and therefore the absolute amplitudes of the signals are one to two orders of magnitude lower than for the inter-fiber failure cases of similar crack surface dimensions.

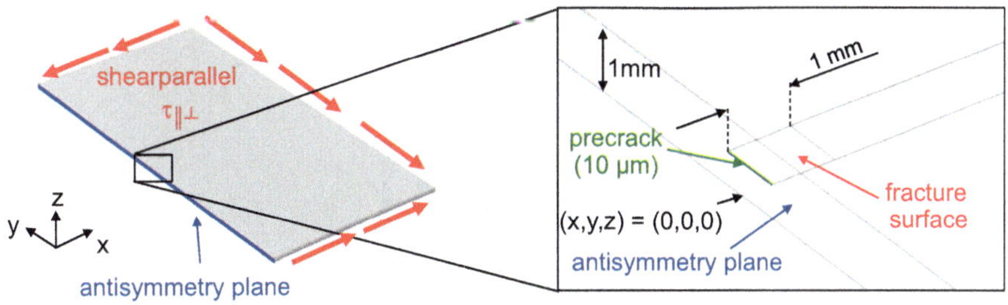

Fig. 4.30 Model configuration including position of antisymmetry plane and details of crack region

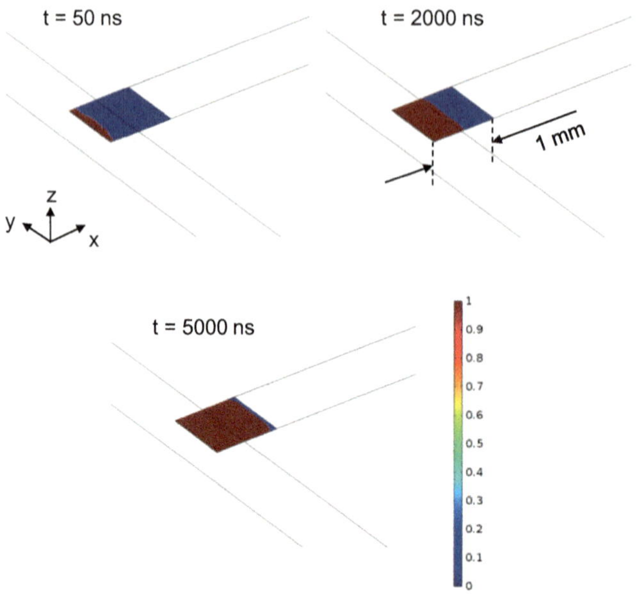

Fig. 4.31 Extension of fracture surface during crack growth at $t = 50$ ns, $t = 2000$ ns, and $t = 5000$ ns after initiation of crack propagation

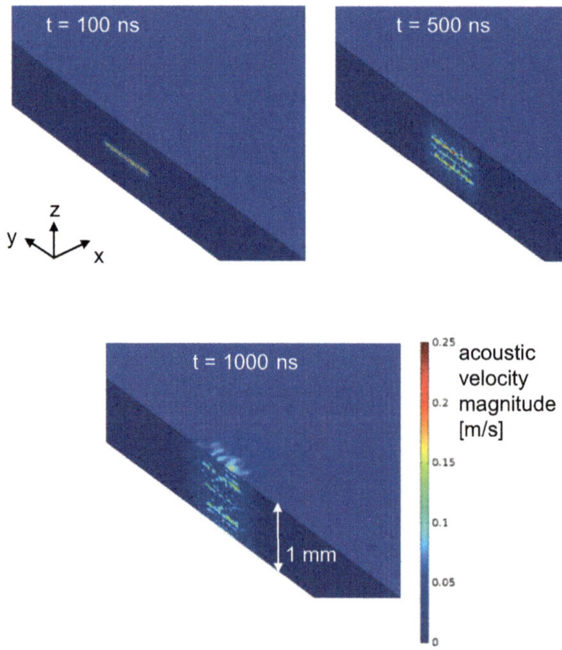

Fig. 4.32 Plot of acoustic velocity magnitude during crack growth at $t = 100$ ns, $t = 500$ ns, and $t = 1000$ ns after initiation of crack propagation

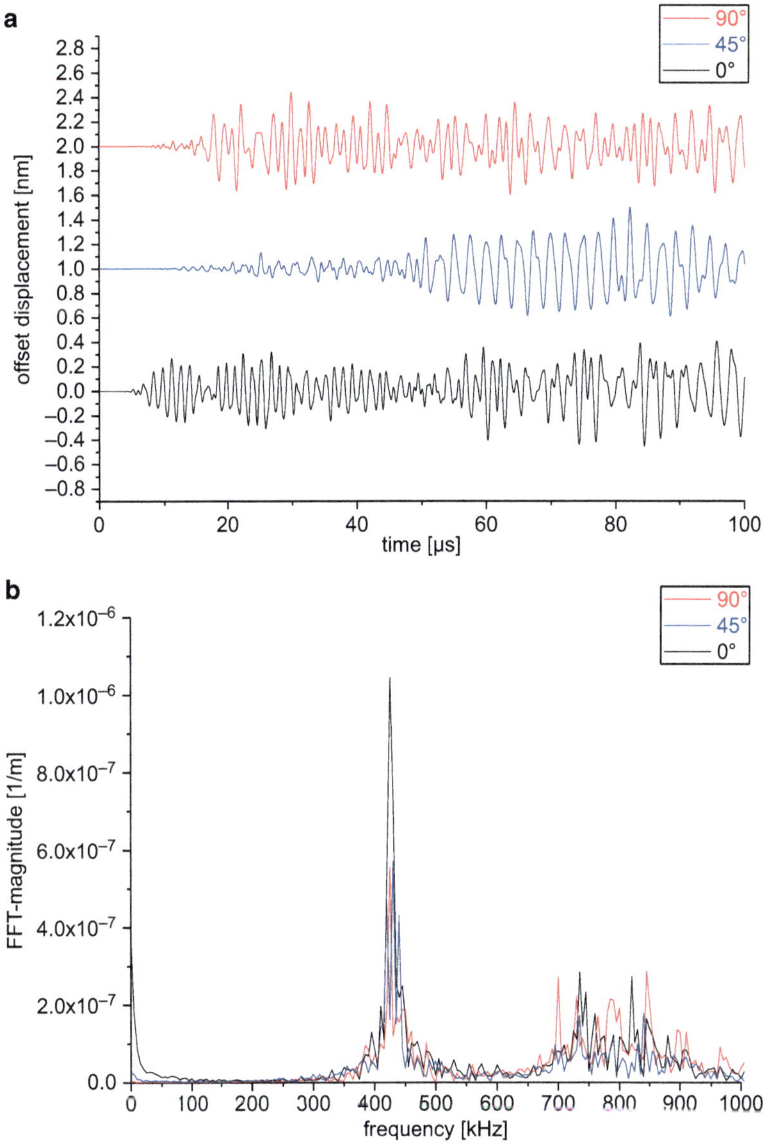

Fig. 4.33 Acoustic emission signals of inter-ply delamination due to $\tau_{\parallel\perp}$ detected in unidirectional plate at 0°, 45°, and 90° relative to fiber orientation at 25 mm distance to source position in time (**a**) and frequency domain (**b**)

Bending σ_n

Another configuration encountered in fiber reinforced materials, which causes initiation of inter-ply delamination, is the presence of an inter-fiber crack in off-axis plies relative to the load axis. As representative example for this scenario,

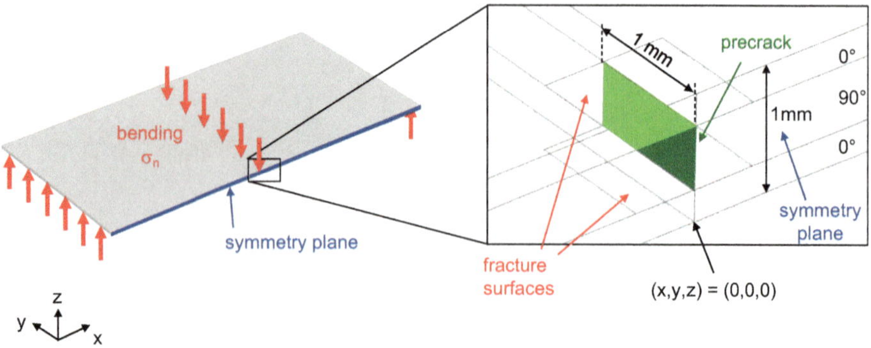

Fig. 4.34 Model configuration including position of symmetry plane and details of crack region

the load case bending σ_n in conjunction with a $[0/90]_{\mathrm{sym}}$ laminate is investigated. The schematically arrangement of this source type is shown in Fig. 4.34. As precrack a fully grown inter-fiber crack spanning the thickness of the two 90° plies is chosen. The extent of this precrack is identical to the previously studied case of inter-fiber failure due to bending load σ_n. In such a case, inter-ply delamination is expected to occur at the interface between 0° and 90° plies at the top and at the bottom, respectively. The size of each of the delamination areas was chosen to be 1.0 mm × 1.0 mm.

At an applied bending load $F_n = 750\,\mathrm{N}$, Puck's failure criterion indicated initiation of crack growth at both of the interfaces. As seen from the evaluation of the degradation function as false-color plot at three distinct time steps in Fig. 4.35, the crack front advances along the x-axis starting at the position of the precrack. Due to the distinctly different load conditions at the top and at the bottom interface, the crack front at the top interface spreads out first and advances faster than at the bottom. However, both cracks are active at the same time, thus yielding a combined contribution to the near-field zone and thus effectively yielding one acoustic emission signal.

The combined activity of both delamination fronts is also seen in the acoustic velocity magnitude shown in Fig. 4.36 at three distinct time steps. Here the zone at the top interface appears larger, which is due to the delamination at the top interface preceding the delamination at the bottom interface. The near field is shaped by the direction of crack propagation along the x-axis but is also facilitated by the acoustic anisotropy of the 0° layers causing faster wave propagation along the x-axis than in the perpendicular directions. Therefore the near-field zone appears rather elliptical at the top and at the bottom position.

Due to the close physical distance and the close time-wise distance, the activity of the two delamination fronts causes only one acoustic emission signal. This is demonstrated by the calculated signals in 25 mm distance to the source positions for three different angles as shown in Fig. 4.37. The acoustic emission signals do not reveal any indications of two merged signals but rather appear as typical transient signal with one distinct onset and according activity after that. The

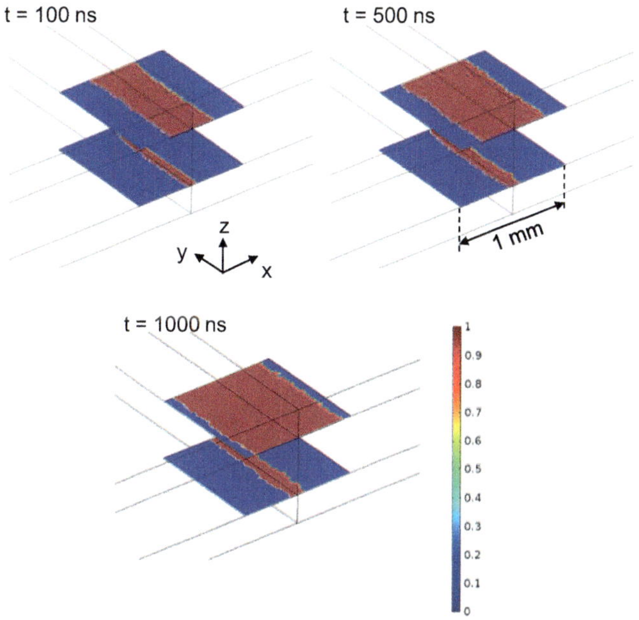

Fig. 4.35 Extension of fracture surface during crack growth at $t = 100$ ns, $t = 500$ ns, and $t = 1000$ ns after initiation of crack propagation

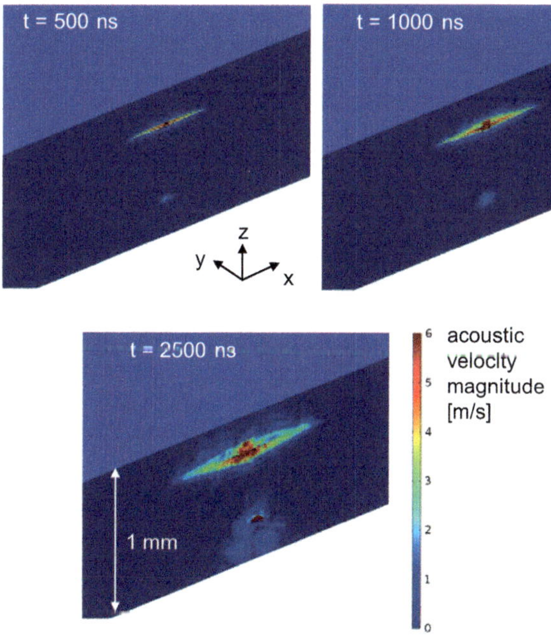

Fig. 4.36 Plot of acoustic velocity magnitude during crack growth at $t = 500$ ns, $t = 1000$ ns, and $t = 2500$ ns after initiation of crack propagation

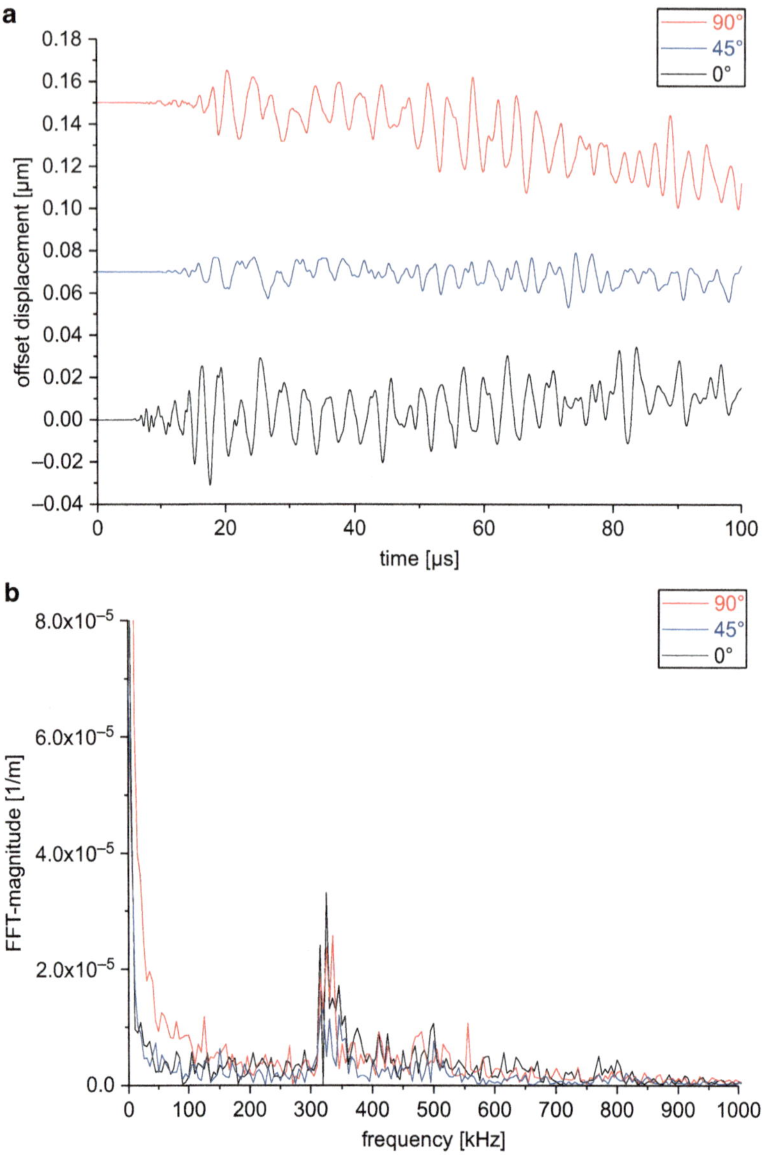

Fig. 4.37 Acoustic emission signals of inter-ply delamination due to σ_n detected in $[0/90]_{sym}$ plate at 0°, 45°, and 90° relative to fiber orientation at 25 mm distance to source position in time domain (**a**) and frequency domain (**b**)

dominant contributions seen in the signals can be attributed to similar origins as before. The high-frequency portion is mostly caused by the elementary crack arrest and crack advance of the fracture surface, whereas the low-frequency portion is again due to a settlement of the damage zone induced by the change in stiffness reduction.

4.2.3.3 Fiber Failure

Distinctly different to the two previous categories of failure mechanisms, fiber failure is a type of crack propagation associated with several challenges in source modeling. As evident from experimental findings, fiber failure is a mechanism which is usually not found on the macroscopic scale (e.g., crack zones with mm^2) prior to ultimate failure. Instead composite failure was observed to be caused by accumulation of clusters of less than 20 single fiber filament failures [41]. With some exceptions, rupture of a larger number of loaded fibers will initiate ultimate failure, and the according acoustic emission signal will be useless in the current context, since it will overlap with many other mechanisms and therefore cannot be related to a particular microscopic mechanism. Consequently, the following modeling work aims for the failure of single filaments and small fiber bundles in relation to their acoustic emission signals. Consequently, the RVE approach was used to model a microscopic part of the unidirectional fiber reinforced plate embedded in the macroscopic plate. Distinction is made between two load cases typically causing failure of fiber filaments in unidirectional laminates. One is the tensile parallel load case $\sigma_{\parallel}$ and one is the compression parallel load case $-\sigma_{\parallel}$.

Tension Parallel $\sigma_{\parallel}$

For initiation of fiber breakage, tensile loading parallel $\sigma_{\parallel}$ to the fiber axis is the only relevant situation for a unidirectional laminate. This model configuration is shown in Fig. 4.38 illustrating the macroscopic load configuration and then presenting details of the RVE dimensions and position within the plate. For the crack position, a further magnification of the RVE provides some more details with respect to the chosen model configuration. In the present RVE configuration, all fibers have identical dimensions and mechanical properties.[2] To initiate failure in one single filament, two modifications to the perfect RVE arrangement were required. First, within the designated yz-plane, the matrix area surrounding the fiber filament was defined as fully cracked, resembling presence of a void or an inter-fiber crack acting as stress concentrator at the surface of the fiber. Second, to avoid failure of other

[2] This assumption is not absolutely justified in reality, since it is well known that fibers show Weibull-type strength distributions, and may have significant variation in their cross-sections and their modulus.

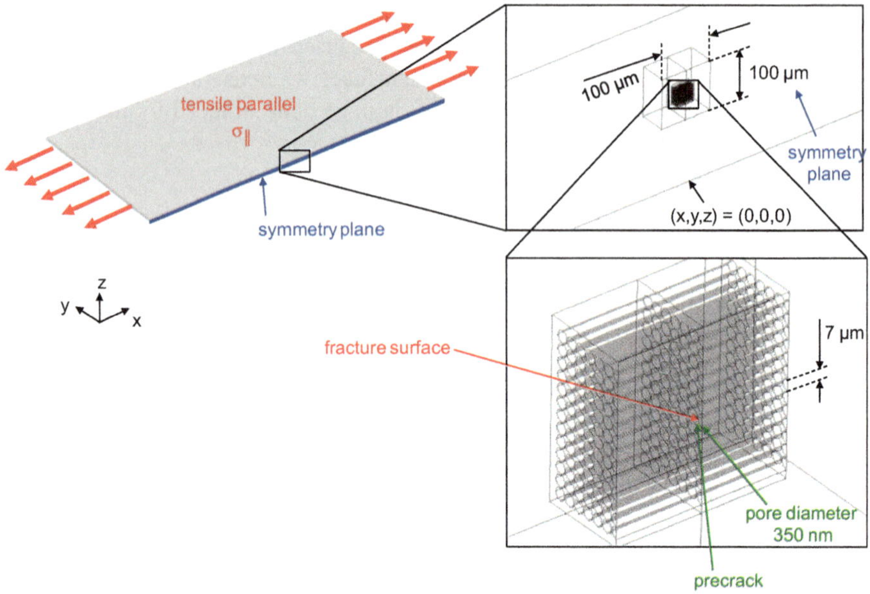

Fig. 4.38 Model configuration including position of symmetry plane, details of RVE position, dimensions and details of crack region

surrounding fibers (subject to the same stress concentration effect), the designated fiber filament was additionally weakened by the presence of a small pore with 350 nm diameter. This is one tenth of the fiber radius and therefore causes enough stress concentration at one particular fiber filament to enable crack propagation only within its cross-section. That way it is possible to systematically vary the filament failure strength between 300 and 4061 MPa to investigate its relation to the detectable acoustic emission signal amplitude. The value of 4061 MPa was taken from the material supplier's datasheet and is interpreted as mean fiber strength.

In order to reach a stress concentration of 4061 MPa at the designated fracture plane, a macroscopic load of $F_{\|} = 263\,\text{kN}$ was necessary. As failure criterion the von Mises equivalent stress as defined in (4.6) was used. Due to the presence of the modeled pore, failure initiates at the bottom of the fiber as seen in the false-color plots of the degradation function in Fig. 4.39. The crack propagation takes about 1.3 ns until the fracture surface is finally established. This duration seems quite reasonable when compared to the value of 1.2 ns calculated for a free fiber filament in [13] which has been validated against experimental data. As distinct difference to the case of a free fiber filament, the crack front seems to initiate at the position of the pore, but later starts to advance along the radial direction of the fiber starting at all directions reducing the cross-section of the fiber upon crack growth.

The near field of the single fiber filament failure is visualized as acoustic velocity magnitude in Fig. 4.40 for three distinct time steps after fiber failure. The shape of the near-field zone resembles close similarity to a dipole radiation pattern.

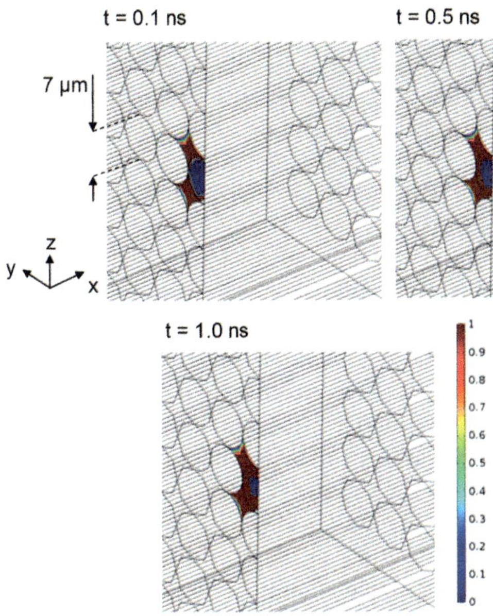

Fig. 4.39 Extension of fracture surface during crack growth at $t = 0.1$ ns, $t = 0.5$ ns, and $t = 1.0$ ns after initiation of crack propagation

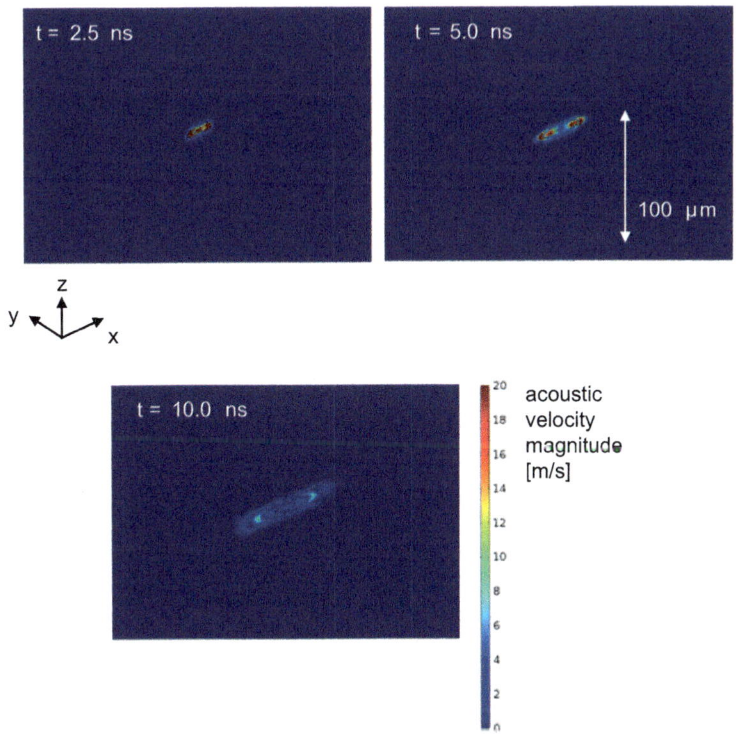

Fig. 4.40 Plot of acoustic velocity magnitude during crack growth at $t = 2.5$ ns, $t = 5.0$ ns, and $t = 10.0$ ns after initiation of crack propagation

This pattern is slightly distorted over time, because of the high acoustic velocity along the fiber axis causing a faster spreading along the fiber axis than in the surrounding matrix material. The dominant radiation direction is found to be in the in-plane direction which is in good accordance with results from previous source models [24, 35].

Despite of the high macroscopic failure load and the intense acoustic velocity magnitudes seen in Fig. 4.40, the amplitudes of the detected acoustic emission signals are substantially lower than for inter-fiber failure or delamination (cf. Fig. 4.41). For the three acoustic emission signals detected at 25 mm to the source position at three different angles, the values are found to be three to four orders of magnitude lower than for the previous cases. This is caused by the substantially lower fracture surface area in the present case. Compared to the previous fracture surface area of the order of $1 \times 10^{-6}\,\mathrm{m}^2$ in the present case, the fiber cross-section is of the order of only $1 \times 10^{-10}\,\mathrm{m}^2$ having a difference of about four orders of magnitude. The main cause for the difference in intensity for the 0° orientation and the 45° and 90° orientation is the unidirectional arrangement of the fiber and the according radiation pattern having a dominant radiation in parallel to the fiber axis. Independent of the angle, the detected acoustic emission signals are dominated by high-frequency contributions, which are caused by the short rise-time of the source.

Compression Parallel $-\sigma_{\|}$

As further load case for single filament failure, the compression load parallel $-\sigma_{\|}$ to the fiber axis is investigated. As seen from the micromechanical representation in Fig. 4.42, this type of failure usually causes fragmentation of the fiber filaments due to the occurrence of band kinking prior to failure. Therefore the resulting AE signal is caused by failure of more than one material and a mixed mode fracture of the individual microscopic components. Being a representative case of this class of failure, the kinking of one single fiber filament was investigated.

As seen from Fig. 4.42, the RVE was embedded in the laminate at the center position of the plate. To facilitate the visibility of the relevant modifications, the fibers of the RVE surrounding the failing filament are not shown in Fig. 4.42. The single fiber filament subject to failure was modified to include two pores at opposite edges along the z-axis. Due to this pore configuration, a skew microscopic load situation occurs, which causes kinking of the designated fiber filament.

Using a macroscopic stress level of 800 MPa at the designated fracture planes, a microscopic load of $-\sigma_{\|} = 1800\,\mathrm{MPa}$ was observed. As failure criterion in this case, the von Mises equivalent stress as defined in (4.6) was used for this case. Similar to the previous case, the presence of the modeled pores initiates crack growth at these positions as seen from the false-color plots of the degradation function in Fig. 4.43. The crack propagation takes about 1.3 ns until the full extent of the fracture surface is reached, which is the identical duration as observed for the

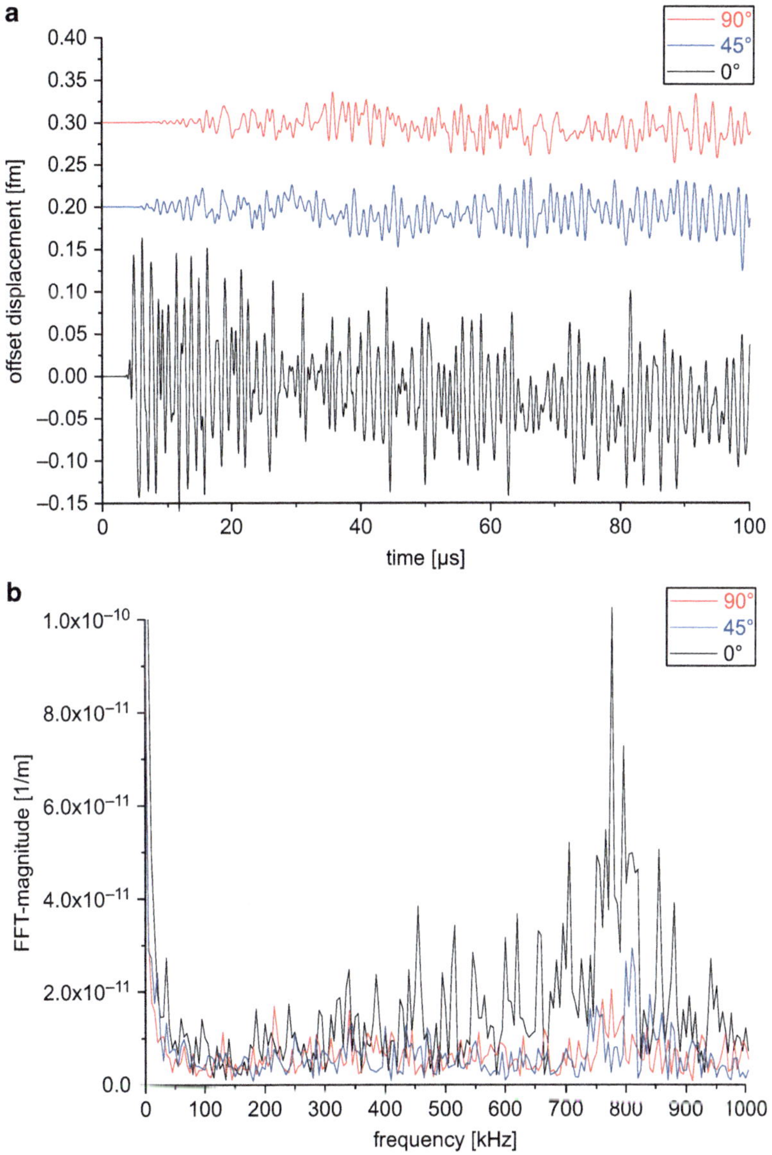

Fig. 4.41 Acoustic emission signals of fiber failure due to $\sigma_{\parallel}$ detected in unidirectional plate at 0°, 45°, and 90° relative to fiber orientation at 25 mm distance to source position in time domain (**a**) and frequency domain (**b**)

tensile fiber failure case. This seems well applicable since the crack speed is dominated by the carbon fiber properties in both cases.

The signal propagation after filament failure is visualized as acoustic velocity magnitude in Fig. 4.44 for three distinct time steps after failure initiation. Different

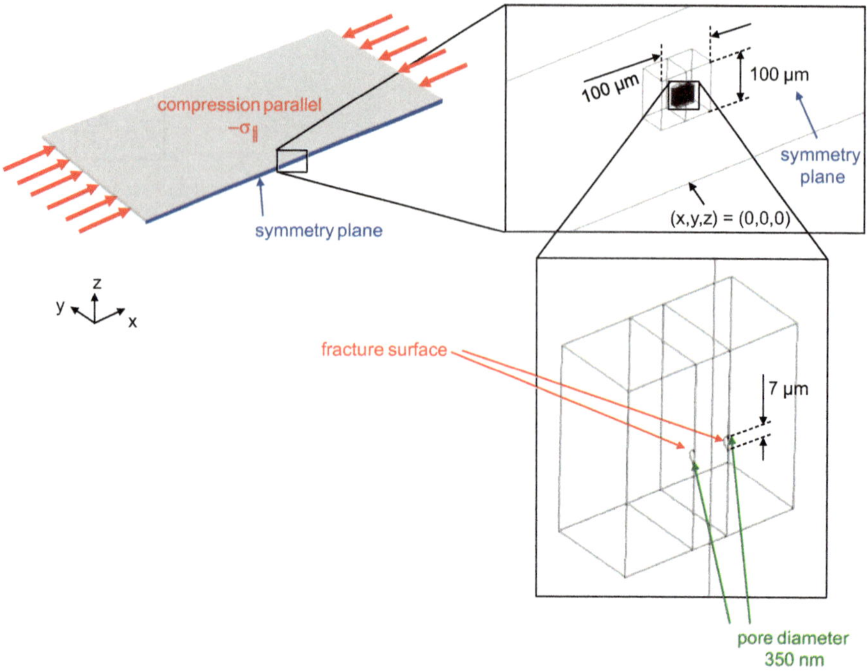

Fig. 4.42 Model configuration including position of symmetry plane, details of RVE position, dimensions and details of crack region

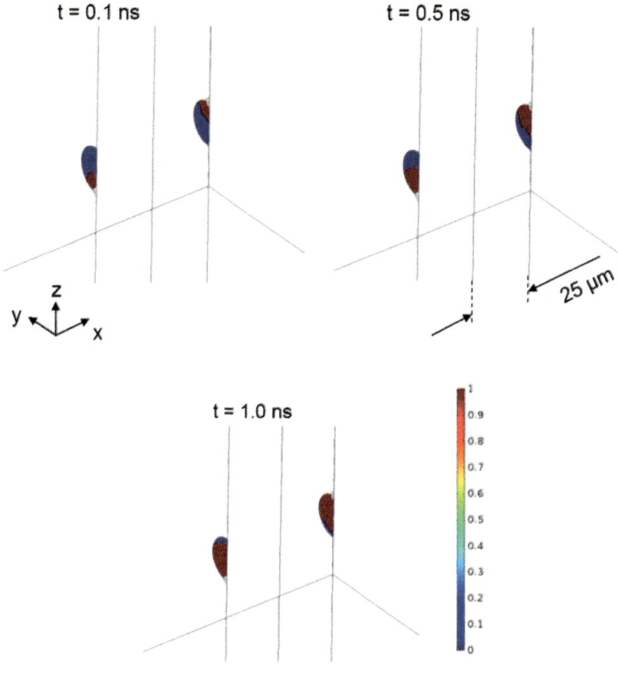

Fig. 4.43 Extension of fracture surface during crack growth at $t = 0.1$ ns, $t = 0.5$ ns, and $t = 1.0$ ns after initiation of crack propagation

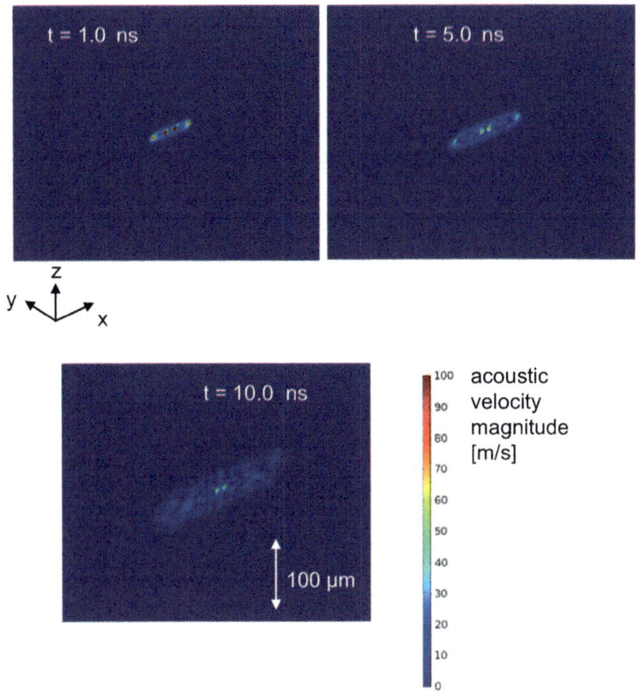

Fig. 4.44 Plot of acoustic velocity magnitude during crack growth at $t = 1.0$ ns, $t = 5.0$ ns, and $t = 10.0$ ns after initiation of crack propagation

to the tensile load case, the presence of two moving fracture surfaces causes formation of two dipole-like radiation patterns instead of one. Their superposition causes the wave propagation in the near field seen in Fig. 4.44. The dominant radiation direction is found to be in the in-plane direction with stronger intensity than for the tensile load case.

The resulting acoustic emission signals of compressive fiber failure are shown in Fig. 4.45. They obey the same trend as function of angle like the signals of tensile fiber failure, but tend to be larger in amplitude by almost one order of magnitude. This can be partially attributed to the formation of two times the fracture surface when compared to the tensile load case. Another factor is the tilted movement of the fracture zone (including the fractured filament and the surrounding matrix) which turns into larger microscopic displacements than for the tensile fiber failure case.

4.2.3.4 Fiber Bridging

Beyond the individual failure of the matrix material, the fibers, or the interface, there are several failure mechanisms involving a specific arrangement of micromechanical processes. As discussed in the previous section, compressive

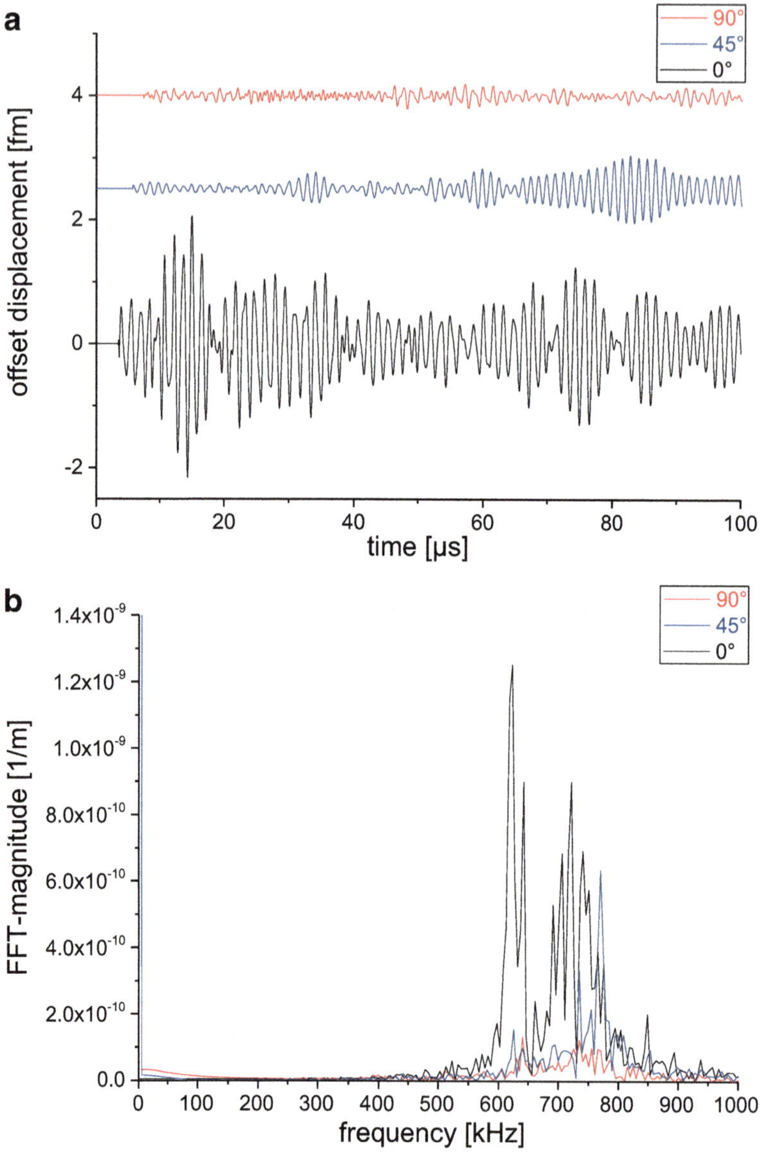

Fig. 4.45 Acoustic emission signals of fiber failure due to $-\sigma_{\|}$ detected in unidirectional plate at 0°, 45°, and 90° relative to fiber orientation at 25 mm distance to source position in time domain (**a**) and frequency domain (**b**)

fiber failure is such a process. Another failure mechanism in this context is fiber bridging. Here the crack propagates transverse to the fiber axis, but the individual filaments remain intact during crack propagation. Instead of fiber failure, a combination of fiber-matrix debonding, matrix cracking, and interfacial failure occurs

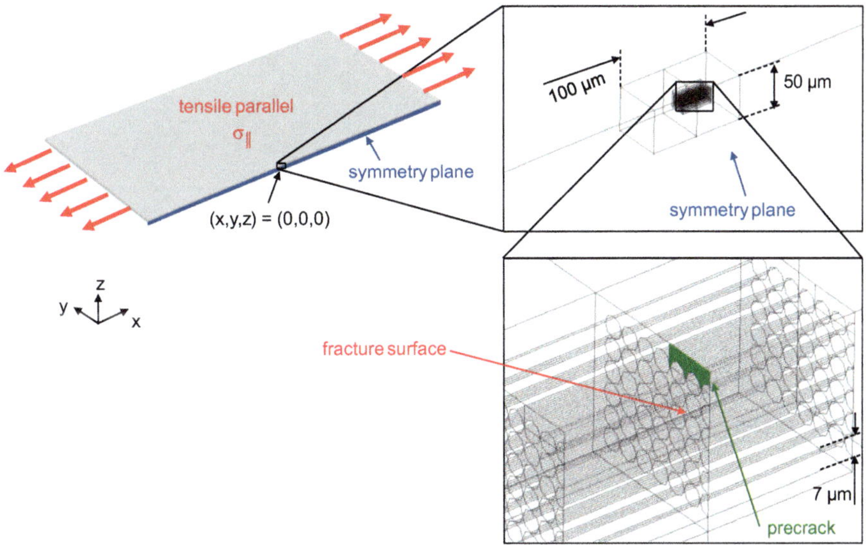

Fig. 4.46 Model configuration including position of symmetry plane, details of RVE position, dimensions and details of crack region

(cf. Sect. 2.1). To model this phenomenon on the microscopic scale, the RVE approach was used, and the von Mises criterion was applied to evaluate failure of the matrix material within the RVE. As macroscopic load configuration, the tensile load parallel case $\sigma_{\parallel}$ was studied. To initiate the crack used for fiber bridging, a precrack starting at the surface position of the laminate was chosen as indicated in Fig. 4.46.

The selected position and dimensions of the precrack assure that due to a macroscopic stress level of 2000 MPa, a microscopic load of $\sigma_{\parallel} = 52\,\text{MPa}$ is reached at the edge of the precrack. This fulfills the fracture condition for matrix cracking and thus drives the crack propagation plotted in Fig. 4.47. As seen for three distinct time steps after failure initiation, the growth of the fracture plane is not straightforward along the *z*- or *y*-axis, but is influenced by the stress concentration effect of the nearby fibers. Accordingly, the full fracture surface is finally established by merging of three individual matrix areas between the surrounding fibers after 5 ns. Since the model is calculated with symmetry conditions at the *xz*-plane, this configuration would approximate an elliptical crack front, which includes the precrack area.

After crack growth, the signal spreads out and propagates along the free surface and into the volume of the laminate. As seen in the acoustic velocity magnitude from Fig. 4.48, the intensity of the wave is comparatively low, and the corresponding acoustic emission signals shown in Fig. 4.49 are among the lowest intensities found for the cases studied herein. This is readily understood from the small size of the fracture zone and the reduced mobility of the crack wall movement. The latter is bridged by the fibers and these in turn limit the deflection of the

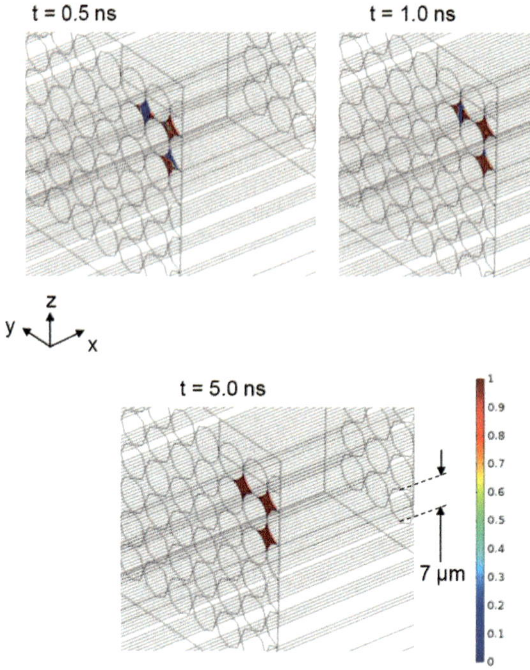

Fig. 4.47 Extension of fracture surface during crack growth at $t = 0.5$ ns, $t = 1.0$ ns, and $t = 5.0$ ns after initiation of crack propagation

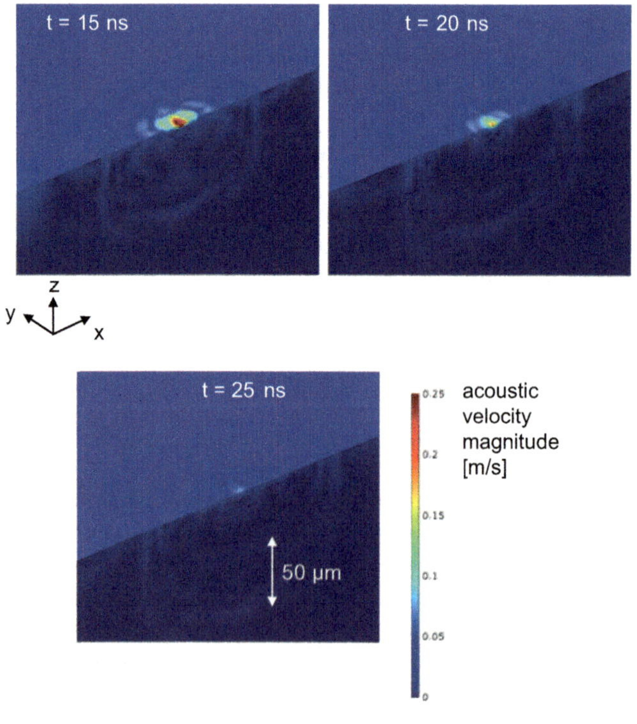

Fig. 4.48 Plot of acoustic velocity magnitude during crack growth at $t = 15$ ns, $t = 20$ ns, and $t = 25$ ns after initiation of crack propagation

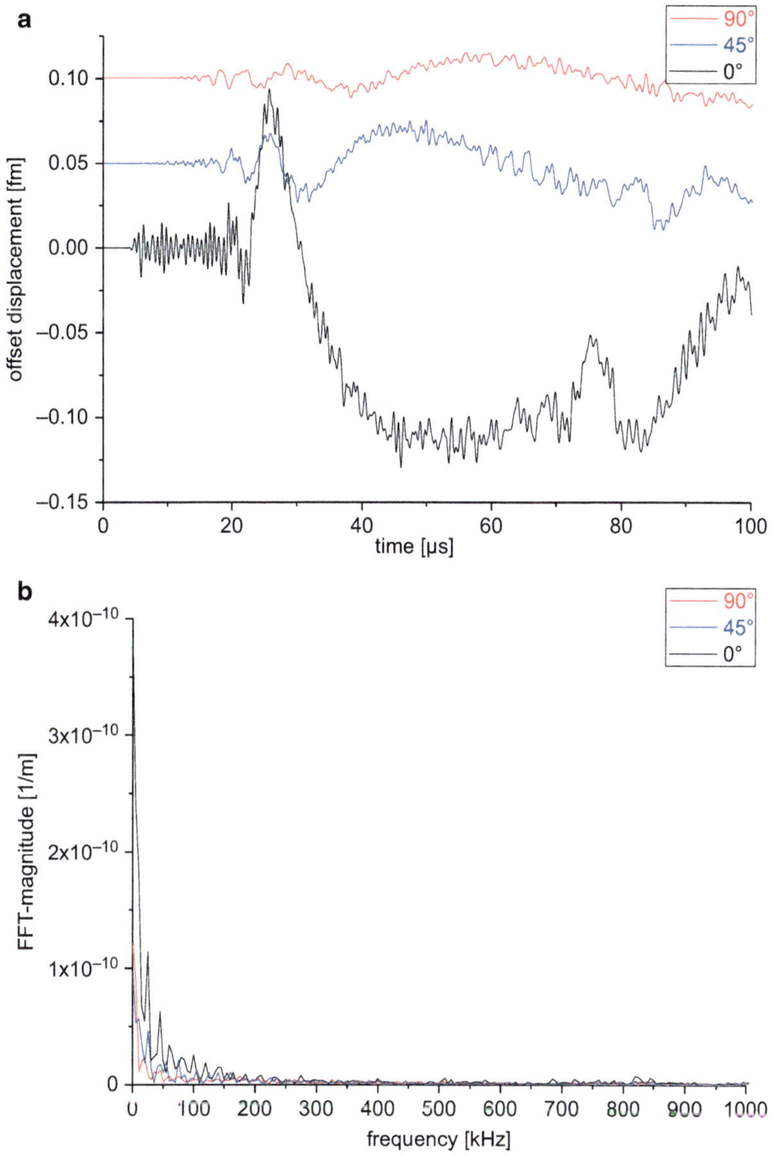

Fig. 4.49 Acoustic emission signals of fiber bridging failure due to $\sigma_{\parallel}$ detected in unidirectional plate at 0°, 45°, and 90° relative to fiber orientation at 25 mm distance to source position in time domain (**a**) and frequency domain (**b**)

crack walls along the fiber axis. Hence, the acoustic emission magnitudes are expected to be rather low.

Especially for the acoustic emission signal detected along the 0° axis, some high-frequency contributions are visible for the S_0 mode at the beginning of the signal shown in Fig. 4.49. These are similar to the frequencies seen for the fiber breakage cases, which are in turn linked to the size of the fracture zones. But the dominant contribution in the present case are lower frequencies originating from the A_0 mode contributions starting after approximately 10 µs. These are predominantly due to the chosen source position at the surface of the laminate. Thus a decrease of their intensity is expected if fiber bridging would occur inside the laminate (cf. Sect. 4.2.3.7).

4.2.3.5 Influence of Crack Length

Beyond the distinct relationship of individual source types and their acoustic emission signals, it is worthwhile to consider some further factors of influence. First a comparison is made between the motions of the crack surface caused by three different crack propagation lengths for inter-fiber failure in the $\sigma_\perp$ load case. Therefore the y-displacement of the crack surface position is evaluated following the procedure described in Sect. 4.2.3.1. To this end, the length of the thin elastic layer is adjusted to allow crack propagation only to the designated maximum length. The respective evaluation of the y-displacement of the crack surface is shown in Fig. 4.50 as comparison of the three cases. During the duration of crack propagation, the y-displacement increases almost linearly. However, at the moment of crack arrest, the y-displacement does not stop or cease, but increases further until a maximum value is reached as discussed previously in Sect. 4.2.3.1. Subsequently, the crack surface starts to vibrate and finally settles at a new equilibrium position. The predictions of the y-displacement derived from the theory of Green [40] are

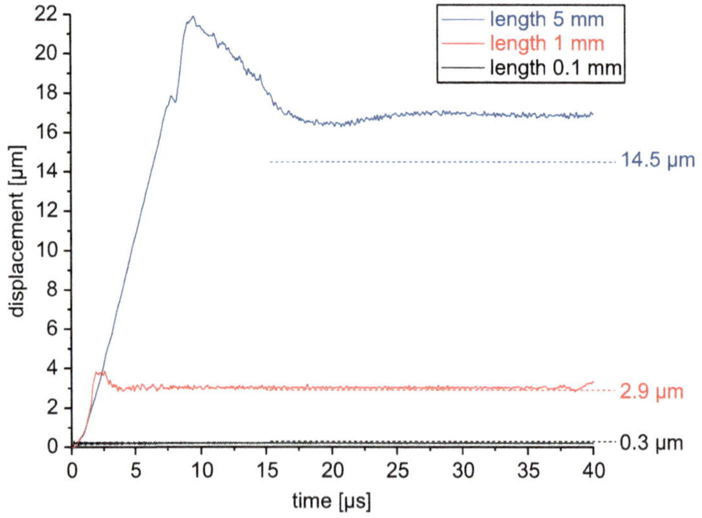

Fig. 4.50 Evaluation of source displacement in y-direction for $\sigma_\perp$ load case with three different maximum propagation lengths and dashed lines to mark predictions from analytical theories

marked as dashed lines in Fig. 4.50. As can be seen from the comparison, in the initial part, these values are systematically lower than the present model predictions. This originates from the underlying assumptions in the theory. In [40] static crack opening is assumed, i.e., the length of the crack is already present as a flaw in the material and merely opens due to an external force. If the same assumptions are made in the present model, the achieved deformation state is in good agreement to the values predicted by [40]. However, the dynamic crack propagation seems to generate initial acoustic emission amplitudes larger than predicted by analytical theories, which has also been reported for the case of isotropic materials before [36].

4.2.3.6 Influence of Fracture Surface

Another factor of influence contributing to the intensity of the acoustic emission signal is the size of the fracture surface. To demonstrate this effect, the tensile parallel load case $\sigma_{\|}$ for fiber failure is chosen. As major difference to the failure model shown in Fig. 4.38, the position of the pore acting as precrack and the number of failing filaments were varied. Studies are carried out for failure of one, three, and five filaments. In this context, the number of filaments failing simultaneously determines the effective fracture surface as shown in blue in Fig. 4.51.

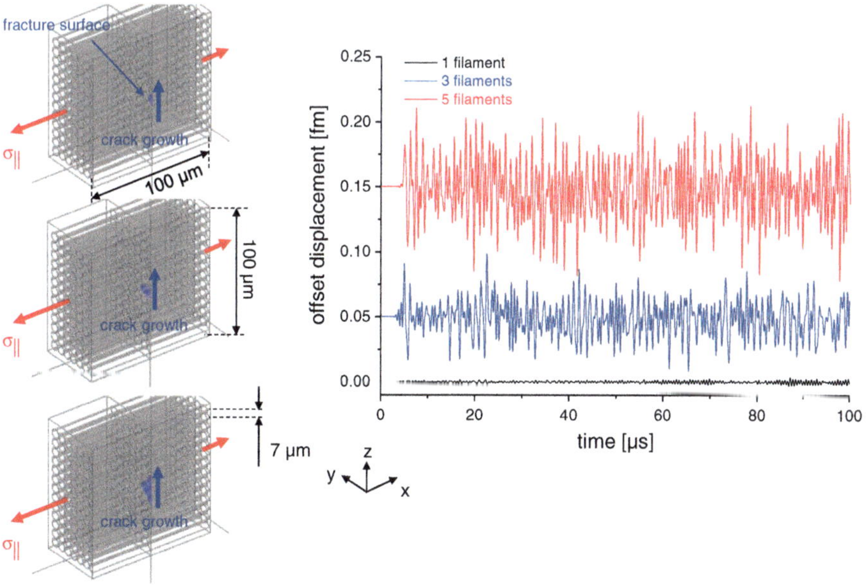

Fig. 4.51 Schematically representation of fracture surface within the RVE marked in *blue* (*left side*) and evaluation of acoustic emission signals due to fiber failure with one, three, and five filaments in unidirectional laminate at 25 mm distance to source position at 0° angle relative to the fiber orientation (*right side*)

The position of the pore was chosen at the bottom of the lowest fiber filament, and crack propagation was found to be mostly oriented along the z-axis direction as indicated by the arrow in Fig. 4.51. Based on the generalized theory of acoustic emission [21, 22], a larger fracture surface translates into larger vibrating volume ΔV if the same stress level is applied to the fracture surface. Since the external load level was kept identical for all model configurations at $F_{\parallel} = 263\,\text{kN}$, the dominant change is solely due to the change in fracture surface. As seen from the detected acoustic emission signals in Fig. 4.51, this change in fracture surface directly translates into a change in signal amplitude. Here the higher the numbers of failing filaments, the higher the signal amplitudes were calculated. The observed frequencies of the signals are still comparable, since the rise-time is still in the same order of magnitude.

4.2.3.7 Influence of Depth Position

So far all crack positions were chosen symmetric with respect to the neutral axis of the plate in order to avoid any influences due to asymmetric source positions as described in Sect. 4.2.1. However, for the unidirectional fiber reinforced composite used in this study, there is no preferential position for fiber breakage along the z-axis. Therefore, various depth positions below the surface were studied to compare the resulting out-of-plane displacement signals as seen in Fig. 4.52. Except for the position of the fiber breakage sources, all model parameters are chosen identical to the description given in Sect. 4.2.3.3 for single fiber filament failure.

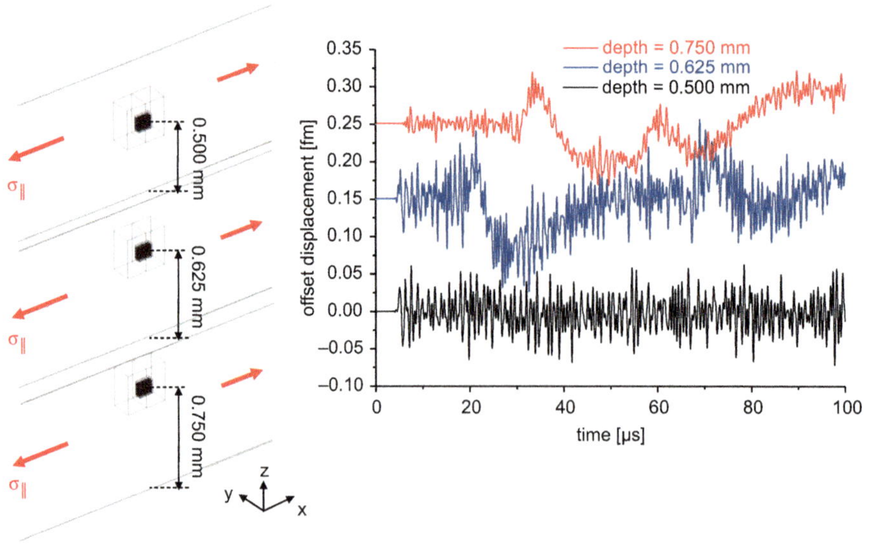

Fig. 4.52 Evaluation of acoustic emission signals due to fiber failure at three different depth positions in unidirectional laminate at 25 mm distance to source position at 0° angle relative to the fiber orientation

Starting at the midplane of the plate ($z = 0.50$ mm), the RVE region was gradually moved to a position shortly below the surface with the fiber breakage being at $z = 0.75$ mm (cf. Fig. 4.52). For better visibility only one half of the RVE region is shown. For the tensile parallel load case $\sigma_{\|}$, failure of one fiber filament results in the out-of-plane displacement signals shown in Fig. 4.52. As expected by the principles of modal AE, the gradual shift from the midplane to the surface causes a noticeable increase of the A_0 mode. However, there is still noticeable contribution of the high-frequency components propagating as S_0 mode. In fact, there is no reduction of bandwidth for the three signals. But there is a distinct shift of the relative intensity of these contributions, causing higher contributions at lower frequencies with increased distance to the midplane. Here the A_0 mode also causes increased absolute signal amplitudes, which is beneficial from the detection point of view.

4.2.4 Detectability of Failure Mechanisms

Based on the multitude of calculated signals for different failure mechanisms presented in the previous section, it is possible to assess the detectability of different failure mechanisms by AE measurements. As will be demonstrated in Sect. 4.4, the sensitivity of AE sensors is extraordinary high, but still there are physical limitations. Various authors claim the expected out-of-plane sensitivity of acoustic emission sensors to be in the order of 10^{-13}–10^{-14} m [42, 43] and thus will act as ultimate detection limit in the following. Using the investigations of the previous section, all signals are evaluated with respect to their maximum absolute amplitude within the first 100 μs. This procedure is identically carried out for signals of all source mechanisms and for all detection angles to account for these effects as scatter of the obtained values.

The aim of this subsection is twofold. First, the general relationship between the detectable acoustic emission signal amplitude and the physical size of the failure mechanism is established. Second, the aspect of detectability in terms of the distance and orientation relative to the source position is discussed.

Figure 4.53 presents the results of the calculated out-of-plane displacement of the acoustic emission signal plotted as function of the "crack area." For the quantification of the acoustic emission signal, the absolute value of the out-of-plane displacement is used, since detectability by acoustic emission sensors does not distinguish between positive or negative amplitude values. The "crack area" is quantified as area of the fracture surface ΔA finally reached by the respective failure type (i.e., for single fiber filament failure, this is equivalent to the cross-section of one filament). In general, the vibrating volume ΔV is the acoustically relevant quantity. The vibrating volume is related to ΔA by considering the respective displacement normal to the surface A. However, the quantity ΔV cannot be readily derived from simple geometrical considerations and therefore does not constitute a

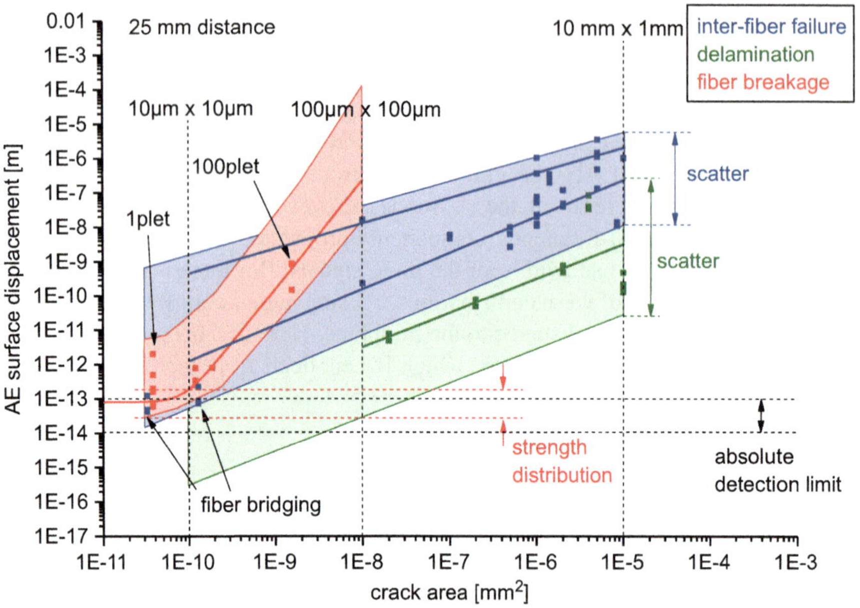

Fig. 4.53 Out-of-plane component of acoustic emission signal detected at 25 mm distance to the source for different failure types as function of the crack area in 1 mm thick T700/PPS plate

useful quantity to assess detectability. Instead, the size of the fracture zone ΔA can be understood in close analogy to the flaw size as typically used in NDT standards and therefore allows better interpretability of Fig. 4.53. To facilitate reading further, some exemplary crack dimensions representative of the scales are noted in Fig. 4.53 as black vertical lines.

As expected from the physical dimensions of the fracture zones modeled, the detectable crack areas span five orders of magnitude. Similarly, the detected signal amplitudes cover eight orders of magnitude. Hence, both axes are shown in a logarithmic scale to allow for better visibility of the individual signals. There is a general trend common to all failure mechanisms showing larger signal amplitudes with larger crack areas. Based on the fundamental considerations made in (4.1) and (4.2), this finding is not surprising.

For the failure mechanism of fiber breakage, the variation of fiber filament strength causes a characteristic distribution of expected AE signal amplitudes for one particular crack area since the latter was kept constant. This is indicated exemplarily for the case of one fiber filament. Additional scatter arises from the intensity radiated to the different angles relative to the source. Both effects define a mean value of fiber filament failure for the present case and a respective scatter expected due to the underlying Weibull strength distribution and the extreme cases of observation angles relative to the source position. For an increasing number of fiber filaments failing simultaneously, the detected AE signal amplitude increases as does the crack area. Given the applied macroscopically tensile stress is kept constant, a power law fit indicated by the solid red line was found to describe the relationship between these two quantities best. This allows a slight extrapolation of this relationship beyond the case of 100 fibers failing simultaneously (which was the largest number studied by numerical methods). Accordingly, the extrapolation of the scatter band establishes an approximate area in Fig. 4.53 where signal amplitudes of all fiber failures at a detector distance of 25 mm are expected regardless of the orientation of the detector position relative to the source. It is worth noting that at the distance of 25 mm, some of the weak fiber filament failures are approaching the absolute detection limit of 10^{-14} m and thus may not be detected in an experimental setup.

Signals of fiber bridging are found close to the intensities of single fiber failure and due to the crack areas studied herein also have closely related crack areas. In average these signal amplitudes are even lower, which is mostly due to the lower macroscopic tensile load required to generate this failure mechanism and the constrained movement due to the fibers bridging the crack walls. Hence these signals also partially fall below the detection limit at 25 mm distance.

The different cases of inter-fiber failure studied in the previous section typically show larger crack areas and, consequently, also larger signal amplitudes. For a particular source mechanism, there is an almost linear relationship in the log-log scale as indicated by the solid blue lines as a result from two power law fits for the cases of inter-fiber failure due to $\sigma_{\perp}$ and the inter-fiber failure due to $\tau_{\parallel\perp}$, both with varying crack lengths. However, it is difficult to spot a general relationship covering all of the different load cases. Also, data points in Fig. 4.53 originate from all angles of orientation relative to the source position as well as results from the unidirectional plate and the cross-ply plate. Therefore the blue rectangle is sufficient to cover all cases of inter-fiber failure studied herein and can thus be used as areal representation of inter-fiber failure. Within this area all AE signals of inter-fiber failure detected at 25 mm to the source should locate, regardless of their specific type and angular position relative to the detector. The lower and upper boundary of the rectangle in terms of the crack area are chosen arbitrarily at dimensions approaching the range of the microscopic constituents of the composite on one end and at dimensions approaching the cross-section of the plate investigated on the other end. Hence these boundaries do not constitute absolute limits of the relationship, but are limits dictated by the chosen size of the model.

For the case of inter-ply delamination, results of the unidirectional plate and the cross-ply plate are shown in Fig. 4.53. Marked in green color, these seem to generally exhibit lower AE signal amplitudes for the same crack area when compared to inter-fiber failure. The result of a power law fit for the case of inter-ply delamination due to $\tau_{\parallel\perp}$ with varying crack area is shown as solid green line. Based on the same considerations as for inter-fiber failure, the green rectangle is established as area of largest likelihood for the location of AE signals due to inter-ply delamination.

To summarize the findings shown in Fig. 4.53, the marked areas related to the occurrence of inter-fiber failure, delamination, and fiber breakage include all extreme cases of signals as arising due to:

(a) Variation of detector position by an angle from 0° to 90°
(b) Signals of a unidirectional and a cross-ply laminate
(c) Variation of source strength at identical crack area
(d) Variation of source depth position within the laminate

As seen from Fig. 4.53, reasonable overlap between the calculated AE signal amplitudes for the different failure mechanisms occurs. So based on this investigation, there is a general trend for the intensity range of the individual mechanisms, but there is no decisive boundary to separate failure mechanisms based on their intensity for the types of failure mechanisms studied herein. Such classification can instead be achieved by means of suitable frequency analysis as discussed in Sect. 4.5.

To allow more general conclusions on detectability of AE signals in fiber reinforced materials, the results of Fig. 4.53 may be further generalized with respect to the attenuation of signals with distance to the source position. Considering the signal attenuation, this allows establishing a 3D detectability diagram as derivation of the 2D cross-section at 25 mm distance shown in Fig. 4.53. In the following a value of 0.1 dB/mm was chosen as extreme case found for the T700/PPS plates investigated.

In Figs. 4.54, 4.55, and 4.56, an extrapolation of the envelopes of each failure type presented in Fig. 4.53 is extrapolated to a distance of 500 mm between source and detector. For comparison the gray plane shows the upper level of the absolute

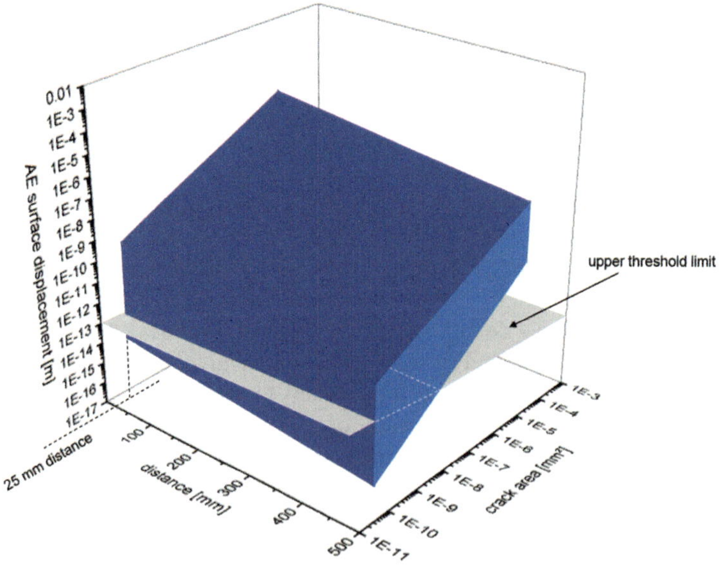

Fig. 4.54 Extrapolation of envelope of out-of-plane component of acoustic emission signal due to inter-fiber failure to 500 mm detection distance

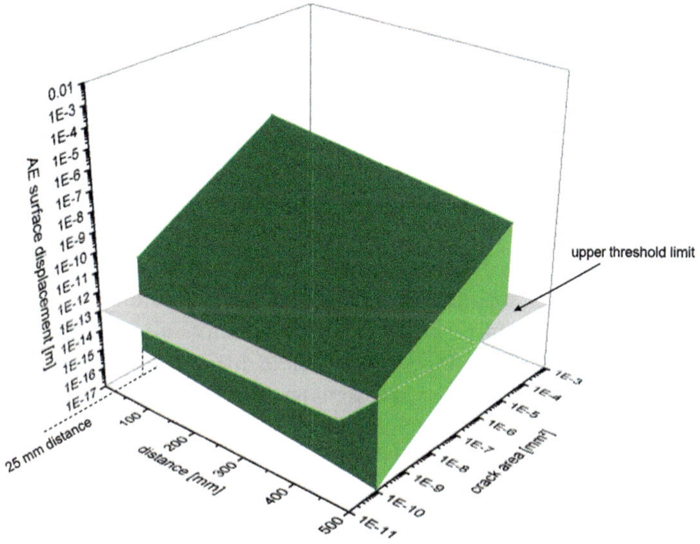

Fig. 4.55 Extrapolation of envelope of out-of-plane component of acoustic emission signal due to interfacial failure to 500 mm detection distance

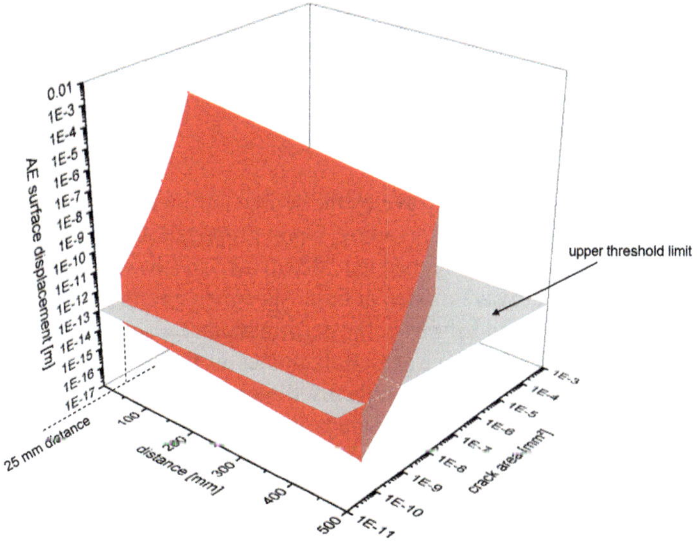

Fig. 4.56 Extrapolation of envelope of out-of-plane component of acoustic emission signal due to fiber breakage to 500 mm detection distance

threshold limit at 10^{-13} m. As seen in Fig. 4.54 for the case of inter-fiber failure, a part of the signals falls already below this limit at 25 mm distance. Accordingly, due to attenuation a larger fraction of signals falls below this limit at higher distances. The 3D envelope shown in Fig. 4.54 may be used to estimate if a certain crack size will be detectable by means of AE measurements. As mentioned previously, the envelope was found to be applicable for all angles of detection relative to the source and unidirectional and cross-ply stackings. However, results are based on the material properties of T700/PPS, and the attenuation coefficient was assumed to be linear and isotropic and constitutes an extreme value. Hence, the envelope may be considered as conservative approach for detection probability of the AE method.

Similar diagrams are shown in Figs. 4.55 and 4.56 for the cases of interfacial failure and fiber breakage. The case of interfacial failure may be readily understood in analogy to the matrix cracking case, yet the overall number of signals that may fall below the detection limit was assessed to be higher. Accordingly, less PoD may be expected at higher distances as well. For the extrapolation of the envelope of fiber breakage shown in Fig. 4.56, one remarkable difference arises. It was already shown in Fig. 4.53 that a part of the signals corresponding to failure of weak filaments falls below the detection limit already at 25 mm distance. Thus Fig. 4.56 predicts the loss of all signals of single fiber filament failure after a certain distance to the source. Since the occurrence of fiber breakage is a crucial indicator to the structural health of a fiber reinforced composite, this is of vital importance.

For the T700/PPS material investigated here, this distance was evaluated at 400 mm distance. Keeping in mind the conservative assumptions made regarding the attenuation coefficients, this distance might be estimated slightly too short but is certainly within a realistic range.

4.3 Wave Propagation

After the emission of an acoustic wave by the source, the signal propagates into the solid and is subject to geometric spreading. Upon propagation the wave is affected by dispersion and attenuation effects due to dissipative mechanisms in the material and formation of distinct wave modes in finite specimen geometries.

According to Landau and Lifschitz, the deformation of a solid causes heat to be generated at the deformation zone which fluctuates in time and space. But if the spatial heat propagation is slower than the oscillatory movements within the solid, the deformations can assumed to be adiabatic [16]. For this case, the theory of elastic waves describes the propagation of deformations within the solid for small initial deformations. This yields the basic theoretical formulation for the understanding of acoustic emission given by the theory of elastic waves, which in general requires solving the momentum balance equation for the case of negligible Lagrangian inertia:

$$\rho \frac{\partial^2 \boldsymbol{u}}{\partial t^2} = \frac{\partial \boldsymbol{\sigma}}{\partial \boldsymbol{r}} \tag{4.11}$$

Here ρ is the material density, $\boldsymbol{u}$ is the displacement, and $\boldsymbol{\sigma}$ is the stress tensor derived from the generalized formulation of Hooke's law:

$$\sigma = \mathbf{C} \cdot \varepsilon \tag{4.12}$$

This equation, also known as constitutive relation, couples the stress tensor $\boldsymbol{\sigma}$ with the deformation tensor $\boldsymbol{\varepsilon}$ using the elasticity tensor $\mathbf{C}$ for the case of small elastic deformations in a homogeneous medium. In general $\mathbf{C}$ is a fourth-order tensor with 81 elastic coefficients. Since the deformation tensor $\boldsymbol{\varepsilon}$ is symmetric, the elasticity tensor $\mathbf{C}$ exhibits the same symmetry constraints regarding index permutation, which reduces the number of independent coefficients for a solid material to 21 [16]. The occurrence of symmetries within the solid reduces the number of independent coefficients further. For the case of orthotropic materials like a composite laminate, the elasticity tensor $\mathbf{C}$ can be reduced to a 6×6 matrix with nine independent elasticity coefficients and one angle defining the axis orientation in the xy-plane. Using the Voigt notation, the stress tensor $\boldsymbol{\sigma}$ can be reduced to a vector with six independent components, and the deformation tensor $\boldsymbol{\varepsilon}$ is also written as a vector with six independent components. Hence, (4.12) may be rewritten in matrix form as

$$\begin{pmatrix} \sigma_1 \\ \sigma_2 \\ \sigma_3 \\ \sigma_4 \\ \sigma_5 \\ \sigma_6 \end{pmatrix} = \begin{pmatrix} C_{11} & C_{12} & C_{13} & 0 & 0 & 0 \\ C_{12} & C_{22} & C_{23} & 0 & 0 & 0 \\ C_{13} & C_{23} & C_{33} & 0 & 0 & 0 \\ 0 & 0 & 0 & C_{44} & 0 & 0 \\ 0 & 0 & 0 & 0 & C_{55} & 0 \\ 0 & 0 & 0 & 0 & 0 & C_{66} \end{pmatrix} \cdot \begin{pmatrix} \varepsilon_1 \\ \varepsilon_2 \\ \varepsilon_3 \\ \varepsilon_4 \\ \varepsilon_5 \\ \varepsilon_6 \end{pmatrix} \tag{4.13}$$

If the solid is isotropic, (4.13) is reduced to

$$\begin{pmatrix} \sigma_1 \\ \sigma_2 \\ \sigma_3 \\ \sigma_4 \\ \sigma_5 \\ \sigma_6 \end{pmatrix} = \begin{pmatrix} \kappa + 2\varsigma & \kappa & \kappa & 0 & 0 & 0 \\ \kappa & \kappa + 2\varsigma & \kappa & 0 & 0 & 0 \\ \kappa & \kappa & \kappa + 2\varsigma & 0 & 0 & 0 \\ 0 & 0 & 0 & \varsigma & 0 & 0 \\ 0 & 0 & 0 & 0 & \varsigma & 0 \\ 0 & 0 & 0 & 0 & 0 & \varsigma \end{pmatrix} \cdot \begin{pmatrix} \varepsilon_1 \\ \varepsilon_2 \\ \varepsilon_3 \\ \varepsilon_4 \\ \varepsilon_5 \\ \varepsilon_6 \end{pmatrix} \tag{4.14}$$

The remaining two independent elements κ and ς are named Lame's constants. These completely describe all elastic parameters for isotropic materials like Young's modulus E_{Y}, compressive modulus E_{C} , shear modulus E_{S}, and Poisson's ratio ν.

$$E_{\mathrm{Y}} = \varsigma \frac{3\kappa + 2\varsigma}{\kappa + \varsigma} \tag{4.15}$$

$$E_{\mathrm{C}} = \kappa + \frac{2}{3}\varsigma \tag{4.16}$$

$$E_{\mathrm{S}} = \frac{\kappa(1-2\varsigma)}{2\varsigma} \tag{4.17}$$

$$\nu = \frac{\kappa}{2(\kappa+\varsigma)} \tag{4.18}$$

For such isotropic materials, (4.11) can be simplified using the Lame's constants to

$$\rho\frac{\partial^2 \boldsymbol{u}}{\partial t^2} = (\lambda+\varsigma)\nabla(\nabla\cdot\boldsymbol{u}) + \mu\boldsymbol{\Delta u} \tag{4.19}$$

The solution of (4.19) can then be formulated with scalar and vector potentials as

$$\boldsymbol{u} = \nabla \mathrm{X} + \nabla\times\boldsymbol{\psi}. \tag{4.20}$$

This yields the two independent wave equations in solid isotropic media:

$$\Delta \mathrm{X} = \frac{1}{c_{\mathrm{L}}^2}\frac{\partial^2\varsigma}{\partial t^2} \tag{4.21}$$

$$\boldsymbol{\Delta\psi} = \frac{1}{c_{\mathrm{T}}^2}\frac{\partial^2\boldsymbol{\psi}}{\partial t^2} \tag{4.22}$$

Equations (4.21) and (4.22) use the definition of the longitudinal and transversal propagation velocities c_{L} and c_{T} for an elastic wave given as

$$c_{\mathrm{L}} = \sqrt{\frac{E_{\mathrm{Y}}(1-\nu)}{\rho(1+\nu)(1-2\nu)}} \tag{4.23}$$

$$c_{\mathrm{T}} = \sqrt{\frac{E_{\mathrm{Y}}}{2\rho(1+\nu)}} \tag{4.24}$$

With respect to the boundary conditions, (4.21) and (4.22) can be used to obtain solutions for wave propagation in arbitrary geometries. In infinite isotropic and homogeneous media equations, (4.21) and (4.22) have solutions of the form of a monochromatic plane wave with an amplitude u, wave vector $\boldsymbol{k}$, and angular frequency ω:

$$\boldsymbol{u}(\boldsymbol{r},t) = u\cdot \mathrm{e}^{-i(\boldsymbol{kr}-\omega t)}. \tag{4.25}$$

Here the propagation velocities c_{L} and c_{T} are identical to the respective phase velocities c_{P} of the plane waves. The propagation of an elastic wave in this case can thus be described by independent longitudinal wave and transversal wave propagation.

In general, for non-monochromatic waves and anisotropic or finite media, the propagation velocity of the wave is the group velocity c_G defined as

$$c_G = \frac{\partial \omega(\mathbf{k})}{\partial \mathbf{k}} \tag{4.26}$$

Similarly, the phase velocity c_P is defined by the relationship between $\boldsymbol{k}$ and ω, which is called dispersion relation:

$$c_P = \frac{\omega(\mathbf{k})}{\mathbf{k}} \tag{4.27}$$

For geometries of finite extent, acoustic wave propagation has to take into account the respective boundary conditions. This ultimately results in wave propagation which is "guided" by the geometry of the propagation medium. The most prominent types of guided waves are the surface waves, which occur for a solid with free surface. These were first predicted by Lord Rayleigh in 1885 for the special case of an infinite solid with free surface. Later Stoneley and Love generalized Rayleigh's description for the case of a solid with an adjacent liquid half-space and planar interfaces located between two half-spaces [44, 45]. The guided waves in such structures with liquid-solid interfaces are hence often called Stoneley or Love waves. For infinitely extended solids with two parallel free surfaces, H. Lamb published the respective solutions for the wave equation in 1917 [46]. According to the geometry of the propagation medium, the respective guided elastic waves are commonly called "plate waves." These guided waves are the dominant mode of wave propagation in fiber reinforced structures, since these are often designed in the geometry of shell-like structures. In other types of geometries like rods or cylinders, similar types of guided waves exist. Due to the planar structure of the experimental specimens used for the present study, the propagation behavior of guided waves is very important and will be discussed in more detail in the following. For a closer review about Rayleigh waves, the reader is referred to [47], while other types of guided waves are comprehensively discussed in [48–50].

In order to understand guided waves, it is necessary to understand the behavior of elastic waves at interfaces. The relevant physical quantity for this case is the acoustic impedance Z, which is defined using the density ρ and the respective sound velocity c_i as

$$Z = \rho c_i. \tag{4.28}$$

If a planar acoustic wave arrives with orthogonal incidence to an interface, the wave is partially reflected and transmitted dependent on the acoustic impedance of the two materials. When the wave propagates from medium one with impedance Z_1 to medium two with impedance Z_2 as shown in Fig. 4.57, the coefficients for reflection Rc and transmission Tr are given by

$$\text{Re} = \frac{Z_2 - Z_1}{Z_2 + Z_1} \tag{4.29}$$

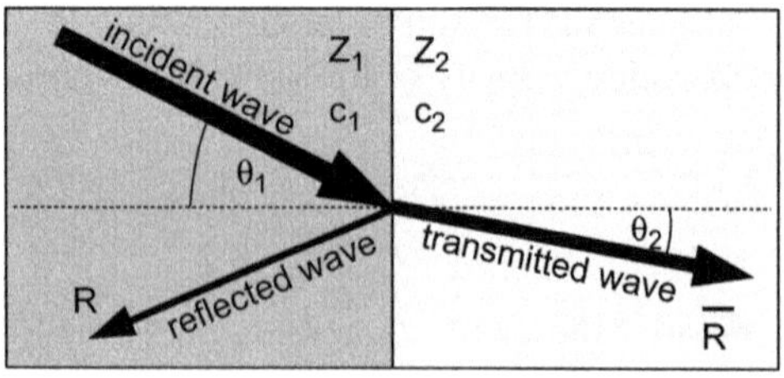

Fig. 4.57 Schematic representation of law of refraction for acoustic waves (based on [35])

$$\mathrm{Tr} = \frac{2Z_2}{Z_2 + Z_1} \tag{4.30}$$

For incidence angles other than $\theta_1 = 0°$, the law of refraction is applied. In contrast to electromagnetic waves, elastic waves usually undergo modal conversion during transmittance; thus, longitudinal waves can be completely or partially converted to transversal (shear) waves dependent on the incidence angle. In the most general case, the law of refraction can be written for arbitrary combinations of longitudinal and transversal sound velocities as

$$\frac{\sin(\theta_1)}{\sin(\theta_2)} = \frac{c_1}{c_2}. \tag{4.31}$$

For certain incident angles, the transmitted parts of the elastic waves are of very special nature. The most important among them are the surface waves which occur for elastic wave incidence from gaseous or liquid media to solids under certain angles. The total reflection condition for the longitudinal wave is termed first critical angle and causes excitation of a creep wave propagating along the surface of the solid having a propagation velocity identical to the longitudinal wave and a fast energetic decay. More important for practical applications is the Rayleigh wave which occurs under the second critical angle given by the total reflection condition of the transversal wave. L. Bergman proposed an approximation formula for the sound velocity of Rayleigh waves [51]:

$$c_R \approx \frac{0.874 + 1.12\nu}{1 + \nu} c_T \tag{4.32}$$

The velocity of the Rayleigh wave is generally slower than the transversal wave velocity of the solid. The amplitude of the Rayleigh wave decays exponentially with depth position below surface level. This is the reason why these types of waves are called "surface" waves as they propagate and interact only close to the solid surface.

If the dimension of the solid is reduced and approaches the shape of a plate with thickness z_d, the surface wave is further influenced and another type of guided wave propagation results. Such a plate with infinite extensions in the x and y direction and a finite thickness in z-direction defines boundary conditions for the solution of the acoustic wave equation which result in a classic eigenvalue problem. The respective solution of the acoustic wave equation is given by:

$$\begin{aligned} u_x &= f_x(z)\mathrm{e}^{-i(kx-\omega t)} \\ u_y &= 0 \\ u_z &= f_z(z)\mathrm{e}^{-i(kx-\omega t)} \end{aligned} \tag{4.33}$$

Equation (4.33) describes the propagation of an elastic wave in x-direction as function of wave number k, frequency ω, and as function of the z-position $f_i(z)$. The (x, y) origin is located in the medial plane of the plate as shown in Fig. 4.58. The specific type of guided plate waves described by (4.33) is referred to as Lamb waves which exhibit no displacement in y-direction for an infinite plate. Complementary to the Lamb waves, the so-called shear-horizontal waves exist, which have no displacement in the x- or z-directions and thus propagate only in y-direction.

If the boundary condition $\sigma = 0$ is applied at the surface level $\pm 0.5\,h$ of a plate with thickness h, this results in two characteristic periodic equations for symmetric and antisymmetric motion with respect to the plane $z = 0$ as schematically represented in Fig. 4.58:

$$\frac{\tan\left(0.5\,\widetilde{\mathrm{B}}h\right)}{\tan\left(0.5\,\widetilde{\mathrm{A}}h\right)} = -\frac{4\,\widetilde{\mathrm{A}}\widetilde{\mathrm{B}}k^2}{\left(k^2 - \widetilde{\mathrm{B}}^2\right)^2} \tag{4.34}$$

$$\frac{\tan\left(0.5\,\widetilde{\mathrm{B}}h\right)}{\tan\left(0.5\,\widetilde{\mathrm{A}}h\right)} = -\frac{\left(k^2 - \widetilde{\mathrm{B}}^2\right)^2}{4\,\widetilde{\mathrm{A}}\widetilde{\mathrm{B}}k^2} \tag{4.35}$$

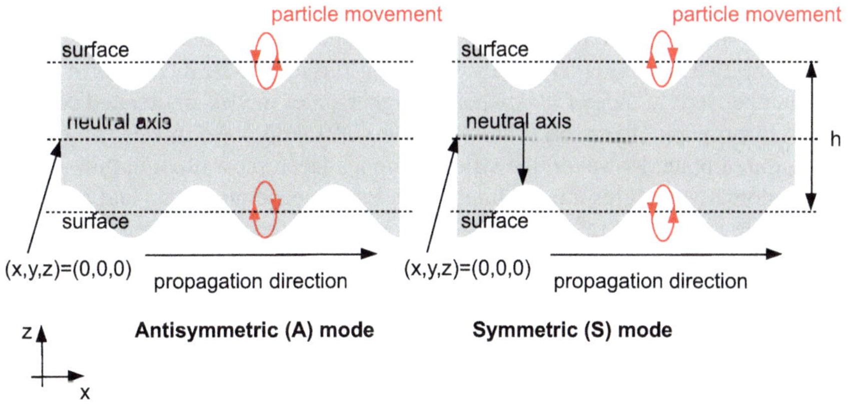

Fig. 4.58 Schematic representation of symmetric and antisymmetric Lamb wave modes propagating in x-direction with strongly emphasized surface motion (based on [35])

Using the abbreviations $\tilde{A}$ and $\widetilde{\mathrm{B}}$:

$$\widetilde{\mathrm{A}}^2 = \frac{\omega^2}{c_\mathrm{L}^2} - k^2 \tag{4.36}$$

$$\widetilde{\mathrm{B}}^2 = \frac{\omega^2}{c_\mathrm{T}^2} - k^2 \tag{4.37}$$

The equations (4.34) and (4.35) implicitly relate the wave number k and the frequency ω to the plate thickness h. Because the velocity of propagation is dependent on the frequency, this results in dispersion. For Lamb waves the relation between phase velocity c_P and group velocity c_G is thus given by

$$c_\mathrm{G} = c_\mathrm{P}(\omega h)\left(1 - \frac{1}{1 - \dfrac{c_\mathrm{P}}{\omega h \dfrac{\mathrm{d}c_\mathrm{P}}{\mathrm{d}(\omega h)}}}\right). \tag{4.38}$$

The relationship between frequency, plate thickness, and phase or group velocity is typically visualized within a set of dispersion curves for symmetric (S_i) and antisymmetric (A_i) Lamb wave modes as well as the shear-horizontal modes (SH_i). An example of such dispersion curves for a 1.0 mm thick unidirectional T800/913 composite plate using the material properties of Table B.1 in Appendix B is shown in Fig. 4.59. Since the wave propagation is dependent on the orientation, Fig. 4.59a displays the dispersion curves for a propagation direction parallel to the fiber orientation. For a propagation direction perpendicular to the fiber orientation, the dispersion curves appear differently as seen in Fig. 4.59b. This is due to the acoustical anisotropy introduced by the orientation of the fibers.

The dispersion curves show the characteristic dependency of the phase velocity and group velocity on the product $\omega \cdot h$ for the symmetric and antisymmetric wave modes. The two fundamental modes exhibiting the lowest frequency for a given wave number start at index $i = 0$, whereas higher-order modes are iterated consecutively. In principle, there are infinite numbers of Lamb wave modes, which can exist within a plate. However, the types of modes faced most often in thin-walled fiber reinforced structures are the fundamental symmetric mode (S_0) and antisymmetric mode (A_0), often referred to as extensional and flexural mode.

In guided wave testing, Lamb waves are typically excited by a plane elastic wave with appropriate angle of incidence ϕ as shown in Fig. 4.60. The excited wave modes in turn depend on the used frequency and plate thickness. If the wavelength λ of the used frequency is chosen inappropriately, the incident elastic wave is reflected at the top and bottom boundaries and propagates in the same way as in a bulk medium (Fig. 4.60, top). If the wavelength is sufficiently high or the thickness

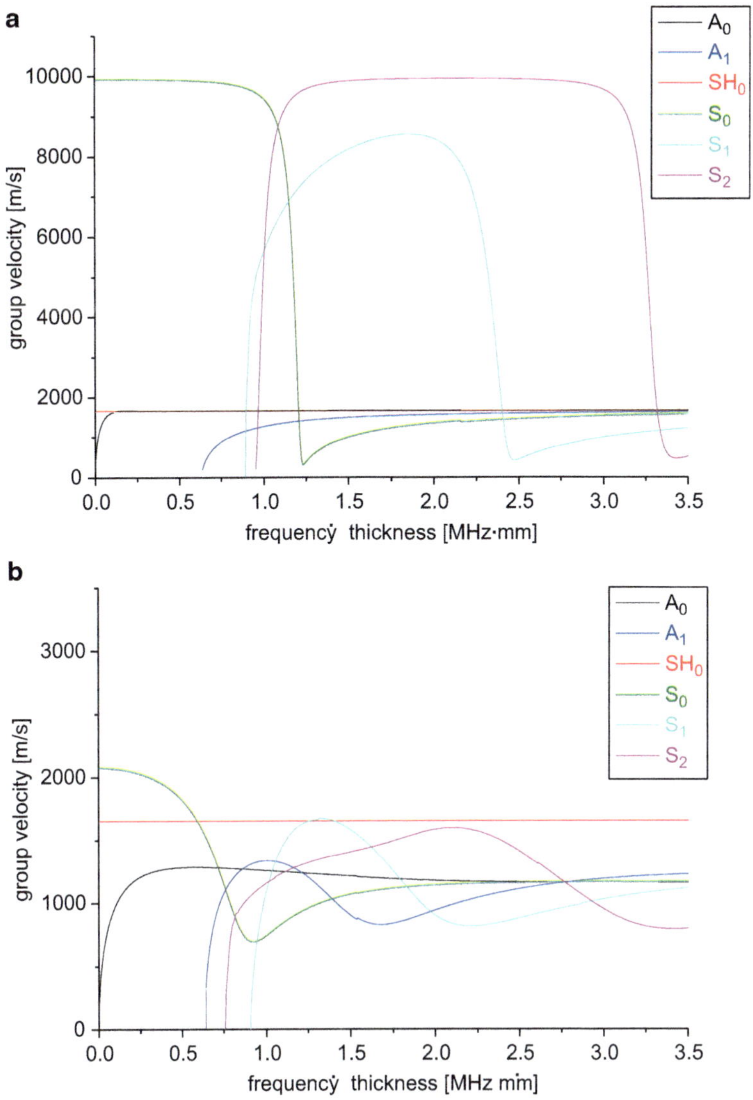

Fig. 4.59 Set of dispersion curves calculated for a 1.0 mm thick unidirectional fiber reinforced material for propagation direction parallel to the fiber orientation (**a**) and perpendicular to the fiber orientation (**b**)

z_d is small enough, at the same angle of incidence, interference occurs and causes excitation of Lamb waves (Fig. 4.60, bottom).

As discussed in Sect. 4.2, the excitation process of Lamb waves due to acoustic emission sources is different and is affected by various additional factors.

One particular difference to conventional guided wave testing is the bandwidth of AE sources. In guided wave testing, usually a center frequency is chosen, which

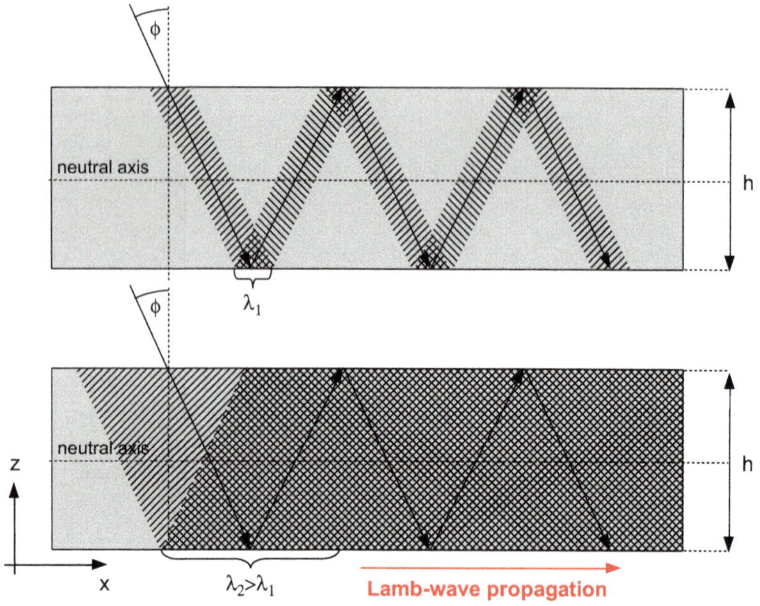

Fig. 4.60 Excitation by planar acoustic wave with angle of incidence ϕ and small wavelength λ_1 (*top*) and larger wavelength λ_2 causing Lamb wave excitation due to interference (*bottom*) (based on [35])

is enveloped by a Gaussian bell to yield a finite duration wave package. Although the frequency content of such an enveloped center frequency is theoretically infinite, still the dominating component is the center frequency. Therefore, one can select a particular propagation velocity of a wave mode at a particular frequency. Compared to that, AE sources typically show broad bandwidth, which is not controlled by the operator, but controlled by the AE source. Consequently, multimode excitation and variation of intensity as function of frequency may occur.

But even for equal intensity within a certain bandwidth, unequal excitation of the according wave modes is found (cf. [31, 34]). As seen in Fig. 4.61 for the calculations of a dipole source in aluminum plate as discussed in Sect. 4.2.1, the source bandwidth falls in the range up to several MHz. Despite of the source spectrum in Fig. 4.61a being smooth with frequency in the range up to 1 MHz, the branches of the S_0 mode and A_0 mode are not excited equally, but predominantly in local regions (see Fig. 4.61b). This translates in characteristic frequencies of the final AE signal as seen in the respective frequency spectrum of the AE signal (see Fig. 4.61a).

For other geometries as rods or cylinders, other types of guided waves exist, which can also be excited by acoustic emission sources. Common to all guided waves is the fact that these need a certain distance to fully develop. Typically this is dominated by the relevant geometrical boundary constraint such as the plate thickness or the rod diameter. Thus guided waves may only be interpreted after

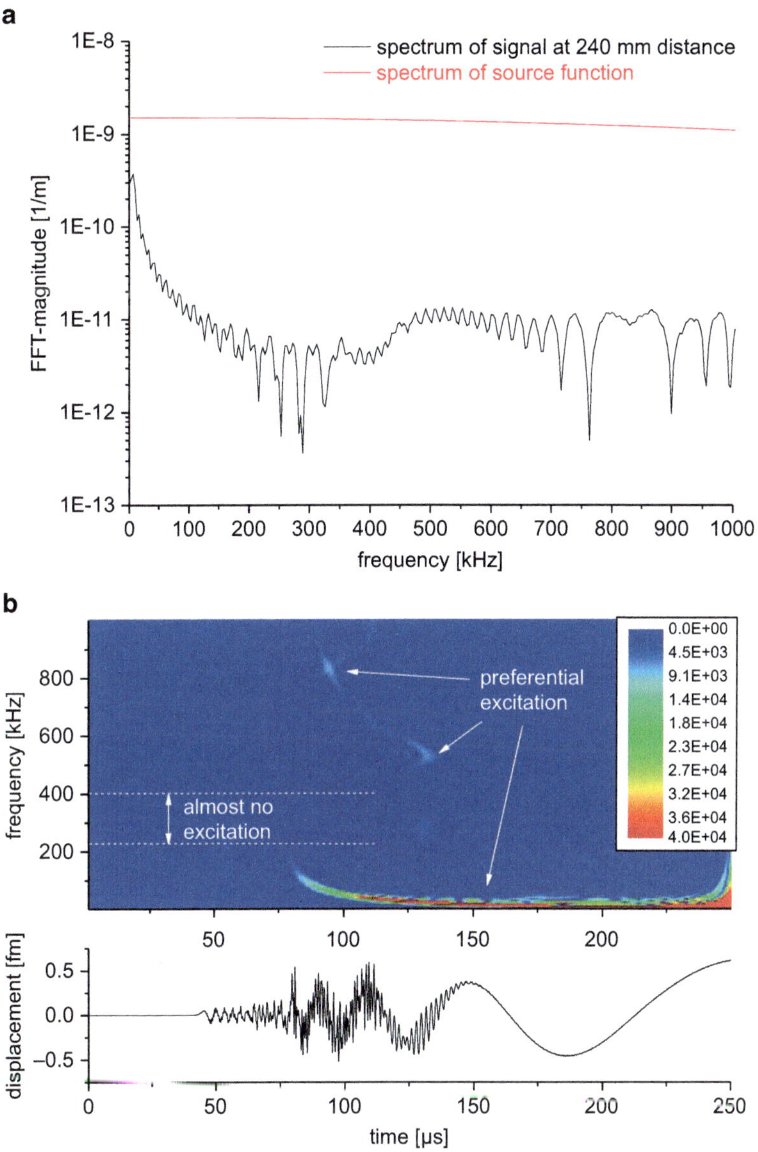

Fig. 4.61 Source spectrum of a buried dipole in an aluminum plate (**a**) and according Choi-Williams distribution at 234 mm distance to the source showing preferential excitation of Lamb wave modes (**b**)

some distance to the source. More information on this topic can be found, e.g., in [48, 49, 52].

Lamb wave propagation in undamaged carbon fiber reinforced polymers (CFRP) has been investigated extensively before [53–56]. Scattering of Lamb waves at internal damage, like cracks or delamination, is the key principle for structural health monitoring of CFRP by guided wave testing [57]. The impact of such discontinuities on acoustic emission analysis has been investigated less and is therefore reviewed in more detail in Sect. 4.3.3. Since the guided wave propagation is the carrier of information of AE source activity in the material, distortion of this information due to interaction of Lamb waves with internal discontinuities is closely related to the question of reliability of the information. Changes of modal intensity or occurrence of alternative propagation paths due to scattering can readily affect source localization accuracy and/or complicate source identification procedures.

In principle, numerical methods for analytical solutions to describe wave propagation in multilayered structures were already given by Thomson and Haskell [58, 59]. However, many numerical challenges like the problem of ill-conditioned matrices as used for the transfer-matrix method had to be overcome. A good review on those developments is found in [54]. Also, analytical predictions for wave propagation in multilayered structures are given by Mal and Kundu for the case of infinite half-spaces [60]. Compared to analytical solutions, a finite element approach to model wave propagation in fiber reinforced materials is advantageous in two ways. First, arbitrary geometries can be implemented without substantial modifications of the chosen approach. In comparison, for analytical approaches boundary conditions along with complex geometries are often difficult to implement. Second, three-dimensional FEM takes into account the boundary reflections of edges and additional interfaces, which are rarely captured adequately by analytical approaches. As has been demonstrated by measurements using a laser vibrometer, the accuracy of the FEM approach used in the following is sufficient to describe acoustic emission signal propagation in a variety of cases [53, 61, 62].

In general, the propagation of the acoustic waves is subject to the boundary conditions formed by the propagation medium geometry. In addition, the propagation of acoustic waves is subject to attenuation which significantly alters the appearance of the waves as function of propagation distance and is therefore discussed in the subsequent section.

4.3.1 Attenuation

The propagation of acoustic waves in an infinite isotropic medium exhibits no loss of energy. However, for real materials, the elastic waves experience several attenuation effects. As demonstrated by Auld, attenuation of acoustic waves propagating in solids close to room temperature can be formulated by a viscous damping term [63]. As shown in [55, 63–65], the acoustic wave attenuation is sufficiently described by introducing a complex stiffness tensor $\boldsymbol{C}^*$ linking the stress tensor $\boldsymbol{\sigma}$ to the deformation tensor $\boldsymbol{\varepsilon}$:

$$\boldsymbol{\sigma} = \boldsymbol{C}^* \boldsymbol{\varepsilon} \tag{4.39}$$

This formulation can be rewritten using complex elastic coefficients if the respective acoustic wave field has $e^{i\omega t}$ time dependence:

$$C^* = (C + i\omega C'') \tag{4.40}$$

Acoustic attenuation is then defined as decay of amplitude of an acoustic wave between two observation positions 1 and 2:

$$\text{Attenuation} = 10\log\left(e^{-\alpha(u_1-u_2)}\right)^2 \text{dB} \tag{4.41}$$

This introduces the attenuation coefficient α used as measure for the damping properties of a material. The attenuation is typically expressed in the dB scale to cover the large range of attenuation effects. As discussed by Pollock [66] and Prosser [67] from the viewpoint of acoustic emission, there are five contributions for attenuation of acoustic emission signals, which are discussed in more detail in the following.

4.3.1.1 Geometric Spreading

In the near field, close to the acoustic emission source, geometric spreading is the main reason for attenuation. This is caused by the spherical radiation of energy into the solid volume, which results in an energetic decay per dihedral angle with inverse propagation distance r. When Lamb waves are excited, the attenuation in the near-field range will even be more pronounced, since the elastic wave energy is separated into the distinct symmetrical and antisymmetrical modes. In addition, the Lamb waves suffer from velocity dispersion as will be described below. According to [67] for the case of Lamb waves, the distance of dominating attenuation by geometric spreading was estimated as $r < 4.34/\alpha$.

4.3.1.2 Thermoelastic and Akhiezer Dissipation

In the far field of the acoustic emission source, the main contribution to attenuation at room temperature originates from the thermoelastic mechanism and the Akhiezer mechanism. The first one results from irreversible heat conduction originating from compression during propagation of the longitudinal wave. The second mechanism arises from the disturbance of the equilibrium distribution of thermal acoustic waves (phonons). The propagation of a coherent acoustic wave disturbs the equilibrium positions of the phonons and thus leads to a dissipative energy contribution. For homogeneous isotropic solids, the thermoelastic and the Akhiezer mechanism result in attenuation for elastic waves dependent on the square of the oscillation frequency ω^2.

4.3.1.3 Dispersion

For guided waves in anisotropic solids, even more effects contribute to the total acoustic attenuation. Specifically for the case of Lamb waves, the total attenuation is affected by spatial dispersion and frequency dispersion [68]. Attenuation from spatial dispersion is caused by the dependency between phase velocity and wave vector inherent in (4.34) and (4.35). As discussed by Ward et al. for dispersive media, the phase velocity is directly linked to the attenuation by Kramers-Kronig relations deduced from the principle of causality [69]. The effect of attenuation due to frequency dispersion arises in propagation of non-monochromatic wave packages. Since each frequency component travels at a distinct velocity, an initially short pulse begins to spread in time during propagation. This causes attenuation with propagation length. For the case of Lamb waves, this is of great importance since the different modes exhibit different α-coefficients. For CFRP plates the attenuation of the antisymmetrical modes is still higher than the attenuation of the symmetrical modes. This effect superimposes the thermoelastic effect, since the S_0 Lamb wave mode typically shows higher frequencies than the A_0 Lamb wave mode and consequently is attenuated more [67].

4.3.1.4 Scattering

In addition, for wavelengths approaching the dimensions of included inhomogeneities, strong scattering of the elastic waves can occur. Typical examples are the grain structure within metals, fibers in composites, or voids. This can also result in severe attenuation and has to be taken into account when discussing attenuation effects.

4.3.1.5 Dissipation into Adjacent Media

In experimental environments acoustic emission signals propagate within a material under testing, which typically is in contact with adjacent media like, e.g., fixtures or supply pipes. Because a partial transmission of the acoustic emission signal into adjacent media results in a respective loss of the reflected wave, such losses can be fairly large and thus can even dominate the acoustic attenuation.

4.3.1.6 Modeling Attenuation

Moreover, attenuation in fiber reinforced structures also depends on the orientation of the fibers in the laminates. To demonstrate this effect of acoustic anisotropy, Ono and Gallego conducted measurements of the signal attenuation in AS4/3502 plates (see Table B.1 of Appendix B) using piezoelectric pulsers (see [70] for details).

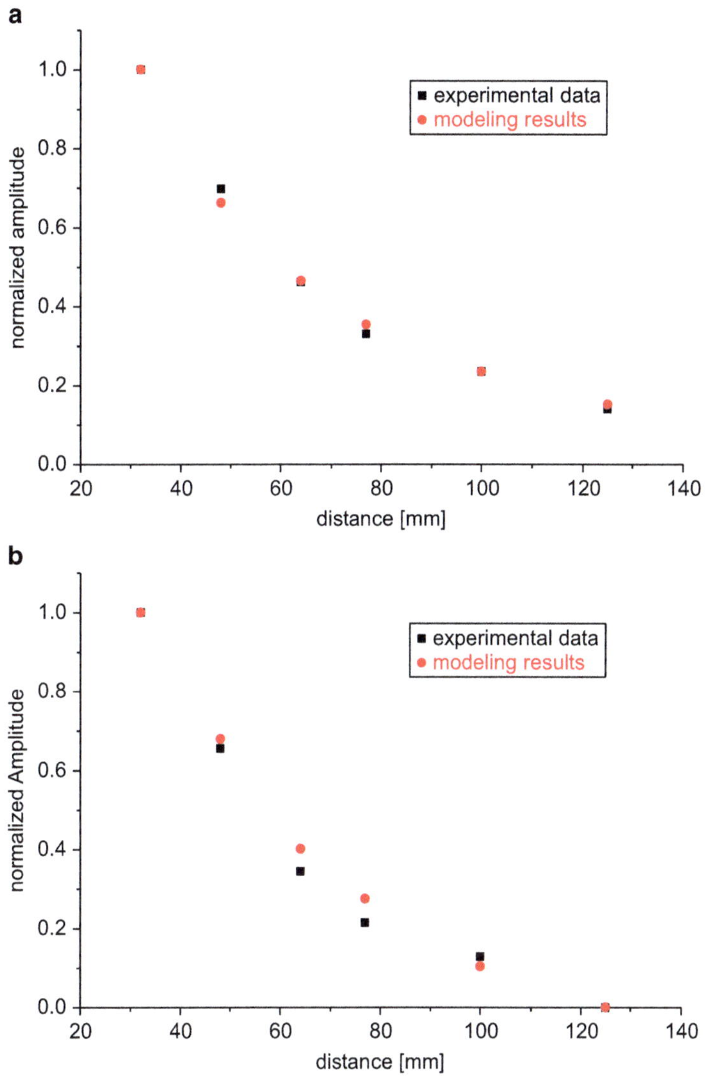

Fig. 4.62 Comparison between normalized experimental attenuation measurements and simulation result using complex elastic coefficients for the S_0 mode (**a**) and the A_0 mode (**b**)

Figure 4.62 shows a comparison of these measurements of a unidirectional plate to respective results of modeling work using the complex elastic coefficients given in Appendix B. As seen from Fig. 4.62 as a function of the distance to the source, the detected amplitudes decrease significantly. In the model the total attenuation is due to geometric spreading (result of using real part of stiffness tensor) but also due to a contribution of thermoelastic dissipation (taken into account by the complex elastic coefficients). Moreover, there is a difference in the attenuation of the two fundamental modes as seen by comparison of Fig. 4.62a, b.

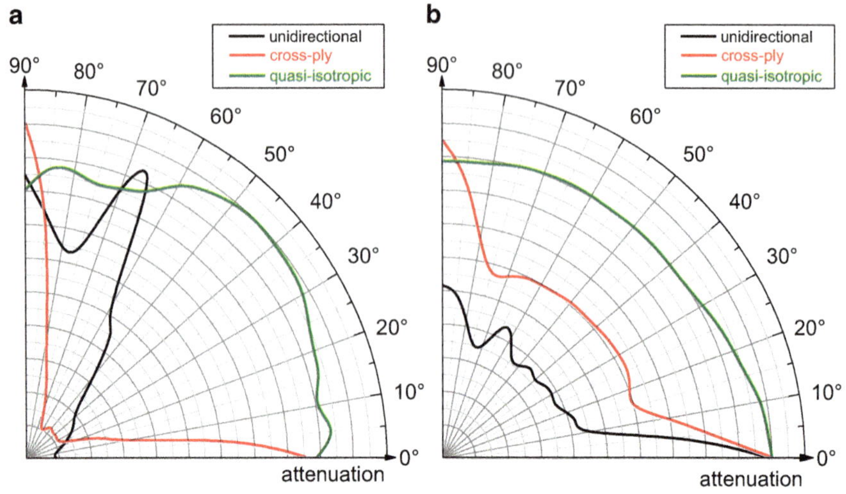

Fig. 4.63 Comparison between normalized attenuation levels of unidirectional, cross-ply, and quasi-isotropic stacking for the S_0 mode (**a**) and the A_0 mode (**b**)

The specific challenges arising in arbitrary stackings of laminates are shown in Fig. 4.63a, b. As seen in the comparison of the three different stackings in one polar diagram, each layup causes different attenuation levels as a function of distance to the source. Also, the ratio of attenuation between S_0 mode and A_0 mode is influenced by the layup investigated. In particular for the S_0 mode, the effect of the change in in-plane stiffness (dominated by the fiber orientation) is clearly seen.

In general, these attenuation effects will cause difficulties to interpret the signals after a certain distance of propagation, since the information included in the signals is altered, and in polymeric materials, the higher-frequency components are attenuated faster due to the thermoelastic dissipation. This will render signals useless for frequency interpretation after a certain distance and has to be included in predictive tools to account for this effect. For the modeling part, the usage of complex elastic coefficients according to the approach by [55, 64] is found to be a suitable approach to account for this effect in FEM. However, care has to be used for those frequency ranges, where strong changes to the complex elastic coefficients occur. As reported by [64], the values of the complex stiffness tensor are usually a function of the frequency of propagation and therefore have to be implemented as such in a modeling environment if not constant within the investigated bandwidth.

4.3.2 Influence of Geometry

Similar as the propagation within an attenuating medium, a change of the boundary conditions of the propagation medium can have a substantial impact on the acoustic emission signal. For the case of fiber reinforced materials, this mostly refers to changes in the specimen geometry, changes of the stacking sequence, and changes to the material configuration.

In Fig. 4.64a, b, modeled signals for a fiber breakage source are shown as calculated for a unidirectional T800/913 material using the material properties

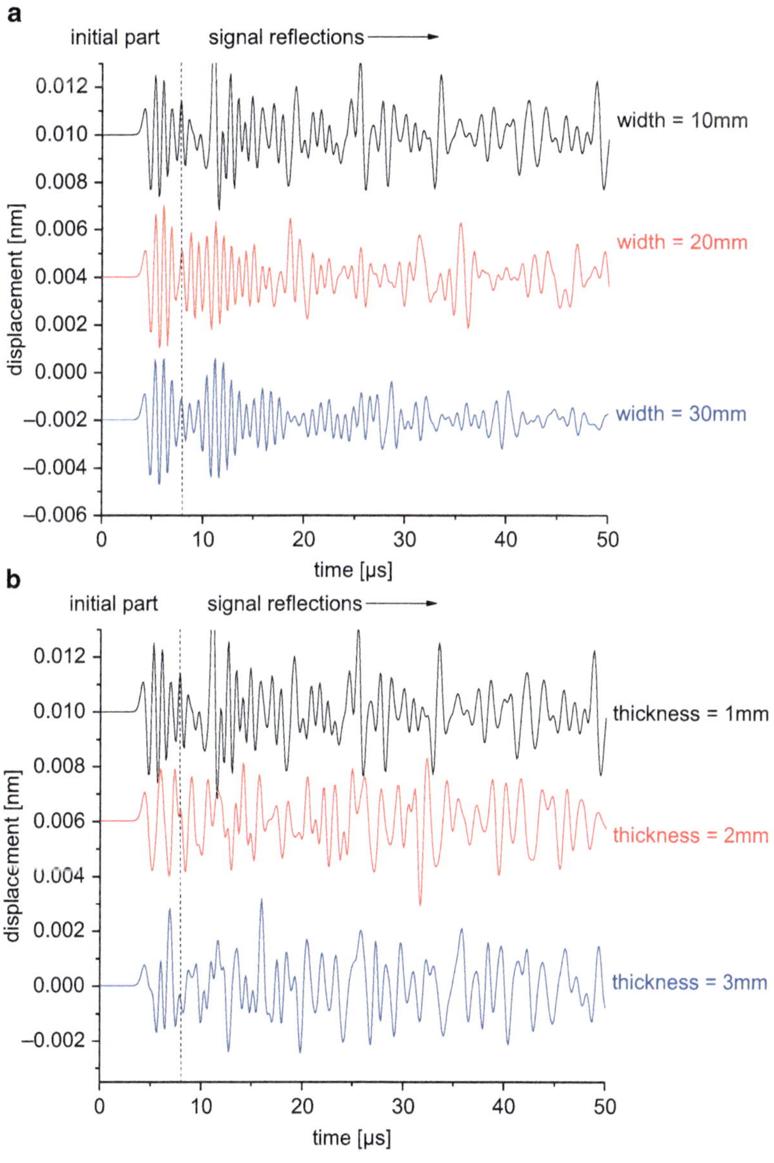

Fig. 4.64 Influence of specimen width at 1 mm thickness (**a**) and specimen thickness at 10 mm width (**b**) on detected signals of a fiber breakage source model in 30 mm distance

from Table B.1 in Appendix B. The acoustic emission signals are evaluated as out-of-plane displacement in 30 mm distance to the source position. The width of the specimen and the thickness of the specimen are varied as noted in the Fig. 4.64a, b, respectively. While changes of the specimen width mainly affect the presence and intensity of the edge reflections, changes of the thickness apparently have larger impact on the signal frequency content. Due to the nature of plate waves, their frequency content is implicitly related to the thickness of the plate. This causes a substantial change of the AE signal frequency, being the highest for the thinnest specimen thickness.

Similar changes to the frequency content are observed, when the acoustic wave passes a plate section with a gradual increase of the thickness. Such gradual changes of the specimen thickness have similar impact on the signal frequency content as a propagation medium with different thickness. However, additional attenuation effects occur due to the change in propagation volume. Figure 4.65 presents a

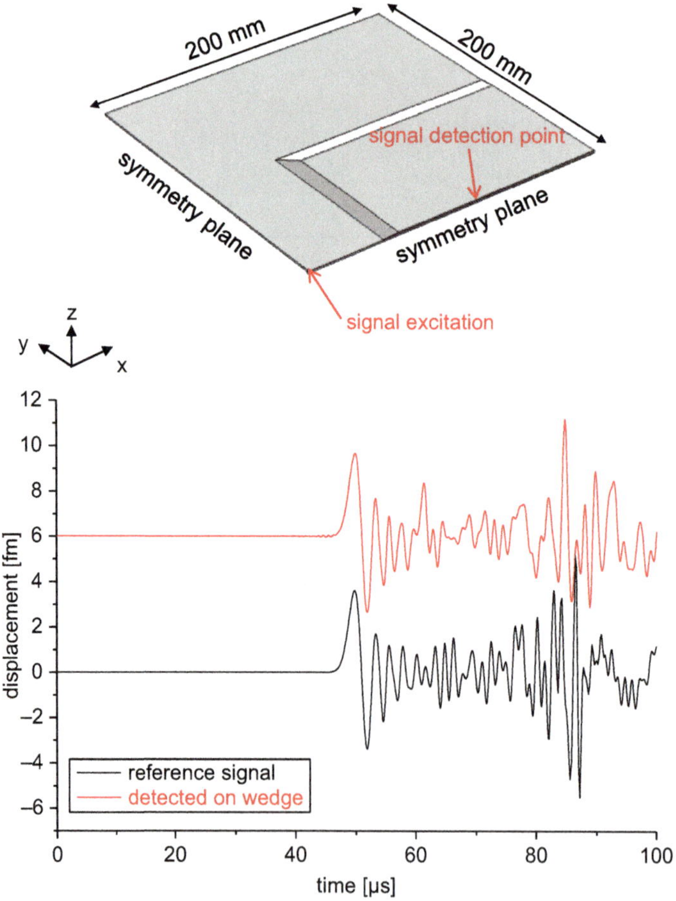

Fig. 4.65 Model setup showing wedge profile and comparison of signals after propagating 100 mm distance with and without presence of a wedge to increase the plate thickness

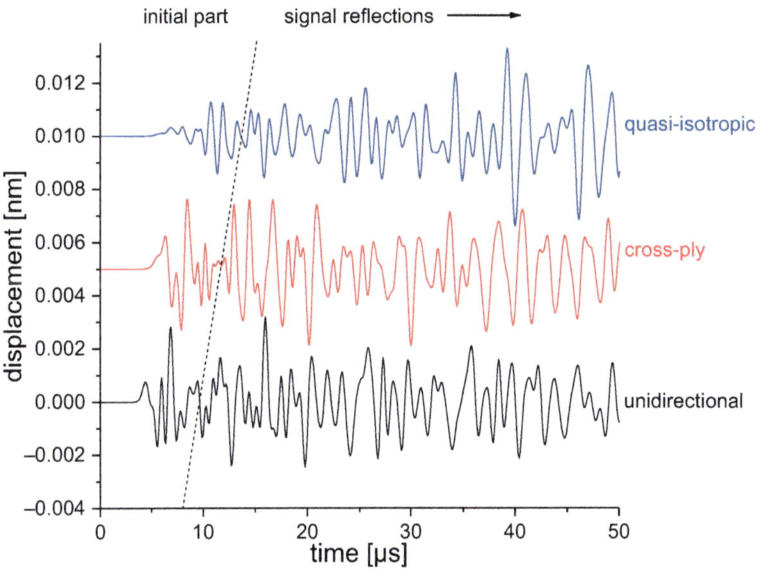

Fig. 4.66 Influence of stacking sequence on detected signals of a fiber breakage source model in 30 mm distance

comparison of signals detected in a 1 mm thick T800/913 unidirectional fiber reinforced polymer plate with properties of Table B.1 in Appendix B. One curve shows the reference signal detected as out-of-plane displacement in 100 mm distance to a dipole source buried in the plate. The other curve is the signal detected in the same distance after passing through a wedge section increasing the thickness to 2 mm using a wedge length of 10 mm.

As further parameter, the stacking sequence is varied, and its influence on the signal emitted by the fiber breakage source model is shown in Fig. 4.66. Three different configurations of identical plate thickness of 3 mm were chosen, comprising a unidirectional, a cross-ply $[0/90/90/0]_{sym}$, and a quasi-isotropic $[90/\text{-}45/45/0]_{sym}$ stacking sequence. As seen in Sect. 4.3.1, these different stacking sequences translate into different acoustic properties, which have a significant impact on the visual appearance and frequency content of the AE signals, as well as on the arrival time of the signals.

Further complications arise, when the material is not only anisotropic but also inhomogeneous as for the case of metallic liners attached to the composite or hybrid plates of fiber reinforced materials interspersed with metallic sheets. First, the influence of the thickness ratio between layers of aluminum and a fiber reinforced material is investigated as demonstrated in [53]. Using the results of aluminum and T800/913 unidirectional CFRP with values as given in Table B.1 of Appendix B, a combination of both materials is investigated as a two-layer system. The thickness ratio between aluminum and CFRP is systematically varied while keeping the total thickness of 1 mm. Simulated signals shown in Fig. 4.67 are for an AE dipole

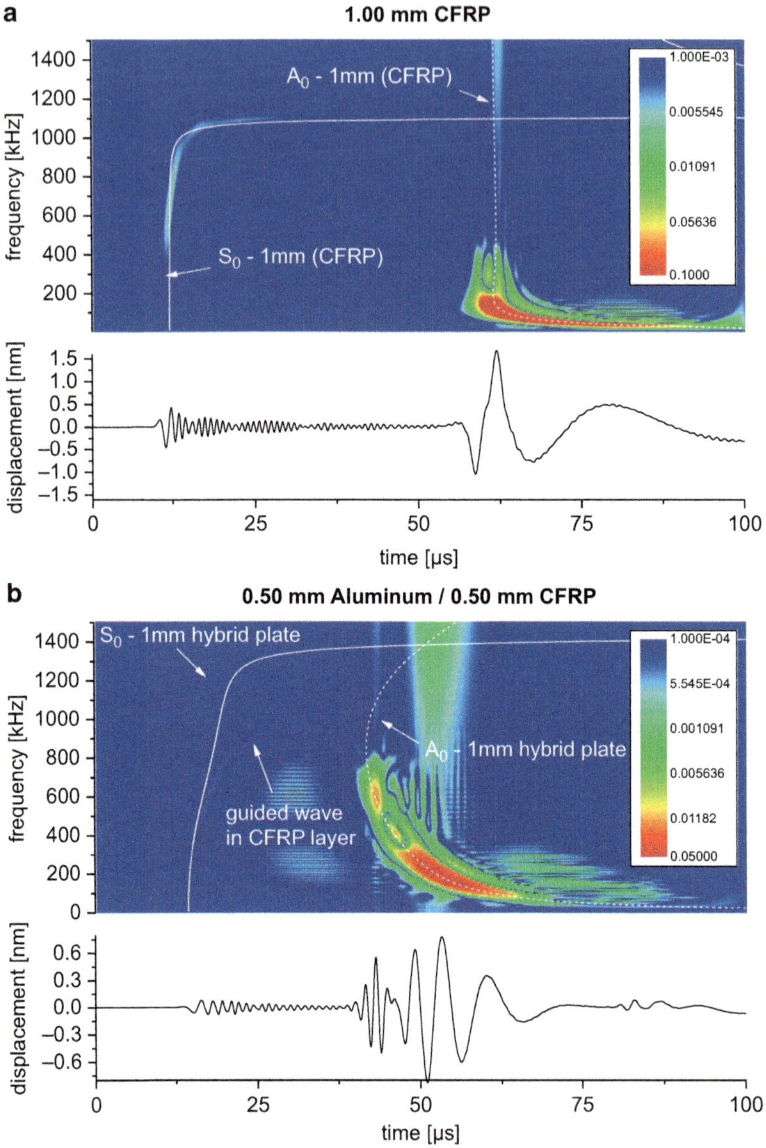

Fig. 4.67 Simulated signals of AE dipole source in two-layer system with varying thickness ratio between aluminum and T800/913 unidirectional CFRP obtained at 100 mm distance for 90° propagation angle. (**a–c**) show Choi-Williams distribution (logarithmic scale) of selected configurations with superimposed dispersion curves; (**d**) shows comparison of all configurations (based on [53])

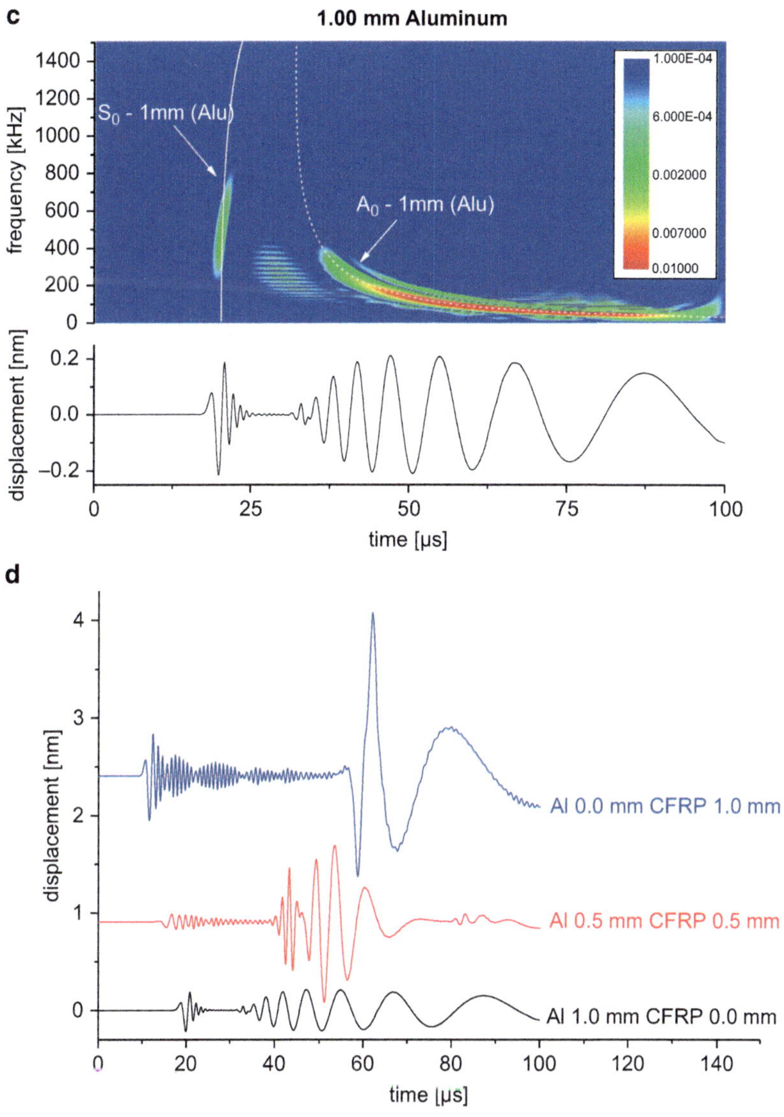

Fig. 4.67 (continued)

source buried in the CFRP part with axis orientation along the *y*-axis that is detected at 100 mm distance under 90° propagation angle on the top surface (composite side) of the plate. To identify particular Lamb wave modes, the CWDs shown in Fig. 4.67 are superimposed by results of dispersion curve calculations following reference [71].

The separation in time of the S_0 mode and A_0 mode is observed strongest in homogeneous CFRP and weakest in homogenous aluminum. The modes of the signal shown in Fig. 4.67b are identified as symmetric and antisymmetric motion relative to the midplane of the plate. However, there is no match between the detected Lamb wave modes and the dispersion curve calculations for the individual materials. Therefore, the Lamb wave propagation seems to occur as a motion of the complete plate for this case.

For varying ratio between aluminum and CFRP, a systematic transition of signal shape and propagation velocities is found. Similar to the S_0 mode, the signal shape of the A_0 mode is changed continuously between the two extreme values of only CFRP and only aluminum, while the amplitude is subject to a nonlinear transition. There is significant influence on the propagation velocity of the S_0 mode and the A_0 mode when the thickness ratio between aluminum and CFRP is changed. As demonstrated in [53], the agreement between the simulated velocities and those obtained from the rule of mixture of the velocities of both materials indicates that the velocity of Lamb wave propagation of the S_0 mode is governed by the average elastic properties of the materials involved. This is of particular interest for AE source localization approaches that are based on the initial threshold crossing. A similar transition of propagation velocities is found for the A_0 mode. However, a direct application of the rule of mixture does not capture the changes in propagation velocity. This is due to the correlation of the A_0 mode velocity with the overall bending stiffness of the plate, which cannot be obtained from a direct rule of mixture calculation of the individual sound velocities of the materials involved.

Also, the ratio between anisotropic and isotropic material seems to be the key factor for the changes in signal propagation. Thus for thickness ratios as typically faced for thin liner materials (<0.125 of the total thickness), the influence of the liner on signal propagation can assumed to be minimalistic.

For an increased total thickness of the plates, more than the two fundamental Lamb wave modes can be identified in the CWD. For a case with 2.00 mm thickness, two propagation modes are found at the beginning of the signal. These originate from a S_0 mode Lamb wave of the hybrid plate and an additional guided wave propagating in the CFRP layer [53].

This effect is seen best in the signals detected at 0° propagation angle, as shown in Fig. 4.68. The comparison of the signal detected on the aluminum side (Fig. 4.68a) and detected on the composite side (Fig. 4.68b) demonstrates the distinct difference of the signals. While both signals exhibit the same intensity of A_0 mode and S_1 mode, the additional propagation of the S_0 mode is seen only for the detection on the composite side.

This demonstrates that Lamb wave propagation in such multilayered plates can take place as motion of the complete plate, but also as a guided wave in only one of the layers.

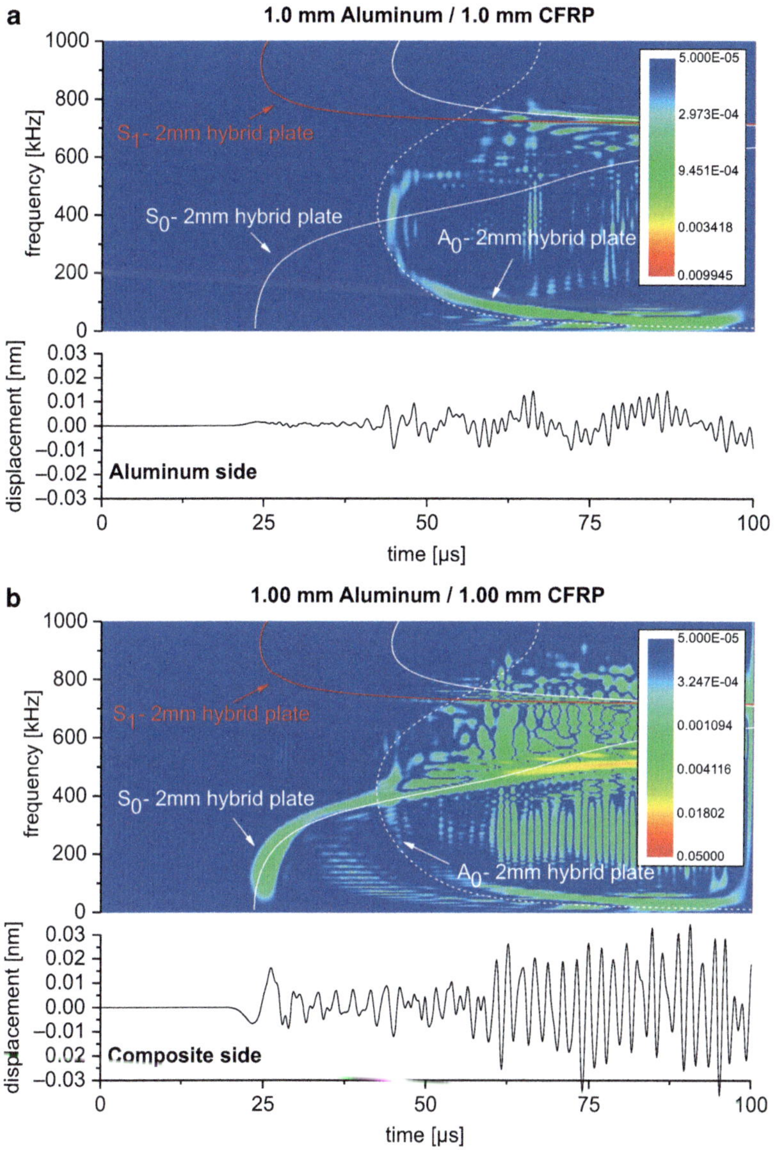

Fig. 4.68 Choi-Williams distribution (logarithmic scale) of simulated signals of AE dipole source in a two-layer system with total thickness of 2.00 mm obtained at 100 mm distance for 0° propagation angle on aluminum side (**a**) and composite side (**b**) (based on [53])

4.3.3 External and Internal Obstacles

So far the acoustic wave propagation was discussed for undisturbed media, which is certainly not the case for a loaded composite structure with a continuously increasing amount of damage. Therefore, the influence of pre-damaged structures on wave propagation has to be considered for valid interpretation of acoustic emission signals from composites. Due to the controlled situation, the previously validated finite element method is further applied to account for the presence of several types of external and internal obstacles. Within this section only aspects of wave propagation are discussed. The influence on source localization methods will be presented in Sect. 4.6 and the impact on source identification in Sect. 4.5.

The model setup used in the following is shown in Fig. 4.69. A rectangular plate of 200 mm × 200 mm size and 1 mm thickness is used as propagation medium. For evaluation of the signals, a unidirectional CFRP plate was selected as propagation medium, since this exhibits a maximal elastic anisotropy.

The elastic properties of a unidirectional T800/913 CFRP plate are used as given in Table B.1 of Appendix B. The fiber direction of the unidirectional laminate is oriented along the 0° axis as noted in Fig. 4.69. Two symmetry planes were chosen to reduce the volume modeled to one quarter of the overall volume. The symmetry planes are the *yz*- and *xz*-plane with respect to the origin of the coordinate system.

As acoustic emission source, a point couple was defined following [61, 62]. The position of the source was at $(x, y, z) = (0.10, 0.00, 0.53)$ mm, slightly asymmetric with respect to the midplane of the plate, to excite a reasonable amount of S_0 and A_0 Lamb wave modes at the same time. A linear ramp function with excitation time $t_e = 1\,\mu s$ and maximum force $F_{max} = 3\,N$ is used:

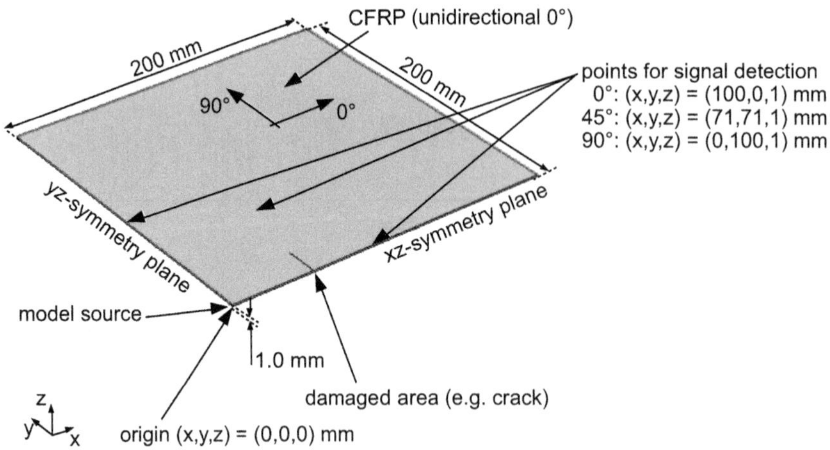

Fig. 4.69 Three-dimensional model setup used for simulation of Lamb wave propagation including details for signal detection points and position of obstacle (e.g., crack position)

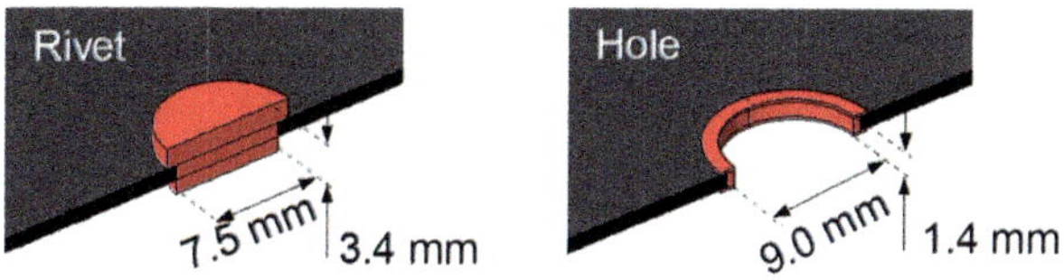

Fig. 4.70 Three-dimensional model geometry used for simulation of bolts and holes

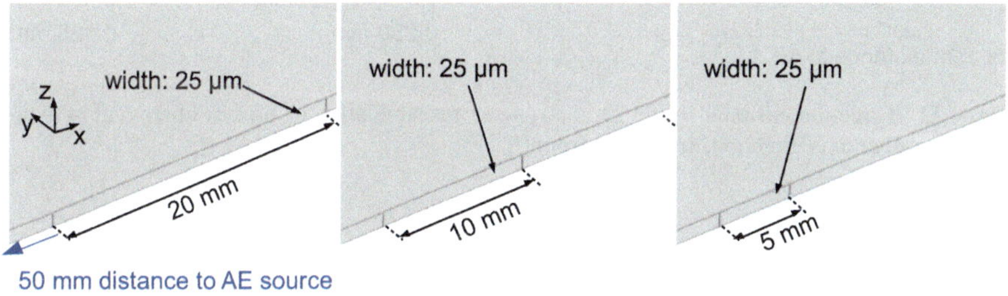

Fig. 4.71 Three-dimensional model geometry used for simulation of inter-fiber cracks (cut parallel to fiber axis direction) (reprinted from [62])

$$F(t) = \begin{cases} F_{\max} \cdot (t/t_{\mathrm{e}}) & t \leq t_{\mathrm{e}} \\ F_{\max} & t > t_{\mathrm{e}} \end{cases} \tag{4.42}$$

The obstacle was placed with midpoint at a distance of 50 mm in x-direction to the source. In the following, the influence of the presence of rivets, bolts, holes, inter-fiber cracks, broken fibers, and delamination is discussed. Each of these obstacles refers to one prototype of discontinuity. Rivets and bolts are through-thickness metallic enclosures that allow direct transmission of elastic waves, but are expected to cause wave scattering due to their acoustic impedance mismatch. All signals are evaluated at 100 mm distance to the AE source in 0°, 45°, and 90° propagation direction as z-displacement of the plate. For further details of this study, the reader is referred to [61, 62].

Details of the geometries chosen for rivets and holes are given in Fig. 4.70. All metallic parts marked in red are modeled as high-strength alloy steel parts, with elastic properties as listed in Table B.1 in Appendix B.

Inter-fiber cracks (also called transverse cracks) weaken the link between neighboring fibers and can thus affect the transmission of acoustic waves. In the chosen configuration, the growth direction of such cracks is along the fiber axis direction (0°). From the perspective of wave propagation, the worst case is a growth of such cracks through the complete thickness. This configuration is studied here as cut of 25 µm width ranging from the top to the bottom of the plate and various lengths as shown in Fig. 4.71.

As further prototype, rupture of fibers can accumulate locally and can become a macroscopic crack. This is modeled as a 100 µm thick cut perpendicular to the fiber axis direction as shown in Fig. 4.72. Since such crack growth typically initiates ultimate failure of the load-carrying structure, three depths of the cut were

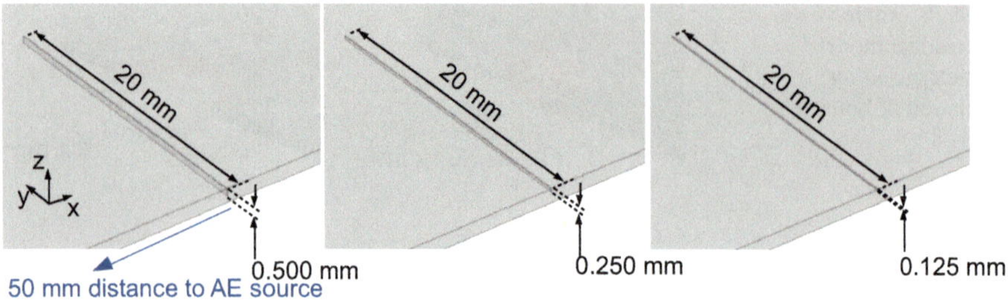

Fig. 4.72 Three-dimensional model geometry used for simulation of broken fibers (cut perpendicular to fiber axis direction) (reprinted from [62])

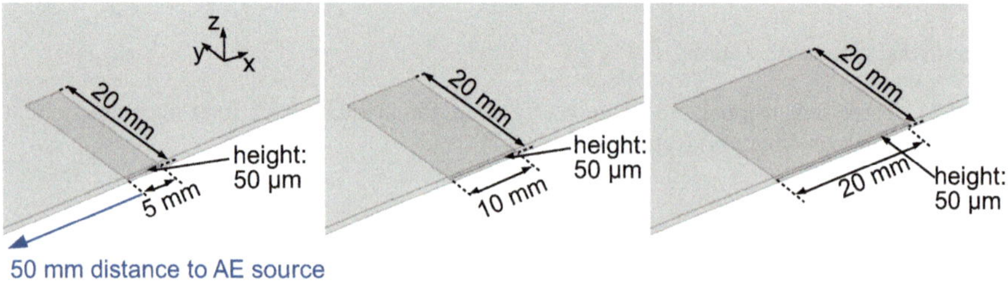

Fig. 4.73 Three-dimensional model geometry used for simulation of inter-ply delamination (cut parallel to fiber axis direction) (reprinted from [62])

investigated to study the influence on wave propagation. The depth of 0.5 mm already refers to a significant amount of damage, reflecting a rupture of more than 800,000 fiber filaments. Also, for bending loads, the initiation of these cracks is typically found on the surface of the structure. Therefore, the current geometry of broken fibers starting at the surface exhibits an asymmetric location of the crack with respect to the midplane.

Finally, inter-ply delamination is an in-plane discontinuity that weakens the local bending stiffness of the plate. This is often encountered in fiber reinforced structures as a consequence of impact damage, excessive loading, or even as residue of the manufacturing process. Since this can occur in various extents, three different sizes were investigated to reflect reasonable variability of the size of the delaminated area. For all configurations a thickness of the delamination of 50 µm is chosen centered within the plate (Fig. 4.73).

4.3.3.1 Reference Case

In the following the influence of the modeled obstacles is demonstrated. Figure 4.74 shows the CWD of the calculation result for signal propagation along the 0° axis of the unidirectional CFRP plate. The CWD in Fig. 4.74a uses a different color,

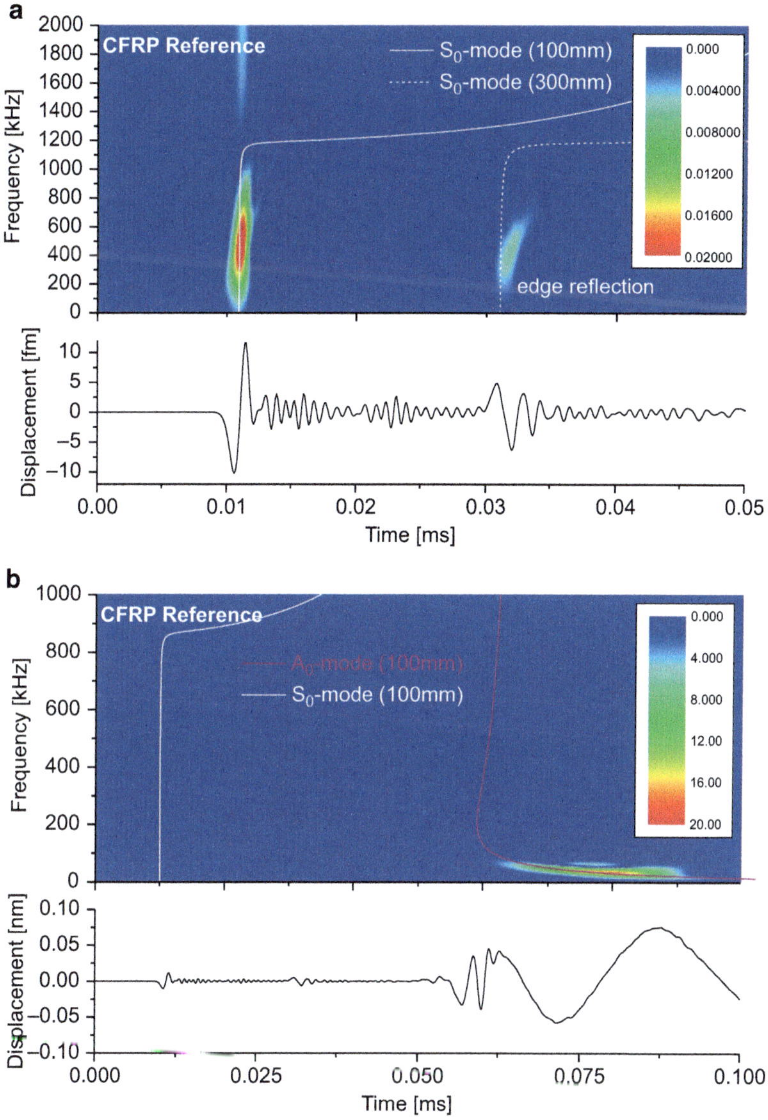

Fig. 4.74 Simulation results of signal propagation at 100 mm distance along 0° direction in unidirectional CFRP. Truncated time scale shows S_0 mode (**a**) and full time scale shows A_0 mode (**b**) (reprinted from [62])

frequency, and time range as Fig. 4.74b, which is necessary to visualize the S_0 mode at the initial part of the signal. To identify particular Lamb wave modes, the dispersion curves of the fundamental modes were calculated.

As indicated by the superimposed dispersion curves for the S_0 mode at 100 and 300 mm source distance, the S_0 mode is reflected at the edge of the plate in

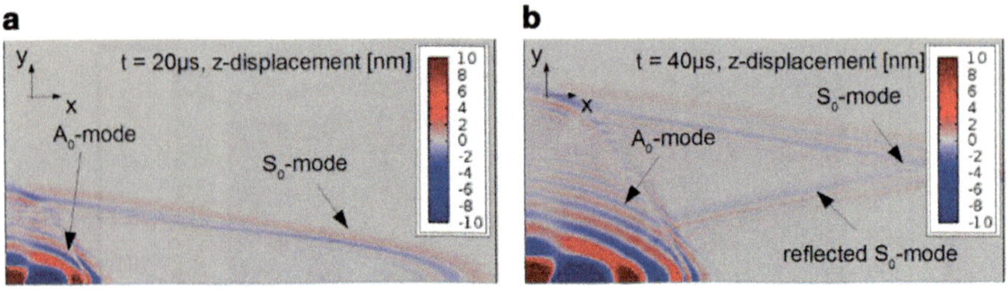

Fig. 4.75 Comparison of simulated wave fields z-displacement for reference case at $t = 20$ μs (**a**) and at $t = 40$ μs (**b**) after signal excitation (reprinted from [62])

x-direction and is detected twice. The shape of the A_0 mode shown in Fig. 4.74b agrees well with the calculated result of the dispersion curve for 100 mm distance. The small dips in the curve seen at 52 and 62 μs are caused by multiple reflections of the S_0 mode, which has been reflected at all edges of the plate at this time. The calculation results for the CFRP plate without defects in Fig. 4.74 will act as a reference case to evaluate the influence of any obstacles modeled within the propagation path.

The wave fields obtained for the reference case without obstacles are shown in Fig. 4.75a, b for two distinct times $t = 20$ μs and $t = 40$ μs after signal excitation. Effects of edge reflections and modal conversions are easy to spot in these stills of the wave field.

4.3.3.2 Rivets and Bolts

Bolts and rivets are commonly used as fasteners in CFRP structures and are frequently encountered during testing of real structures. The calculation result for signal propagation in the presence of such obstacles is shown in Fig. 4.76 with time, frequency, and coefficient range settings slightly different to Fig. 4.74. As observed from Fig. 4.76a, shape and intensity of the S_0 mode are influenced by the presence of the rivet within the propagation path. The intensity is less than for the reference structure. On the one hand, this is caused by the acoustic transmission coefficient (0.75) of the CFRP-metal and metal-CFRP interface in 0° direction calculated under the assumption of planar wave incidence. On the other hand, interfaces are likely to cause modal conversion. For the current case, a significant part of the incident S_0 mode is converted to an A_0 mode radiating away from the rivet (cf. also Fig. 4.77). This causes an additional emission of an A_0 mode that is ahead of the A_0 mode originally emitted by the point source. Thus, determination of the correct arrival time of the A_0 mode with respect to the point source is almost impossible.

The presence of the rivet causes a strong interaction with the incident Lamb waves. As seen in Fig. 4.77a, the interaction of the rivet with the S_0 mode causes excitation of a strong secondary A_0 mode with almost circular radiation pattern around the center of the rivet. The only explanation for this is a modal conversion of

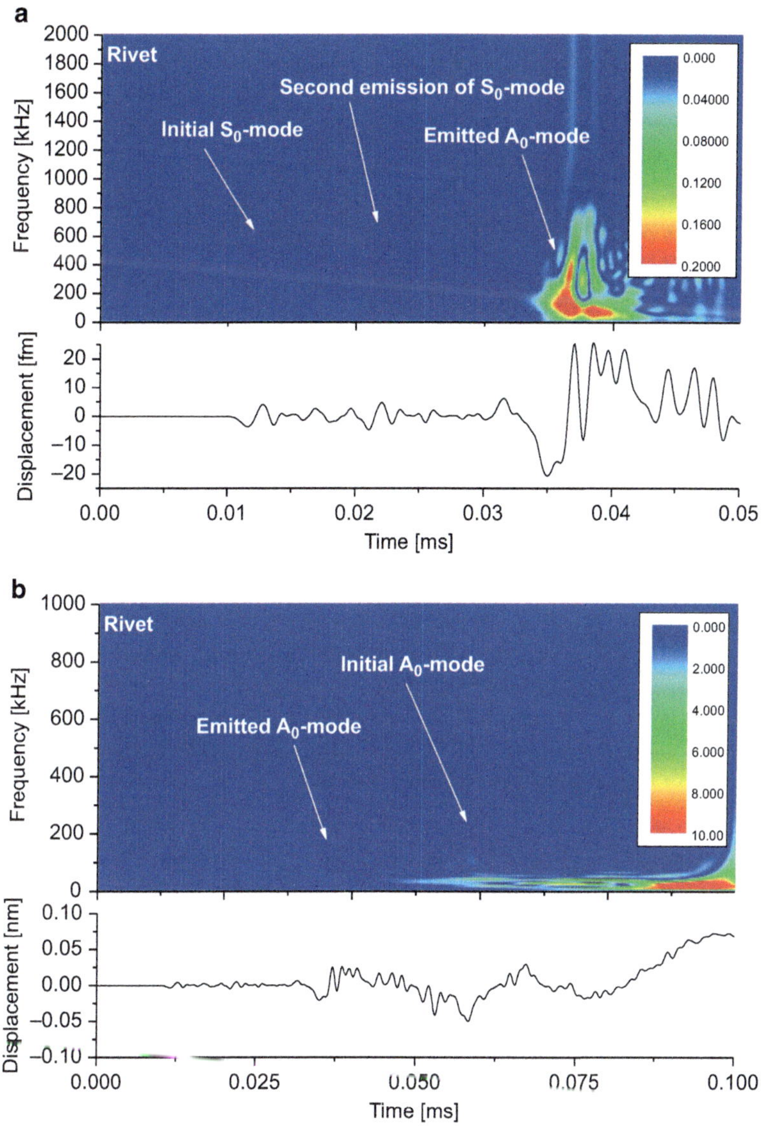

Fig. 4.76 Simulation results of signal propagation at 100 mm distance along 0° direction in unidirectional CFRP with included rivet. Truncated time scale shows S_0 mode (**a**) and full time scale shows A_0 mode (**b**)

the primary S_0 mode to a secondary A_0 mode, since the primary A_0 mode did not arrive at the rivet for $t < 20$ μs. In Fig. 4.77b, the wave field for $t = 40$ μs shows the interaction of the A_0 mode with the rivet. Similar as for the incident S_0 mode, the A_0 mode is partially reflected at the rivet.

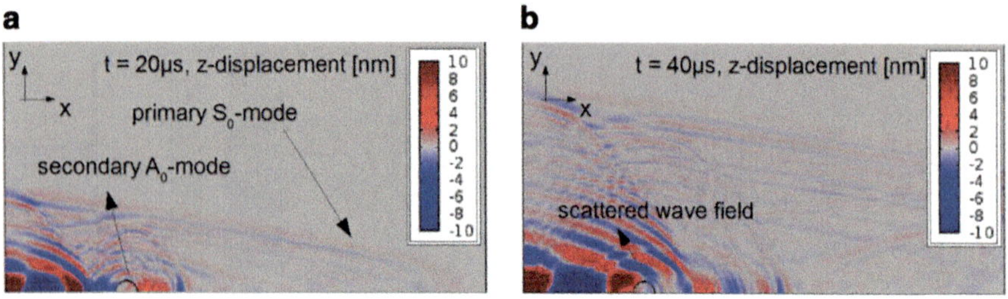

Fig. 4.77 Comparison of simulated wave fields *z*-displacement for rivet obstacle at $t = 20$ μs (**a**) and at $t = 40$ μs (**b**) after signal excitation (reprinted from [61])

4.3.3.3 Holes

Holes in CFRP are another type of acoustic obstacle often encountered in technical CFRP structures. These can originate from drilling processes or represent seriously damaged areas, e.g., after high-velocity impacts. As shown in the CWDs in Fig. 4.78 with settings identical to Fig. 4.74, the influence of this kind of obstacle is more severe than the influence of the rivet. As seen in Fig. 4.78a, the presence of the through-thickness hole causes less-intense transmission of the initial S_0 mode as compared to the reference case. Also, the shape of the hole results in a second emission of a S_0 mode with a virtual source position located at the center of the hole. This process is observed in more detail in the wave fields of Fig. 4.79a, b. As soon as the wavelength of the A_0 mode becomes sufficiently larger than the diameter of the hole, there is no interaction with the hole. Thus, only the short wavelengths of the A_0 mode (i.e., the initial part) are affected by the presence of the hole and are scattered at the obstacle.

4.3.3.4 Inter-fiber Cracks

Inter-fiber cracks are one of the basic damage types found in fiber reinforced composites. These are cracks with propagation direction perpendicular to the fiber axis direction. Since the strength of the direction perpendicular to the fiber axis is the weak link between the fibers, this type of damage is observed frequently in composite structures. In the current configuration, the inter-fiber crack is modeled as crack-through process, i.e., the crack reaches from the top to the bottom of the plate. The length of the inter-fiber crack is varied between 5 and 20 mm to cover a broad range of macroscopic crack sizes.

As seen in Fig. 4.80, the S_0 mode detected in these damaged structures has only negligible differences when compared to the reference case (black line). But there is a detectable influence of the presence of the inter-fiber cracks on the initial part of the A_0 mode. For the various lengths of 5, 10, and 20 mm of the inter-fiber crack, no

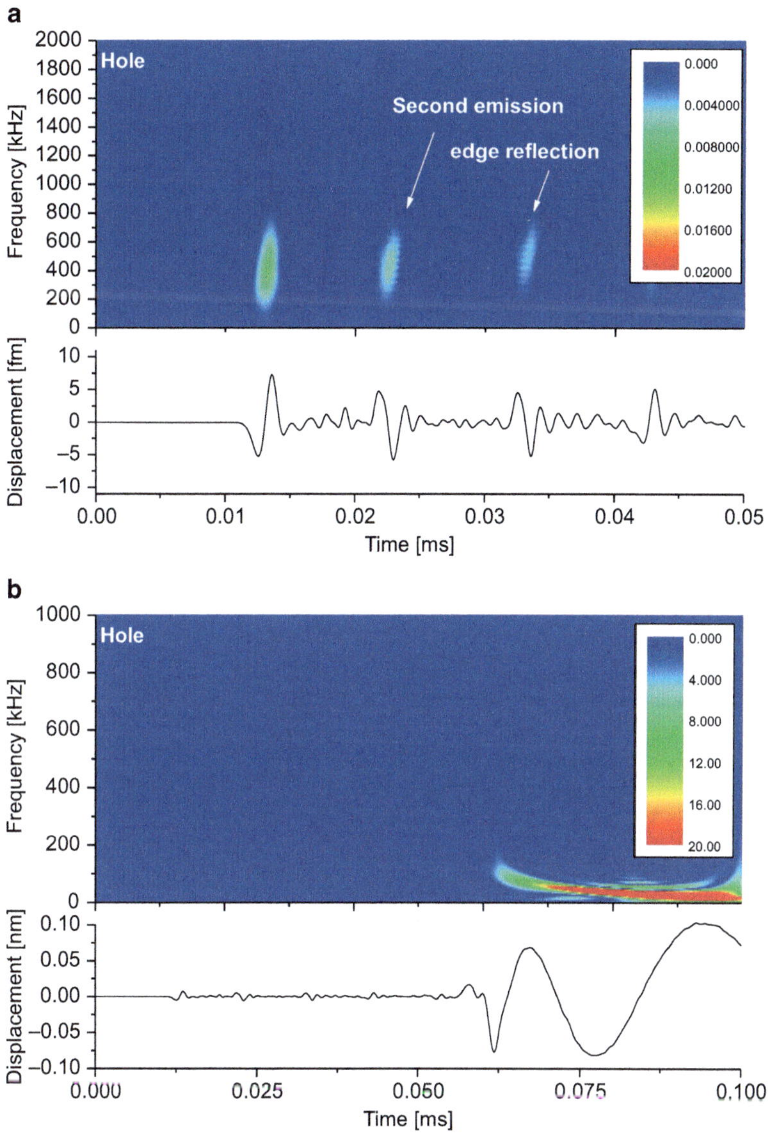

Fig. 4.78 Simulation results of signal propagation at 100 mm distance along 0° direction in unidirectional CFRP with hole. Truncated time scale shows S_0 mode (**a**) and full time scale shows A_0 mode (**b**) excitation (reprinted from [61])

significant differences are observed relative to each other besides a small signature at 63 μs.

For the inter-fiber cracks, only weak interaction with the S_0 mode and A_0 mode is found as seen in Fig. 4.81a, b.

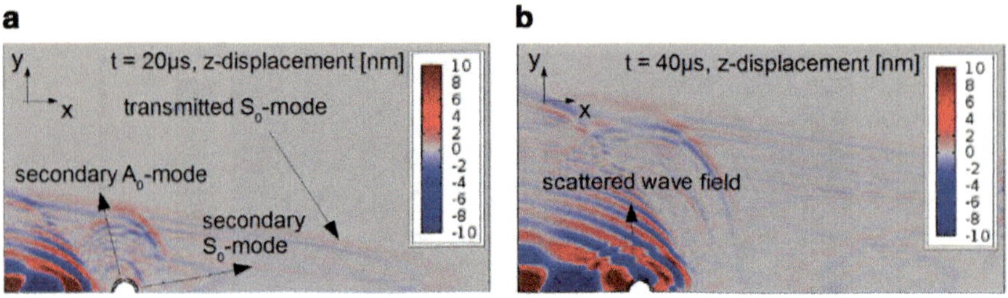

Fig. 4.79 Comparison of simulated wave fields *z*-displacement for hole obstacle at $t = 20$ µs (**a**) and at $t = 40$ µs (**b**) after signal excitation (reprinted from [61])

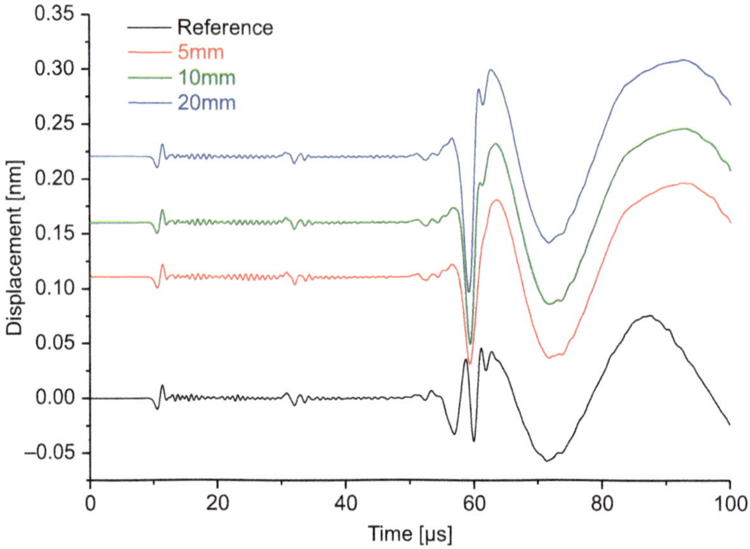

Fig. 4.80 Simulation results of signal after propagation of 100 mm along 0° direction in unidirectional CFRP with inter-fiber cracks of various lengths (reprinted from [62])

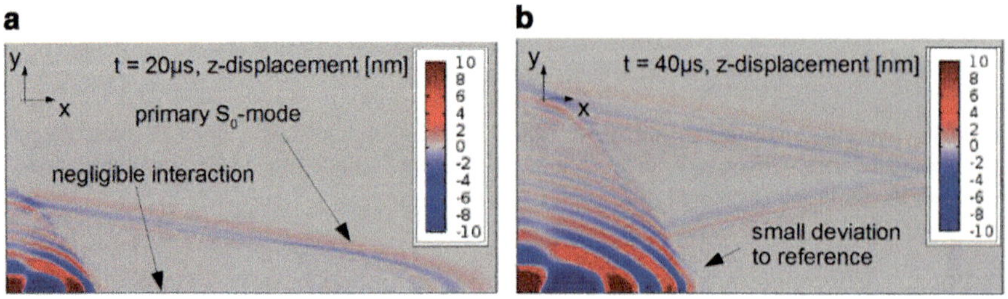

Fig. 4.81 Comparison of simulated wave fields *z*-displacement for inter-fiber crack at $t = 20$ µs (**a**) and at $t = 40$ µs (**b**) after signal excitation (reprinted from [62])

4.3.3.5 Broken Fibers

Among the damage types found in fiber reinforced composites, a rupture of the load-bearing constituents is the most severe damage. In the model, the presence of such discontinuous fiber filaments is taken into account by an air gap with extension perpendicular to the fiber axis direction (cf. Fig. 4.72). To consider the various degrees of damage for this type of failure, the depth of the air gap in the plate thickness direction is varied between 0.125 and 0.500 mm.

As seen in Fig. 4.82, the influence of the discontinuity on the first arrival of the S_0 mode ($t < 20$ μs) is negligible for all cases investigated. Dependent on the through-thickness dimension of the modeled discontinuity, a secondary peak arises around 35 μs. This is caused by modal conversion of the incident primary S_0 mode at the discontinuity into a secondary A_0 mode and into a secondary S_0 mode (see Fig. 4.83a, b). The later parts of the signal ($t > 40$ μs) are also affected by this

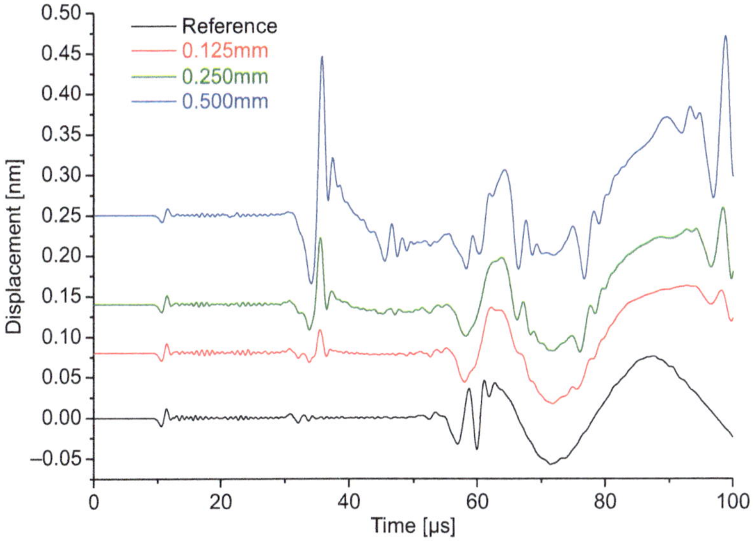

Fig. 4.82 Simulation results of signal after propagation of 100 mm along 0° direction in unidirectional CFRP with broken fibers modeled with various dimensions (reprinted from [62])

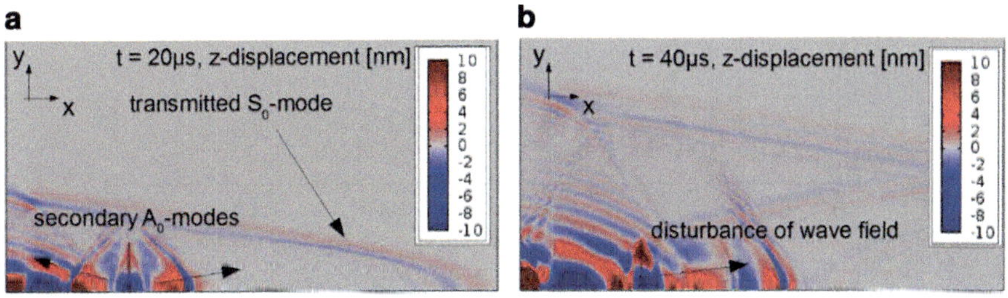

Fig. 4.83 Comparison of simulated wave fields z-displacement for fiber breakage at $t = 20$ μs (**a**) and at $t = 40$ μs (**b**) after signal excitation (reprinted from [62])

interaction, which results in a superposition of the incident A_0 mode with the multiple reflections of the secondary modes. It is worth noting that the model is energy conservative. Still, the calculated amplitude of the signals is distinctly different. This is solely caused by the presence of the broken fibers, which are responsible for a preferential distortion of the displacement field resulting in a transfer of energy from the in-plane displacement to the out-of-plane displacement.

4.3.3.6 Delamination

Inter-ply delamination is one of the most common types of damage found in fiber reinforced composites, since it is often caused by local impact or as residue of a manufacturing error. During mechanical testing of fiber reinforced structures, delamination can evolve step by step and can affect the elastic properties along the signal propagation path. Thus, the influence of delamination on the detected AE signals is of considerable interest. To resemble the variety of dimensions of inter-ply delamination, the size of the delaminated area in the in-plane direction was varied between 5 and 20 mm (cf. Fig. 4.73).

Figure 4.84 shows the calculated signals for all three inter-ply delamination sizes investigated and the reference signal for comparison. For the S_0 mode, there are only negligible differences observed to the reference case. Since propagation of the S_0 mode is dominated by the in-plane stiffness, this is explained by considering a multilayer specimen composed of 950 μm CFRP and 50 μm air. The in-plane stiffness of such a plate is almost 95 % of the stiffness of a 1000 μm CFRP plate.

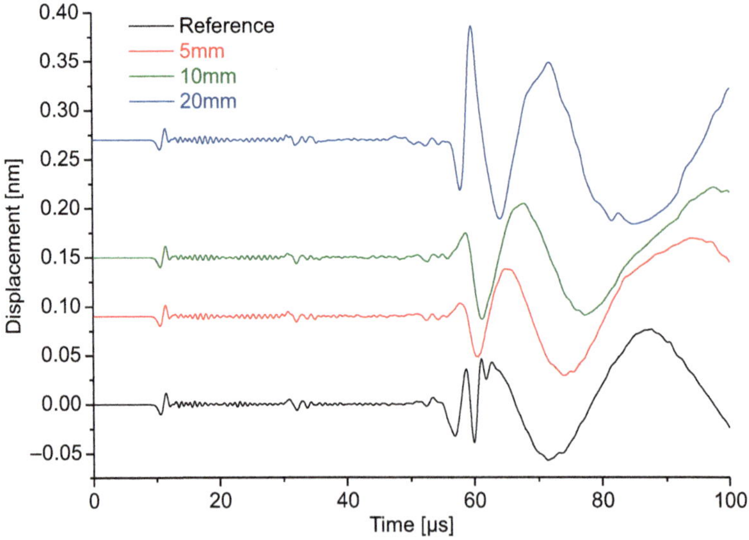

Fig. 4.84 Simulation results of signal after propagation of 100 mm along 0° direction in unidirectional CFRP with inter-ply delamination of various dimensions (reprinted from [62])

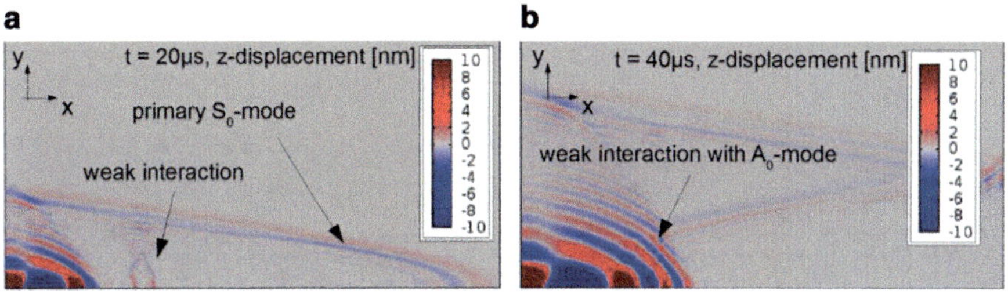

Fig. 4.85 Comparison of simulated wave fields z-displacement for delamination at $t = 20$ μs (**a**) and at $t = 40$ μs (**b**) after signal excitation (reprinted from [62])

A larger influence is observed for the propagation behavior of the A_0 mode. The shape and intensity of the A_0 modes differ from the reference signal due to the change of local bending stiffness as introduced by the inter-ply delamination. With increasing size of the inter-ply delamination, the deviation compared to the reference case increases as well.

As seen in the wave field in Fig. 4.85a, b, for the delamination case, only weak interaction with the S_0 mode and the A_0 mode is found.

4.3.3.7 Influence on Signal Arrival Times

As pointed out in the previous sections, obstacles like rivets, holes, or inclusions as well as the occurrence of damage can have severe impact on the transmittance of acoustic waves. Since damage may evolve during loading of the material, their sudden presence may alter the transmission path between an AE source and the AE sensor. Therefore, the acoustic emission signals calculated for the damage configurations are also evaluated with respect to the arrival time of signal.

In Fig. 4.86, the extracted initial arrival times are shown for all damage configurations (inter-fiber failure, inter-ply delamination, and broken fibers) and the reference case in 0°, 45°, and 90° propagation direction. The largest influence of the internal damage types is found for the 90° propagation direction, which demonstrates that the complete wave field is affected by the presence of the damage and not solely the part of the wave propagating through the obstacle (0° direction).

Based on the maximum deviation of 4.2 μs to the reference case of the calculated arrival time of the S_0 mode in 90° propagation direction, a corresponding error of localization in the range of several centimeters can be expected. The estimation is based on the calculated group velocity of the S_0 mode in 90° propagation direction. A significantly larger influence was found for the arrival time of the A_0 mode (not shown in Fig. 4.86). Due to modal conversion occurring at the internal damage, the arrival of the first detectable A_0 mode is 48.6 μs ahead compared to the reference case in one configuration. However, this is not the arrival of the primary A_0 mode as discussed before. For the majority of the cases studied, the arrival times of the A_0 mode are within 10 μs difference to the reference case. Consequently,

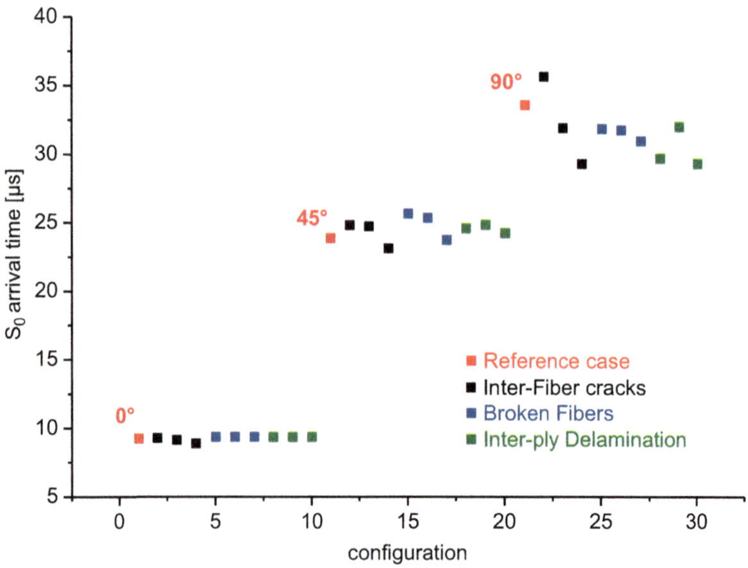

Fig. 4.86 Calculated arrival time of S_0 mode for the various configurations of internal obstacles and internal damage in comparison to the reference case for 0°, 45°, and 90° propagation direction (reprinted from [62])

for localization methods using arrival times of both fundamental Lamb wave modes, the observed difference in arrival time of the A_0 mode might have a large impact as well.

4.4 Detection of Acoustic Emission Signals

The detection of an acoustic emission signal occurs at the surface of a solid. The most common type of measurement uses the sensitivity of piezoelectric materials to detect the wave amplitudes down to the order of 10^{-13}–10^{-14} m [42, 43]. In many cases this piezoelectric element is included within a protective housing and connected to the surface by a viscous coupling medium [72]. There is a large variety of acoustic emission sensors available today, which are almost completely based on piezoelectric detector materials, such as lead zirconate titanate Pb $[Zr_xTi_{1-x}]O_3$ (PZT) or lead metaniobate $PbNb_2O_6$ (PM) [73]. As technical alternative to piezoelectric materials, various other detection methods were proposed. While mostly capacitance sensors were used in the past decades for broadband applications [74, 75], recently other methods using fiber Bragg gratings (resonant) [76–79] or laser interferometers (broadband) have demonstrated their feasibility as approaches to detect AE [80, 81].

The main purpose of the acoustic emission sensor is the conversion of surface motion of the solid to an electric signal. For the piezoelectric sensors, this conversion is based on the piezoelectric effect which is directly related to the occurrence of electric dipole moments within piezoelectric materials. For the case of PZT, these dipole moments result from the asymmetric charge density located within the oxygen octahedron. The magnitude of the dipole moment along the axis of the octahedral is directly proportional to the applied mechanical stress along that direction.

The fundamental equation for the electrical displacement $\boldsymbol{D}$ of a material with electrical permittivity ξ within an electric field with strength $\boldsymbol{E}$ is given by

$$\boldsymbol{D} = \xi \cdot \boldsymbol{E}. \tag{4.43}$$

In combination with Hooke's law, this yields the coupled equations of the piezoelectric effect in stress-charge form:

$$\boldsymbol{D} = \boldsymbol{S} \cdot \boldsymbol{\varepsilon} + \xi \cdot \boldsymbol{E} \tag{4.44}$$

$$\boldsymbol{\sigma} = \boldsymbol{C} \cdot \boldsymbol{\varepsilon} - \boldsymbol{S}^t \cdot \boldsymbol{E} \tag{4.45}$$

The coupling matrix is given by the direct $\boldsymbol{S}$ and inverse $\boldsymbol{S}^t$ piezoelectric constants, which can be written in the stress-charge form for a C_{4v}-crystal class as for the case of PZT using the Voigt notation as

$$\begin{pmatrix} D_1 \\ D_2 \\ D_3 \end{pmatrix} = \begin{pmatrix} 0 & 0 & 0 & 0 & S_{15} & 0 \\ 0 & 0 & 0 & S_{24} & 0 & 0 \\ S_{31} & S_{32} & S_{33} & 0 & 0 & 0 \end{pmatrix} \begin{pmatrix} \varepsilon_1 \\ \varepsilon_2 \\ \varepsilon_3 \\ \varepsilon_4 \\ \varepsilon_5 \\ \varepsilon_6 \end{pmatrix} + \begin{pmatrix} \xi_{11} & 0 & 0 \\ 0 & \xi_{22} & 0 \\ 0 & 0 & \xi_{33} \end{pmatrix} \begin{pmatrix} E_1 \\ E_2 \\ E_3 \end{pmatrix} \tag{4.46}$$

$$\begin{pmatrix} \sigma_1 \\ \sigma_2 \\ \sigma_3 \\ \sigma_4 \\ \sigma_5 \\ \sigma_6 \end{pmatrix} = \begin{pmatrix} C_{11} & C_{12} & C_{13} & 0 & 0 & 0 \\ C_{12} & C_{22} & C_{23} & 0 & 0 & 0 \\ C_{13} & C_{23} & C_{33} & 0 & 0 & 0 \\ 0 & 0 & 0 & C_{44} & 0 & 0 \\ 0 & 0 & 0 & 0 & C_{55} & 0 \\ 0 & 0 & 0 & 0 & 0 & C_{66} \end{pmatrix} \begin{pmatrix} \varepsilon_1 \\ \varepsilon_2 \\ \varepsilon_3 \\ \varepsilon_4 \\ \varepsilon_5 \\ \varepsilon_6 \end{pmatrix} - \begin{pmatrix} 0 & 0 & S_{31} \\ 0 & 0 & S_{32} \\ 0 & 0 & S_{33} \\ 0 & S_{24} & 0 \\ S_{15} & 0 & 0 \\ 0 & 0 & 0 \end{pmatrix} \begin{pmatrix} E_1 \\ E_2 \\ E_3 \end{pmatrix} \tag{4.47}$$

Thus the lattice deformation of a piezoelectric solid results in an electric field, which is called the direct piezoelectric effect and is typically used in sensor applications. Complementary, the application of an electrical field on a piezoelectric material causes a mechanical displacement known as inverse piezoelectric effect which is instead often used in actuator elements.

Although such piezoelectric sensors are very sensitive to detection of surface displacements, they show intrinsically resonant behavior. This resonant frequency response is caused by the piezoelectric material properties and the finite dimensions of the sensing element. For the frequently used geometrical shape of a disk, the possible resonances are for movements along the thickness direction, movements along the radial direction, as well as due to torsional movement. These effects superimpose the frequency characteristic of the detected elastic wave. Consequently, for interpretation of the frequency distribution of the acoustic emission signals, the effect of the sensor has to be taken into account, which is discussed in more detail in Sect. 4.4.1.

A related concept dealing with such characteristic system responses is the convolution theory. The output function $h_A(t)$ of a system is given by the convolution integral of input function $h_E(t)$ and the transfer function $\Xi(t)$:

$$h_A(t) = \int_{-\infty}^{+\infty} h_E(t) \cdot \Xi(t' - t)\mathrm{d}t'. \tag{4.48}$$

The transfer function itself is defined as the response of the system to a δ-function in the time domain as

$$\Xi(t) = \int_{-\infty}^{+\infty} h_E(t) \cdot \delta(t' - t)\mathrm{d}t'. \tag{4.49}$$

For most considerations, (4.48) is transferred to the frequency domain by Fourier or Laplace transformation using the complex frequency ω^*:

$$\widetilde{h}_A(\omega^*) = \widetilde{h}_E(\omega^*) \cdot \widetilde{\Xi}(\omega^*) \tag{4.50}$$

Thus in the frequency domain, the system output is simply given by multiplication of the input function by the transfer function. Moreover, transfer functions of concatenated systems can be written as product of each singular transfer function. With respect to the case of acoustic emission detection, this can be expressed according to [2, 52, 82] as

$$\widetilde{u}_{AE}(\omega^*) = \widetilde{d}_s(\omega^*) \cdot \widetilde{\Xi}_{med}(\omega^*) \cdot \widetilde{\Xi}_{geo}(\omega^*) \cdot \widetilde{\Xi}_{sen}(\omega^*) \cdot \widetilde{\Xi}_{amp}(\omega^*) \tag{4.51}$$

This includes the acoustic emission source-time function $\widetilde{d}_s(\omega^*)$ in the frequency domain, the influence of the propagation medium $\widetilde{\Xi}_{med}(\omega^*)$, its geometry $\widetilde{\Xi}_{geo}(\omega^*)$, the sensors response $\widetilde{\Xi}_{sen}(\omega^*)$, and the characteristic of the amplifier $\Xi_{amp}(\omega^*)$ of the recording equipment. Figure 4.87 shows exemplary frequency spectra of these transfer functions illustrating their influence on the final acoustic emission signal.

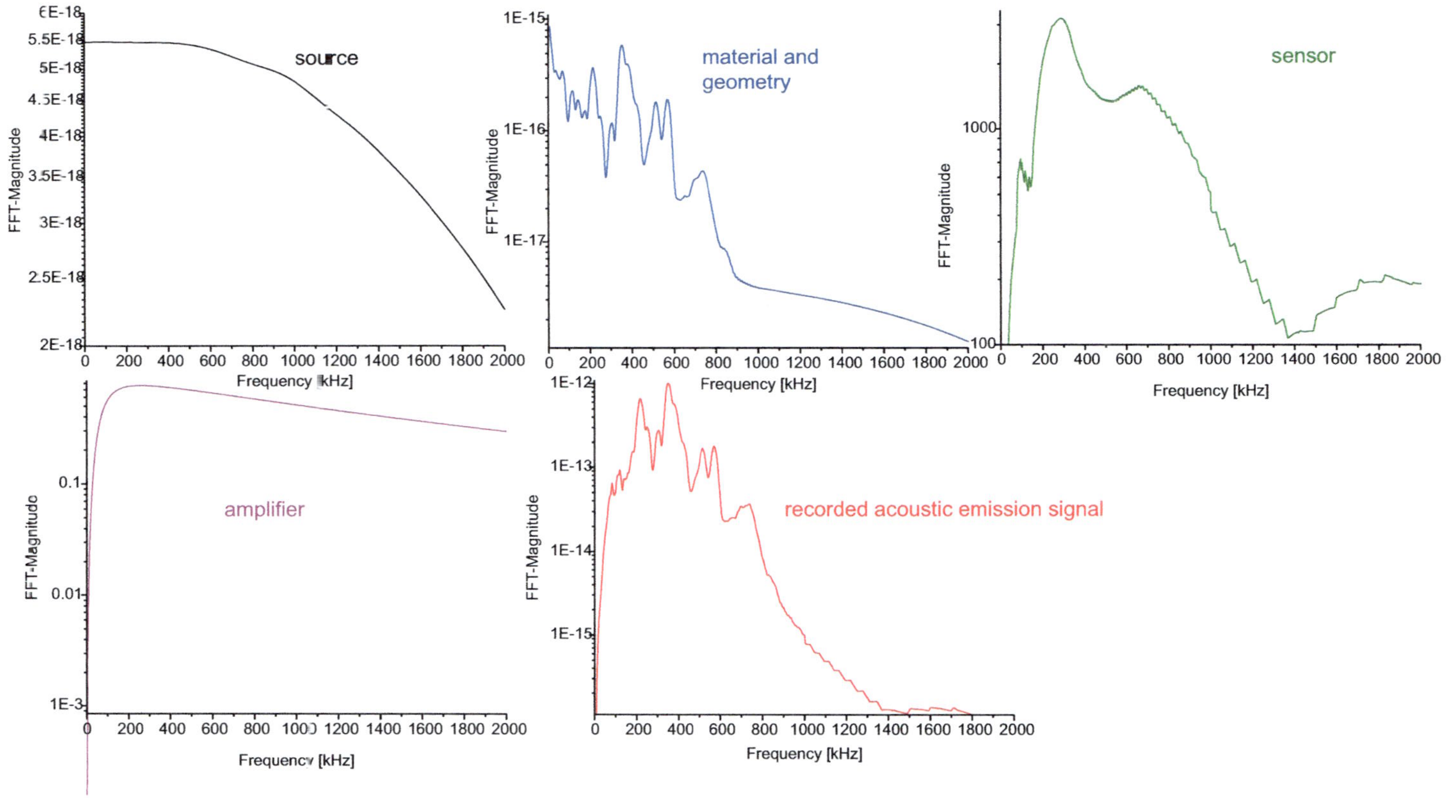

Fig. 4.87 Contributions of the different transfer functions of source, material and geometry, sensor and acoustic emission equipment obtained from FEM-simulations to the final acoustic emission signals, visualized following Bohse [82] (figure reprinted from [35])

As already pointed out by Giordano et al. [2] and Wilcox et al. [10], it is possible to evaluate each transfer function of (4.51) independently. Consequently, an application of the deconvolution analysis according to [27] to (4.51) can deduce the source function $\widetilde{d}_s(\omega^*)$ from the recorded signal $\widetilde{u}_{AE}(\omega^*)$ if all other transfer functions are known.

Usually the influence of the propagation medium $\widetilde{\Xi}_{med}(\omega^*)$ and its geometry $\widetilde{\Xi}_{geo}(\omega^*)$ cannot be separated sufficiently. Moreover their exact appearance is also correlated to the propagation path of the acoustic wave. Therefore analytical solutions for these two contributions are usually approximated by idealistic assumptions on the medium influence and assumptions with respect to the real geometry, typically applying infinite extensions along one or more axis [21, 22, 29, 30].

In contrast, the estimation of the transfer function of the sensor and the amplifier is an experimentally well-established procedure. While the characteristic of the amplifier can easily be obtained by a network analyzer, the measurement of the transfer function of acoustic emission sensors, also known as primary absolute calibration, is more difficult. Several experimental methods are thus established for primary absolute calibration of acoustic emission sensors. The most common are the step-function calibration according to ASTM E 1106 and the reciprocity calibration according to NDIS 2109. Originally, Breckenridge et al. applied a capacitive sensor for primary absolute calibration of piezoelectric sensors [83]. In recent years laser vibrometers are frequently applied to the same purpose [84, 85].

As pointed out by Goujon et al., there is no expectation for an absolute transfer function of the sensor, since the frequency sensitivity characteristic of the sensor also depends on the elastic properties of the medium it is applied on [85]. This effect is caused by the mismatch of acoustic impedance between sensor and medium and is discussed in Sect. 4.4.2.3.

4.4.1 Comparison of Sensor Types

Despite of the common usage of piezoelectric materials as sensing element, there are many different acoustic emission sensor designs available. The selection of the appropriate sensor type follows obvious aspects like operation temperature, but also more complicated parameters like bandwidth and flatness of frequency response. For many applications it is only relevant to detect acoustic emission signals in general, so the frequency characteristics might not be that relevant. Other approaches use the frequency information of the detected signal to perform source identification or modal analysis, which requires sensors with a certain bandwidth. Also, a flat sensor response allows better detection of signal arrival at a certain frequency. This leads to the type of sensor classification into resonant, multi-resonant, and wideband sensor types. Some examples of commercially available

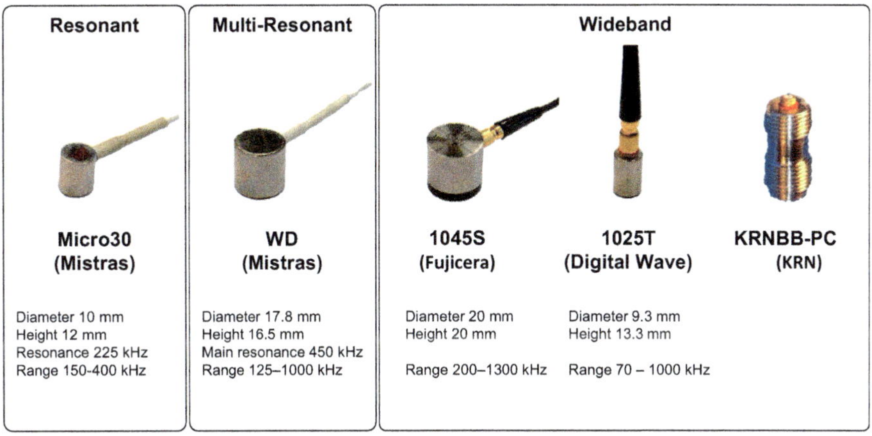

Fig. 4.88 Selection of different acoustic emission sensor types including specifications according to manufacturer's data sheets

sensor types are shown in Fig. 4.88 including their most relevant specifications according to the manufacturer's data sheets.[3]

To compare the response of the different sensors, an aluminum plate of 1500 mm × 1000 mm × 3 mm is chosen with the sensor positioned at the center of the plate. Pencil lead breaks with 0.5 mm diameter and 2H hardness were performed at a distance of 250 mm to the sensor midpoint. Pencil lead breaks were repeated 20 times, and AE signals were detected using a preamplifier with 10 kΩ input impedance and 15 pF input capacitance operating at fixed gain. The mean frequency spectrum was calculated from the 20 detected signals for comparison of sensor response. As seen for the representative data in the time domain, the choice of sensor type has significant impact on the amplitude and shape of the signal. The different sensitivity is owed to the fact that vastly different geometries of sensing elements are used in the type of sensors studied. Clearly seen in the frequency spectra in Fig. 4.89b are the resonances of the Micro30 sensor at around 250 kHz. Here the sensing element is close to its (damped) Eigen resonance and therefore allows for preferential conversion of the incident wave into an electric signal.

However, such resonances limit the effective bandwidth of the sensor and usually cause strong drops of sensitivity outside the resonance range. Improvements of this are given by multi-resonant sensors such as the WD sensor. Using a combination of two sensing elements (cf. [24, 35]), this allows for signal detection with a higher bandwidth, since resonances at 300 and 600 kHz are intrinsic to this geometry. For those sensor types referred to as wideband sensors, a good match to the theoretical frequency content of the pencil lead break source is found. While the 1045S and the 1025T still show some minor resonances, the KRNBB-PC seems to

[3] According to ASTM E976 calibration standard [V/μbar]

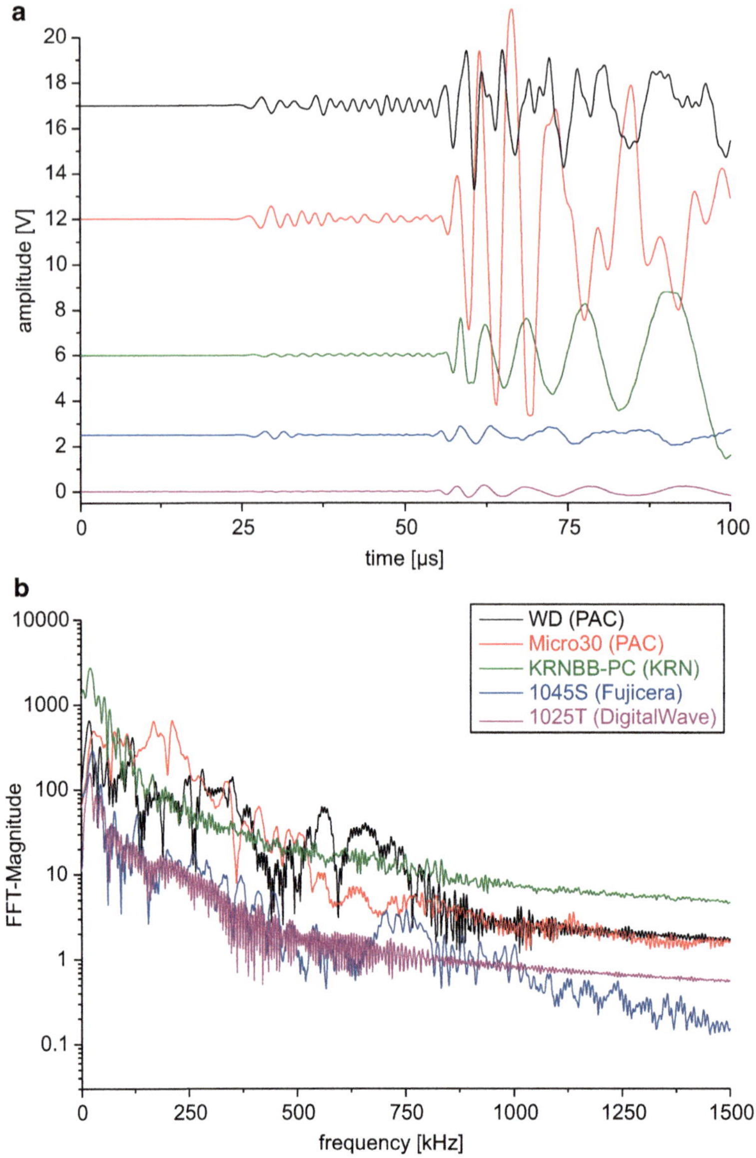

Fig. 4.89 Comparison of detected PLB signals in 250 mm distance in the time domain (**a**) and the respective frequency spectrum averaged over 20 signals (**b**)

match best with the theoretically expected smooth falloff with frequency. The latter sensor type uses a conical sensing element which was found to yield very flat response in the frequency range of interest for acoustic emission analysis [42, 86–89]. However, as seen in Fig. 4.89b within the range of the resonances of other sensor types, the KRNBB-PC has less sensitivity. Therefore, a suitable trade-off has to be made in practice between sensitivity and flatness with frequency when selecting sensor types for a particular application.

4.4.2 Sensor Modeling

To better understand some of the principles of the sensor systems used, several attempts demonstrate the feasibility to model the response of piezoelectric sensors by FEM [33, 87, 90, 91]. For realistic modeling of sensor signals, aspects of signal propagation and sensor modeling have to be combined as demonstrated for a commercial sensor type [35] and a parameter study of conical PZT elements [87]. The modeling strategy presented in this section enables a dedicated analysis of the key factors in sensor design, which are responsible for the different transfer functions observed for the various sensor types [92]. Also, a comprehensive modeling strategy of acoustic emission sensors enhances the possibilities to investigate possible new sensor designs [93–95] and allows new insights into other sensor concepts to detect acoustic emission [95–97].

While acoustic wave propagation is sufficiently described by solving the constitutive equations of structural mechanics, the interaction between acoustic waves and a piezoelectric sensor requires new approaches using coupled partial differential equations [35, 98]. In particular, the generation of electrical charges due to deformation of the piezoelectric material requires simultaneous solving of the electrical and structural mechanics constitutive equations (4.44) and (4.45). Another important aspect in simulation of transfer functions of piezoelectric sensors is the interaction between the sensing material and the attached circuitry. In contrast to conventional open-loop simulations of sensor response, additional input impedances of preamplifiers and cables can change the sensor response significantly and have to be taken into account for realistic prediction of the sensor behavior. To include these effects, a FEM approach is demonstrated in this section following reference [87] with technical details reported therein.

As seen in Fig. 4.90, signal propagation due to a pencil lead break in 250 mm distance is studied in a 3.0 mm thick AlMg3 aluminum plate. All material properties used in the modeling are given in Tables B.1 and B.2 in Appendix B. As a first step, the interaction of the sensor as function of the attached circuitry is discussed. The respective geometric details of the sensor type investigated are shown in Fig. 4.91 with the layout of the attached circuit given in Fig. 4.92.

A comparison of the experimentally detected signal and the modeling result is shown in Fig. 4.93a for the conical sensor type and for the WD sensor type in Fig. 4.93b. Despite of some differences at the beginning of the A_0 mode, the main characteristics of the two different sensor types are well captured by the modeling approach. In particular it is possible to calculate the absolute values of the sensor output since the strength of the pencil lead break source is known (cf. [53, 99]).

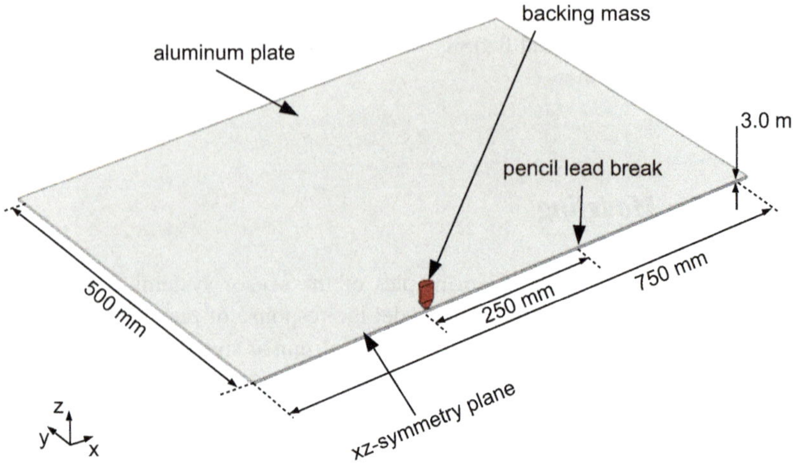

Fig. 4.90 Three-dimensional model setup of experimental configuration (based on [87])

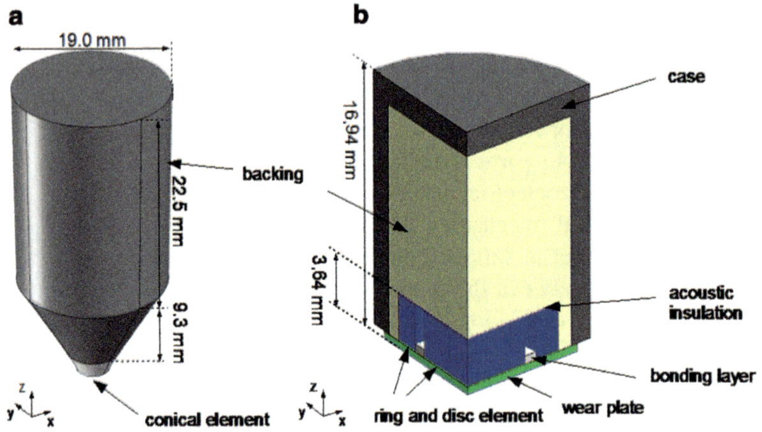

Fig. 4.91 Three-dimensional geometry of conical sensor type used in [87] (**a**) and WD sensor type used in [35] (**b**) (images based on [35] and [87])

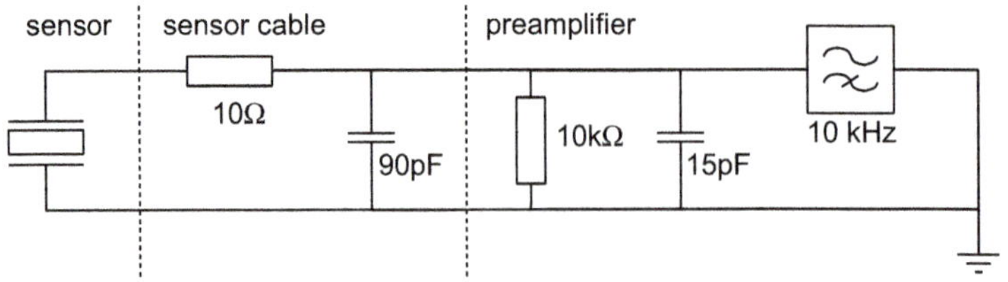

Fig. 4.92 Drawing of electrical circuit used for simulation of connection cable and preamplifier (based on [87])

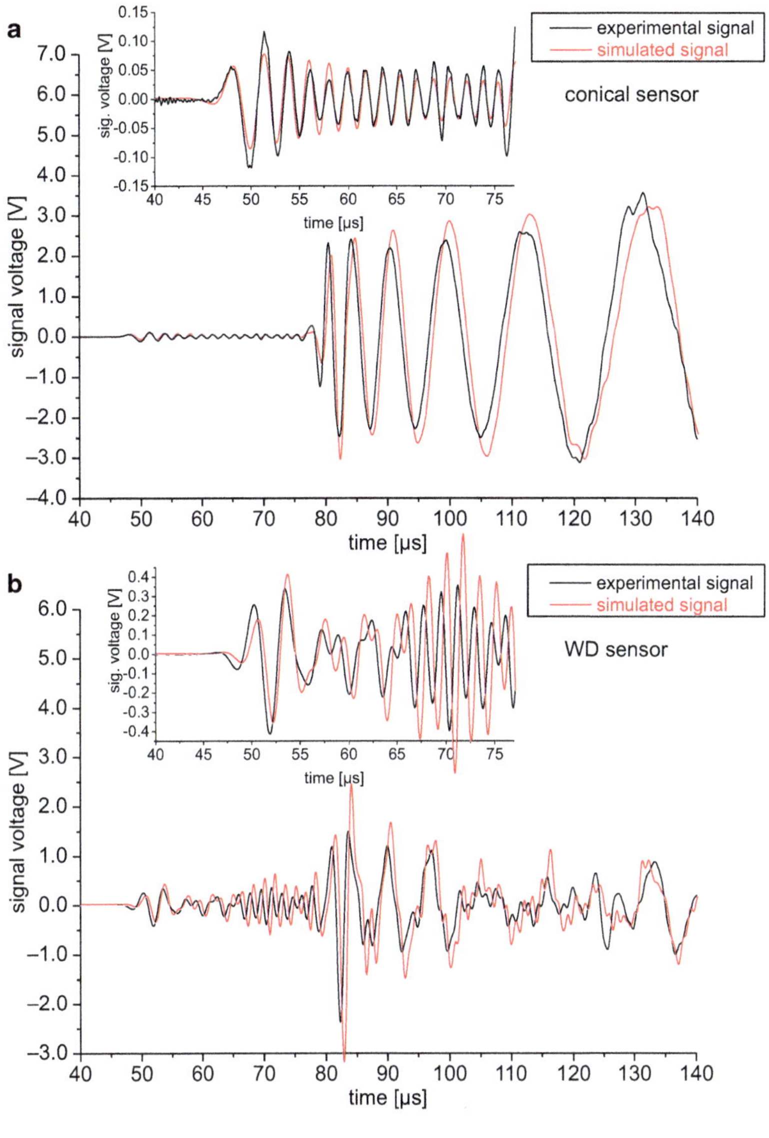

Fig. 4.93 Comparison of experimentally detected pencil lead break signals in 250 mm distance and simulated signals for conical sensor type (**a**) (based on [87]) and WD sensor type (**b**). Insets show S_0 mode arrival between 40 and 75 µs after signal excitation

4.4.2.1 Influence of Attached Circuit

As the first parameter study, the capacitance of the cable model was systematically changed between 10 and 1000 pF, while holding all other circuit parts of Fig. 4.92 constant. The resulting sensor signals of the simulation are shown in Fig. 4.94a.

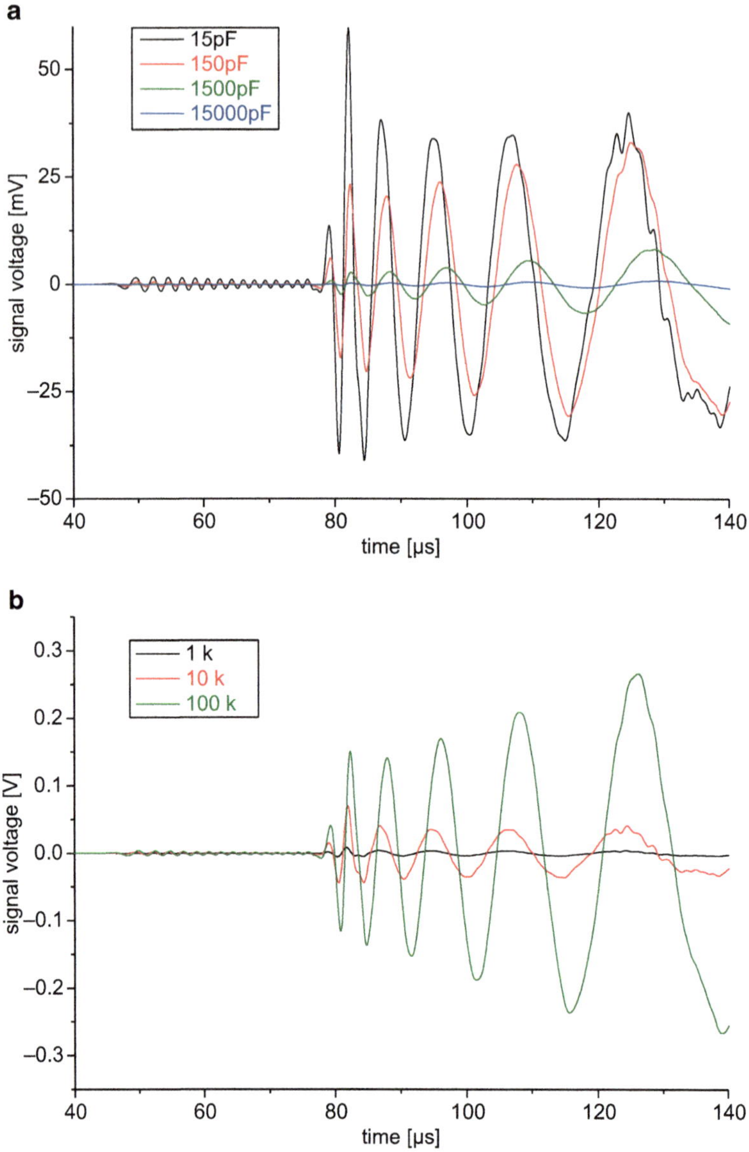

Fig. 4.94 Variation of cable capacitance for 1.50 mm conical element diameter (**a**) and preamplifier impedance (**b**)

In a second study, the resistance value of the preamplifier was changed from 1 to 100 kΩ (see Fig. 4.94b), while the cable capacitance was kept constant. Since the ranges of both cover a broad range of the intrinsic sensor impedance and capacitance (cf. [87]), a strong interaction between the piezoelectric element and the circuitry is found.

The change in signal amplitude due to changes in the resistance and capacitance was of similar magnitude as the geometrical changes of the conical elements. For example, an increase in the capacitance of the piezoelectric element has the same effect as a decrease in the circuit capacitance. This is a consequence of the coupled electrical system of piezoelectric material and the attached circuit. Therefore, a suitable match between the sensor type and the preamplifier has to be assured to avoid a negative impact on the sensitivity of the system and the transmitted bandwidth.

4.4.2.2 Influence of Sensor Aperture

In addition to the attached circuitry, other effects also contribute to the final appearance of the signal. As discussed previously [12, 24, 87, 100, 101], one major influence on the signal shape is the aperture effect due to the wavelengths of the propagating wave. This influence is pointed out in Fig. 4.95. A comparison of a simulated signal without the presence of the sensor reveals the ideal shape of the Lamb wave in the aluminum plate used.

An overlay with the normalized simulated sensor voltage signal for sensor tip diameters of 0.25 and 4.00 mm demonstrates the characteristic differences in the signal shape due to the increase of aperture. The out-of-plane signal is in phase with the sensor signals at the beginning of the A_0 mode, but increasingly deviates at

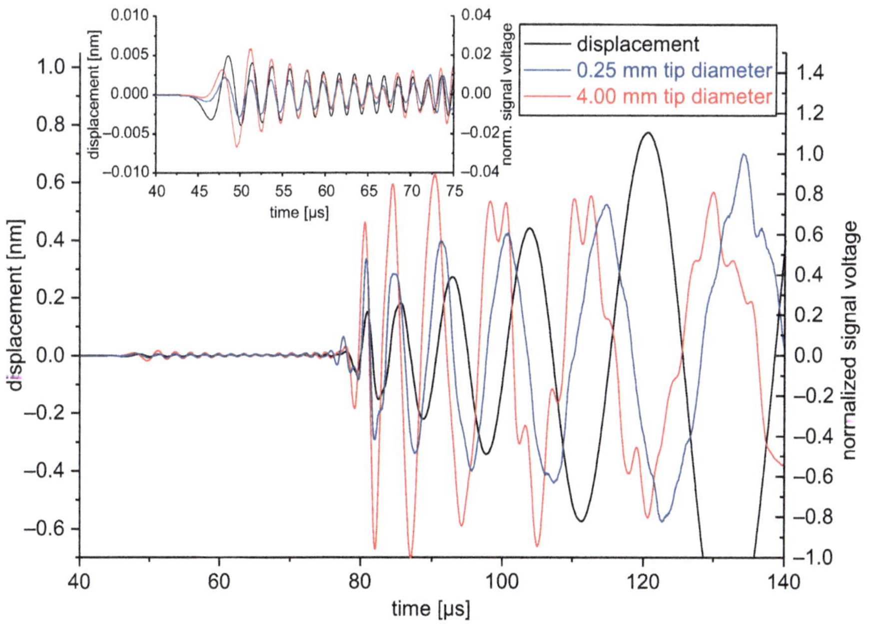

Fig. 4.95 Comparison between out-of-plane displacement at position of sensor midpoint without the presence of sensor model and simulated signal voltage for conical elements with 0.25 and 4.00 mm tip diameter (based on [87])

times larger than about 80 μs. This is caused by the intrinsic phase shift of the piezoelectric conversion and the attached circuit. Despite the phase shift, a better match to the simulated out-of-plane displacement is found for the 0.25 mm tip diameter. Since this diameter approaches a point contact, the effects of aperture become more negligible. In comparison, the 4.00 mm tip diameter suffers from aperture more and thus shows a signal with more difference compared to the original Lamb wave.

4.4.2.3 Influence of Impedance Mismatch

For the detection of an acoustic emission signal, the incident wave needs to be transmitted from the propagation medium to the outermost part of the sensor. This is the casing, the wear plate, or the piezoelectric element. This transmission is subject to the considerations of the transmission and reflections coefficients as given in (4.29) and (4.30). Therefore, the mismatch of acoustic impedance between the sensor and the propagation medium is expected to show a significant role. In addition, the detecting sensor can also be understood as spring-mass system. Consequently, a relation to the stiffness of the propagation medium is also expected. While it is hard to provide a general description of this relationship[4] between the detected signal amplitude, the acoustic impedance, and the stiffness of the structure, it is still worthwhile to assess the magnitude of this effect for an exemplary case to yield the map of expected signal amplitudes shown in Fig. 4.96.

To this end the setup shown in Fig. 4.90 using the conical sensor model is used and the properties of the plate are varied to investigate the difference in detected signal amplitudes. The material properties of the isotropic materials listed in Table B.1 in Appendix B span a broad range of acoustic impedances and Young's modulus values. In addition to the "real" materials, "artificial" materials were used having values of acoustic impedance and Young's modulus with equidistant spacing. This allows interpolating between the different materials and allows computing the full range of possible material properties.

Following [87] the incident acoustic waves are converted into electrical signals using simulation of the piezoelectric conversion process and the attached circuitry. Due to the different sound velocities in the materials, the durations for the individual computations are modified appropriately to allow for efficient computations.

The amplitude evaluated in Fig. 4.96 is the peak value of the A_0 mode propagating in the structure divided by the incident out-of-plane component of the elastic wave. Computations for "real" materials are given by the black dots, and the interpolated field was created from an array of 9×12 "artificial" materials spanning the range between 5.4 and 54.4 MRayl and 2.5–1000.0 GPa.

It can be concluded from the results in Fig. 4.96 that neither the acoustic impedance nor the stiffness of the plate shows strictly linear dependence on the detected signal

[4] Note that [48, 73] discuss the effect of the acoustic impedance but not for incident plate waves.

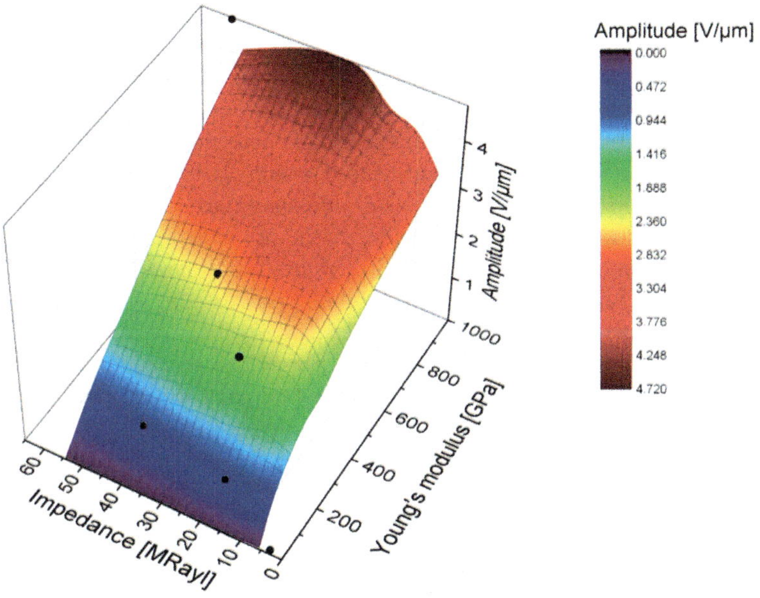

Fig. 4.96 Map of detected signal amplitude for configuration used in [87]. Plate materials are varied using "real" materials (*black dots*), and interpolated sensor response field is computed from "artificial" materials

amplitudes. Moreover, the computed behavior seems to behave nonproportional in some regions. However, there is a strong tendency seen for a change of the Young's modulus of the material. Here, the stiffness effect seems to overrule the effect of acoustic impedance mismatches. For those materials with the highest Young's modulus, the detected signal amplitudes are more than 300 times larger than for the materials with low Young's modulus. In comparison, the largest effect for a variation of the acoustic impedances was observed to be in the range of a factor of 1.3.

These findings strongly indicate that the detectable signal amplitudes for the same type of sensor are unique for every material. Therefore, calibration curves of a sensor are unlikely to be representative for inspection of other materials than the material used for the calibration. Since these calibration routines are usually carried out on metallic materials (i.e., steel or aluminum), the direct transfer to fiber reinforced composites might be questionable.

4.4.3 Waveguides

Another particular challenge arises if test temperatures exceed the operation temperatures of piezoelectric sensors. In the past, suitable alternatives to PZT were investigated to extend the accessible temperature range. One solution to avoid using such PZT sensors at operation temperatures outside their specifications was found by the researchers like Lynnworth et al. and involves the usage of thin rods, clad rods, or hollow tubes as ultrasonic waveguides for signal detection [102]. They demonstrated the use of such waveguides to duplicate acoustic emission signals propagating as guided waves in a plate. Also various short waveguides with different material properties, length, diameter, and detection angles were tested in order to investigate the effects upon pulsed events [102, 103]. However, the selection of appropriate waveguides is still an experimental procedure and relies on the experience to a large extent. Although the principle description of guided wave modes in rods is found in standard literature [104], numerical methods are nowadays also found to be a convenient tool to assess the properties of such waveguides prior to experimental design [105, 106].

Following the established and validated numerical procedure described in [105, 106], the response of a model sensor is evaluated in various geometrical configurations and material pairings. In the following, effects of the waveguide termination, the diameter, material, and temperature gradients are briefly discussed.

As seen in Fig. 4.97, the computations are performed for one reference case with the sensor directly attached to a 3.0 mm thick aluminum plate. As an acoustic emission source, a pencil lead break test in 100 mm distance to the sensor location is used. To reduce the computational intensity of the model, a symmetry condition is chosen along the *xz*-plane. The details of the model sensor are indicated in Fig. 4.97 and confirm with the previous choice in [87] (see also Sect. 4.4.2). As the second model, the propagation path between source and sensor is modified by incorporation of an acoustic waveguide as seen in Fig. 4.97. Based on an experimentally used design, the diameter is selected as 1.59 mm and the length as 306.4 mm.

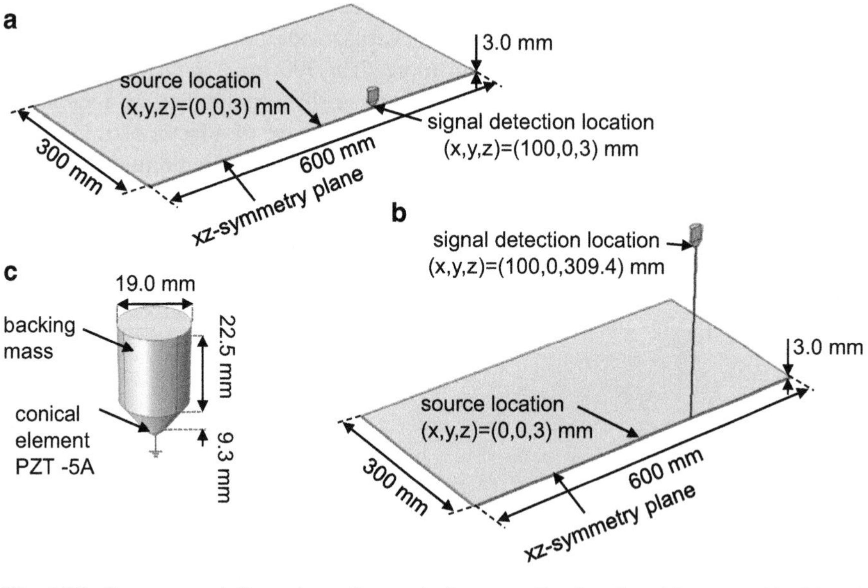

Fig. 4.97 Geometry and dimensions of numerical case studies for plate (**a**), waveguide (**b**) and sensor model (**c**) following [105, 106]

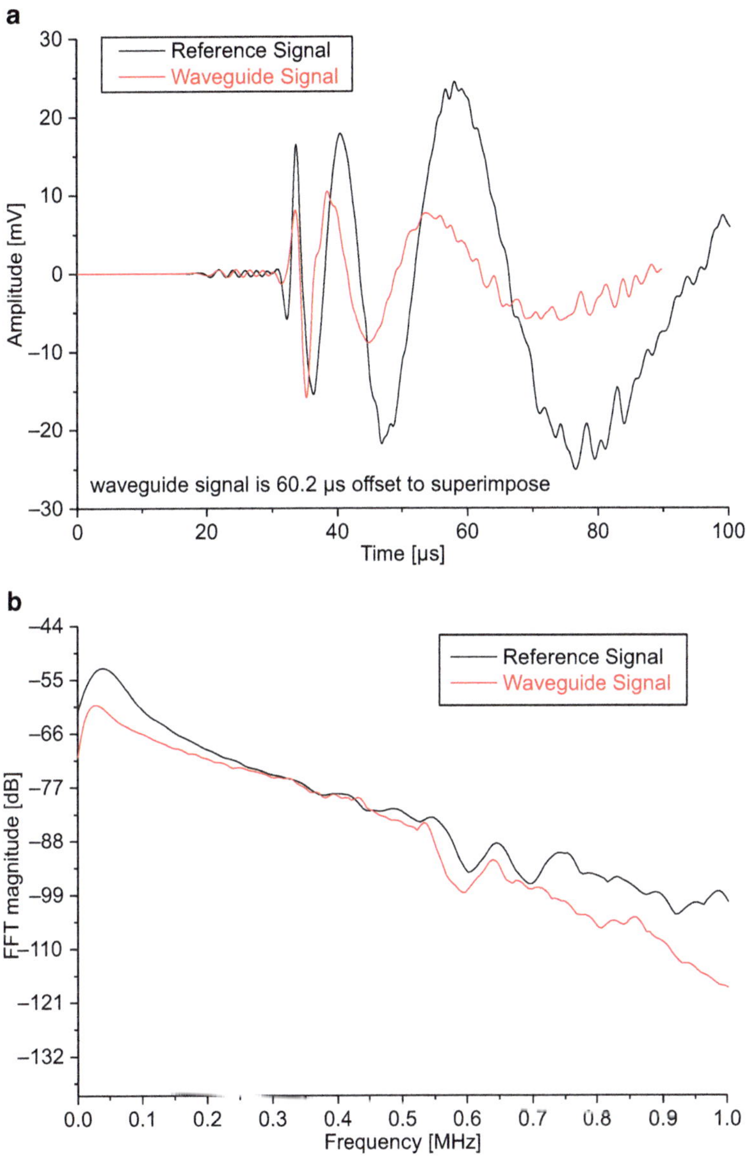

Fig. 4.98 Comparison of detected acoustic emission signals after propagation in aluminum plate and after additional propagation in acoustic waveguide (**a**) and respective frequency spectra (**b**). Signal detected at the top of waveguide in (**a**) is shifted forward by 60.2 μs to compensate for the effect of signal propagation within the waveguide

The result of both computations is shown in Fig. 4.98. In order to allow better comparison of both acoustic emission signals, the result including the computation with the waveguide is shifted forward by 60.2 μs to compensate for the effect of different propagation distances.

Overall, both signals still show some resemblance, but the frequency content and the amplitude of the individual guided wave modes are not fully retained. This is also seen in the comparison of the frequency spectra of both signals in Fig. 4.98b. In the present case, the signals propagate as guided plate waves in the aluminum plate. At the position of the waveguide, the out-of-plane components are transmitted into the waveguide. At this position, the signal transmittance is governed by the impedance match of the waveguide to the plate and the aperture of the waveguide. After transmittance into the waveguide, the wave propagation starts to transform into the guided wave modes of the waveguide geometry (e.g., rod modes). The final signals' time-frequency characteristic at the top of the waveguide thus suffers from two types of guided wave propagation and modal conversions, respectively. This causes characteristic deviations to the acoustic emission signals detected by a sensor mounted directly on the structure of interest, which is further denoted as reference signal.

4.4.3.1 Shape of Waveguide

In order to ensure stable mechanical attachment of the waveguide and the acoustic emission sensor, several terminations are typically used in practice. Therefore it is helpful to understand their impact on the signal transmission when being used in an application. To this end, [105] has investigated several terminations, keeping the diameter and material constant. As seen in the frequency spectra of the acoustic emission signals detected on the top of the waveguide in Fig. 4.99, the various configurations do cause substantial differences in the transmitted signals.

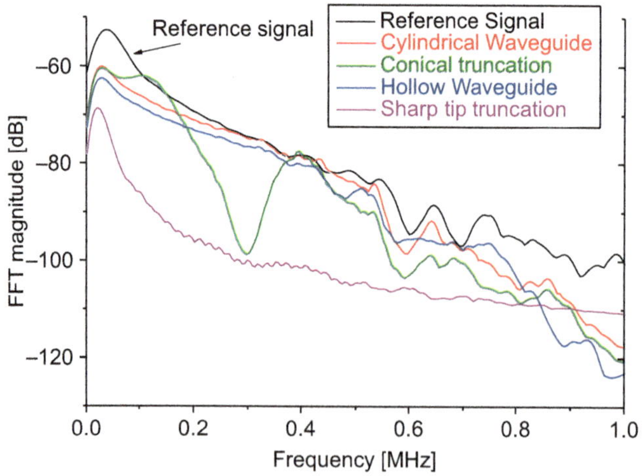

Fig. 4.99 Comparison of frequency spectra of signals detected with different waveguide geometries

Relative to a simple circular waveguide truncations like cones or sharp tips may either cause resonances (especially the conical truncation at the sensor position) or may reduce the detectable signal amplitude by up to 17 dB. In comparison, the use of hollow waveguides or such filled with liquids was found to be of less impact concerning the waveguide transmission characteristics in the investigated diameter and frequency range (for details on dimensions, see [105]).

4.4.3.2 Diameter of Waveguide

Based on the intended application, it might be required to select a particular minimum waveguide diameter to offer a certain mechanical strength or stiffness. On the contrary it might be necessary to select a certain maximum diameter to limit temperature reduction at the contact position with the object or to confirm with certain lead-through dimensions. For cylindrical geometries, the increase of the diameter will directly increase the possible number of guided wave modes (see Sect. 7.2.4.2 for an example). Mixed mode propagation and frequency specific attenuation will thus affect the transmitted acoustic emission signals. Considering the same model geometry as before, Fig. 4.100a presents some signals detected on top of cylindrical waveguides with different diameters. Apart from the shift in intense frequency regions, the signal amplitude is also affected. As seen in the evaluation of the signal amplitude at a constant frequency of 450 kHz in Fig. 4.100b, there is a preferred waveguide diameter at around 2.2 mm, which offers highest signal detectability at the top of the waveguide. However, this is not a simple resonance of the waveguide, but a specific transmission characteristic of the waveguide at 450 kHz. Based on the material of the waveguide and additional effects of the truncation at the end of the waveguide, the regions of preferred frequency transmission may shift and therefore another diameter would be the optimal choice. Thus, for practical applications, it is recommended to perform a sensitivity analysis for typical acoustic emission signals found in the structure under test.

4.4.3.3 Material of Waveguide

As indicated in [106], the impedance mismatch between the structure under test and the waveguide will also cause noticeable differences in the detectable acoustic emission signals. This is owed to the law of refraction as seen in Fig. 4.57 and causes the same consequences as already pointed out in Sect. 4.4.2.3. Thus it is recommended to use waveguide materials with acoustic impedance close to the material under inspection to avoid strong impedance gaps. If this is not feasible, the acoustic impedance of the waveguide should be selected larger than that of the material under investigation to provide good acoustic transmission into the waveguide. However, this can cause additional modal conversion, therefore reducing the effective transmitted amplitude.

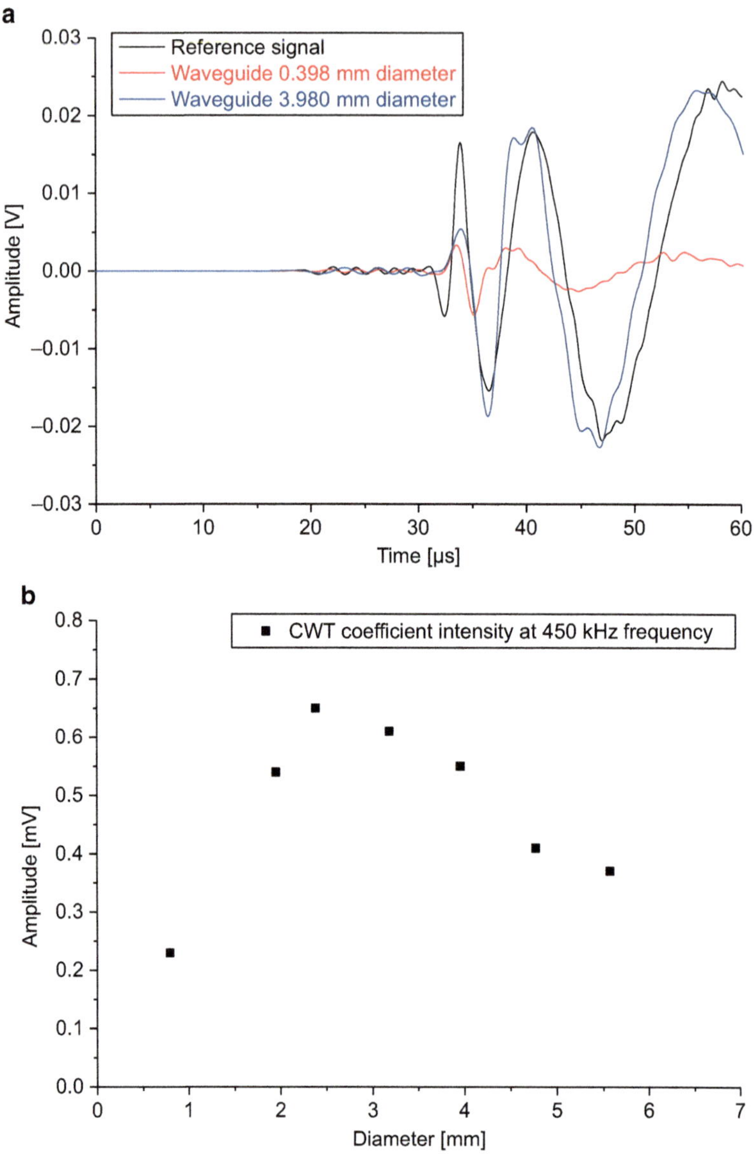

Fig. 4.100 Comparison of different acoustic emission signals detected at the top of waveguide with different diameter (**a**) and evaluation of maximum signal amplitude at constant frequency of 450 kHz using continuous wavelet transformation (**b**)

4.4.3.4 Influence of Temperature Gradient

Many waveguide applications deal with objects operating or being loaded at elevated temperatures. Since typical acoustic emission sensors may not exceed their operational temperature range, the waveguide is selected long enough to reduce the temperature from the elevated temperature to safe operation conditions for the acoustic emission sensor. In some applications this temperature gradient within the waveguide can be in the order of several 100 °C or even up to 2000 °C. Accordingly, the waveguide material changes its material properties as function of the local temperature. Depending on the sensitivity of Young's modulus, Poisson's ratio, and density of the material, the wave propagation is noticeably changed. An exemplary calculation for an UNS K12211 steel alloy with a temperature gradient of 500 °C was previously presented in [106]. The findings did not reveal a strong impact of the waveguide operation due to the existence of the temperature gradient. Except for a delay in signal arrival of 1.65 μs, both signals were almost identical. However, other waveguide materials may exhibit stronger changes of their material properties as function of temperature and therefore may affect signal propagation more than this.

4.4.4 Other Factors Affecting Sensor Sensitivity

Beyond the specific choice of the sensor type, several other factors will affect the detectable frequency range and the sensitivity of the acoustic emission sensor. Apart from aging of the sensor itself and environmental test conditions, these are the choice of the coupling medium and the way of sensor fixation.

4.4.4.1 Coupling Medium

In principle, the purpose of the coupling medium is to avoid air gaps in between the object under test and the acoustic emission sensor. Typically, the test object surface is not fully flat or comes with a certain roughness, so the planar sensor surface would only be coupled at distinct locations. Having a grease or a liquid as coupling medium can thus fill these gaps and ensure a good acoustic transmission. Typically, coupling media are viscoelastic materials causing higher acoustic attenuation at higher frequencies. The choice of coupling media may thus limit the bandwidth of the measurement chain as seen in Fig. 4.101. The spectra shown are averaged out of 20 pencil lead break signals, which were detected by a WD type sensor under identical conditions on an aluminum plate except for the choice of couplant. Obviously the hot glue reduces the sensitivity at higher frequencies when compared to the less viscous Baysilone silicone grease couplant.

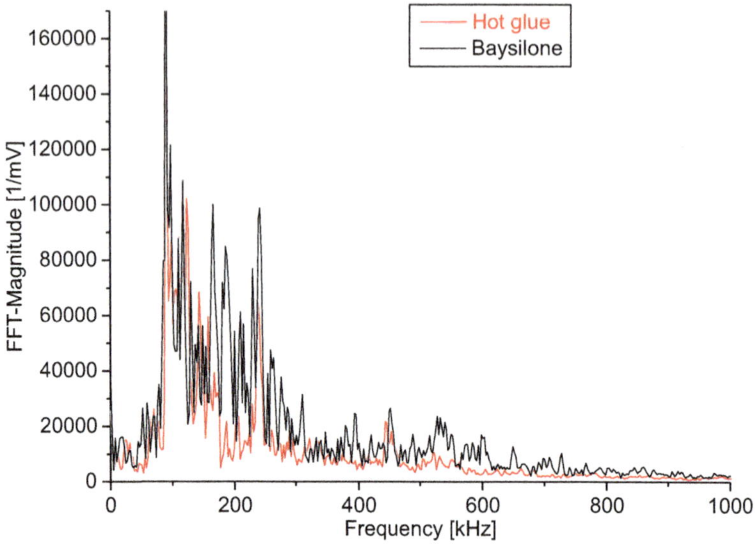

Fig. 4.101 Comparison of frequency spectra recorded under identical conditions except for the choice of couplant

In combination with temperature ranges other than room temperature, several other factors may limit the applicability of coupling media. At temperatures <0 °C included water or the coupling medium itself may freeze, causing microcracking with noticeable acoustic emission. Although this acoustic emission is often with relatively low amplitudes, it may cause a substantial amount of additional signals, which may interfere with acoustic emission signals emitted by the test object. For such applications a coupling medium needs to be selected which does not show intrinsic acoustic emission activity in the intended temperature range.

In the range above room temperature, the primary challenge is to avoid degradation of the coupling medium. This can be assessed prior to the monitoring task by a face-to-face arrangement. Here, one sensor is used as a pulser and another sensor is used as a detector, both coupled together in a face-to-face arrangement using the coupling medium to be tested. While the pulser is emitting a defined broadband pulse at distinct intervals, the whole configuration is subject to continuous temperature variations. The evaluation of the detected signal amplitudes is shown in Fig. 4.102 for three different types of coupling media and one dry contact coupling. As seen from the difference in signal amplitude at room temperature, the different coupling choices come with different initial sensitivities being worst for the dry coupling. During the temperature increase up to 200 °C, the detected signal amplitude decreases, which is mostly due to the thermally induced expansion of the overall face-to-face arrangement and the respective changes in the contact force. Still, the overall loss in signal amplitude is still different for each coupling medium. Upon cooling, the coupling media show hysteresis effects of the detected signal amplitudes. Back at room temperature, only the Apiezon has returned to its previous sensitivity. The other two choices do not get back to their original sensitivity within a range of ±2 dB, thus indicating possible degradation effects of the coupling medium. Such procedures allow to evaluate the performance of the coupling medium prior to any testing.

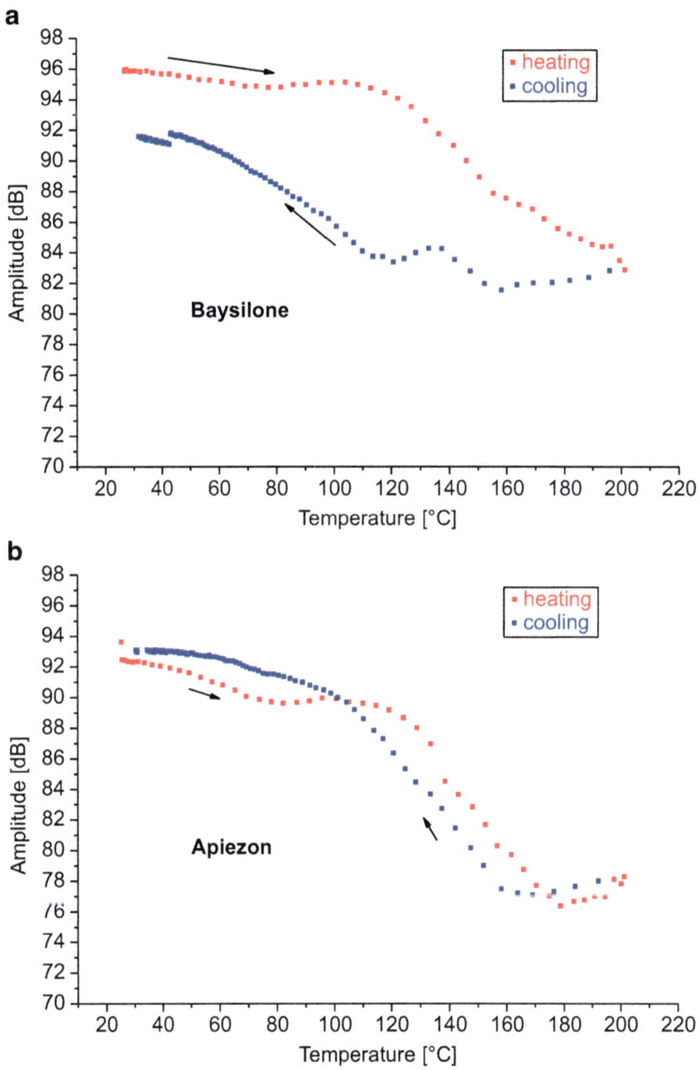

Fig. 4.102 Detectable signal amplitude for various coupling media (**a**, **b**, **c**) as function of temperature in a face-to-face configuration. Reference case without coupling medium in (**d**)

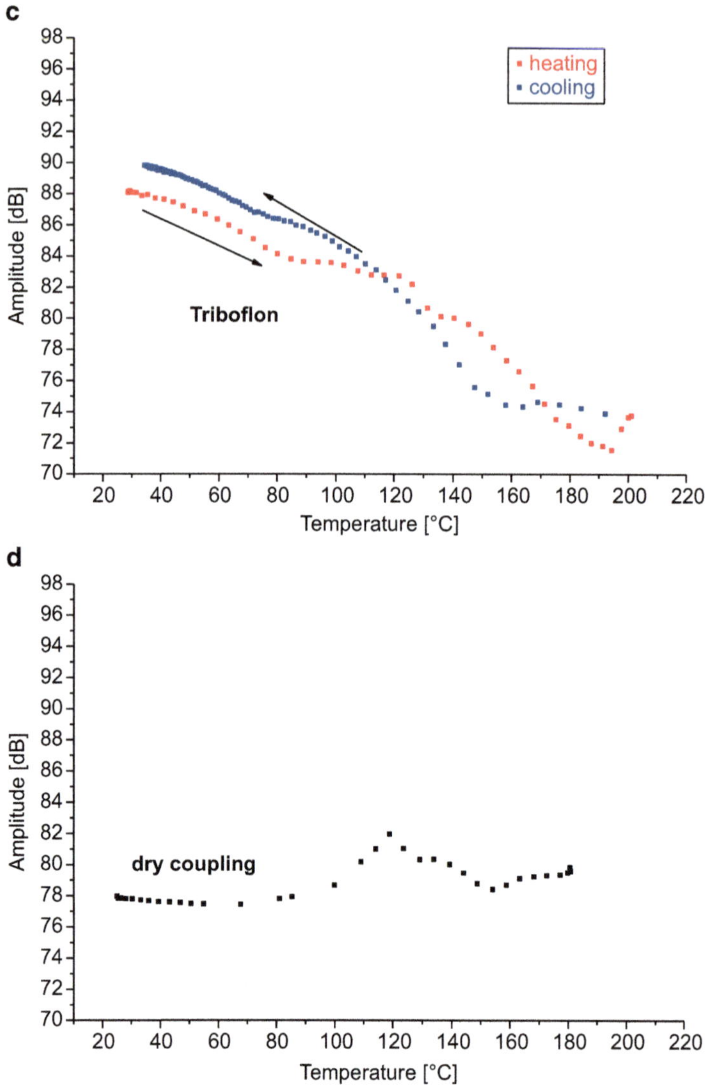

Fig. 4.102 (continued)

4.4.4.2 Sensor Fixation

Apart from the coupling medium, another critical item for acoustic emission sensors is the way of sensor fixation during loading of the test object. Basically, the coupling medium may be categorized in those that bond the acoustic emission sensor to the structure and those which do not allow to carry its weight.

For the bond couplings, all types of permanent adhesives, hot glue, as well as two-sided duct tapes are typically used. Despite of their ease of handling, the main drawback is to assure good bonding quality during loading of the test structure and possibly also during temperature changes and other superimposed environmental factors. Especially for fiber reinforced materials, the adhesive needs to be compatible with the matrix material and the sensor casing material (typically metallic or ceramic). Upon large deformations or superimposed temperatures, this may often lead to failure of the adhesive, thus causing additional acoustic emission signals or drop-down of the attached sensor. Also, the choice of adhesive may affect the frequency-dependent sensitivity as seen in Fig. 4.102.

As alternative, the acoustic emission sensor can be attached to the test object using a viscous or liquid coupling medium, but a mechanical force is required to carry the weight of the sensor and to keep it in its designated position. To this end, several fixation systems are established. Custom-made systems operating from mechanically stable platforms or fixtures can be used as well as clamp systems, rubber bands, (external) tacky tapes, suction cups, or magnetic holders. In Fig. 4.103 some of these concepts are shown in exemplary applications. Note that the magnetic holders are somewhat restricted in terms of composite applications, since these require a ferromagnetic material to be attached to. For many of these systems, their main advantage is the ease of handling and the instant contact of the sensor with the structure (as compared to adhesives which might need time to cure). Also, unmounting and repositioning of the sensor is fairly easy, since no permanent contact is made with the test structure.

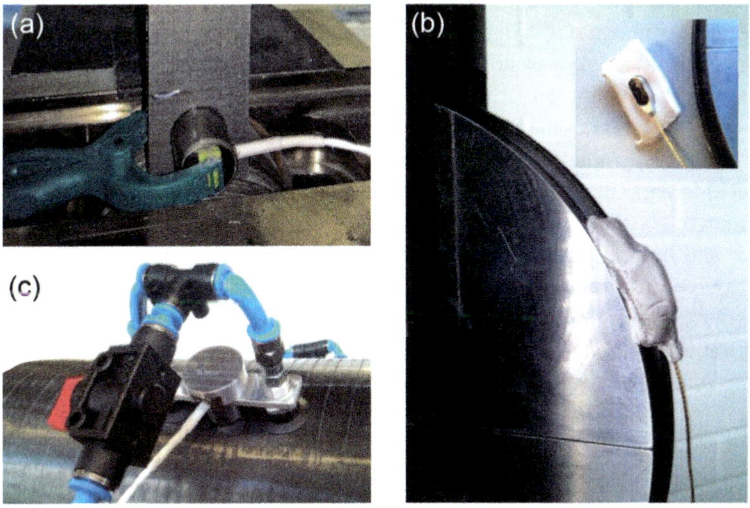

Fig. 4.103 Sensor fixation using clamp system (**a**), (external) tacky tape (**b**), and suction cup holders (**c**)

4.5 Signal Classification

In recent years, many new approaches were proposed that add the possibility of AE source identification to the established approaches of AE signal accumulation and AE source localization. Namely, these are moment tensor inversion, time reversal approaches, guided wave analysis, and pattern recognition techniques.

In volumetric media, one suitable approach to deduce the source type is to use the moment tensor inversion. A comprehensive review on this topic was recently given in [52], so only a brief overview will be provided in the following. Based on the type of source description by the moment tensor as given in Sect. 4.2, the basic measurement principle stems from the acquisition of a set of signals at different angles to the source position. Using these different signals and knowing the Green's function of the medium, the moment tensor components are readily computed by a deconvolution analysis after Wadley [27]. In this field, Ohtsu and Ono demonstrated that the moment tensors from different crack modes in volumetric media are readily distinguishable (see [18, 21, 52]). Green extended this approach used in isotropic media to the needs of acoustic emission sources in composite laminates [29, 30]. Despite these valuable efforts, there are several drawbacks associated with the moment tensor inversion in application to fiber reinforced composites. One challenge stems from the reduced field of view on the source in the typically thin composite materials. Since signals propagate as guided waves, much of the information on source orientation is lost at the first millimeters of propagation. Therefore, even for thick laminates, the different observation angles of the source required for moment tensor inversion are difficult to obtain in real testing of composite materials or structures. Another challenge is the precise knowledge of the Green's functions of the medium. Although it was demonstrated that these can be computed practically for every piece of material, it is still burdensome to achieve in practice.

However, if moment tensor inversion is performed on a composite material, it is a well-suitable technique to classify signal origins in different source radiation types. This yields valuable insight on the material failure at this level.

In recent years, the approach of time reversal acoustic emission has emerged as an alternative source localization strategy [107–109]. Here the detected signals are inversed and basically back projected to determine their signal origin. Beyond the spatial source position, this also yields further information on the radiation pattern of the source. This was successfully demonstrated for measurement of the crack orientation in thin plates [109]. Consequently, the time reversal approach might be used for composites to extract further information on the source origin. But still the approach faces the same challenge as the moment tensor inversion, which is the knowledge of the Green's function of the system required to allow a valid interpretation of the source activity.

Another classification technique more adapted to composite materials is the analysis of guided waves. This approach has a prime focus on the relationship between certain source types and the type of guided wave modes found in platelike structures [6, 67, 110–117]. The aim is to find characteristic ratios of certain guided wave modes, which are characteristic of a particular source type. Although such relations are partially expected, it is still hard to allow for unique assignments of such guided wave ratios to source mechanisms (see Sect. 4.2). The first publications of this approach investigating the relation between source type and guided wave formation attracted much attention and allowed for better understanding of the signal propagation in platelike materials. Also the pattern recognition techniques presented in the following might be understood as more general classification routine following the idea of guided wave acoustic emission analysis.

In the context of acoustic emission analysis, the purpose of pattern recognition techniques is the identification of similarities in the recorded acoustic emission signals. In contrast to conventional feature-based interpretation of the signals, pattern recognition techniques use a multitude of features and consequently can identify distinct signal types even if they cannot be described by fixed feature limits. Basically, pattern recognition methods can be separated into two approaches, namely, supervised and unsupervised pattern recognition.

The task of unsupervised pattern recognition is to separate a set of given objects into distinct groups according to their similarity to each other without any previous knowledge. Various approaches with a focus on detection of characteristic similarities of the recorded signals have been published [7, 24, 77, 118–130].

As distinctly different approach, supervised pattern recognition techniques consist of two subsequent stages. In the supervised stage, a set of objects with known assignment to the respective classes is prepared. This assignment is usually denoted labeling. Here an algorithm is trained to recognize these types of objects based on a given set of features. In the subsequent stage, the algorithm is applied on objects with unknown assignment and classifies them based upon their similarity to the object classes provided in the supervised stage. To this end, discrimination between noise and non-noise acoustic emission signals is often achieved by unsupervised pattern recognition techniques [118, 119]. Acoustic emission signals originating from friction and electromagnetic inductions can easily be identified in this respect, due to their inherent characteristic difference to transient acoustic emission signals. Utilizing suitable experimental considerations and finite element simulations, the respective signal types can also be associated with specific failure mechanisms [7, 24, 120, 122–125, 131–134].

In terms of pattern recognition techniques, the problem faced is the identification of natural clusters of acoustic emission signals. One of the major difficulties in this concern is an adequate evaluation of the classification results of different clustering approaches. There are many ways to choose features, normalization procedures, and algorithms that will influence the result of the classification process.

The first step in every signal classification procedure is a suitable data reduction of the acquired AE signals. This comprises elimination of obvious noise signals and a suitable strategy to focus on the AE signals relevant for material failure (see Sect. 4.5.1). Subsequently, the detected AE signals are reduced to a number of features calculated from the signals as defined in Appendix C. A number of those extracted features are then used as dataset and are investigated by an unsupervised pattern recognition method to yield groups of similar AE signals (AE signal clusters). These basic steps used in the following are schematically shown in Fig. 4.104. As a last step, a suitable validation of the resulting partition is

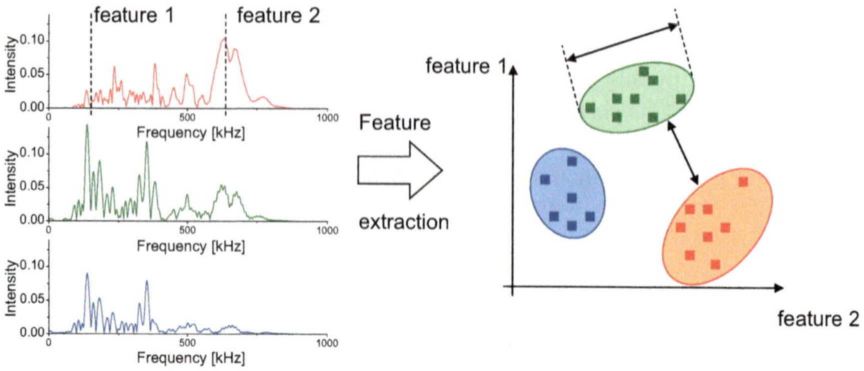

Fig. 4.104 Three exemplary frequency spectra resulting in three data points after feature extraction process. Evaluation of many signals potentially results in cluster formation for frequency intensity at position feature 1 and feature 2. *Ellipsoids* indicate extension of clusters and can be evaluated by outer diameters and relative distances

mandatory. Since a partition is by definition an arbitrary grouping of signals, one has to assure that the grouping is meaningful in a physical sense. For that purpose, cluster validity measures can be used to compare partitions obtained throughout different pattern recognition approaches.

4.5.1 Recommended Practices Before Starting Signal Classification

As seen in Fig. 4.105, generally one can distinguish between continuous acoustic emission and transient acoustic emission (burst) signals. Both types are encountered in fiber reinforced materials, and the transition between the observations of either type is also subject to the loading rate of the experiment, the inspected volume, and the general acoustic emission activity of the inspected material. Ultimately if damage progression in the material is fast and many AE sources are operative, the individual burst signals superimpose and merge into a continuous AE release.

In the following all descriptions refer to the burst-type signals, since they can be attributed to a particular source type, whereas the interpretation of continuous acoustic emission has to use fairly different concepts [52].

Before starting a source discrimination routine in terms of failure discrimination, some practices are suggested in order to allow for better interpretation of the data. The first part consists of a suitable removal of noise signals potentially present in the recorded dataset. Some recommendations are to look at the distribution of feature values as seen, e.g., in Fig. 4.105:

- Signals with durations at the limits of the detection system indicate signals with long durations, potentially from continuous acoustic emission sources, such as friction.

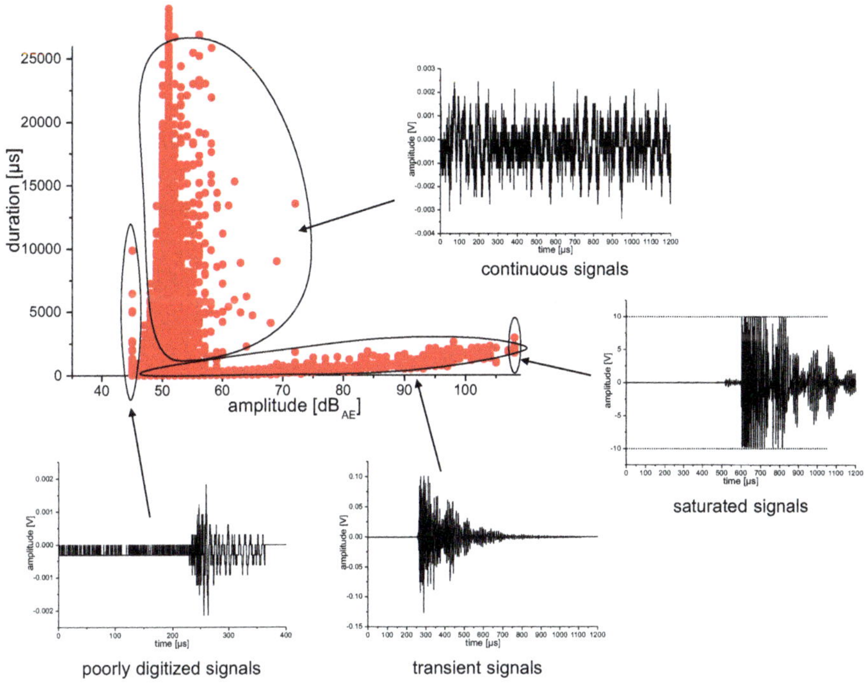

Fig. 4.105 Typical diagram of amplitude versus duration of acoustic emission signals with noise signals

- Signals with very short durations are sometimes just electromagnetically induced noise (spikes).
- Signals with very low amplitudes may fall into the digitization limit and may not yield accurate feature values.

Another recommendation is to use source localization routines to increase the level of relevance of a particular signal origin. If the acoustic emission of a source is recorded by more than one sensor within a certain time interval, this provides additional evidence that the signal is within the region of interest of interpretation (c.g., the gage section of a specimen). This excludes all signals outside the region of interest and consequently is sometimes referred to as Δt-filter. This can inhibit triggering if there is no signal arrival within a certain Δt at two or more sensors (see also Fig. 4.122). Also, the same concept can be used in a postprocessing step, eliminating all signals for further analysis, which are not detected at more than one sensor within a certain time interval Δt. If the source can also be localized[5] in this region of interest, this indicates even higher relevance for the interpretation of the respective acoustic emission signals.

[5] Note that the arrival of a signal at two or more sensors within a certain time interval does not necessarily imply that the source position can be determined by a source localization algorithm.

If there is no possibility to remove "noise" signals, such as spurious signals or continuous signals from transient pulses, one can also specifically train neural networks to detect the "noise" signals based on prototypes of "useful" signals and "noise" signals. For such an application, one can choose a multitude of features for description of the signal types and train a neural network to distinguish the two groups of signals. As seen in Fig. 4.106, reapplication of such a neural network then can be used to automatically classify and select the signal types of interest.

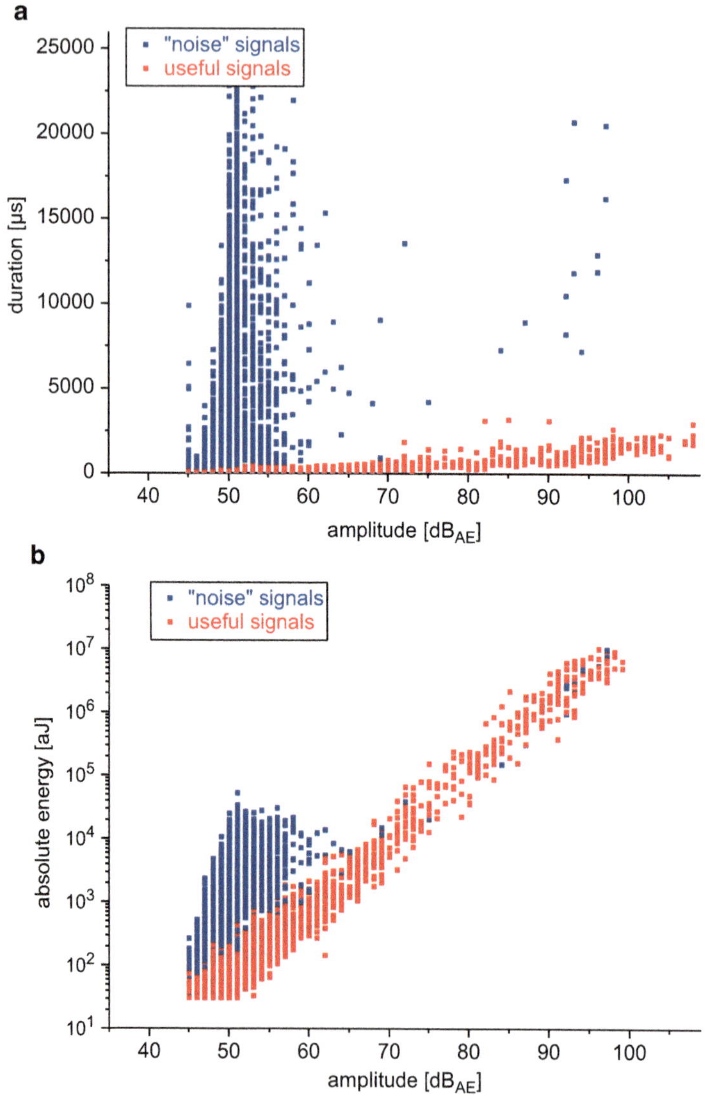

Fig. 4.106 Neural net-based removal of signals attributed to noise sources as plots of signal amplitude vs. duration (**a**) and as signal amplitude vs. absolute energy (**b**)

However, if there are changes to the type of experiment such as type of loading, type of material, stacking sequences, or other parameters, a direct application of such trained neural networks to new datasets is often not possible.

Another challenge arises due to the feature extraction process itself (see Appendix C for mathematical definitions of typical AE features). First of all, feature values are always linked to the acquisition settings of the equipment and the sensor type. Therefore it is hard to compare feature values of different setups in a quantitative fashion. Moreover, the feature values will strictly depend on the selected portion of the signal. As seen in Fig. 4.107 in the time and frequency

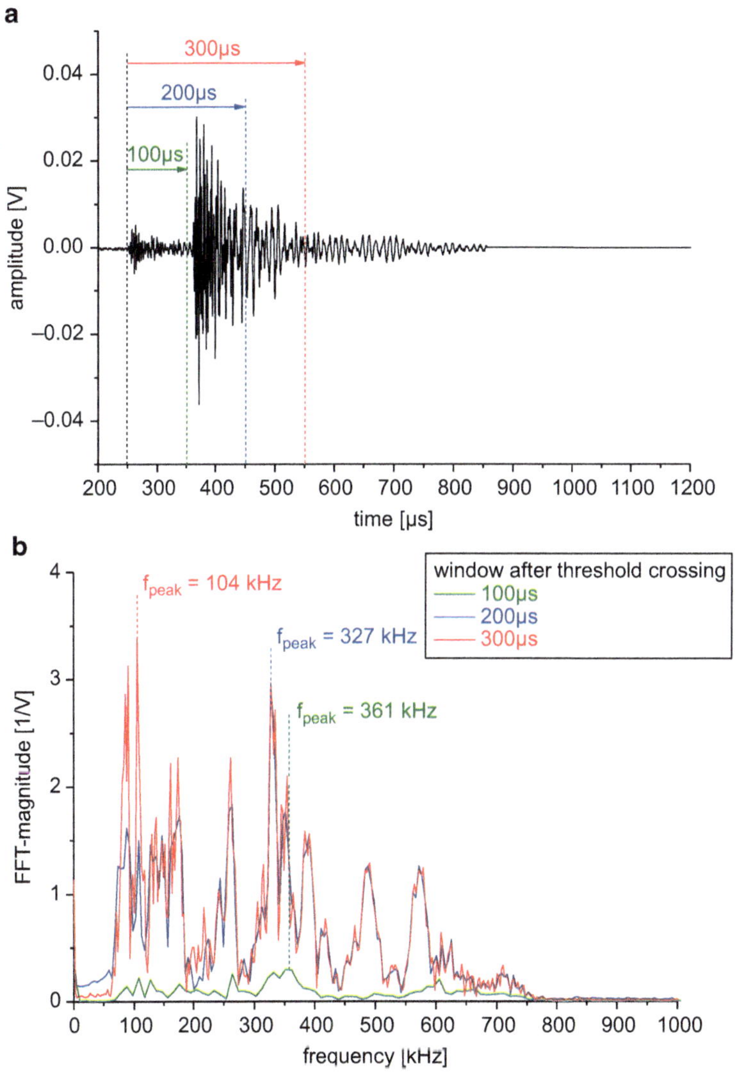

Fig. 4.107 Feature values for feature extraction process using three different window lengths in time domain (**a**) and frequency domain (**b**)

domain, the change of the time window to extract feature values has significant impact. In particular for small specimens with many occurrences of edge reflections, the relevant portion of the signal is at the beginning [35, 56]. Therefore, feature extraction is recommended to start at the time of first arrival and then to extend until a significant part of the signal has arrived at the sensor position. In particular for guided waves, this should imply arrival of all wave modes propagating in the structure to a certain amount. For larger specimens with vast differences in the arrival time of the different modes, a suitable compromise has to be achieved between the length of the feature extraction window and the inclusion of a large amount of reflections of the specimen edges.

Although the method proposed in [125] can potentially deal with arbitrary numbers of features, it is recommended to preselect a number of features for such an investigation. One way to reduce some redundancy of the features is to consider feature correlation dendrograms as shown in [35, 118]. Another approach is to consider the individual feature distributions as recommended in [135, 136]. If the feature distribution does not indicate more than one maximum, it is not likely a feature of large relevance to segment data. However, all of these feature reduction methods should be used carefully as higher-order correlation between different features might be lost in such attempts [135, 136].

4.5.2 Pattern Recognition Method to Detect Natural Clusters

As indicated in the previous section, there have been many attempts to perform pattern recognition in order to allow a classification of AE source mechanisms. In general, such attempts may be distinguished in two phases. The first phase is the statistical data analysis, which is used to form groups of signals. The second phase is the attempt to correlate each group with a particular AE source mechanism. In this section a data analysis algorithm is presented, which aims for the detection of "natural clusters." These are understood as clusters of entities, which provide highest discrimination to the members of other groups. In the context of acoustic emission, it is important to mention that these natural clusters are only potentially related to different AE source types. It may also happen that the natural clusters of an AE dataset are linked to different sensor types, due to propagation effects or similar effects which also alter the detected AE signals. Thus it is worthwhile to consider the occurrence of natural clusters, but this is not necessarily enough to allow full distinguishability of different AE source types. Moreover, the direct relation between particular cluster structures and respective AE source mechanisms needs to be validated in every experimental configuration. However, it has been demonstrated by various groups that the cluster strategy proposed in [125] works for source discrimination of AE signals of composite materials and wood fracture and in plant science [24, 124, 129, 137–143].

In order to find the natural clusters of a dataset without previous assumptions on the number of clusters and previous feature selection, an automated screening of feature combinations is a suitable approach. To evaluate the quality of the partition of a dataset, several cluster validity indices are established in literature [144–147]. Since each cluster validity index shows a different performance depending on the occurrence of outliers and the shape of the clusters, a combined evaluation of multiple cluster validation indices is a possibility to increase the reliability of the cluster identification method. Such combined evaluation was already proposed by Günter and Bunke [148] in the form of a voting scheme to measure the overall performance of the different indices.

The goal of the presented unsupervised pattern recognition methodology is to introduce an automated technique to detect promising feature combinations for clustering of acoustic emission signals as presented in [125] based on the method proposed in [35]. In contrast to feature reduction procedures (e.g., based on correlation dendrograms [118]), an exhaustive search of global optimal combinations of the signal features used is performed. As already pointed out by Polikar, such exhaustive search methods are conceptually simple but computationally devastating since the number of feature combinations to investigate grows combinatorial with the number of available features [136].

Consequently, before application of the presented exhaustive search method, an educated preselection of promising features for description of the respective acoustic emission signals is suggested. In principle the proposed method has no limitation regarding the number of features, but the exhaustive screening of a large number of feature combinations will lead to increased computation times. A visualization of the complete method as described in the following is shown in Fig. 4.108 as a flowchart diagram.

Determine all feature combinations:
The first step consists of selection of all possible feature combinations with a given minimum number of features $K_{\min}$ to be used for clustering. The total number N_{tot} of combinations of K previously selected features is thus

$$N_{\mathrm{tot}} = \sum\nolimits_{X=K_{\min}}^{K} \binom{K}{X} \tag{4.52}$$

Preprocessing:
As a second step, all preprocessing options for each feature combination are applied. In the current investigation, the preprocessing consists solely of feature normalization by dividing by the standard deviation. This is based on preceding investigations on classification of acoustic emission signals [24, 124, 125, 137]. For a generalized approach, the settings should be modified to fit the specific needs of the classification problem investigated.

Clustering:
Subsequently the respective partitions for 2, 3, 4, ... Ω clusters are calculated utilizing a cluster algorithm with Ω as the maximum number of clusters. The k-means algorithm is one of the most widely used algorithms to this end. However, depending on the data structure, the k-means algorithm may fail to detect

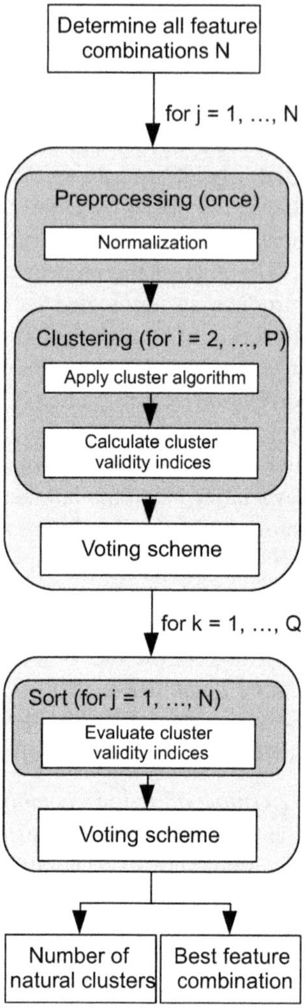

Fig. 4.108 Flowchart of generalized pattern recognition method used in the following (based on [125])

meaningful partitions [126, 149, 150]. Hence, other cluster algorithms, such as the Gustafson-Kessel algorithm, LVQ, SOM, or other computationally intense approaches may yield improved results. To increase the chances of finding the global minimum using the *k*-means algorithm, MacQueen's implementation with ten random cluster center initializations and a maximum number of 100 iterations to convergence was applied [151].

For the present implementation, as a measure of similarity, the Euclidean distance was used. For each partition the Tou index TOU, the Davies-Bouldin index DB, Rousseeuw's silhouette value SIL, and Hubert's Gamma statistic GAMMA were calculated. Since these cluster validity indices are comprehensively described in the respective authors' original publications [144–147], they are not explicitly repeated. It is worth pointing out that the TOU index, the SIL value, and the GAMMA statistics indicate good partitions by value maximization, while the DB index indicates good partitions by value minimization.

Voting scheme:
Based on these four cluster validity measures, the numerical performance of each partition is evaluated in the following scheme adapted from Günter and Bunke [148]:

1. The number of clusters with the best index performance is given Ω points.
2. The number of clusters with second-best performance is given $(\Omega - 1)$ points.
3. The number of clusters with third-best performance is given $(\Omega - 2)$ points.
4. ...

Subsequently the points of all four cluster validity indices are accumulated as a function of the number of clusters and are evaluated for their global maximum in points. The respective number of clusters is defined as numerically optimal and thus selected for the current feature combination. As discussed in [148], this combined evaluation of multiple indices can improve automated determination of the optimal number of clusters. Since cluster validity indices are often affected differently by outliers and the dataset structure, strengths of different indices are effectively combined by the voting strategy, while weaknesses are reduced.

In this context it is worth pointing out that the cluster validity indices chosen have low numerical complexity. There are numerous other cluster separation measures available, but these typically come with increased computational costs. Thus they are less desirable for automated screening of large numbers of feature combinations.

Detection of globally best feature combination:
As a final step, the feature combination with best global performance is determined. By definition, the partition with best global performance is detectable by the extreme values of the associated cluster validity indices. Therefore the 25 best partitions for each cluster validity index are evaluated as follows:

1. The partition (feature combination) with best index performance is given 25 points.
2. The partition with second-best performance is given 24 points.
3. The partition with third-best performance is given 23 points.
4. ...

Subsequently, the individual results of the cluster validity indices are cumulated. Thus a partition which is favored by all cluster validity indices receives exactly 100 points. Similarly, good partitions would get high numbers of points, while less favored or ambiguous partitions get fewer points. Typically, the partition with best global performance has less than 100 points. Thus the individual cluster validity indices would suggest different partitions. This indicates why the voting strategy is better than simple extreme value search, since it takes into account an averaged vote of the individual cluster validity indices.

Since the first publication of the method, it has been adopted by various groups to classify acoustic emission signals [129, 139, 140, 142, 143, 152]. The overall task of the method is to detect the most significant clusters of the entirety of acoustic emission signals with a minimum of initial assumptions on the cluster structure. Therefore, no assumptions are made on the exact number of signal classes or the number of AE features or the type of AE features.

Compared to signal classification methods based on single features, pattern recognition methods are computationally intense. But single AE features like peak-frequency or signal amplitudes can have significant dependency on the type of sensor, the details of the specimen geometry (cf. Sect. 4.5.3), or the propagation distance between source and sensor [153]. Consequently, source classification by static AE feature ranges cannot be generalized beyond certain limits. In contrast, pattern recognition techniques are adaptive to the problem investigated and do not rely on static AE feature ranges.

4.5.3 Uncertainty of Classification

Any classification of acoustic emission ultimately results in a partition. This grouping of signals might show an overlap of some of the feature values to a certain degree. This leads to ambiguity for some of the signals located at the boundaries of the groups, since they could potentially fall into two or more groups. In order to quantify such uncertainty of the classification process, recently a stochastically method was proposed to estimate the overlap of such cluster structures as resulting from pattern recognition [154]. In the following, the pattern recognition process is used as an example to quantify the UoC, but the method is straightforward to apply on any type of classification achieved by other methods.

It is demonstrated how the values of cluster validity measures DB, TOU, SIL, and GAMMA can be used to estimate the UoC. In order to establish an analytical correlation between the cluster validity measures and the UoC, artificially generated datasets were investigated [154]. Following the approach of Milligan [155] and the refinement by Qiu and Joe [156], datasets according to the implementation within the software package "R" by Qiu et al. were generated. As a measure of the degree of separation, J as introduced by [156] is used. The measure J is based on the separation of two clusters generated from two univariate normal distributions $(0, 1)$ and $(0, \tilde{A})$ [156]. For the values of $\tilde{A} = 8$ and $\tilde{A} = 4$, the measure of the degree of separation ranges from "well-separated" $(J = 0.342)$ to "close" $(J = 0.010)$ as shown in the scatterplots in Fig. 4.109.

The proposed pattern recognition method can identify the number of clusters accurately down to values of $J = 0.010$ [125]. Since the internal structure of the artificial datasets is initially known, this allows quantifying the classification error using the Rand index [157]. This index relates the pairwise assignment of objects in two partitions to all pairs of the partitions. In the present case, one partition is the

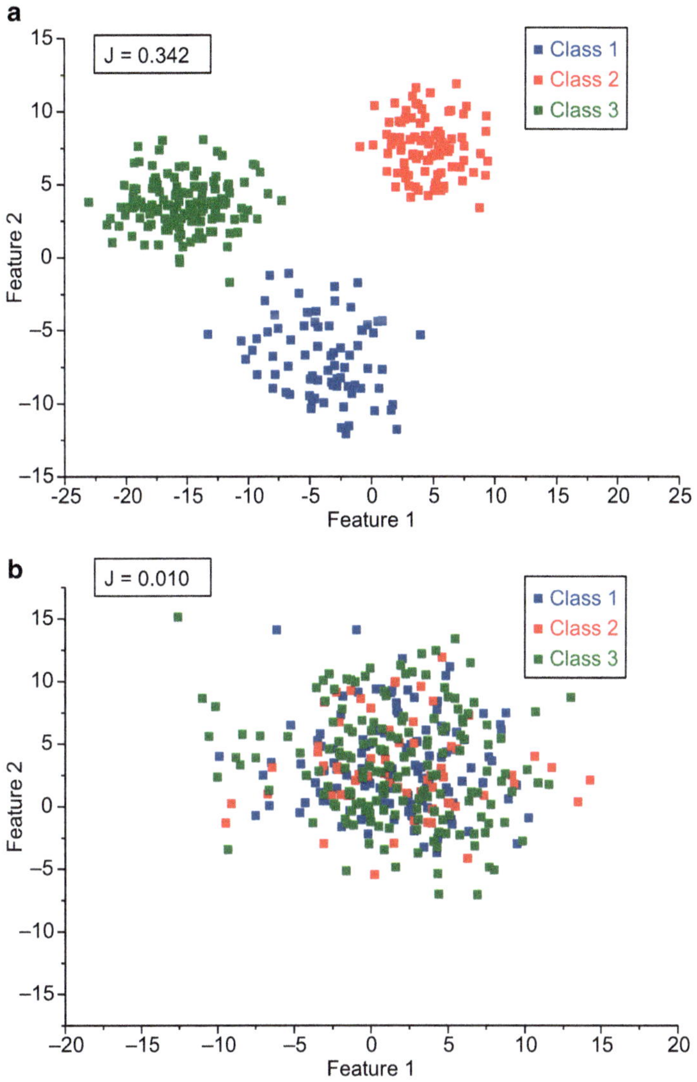

Fig. 4.109 Visualization of cluster structure for degree of separation of $J = 0.342$ (**a**) and $J = 0.010$ (**b**) (based on [154])

initially known assignment of the objects to a particular cluster. The other partition is the resulting assignment of objects of the pattern recognition method. A true positive (TP) decision is a classification of both objects to the same cluster; a true negative (TN) decision assigns two dissimilar objects to dissimilar clusters. These quantities are related to the false-positive (FP) decisions, which assigns two dissimilar objects to the same cluster and false negative (FN) decisions that assign two similar objects to different clusters:

$$\mathrm{RAND} = \frac{\mathrm{TP} + \mathrm{TN}}{\mathrm{TP} + \mathrm{TN} + \mathrm{FP} + \mathrm{FN}} \tag{4.53}$$

The Rand index is a direct measure of the percentage of decisions that are correct and can be employed as direct measure of the UoC. Next the correlation between the calculated cluster validity measures DB, TOU, SIL, and GAMMA and the Rand index is considered. To this end, a number of artificial datasets with varying degree of separation J between -0.45 and 0.45 are investigated. The number of objects in each cluster was randomly chosen within the range [50, 200] which reflects reasonable variation of the cluster sizes.

In the following, the case of three clusters and five features is discussed as an example to demonstrate the relationship between the cluster validity measures and the Rand index. In Fig. 4.110 values of the Rand index are plotted as a function of DB, TOU, SIL, and GAMMA. Since all cluster validity measures are by definition linked to the quality of the partition, a correlation to the Rand index is expected. Here, DB indicates a partition of high quality by a numerical minimum, while TOU, SIL, and GAMMA maximize their values for high cluster separation. The scatter in Fig. 4.110 indicates that there is no unique analytical correlation between a particular cluster validity measure and the Rand index. This is owed to the statistical distribution of clusters elements, which can cause nearly identical values of DB, TOU, SIL, or GAMMA for two partitions, while the respective number of correct classifications is still different for both partitions. In order to bring this behavior down to an analytical relationship, a five-parameter logistic function is used to fit the cluster validity measure c to the Rand index:

$$\mathrm{RAND} = A_{\mathrm{min}} + \frac{A_{\mathrm{max}} - A_{\mathrm{min}}}{\left(1 + \left(\frac{c}{c_0}\right)^{-h}\right)^{s}} \tag{4.54}$$

The boundary conditions of the Rand index with lower limit of $A_{\mathrm{min}} = 0$ and upper limit of $A_{\mathrm{max}} = 1$ reduce the number of independent fit parameters by two. The resulting fit including the 95 % prediction band is shown in Fig. 4.110. Values of the fit parameters c_0, h, and s are shown in the respective inset.

As indicated by the values of the adjusted least square errors (adj. R^2), the various cluster validity measures show different performance in their correlation to the Rand index value. As noted in [154], best fit quality was observed for SIL and GAMMA. A strong correlation was found for DB, while the size of the 95 % prediction band for TOU was found to be insufficient for reliable estimation of the UoC.

As demonstrated in [154], the application of (4.54) to the cluster validity measures of an experimental dataset using the fit parameters from Fig. 4.110 directly yields the UoC given by the estimated Rand index of the experimental partition. The UoC is understood as fraction of the signals, which are potentially classified in the wrong cluster. However, influences of the number of features selected for the process and the number of clusters have to be considered, when using values for the fit parameters c_0, h, and s (cf. [154]).

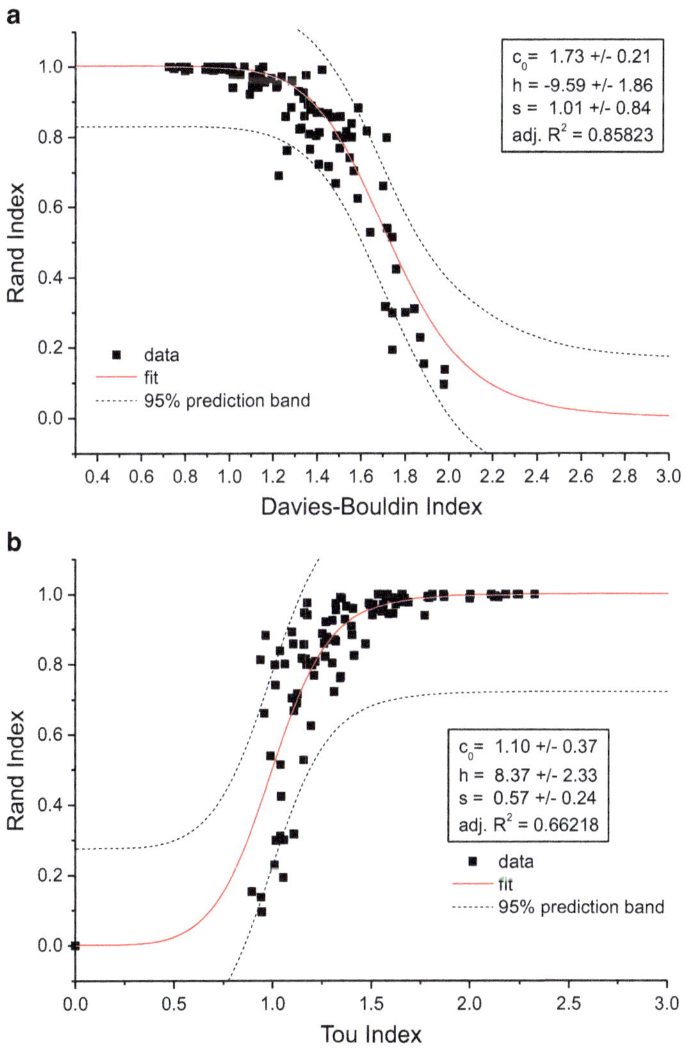

Fig. 4.110 Rand index as a function of cluster validity statistics. Plots include results of fit model and their corresponding 95 % prediction band. Diagrams are shown for Davies-Bouldin index (**a**), Tou index (**b**), Hubert's Gamma coefficient (**c**), and Rousseeuw's silhouette value (**d**) (based on [154])

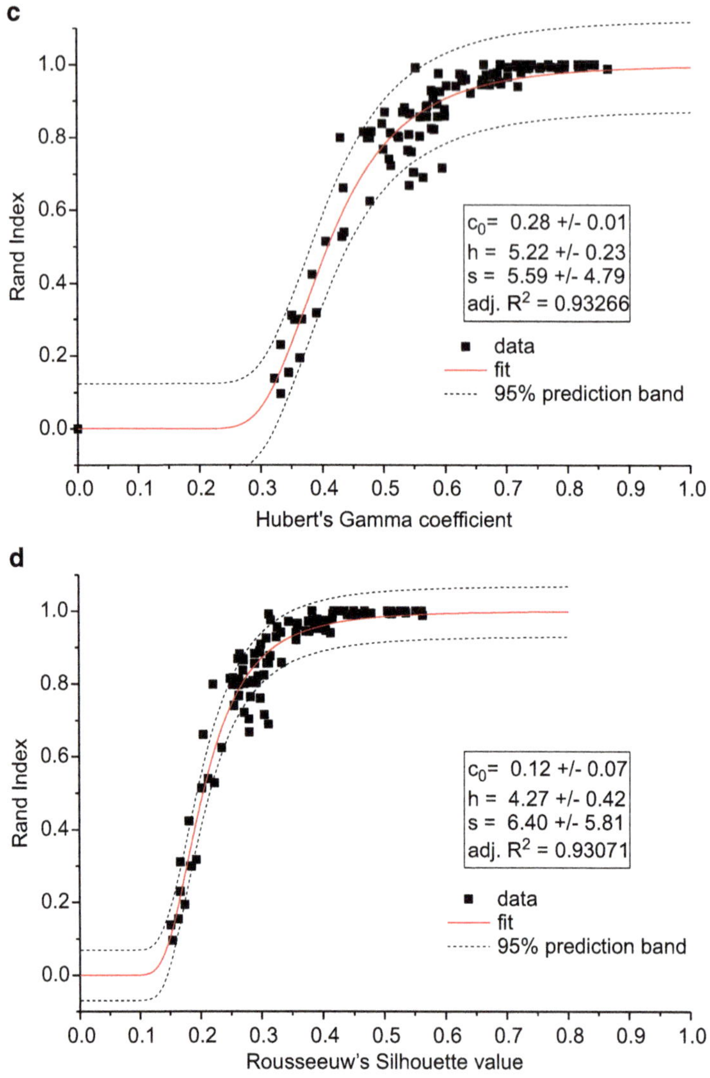

Fig. 4.110 (continued)

4.5.4 Factors of Influence

In the following some of the factors influencing the quality of signal classification procedures are presented. All factors are discussed in terms of the respective origin, and their effect on the pattern recognition approach proposed in references [35, 125] is used as an example to demonstrate their impact. To judge on the different factors, the quantification method from [154] is applied to yield a measure of the classification accuracy in form of the UoC, whenever appropriate.

4.5.4.1 Source-Sensor Distance

Since attenuation and dispersion affect the acoustic emission signal during propagation, these naturally have a significant impact on the possibilities to distinguish different signal sources. The impact of this effect can be experimentally studied in a test specimen using test sources being able to excite acoustic emission signals at various frequency ranges. To demonstrate this effect, reference [154] investigated this effect using large double-cantilever beam specimens. Three types of acoustic emission test sources were applied at source-sensor distances between 80 and 280 mm as seen in Fig. 4.111. In analogy to the signals detected during testing of double-cantilever beams, signal features Partial Power 2 and Weighted Peak Frequency were calculated from the obtained signals as defined in Appendix C. As demonstrated in [154], each of the three acoustic emission test sources excites a unique combination of S_0 mode, A_0 mode, and SH_0 mode. This translates into a unique feature range as seen in Fig. 4.112.

Since the depth position of the pencil lead breaks is constant and the source type is identical, the variation in feature values shown in Fig. 4.112 is solely caused by the change in propagation length to the sensor. Similarly, the feature range of signals of pencil lead breaks at the edge of the plate starts at different feature values and spans a range of feature values in the diagram of Partial Power 2 and Weighted Peak Frequency that coincides well with the cluster attributed to interfacial failure. Here, both feature ranges (Partial Power 2 and Weighted Peak Frequency) are well reflected by the signals detected after different propagation lengths. For the test signals from the WD pulser, the change of propagation length to the sensor also causes a significant shift in the feature values Partial Power 2 and Weighted Peak Frequency. However, the trajectories of the values extracted for the different test sources do not intersect with each other, despite of the change in propagation

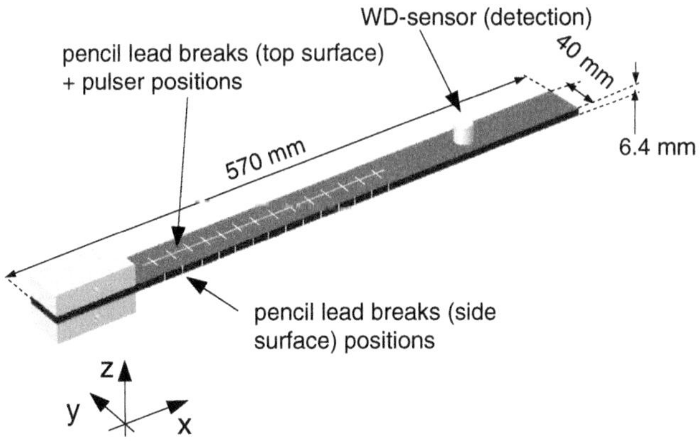

Fig. 4.111 Scheme of experimental setup used for generation of AE test sources including dimensions of DCB specimen used as propagation medium and subsequent mechanical test (based on [154])

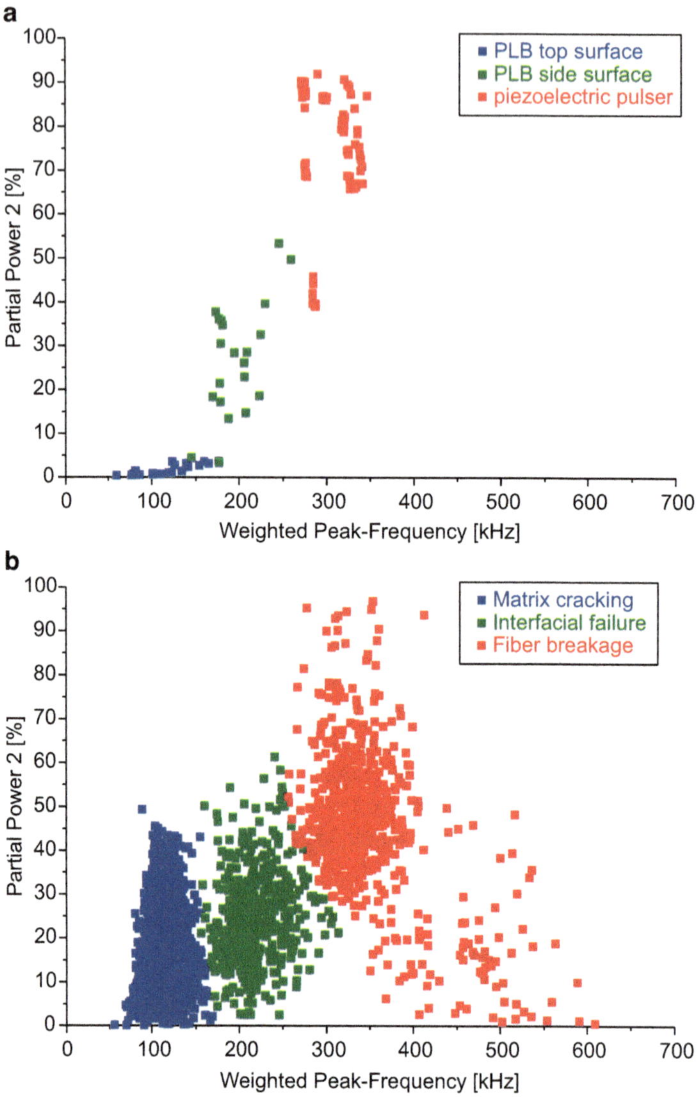

Fig. 4.112 Change of feature values for AE test source measurements with different distance between source and sensor (**a**) and experimental dataset partition as obtained from pattern recognition method from test on DCB specimen shown in Fig. 4.111 (**b**)

distance between 80 and 280 mm. In combination with investigations on typical size geometries for mode I testing (cf. Sect. 4.7.4.1), this indicates that distinction of sources operating at different frequency ranges should be possible for propagation distances up to 280 mm.

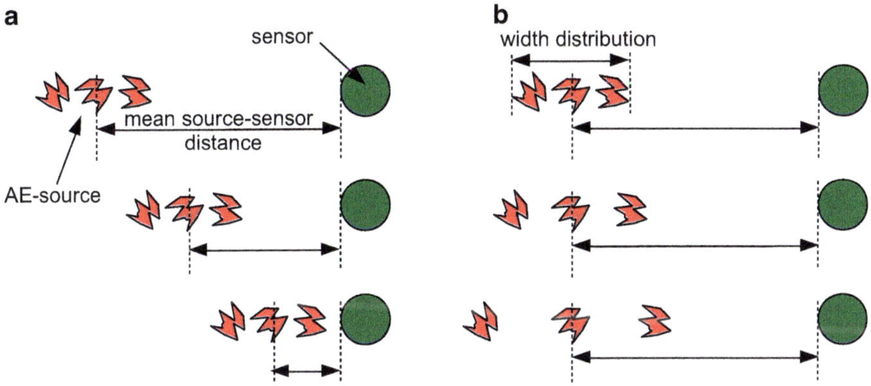

Fig. 4.113 Scheme for selection of source position and distribution to calculate cluster validity indices as function of source-sensor distance (**a**) and as function of width distribution (**b**)

In order to evaluate the UoC error as function of source-sensor distance, the double-cantilever beam shown in Fig. 4.111 was tested as reported in detail in [154]. The x-position of the acoustic emission signals was localized using a Δt localization procedure. In such an experiment, the majority of the acoustic emission signals originate from positions close to the crack tip. Only few signals are localized significantly ahead of or behind the position of the crack tip. This allows calculating the distance to the detecting sensor. In addition, the depth position (z-position) of the acoustic emission sources is expected to be dominantly around the medial plane of the specimen, and the y-position is expected to be randomly distributed.

Two approaches were evaluated to investigate the influence of the source-sensor distance on the UoC. As shown in the scheme in Fig. 4.113a, as the first approach the signals were analyzed at constant width distribution as function of source-sensor distance. As the second approach the source-sensor distance is kept constant and the width distribution is varied. The subsets were chosen in stepwise increments of 500 signals until the full dataset size is reached.

For each subset of the whole dataset, Rousseeuw's silhouette value and Hubert's Gamma coefficient were calculated based on the features Partial Power 1, Partial Power 2, Partial Power 4, Peak Frequency, and Weighted Peak Frequency as defined in Table C.2 in Appendix C. The resulting UoC for both approaches is shown in Fig. 4.114a, b, respectively.

For the shifted approach, no clear trend of the cluster validity measures or their respective UoC is observed as a function of the source-sensor distance. The cumulative approach reveals an increased UoC with increasing width distribution of the source positions. This leads to the conclusion that the source-sensor distance itself is not a critical parameter for valid source identification within the investigated range below 375 mm. In contrast, the width distribution of sources within the specimen seems to influence the accuracy of the source identification procedure significantly. For these experiments, an increased width distribution was found to cause an increased UoC.

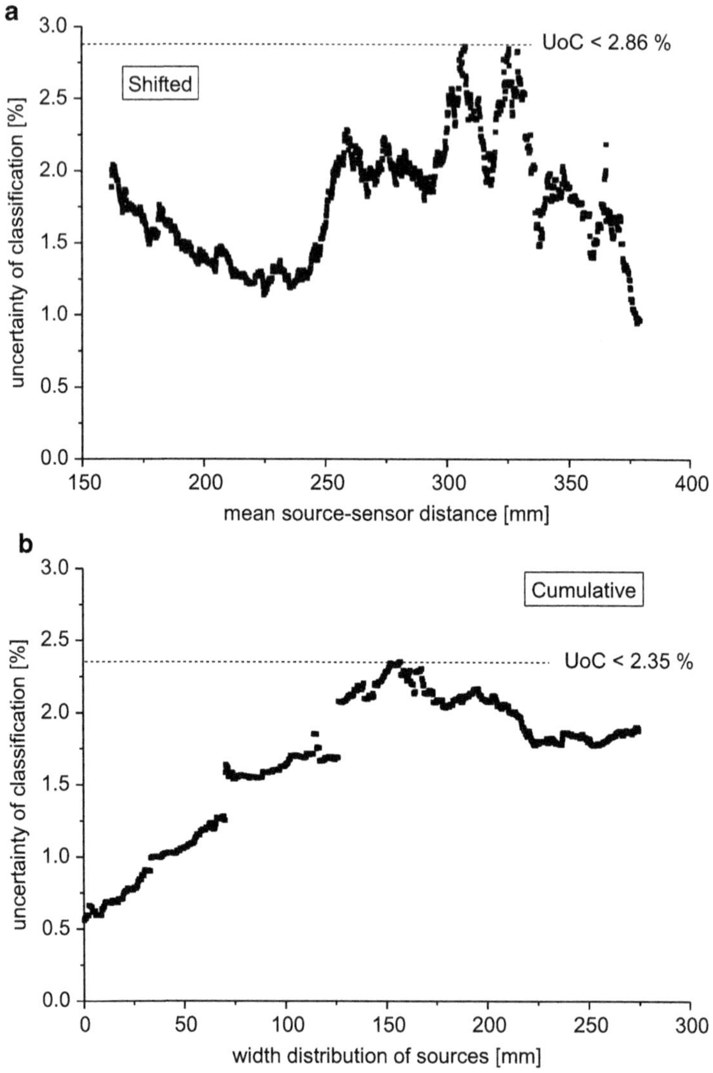

Fig. 4.114 Calculation of uncertainty of classification as function of distance for shifted (**a**) and cumulative (**b**) approach

4.5.4.2 Plate Thickness

Similar as for the source-sensor distance, a change in the propagation medium geometry is likely to affect the UoC. In order to demonstrate the effect of thickness changes, laminates used for bearing strength testing were investigated. The test fixture for the 10.8 mm thick specimens is seen in Fig. 4.115. For the variation of the specimen dimensions, an approximate ratio of 0.5:1:1.5 was selected, which are

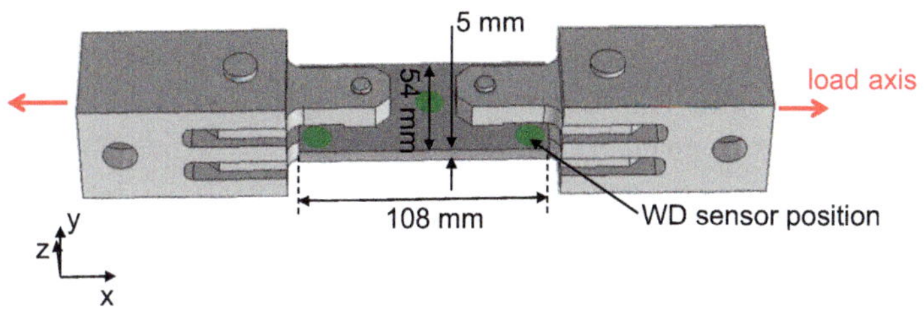

Fig. 4.115 Test fixture for smallest specimen size used for bearing strength tests including AE sensor positions

applied to the lateral dimensions, the thickness of the specimen, as well as to the diameter of the bolts. Thus the smallest specimen is of dimensions of 108 mm × 54 mm × 5 mm (height × width × thickness), the medium size is of dimensions of 220 mm × 110 mm × 10 mm (length × width × thickness), and the largest size is of dimensions of 330 mm × 165 mm × 15 mm (length × width × thickness). This scaling of specimen dimensions inevitably involves a change of the average source-sensor distance. Since for this test, the position of damage spots is well localized behind the pins connecting the laminate to the test fixtures, their width distribution remains rather narrow (<10 mm). Based on the findings in Sect. 4.5.4.1, the lateral scaling is thus unlikely to contribute much to the UoC. To allow for comparable propagation paths, only signals of one of the sensors indicated in Fig. 4.115 are used in the following. Due to the bearing failure mode observed in all of the specimens, it is also likely to generate equal ratios of failure mechanisms for each of the three specimen dimensions. For signal acquisition, a type WD sensor with 20 dB_{AE} preamplification and a band-pass from 20 kHz to 1 MHz is used. Further details on the experimental setup and the data acquisition parameters are found in [158, 159].

One representative partition of each specimen type as achieved by the pattern recognition method following [125] is shown in Fig. 4.116. The respective assignment to failure mechanisms is based on accompanying FEM modeling work following the strategies presented in Sects. 4.2 and 4.7. Clearly, all three thickness types show clusters of acoustic emission signals, which can be partitioned using pattern recognition methods. Remarkably, the UoC is nearly identical for all three specimen types. This gives rise to the conclusion that the thickness of the specimen does not have a substantial impact on the uncertainty involved in the classification procedure. However, larger discrepancies are expected, when such thickness changes arise during signal propagation (see Sect. 4.3.2).

4.5.4.3 Stacking Sequence

As already seen from the studies on wave propagation in Sect. 4.3.2, the stacking sequence has significant impact on the shape of the signal, even for identical sources. Therefore it is necessary to investigate the impact of this parameter on the distinguishability of different AE source types. To this end three different stacking sequences of Sigrafil CE1250-230-39 prepreg material were prepared. Plates of 570 × 570 mm and 1.3 mm thickness were fabricated following the production cycle recommend by the material supplier. As stacking sequence,

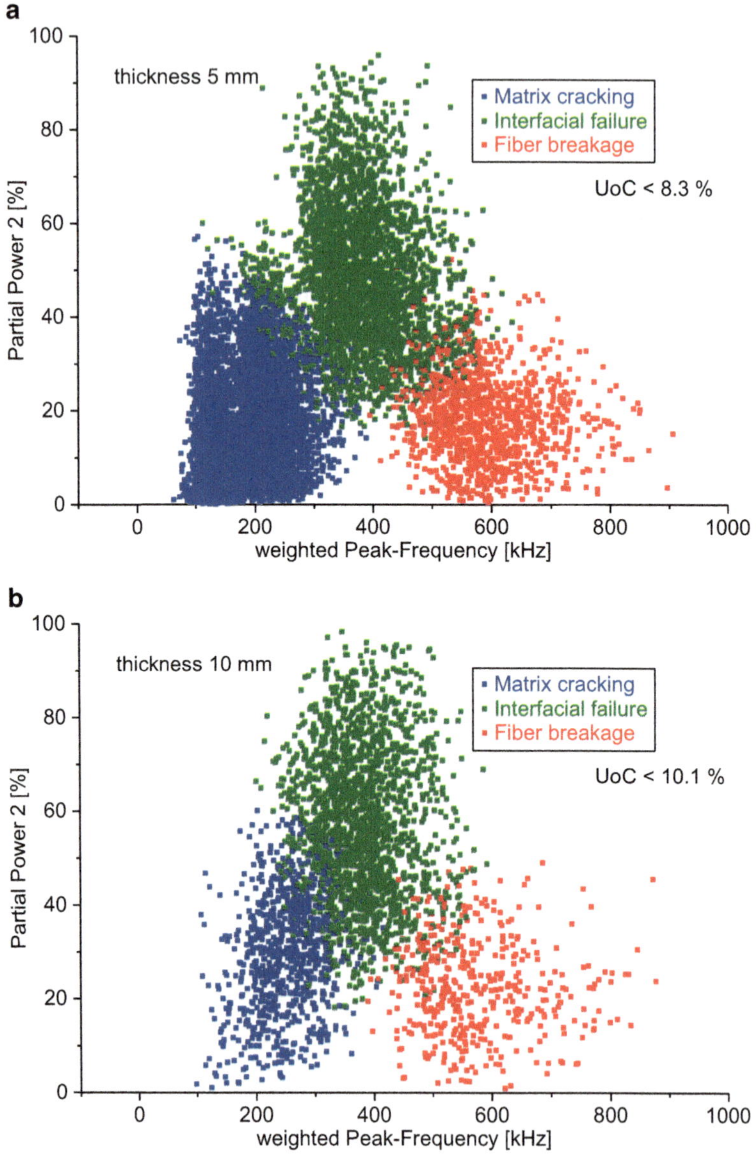

Fig. 4.116 Classification results of thickness 5 mm (**a**), 10 mm (**b**) and 15 mm (**c**) with similar ratios to all axis (nearly quasi-isotropic) including calculated uncertainty of classification

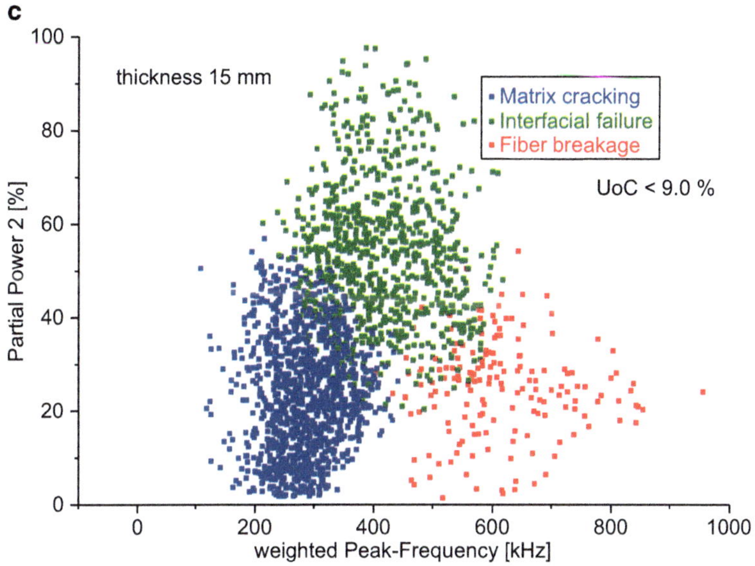

Fig. 4.116 (continued)

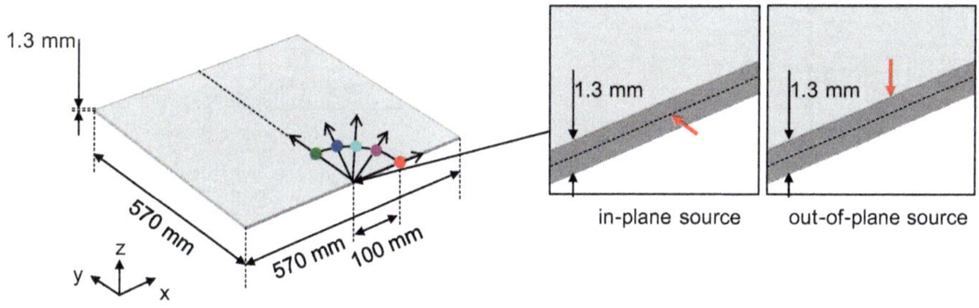

Fig. 4.117 Scheme of test configurations used including test source position and sensor positions

a unidirectional plate $[0_3]_{sym}$, a cross-ply laminate $[0/90_2]_{sym}$, and an quasi-isotropic plate $[0/60/-60]_{sym}$ were used. To study wave propagation in these plates, four different configurations were selected with an arrangement as shown in Fig. 4.117. In every configuration two different test sources are used to generate ten test signals. As the first test source, pencil lead breaks in the out-of-plane (OoP) direction were selected, which cause signals with dominant A_0 mode having mostly low-frequency content. As the second test source type, the pencil lead breaks were carried out at the edge of the plate. This in-plane (IP) direction at the middle plane of the plate causes dominant S_0 modes having substantially higher-frequency content. To avoid influences of distance changes between source and sensor, the test source position was kept constant. Every test source signal was monitored by five type WD sensors arranged in a quarter circle around the source position at a distance of 100 mm. Prior to testing it was assured that the sensors show similar

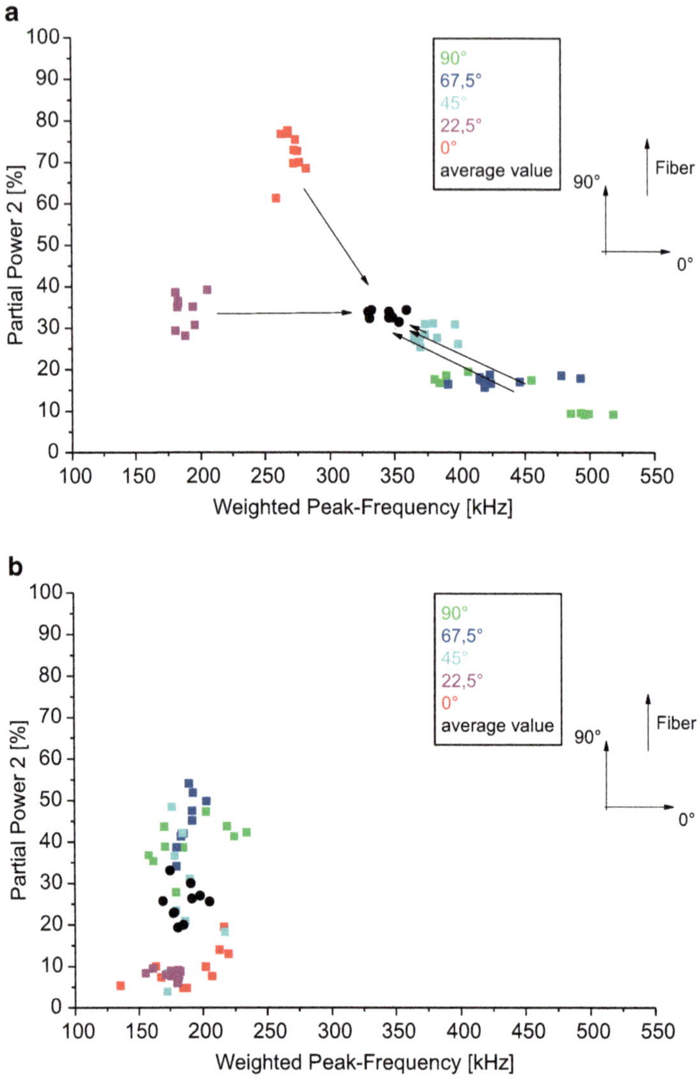

Fig. 4.118 Feature values as extracted for configuration 1 (unidirectional) for the IP-source (**a**) and the OoP-source (**b**)

frequency response and sensitivity. For the unidirectional plate, the two fundamental orientations were investigated, since different test source signals are expected based on the difference in stiffness along these two directions.

Feature extraction using a time window of 100 μs after signal arrival yields the diagrams shown in Figs. 4.118, 4.119, 4.120, and 4.121. The angles noted throughout Figs. 4.118, 4.119, 4.120, and 4.121 refer to the orientation of the sensor relative to the source position, whereas the orientation of the different plies is indicated by

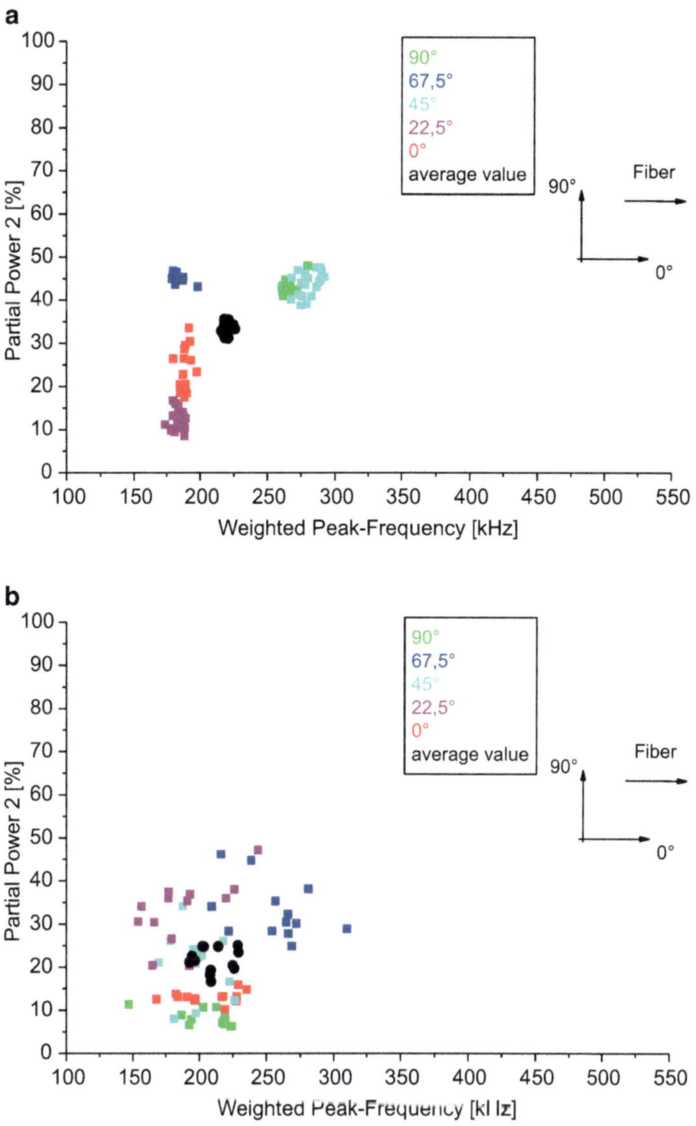

Fig. 4.119 Feature values as extracted for configuration 2 (unidirectional, rotated) for the IP-source (**a**) and the OoP-source (**b**)

the fiber arrows. For the configuration 1 (unidirectional plate, source parallel to fiber axis), the result is given in Fig. 4.118a, b. A first comparison of these two figures clearly indicates the difference as expected from the different test sources. For the case of the IP-source shown in Fig. 4.118a, the detected frequency content expressed by the feature values Partial Power 2 and Weighted Peak Frequency is

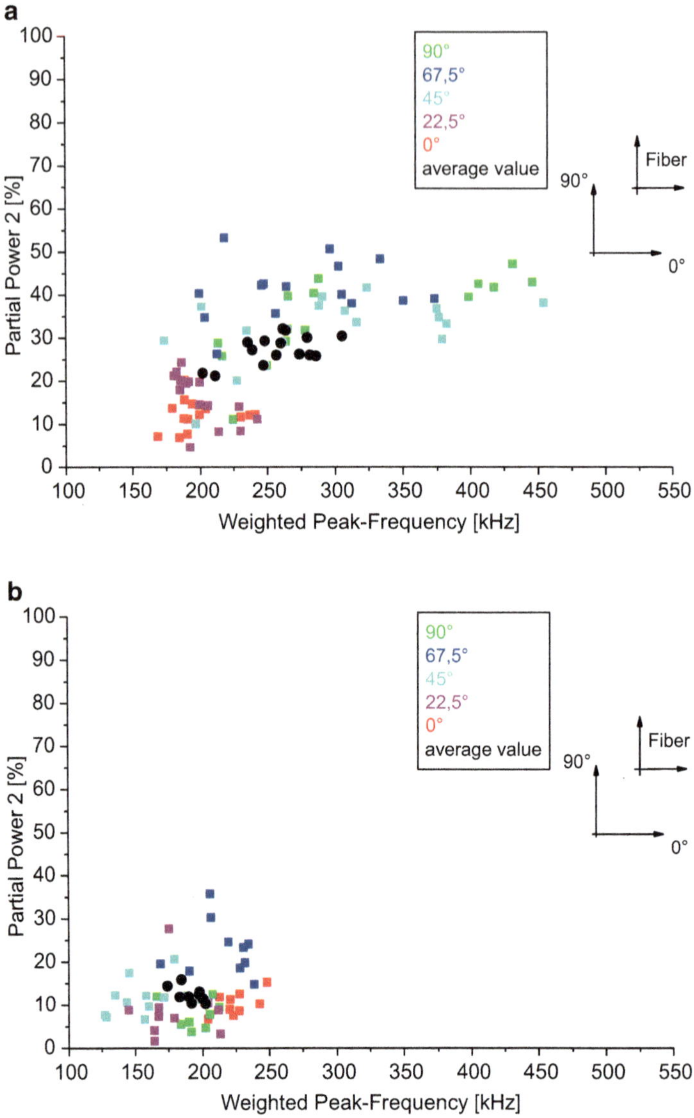

Fig. 4.120 Feature values as extracted for configuration 3 (cross-ply) for the IP-source (**a**) and the OoP-source (**b**)

clearly different for the different sensor positions. Moreover, the feature range covered by the individual sensors spans the typical feature ranges also used to sort out failure types based on previous work [24, 125, 138, 160]. For the OoP-source results seen in Fig. 4.118b, this effect is less pronounced, but still visible due to the formation of individual clouds for each sensor. It is worth noting

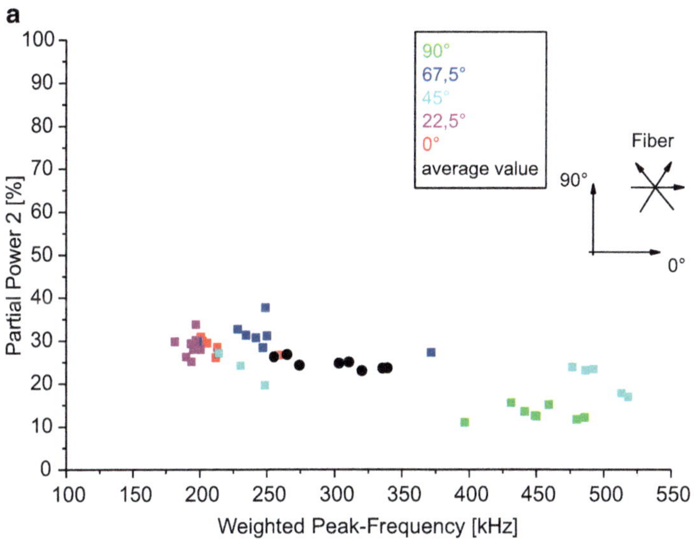

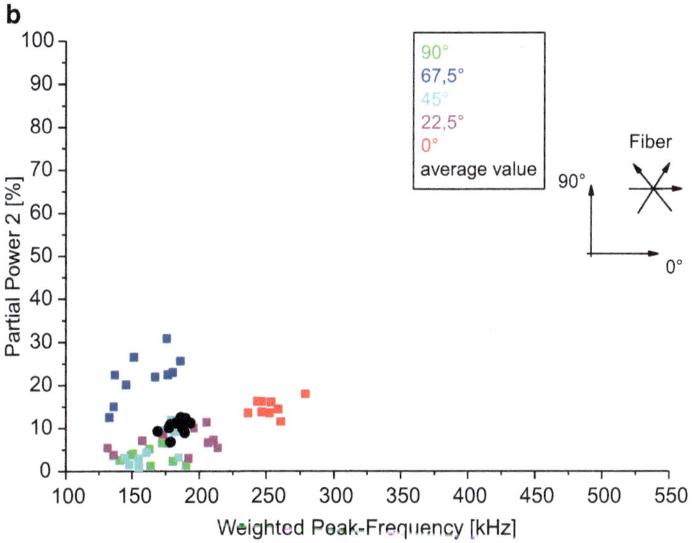

Fig. 4.121 Feature values as extracted for configuration 4 (quasi-isotropic) for the IP-source (**a**) and the OoP-source (**b**)

that for each sensor, there is distinct difference in the cluster position for the IP-source and the OoP-source. Therefore, source discrimination would be still possible if only signals of one sensor are investigated at a time. The problem of overlapping cluster ranges solely arises, if signals of all sensors are mixed and are interpreted at the same time. In order to overcome this problem, it is possible to calculate average feature values for each source event.

If signals are known to originate from the same source and are detected by N sensors, it is straightforward to calculate the average feature value $\langle q \rangle$:

$$\langle q \rangle = \sum_{i=1}^{N} q_i \tag{4.55}$$

These average feature values are superimposed as black dots in Fig. 4.118a, b, respectively. These cluster positions of q are found well separated and therefore will allow meaningful source type distinction.

Identical investigations on the remaining three cases yield Figs. 4.119, 4.120, and 4.121 comparing the results of the IP-source and the OoP-source. For configuration 2 (unidirectional plate, source perpendicular to fiber axis), similar behavior as for configuration 1 is observed. However, the average frequency content of the IP-source is found to be significantly lower, which is caused by the source mechanism rather than the orientation aspect. Therefore, the final separation of the averaged features seems to be less pronounced.

For the configurations 3 (cross-ply) and 4 (quasi-isotropic), the decreasing degree of acoustic anisotropy causes less distinctly separated cluster for each of the detecting sensors. Nevertheless, the basic findings established for configurations 1 and 2 are still found to be valid.

Based on these findings, it seems to be possible to distinguish different failure types in the stacking sequences investigated above. More generally, the distinction based on frequency content should always be possible if sources are monitored at the same propagation angle (i.e., the different sources are subject to the same propagation path). If this condition cannot be fulfilled (i.e., a technical structure with multiple possible source positions), it is recommended to use the average feature values $\langle N \rangle$ of a source event instead of the feature values of the individual sensors. As a second requirement for the latter, a significant portion of orientations relative to the source should be covered (i.e., spanning the range between 0° and 90°).

4.5.4.4 Internal Damage

As discussed in Sect. 4.3.3 and demonstrated in Sect. 7.2.7, the interaction of propagating waves with external or internal obstacles may have substantial influence on the detectable AE signals. Therefore it is straightforward to expect a respective change in the feature values extracted from such signals. For acoustic emission sources detected during mechanical loading of composites, this influence is quite relevant, since there are lots of microscopic and mesoscopic failure process within the propagation path between potential sources and the sensor positions until final failure.

As demonstrated in [61, 62], it is useful to evaluate the changes of feature values due to the presence of such internal damage. One example is the signal amplitude shown in Fig. 4.122 for each configuration described in Sect. 4.3.3, evaluated in 0°, 45°, and 90° propagation directions. Compared to the reference case (red), the

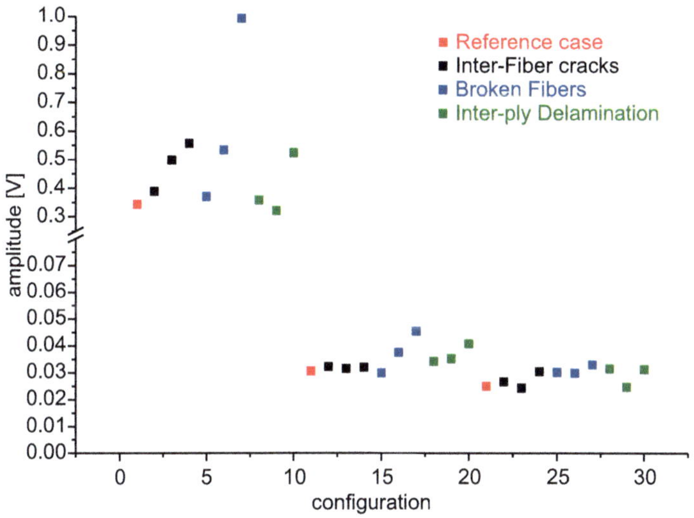

Fig. 4.122 Calculated signal amplitude of signals for the various configurations of internal damage in comparison to the reference case for 0°, 45°, and 90° propagation direction (reprinted from [62])

signals passing through the damaged region experience significant changes of their signal amplitude. This is due to the modal conversion at the damage position and the scattering introduced by some of the damage types. Except for one of the models with broken fibers, the overall deviation of amplitude values was found to be within ±0.25 V (i.e., ±5 dB_{AE}) to the reference case. Compared to the ±3 dB_{AE} recommendation of ASTM E 2191 for sensor replacement, this amplitude range is well within typically encountered uncertainties of measurement setups used.

The calculated Weighted Peak Frequency for the signals of different damage configurations and the reference case are shown in Fig. 4.123 for 0°, 45°, and 90° propagation directions. Dependent on the propagation angle, different values for the Weighted Peak Frequency are observed. This is caused by the anisotropic propagation of Lamb wave modes in the unidirectional plate used in this modeling work [61, 62]. Similar to the signal amplitude, deviations compared to the reference case are up to 100 %. Again, features evaluated in all propagation directions are affected by the presence of the modeled damage types.

As discussed in Sect. 4.5.4.3, the difference in feature values to the different angles is expected from the anisotropy of the unidirectional plate studied in the modeling work. For valid source identification, one can use the average frequency features detected by a set of sensors as described in Sect. 4.5.4.3. The distribution range of feature values as caused by the modeled damage types within the propagation path is within the typical experimentally encountered distribution range for an individual feature. Therefore, the proposed pattern recognition methods are likely to deal with these changes [61, 62]. However, the presence of such internal obstacles is expected to increase the mean distribution of feature values of the

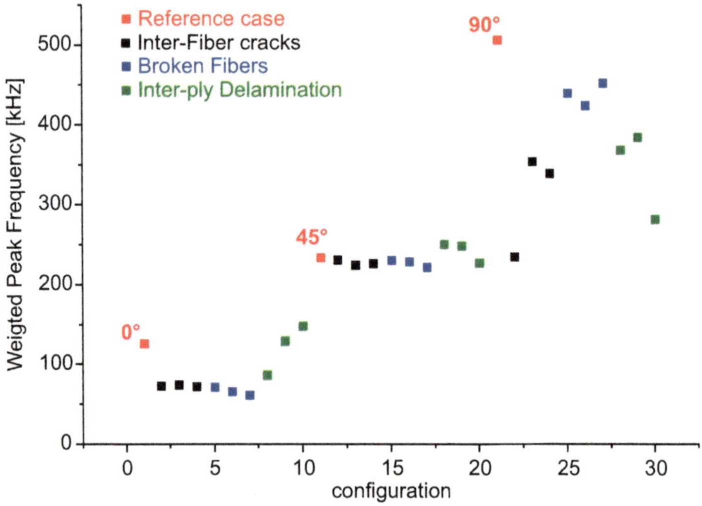

Fig. 4.123 Calculated Weighted Peak Frequency of signals for the various configurations of internal damage in comparison to the reference case for 0°, 45°, and 90° propagation direction (reprinted from [62])

cluster belonging to a particular failure type. Hence, with increasing amount of damage, cluster structures may overlap to a significant amount and may cause an increase in the UoC.

4.5.4.5 Sensor Type

As discussed in Sect. 4.4, the detection process of AE signals by an AE sensor may have significant impact on the frequency content of the signal. The commercially available types of AE sensors can be categorized as resonant, multi-resonant, or wideband types. Naturally, resonant sensor types reduce the detectable frequency spectrum, while broadband types are sensitive in a wide frequency range. Some sensor designs show almost flat frequency response, minimizing the influence of the sensor on the detection process [87–89].

To estimate the influence of the chosen AE sensor type on the outcome of frequency-based signal classification, three different sensor types are chosen and their results are compared. As typical representatives, the Digital-Wave 1025T (wideband), the Physical Acoustics WD (multi-resonant), and the Physical Acoustics Micro30 (resonant) were chosen. The sensors were mounted during mode I testing of Sigrafil CE1250-230-39 prepreg at identical locations. The according experimental setup and data reduction routine to yield the graphs in Fig. 4.124 are described in more detail in [154].

Figure 4.124 shows the results of the proposed pattern recognition procedure as scatterplots of Partial Power 2 over Weighted Peak Frequency with definition of these features found in Appendix C. AE signals are assigned to matrix cracking, interfacial failure, and fiber breakage following the FEM-based strategy as outlined in Sects. 4.2–4.4 and as described in detail in [138].

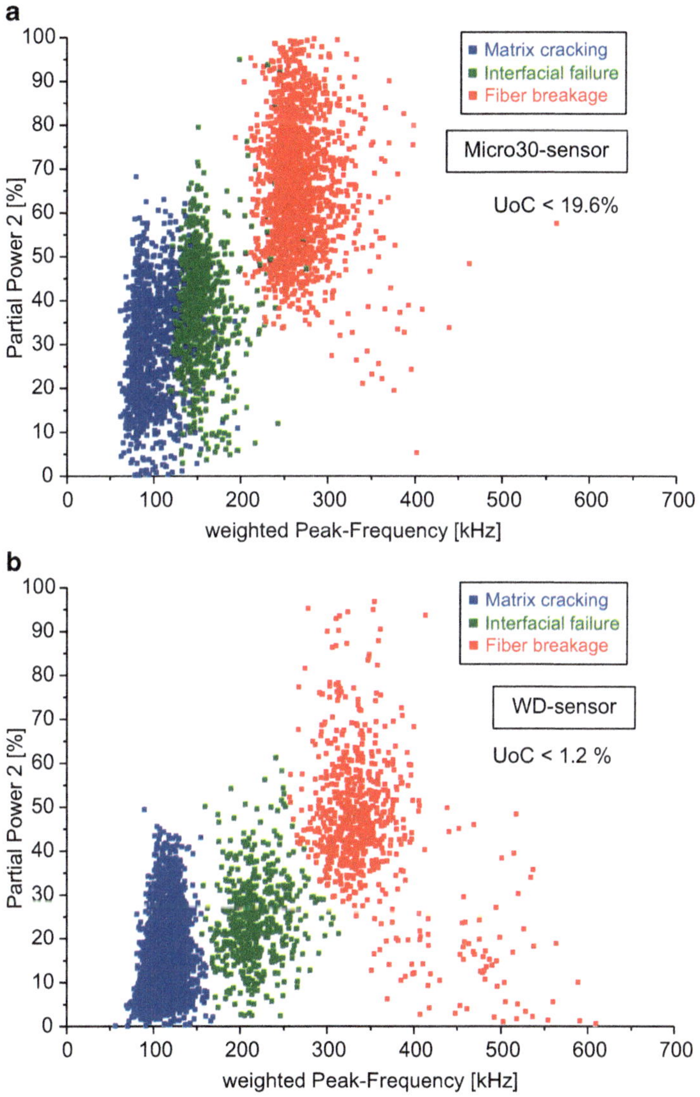

Fig. 4.124 Positions of signal clusters achieved by pattern recognition approach for three types of AE sensors used with same type of experiment. Figures show resonant sensor type (**a**), multi-resonant sensor type (**b**), and broadband sensor type (**c**) including calculated uncertainty of classification

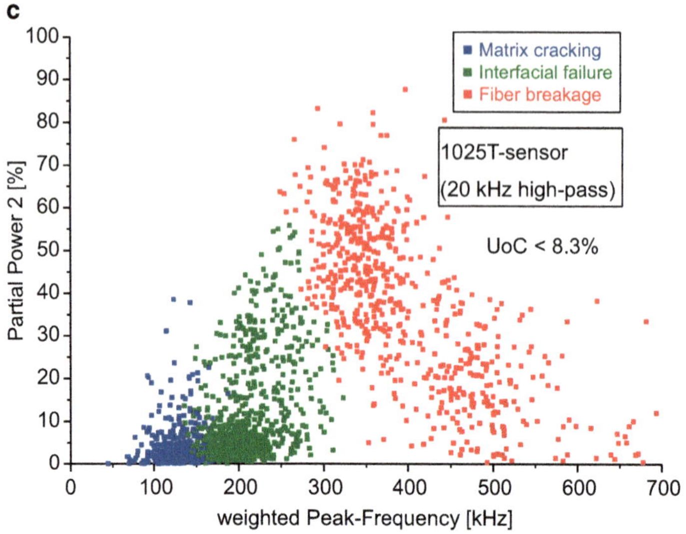

Fig. 4.124 (continued)

The AE signals detected by the Micro30 sensor shown in Fig. 4.124a are centered in their Weighted Peak Frequency around the sensor resonant frequency of 200 kHz. Still, the pattern recognition algorithm is able to distinguish between three types of AE signals, but the clusters associated with matrix cracking and interfacial failure show substantial overlap. Thus, the AE signals lose their discriminative frequency information by the detection process with a narrow bandwidth sensor type, which is expressed in a high UoC of 19.6 %.

In Fig. 4.124b the results of the same experiment using the WD sensor are shown. Compared to the results in Fig. 4.124a, the values of the Weighted Peak Frequency are spread out over a range of 500 kHz. Although the WD sensor exhibits multiple resonant frequencies, these do not directly correspond to the positions or number of the clusters. Instead, the concentration of AE signals around distinct Weighted Peak Frequencies is influenced mostly by the specimen geometry and the stacking sequence (cf. also [35]). The overlap of the clusters is found to be less than for the resonant-type sensor yielding an UoC of only 1.2 %.

Initially, the AE signals detected by the broadband-type sensor 1025T did not show any cluster structure. All feature values were located within a small bandwidth of (150 ± 100) kHz Weighted Peak Frequency and (5 ± 5) % Partial Power 2. The result shown in Fig. 4.124c is the same dataset after postprocessing by a 20 kHz Butterworth high-pass filter of eighth order. Using this step, high sensitivity of this sensor type at frequencies below 20 kHz is suppressed. The data structure revealed is similar to that achieved by the other two sensors and can be classified accordingly with an UoC of 8.3 %.

It was found that the bandwidth of the sensor type used has significant influence on the separability degree of the cluster structure. Thus the according error of assignment to a failure mechanism will depend on the choice of sensor type as well as on the suitability of the postprocessing steps.

4.5.4.6 Signal-to-Noise Ratio

AE signals of composite materials usually cover a broad range of signal amplitudes. Even acquisition equipment with substantially high digitalization depth cannot assure an equal signal-to-noise (*S*/*N*) ratio for every AE signal. This raises the question if feature values extracted from AE signals with poor *S*/*N* ratio are different to those extracted from AE signals with good *S*/*N* ratio.

Therefore typical AE signals of each failure mechanism were chosen to investigate the influence of *S*/*N* ratio on the cluster structure. White noise U_{noise} was superimposed to the AE signal U_{AE} with increasing intensity as shown in Fig. 4.125a:

$$U'_{\text{AE}}(t) = \widetilde{\alpha} \cdot U_{\text{noise}}(t) + U_{\text{AE}}(t) \tag{4.56}$$

The noise level α was chosen relative to the maximum signal amplitude $U_{\max}$ of the AE signal:

$$\widetilde{\alpha} = U_{\text{noise}} / U_{\max} \tag{4.57}$$

The resulting AE signals based on one matrix cracking signal are shown in Fig. 4.125a for five different noise levels $\widetilde{\alpha}$. The noise levels were selected to span the full range of *S*/*N* ratios encountered in AE testing. The resulting extracted feature values are plotted in Fig. 4.125b as scatterplot of Partial Power 2 over Weighted Peak Frequency. The dependency of the feature values on α is indicated by an arrow starting at the original AE signal's feature value.

One observation is the tendency of high $\widetilde{\alpha}$-values to cause shifts of the feature values to higher frequencies. For the Weighted Peak Frequency, this is observed directly from Fig. 4.125, while the influence of the *S*/*N* ratio on Partial Power 2 originates from the negative correlation to the other Partial Powers. Thus for low-frequency dominated AE signals (e.g., matrix cracking), Partial Power 2 values increase with $\widetilde{\alpha}$. Contrary, AE signals dominated by intermediate or high frequencies (e.g., interfacial failure, fiber breakage) decrease their Partial Power 2 values with $\widetilde{\alpha}$, because Partial Powers 3 and 4 increase instead.

Another observation is the path of the trajectories for increasing $\widetilde{\alpha}$-values. Even for the extreme values of *S*/*N* ratio considered here, the feature values shown in Fig. 4.125b reside inside the typical extent of respective signal clusters.

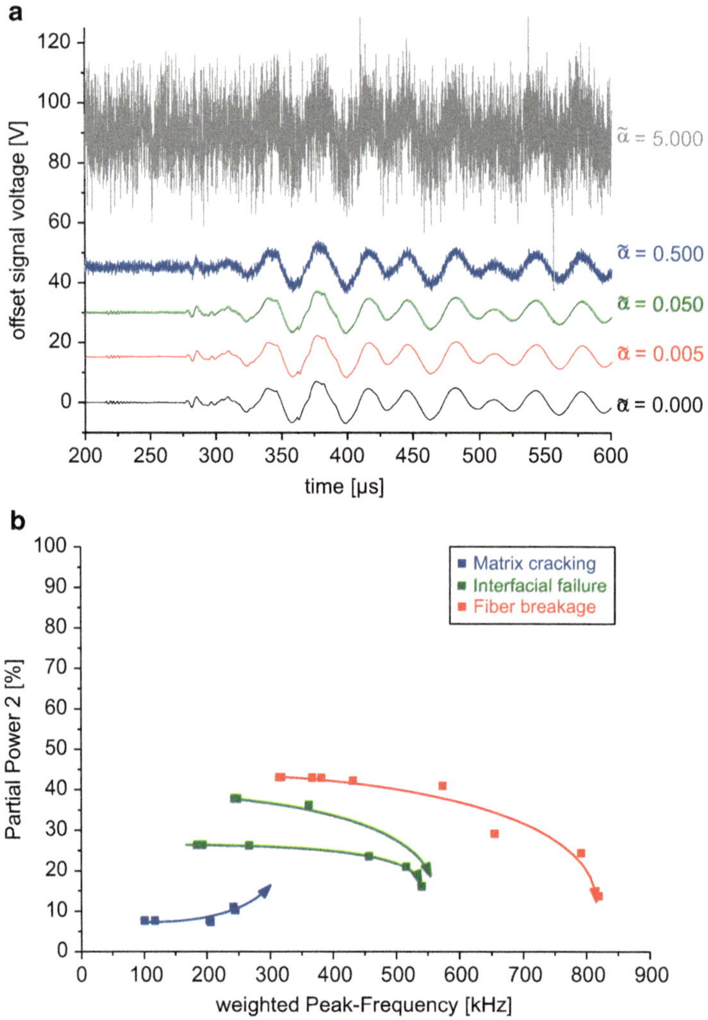

Fig. 4.125 Increase of *S/N* ratio for AE signal of matrix cracking (**a**). Influence of $\widetilde{\alpha}$-values on extracted frequency feature values (**b**)

4.6 Source Localization

Another important concept for interpretation of acoustic emission signals is the localization of the acoustic emission source position. Once the acoustic emission source is localized, its origin can be correlated with microscopic observations and usually is called an acoustic emission event. For most of the algorithms used for source localization, the calculation of the source position is based on the differences of the arrival times between several sensor positions. As shown in Fig. 4.126,

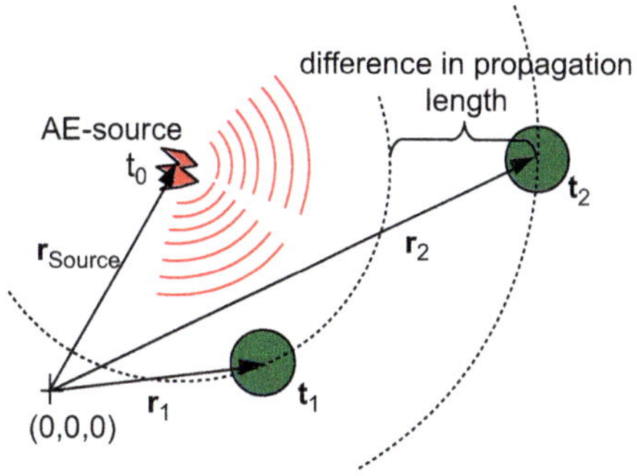

Fig. 4.126 Scheme for localization of acoustic emission source positions based on differences in signal arrival at different sensor positions (reprinted from [35])

the elastic wave originating from an acoustic emission source radiates into the solid volume and "arrives" at the respective sensor positions with distinct differences in arrival time $\Delta t = t_2 - t_1$.

A comprehensive review on state-of-the-art source localization procedures was recently given by [52]. The purpose of this section is thus not to repeat the substantial developments that have been carried out previously. Instead, the focus of this section rests on the specific challenges that arise for the application of source localization procedures in fiber reinforced materials and some recent developments to address these needs.

A first challenge arises due to the type of wave propagation faced in most of the composite applications. For source localization in volumetric media, the relevant sound velocities are the longitudinal c_L and transversal c_T bulk wave velocities. Although these velocities suffer from dispersion, their values can be assumed to be reasonably constant within typical AE applications. But as extensively discussed in Sect. 4.3, the acoustic waves found in thin platelike structures are guided waves. For these Lamb-type guided waves, the propagation velocities are strongly dependent on frequency (cf. Sect. 4.3) and therefore need to be considered explicitly. Moreover, for these guided waves, every wave mode has an individual dependency of propagation velocity as a function of frequency. Finally, the formation of such guided waves reduces a 3D localization problem to a 2D localization problem, since the depth position of the AE source cannot be localized based on differences in arrival time at the AE sensors.

Especially in the case of composite plates, an acoustic anisotropy is introduced by the orientation of the fiber reinforcements. Their dominating elastic modulus results in highly anisotropic sound velocities and therefore is another significant challenge for source localization. For the nonuniform stacking sequences typically used in composites, these acoustic anisotropies can result in complex propagation behavior of the acoustic emission signals and have to be taken into account for localization and interpretation. An example of the acoustic anisotropy faced in such fiber reinforced composites is given in Fig. 4.127.

Here the angular dependency of the group velocity of the fundamental Lamb wave modes and the fundamental shear-horizontal mode is plotted for a frequency of 250 kHz. The calculated group velocities of the S_0 mode show strongest

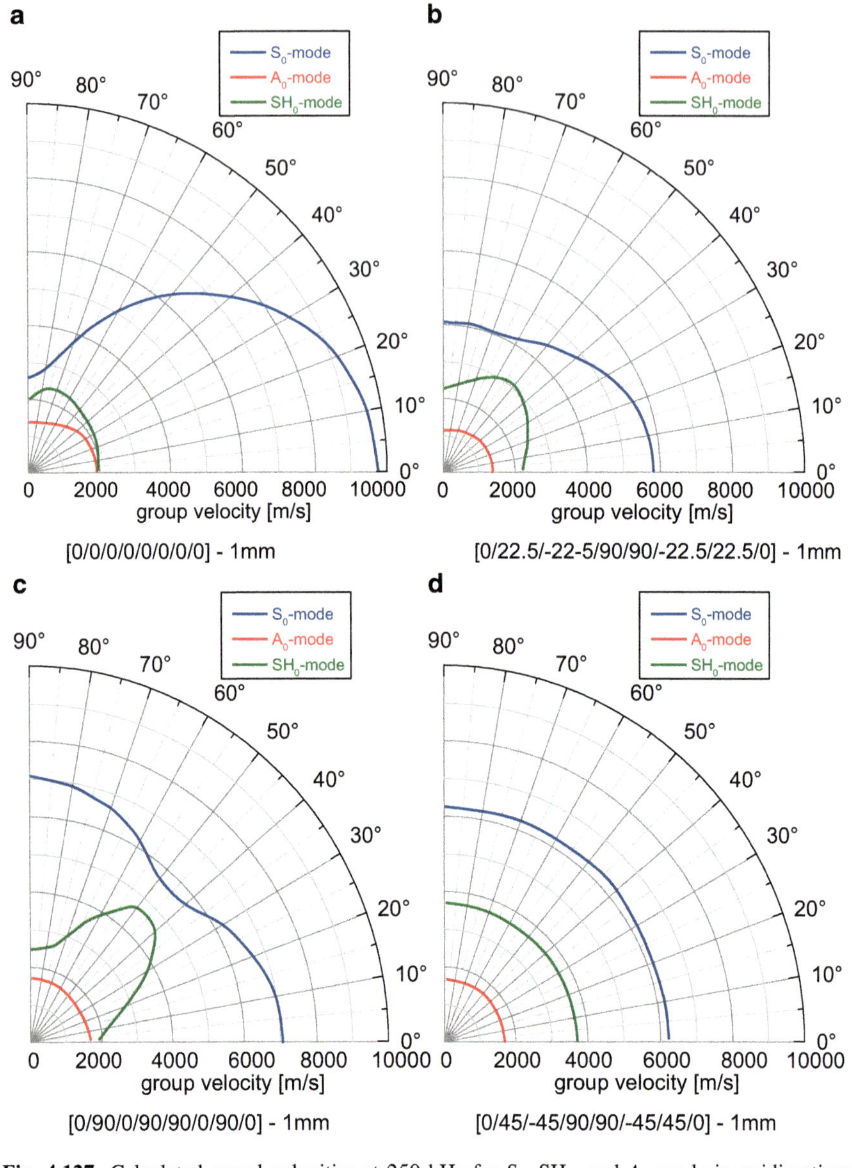

Fig. 4.127 Calculated sound velocities at 250 kHz for S_0, SH_0, and A_0 mode in unidirectional stacking (**a**), oriented cross-ply stacking (**b**), cross-ply stacking (**c**) and quasi-isotropic stacking (**d**) of 1 mm thick CFRP plates

anisotropy for the unidirectional stacking sequence as seen in Fig. 4.127a. The degree of anisotropy in the group velocities tends to vanish as more layers are orientated in different directions as seen in Fig. 4.127b, c. For the quasi-isotropic stacking sequence, the group velocity is almost equal to every angle as seen in

Fig. 4.127d. The A_0 mode is less affected by the stacking sequence than the S_0 mode and shows almost isotropic nature for all of the stacking sequences. Only for the unidirectional stacking sequence a slight elliptical group velocity profile is visible in Fig. 4.127a.

While acoustic anisotropy and guided wave propagation are challenges closely linked to fiber reinforced materials, even further challenges arise in realistic structures made from these materials. These challenges comprise complex propagation paths including reflections at boundaries and interfaces as well as changing acoustic properties due to external stresses or due to growth of damage.

In the following, first the important step of onset picking is reviewed, since this comprises one of the fundamental tasks in AE source localization. Subsequently, the traditional way of Δt-based source localization is briefly introduced to act as reference for new developments. In the last section, a recently developed approach of AE source localization using neural networks is presented in some applications.

4.6.1 Determination of Signal Onset

The main experimental challenge in AE source localization is the exact determination of the arrival time of a transient AE signal. As depicted in Fig. 4.128a, the first occurrence of a threshold crossing can be used as onset of the signal. The so-obtained arrival time is obviously linked to the chosen threshold level and consequently can falsify the calculated source coordinates. As alternative approaches other methods for determination of the arrival time of elastic waves were adopted from seismology in the past. The most important techniques to mention are the autoregressive methods. The modified Hinkley criteria use the position of the energetic minimum of transient signals at the onset [161]. An application of the Hinkley criterion using two different scale parameters a is shown in Fig. 4.128b. In contrast to the threshold-based techniques, this method yields a more reliable determination of the signal onset. Another alternative is the Akaike information criterion (AIC), which is obtained from the signals envelope after applying a Hilbert transformation [162]. The result for the AIC criterion in application to an acoustic emission signal is shown in Fig. 4.128c.

In comparison to the previous techniques, the accuracy for determination of the signal onset is substantially increased. However, both autoregressive methods benefit from previous knowledge of the approximate region of signal onset to provide stable results. In this case, these methods yield the most reliable results to determine the signal onset [161]. Further techniques focus on direct determination of time of arrival differences by inverse calculation of attenuation effects, by beam forming [163] or by extracting arrival times at constant frequencies using time-frequency analysis [164, 165]. Beyond these standard methods, various approaches have been proposed, which are found in a comprehensive review on the different established techniques in [52].

The difference in arrival time between two sensor positions can also be used to filter signals based on the Δt-value. As an example of its usage, the frequently used test configuration of a tensile test is shown in Fig. 4.130. Here the goal is to interpret AE signals originating from the reduced section of the specimen. In such test configurations, failure of the grips and additional damage at the reinforcement section may cause a substantial amount of acoustic emission signals. To avoid occupying system resources in detecting these unwanted signals, a Δt-filtering method can be used. As indicated in Fig. 4.129, all sources originating from a certain part of the specimen fulfill the condition $|\Delta t| \leq 20\,\mu s$, which can be used to

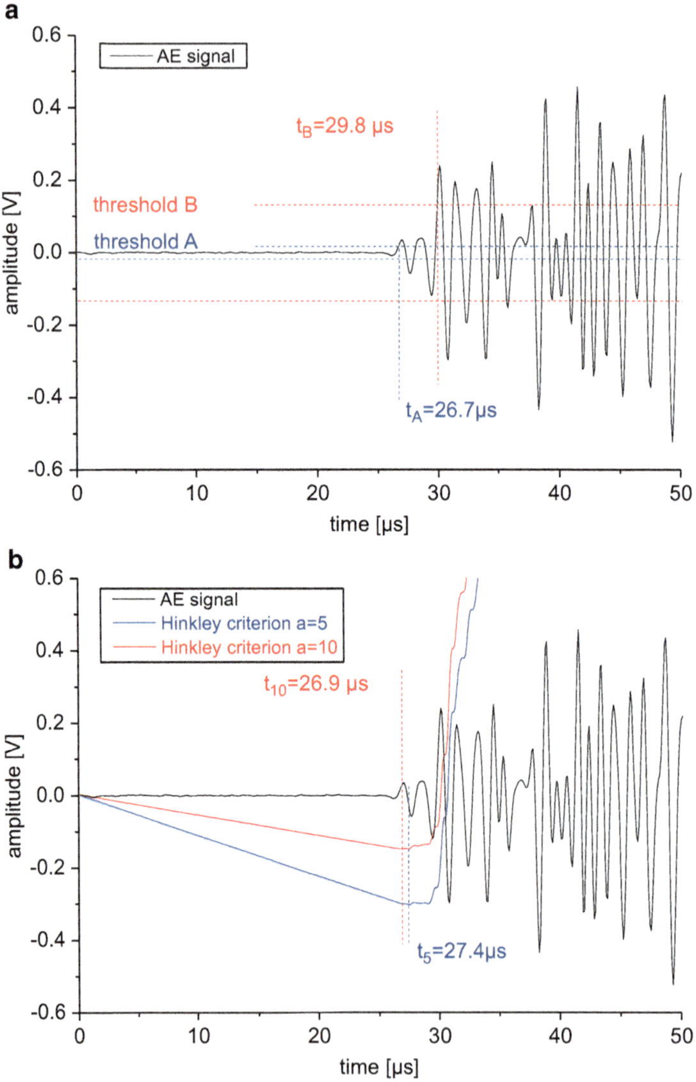

Fig. 4.128 Determination of arrival time showing influence of threshold level (**a**) results of modified Hinkley criterion (**b**) and results of AIC criterion (**c**)

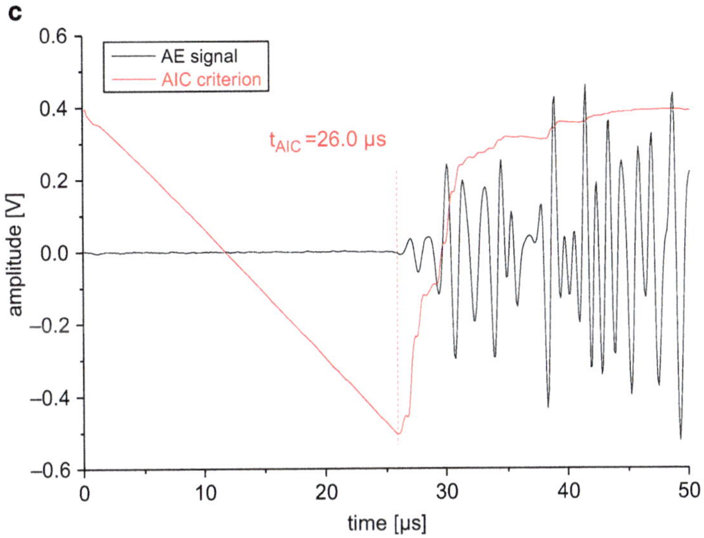

Fig. 4.128 (continued)

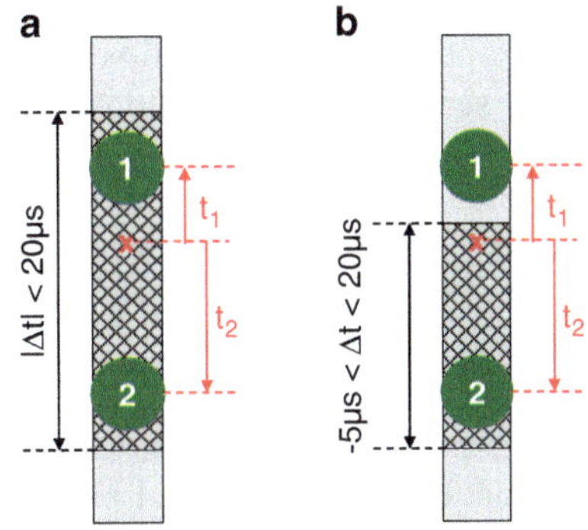

Fig. 4.129 Application of Δt-filter to tensile test setup detecting signals symmetrically between the sensors (**a**) and unsymmetrically between the sensors (**b**)

acquire only signals out of the cross-hatched area. Using the same technique, one can define areal sections for configurations with more than two sensors or define unsymmetrical portions between sensors, if different Δt limits are defined in the positive and negative direction (see Fig. 4.129b).

4.6.2 *Classical Localization Methods*

Based on the measured arrival time differences as discussed in Sect. 4.6.1, the simplest approach to perform source localization is the zonal localization. For undisturbed signal spreading, the sensor showing first arrival of the wave is closest to the source position. Therefore the vicinity of the sensor is selected as according source "zone." For larger objects, this may already give useful indications of areas with high source activity and might be used to identify such spots. However, the source position can rarely be determined with an accuracy higher than a few centimeters. Typically the accuracy is expected to be roughly around half the sensor spacing. Based on a set of arrival time values of different sensors, the source position $\boldsymbol{r}_0$ and time of acoustic emission t_0 can be calculated as:

$$|\boldsymbol{r}_i - \boldsymbol{r}_0| = \boldsymbol{c} \cdot (t_i - t_0). \tag{4.58}$$

This equation relates the positions of the different sensors $\boldsymbol{r}_i$ and their signal arrival times t_i to the acoustic emission source location $\boldsymbol{r}_0$ using the sound velocity $\boldsymbol{c}$. The localization of acoustic emission signals in isotropic media is closely related to the problems faced in GPS (Global Positioning System) applications. Consequently, the corresponding algorithm of Bancroft is typically applied to obtain the acoustic emission source position in a direct algebraic approach [166]. As an alternative, algorithms adapted from seismology science are often applied, which use an inverse approach. Here the specimen is subdivided into elements with finite resolution and hypothetical runtimes are calculated for each element. The element showing sufficient agreement with the measured runtimes is chosen as source position [52].

In general, (4.58) applies to arbitrary dimensions. Since the unknowns are the source position $\boldsymbol{r}_0$ and the time of occurrence t_0 for localization in N-dimensions, $N+1$ sensors are required to solve (4.58). Thus, for a three-dimensional localization, at least four sensors are necessary to obtain solutions. Typically more than $N+1$ sensors are used to localize the source positions which results in overdetermined equation systems. This additional information can be used to estimate the source localization error or to improve the accuracy of localization by multiple regression analysis.

In practice the localization accuracy depends on the sampling rate ΔT^{-1} used to record the acoustic emission signals, the aperture of the sensors, the validity of the assumptions made on signal propagation, and the uncertainty in determining the signal arrival. The basic inaccuracy of the measured signal runtime can be calculated from the inverse sampling rate. This is directly correlated to a metric localization error via multiplication of the sampling interval ΔT with the sound velocity. The physical size of the sensor results in an additional inaccuracy of the determination of the arrival time caused by aperture effects. If the sensor radius is small in comparison to the signal propagation length (e.g., if the sensor can be assumed as point sensor), this error contribution can be neglected. For complex geometries and anisotropic media, the largest errors arise from deviations in the assumed signal propagation path. If the used algorithm does not take into account the reflections at specimen boundaries, the assumed signal propagation velocity might be over- or underestimated. In the case of planar specimens, two-dimensional localization of acoustic emission sources generating Lamb waves can be treated in this respect by using the propagation velocity of the S_0 mode for localization. As depicted in Fig. 4.127, for the frequency range of acoustic emission signals and reasonable plate thickness, the S_0 mode shows higher group velocity than the other Lamb wave modes and thus arrives first at the sensor position.

When the propagation medium exhibits elastic anisotropy, (4.58) is no longer valid and should be replaced by a respective approach for anisotropic sound velocities. In general this boosts the complexity of the equations. For simple two-dimensional localization problems like in CFRP plates, the varying ply layups will always result in elastic anisotropy to a certain degree. A common approach is to scale the geometry to the ratio of anisotropy in the two principal sound velocities. Afterwards the common algorithms are used to infer the source location based upon equation (4.58). A direct implementation for cases where a bidirectional velocity characteristic is derived by rewriting (4.58) to yield:

$$t_i - t_0 = \sqrt{\frac{(x_0 - x_i)^2}{c_x^2} + \frac{(y_0 - y_i)^2}{c_y^2}} - \sqrt{\frac{(x_0 - x_j)^2}{c_x^2} + \frac{(y_0 - y_j)^2}{c_y^2}} \quad (4.59)$$

Here x_i, y_i and x_j,y_j denote the x- and y-coordinates of the sensor positions on the surface of the test object. The values x_o and y_o are the x- and y-coordinate of the source position, and c_x and c_y are the sound velocity in the respective directions. As a solver for this equation system, usually the Nelder-Mead-based algorithm is used.

One of the major uncertainties entering the calculation of the source position is given by the extraction of the arrival time value from the signal as mentioned in Sect. 4.6.1. For example, constant threshold values to detect the signal arrival time depend substantially on the slope of the initial part of the signal as well as on its amplitude and, therefore, may cause scatter of the so-obtained Δt-values. If the signal amplitude is sufficiently high above the threshold level, the scatter in Δt-values is dominated by the slopes of the acoustic emission signal. For signal amplitudes close to the noise level, it is hardly possible to determine the correct arrival time, which may cause even stronger scatter of the extracted values. Other factors to take into account for the calculation of the uncertainty of localization are the aperture of the sensor and the sampling rate of the system. Further factors governing the accuracy of the method are the system sensitivity and the chosen couplant. These are known to be relevant factors that may have a systematic influence on the quality of the localization results, but are neglected in the following, since these are difficult to quantify and are subject to the individual experiment.

The uncertainty of localization is also dependent on the sensor positions and the position of the acoustic emission source. Source localization based on Δt-values leads to a spatial variation of the localization sensitivity. This means that an identical shift in position Δr at different locations on the test object will lead to different shifts in the respective Δt-values.

The Δt-values of two sensors are directly obtained using formula (4.58). To determine the source localization error, the gradient of the Δt-values may be

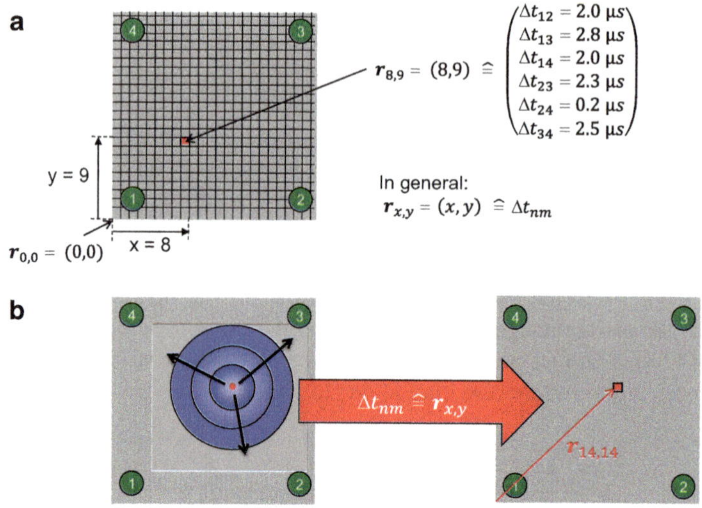

Fig. 4.130 Scheme of Δt-mapping source localization technique to establish Δt-map (**a**) and during application (**b**)

calculated. This vector field is defined as gradient of formula (4.58) in all spatial directions:

$$\nabla(t_i - t_0) = \nabla\left(\frac{|\boldsymbol{r}_i - \boldsymbol{r}_0|}{c}\right) \tag{4.60}$$

This gradient of Δt-values also determines the maximum source localization accuracy that can be achieved for a particular source position since it directly affects the accuracy of the localization procedure. Because the errors discussed so far are mutually uncorrelated, the resulting temporal and spatial errors may be derived using the Gauss formalism to yield a final estimate for the uncertainty of localization.

One strategy to overcome the problem of acoustic anisotropy in source localization procedures was proposed by Pullin, Baxter, and Eaton and is referred to as Δt-mapping [167, 168] schematically shown in Fig. 4.130.

As schematically shown in Fig. 4.130a, for a structure with arbitrary geometry and acoustic properties, first a subdivision into discrete zones is made. Given a sensor network at fixed positions, for every discrete zone, a theoretical Δt-value is calculated for every sensor combination. This effectively yields a Δt-map of the structure. For the Δt-values of the experimentally detected signals, the closest combination of Δt-values can then be estimated by automatic search algorithms [167, 169, 170]. This provides an effective localization procedure for fiber reinforced materials. However, the drawback of this approach is the necessity to rely on theoretical Δt-values. Material imperfections, erroneous sound velocities, attenuation, and reflections may cause distinct differences between theoretical values and those Δt-values measured in a real structure.

4.6.3 Source Localization Methods Based on Neural Networks

First recommendations have been made to use neural networks for source localization in geometrically complex metallic structures [171]. For the neural network approaches proposed so far, the basic strategy involves two subsequent stages (training and application) as shown schematically in Fig. 4.131 and is explained in more detail in the next passages.

Similar to the Δt-mapping concept, artificial neural networks can be used to establish a symbolic relationship between input data (Δt-values) and output values (source position coordinates).

By conception, the application of a neural network can be subdivided in two stages:

1. Training of a functional relationship based on Δt-values of known source positions
2. Application of this trained relationship to Δt-values with unknown source position

For the first part (the training stage), Δt-values need to be extracted from signals detected for known source positions. This can be achieved by experimental test sources (see Sect. 4.6.3.1) or by respective modeling of the signals (see [172]). While for the modeling approaches, a precise knowledge of the acoustical properties is necessary, the experimental approach using test sources does not even require an explicit knowledge of the sound velocities in the material.

Based on either of these input data, a neural network can be used to approximate a functional relationship to the source coordinates. To achieve such training of

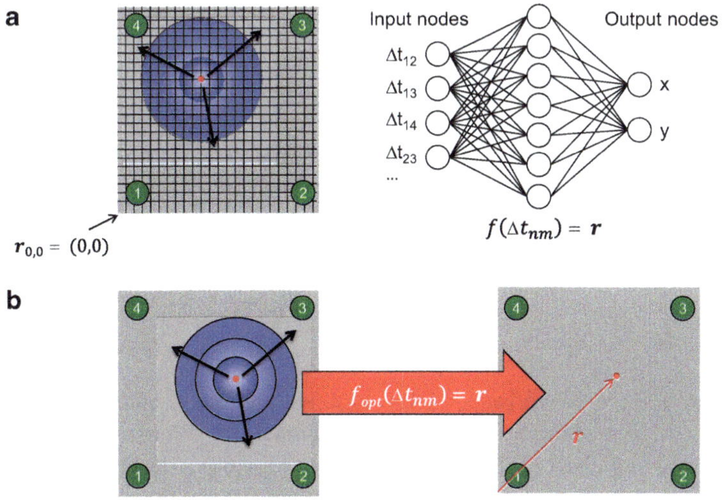

Fig. 4.131 Scheme of source localization technique applying neural networks including training stage (**a**) and application stage (**b**)

functional relationships, typically back-propagating algorithms such as the back-propagation multilayer perceptron (MLP) or the SuperSAB are applied [136]. In both cases, the training error is minimized by a gradient method using a feedback of the resultant training error. In the following a review of the resilient back-propagating MLP is given, since this was used for the work presented in [158, 159, 173, 174].

The architecture of a MLP is a neural network as schematically shown in Fig. 4.131a. The training of such neural network structures can be understood as function approximation problem, where the "function" maps values given at the input nodes to values at the output nodes. Since this correlation is generally a nonlinear problem, multiple artificial neurons are typically used, arranged in a number of hidden layers between input and output nodes (see Fig. 4.131a), powerful enough to allow approximation of arbitrary functions.

The response of a single neuron node h_{A} to various inputs h_{E_i} with weights w_i is formulated using the activation function $f_{\mathrm{activation}}$ as

$$h_{\mathrm{A}} = f_{\mathrm{activation}}\left(\sum_i w_i h_{E_i}\right). \tag{4.61}$$

The activation function is used to modify the linear decision boundary given by $\sum_i w_i h_{E_i}$ and operates as decision criteria if the neuron is active or inactive. This can either be realized by a threshold function (hard-limit function) or by continuous (linear or sigmoidal) functions. The functionality of the neural network now is the adaptive adjustment of the node weights w_i to generate an optimized mapping between the input layer and the output layer. This is done by adjusting the weights of the nodes in order to find a global minimum of the quadratic error surface function $\widetilde{L}$:

$$\widetilde{L}(\boldsymbol{w}) = \frac{1}{2}\sum_{i=1}^{O_{\mathrm{train}}}\sum_{j=1}^{O}\left(q_{\mathrm{ref},j} - q_j\right)^2 \tag{4.62}$$

Here $\boldsymbol{w}$ represents the vector of all node weights, $q_{\mathrm{ref},j}$ the output value of the reference objects, and q_j the present output value. The dependency of the error surface function on $\boldsymbol{w}_{\mathrm{tot}}$ is implicitly given in (4.62), since the node weights strictly depend on the values of q_j. The back-propagating multilayer perceptron updates these weights $\boldsymbol{w}$ iteratively based upon a gradient search of the global minimum with a learning rate ι:

$$\boldsymbol{w}(i+1) = \boldsymbol{w}(i) - \iota\frac{\partial\widetilde{L}(\boldsymbol{w})}{\partial\boldsymbol{w}}. \tag{4.63}$$

In order to achieve convergence of this iterative process, a minimum in $\widetilde{L}(\boldsymbol{w})$ is necessary. As pointed out by Polikar, such minima of the error surface function can either be local or global [136]. Thus, convergence of (4.62) is dependent on the initial settings of the weights and does not necessarily converge to the global minimum.

Other typical problems faced in back-propagating MLPs are the appropriate choices of the learning rate, the network structure, and the threshold level of the error function. The latter one is the termination criterion which is used as the lower boundary assuming that the network is sufficiently trained when this threshold level is reached. While the learning rate and the network structure can often be determined empirically, a low threshold level of the error function can result in overtraining of neural networks. Then the neural network tries to capture very detailed the characteristics of the input objects.

After successful training of such neural networks, their ultimate strength is the computationally fast reapplication to new input data. This is the second stage and is known as application stage. Here the neural network is able to approximate source positions based on new Δt-value combinations.

In the following a review is made of the implementation introduced in [158, 159, 173, 174] to yield a source localization technique particularly useful for fiber reinforced materials. To this end, the next subsections present some aspects of experimental test sources to generate input data. Subsequently, it is demonstrated how source localization can be performed in highly anisotropic media such as unidirectional fiber reinforced composites and how the influence of obstacles in the propagation path is compensated. The section closes with some examples of source localization in composite structures subject to mechanical loading.

4.6.3.1 Training Stage Using Test Sources

For the training stage, various experimental test sources can be used. These test sources are applied at different positions on the test object to generate a dataset of representative Δt-values. Therefore test sources need to be placed at sufficient spatial density. Moreover, the test sources need to fulfill two further requirements:

1. At every test position, the signal of the test source needs to be detected by a sufficient number of sensors.
2. The test source signals' modal composition and frequency content need to be representative for the experimentally detected signals in the application stage.

The most frequently used test source in AE is the pencil lead break source (Hsu-Nielsen source) according to standard ASTM E976. The signals of pencil lead breaks are typically of high intensity and are therefore useful to allow signal detection also at sensors with far distance to the source position. However, pencil lead breaks are usually performed in the out-of-plane direction and their bandwidth is dominantly below 200 kHz. This causes a preferential stimulation of the fundamental antisymmetric mode in platelike structures as seen in a CWD in Fig. 4.132a.

Since various sources in fiber reinforced materials tend to stimulate signals with dominant contributions of other than A_0 modes, it is necessary to also take into account their propagation behavior. As seen in the CWD in Fig. 4.132b, piezoelectric pulsers can be used to excite fairly different signals in the same test structure. Although being placed in the out-of-plane direction, they can stimulate also symmetric modes, as seen by the presence of the S_0 mode in Fig. 4.132b.

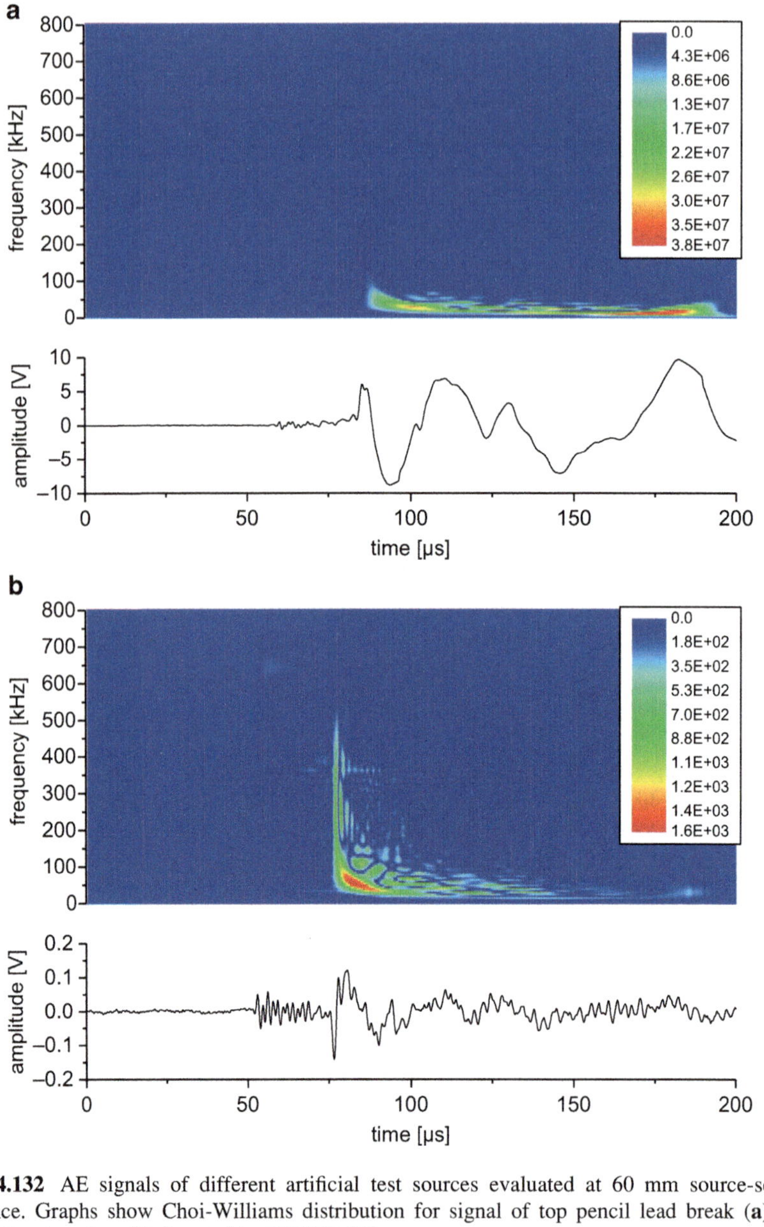

Fig. 4.132 AE signals of different artificial test sources evaluated at 60 mm source-sensor distance. Graphs show Choi-Williams distribution for signal of top pencil lead break (**a**) and piezoelectric pulser (**b**) detected by KRNBB-PC sensor

An alternative approach to the use of physically different test sources is to have a type of test source, which can generate a broad range of frequencies. To this end, piezoelectric pulsers operating within a reasonably broad frequency range can be applied. Using a piezoelectric pulse with 20 ns rise-time, 1 μs width, and 10 V amplitude applied to the conical type sensor described in Sect. 4.4.2, a KRNBB-PC sensor is able to detect both fundamental Lamb wave modes with reasonable intensities as shown in the CWD in Fig. 4.132b. After detection, it is possible to use high- or low-pass filters to select particular frequency ranges of the signals to generate different input signals as basis for the neural network training. As seen in Fig. 4.133a, a seventh-order Butterworth high-pass filter of 400 kHz generates a

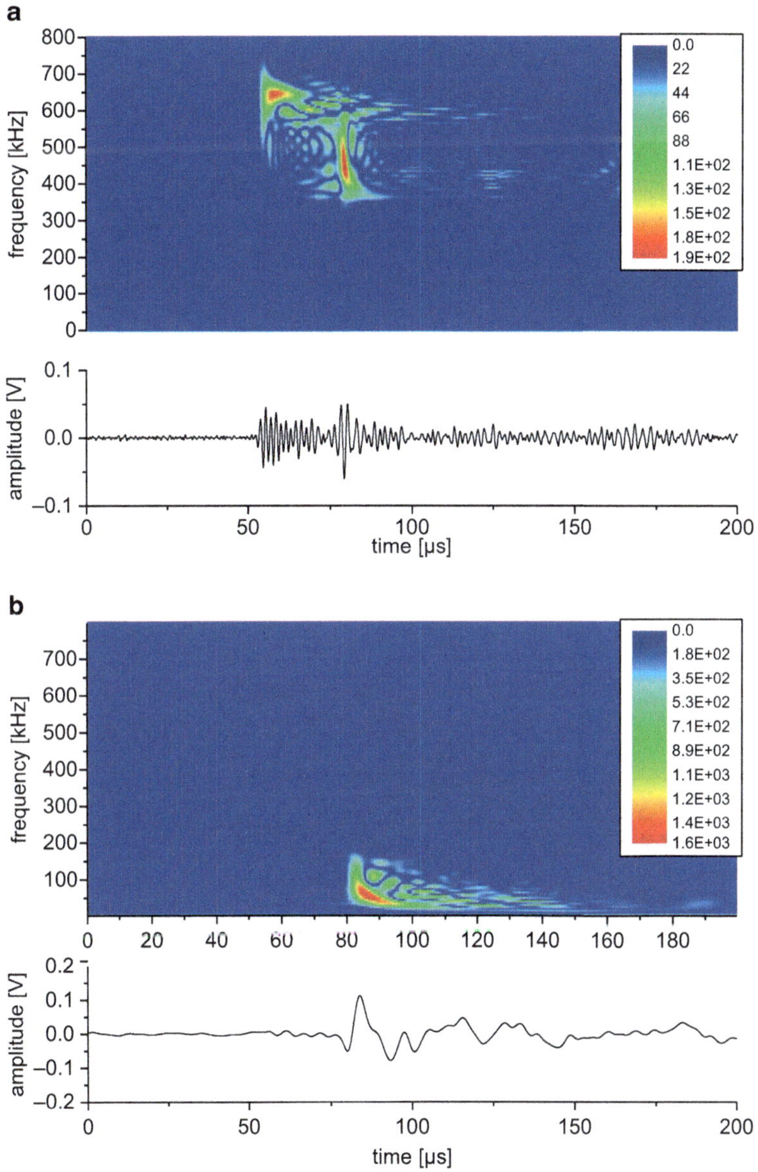

Fig. 4.133 AE signals of piezoelectric pulser filtered by seventh-order Butterworth high-pass of 400 kHz (**a**) and low-pass of 200 kHz (**b**)

signal dominated by the S_0 mode showing a signal arrival at 52.6 μs, whereas a seventh-order Butterworth low-pass filter of 200 kHz yields a signal dominated by the A_0 mode (cf. Fig. 4.133b), which has a signal arrival at 77.3 μs due to the lower propagation velocity of this mode. Using such filtering, this effectively allows to generate signals with frequency contents being representative for a broad range of source types as faced in fiber reinforced materials and therefore allows to train the neural network with a variety of input signals without the need for repetitive testing at a single test source position.

An obvious alternative to test sources with broad frequency range is to filter the detected signals of the application stage to yield a characteristic similar to those of pencil lead break signals. Practically this means using low-pass filtering to yield only the low-frequency portion of the signal. However, this may yield poor signal-to-noise ratios for the signals that are dominated by frequencies outside the filtering range.

Further considerations are required with respect to the density of training points. For a simple example as a plate structure, a homogeneous distribution of training points might be expected to yield appropriate results. Judgment on the suitability of the density and position of training points can be made based on the appearance of Δt-maps. Here the Δt-values of a particular sensor pair is plotted as function of the spatial coordinates. This is exemplarily shown for a set of laminates ranging from a unidirectional stacking sequence to a cross-ply stacking and a quasi-isotropic stacking in Fig. 4.134. Due to the different fiber orientations, a corresponding change in the acoustic anisotropy may be expected as shown in the diagrams in

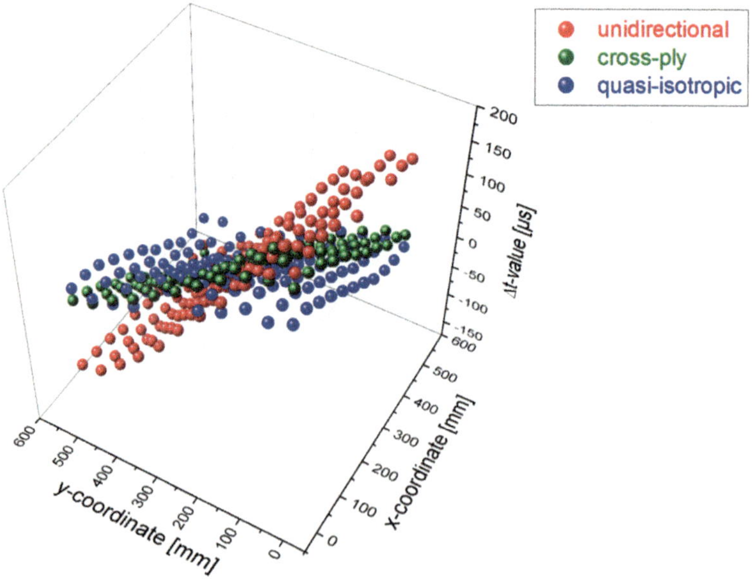

Fig. 4.134 Example of Δt-map for one sensor pair mounted on laminate plates with 400 mm edge length as discussed in Sect. 4.6.3.2. Comparison is made between unidirectional stacking, cross-ply stacking, and quasi-isotropic stacking

Fig. 4.127. This directly translates into the expected variance of Δt-values, which are governed by the wave velocities of the material. As seen from the examples in Fig. 4.134, the individual Δt-maps are reasonably smooth, and their steepness and curvature is just owed to the degree of acoustical anisotropy in the laminate. Hence the full character of the laminates may already be suitably approximated by subsets of the training data points shown. This allows the reduction of the overall number of training data points in these cases as will be shown in Sect. 4.6.3.2.

However, for a geometrically more complex structure, the ideal distribution and position of training points is not so straightforward to conclude. In such cases, the Δt-map may be used to detect divergences or areas with less smooth structures, which might benefit from refinement. An example for such an object made from fiber reinforced composites is shown in Fig. 4.135. This structural CFRP part contains several sinks, holes, and steep curvatures as well as local changes in the stacking sequence and the thickness.

As discussed in Sect. 4.3, all of these aspects influence the signal propagation and thus will add to substantial deviations in the signal arrivals at the sensor position. After applying a test source at several points on a test grid, the respective Δt-map of the structure is obtained as shown in Fig. 4.135. Here only the Δt-value of sensors 3 and 12 was used for visualization of the Δt-map. Due to the size of the component, the overall arrival time difference spans a range of 250 μs. The Δt-map does not resemble a topological map of the part, but instead it reveals the degree of acoustic anisotropy as well as regions with strong discontinuities as seen close to

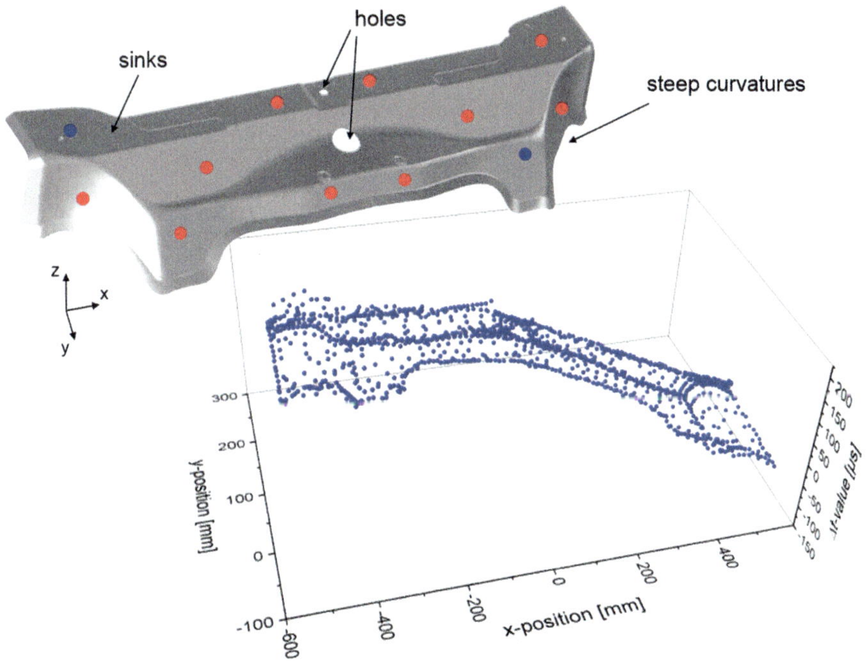

Fig. 4.135 Example of complex structural part made from CFRP with respective Δt-map

the position of the holes. This allows to identify regions, which benefit from a higher number of training points to be used for the neural network approach. Also, the scatter of Δt-values at x-positions below -200 mm indicates that the detected signals approach the level of the background noise, thus making the correct evaluation of the signal arrival time more difficult.

4.6.3.2 Compensation of Acoustic Anisotropy

To demonstrate the capabilities of neural network-based source localization, an investigation on carbon fiber reinforced laminates is presented. Plates of 570×570 mm edge length and 1.3 mm thickness were fabricated using three stacking sequences. All plates are made from unidirectional epoxy prepreg Sigrafil CE1250-230-39 and were cured according to the material supplier recommendations at 120 °C. The first plate is a quasi-isotropic laminate with $[0/60/\text{-}60]_{\text{sym}}$ stacking sequence, the second is a cross-ply laminate with $[0/90_2]_{\text{sym}}$ stacking, and the third is a unidirectional laminate with $[0_3]_{\text{sym}}$ stacking. Therefore, the three plates show an increasing amount of acoustical anisotropy due to the orientation of the fibers in the individual layers.

Using a rectangular training point grid of 11×11 points, a total number of 121 source positions is distributed across the plate in equal mutual distances of 5 cm. At every position a pencil lead break is performed to yield the training data (i.e., to generate signals at known source positions). Four WD sensors mounted at the corner of the plates are used to detect the AE signals using an acquisition rate of 10 MSP/s, 40 dB_{AE} preamplification, and a band-pass filter ranging from 100 kHz to 1 MHz. The arrival time of the AE signals is determined using the AIC.

In Fig. 4.136 the localization error of the classical methods based on (4.58) is shown. The source localization error Δr is obtained from the calculated source position $\boldsymbol{r}_0$ and the known test source position $\boldsymbol{r}_{\text{test}}$ as

$$\Delta r = |\boldsymbol{r}_0 - \boldsymbol{r}_{\text{test}}| \tag{4.64}$$

The false-color range in Fig. 4.136 is used to represent the source localization error.

As clearly seen in the figures, the classical method suffers from the increasing amount of acoustical anisotropy. While the quasi-isotropic laminate still allows a successful localization of all test signals, this is not the case for the cross-ply laminate and the unidirectional laminate. Here a decreasing number of only 37 out of 121 source positions can be calculated within the dimensions of the cross-ply laminate. For the unidirectional laminate, this is reduced even further to 6 out of 121 source positions. Also the average source localization error increases substantially. Especially in the areas outside the rectangle spanned by the four sensors, the source localization accuracy is very poor.

For the neural network-based approach, 62 test positions are chosen as training dataset. An artificial neural network using two hidden layers and ten neurons is applied. As activation function of the neurons, a sigmoidal hyperbolic tangent function is defined. For every laminate the trained neural network allows a localization of all

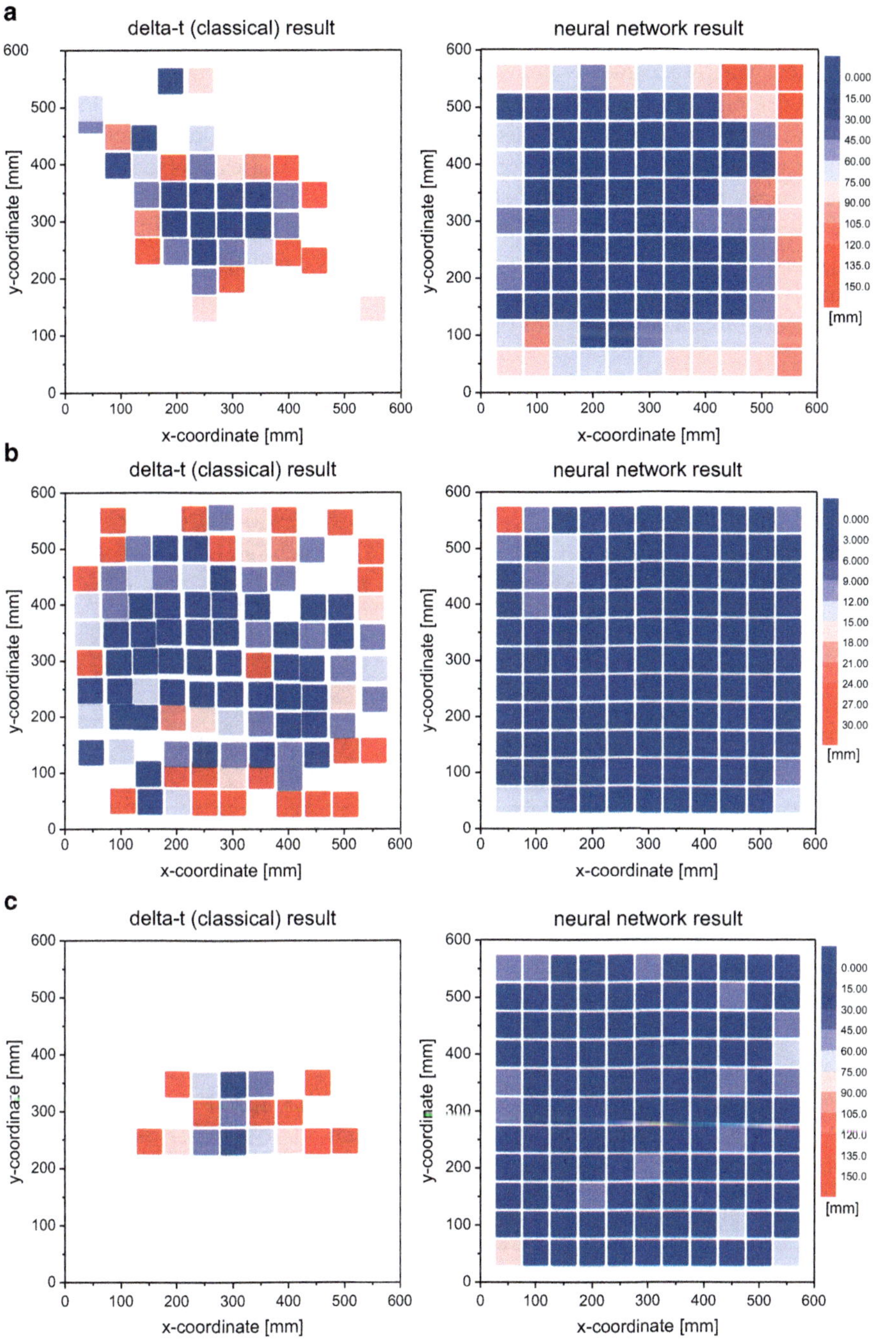

Fig. 4.136 Comparison of AE source localization accuracy for classical localization (*left*) and neural network-based approach (*right*) for quasi-isotropic laminate (**a**), cross-ply laminate (**b**), and unidirectional laminate (**c**)

121 source positions within the dimensions of the plate. Moreover, the source localization error is four to five times less than for the classical localization method, and the source localization accuracy is almost identical in all parts of the plate.

This demonstrates the high flexibility of neural network-based source localization with respect to acoustically anisotropic media. However, from a practical point of view, the number of 62 test source positions used for the area of 60 cm × 60 cm is unnecessarily high, since the Δt-fields of these plates do not exhibit strong gradients. Hence, similar source localization accuracy should be expected for systematically less number of training data positions. To demonstrate the impact of a reduced number of training points, Fig. 4.137 shows the false-color representation of the source localization error of the quasi-isotropic plate.

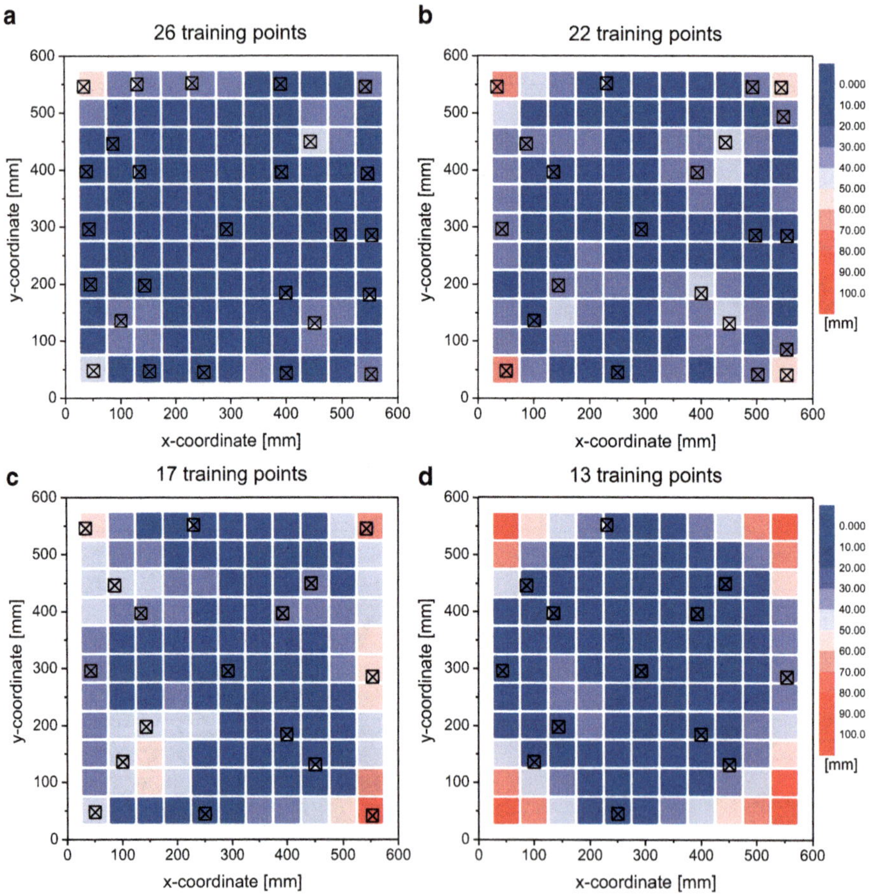

Fig. 4.137 AE source localization accuracy for systematic reduction of training data positions to 26 (**a**), 22 (**b**), 17 (**c**), and 13 (**d**)

As seen in the source localization error indicated in the figures, a systematic reduction of the number of training data points from 62 (cf. Fig. 4.129a) to 26, 22, 17, and 13 in Fig. 4.137 still yields a reasonable source localization accuracy. The average source localization error was found to increase from 3.5 to 25.4 mm in the case with only 13 training data positions.

So far, the results were obtained only for test source signals using pencil lead breaks. But as previously discussed in Sect. 4.6.3.1, the typical AE signals of fiber reinforced materials may show distinct differences in their modal composition and frequency range, which is not sufficiently represented by pencil lead break sources.

A suitable alternative is the usage of two different test sources as the piezoelectric pulser and the pencil lead break at identical test source positions. Subsequently the neural network is trained using the same approach as before but with one additional input node. This input node is set to 0 for the pencil lead break input data and to 1 for the input data of the piezoelectric pulser. As a consequence, the neural network virtually forms two branches to deal with the source localization task of the one or the other set of input data. Figure 4.138 presents the resulting source localization error using this approach. In both cases, a combined set of two times 45 training data points was used. Figure 4.138a shows the application of this trained neural network to the signals of the piezoelectric pulser only and Fig. 4.138b to the signals of the pencil lead breaks only. Clearly in both cases, the neural network provides similar source localization accuracy being mostly dominated by the number of training data points rather than the presence of two different sets of input data.

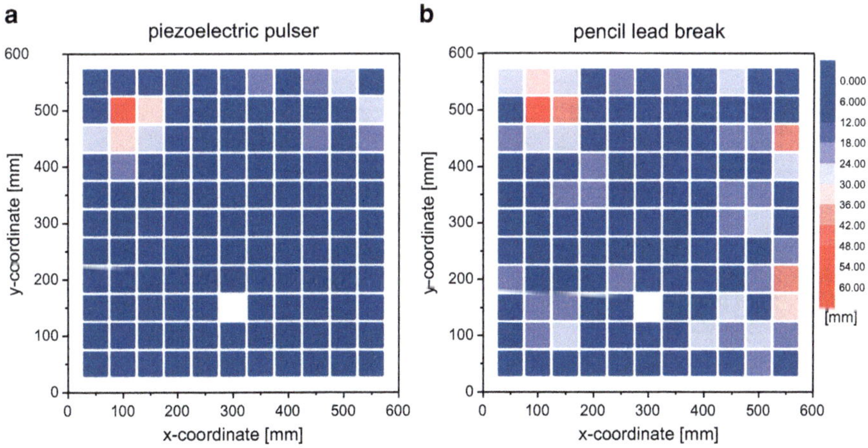

Fig. 4.138 AE source localization accuracy for quasi-isotropic laminate with two times 45 training data points based on simultaneous training data of piezoelectric pulser and pencil lead break source applied to data of piezoelectric pulser (**a**) and applied to data of pencil lead break (**b**)

4.6.3.3 Discontinuous Δt-Fields

Other than the plates studied in Sect. 4.6.3.2, realistic structures usually show more geometrical complexity. This may involve curvatures, internal interfaces, holes, supports, or other factors, which affect the way of acoustical propagation. All these disturbances in acoustical propagation can also be interpreted as disturbances in the arrival time differences Δt. Therefore, the Δt-field may start to exhibit steep gradients or may start to show discontinuities. Overall, the quality of the neural network-based source localization approach is directly linked to the approximation quality of this Δt-field.

To introduce some further complexity to the source localization problem, the specimen geometry typically used for bearing strength test of bolted laminates is a suitable candidate. T700/PPS specimens of 220 mm × 110 mm × 10 mm (length × width × thickness) with 18 mm diameter bolts were used to this end. During the test, four type WD AE sensors were mounted on the specimens using medium viscosity couplants and clamp systems. AE signals were acquired using 10 MSP/s acquisition rate at 40 dB_{AE} preamplification and a band-pass ranging from 20 kHz to 1 MHz. The complex situation of AE signal propagation encountered in this test environment is shown in Fig. 4.139 in an exemplary calculation result using finite element simulations based on the implementation described in Sect. 4.2.

In this test configuration, the presence of a bolt in the laminate causes a pronounced discontinuity in the Δt-field. Following the previously outlined method, a neural network with four neurons and three hidden layers with sigmoidal activation functions is used. Therefore, all effects of the disturbed wave propagation are accounted for in the training stage of the neural network.

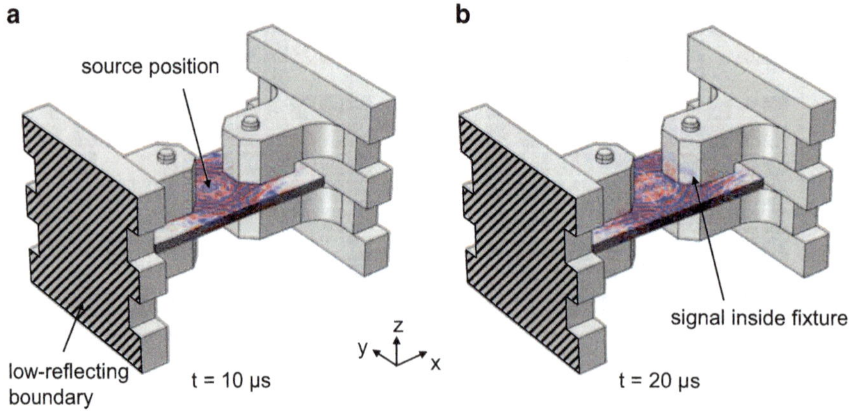

Fig. 4.139 FEM result of acoustic emission wave field of modeled fiber breakage failure at $t = 10$ µs (**a**) and at $t = 20$ µs after source excitation

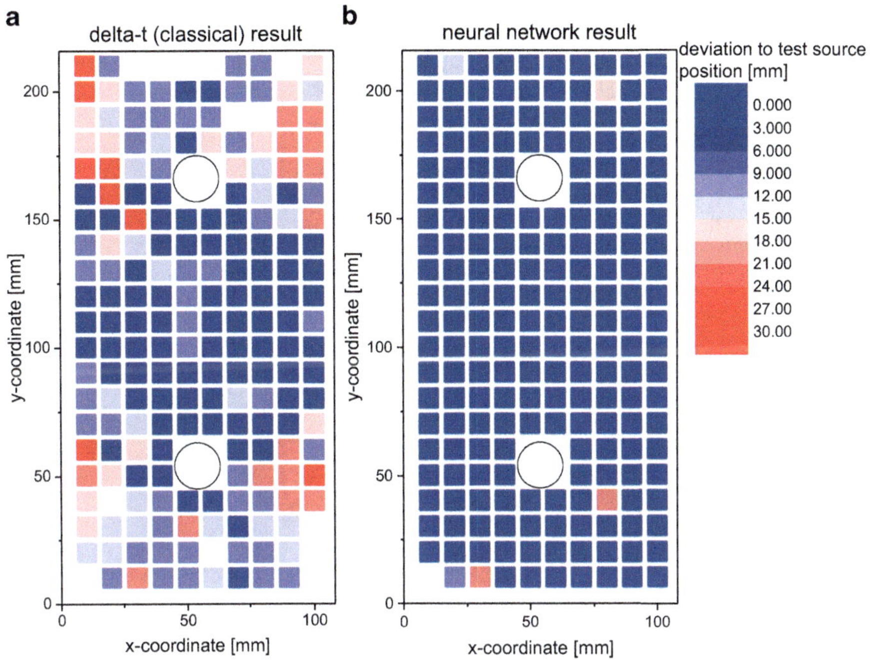

Fig. 4.140 Comparison of AE source localization accuracy for classical localization (**a**) and neural network-based approach (**b**)

A comparison between the source localization error of the classical method and the source localization error of the neural network-based approach (based on 208 training positions) is shown in Fig. 4.140. As average of the whole specimen, the classical method has a source localization accuracy of ±14.5 mm. Out of the 210 test source positions, only 196 were localized within the geometry of the specimen. Compared to that, the neural network approach has a mean source localization accuracy of only 3.3 mm, and 209 out of 210 source positions were localized within the specimen geometry.

4.6.3.4 Scale Invariance and Portability

Another interesting aspect of the neural network-based source localization is the ability to scale existing training data to similar objects of larger size. To demonstrate these capabilities, the specimens of the bearing strength test are used. Here, three types of test specimens with different dimensions are investigated. The smallest specimen investigated has dimensions of 108 mm × 54 mm × 5 mm (length × width × thickness), the medium size dimensions of 220 mm × 110 mm × 10 mm (length × width × thickness), and the largest size dimensions of 330 mm × 165 mm × 15 mm (length × width × thickness). Starting with the

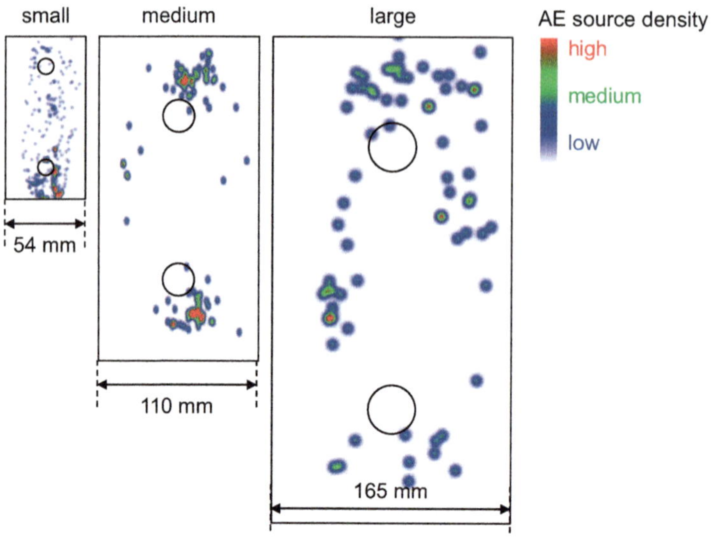

Fig. 4.141 Exemplary localization results by neural network-based technique using training data of the medium-sized specimen

smallest size, the diameter of the bolt is 9 mm with an increase of a factor of 2 and 3 for the medium and large size, respectively. Test signals of pencil lead breaks are detected as described in Sect. 4.6.3.3.

To investigate the scalability and portability of trained neural networks, only one database will be used for the results shown in the following. The training data was generated using the medium size plate. Since the three different specimen types vary in all dimensions by a factor of 2 and 3 relative to the smallest plate, the Δt-values of the training dataset are multiplied by a factor of 0.5 and 1.5, respectively.

The comparison in Fig. 4.141 shows the AE source positions as localized during a bearing strength test. For the small and large cases in Fig. 4.141, the localization results are obtained for a neural network based on the training data obtained on the medium specimen size. The obtained source localization error for the individual training datasets of each specimen type was found to be almost identical to the source localization error of the scaled training datasets.

Similar to the scaling procedure of the training dataset, it is also feasible to transfer the trained neural networks from specimen to specimen with identical geometries. This is particularly relevant for testing of specimen series, where no substantial changes of the material, the geometry, the sensor positions, and the types occur. In such cases a direct reapplication of a trained neural network is an efficient way to perform fast source localization.

4.7 Application to Composites

Based on the possibility to detect the occurrence of all sorts of microscopic failure mechanisms in composite materials, there is a broad range of applications for the acoustic emission method. In this section, a focus to monitoring of materials testing under laboratory conditions is given. However, it is attractive to extend the presented approaches to structural health-monitoring applications. This comprises a field of research on its own and is therefore not covered in this book. For more details on this part, the reader is referred to other literature dealing with this topic [52, 175, 176].

4.7.1 AE Source Identification Using FEM Results

In this section, a methodology to compare experimental data to FEM results is reviewed. The modeling procedure to compute acoustic emission signals used for this purpose is extensively described in Sects. 4.2–4.4 and is not repeated in detail here.

In general, such FEM investigations can be performed for different reasons. In any case it is required to have a validated modeling environment as presented in the previous sections. To allow source identification based on modeling results, one can use different approaches. As schematically shown in Fig. 4.142, the intuitive way is

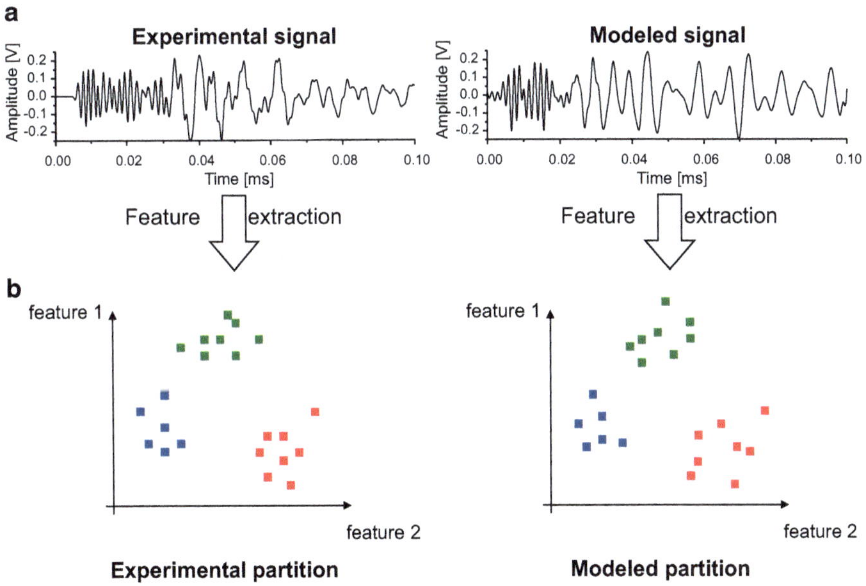

Fig. 4.142 Scheme of data processing to compare experimental AE data and FEM-based AE data based on signals (**a**) and partitions (**b**)

to assume a certain source type at a position within a solid and directly compute the respective signal in a forward approach. Being demonstrated in [13, 35, 87], this approach can yield quantitative prediction of AE signals if all details of the experiment are known. In particular, the accurate source position of an experimental signal is usually hard to obtain, since source localization approaches show some uncertainty. Moreover, for platelike structures, the depth position of the source can usually not be retrieved by the source localization approach (see Sect. 4.6).

Because for composite materials, many acoustic emission signals are detected during the experiment, one can also address the problem of source identification by a statistical approach.

As first presented in [177], the procedure is carried out in three subsequent steps. The volume of interest is defined and modeling of different source types is carried out at representative positions. With reference to a simple tensile test of a unidirectional specimen discussed in [178], the consideration would be as follows:

- The volume of interest is given by the tapered area of the specimen.
- Failure mechanisms are assumed to occur everywhere within the volume of interest.

For cross-ply specimens on the other hand, preferential depth positions of inter-ply delamination arise at the interfaces between the plies. Therefore modeling of this failure type should be considered focused at those depth positions. If notches are present, these likely act as stress concentrator and may initiate all sorts of failure mechanisms in their vicinity. Then the volume of interest has to be adapted accordingly. After the modeling work is completed, a set of waves due to a variety of source positions and different failure types is available.

The second step is to apply the same feature extraction procedure to the modeled signals as for the respective experimental signals. This yields a partition of acoustic emission signal features with known origin and is indicated as modeled partition in Fig. 4.142b.

Accordingly, the third step then consists in comparison of the feature value positions of the modeled signals and the experimental signals. This procedure has been demonstrated in various publications [24, 35, 125, 138, 177, 179] and is demonstrated in the following for one example of a tensile test as described in [178]. All acoustic emission signals were acquired using the settings reported in Table 4.2 with sensor positions at the end of the tapered area of the specimen.

Figure 4.143a shows the partition of acoustic emission signals obtained by unsupervised pattern recognition for one representative specimen. To visualize the position of the signal clusters, a projection to the features Weighted Peak Frequency and Partial Power 2 was used. For the experimental data, the clusters are well defined, but their edges are close together. As discussed in Sect. 4.5, this may cause an uncertainty in the assignment of the signals to the respective cluster. Accordingly, simulations for the failure mechanisms as described in Sect. 4.2.2.2 were performed for the type of tensile specimen considered. For the simulated AE signals, the extracted feature values are plotted in Fig. 4.143b.

Table 4.2 Acquisition settings and postprocessing parameters used for tensile tests

Acquisition settings	Value	Postprocessing parameters	Value
Preamplification	20 dB_{AE}	Feature extraction window	100 μs after threshold
Threshold	35 dB_{AE}	Δt-filter	(7 ± 2) μs
Triggering	Individual channel		
Acquisition rate	10 MS/s		
Band-pass range	20 kHz to 1 MHz		
Couplant	Medium viscosity silicone grease		
Mounting system	Clamp		
Sensor type	WD		
Number of sensors	2		

The signals simulated for fiber breakage are well separated from the rest of the signals. Compared to the experimental data, the simulated signals show slightly higher frequency contributions, but overall coincide with the experimental feature range.

The feature values extracted from modeled signals for inter-fiber failure and inter-ply delamination are observed within similar ranges as for the experimental data. Signals originating from inter-fiber failure with dominant crack opening (mode I) components are found with lowest "Partial Power 2." The simulations of inter-fiber failure with significant shear (mode II) contributions are found with higher "Partial Power 2" values up to 40 %. The variation of the orientation of the fracture plane causes variability in the absolute values as seen in Fig. 4.143b, but does not substantially change the extracted feature values.

For inter-ply delamination, a small overlap to the feature value ranges of inter-fiber failure is found. This is in very good agreement to the experimental observations. However, the separation to signals associated with fiber breakage is much more pronounced in the simulation data than in the experimental data. One likely explanation is the exclusion of the signal-to-noise effects as described in Sect. 4.5.4.6. Since the signals of 1-plet fiber breakage is of small amplitude, the presence of noise will likely add to a larger variability of feature values in the experimental data than currently observed in the modeled data.

As example to demonstrate the relation between source position and feature ranges, the feature trajectory for a change in source-sensor distance is marked for one case of inter-ply delamination. Here the same source is modeled at different source-sensor distances, and the feature values are extracted consistently. This evidently causes a substantial shift in the values of Partial Power 2 and Weighted Peak Frequency. However, the length and shape of the trajectories are typically responsible for the cluster dimensions and do not result in crossing between clusters for the example of Fig. 4.143.

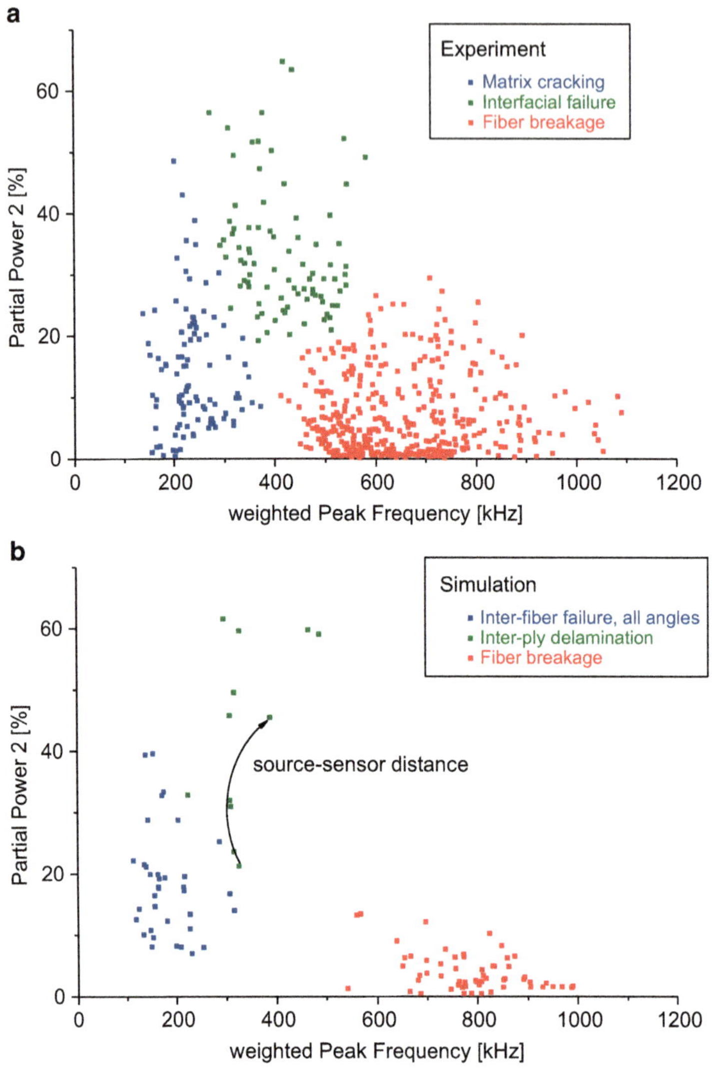

Fig. 4.143 Comparison between feature values extracted from experimental (**a**) and simulated (**b**) AE signals

Based on these findings, it is thus possible to conclude that the clusters detected by the pattern recognition approach allow meaningful distinction between the occurrences of three types of signal classes. In the modeling part, these were strictly termed as fiber breakage, inter-fiber failure, and inter-ply delamination. In previous work it was demonstrated that effects such as crack formation in pure matrix material also fall within the feature ranges observed for the class of inter-fiber failure. Moreover, other interfacial failure types such as fiber pullout and intralaminar delamination were observed to fall within the feature range modeled for inter-ply delamination. To account for this difference, the experimentally derived signal classes are termed as fiber breakage, matrix cracking, and interfacial failure in Fig. 4.143 and in the other parts of this chapter.

Beyond the nature of the AE source, other factors were previously discussed to affect the feature range of the clusters and their overlap. In addition to the source-sensor distance and the signal-to-noise ratio mentioned above, the ply layup, complex 3D-geometries of the individual plies (e.g., fabrics), and the AE sensor type will influence the quality of the partition. In many cases, the accumulation of these effects will cause significant overlap of the clusters. For such cases, any attempts to distinguish AE signals based on the proposed feature values using unsupervised pattern recognition strategies are unlikely to yield meaningful partitions of clusters. However, using the modeling strategies presented in Sects. 4.2–4.4, it is possible to evaluate the severity of these effects. In particular, the presented approaches were found to be sufficient to predict cluster overlaps and merging of clusters, which constitutes the ultimate limit to perform source discrimination using signal features.

4.7.2 *Detection of Failure Onset*

As already discussed for the application of DIC in Chap. 3, one of the main tasks in the context of fracture mechanics is the detection of first failure onsets. The following section demonstrates some applications, where acoustic emission monitoring is used to provide better input quantities for mechanical testing in this context. It seems natural to associate the onset of acoustic emission signals with the onset of damage in a material. However, this direct association is only valid if no noise sources are present in the experimental setup. But even for such ideal laboratory conditions, detection of first acoustic emission is not identical to detection of first damage onset. Depending on the type of testing, it is also required to detect the initiation of the failure mode relevant for the given mechanical test procedure.

4.7.2.1 Apparent Interlaminar Shear Strength Tests

The measurement of the apparent interlaminar shear strength using 3-point bending is one typical application requiring the detection of failure initiation. As previously described in Sect. 3.4.2.1 at the moment of onset of interlaminar crack growth, the maximum shear force occurs at the neutral axis of the beam. According to the test standards, this onset is reached at the first load drop or at the maximum force recorded during the test. The value for the apparent interlaminar shear strength is then deduced from the force at onset F_{onset}, the thickness h of the specimen, and the width t according to (3.20).

However, there are several cases, where these assumptions are not fulfilled. One possibility is to have failure at the load introduction points rather than at the neutral axis. Accordingly, the load drop may not be interpreted as interlaminar failure. Another possibility is plastic deformation prior to interlaminar failure. This also causes signatures in the load-displacement curve, which may not be interpreted as failure onset. Moreover, the real onset of interlaminar failure may likely precede any signatures visible in load-displacement curves.

As demonstrated by [180], a combination of the short-beam shear test with acoustic emission monitoring is a suitable approach to increase the reliability of the test. Figure 4.144 shows test results of a specimen made from Sigrafil CE1250-230-39 prepreg subject to a short-beam shear test according to DIN EN 2563. For acoustic emission monitoring, a type WD sensor was mounted on the load nose fixture, and all signals were acquired using the acquisition settings given in Table 4.3. As seen in Fig. 4.144, there is a clear signature in a load-displacement curve indicated by F_{visible}. The maximum of the curve appears at later stages and is obviously not the onset of damage. As revealed by the superimposed curve of

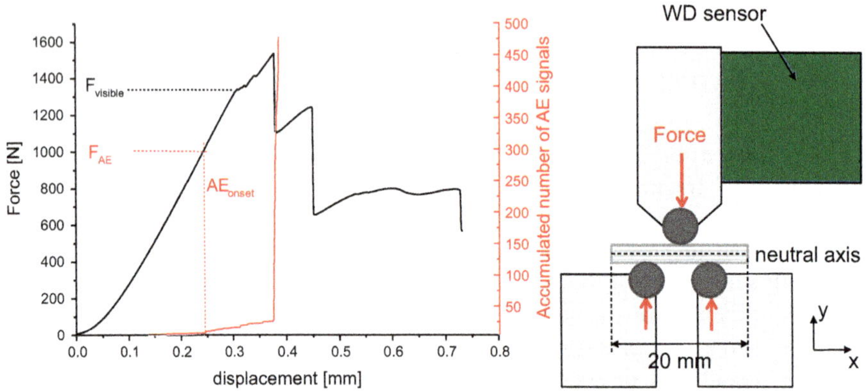

Fig. 4.144 Exemplary load-displacement curve superimposed by detected acoustic emission signals (*left*) and according scheme of experimental setup (*right*)

Table 4.3 Acquisition settings used for short-beam shear tests

Acquisition settings	Value
Preamplification	40 dB_{AE}
Threshold	35 dB_{AE}
Triggering	Individual channel
Acquisition rate	10 MS/s
Band-pass range	20 kHz to 1 MHz
Couplant	Medium viscosity silicone grease
Mounting system	Clamp
Sensor type	WD
Number of sensors	1

accumulated acoustic emission signals, a first onset of damage is detected at a much lower load level F_{AE}. Compared to the possibility of DIC monitoring as described in Sect. 3.4.2.1, the possibility to leave the acoustic emission sensor permanently on the test fixture seems advantageous.

4.7.2.2 End-Notched Flexure Tests

A test configuration frequently used in fracture mechanics of fiber reinforced polymers is the end-notched flexure test to measure the mode II fracture toughness. As described previously in Sect. 3.4.2.2, the aim of this test is also to determine the load at the moment of onset of interlaminar crack growth. It has already been demonstrated in Sect. 3.4.2.2 that it is not absolutely necessary to accompany tests of fiber reinforced polymers with brittle epoxy matrix systems. These materials tend to fail very brittle in mode II loading, and the first load drop was found to coincide well with the first occurrence on interlaminar crack growth. However, this test procedure becomes particularly more challenging, when ductile matrix materials, such as polyphenylensulfide (PPS), are tested.

To demonstrate the deviations to the failure behavior expected by the established test standards, ENF tests using carbon fiber reinforced PPS laminates were performed [181]. Specimen geometries in accordance with ASTM D7905 were prepared and tested including acoustic emission monitoring as seen in Fig. 4.145.

All acoustic emission signals were acquired using the acquisition setting given in Table 4.4 using the sensor position indicated in Fig. 4.145.

The detected signals are analyzed by the pattern recognition method described in [125], which results in three natural clusters that can be assigned to three types of failure mechanisms as indicated in Fig. 4.146.

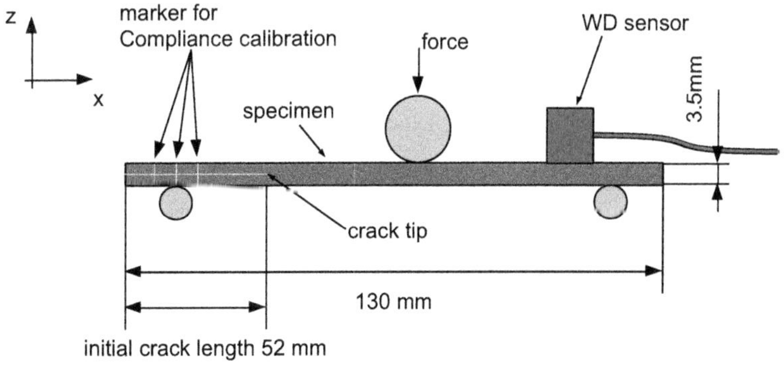

Fig. 4.145 Test configuration used for mode II ENF testing in combination with acoustic emission detection

Table 4.4 Acquisition settings and postprocessing parameters used for mode II ENF tests

Acquisition settings	Value	Postprocessing parameters	Value
Preamplification	40 dB_{AE}	Feature extraction window	100 µs after threshold
Threshold	35 dB_{AE}		
Triggering	Individual channel		
Acquisition rate	10 MS/s		
Band-pass range	20 kHz to 1 MHz		
Couplant	Medium viscosity silicone grease		
Mounting system	Clamp		
Sensor type	WD		
Number of sensors	1		

Interpreting the load-time curves in accordance with standard ASTM D7905 using the force maximum as value for F_{onset} yields the values of G_{IIc} as noted in Table 4.5. These differ to existing values in literature of 802–933 J/m^2 [182, 183] by a factor of 4. However, clear nonlinearity is observed prior to the load maximum. At the same time, a significant amount of acoustic emission signals are detected as seen in Fig. 4.146b. This indicates the onset of subcritical crack growth and therefore suggests the usage of the acoustic emission onsets to deduce the onset of crack growth.

If no distinction is made between different failure types, the evaluation of the overall onset of all acoustic emission signals yields a very conservative value of 647 J/m^2 having a coefficient of variation of 36.1 %. However, the significance of the acoustic emission analysis can be increased substantially by distinction of different failure types.

For the onset of interlaminar crack growth, the relevant class of acoustic emission signals is interfacial failure. Evaluation of the specific onset of this class of signals yields a value of 980 J/m^2 with a significantly reduced coefficient of variation of 18.8 %. This value was found to be more reasonable given the reports in literature [182, 183] and the typically expected range of mode II fracture toughness values for fiber reinforced polymers.

4.7.2.3 Transverse Crack Tension Tests

A further example of a test procedure based on valid detection of damage onsets is the transverse crack tension test. Standard tensile test specimen geometries with a tapered area and tab reinforcement are usually used. As a major modification, one of the center plies is fabricated discontinuous, i.e., exhibits a cut transverse to the load direction. Due the applied tensile load, this translates into a shear component acting between the inner plies at the discontinuity. Therefore, this refers to a mode II load situation. Using the relation between the force at onset F_{onset}, the thickness

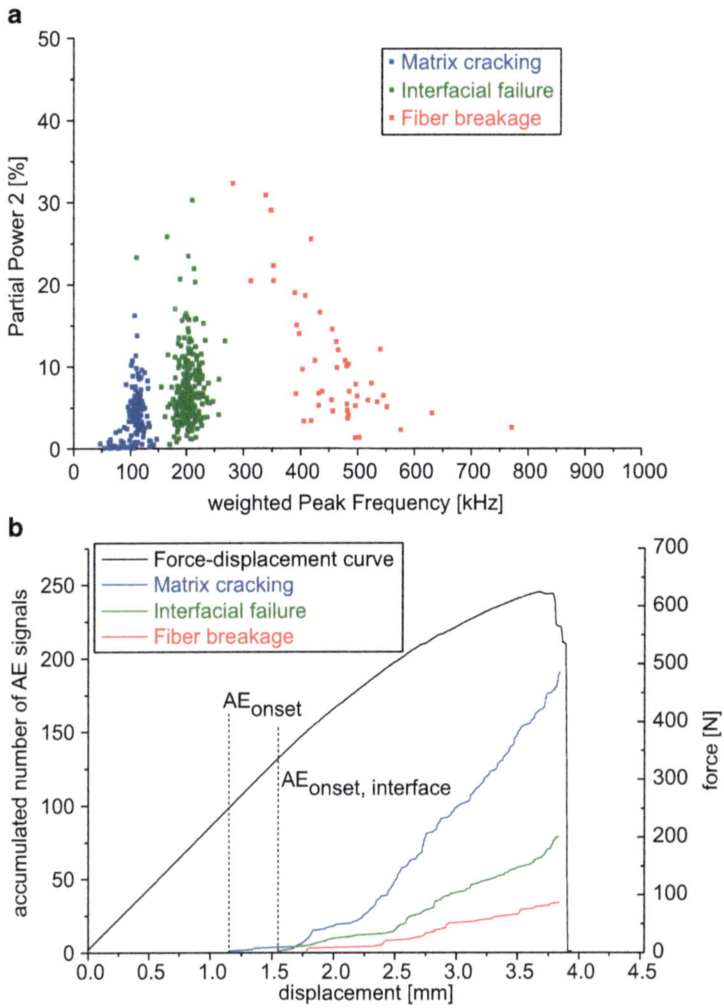

Fig. 4.146 Partition of signals as obtained by pattern recognition method applied to signals detected in ENF tests (**a**) and accumulated number of acoustic emission signals with superimposed load-time curve (**b**)

h of the specimen, the thickness h_t of one ply, the width b, and the modulus of the specimen E, it is possible to calculate the mode II fracture toughness G_{IIc} according to

$$G_{\mathrm{IIc}} = \frac{F_{\mathrm{onset}}^2 \cdot h_t}{4b^2 E \cdot h(h - h_t)} \tag{4.65}$$

As an example for this test procedure in combination with acoustic emission analysis, test results of a unidirectional carbon fiber fabric made from Toray T700 50C, 24k are presented in [143]. The transverse crack in the center ply was

Table 4.5 Evaluation results of mode II ENF test using standard procedure and using acoustic emission analysis

Energy release rate	ASTM D7905	AEonset, total	AEonset, interface
Mean [J/m^2]	3722	647	980
Standard deviation [J/m^2]	1227	234	185
Coefficient of variation [%]	32.9	36.1	18.8

Table 4.6 Acquisition settings and postprocessing parameters used for transverse crack tension tests

Acquisition settings	Value	Postprocessing parameters	Value
Preamplification	20 dB_{AE}	Feature extraction window	100 μs after threshold
Threshold	35 dB_{AE}	Δt-filter	(15 ± 2) μs
Triggering	Individual channel		
Acquisition rate	10 MS/s		
Band-pass range	20 kHz to 1 MHz		
Couplant	Medium viscosity silicone grease		
Mounting system	Clamp		
Sensor type	WD		
Number of sensors	2		

Table 4.7 Acquisition settings of DIC system for transverse crack tension tests

Measurement setup	Value
Camera	Toshiba CMOS camera CSC12M25BMP19-01B with 4096 × 3072 pixels
Lenses	Titanar, focal length 2.8/50 mm
Lighting	KSP 0495-0001A LED 20W/24V white 30°
Filter	Polarization-filter Schneider-Kreuznach
Field of view	(113.4×86.7) mm^2
Scale factor	0.028 mm/pixel
Subset size	20 × 10 pixel
Acquisition rate	2 Hz
Software	ARAMIS
Configuration	3D

introduced by a standard cutter. Mechanical testing is carried out with a 250 kN load cell and displacement rate of 1 mm/min. All acoustic emission signals were acquired using the parameters given in Table 4.6 using two AE sensors attached to the specimen in 125 mm distance to each other. For digital image correlation, the edge of the specimen was monitored. Acquisition is done by 2D-measurements applying the parameters found in Table 4.7.

An application of the pattern recognition method described in [125] yields three signal classes for all specimens investigated. Based on the method proposed in [154], the uncertainty of the classification was evaluated to be less than 0.20 % for all specimens investigated. Assignment of the signal classes to respective failure mechanisms is carried out as described in Sect. 4.7.1.

This allows comparing different data reduction methods in one test. As the first technique, the classical approach to deduce the onset level of the propagation load, which is normally used for the quantification of mode II interlaminar fracture toughness, was evaluated [184, 185]. This classical approach is essentially based on signatures of the recorded force-displacement curves. As the second technique, DIC was used to measure the load corresponding to first initiation of delamination. To obtain a quantitative measure of the strain field, the displacement in the through-thickness direction (z-direction) was evaluated at several x-positions in the vicinity of the discontinuous center ply. The distance between two points was measured at positions as indicated in the top image in Fig. 4.147. However, as seen in the exemplary evaluation of the measured distance of two positions as a function of loading in Fig. 4.147, there were no clear signatures found in the curves. Instead, the change in slope of the signals was mostly gradual, which makes picking of a trend change challenging. Also, there was no mutual agreement in the shape of the individual curves recorded for different specimens. Therefore, the optical detection of the delamination onset was found to yield unreliable results for this kind of application, although the process of damage progression was easily visible in the DIC images.

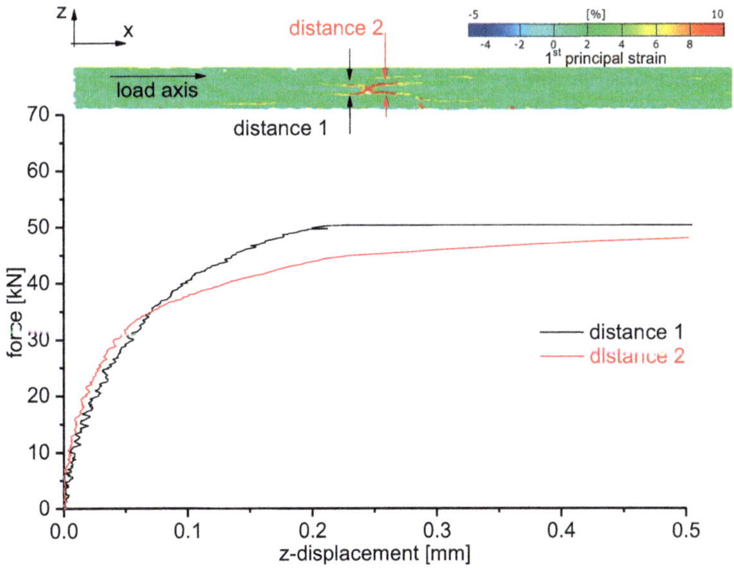

Fig. 4.147 Evaluation of distance between points in z-direction as function of loading at two exemplary positions in the vicinity of the discontinuous center ply

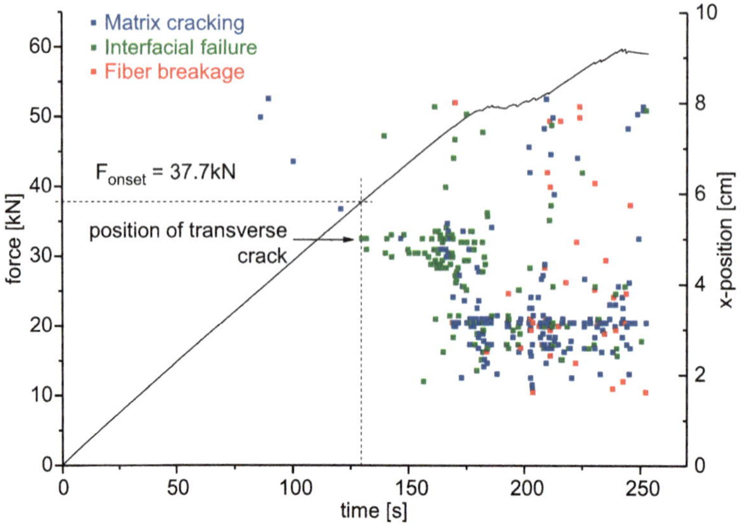

Fig. 4.148 Exemplary evaluation results of acoustic emission source positions as function of time with superimposed load-time curve for a transverse crack tension specimen

As third method, the onset of localized acoustic emission signals was evaluated. At the time of acoustic emission onset (see Fig. 4.148), no evident signature could be detected in the DIC data or in the force-time curve. Based on previous results from tensile specimens [177, 186], the detected acoustic emission signals could be assigned to the occurrence of matrix cracking, interfacial failure, and fiber breakage. One representative classification result is shown in Fig. 4.148. With increased load, acoustic emission signals initiate at weak spots inside the laminate. Although the discontinuous center ply is intentionally designed as a weak spot, the first acoustic emission signals are localized at various positions along the x-axis of the specimen. Since these signals were classified as matrix cracks, it is likely that they originate from failure at the specimen edges, failure at resin rich areas, or failure at randomly distributed minor fabrication defects. At a distinct load level, signals classified as interfacial failure initiate specifically at the position of the discontinuous center ply (x-position 5 cm).

Subsequently, acoustic emission signals are localized in the vicinity of the discontinuous center ply, which densify around $x = 50$ mm and spread outward with increased load level. This is indicative of interlaminar crack growth, starting at the interrupted center ply. At increased load levels, acoustic emission signals are localized in a broad range between the two sensors. This reveals the occurrence of additional damage occurring at several positions inside the specimen. Among these signals some are classified as fiber breakage signals. Since the tensile strength of carbon fibers typically shows Weibull-type distributions, failure of individual fiber filaments is well expected before the ultimate load level of the laminate is reached [187]. Therefore occurrence of sporadic single fiber breakage long before ultimate failure is a likely source for these acoustic emission signals.

Remarkably, the relevant acoustic emission signals occur long before significant signatures are visible in the load-displacement curves. Since initiation of interfacial failure at the position of the discontinuous center ply is of key interest for further analysis, the load level of initiation was also quantified by acoustic emission measurements. Therefore, the first occurrence (onset) of the acoustic emission signals, classified as interfacial failure within the range of the discontinuous center ply at (50 ± 5) mm x-coordinate, was used to obtain the respective load level of failure initiation F_{onset}.

The occurrence of early acoustic emission indicates that the initiation of delamination is not a uniform process over the width of the transverse crack tension specimens, but may also initiate within the specimen volume. Consequently, detection of the delamination onset based solely on the DIC information obtained at the specimen surface may be error prone.

In summary, the traditional approach based on signatures in force-displacement curves allows to deduce the fracture toughness for crack propagation. However, more conservative values for the fracture toughness values for crack initiation can be obtained by the acoustic emission method proposed above. As has been shown in [143], the sensitivity of the acoustic emission method is sufficient to assess the quality of different dry fiber-cutting processes such as mechanical cutting or laser cutting. In this application no differences were found by the traditional approach, and therefore this data reduction would significantly underestimate the influence of the cutting process [143].

4.7.3 Comparison to Failure Criteria Predictions

Beyond the measurement of first onset of damage in specimens explicitly designed for fracture mechanics testing, it would also be interesting to have methods capable of visualizing the onset of specific failure modes, e.g., as predicted by failure criteria. In the following, some examples are presented, which compare measurement results using acoustic emission analysis to the predicted onsets of inter-fiber failure and fiber failure using Puck's failure criterion. Simple tensile test specimens were selected for this purpose, since this load scenario is expected to yield a high predictive capability by the theory. However, the application of the experimental method to more complicated load scenarios is straightforward and will be demonstrated in Sect. 4.7.4.3 for a structural component.

4.7.3.1 Unidirectional Laminates

As the first application, the case of tensile test of unidirectional laminates with load axis in parallel to the fiber axis is discussed. According to the failure theory by Puck, no inter-fiber failure is expected prior to ultimate failure due to fiber failure.

Table 4.8 Acquisition settings and postprocessing parameters used for tensile tests

Acquisition settings	Value	Postprocessing parameters	Value
Preamplification	20 dB_{AE}	Feature extraction window	100 μs after threshold
Threshold	35 dB_{AE}	Δt-filter	(7 ± 2) μs
Triggering	Individual channel		
Acquisition rate	10 MS/s		
Band-pass range	20 kHz to 1 MHz		
Couplant	Medium viscosity silicone grease		
Mounting system	Clamp		
Sensor type	WD		
Number of sensors	2		

However, a comment is made in [39] that experimental findings indicate the occurrence of inter-fiber failure at about 70 % of the ultimate tensile strength of the specimen.

To evaluate the acoustic emission release of specimens under this load scenario, tensile tests according to DIN EN 2561 using specimens made from T700/PPS were performed [178]. Acoustic emission signals were detected using the acquisition settings given in Table 4.8. All signals were localized within the tapered area of the specimen and were subject to the pattern recognition method described in Sect. 4.5.

Relevant for confirmation of failure theories is the experimental detection of initiation of specific failure modes. As seen in Fig. 4.149, contributions of all failure mechanisms are detected during loading. The usual sequence of their occurrence starts with signals associated with matrix cracking, followed by interfacial failure and fiber breakage. However, for the present case, distinction between the occurrence of interfacial failure and matrix cracking is hardly justified, since both of these types are associated with inter-fiber failure in the sense of Puck's failure criterion. Therefore, distinction is made between the average onset of signals from matrix cracking and interfacial failure and the onset of acoustic emission signals of fiber breakage.

The ratio of the load level at acoustic emission onset and the ultimate failure load is shown in Fig. 4.149a. The average onset of acoustic emission signals of matrix cracking is found at load levels between 5 and 30 % of the ultimate load. Based on observations of simultaneously recorded videos, some of these first onsets of acoustic emission signals correspond to the onset of splitting at the edges of the specimen. Subsequently, damage progression can be expected, which is likely responsible for the early onset of acoustic emission. However, as seen in Fig. 4.149b, many of the unidirectional specimens show a certain change in acoustic emission activity at (76.0 ± 4.1) % load to failure. This is likely due to the formation of inter-fiber cracks in parallel to the load axis as predicted by [39].

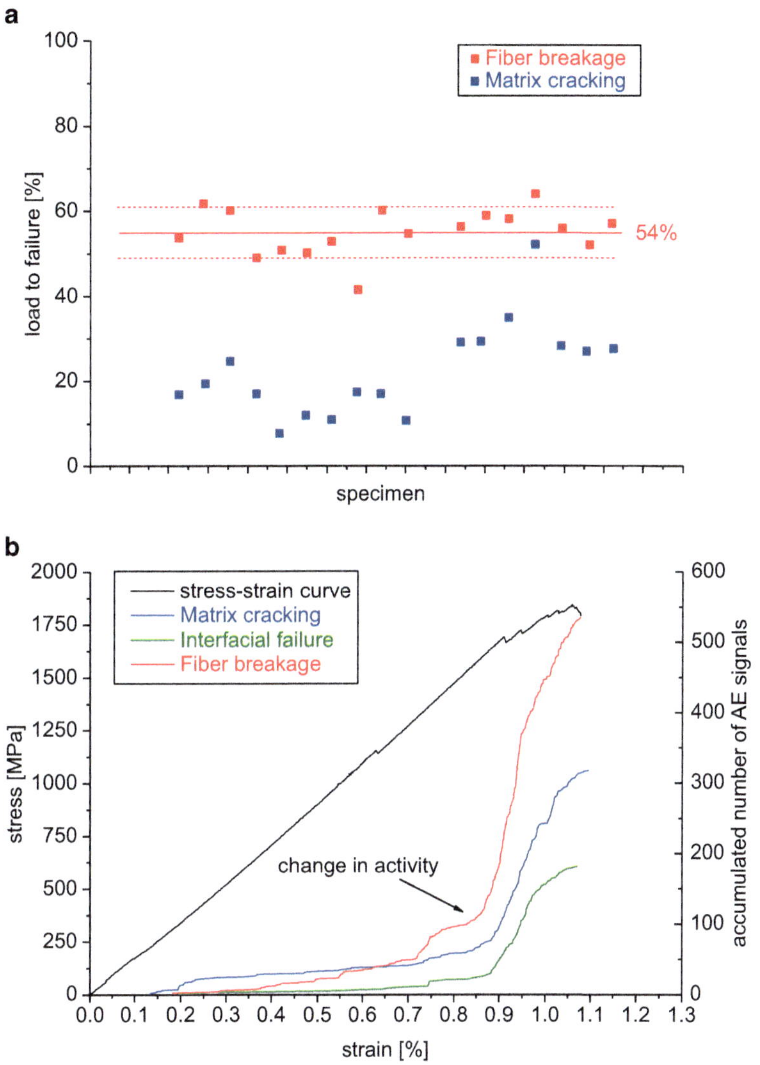

Fig. 4.149 Evaluation results of acoustic emission onsets for different specimens for unidirectional tensile specimens (**a**) and exemplary result of change in activity of matrix cracking signals at 75 % of ultimate load level (**b**)

The average onset of fiber breakage signals is quantified to be at around 54 % of the ultimate failure load. Based on the theory of Zhou et al. [188] and the values for the T700 filaments from [189], it is quite probable to have fiber failure at 54 % of the ultimate strength. In fact, the typical Weibull strength distribution of filaments would even expect fiber failure to occur at lower load levels as recently demonstrated by Lomov et al. [190]. But as demonstrated in Sect. 4.2.3, the energy release of weak filaments is likely to fall below the detection limit of the sensors used. This may therefore explain the general discrepancy between expected onset of fiber breakage signals and AE measurements. However, the constant onset level of fiber failure signals indicates that first measurable fiber breakage is taking place in the loaded plies at a systematic load level prior to ultimate failure of the specimens.

4.7.3.2 Cross-Ply Laminates

As the second application, investigations of tensile tests performed on cross-ply laminates are presented in this subsection. To this end six different cross-ply stackings were fabricated using Sigrafil CE1250-230-39 prepreg material with a nominal ply thickness of 220 μm. In the tapered area, the stacking sequences were chosen as $[0_2/90/0_2]_{sym}$, $[0_2/90_2/0]_{sym}$, $[0/90_3/0]_{sym}$, $[0/90]_{sym}$, $[(0/90)_2]_{sym}$, and $[(0/90)_3]_{sym}$. For production of the reinforcements, the ply-drop method was used. The respective interspersed plies are shown exemplarily in Fig. 4.150. The data reduction follows reference [186], and data interpretation uses the set of quantities described in [35, 138]. All acoustic emission signals were detected using the acquisition settings given in Table 4.9, and only signals localized in the tapered area were used for further analysis.

For each specimen of each stacking sequence, the pattern recognition method as described in Sect. 4.5 was applied. Representative results for four exemplary laminate configurations are shown in Fig. 4.151 as a diagram of the signal features Partial Power 4 over Weighted Peak Frequency.

Based on the different ratio of 0° and 90° plies of the four stacking sequences, a different ratio of microscopic failure mechanisms is expected. As seen from Fig. 4.151, qualitatively the total number of signals correlated with matrix cracking increases for an increasing number of 90° plies. Similarly, the absolute number of signals correlated with fiber breakage increases for an increasing number of

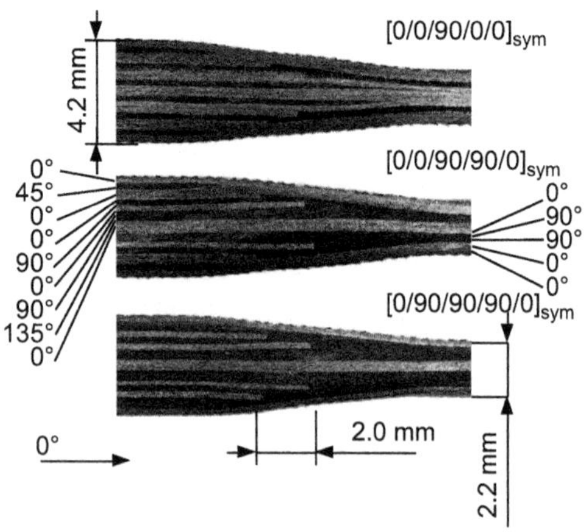

Fig. 4.150 Cross-sectional images of reinforcement sections with interspersed plies for three out of six stacking sequences investigated (reprinted from [186])

Table 4.9 Acquisition settings and postprocessing parameters used for tensile tests

Acquisition settings	Value	Postprocessing parameters	Value
Preamplification	20 dB_{AE}	Feature extraction window	100 μs after threshold
Threshold	35 dB_{AE}		
Triggering	Individual channel		
Acquisition rate	10 MS/s		
Band-pass range	20 kHz to 1 MHz		
Couplant	Medium viscosity silicone grease		
Mounting system	Clamp		
Sensor type	WD		
Number of sensors	2		

0° plies. The absolute number of signals correlated with interfacial failure was found to be similar for all stacking sequences with identical number of 0°/90° interfaces. This type of signals is attributed to the occurrence of inter-ply delamination. The different laminate configurations produce a noticeable change of the cluster positions in the diagrams for each failure mechanism. This is caused by changes of the frequency spectra of the excited Lamb waves and originates from the different elastic properties of the propagation medium and the characteristic depth changes of the acoustic emission sources.

In the following, the pattern recognition results of all specimens are evaluated using the following quantities as introduced in [35, 138, 186]:

The relative number of signals $N_{\mathrm{rel},i}$ and the relative amplitude of signals $U_{\mathrm{rel},i}$, for the ith of the I_{tot} failure mechanisms with N_i of the total signals N_{tot}, are given as

$$N_{\mathrm{rel},i} = \frac{N_i}{N_{\mathrm{tot}}} \tag{4.66}$$

$$U_{\mathrm{rel},i} = \sum_{j=1}^{N_i} U_{ij} \cdot \left(\sum_{i=1}^{I_{\mathrm{tot}}} \sum_{j=1}^{N_i} U_{ij} \right)^{-1} \tag{4.67}$$

Thus, $N_{\mathrm{rel},i}$ and $U_{\mathrm{rel},i}$ express the contribution of a particular failure type relative to all the recorded signals N_{tot} with accumulated signal amplitudes $\sum_{i=1}^{I_{\mathrm{tot}}} \sum_{j=1}^{N_i} U_{ij}$. In addition, the average signal amplitude $\langle U_i \rangle$ of the ith failure mechanism is defined as

$$\langle U_i \rangle = \sum_{j=1}^{N_i} U_{ij} \cdot (N_i)^{-1} \tag{4.68}$$

In comparison to these normalized measures, the absolute values usually lack in comparability due to differences in sensor sensitivities, differences in coupling quality, and numerous other factors affecting the strength and PoD of acoustic emission signals. Therefore the quantities derived in (4.66)–(4.68) are meant to provide an improved possibility to compare and interpret the failure of fiber reinforced composites.

An evaluation of the cross-ply specimens with stacking sequences $[0_2/90/0_2]_{sym}$, $[0_2/90_2/0]_{sym}$, and $[0/90_3/0]_{sym}$ in terms of (4.66)–(4.68) is shown in Fig. 4.152. The design of these three laminate configurations gradually substitutes the 0° plies by 90° plies forming 90° layers with a thickness of one, two, and finally three plies.

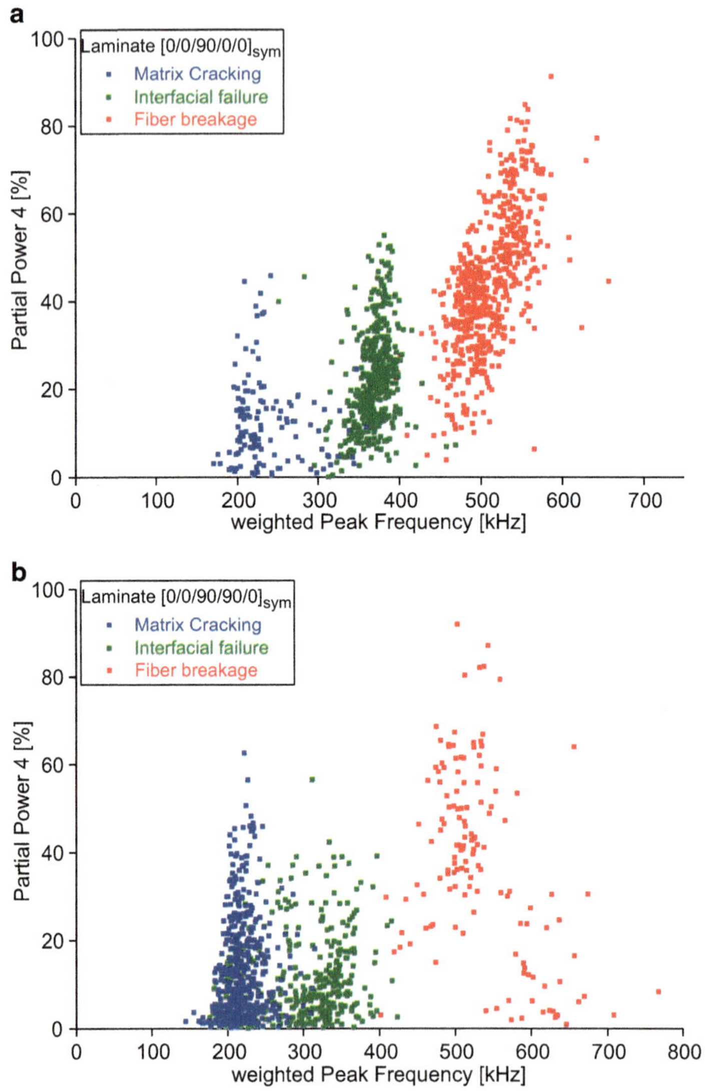

Fig. 4.151 Representative results for the identified natural clusters for different laminate configurations $[0_2/90/0_2]_{sym}$ (**a**), $[0_2/90_2/0]_{sym}$ (**b**), $[0/90_3/0]_{sym}$ (**c**), and $[0/90]_{sym}$ (**d**) (partially based on [186])

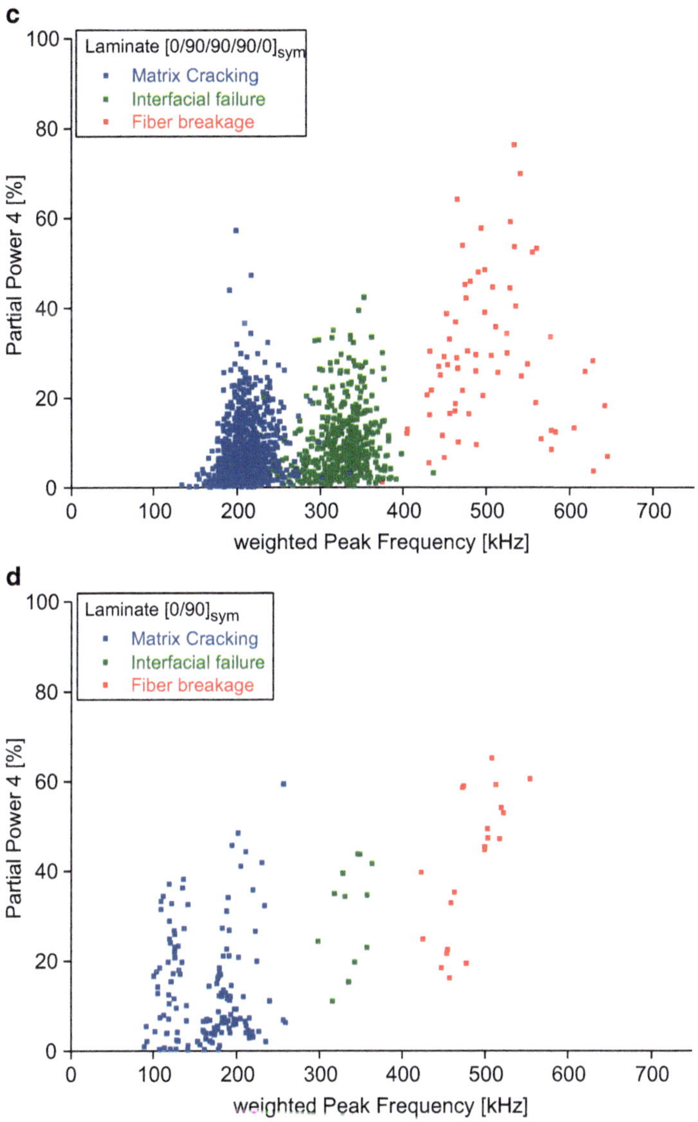

Fig. 4.151 (continued)

As discussed before, $N_{\mathrm{rel},i}$ and $U_{\mathrm{rel},i}$ show quite similar trends [35]. In view of the physical correlation of $U_{\mathrm{rel},i}$ with the source strength, $U_{\mathrm{rel},i}$ is the relevant quantity to discuss. In both diagrams (Fig. 4.152a, b), the contribution of matrix cracking increases as the number of 90° plies increases. Generally, a higher critical crack density is expected for $[0/90_n/0]_{\mathrm{sym}}$ laminates for lower n before the initiation of the delamination at the interface between the 0° and 90° plies [191, 192]. This seems to be in contrast to the quantification of matrix cracking as seen from Fig. 4.152a, b. However, the major amount of signals attributed to matrix cracking for $n = 2$ and $n = 3$ occurs after the onset of delamination and is thus not included in Nairn's theory [191, 192]. The changes observed are thus attributed to damage occurring after the onset of inter-ply delamination.

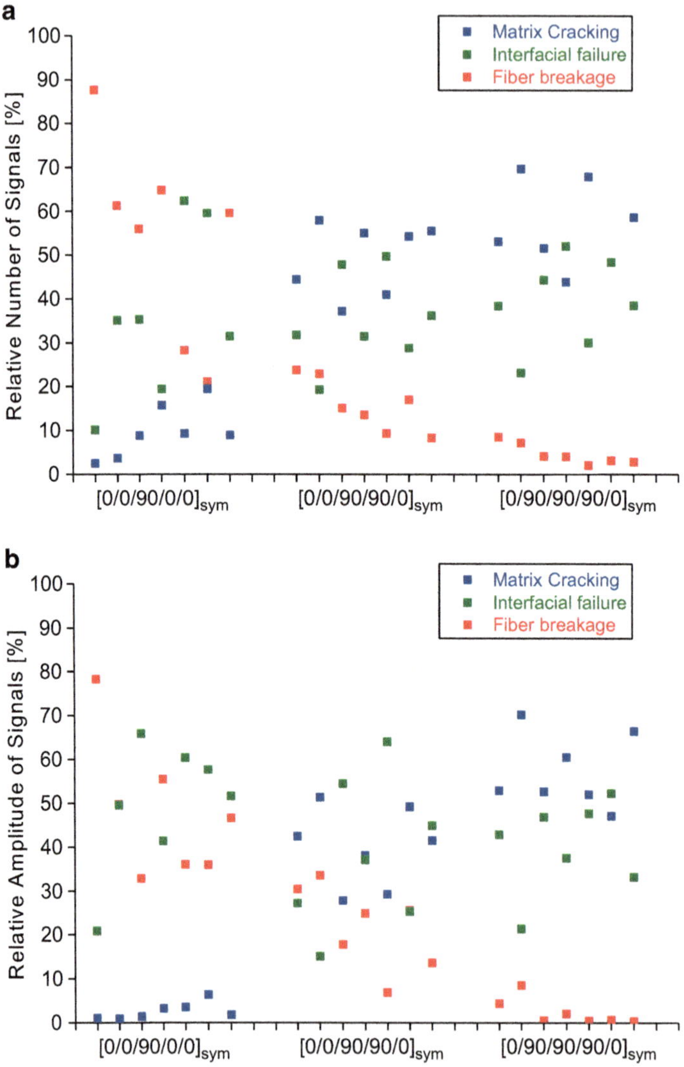

Fig. 4.152 Quantification of relative number of signals (**a**), relative amplitude of signals (**b**), and average amplitude per signal (**c**) for laminate types $[0_2/90/0_2]_{sym}$, $[0_2/90_2/0]_{sym}$, and $[0/90_3/0]_{sym}$ (based on [186])

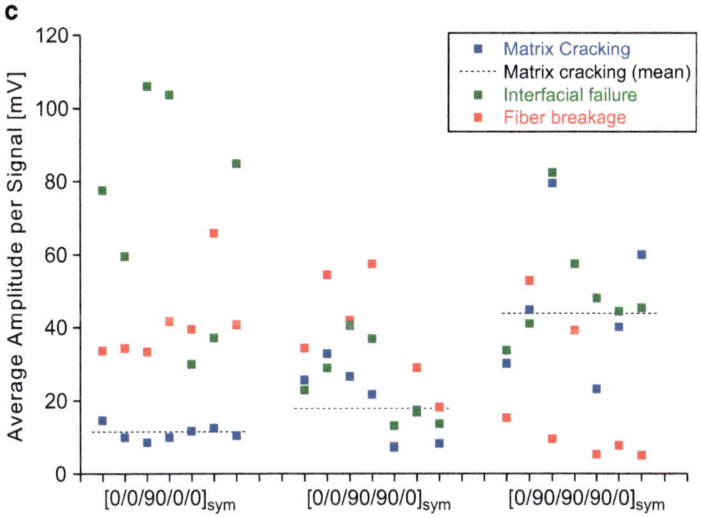

Fig. 4.152 (continued)

The number of fiber breakages increases with increasing number of 0° plies. Based on probabilistic considerations, the frequency of the occurrence of fiber breakage (i.e., single filament failure) before the ultimate failure of the specimen is expected to be proportional to the cross-section of 0° plies.

Contributions of interfacial failure scatter between 10% and 70 %, but exhibit no significant trend as a function of stacking sequence. The large scatter of signals attributed to interfacial failure is consistent with optical observations of the macroscopic failure mode. Here the specimen exhibits vast differences in the visible delamination lengths as will be further discussed in the following.

As predicted by the generalized theory of acoustic emission [21, 22] and also demonstrated with finite element simulations [35], the amplitude U_{AE} of an acoustic emission signal detected at a distance r is proportional to the vibrating crack volume ΔV:

$$U_{AE} = \beta_U \left(c_L^2\right) \cdot \frac{\Delta V}{r} \tag{4.69}$$

The proportionality constant β_U is a linear function of the squared longitudinal sound velocity c_L^2 of the cracking medium and the orientation of the crack surface movement (not taken into account in (4.69)). For the inter-fiber failure case, the vibrating crack volume ΔV is expected to be proportional to the cross-sectional area ΔA of the adjoining 90° plies. Consequently, based on the difference in stacking sequence, the signal amplitudes should increase by a factor of 2 for the $[0_2/90_2/0]_{sym}$ stacking sequence and by a factor of 3 for the $[0/90_3/0]_{sym}$ stacking sequence, compared to the $[0_2/90/0_2]_{sym}$ stacking sequence. The mean U_i for matrix cracking (indicated by the dashed line in Fig. 4.152c) amounts to 11.5 mV (81 dB_{AE}),

17.9 mV (85 dB_{AE}), and 43.9 mV (92.8 dB_{AE}) for the three stacking sequences, respectively. Although strong scattering is observed in U_i of the $[0/90_3/0]_{sym}$ stacking sequence, overall the values are found to be in good agreement with the expected relation derived from (4.69). The average amplitude of the fiber breakage also exhibits a characteristic dependency on the stacking sequence. The mean U_i decreases as the number of 0° plies decreases. Because the majority of the signals belonging to this class originate from single filament failure, the vibrating crack volume ΔV is not proportional to the thickness of the 0° plies. Instead, the amplitude U_{AE} is influenced by the strain energy release, which is higher in the stacking sequences with higher number of 0° plies.

While the diagrams of Fig. 4.152 are mostly useful to visualize the accumulated damage causing final failure of the specimens, the evolution of microscopic failure mechanisms is another relevant factor for comparison to predictions from analytical failure models. Based on the values from Table B.1 and Table B.4 in Appendix B, the stress and strain values for first-ply failure and last-ply failure (ultimate failure) were calculated for the six stacking sequences investigated according to Puck's nonlinear laminate failure model [193–196]. The stress and strain values for the onset of the three failure mechanisms were obtained from the acoustic emission measurements, and their mean value was calculated from all samples of each stacking sequence. A graphical comparison of the obtained values including their margin of error is shown in Fig. 4.153.

The calculated stress and strain values for last-ply failure are found to be in good agreement with the measured values for maximum stress. Only for the $[(0/90)_3]_{sym}$ stacking sequence a laminate strength lower than the predicted value was measured.

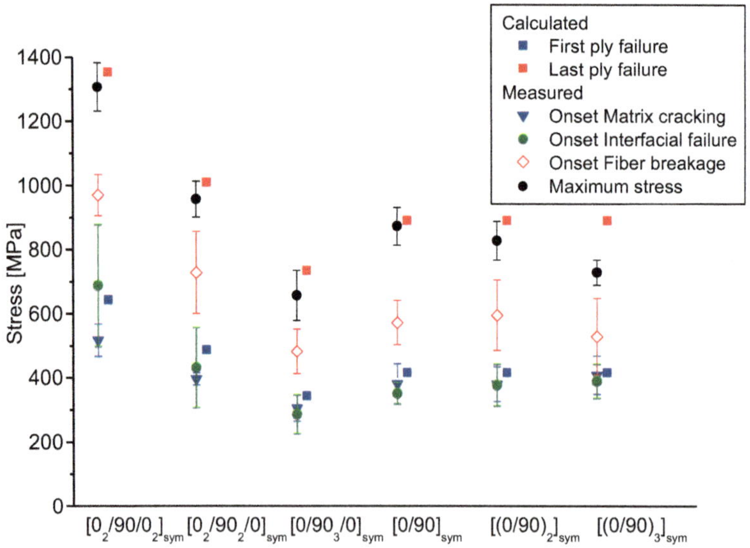

Fig. 4.153 Calculated first-ply failure and last-ply failure based on Puck's nonlinear failure model and respective values derived from acoustic emission measurements

The systematic decrease of the laminate strength for the $[(0/90)_n]_{sym}$ laminates is likely caused by the volume effect [197], since the stacking sequence design is expected to yield identical failure strength based on Puck's nonlinear failure model. Generally, last-ply failure is caused by fiber breakage in the 0° plies of the stacking sequences. In this study, the quantified onset of fiber breakage is found to be systematically lower than the stress values calculated for ultimate failure. This is explained by the fact that acoustic emission measurements are sensitive enough to detect the occurrence of single filament failure. Thus, detection of fiber breakage by pattern recognition can effectively act as early indicator for imminent structural failure down to a range of 70 % of ultimate failure (cf. also Sect. 4.7.3.1).

A good agreement is also found between the calculated stress values for first-ply failure and the initiation of matrix cracking and interfacial failure. Overall, the mean onset of matrix cracking is systematically lower than the values calculated for first-ply failure, while the onset of interfacial failure sometimes exceeds the calculated value. This might be caused by acoustic emission signals originating from the edges of the specimen due to inhomogeneities and improper specimen preparation. Although this kind of initial type of failure is not covered by the analytical description [193–196], it is easily detectable by acoustic emission measurements. Similar to the case of fiber breakage, detection of matrix cracking and interfacial failure can be used to quantify the onset of failure in composites by acoustic emission analysis as a nondestructive measurement technique.

4.7.4 Tracking Failure Evolution

After the detection of first failure onsets, one of the key tasks for monitoring methods is to keep track of the further damage evolution. In the following sections, some typical applications are presented, which have used acoustic emission analysis for this purpose.

4.7.4.1 Double-Cantilever Beam Testing

In general, double-cantilever beam tests according to standards like ISO 15024 or ASTM D 5528 are used to evaluate the mode I interlaminar fracture toughness value of fiber reinforced laminates. A schematic drawing of such a typical test setup is shown in Fig. 4.154. As has been demonstrated previously, acoustic emission analysis can significantly improve the interpretation of composite material failure [35, 131, 138, 181, 198]. In this section, two routines using acoustic emission are demonstrated, which improve the testing procedure in the double-cantilever beam setup.

Typically traveling microscopes or high-resolution camera systems are used to follow the position of the crack tip. The progress of delamination is then obtained via measurement of the crack progress between subsequent images. Evaluating the

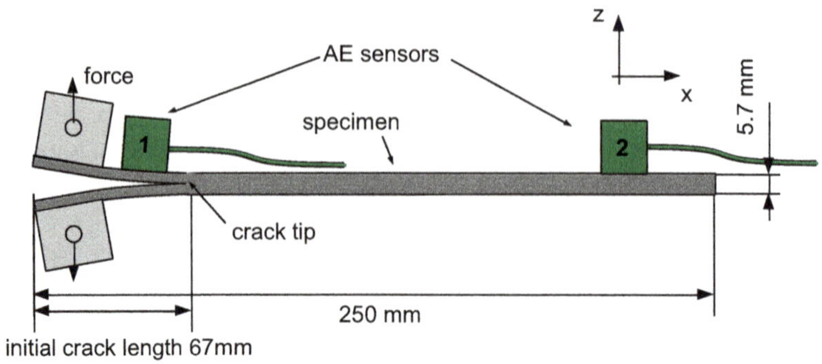

Fig. 4.154 Experimental setup used for double-cantilever beam tests in combination with acoustic emission acquisition

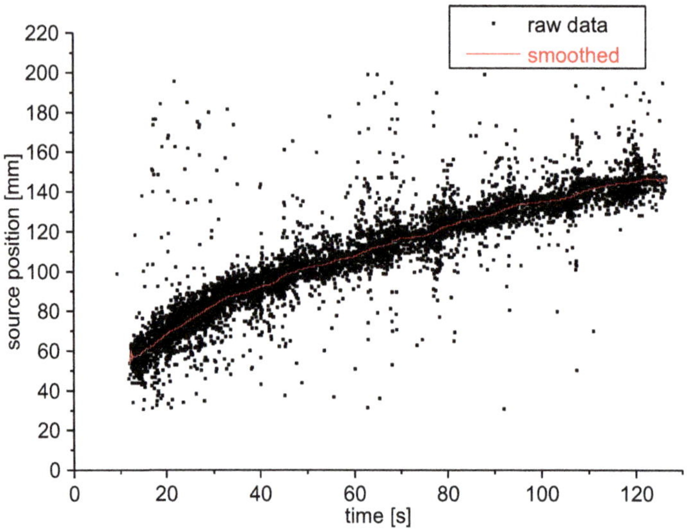

Fig. 4.155 Acoustic emission source positions during double-cantilever beam test as function of time (reprinted from [198])

delamination length as a function of loading, it is thus possible to calculate the delamination resistance curve (*R*-curve). However, the number of images used to quantify the delamination lengths is usually limited, since this involves manual processing of each image.

In the following, acoustic emission source localization is applied to track the position of the crack tip. This approach follows previous investigations by [181, 199] and can be used to generate continuous *R*-curves. In Fig. 4.155 the result from a measurement of a HTA/RTM6 specimen as reported in [198] is shown.

Table 4.10 Acquisition settings and postprocessing parameters used for double-cantilever beam tests

Acquisition settings	Value	Postprocessing parameters	Value
Preamplification	40 dB_{AE}	Feature extraction window	200 µs after threshold
Threshold	35 dB_{AE}		
Triggering	Individual channel		
Acquisition rate	10 MS/s		
Band-pass range	20 kHz to 1 MHz		
Couplant	Medium viscosity silicone grease		
Mounting system	Clamp		
Sensor type	WD		
Number of sensors	2		

All acoustic emission signals were detected according to the parameters reported in Table 4.10. Subsequent to testing, the acoustic emission signals were localized in one dimension between the two sensors and are shown as function of time in Fig. 4.155. In order to obtain an effective crack length as function of time, the raw data was smoothed by a percentile filter with 1500 points window size. Compared to other averaging techniques (adjacent averaging, Savitzky-Golay method), this filter is less affected by outliers and better reflects the achieved crack length. In particular, for those cases with sudden crack progress, steps in the delamination length occur. The filter setting chosen is able to preserve these sharp jumps and therefore resembles a reasonable method to obtain effective crack lengths of the damage zone surrounding the crack tip.

Utilizing an interpolation scheme on the smoothed acoustic emission source position, it is possible to obtain a continuous representation of the delamination length a. In combination with the recorded force F, the cross-head displacement $\widetilde{\delta}$, and specimen width b, this allows for continuous calculation of the G_{I} value by the modified beam theory method recommended in ASTM D5528. The corresponding fit quality to calculate the offset value Δ used in (4.70) is shown in Fig. 4.156:

$$G_{\mathrm{I}} = \frac{3F\widetilde{\delta}}{2b(a + |\Delta|)} \tag{4.70}$$

By knowing the onset of acoustic emission, it is also possible to define a measure for the onset of crack growth. During the experiments it was observed that this definition for the crack onset precedes the visible crack onset and can thus be understood as the most conservative estimation for G_{Ic}. Subsequently the continuous R-curve (strain energy release as a function of delamination length) as seen in Fig. 4.157 can be further used to evaluate the interlaminar delamination process.

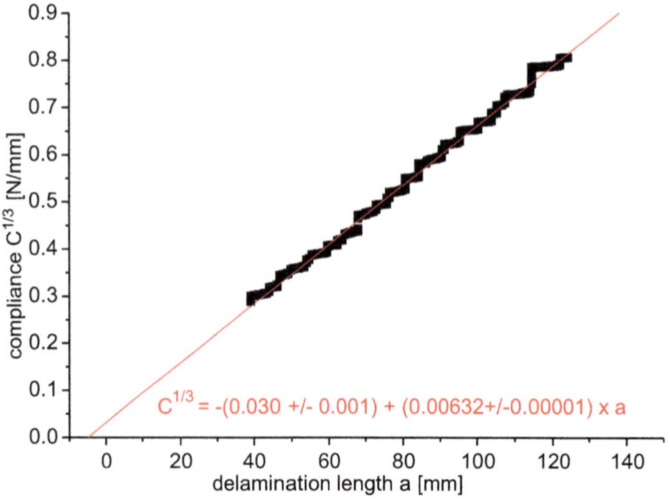

Fig. 4.156 Compliance data and linear fit to obtain shift of effective delamination length Δ (reprinted from [198])

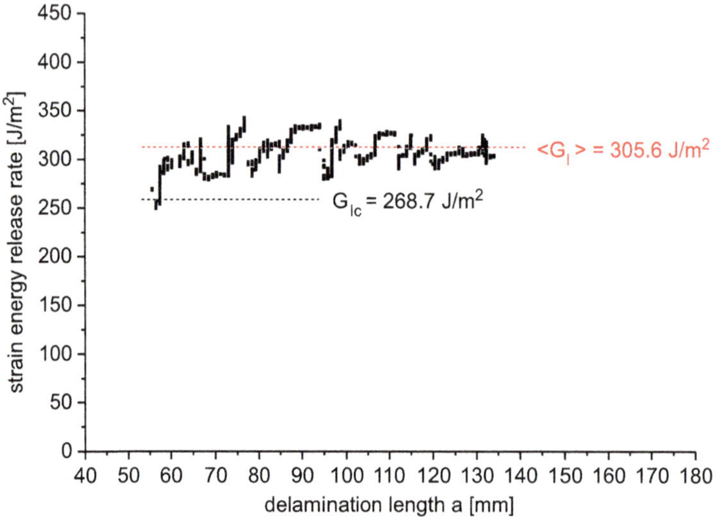

Fig. 4.157 *R*-curve of sample including G_{Ic} onset value and G_{Ic} mean value (reprinted from [198])

The minimum value of the strain energy release rate at the onset of acoustic emission is then used as fracture toughness value G_{Ic} and can be estimated as 268.7 J/m^2 for the example in Fig. 4.157. Accordingly it is feasible to define a mean G_{Ic} value occurring during crack propagation as arithmetic average of all values of the *R*-curve. This mean G_{Ic} value amounts to 305.6 J/m^2.

Table 4.11 Results of mode I energy release rate evaluation

Energy release rate	NL [J/m²]	VIS [J/m²]	5%/MAX [J/m²]	AE_{onset} [J/m²]	AE_{stable} [J/m²]
Mean [J/m²]	222	700	968	289	1028
Standard deviation [J/m²]	22	93	88	80	–
Coefficient of variation [%]	9.9	13.3	9.1	27.7	–

Being typical for a brittle matrix system as RTM6, the fracture toughness value for crack initiation is lowest, followed by a region with stable crack propagation of higher fracture toughness.

However, this behavior is not necessarily identical for ductile matrix materials as will be demonstrated for T700/PPS specimens in the following. Such ductile thermoplastic materials may exhibit a transition between subcritical failure during crack initiation mixed with stable and unstable crack growth. Using the test setup of Fig. 4.154 and acoustic emission acquisition parameters as given in Table 4.10, such behavior can be measured by acoustic emission analysis. To this end, the recorded acoustic emission signals were localized in one dimension, and the pattern recognition method described in Sect. 4.5 was applied. According to the ASTM D 5528 standard, the energy release rate is calculated for the onset of the first nonlinear (NL) force-displacement behavior, for the onset of the first visible crack growth (VIS), and for 5 % increased compliance or the force maximum (5 %/MAX). The corresponding evaluation of these values is given in Table 4.11 for comparison.

Based on the comparison in Fig. 4.158b, the onset of acoustic emission signals coincides well with the first onset of nonlinear force-displacement behavior. This indicates that prior to the force maximum, significant damage occurs within the specimen. With increased loading it is observed that the contributions of signal classes of matrix cracking and interfacial failure dominate, while fiber breakage signals show only minor contributions.

Looking more precisely into the contributions of the individual failure mechanisms, Fig. 4.159a reveals some interesting details. Here the relative contribution of the individual mechanisms was quantified according to (4.67). There are two distinct ranges of substantially different contributions of the failure mechanisms found in all of the tested specimens. At the initiation of crack growth, the contributions of the individual signal classes tend to scatter by up to 30 %. After reaching the force maximum, their contributions stabilize and remain constant within a range of scatter of ±5 %. This transition between the two ranges is hardly visible in the respective force-displacement curve.

After carrying out a series of measurements, this allows to combine measurement results of different specimens at different times during the experiment. A compilation of such results is shown in Fig. 4.159b. Here fracture toughness values were obtained for each specimen at these times, where the VIS criterion is fulfilled.

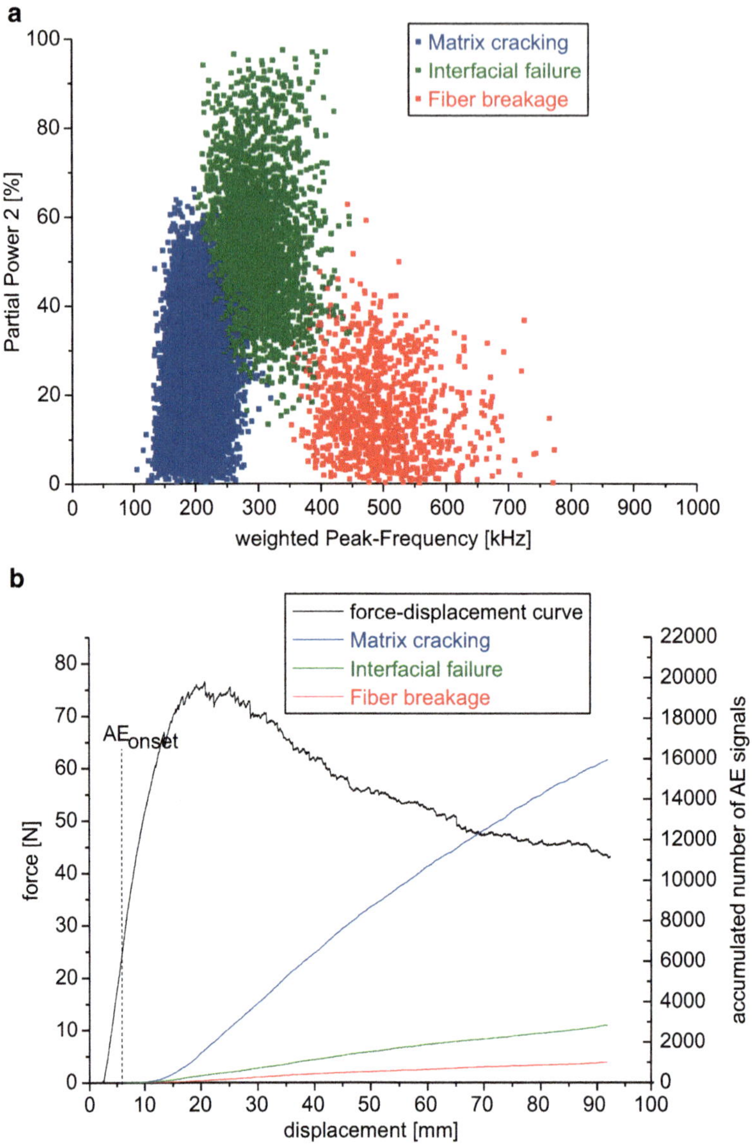

Fig. 4.158 Result of pattern recognition result for one exemplary specimen (**a**) and accumulated number of acoustic emission signals superimposed to force-displacement curve (**b**)

The fracture toughness values are used as *x*-axis in this diagram. As *y*-axis, the corresponding value of the relative amplitude at the time of reaching the VIS criteria is plotted. In addition, the mean values of the fracture toughness as computed from the *R*-curves in the region of stable crack propagation were

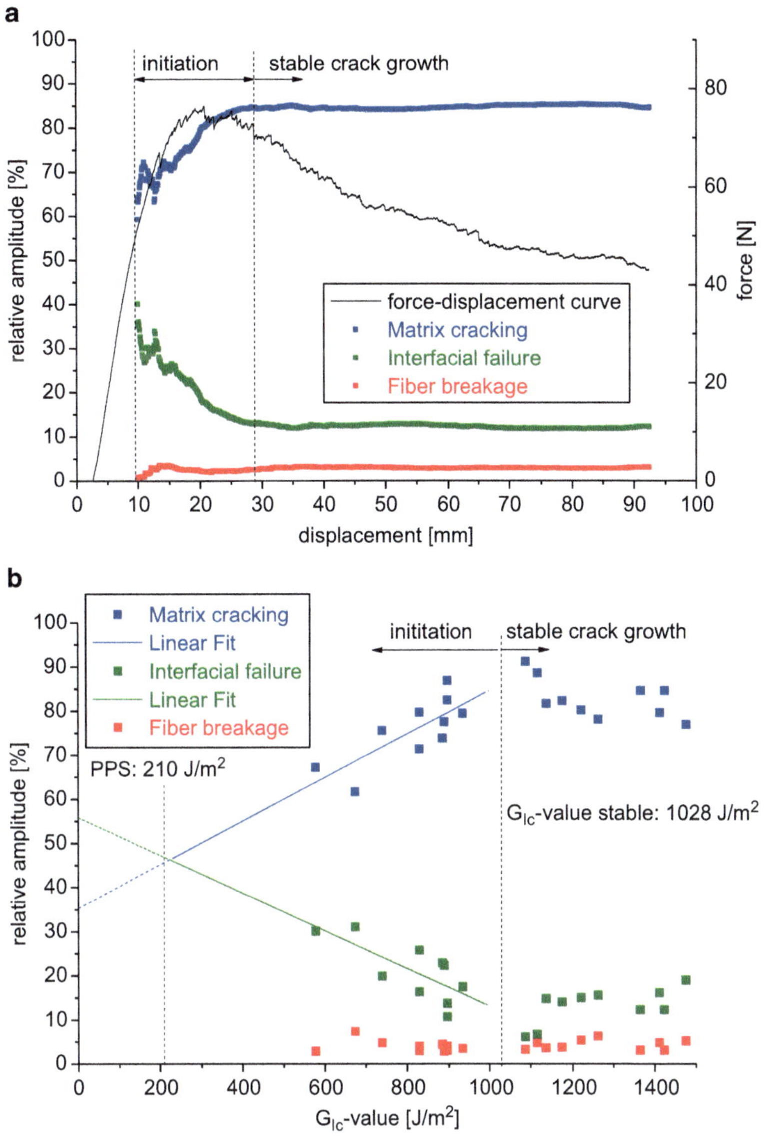

Fig. 4.159 Relative amplitude of failure mechanisms for one representative specimen (**a**) and combined evaluation of all specimens of a series of measurements (**b**)

determined for each specimen. Accordingly, the relative amplitude value of each of the specimens as found in the region of stable crack propagation is plotted on the *y*-axis.

The plot of data points in Fig. 4.159b allows several conclusions for the material system investigated. As reported previously in [35, 138], the back extrapolation of the contributions of matrix cracking and interfacial failure intersects at around 50 % contribution and at an abscissa of the fracture toughness value of the pure matrix material. For PPS this is expected around 210 J/m^2 [182]. This value is also in good agreement with the most conservative fracture toughness value obtained by the NL criterion and the value obtained by the onset of acoustic emission AE_{onset} as reported in Table 4.11. This allows concluding that the onset of nonlinear behavior is merely due to subcritical crack growth, which precedes the macroscopic crack growth [200].

As the second conclusion, it is possible to deduce a lower boundary for the fracture toughness value for stable crack growth. Based on the significant trend change in Fig. 4.159b, this can be evaluated as (1028 ± 50) J/m^2. This value is also in good agreement with the values obtained from the 5 %/MAX criterion (see Table 4.11). Therefore, the acoustic emission analysis allows for reinterpretation of the values of ASTM D 5528 in terms of fracture toughness values for subcritical crack initiation and fracture toughness values for stable crack growth.

4.7.4.2 Tensile Testing

While the double-cantilever beam test is specifically designed to allow for monitoring of crack propagation, this is not the case for most of the tensile specimens. In many cases, without secondary methods, the only noticeable observation of damage occurs at the time of rupture of the specimen. Sometimes preliminary failure is observable due to failure of filaments at the edges or due to audible cracking of the specimen. However, none of these phenomena can be interpreted in terms of quantitative values. As has been demonstrated in Sect. 4.7.3, acoustic emission detected during loading of tensile specimens can be used to detect failure mode specific onsets. Beyond that, the continuous recording of acoustic emission signals also allows to record the evolution of damage. This possibility is demonstrated in the following using the measurements previously described in Sect. 4.7.3.2.

In Fig. 4.160 the accumulated number of signals corresponding to each of the three failure mechanisms is plotted together with the specimen's stress-strain curve. Although the specimens in Fig. 4.160a, b originate from the same sample plate and have the same $[0_2/90/0_2]_{sym}$ stacking sequence, they show a significant difference in their failure behavior. While the specimen in Fig. 4.160a exhibits the theoretically expected order of failure mechanism onsets (matrix cracking, interfacial failure, fiber breakage), this is different for the specimen results in Fig. 4.160b (matrix cracking, fiber breakage, interfacial failure). However, a significant increase in the activity rate of fiber breakage signals is observed above 1 % strain in both specimens in Fig. 4.160. This is expected to be the onset of fiber failure that ultimately causes last-ply failure.

Also, the weight of contributions of fiber breakage and interfacial failure is contrary in both specimens. This is found to be consistent with the following macroscopic observations. Specimen 1 shows large areas of delamination (bottom Fig. 4.161), while specimen 2 fails predominantly by fiber breakage (top

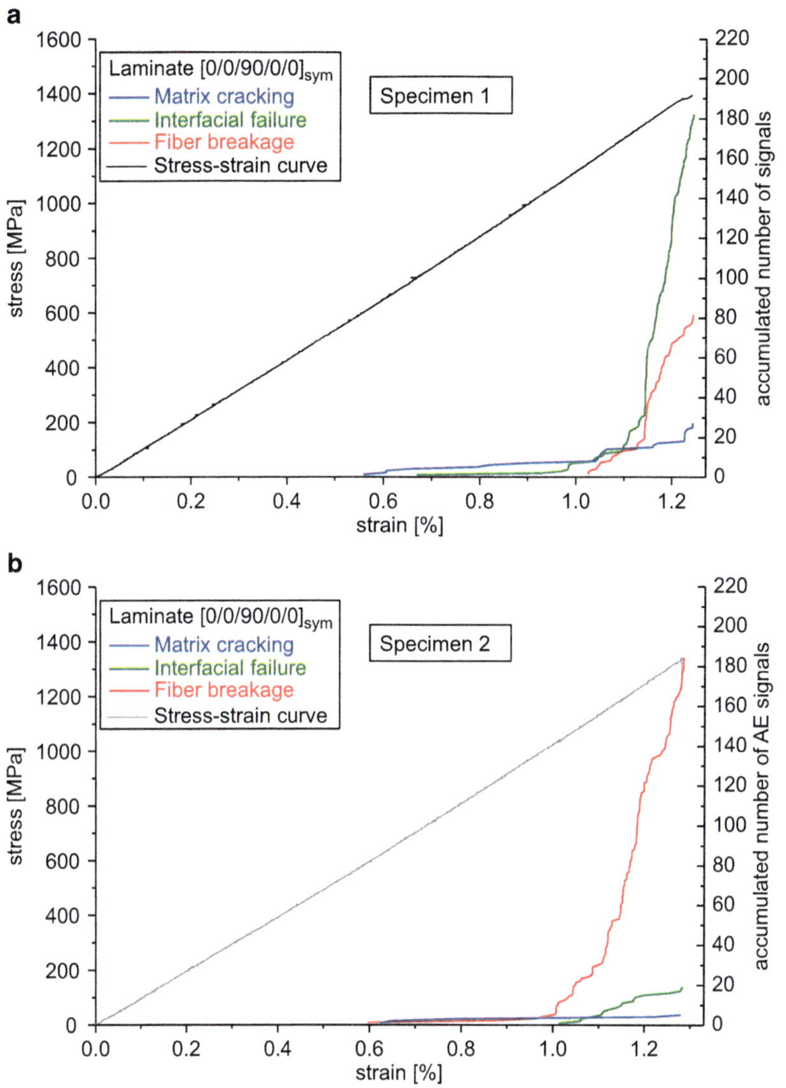

Fig. 4.160 Comparison between stress-strain curves and accumulated number of acoustic emission signals for two specimens with $[0_2/90/0_2]_{sym}$ stacking sequence (specimen 1 in **a**, specimen 2 in **b**)

Fig. 4.161). These differences also account for the substantial differences in the respective contributions of failure mechanisms as seen in Fig. 4.160.

Apart from monotonically increasing the applied load, repetitive loading and unloading cycles are a popular concept used for acoustic emission testing. This testing scheme is shown in Fig. 4.162. Here the specimen is subject to a monotonic

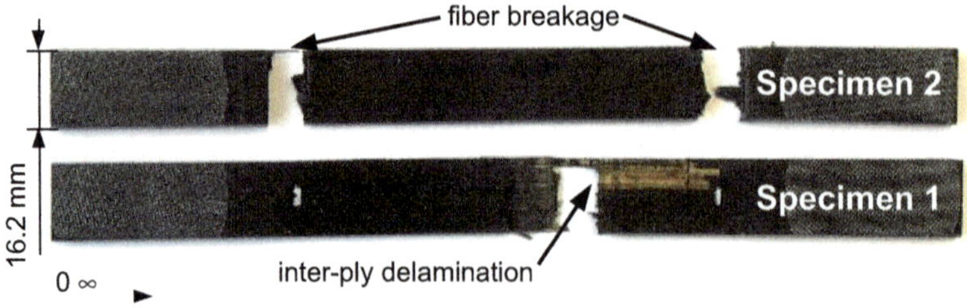

Fig. 4.161 Macroscopic observed failure mode for two different specimens with identical $[0_2/90/0_2]_{sym}$ stacking sequence cut from the same plate

Fig. 4.162 Scheme for definition of Felicity-ratio

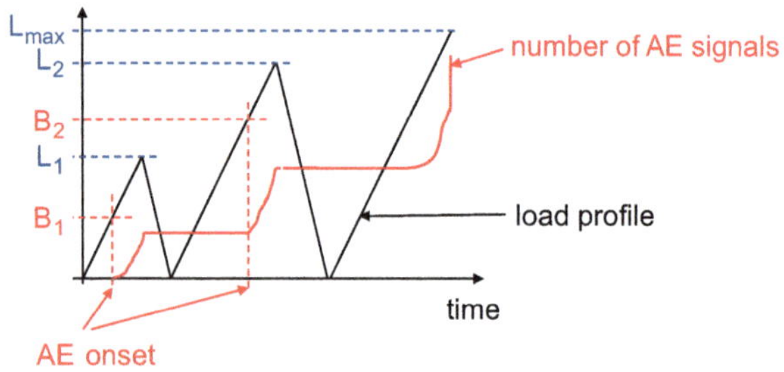

increasing load until a certain load level L_1 is reached. Then the specimen is fully (or partially) unloaded. Subsequently the load is again increased monotonically until a certain load level L_2 is reached, which exceeds the previous load level. This process is repeated N-times up to a certain maximum load level L_{max} or up to specimen failure. During the whole process, acoustic emission is recorded to evaluate the progression of damage. Focus is given on the correct evaluation of the onset of a significant amount of acoustic emission signals.

For the scheme shown below, the onset of signals in the second cycle is at the load level L_2, which allows calculating the Felicity[6]-ratio FR of cycle 2 as follows:

$$\mathrm{FR} = B_2 \,/\, L_1 \tag{4.71}$$

Accordingly, the load-ratio LR of cycle 2 can be calculated using the failure load L_{max}:

$$\mathrm{LR} = L_2/L_{max} \tag{4.72}$$

[6] The term "Felicity" was established by Timothy Fowler, honoring the contributions of his daughter Felicity to his scientific work [218].

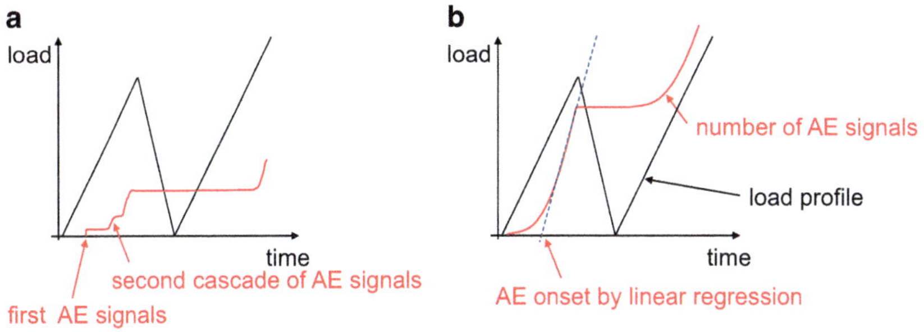

Fig. 4.163 Approaches to establish significant onset of acoustic emission signals at repetitive loading based on first AE signal or cascades (**a**) or based on linear regression (**b**)

As Kaiser has demonstrated for the case of metallic materials, the acoustic emission usually starts after exceeding the previous load level, which implies FR ≥ 1 [201]. This behavior of materials has been termed Kaiser effect to honor his pioneering work in the field of acoustic emission. But in most of the cases for fiber reinforced materials, the acoustic emission signals initiate at lower load levels than previously reached. This fundamental observation goes back to the work of Timothy Fowler providing the definition of (4.71) in [202, 203]. Since the first publication in 1977, the evaluation of the Felicity-ratio has become an important concept to understand the progression of failure in fiber reinforced materials using acoustic emission.

One particular challenge arises from the correct interpretation of the "significant" onset of acoustic emission. As the term significant is also used in standards like ASTM E 1067, it is important to point out that there is no unique definition of significant acoustic emission. Generally, the acoustic emission onset may not necessarily be identical to the occurrence of the first signal, since this may likely be a noise signal or a premature (unique) failure on the microscale, which is not representative for the full specimen or structure under inspection. Especially for cases with large gaps between the occurrence of the first signal and the occurrence of the next cascade of acoustic emission signals (cf. Fig. 4.163a), the assignment of the signal onset based on the first signal is sometimes doubtful.

For cases, where there are a lot of acoustic emission signals starting already at the moment of load application, other aspects might need to be considered. Here trend monitoring of the activity curves may yield an improved approach to detect the onset of "significant" acoustic emission. Especially for cases, where the first signals show only negligible energy compared to the rest of the detected signals, this may yield more meaningful signal onset values. Such can be done by backward extrapolation of linear regression if the activity is rather constant or by other approaches to systematically establish a measure for the trend change in the knee of an acoustic emission activity curve (cf. Fig. 4.163b).

For interpretation of the failure progression in fiber reinforced materials, an exemplary evaluation of a unidirectional T700/PPS tensile specimen is shown in

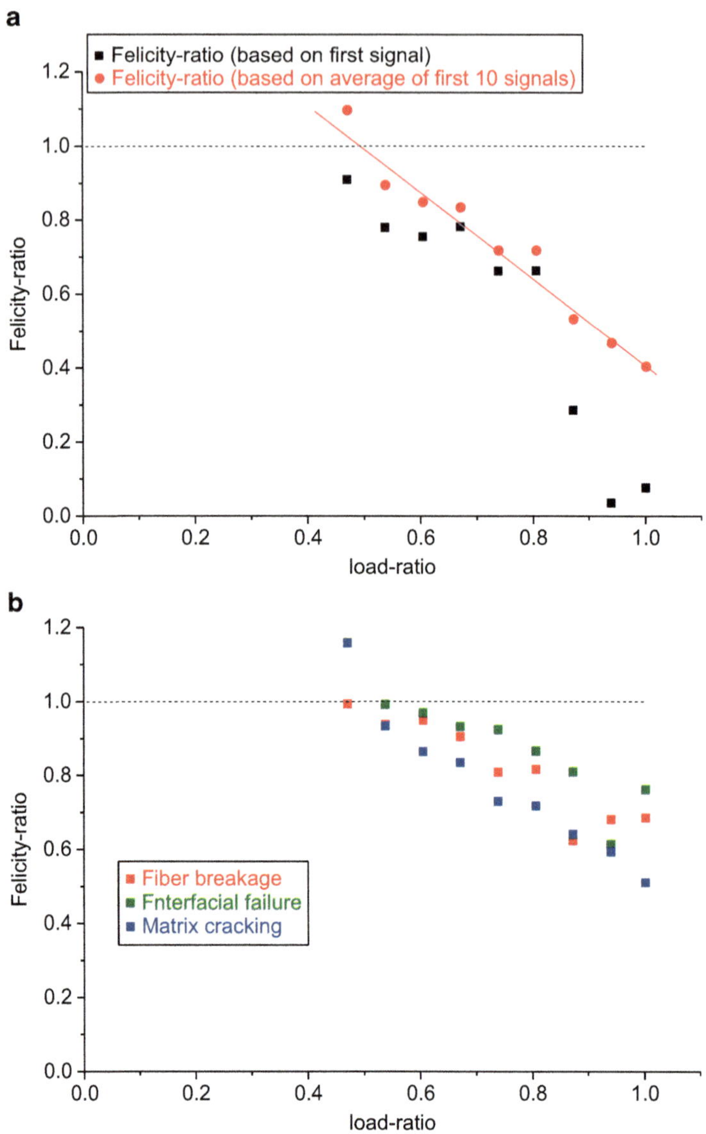

Fig. 4.164 Felicity-ratio evaluation of tensile specimens (UD-T700/PPS) based on first signal and average of first ten signals (**a**) and evaluated according to individual failure mechanisms (**b**)

Fig. 4.164. Except for the load scheme, the acquisition setup confirms with the configuration reported in Sect. 4.7.1. The specimen has been tested until failure, so L_{max} is known and the load-ratio of the respective cycles can be calculated according to (4.72). Figure 4.164a holds an evaluation of the Felicity-ratio based on the occurrence of the first signal (black dots) and based on the average load level of the first ten signals (red dots). Only signals localized in the test section of the specimen have been taken into account for the evaluation. At the cycle of first occurrence of acoustic emission, the calculated Felicity-ratio based on the first signal is already at 0.9. However, the respective Felicity-ratio based on the average of the first ten signals is at 1.1 indicating an undamaged specimen. For both evaluation types, there is a continuous decrease of Felicity-ratios with increased loading. This is indicative of a growing damage state and can be directly related to the increased load-ratio. For the Felicity-ratio of the average of the first ten signals, this is almost a direct linear correlation as indicated by the linear regression. However, the example demonstrates that the onset of the first signal is not necessarily the most stable means of evaluating the Felicity-ratio. Especially for the higher load-ratios, these single signal-based Felicity-ratios drop to very low levels (indicating early onsets), which is due to the presence of many damaged areas in the material. Also, from the comparison of both types of evaluation, it is obvious that the definition of the onset will have significant impact on the absolute values of the Felicity-ratio for the same type of experiment.

In Fig. 4.164b the evaluation of the same specimen is shown using the results of a pattern recognition process as described in Sect. 4.7.1. Here the Felicity-ratios are based on the signal onsets of the individual classes related to fiber breakage, interfacial failure, and matrix cracking. For convenience only the evaluation results using the average load level of the first ten signals per class are plotted. As expected from failure theory, the first onset is usually given by the occurrence of matrix cracking, yielding lowest Felicity-ratios. This is followed by the class of (single) fiber breakage and then by the class of interfacial failure. However, there are some cycles, where the order of occurrence is different. For the first load cycle, the first signal onset is given by fiber breakage, which has been visually observed during the test procedure to originate from failure of filaments at the edge of the specimen. Similar observation was made for the load-ratio 0.85, which explains the early onset of fiber breakage signals in these cycles.

4.7.4.3 Structural Components

So far, the application of acoustic emission analysis was mostly demonstrated for experimental setups used for material testing. However, the method is by far not limited to such applications. Therefore, a representative example of a more complex structure made from fiber reinforced polymers such as an angle-ply torsion shaft as shown in Fig. 4.165 is presented. The aim is to demonstrate the possibilities of the methods introduced in the previous sections in a structural component and to provide an example on how to visualize the evolution of failure in such a structure.

The torsion shafts were fabricated by a wet filament winding process based on a Sigrafil C30 carbon fiber and SIKA CR132/CH132-7 resin as matrix polymer. All specimens are wound as a $\pm 30°$ layup with fibers in circumferential direction as reinforcement. At the load introduction points and at two further locations, which

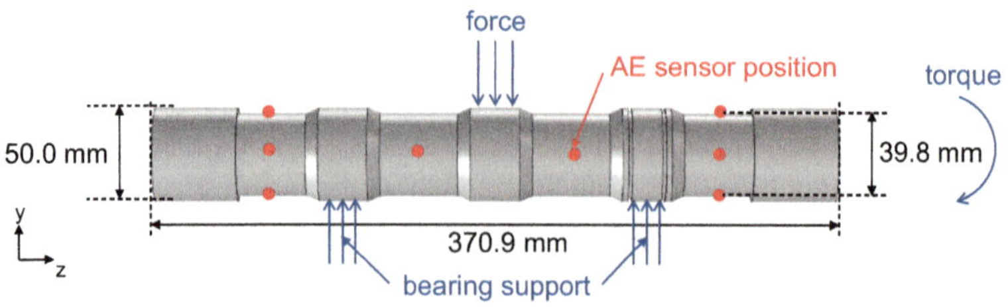

Fig. 4.165 Scheme of torsion shaft experiment including AE sensor positions and load introduction concept

Table 4.12 Values of torque and resultant bending moment as applied to the torsion shafts

Cycle	Torque [Nm]	Bending moment [Nm]	Ratio
1	500	323	1.55
2	750	482	1.56
3	1000	647	1.55
4	1250	798	1.57

are used as support position for a roller bearing, additional 90° layers were added as reinforcement. For load introduction a superimposed bending/torsion load profile as reported in Table 4.12 was chosen.

To apply a static transversal force, a special load frame was used as introduced in [204]. The transverse load is introduced perpendicular to the cylinder axis and is supported by two roller bearings mounted on the torsion shaft. In addition to the transverse load, a torque is superimposed to the full shaft by means of a universal testing machine. In four subsequent cycles, bending force and maximum torque were increased following the levels reported in Table 4.12. Testing was carried out in three steps for each load cycle. First the static bending force was manually increased to the force level corresponding to the bending moment of Table 4.12. As the second step, the torque was automatically increased until the maximum value was reached and subsequently unloaded to zero torque. As final step, the bending force was manually released to reach a zero-stress state for the start of the next load cycle.

During all steps the specimen was monitored by eight acoustic emission sensors mounted at the positions marked in Fig. 4.165 applying the acquisition parameters given in Table 4.13. Given the geometrical arrangement, no mounting of sensors by clamp systems or suction cup holders was feasible. Hence, hot glue was used as couplant and mounting system. This causes somewhat higher attenuation than for silicone grease and also causes a slight shift in frequency sensitivity. Therefore reference measurements on laboratory-scale specimens using different couplant systems are recommended to check the performance of the sensing system with respect to sensitivity and signal distinguishability.

Figure 4.166 shows the classification result of the acoustic emission measurements. The pattern recognition method is able to classify the signals as matrix cracking, interfacial failure, and fiber breakage based on the method proposed in

Table 4.13 Acquisition settings and postprocessing parameters used for torsion shaft test

Acquisition settings	Value	Postprocessing parameters	Value
Preamplification	40 dB_{AE}	Feature extraction window	100 μs after threshold
Threshold	35 dB_{AE}		
Triggering	Individual channel		
Acquisition rate	10 MS/s		
Band-pass range	20 kHz to 1 MHz		
Couplant	Hot glue		
Mounting system	Hot glue		
Sensor type	WD		
Number of sensors	8		

Table 4.14 Acquisition settings and postprocessing parameters used for large-scale pressure vessel testTable 4.14 Citation is missing. Kindly check and provide.was added

Acquisition settings	Value	Postprocessing parameters	Value
Preamplification	60 dB_{AE}	Feature extraction window	250 μs after threshold
Threshold	40 dB_{AE}		
Triggering	Individual channel		
Acquisition rate	10 MS/s		
Band-pass range	100 kHz to 1 MHz		
Couplant	Hot glue		
Mounting system	Hot glue		
Sensor type	WD		
Number of sensors	16		

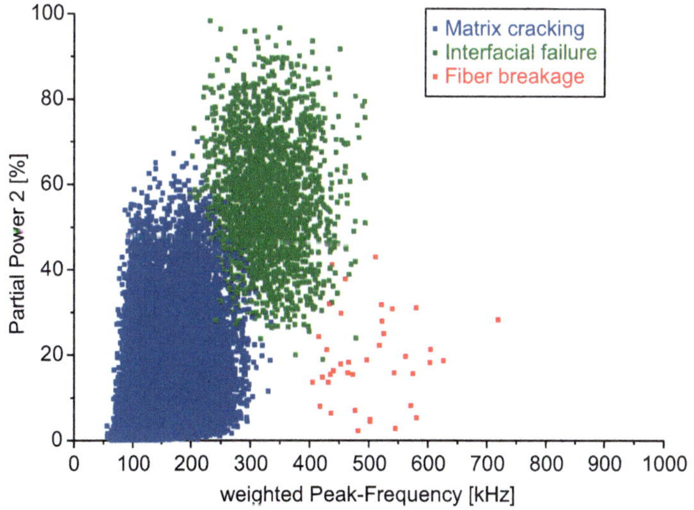

Fig. 4.166 Diagram of Partial Power 2 over Weighted Peak Frequency for one representative specimen (signals shown are result of all load cycles)

Sect. 4.4. The according UoC was found to be below 1.9 % for all specimens investigated. Due to the partitions found for other specimens and the relatively short source-sensor distances, the assignment is expected to be valid.

In Fig. 4.167, the classification results are shown as function of time of the experiment superimposed to the torque as function of time. First acoustic emission signals are observed at the application of the static bending force as well as after some additional application of torque. Hence, first occurrence of damage on the microscale should occur, but as indicated by the few signals during the first three

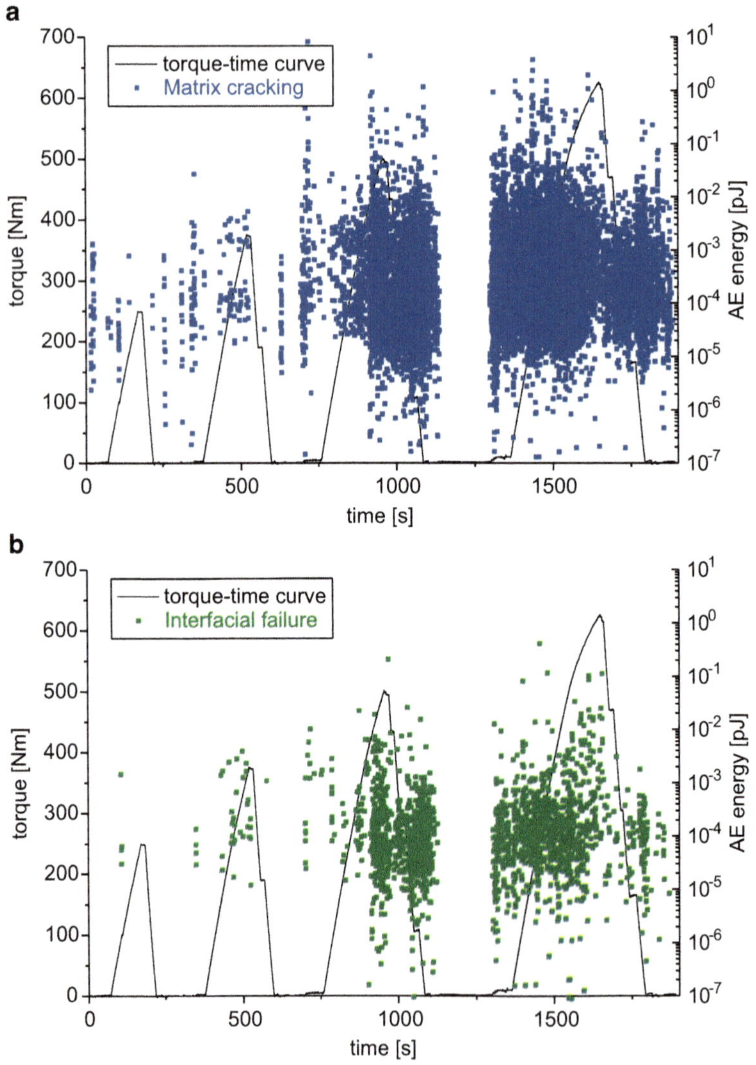

Fig. 4.167 Evolution of failure mechanisms classified as matrix cracking (**a**), interfacial failure (**b**), and fiber breakage (**c**) as function of load cases and accumulated AE energy of all failure mechanisms as function of load cases (**d**)

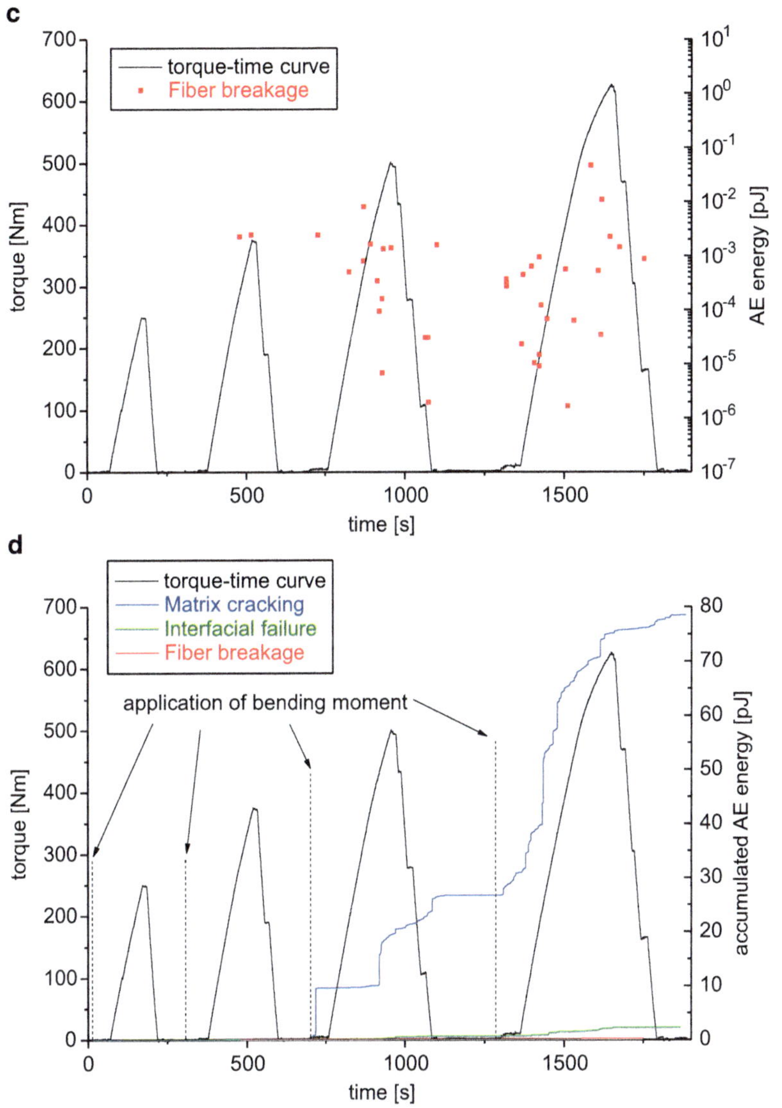

Fig. 4.167 (continued)

load cycles, there is not much damage expected to be present at these load levels. The source positions of these load cycles are plotted as density diagrams superimposed to the torque shaft structure in Fig. 4.168a for matrix cracking sources and in Fig. 4.168b for interfacial failure sources. As seen from these diagrams, no acoustic emission sources are localized in the areas of torque introduction but preferentially within the area of applied transverse load.

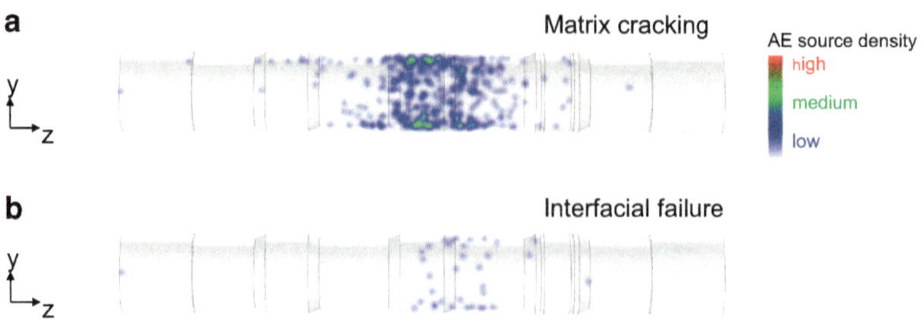

Fig. 4.168 Density diagrams of failure mechanisms classified as matrix cracking (**a**) and interfacial failure (**b**)

At load cycle four, the number and energy of acoustic emission signals increase substantially during the application of the static bending moment. Subsequently the application of the torque causes further increase of the acoustic emission activity as seen from the accumulated acoustic emission energy in Fig. 4.167d. Accordingly, first acoustic emission sources are localized in the area of load introduction of the static bending moment. A further increase of acoustic emission activity is then observed at the beginning of load cycle five causing numerous localized source positions dominantly in the central part of the torsion shaft, but also at the locations of the roller bearings acting as structural supports for the static bending force. Here, Fig. 4.168a, b can be used to visualize the preferential positions of matrix cracking and interfacial failure, which can then be applied to optimize the chosen layup for damage tolerance or to modify the torsion shaft geometry to withstand a higher load level. Finally, the accumulated acoustic emission energies can also be used for comparative purpose to estimate the mechanical energy release due to crack growth (cf. [138]). However, the latter requires comparable acquisition conditions for the experiments including the geometry of the specimen, the sensor types and position, digitization settings, and further parameters (also see Appendix A for a comprehensive list).

Another type of application, which has been investigated using acoustic emission to a substantial extent, are composite pressure vessels as being used for storage of pressurized gases such as compressed natural gas (CNG) and liquefied petroleum gas (LPG) for energy storage, as compressed air cylinder, or as low-pressure tank system for lightweight applications. For filament winding as fabrication technique, one can distinguish between five different categories of pressure vessels as shown schematically in Fig. 4.169:

- Type I: Fully metallic pressure vessel, which does not contain fiber reinforced composite materials as structural reinforcement
- Type II: Mostly metallic vessel with some fiber reinforced material applied along the hoop direction
- Type III: Metallic liner material with full overwrap of fiber reinforced material to carry the structural load (inside and outside pressure, impacts, etc.)

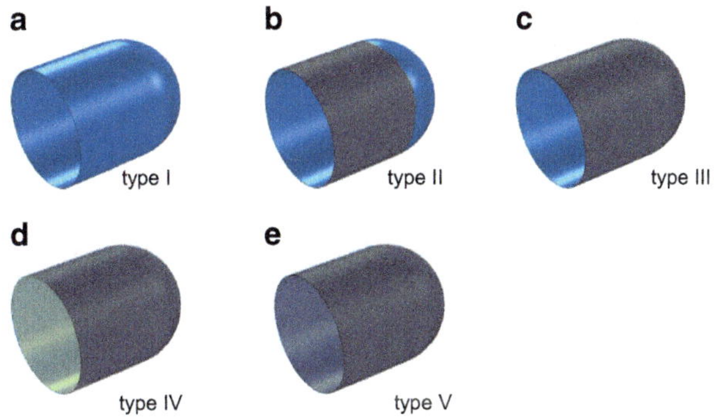

Fig. 4.169 The five basic type categories of composite pressure vessels as seen in (**a**-**e**)

- Type IV: Polymeric liner material with full overwrap of fiber reinforced material to carry the structural load (inside and outside pressure, impacts, etc.)
- Type V: No liner material, but solely fiber reinforced material carrying all structural loads and having sufficient tightness and matrix crack resistance to avoid leakage of contained gases or liquids

Due to the safety criticality of pressure vessels, there is a long history of methods evaluating the performance of these structures. The application of acoustic emission to pursue quality control, to monitor aging, or to predict burst pressures has been followed for decades [205–212]. For some cases, the monitoring of acoustic emission has been covered by standards like ASTM E 2863, ASTM E 1067, ASTM E 1419, ISO 16148, and ISO 19016, whereas in other fields of application, there is still a lack of established standards.

From the viewpoint of acoustic emission monitoring of composite pressure vessels, there are several factors governing the potential success of the measurement. First it is helpful to note that for cylindrical pressure vessels, the usual ply layout chosen to fabricate the vessels is different compared to the typical laminates since the hoop load and the axial load are not identical. Hence, the number of filaments to the axial and hoop directions differs, which causes a pronounced acoustic anisotropy. Second, the filament winding process typically uses a layup sequence, which is not balanced, i.e., is not symmetric around the midplane. Both effects cause a complex type of signal propagation, which is hardly captured by analytical considerations and sometimes counterintuitive to understand. However, numerical methods may be used to visualize and interpret signals of different source mechanisms even in such complex structures [53, 213–215]. Also the possibility of propagation paths through the contained liquid may lead to additional complexity in the interpretation of the detected acoustic emission signals [215, 216].

In the following an example of a large-scale burst pressure test is shown to demonstrate the possibilities of acoustic emission analysis in such an application. The type IV pressure vessel inspected is shown in Fig. 4.170, including the positions of the 16 mounted acoustic emission sensors using the settings of Table 4.14. The pressure vessel was fabricated using a T700/PPS unidirectional

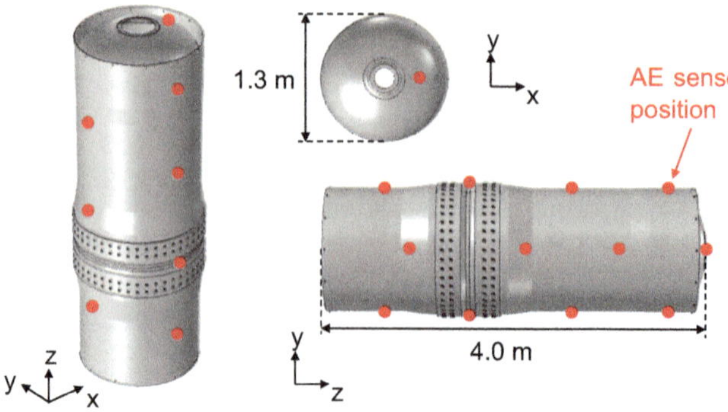

Fig. 4.170 Scheme of burst pressure configuration including 16 AE sensor

tape in situ consolidated using a high-power laser in a filament winding process and is reported in detail in [212, 217]. The axial dimension of the vessel is 4.0 m, with an outer diameter of 1.3 m as seen in Fig. 4.170. The full vessel is composed out of two composite parts joint by a two-sided clevis fastener with two rows of bolts. For the chosen sensor positions, a maximum distance of 0.5 m between the source and sensor is expected for every point on the vessel.

Based on the typical amplitude distribution of sources generated by such a vessel, the PoD of the measurement chain can be calculated. This procedure is based on the measured attenuation curves along the hoop direction and along the axial direction of the vessel. The results of these calculations are reported in Fig. 4.171 based on the 60 dB_{AE} gain of the preamplifier and the acquisition threshold of 40 dB_{AE}. To estimate the lower bound of the PoD, the attenuation curves of the higher-frequency portions of the signals are considered, since these are subject to higher attenuation as pointed out in Sect. 4.3. As seen from Fig. 4.171a, b, respectively, this allows estimation of the PoD in the vessel to be larger than 60 %. Accordingly, the maximum distance between source and sensor to allow a PoD > 90 % is below 150 mm. However, the application of such a dense sensor network is usually not economical in a field testing application; the acquired acoustic emission signals need to be interpreted from a statistical point of view. Since fiber reinforced composites generate numerous acoustic emission signals prior to failure and the signals with highest intensity have a much higher PoD than the signals at the detection limit of the acquisition system, approaches with an overall PoD down to 60 % are sometimes justified.

There are various established protocols to test pressure vessels in combination with acoustic emission. The loading scheme being used depends on the aim of test, such as quality control, repeated testing after being in service, proof testing, or burst pressure evaluation.

As recommended by ASTM E 1067, the load scheme for the pressure vessel shown in Fig. 4.170 was based on subsequent cycles with incremental peak load and intermediate unload to 70 % of the peak load. The peak load was increased until

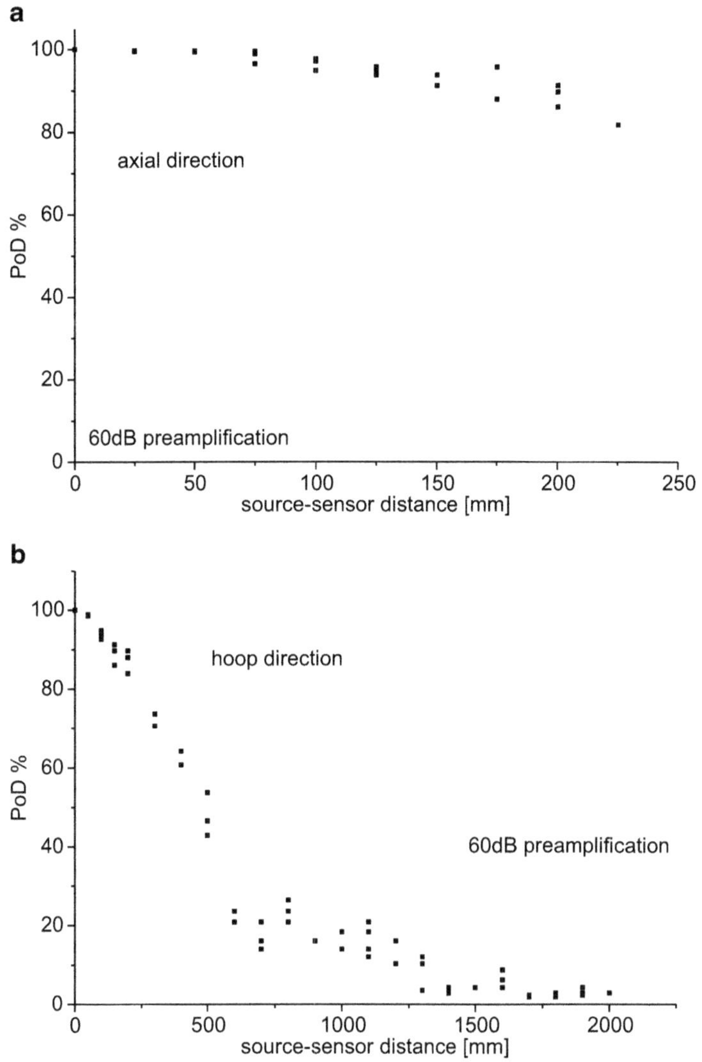

Fig. 4.171 Calculated probability of detection along axial direction (**a**) and along hoop direction (**b**)

final burst failure of the vessel. Using the sensor network mounted on the pressure vessel, it is possible to localize acoustic emission source positions and visualize them as function of the applied pressure. As seen from the density diagram of localized acoustic emission sources in Fig. 4.172 at the moment of vessel rupture, there are numerous sources active at the wedge sections of the composite parts of the pressure vessel. The region of highest acoustic emission density is the actual position of failure of the whole vessel, which was already indicative in the first pressure cycle (24.7 % of burst pressure) as shown in Fig. 4.172.

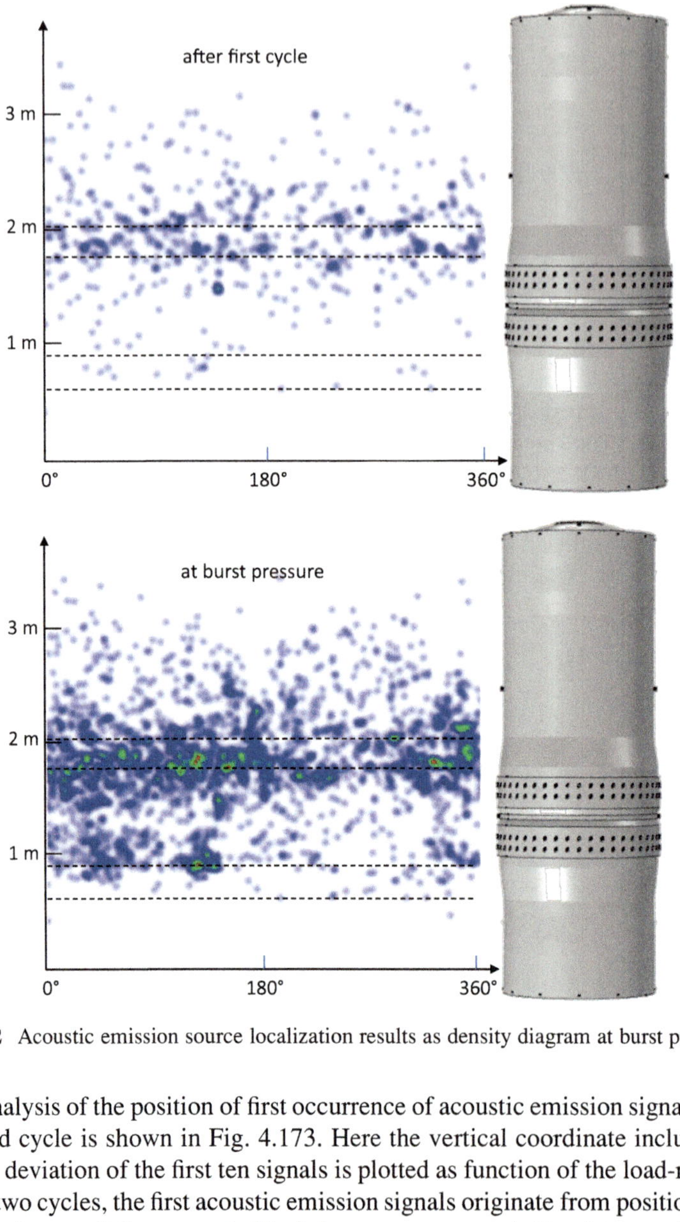

Fig. 4.172 Acoustic emission source localization results as density diagram at burst pressure

An analysis of the position of first occurrence of acoustic emission signals within each load cycle is shown in Fig. 4.173. Here the vertical coordinate including the standard deviation of the first ten signals is plotted as function of the load-ratio. For the first two cycles, the first acoustic emission signals originate from positions in the cylindrical part of the upper half of the pressure vessel. For later cycles, their position changes and stays within the area of locally increased thickness around 1500 mm. For the last load cycles, the position moves to the range covered by the clevis fastener. At this stage, the damage has grown within the laminate and causes increased stress concentration at the bolts. Within the last cycle, the position of occurrence of the first acoustic emission signals precisely indicates the failure location.

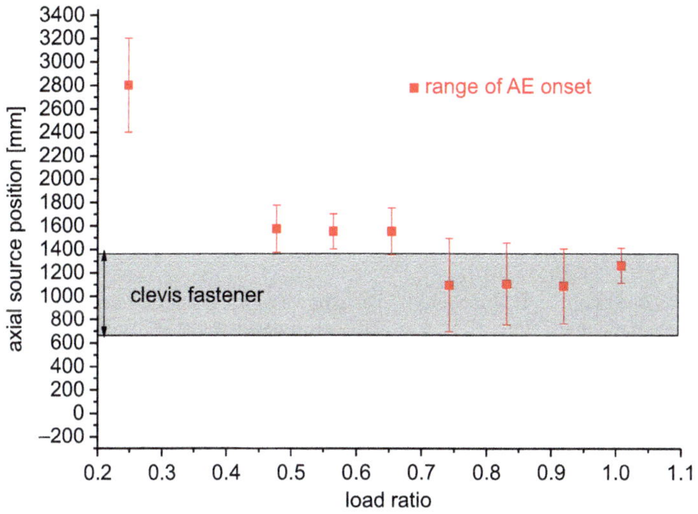

Fig. 4.173 Axial acoustic emission source position as function of the load-ratio indicating shift toward failure location for the eight load cycles with incremental peak load (*left* to *right*)

References

1. Lacidogna, G., Carpinteri, A., Manuello, A., Durin, G., Schiavi, A., Niccolini, G., Agosto, A.: Acoustic and electromagnetic emissions as precursor phenomena in failure processes. Strain **47**, 144–152 (2011)
2. Giordano, M., Condelli, L., Nicolais, L.: Acoustic emission wave propagation in a viscoelastic plate. Compos. Sci. Technol. **59**, 1735–1743 (1999)
3. Hamstad, M.A.: Thirty years of advances and some remaining challenges in the application of acoustic emission to composite materials. In: Kishi, T., Ohtsu, M., Yuyama, S. (eds.) Acoustic Emission Beyond the Millennium, pp. 77–91. Elsevier Science, Amsterdam (2000)
4. Ono, K., Gallego, A.: Research and applications of AE on advanced composites. J. Acoust. Emiss. **30**, 180–229 (2012)
5. Reinhardt, H.W., Grosse, C.U., Kurz, J.H.: Localization and mode determination of fracture events by acoustic emission. In: Carpinteri, A., Lacidogna, G. (eds.) Acoustic Emission and Critical Phenomena. Taylor & Francis Group, London (2008)
6. Prosser, W.H., Jackson, K.E., Kellas, S., Smith, B.T., McKeon, J., Friedman, A.: Advanced waveform-based acoustic emission detection of matrix cracking in composites. Mater. Eval. **53**, 1052–1058 (1995)
7. Huguet, S., Godin, N., Gaertner, R., Salmon, L., Villard, D.: Use of acoustic emission to identify damage modes in glass fibre reinforced polyester. Compos. Sci. Technol. **62**, 1433–1444 (2002)
8. Scholey, J.J., Wilcox, P.D., Wisnom, M.R., Friswell, M.I.: Quantitative experimental measurements of matrix cracking and delamination using acoustic emission. Compos. Part A Appl. Sci. Manuf. **41**, 612–623 (2010)
9. Giordano, M., Calabro, A., Esposito, C., D'Amore, A., Nicolais, L.: An acoustic-emission characterization of the failure modes in polymer-composite materials. Compos. Sci. Technol. **58**, 1923–1928 (1998)

10. Wilcox, P.D., Lee, C.K., Scholey, J.J., Friswell, M.I., Wisnom, M.R., Drinkwater, B.W.: Progress towards a forward model of the complete acoustic emission process. Adv. Mater. Res. **13–14**, 69–75 (2006)
11. Prosser, W.H., Hamstad, M.A., Gary, J., Gallagher, A.O.: Finite element and plate theory modeling of acoustic emission waveforms. J. Nondestruct. Eval. **18**, 83–90 (1999)
12. Sause, M.G.R., Horn, S.: Simulation of Lamb wave excitation for different elastic properties and acoustic emission source geometries. J. Acoust. Emiss. **28**, 142–154 (2010)
13. Sause, M.G.R., Richler, S.: Finite element modelling of cracks as acoustic emission sources. J. Nondestruct. Eval. **34**, 1–13 (2015)
14. Livne, A., Bouchbinder, E., Svetlizky, I., Fineberg, J.: The near-tip fields of fast cracks. Science **327**, 1359–1363 (2010)
15. Livne, A., Bouchbinder, E., Fineberg, J.: Breakdown of linear elastic fracture mechanics near the tip of a rapid crack. Phys. Rev. Lett. **101**, 1–4 (2008)
16. Landau, L.D., Lifschitz, E.M.: Elastizitätstheorie. Akademie Verlag GmbH, Berlin (1987)
17. Freund, L.B.: The initial wave front emitted by a suddenly extending crack in an elastic solid. J. Appl. Mech. **39**, 601–602 (1972)
18. Aki, K., Richards, P.G.: Quantitative Seismology, Theory and Methods. University Science, Sausalito (1980)
19. Ohtsu, M.: Source mechanism and waveform analysis of acoustic emission in concrete. J. Acoust. Emiss. **2**, 103–112 (1982)
20. Lysak, M.V.: Development of the theory of acoustic emission by propagating cracks in terms of fracture mechanics. Eng. Fract. Mech. **55**, 443–452 (1996)
21. Ohtsu, M., Ono, K.: A generalized theory of acoustic emission and Green's function in a half space. J. Acoust. Emiss. **3**, 27–40 (1984)
22. Ohtsu, M., Ono, K.: The generalized theory and source representation of acoustic emission. J. Acoust. Emiss. **5**, 124–133 (1986)
23. Hamstad, M.A., O'Gallagher, A., Gary, J.: Modeling of buried monopole and dipole sources of acoustic emission with a finite element technique. J. Acoust. Emiss. **17**, 97–110 (1999)
24. Sause, M.G.R., Horn, S.: Simulation of acoustic emission in planar carbon fiber reinforced plastic specimens. J. Nondestruct. Eval. **29**, 123–142 (2010)
25. Hamstad, M.A.: Frequencies and amplitudes of AE signals in a plate as a function of source rise time. In: 29th European Conference on Acoustic Emission Testing, Vienna, Austria, 2010, pp. 1–8.
26. Scruby, C.B., Buttle, D.J.: Quantitative fatigue crack measurement by acoustic emission. In: Marsh, K.J., Smith, R., Ritchie, R.O. (eds.) Crack Measurement: Techniques and Applications, pp. 207–287. Engineering Materials Advisory Services Ltd., West Midlands (1992)
27. Wadley, H.N.G., Scruby, C.B.: Acoustic Emission Source Characterization. Advances in Acoustic Emission. Dunhart, Knoxville (1981)
28. Scruby, C.B.: Quantitative acoustic emission techniques. Nondestruct. Test. **8**, 141–208 (1985)
29. Green, E.R.: Acoustic emission sources in a cross-ply laminated plate. Compos. Eng. **5**, 1453–1469 (1995)
30. Green, E.R.: Acoustic emission in composite laminates. J. Nondestruct. Eval. **17**, 117–127 (1998)
31. Hamstad, M.A., O'Gallagher, A., Gary, J.: A wavelet transform applied to acoustic emission signals: part 1: source identification. J. Acoust. Emiss. **20**, 39–61 (2002)
32. Downs, K.S., Hamstad, M.A., O'Gallagher, A.: Wavelet transform signal processing to distinguish different acoustic emission sources. J. Acoust. Emiss. **21**, 52–69 (2003)
33. Hora, P., Cervena, O.: Acoustic emission source modeling. Appl. Comput. Mech. **4**, 25–36 (2010)
34. Hamstad, M.A., O'Gallagher, A., Gary, J.: A wavelet transform applied to acoustic emission signals: part 2: source location. J. Acoust. Emiss. **20**, 62–82 (2002)

35. Sause, M.: Identification of Failure Mechanisms in Hybrid Materials Utilizing Pattern Recognition Techniques Applied to Acoustic Emission Signals. mbv-Verlag, Berlin (2010)
36. Sause, M.G.R.: Modelling of crack growth based acoustic emission release in aluminum alloys. In: 31st Conference of the European Working Group on Acoustic Emission, pp. 1–8. Dresden, Germany (2014)
37. Kalafat, S., Zelenyak, A.-M., Sause, M.G.R.: In-situ monitoring of composite failure by computing tomography and acoustic emission. In: 20th International Conference on Composite Materials, pp. 1–8. Copenhagen, Denmark (2015)
38. Juhasz, T.: Ein neues physikalisch basiertes Versagenskriterium für schwach 3D-verstärkte Faserverbundlaminate. PhD-thesis, University Carolo-Wilhelmina Braunschweig (2003)
39. Puck, A.: Festigkeitsanalyse von Faser-Matrix-Laminaten Modelle für die Praxis. Carl Hanser Verlag, Munich (1996)
40. Green, A.E., Zerna, W.: Theoretical Elasticity. Oxford University Press, New York (2002)
41. Scott, A.E., Sinclair, I., Spearing, S.M., Mavrogordato, M., Bunsell, A.R., Thionnet, A.: Comparison of the accumulation of fibre breaks occurring in a unidirectional carbon/epoxy composite identified in a multi-scale micro-mechanical model with that of experimental observations using high resolution computed tomography. In: Matériaux 2010, pp. 1–9. Nantes, France (2010)
42. Boltz, E.S., Fortunko, C.M., Hamstad, M.A., Renken, M.C.: Absolute sensitivity of air, light and direct-coupled wideband acoustic emission transducers. In: Thompson, D.O., Chimenti, D.E. (eds.) Review of Progress in Quantitative Nondestructive Evaluation, pp. 967–974. Springer, Boston (1995)
43. Tatro, C.A.: Design criteria for acoustic emission experimentation. In: Acoustic Emission ASTM STP 505, pp. 84–99 (1972)
44. Stoneley, R.: Elastic waves at the surface of separation of two solids. Proc. R. Soc. Lond. A. **106**, 416–428 (1924)
45. Love, A.E.H.: Some Problems of Geodynamics. University Press, Cambridge (1911)
46. Lamb, H.: On waves in an elastic plate. Proc. R. Soc. Lond. A. **93**, 114–128 (1917)
47. Rayleigh, L.: On waves propagated along the plane surface of an elastic solid. Proc. Lond. Math. Soc. **s1–17**, 4–11 (1885)
48. Krautkrämer, J., Krautkrämer, H.: Ultrasonic Testing of Materials. Springer, Berlin (1983)
49. Cremer, L., Heckl, M.: Körperschall Physikalische Grundlagen und technische Anwendungen. Springer Verlag, Berlin (1996)
50. Redwood, M.: Mechanical Waveguides; The Propagation of Acoustic and Ultrasonic Waves in Fluids and Solids with Boundaries. Pergamon Press, New York (1960)
51. Bergman, E.H., Shahbender, R.: Effect of statically applied stresses on the velocity of propagation of ultrasonic waves. J. Appl. Phys. **29**, 1736–1738 (1958)
52. Grosse, C.U., Ohtsu, M.: Acoustic Emission Testing. Springer, Berlin (2008)
53. Sause, M.G.R., Hamstad, M.A., Horn, S.: Finite element modeling of Lamb wave propagation in anisotropic hybrid materials. Compos. Part B Eng. **53**, 249–257 (2013)
54. Lowe, M.J.S.: Matrix techniques for modeling ultrasonic waves in multilayered media. IEEE Trans. Ultrason. Ferroelectr. Freq. Control **42**, 525–542 (1995)
55. Castaings, M., Bacon, C., Hosten, B., Predoi, M.V.: Finite element predictions for the dynamic response of thermo-viscoelastic material structures. J. Acoust. Soc. Am. **115**, 1125 (2004)
56. Heidary, Z., Ozevin, D.: On the influences of boundary reflections and piezoelectric sensors to the characteristics of elastic waves for pattern recognition methods. J. Nondestruct. Eval. **34**, 271 (2014)
57. Raghavan, A., Cesnik, C.E.S.: Review of guided-wave structural health monitoring. Shock Vib. Dig. **39**, 91–114 (2007)
58. Thomson, W.T.: Transmission of elastic waves through a stratified solid medium. J. Appl. Phys. **21**, 89 (1950)

59. Haskell, N.A.: The dispersion of surface waves on multilayered media. Bull. Seismol. Soc. Am. **43**, 17–34 (1953)
60. Kundu, T., Mal, A.: Elastic waves in a multilayered solid due to a point source. Wave Motion **7**, 459–471 (1985)
61. Sause, M.G.R., Horn, S.: Influence of internal discontinuities on ultrasonic signal propagation in carbon fiber reinforced plastics. In: 30th European Conference on Acoustic Emissionm pp. 1–11. Granada, Spain (2012)
62. Sause, M.G.R.: Acoustic emission signal propagation in damaged composite structures. J. Acoust. Emiss. **31**, 1–18 (2013)
63. Auld, B.A.: Acoustic Fields ans Waves in Solids. Krieger, Malabar (1990)
64. Hosten, B.: Heterogeneous structure of modes and Kramers-Kronig relationship in anisotropic viscoelastic materials. J. Acoust. Soc. Am. **104**, 1382–1388 (1998)
65. Calomfirescu, M., Herrmann, A.: Attenuation of Lamb waves in composites: models and possible applications. In: Proceedings of the 6th International Workshop on Structural Health Monitoring, pp. 1–8. Stanford, CA, USA (2007)
66. Pollock, A.A.: Classical wave theory in practical AE testing. In: Proceedings of the 8th International AE Symposium, pp. 708–721. Tokyo, Japan (1986)
67. Prosser, W.H.: Advanced AE techniques in composite materials research. J. Acoust. Emiss. **14**, 1–11 (1996)
68. Neau, G., Deschamps, M., Lowe, M.J.S.: Group velocity of Lamb waves in anisotropic plates: comparison between theory and experiments. In: AIP Conference Proceedings, pp. 81–88 (2001)
69. Ward, I.M.: Mechanical Properties of Solid Polymers. Wiley, New York (1971)
70. Gallego, A., Ono, K.: An improved acousto-ultrasonic scheme with Lamb wave mode separation and damping factor in CFRP plates. J. Acoust. Emiss. **30**, 109–123 (2012)
71. Choi, H.-I., Williams, W.: Improved time-frequency representation of multicomponent signals using exponential kernels. IEEE Trans. Acoust. Speech Signal Process. **37**, 862–872 (1989)
72. Beattie, A.G.: Acoustic emission, principles and Instrumentation. J. Acoust. Emiss. **2**, 95–128 (1983)
73. Gautschi, G.: Piezoelectric Sensorics. Springer, Berlin (2002)
74. Breckenridge, F.R.: Acoustic emission: some applications of Lamb's problem. J. Acoust. Soc. Am. **57**, 626 (1975)
75. Scruby, C.B., Wadley, H.N.G.: A calibrated capacitance transducer for the detection of acoustic emission. J. Phys. D Appl. Phys. **11**, 1487–1494 (1978)
76. Read, I., Foote, P., Murray, S.: Optical fibre acoustic emission sensor for damage detection in carbon fibre composite structures. Meas. Sci. Technol. **13**, N5–N9 (2002)
77. de Oliveira, R., Frazão, O., Santos, J.L., Marques, A.T.: Optic fibre sensor for real-time damage detection in smart composite. Comput. Struct. **82**, 1315–1321 (2004)
78. Wild, G., Hinckley, S.: Acousto-ultrasonic optical fiber sensors: overview and state-of-the-art. IEEE Sens. J. **8**, 1184–1193 (2008)
79. Wild, G., Hinckley, S.: Fiber Bragg grating sensors for acoustic emission and transmission detection applied to robotic NDE in structural health monitoring. In: 2007 I.E. Sensors Applications Symposium, pp. 1–6. IEEE (2007)
80. Watanabe, M., Enoki, M., Kishi, T.: Fracture behavior of ceramic coatings during thermal cycling evaluated by acoustic emission method using laser interferometers. Mater. Sci. Eng. A **359**, 368–374 (2003)
81. Enoki, M., Watanabe, M., Chivavibul, P., Kishi, T.: Non-contact measurement of acoustic emission in materials by laser interferometry. Sci. Technol. Adv. Mater. **1**, 157–165 (2000)
82. Bohse, J.: Damage analysis of polymer matrix composites by acoustic emission testing. In: EWGAE 2004—26th European Conference on Acoustic Emission Testing, pp. 339–348. Berlin, Germany (2004)

83. Breckenridge, F.R., Greenspan, M.: Surface-wave displacement: absolute measurements using a capacitance transducer. J. Acoust. Soc. Am. **69**, 1177–1185 (1981)
84. Grosse, C.U.: Quantitative non-destructive testing of construction materials using acoustic emission technique and ultrasound. PhD-thesis, University of Stuttgart (1996)
85. Goujon, L., Baboux, J.C.: Behaviour of acoustic emission sensors using broadband calibration techniques. Meas. Sci. Technol. **14**, 903–908 (2003)
86. Hamstad, M.A.: Re-examination of NIST acoustic emission absolute sensor calibration: part II: finite element modeling of acoustic emission signal from glass capillary fracture. J. Acoust. Emiss. **29**, 175–183 (2011)
87. Sause, M.G.R., Hamstad, M.A., Horn, S.: Finite element modeling of conical acoustic emission sensors and corresponding experiments. Sens. Actuators A Phys. **184**, 64–71 (2012)
88. Hamstad, M.A.: Improved signal-to-noise wideband acoustic/ultrasonic contact displacement sensors for wood and polymers. Wood Fiber Sci. **29**, 239–248 (1997)
89. McLaskey, G.C., Glaser, S.D.: Acoustic emission sensor calibration for absolute source measurements. J. Nondestruct. Eval. **31**, 157–168 (2012)
90. Huang, Y.-H., Ma, C.-C.: Forced vibration analysis of piezoelectric quartz plates in resonance. Sens. Actuators A Phys. **149**, 320–330 (2009)
91. Ohtsu, M., Ono, K.: Resonance analysis of piezoelectric transducer elements. J. Acoust. Emiss. **2**, 247–260 (1983)
92. Ono, K., Cho, H., Matsuo, T.: Transfer functions of acoustic emission sensors. J. Acoust. Emiss. **26**, 72–90 (2008)
93. Feng, G.-H., Tsai, M.-Y.: Acoustic emission sensor with structure-enhanced sensing mechanism based on micro-embossed piezoelectric polymer. Sens. Actuators A Phys. **162**, 100–106 (2010)
94. Or, S.W., Chan, H.L.W., Choy, C.L.: P(VDF-TrFE) copolymer acoustic emission sensors. Sens. Actuators A Phys. **80**, 237–241 (2000)
95. Barbezat, M., Brunner, A.J., Flüeler, P., Huber, C., Kornmann, X.: Acoustic emission sensor properties of active fibre composite elements compared with commercial acoustic emission sensors. Sens. Actuators A Phys. **114**, 13–20 (2004)
96. Marin-Franch, P., Martin, T., Tunnicliffe, D.L., Das-Gupta, D.K.: PTCa/PEKK piezocomposites for acoustic emission detection. Sens. Actuators A Phys. **99**, 236–243 (2002)
97. Zheng, S.X., McBride, R., Barton, J.S., Jones, J.D.C., Hale, K.F., Jones, B.E.: Intrinsic optical fibre sensor for monitoring acoustic emission. Sens. Actuators A Phys. **31**, 110–114 (1992)
98. Nieuwenhuis, J.H., Neumann, J.J., Greve, D.W., Oppenheim, I.J.: Generation and detection of guided waves using PZT wafer transducers. IEEE Trans. Ultrason. Ferroelectr. Freq. Control **52**, 2103–2111 (2005)
99. Sause, M.G.R.: Investigation of pencil-lead breaks as acoustic emission sources. J. Acoust. Emiss. **29**, 184–196 (2011)
100. Greenspan, M.: The NBS conical transducer: analysis. J. Acoust. Soc. Am. **81**, 173 (1987)
101. McLaskey, G.C., Glaser, S.D.: Hertzian impact: experimental study of the force pulse and resulting stress waves. J. Acoust. Soc. Am. **128**, 1087–1096 (2010)
102. Lynnworth, L.C., Umina, J.A.: Extensional bundle waveguide techniques for measuring flow of hot fluids. IEEE Trans. Ultrason. Ferroelectr. Freq. Control **52**, 538–544 (2005)
103. Hamstad, M.A.: Small diameter waveguide for wideband acoustic emission. J. Acoust. Emiss. **24**, 234–247 (2006)
104. Rose, J.L.: Ultrasonic Waves in Solid Media. University Press, Cambridge (2004)
105. Zelenyak, A.-M., Hamstad, M.A., Sause, M.G.R.: Finite element modeling of acoustic emission signal propagation with various shaped waveguides. In: 31st Conference of the European Working Group on Acoustic Emission, pp. 1–8. Dresden, Germany (2014)
106. Zelenyak, A.-M., Hamstad, M., Sause, M.: Modeling of acoustic emission signal propagation in waveguides. Sensors **15**, 11805–11822 (2015)

107. Ernst, R., Dual, J.: Acoustic emission source detection using the time reversal principle on dispersive waves in beams. In: Proceedings of the 2013 International Congress on Ultrasonics (ICU 2013), pp. 87–92. Singapore (2013)
108. Ciampa, F., Meo, M.: Acoustic emission source localization and velocity determination of the fundamental mode A0 using wavelet analysis and a Newton-based optimization technique. Smart Mater. Struct. **19**, 045027 (2010)
109. Ozevin, D., Heidary, Z.: Acoustic emission source orientation based on time scale. J. Acoust. Emiss. **29**, 123–132 (2011)
110. Gorman, M.R., Prosser, W.H.: AE source orientation by plate wave analysis. J. Acoust. Emiss. **9**, 283–288 (1991)
111. Gorman, M.R.: Plate wave acoustic emission. J. Acoust. Soc. Am. **90**, 358 (1991)
112. Gorman, M.R., Ziola, S.M.: Plate waves produced by transverse matrix cracking. Ultrasonics **29**, 245–251 (1991)
113. Prosser, W.H.: The Propagation Characteristics of the Plate Modes of Acoustic Emission Waves in Thin Aluminum Plates and Thin Graphite/Epoxy Composite Plates and Tubes (1991)
114. Morscher, G.N.: Modal acoustic emission of damage accumulation in a woven SiC/SiC composite. Vacuum **59**, 687–697 (1999)
115. Surgeon, M., Wevers, M.: Modal analysis of acoustic emission signals from CFRP laminates. NDT E Int. **32**, 311–322 (1999)
116. Prosser, W.H., Jackson, K.E., Kellas, S., Smith, B.T., McKeon, J., Friedman, A.: Evaluation of damage in metal matrix composites by means of acoustic emission monitoring. NDT E Int. **30**, 108 (1997)
117. Gorman, M.R.: Modal AE analysis of fracture and failure in composite materials, and the quality and life of high pressure composite pressure cylinders. J. Acoust. Emiss. **29**, 1–28 (2011)
118. Anastassopoulos, A.A., Philippidis, T.P.: Clustering methodology for the evaluation of acoustic emission from composites. J. Acoust. Emiss. **13**, 11–21 (1995)
119. Philippidis, T., Nikolaidis, V., Anastassopoulos, A.: Damage characterisation of C/C laminates using neural network techniques on AE signals. NDT E Int. **31**, 329–340 (1998)
120. Richardson, J.M., Elsley, R.K., Graham, L.J.: Nonadaptive, semi-adaptive and adaptive approaches to signal processing problems in nondestructive evaluation. Pattern Recognit. Lett. **2**, 387–394 (1984)
121. Vi-Tong, E., Gaillard, P.: An algorithm for non-supervised sequential classification of signals. Pattern Recognit. Lett. **5**, 307–313 (1987)
122. Ramirez-Jimenez, C.R., Papadakis, N., Reynolds, N., Gan, T.H., Purnell, P., Pharaoh, M.: Identification of failure modes in glass/polypropylene composites by means of the primary frequency content of the acoustic emission event. Compos. Sci. Technol. **64**, 1819–1827 (2004)
123. Marec, A., Thomas, J.-H., Guerjouma, R.: Damage characterization of polymer-based composite materials: multivariable analysis and wavelet transform for clustering acoustic emission data. Mech. Syst. Signal Process. **22**, 1441–1464 (2008)
124. Sause, M.G.R., Haider, F., Horn, S.: Quantification of metallic coating failure on carbon fiber reinforced plastics using acoustic emission. Surf. Coat. Technol. **204**, 300–308 (2009)
125. Sause, M.G.R., Gribov, A., Unwin, A.R., Horn, S.: Pattern recognition approach to identify natural clusters of acoustic emission signals. Pattern Recognit. Lett. **33**, 17–23 (2012)
126. Doan, D.D., Ramasso, E., Placet, V., Boubakar, L., Zerhouni, N.: Application of an unsupervised pattern recognition approach for AE data originating from fatigue tests on CFRP. In: 31st Conference of the European Working Group on Acoustic Emission, pp. 1–8. Dresden, Germany (2014)
127. Anastassopoulos, A.A., Nikolaidis, V.N., Philippidis, T.P.: A comparative study of pattern recognition algorithms for classification of ultrasonic signals. Neural Comput. Appl. **8**, 53–66 (1999)

128. Yu, P., Anastassopoulos, V., Venetsanopoulos, A.N.: Pattern recognition based on morphological shape analysis and neural networks. Math. Comput. Simul. **40**, 577–595 (1996)
129. Baensch, F., Sause, M.G.R., Brunner, A.J., Niemz, P.: Damage evolution in wood—pattern recognition based on acoustic emission (AE) frequency spectra. Holzforschung **69**, 1–9 (2015)
130. Kostopoulos, V., Loutas, T., Kontsos, A., Sotiriadis, G., Pappas, Y.: On the identification of the failure mechanisms in oxide/oxide composites using acoustic emission. NDT E Int. **36**, 571–580 (2003)
131. Bohse, J., Chen, J.: Acoustic emission examination of mode I, mode II and mixed-mode I/II interlaminar fracture of unidirectional fiber-reinforced polymers. J. Acoust. Emiss. **19**, 1–10 (2001)
132. Haselbach, W., Lauke, B.: Acoustic emission of debonding between fibre and matrix to evaluate local adhesion. Compos. Sci. Technol. **63**, 2155–2162 (2003)
133. Li, L., Lomov, S.V., Yan, X., Carvelli, V.: Cluster analysis of acoustic emission signals for 2D and 3D woven glass/epoxy composites. Compos. Struct. **116**, 286–299 (2014)
134. Li, L., Lomov, S.V., Yan, X.: Correlation of acoustic emission with optically observed damage in a glass/epoxy woven laminate under tensile loading. Compos. Struct. **123**, 45–53 (2015)
135. Bishop, C.M.: Neural Networks for Pattern Recognition. Clarendon, Oxford (1995)
136. Polikar, R.: Pattern recognition. In: Akay, M. (ed.) Wiley Encyclopedia of Biomedical Engineering, pp. 1–22. Wiley, Hoboken (2006)
137. Sause, M.G.R., Schultheiß, D., Horn, S.: Acoustic emission investigation of coating fracture and delamination in hybrid carbon fiber reinforced plastic structures. J. Acoust. Emiss. **26**, 1–13 (2008)
138. Sause, M.G.R., Müller, T., Horoschenkoff, A., Horn, S.: Quantification of failure mechanisms in mode-I loading of fiber reinforced plastics utilizing acoustic emission analysis. Compos. Sci. Technol. **72**, 167–174 (2012)
139. Ritschel, F., Sause, M.G.R., Brunner, A.J., Niemz, P.: Acoustic emission (AE) signal classification from tensile tests on plywood and layered wood. In: 31st Conference of the European Working Group on Acoustic Emission, pp. 1–7. Dresden, Germany (2014)
140. Vergeynst, L.L., Sause, M.G.R., Steppe, K.: Acoustic emission signal detection in drought-stressed trees: beyond counting hits. In: 31st Conference of the European Working Group on Acoustic Emission, pp. 1–8. Dresden, Germany (2014)
141. Njuhovic, E., Bräu, M., Wolff-Fabris, F., Starzynski, K., Altstädt, V.: Identification of interface failure mechanisms of metallized glass fibre reinforced composites using acoustic emission analysis. Compos. Part B Eng. **66**, 443–452 (2014)
142. Kempf, M., Skrabala, O., Altstädt, V.: Acoustic emission analysis for characterisation of damage mechanisms in fibre reinforced thermosetting polyurethane and epoxy. Compos. Part B Eng. **56**, 477–483 (2014)
143. Priess, T., Sause, M.G., Fischer, D., Middendorf, P.: Detection of delamination onset in laser-cut carbon fiber transverse crack tension specimens using acoustic emission. J. Compos. Mater. **49**, 2639–2647 (2015)
144. Tou, J.T.: DYNOC—a dynamic optimal cluster-seeking technique. Int. J. Comput. Inf. Sci. **8**, 541–547 (1979)
145. Davies, D.L., Bouldin, D.W.: A cluster separation measure. IEEE Trans. Pattern Anal. Mach. Intell. **1**, 224–227 (1979)
146. Rousseeuw, P.J.: Silhouettes: a graphical aid to the interpretation and validation of cluster analysis. J. Comput. Appl. Math. **20**, 53–65 (1987)
147. Hubert, L.J., Arabie, P.: Comparing partitions. J. Classif. **2**, 193–218 (1985)
148. Günter, S., Bunke, H.: Validation indices for graph clustering. Pattern Recognit. Lett. **24**, 1107–1113 (2003)
149. Placet, V., Ramasso, E., Boubakar, L., Zerhouni, N.: Online segmentation of acoustic emission data streams for detection of damages in composites structures in unconstrained

environments. In: 11th International Conference on Structural Safety & Reliability, pp. 1–8. New York, USA (2013)
150. Serir, L., Ramasso, E., Zerhouni, N.: Evidential evolving Gustafson–Kessel algorithm for online data streams partitioning using belief function theory. Int. J. Approx. Reason. **53**, 747–768 (2012)
151. MacQueen, J.B.: Some methods for classification and analysis of multivariate observations. In: Proceedings of 5th Berkeley Symposium on Mathematical Statistics and Probability, pp. 281–297 (1967)
152. Baensch, F., Zauner, M., Sanabria, S.J., Sause, M.G.R., Pinzer, B.R., Brunner, A.J., Stampanoni, M., Niemz, P.: Damage evolution in wood: synchrotron radiation micro-computed tomography (SRμCT) as a complementary tool for interpreting acoustic emission (AE) behavior. Holzforschung **69** (2015)
153. Eaton, M.J., Holford, K.M., Featherston, C.A., Pullin, R.: Damage in carbon fibre composites: the discrimination of acoustic emission signals using frequency. J. Acoust. Emiss. **25**, 140–148 (2007)
154. Sause, M.G.R., Horn, S.: Quantification of the uncertainty of pattern recognition approaches applied to acoustic emission signals. J. Nondestruct. Eval. **32**, 242–255 (2013)
155. Milligan, G.W.: An algorithm for generating artificial test clusters. Psychometrika **50**, 123–127 (1985)
156. Qiu, W., Joe, H.: Generation of random clusters with specified degree of separation. J. Classif. **23**, 315–334 (2006)
157. Rand, W.M.: Objective criteria for the evaluation of clustering methods. J. Am. Stat. Assoc. **66**, 846–850 (1971)
158. Kalafat, S., Sause, M.G.R.: Localization of acoustic emission sources in fiber composites using artificial neural networks. In: 31st Conference of the European Working Group on Acoustic Emission, pp. 1–8. Dresden, Germany (2014)
159. Sause, M.G.R., Kalafat, S., Zelenyak, A., Hoeck, B., Horn, S.: Acoustic emission source localization in bearing tests of fiber reinforced polymers by neural networks. In: 16th International Conference on Experimental Mechanics, pp. 1–3. Cambridge (2014)
160. Plöckl, M., Sause, M.G.R., Scharringhausen, J., Horn, S.: Failure analysis of NOL-ring specimens by acoustic emission. In: 30th European Conference on Acoustic Emission, pp. 1–12. Granada, Spain (2012)
161. Kurz, J.H.: Verifikation von Bruchprozessen bei gleichzeitiger Automatisierung der Schallenmissionsanalyse an Stahl- und Stahlfaserbeton. University of Stuttgart, Stuttgart (2006)
162. Akaike, H.: Markovian representation of stochastic process and its application to the analysis of autoregressive moving average processes. Ann. Inst. Stat. Math. **26**, 363–387 (1974)
163. Kühnicke, H., Schulze, E., Voigt, D.: Verbesserte Lokalisation mittels Signalformanalyse. In: DGZfP-BB, pp. 44–51 (2007)
164. Hamstad, M.A.: Comparison of wavelet transform and Choi-Williams distribution to determine group velocities for different acoustic emission sensors. J. Acoust. Emiss. **26**, 40–59 (2008)
165. Hamstad, M.A., O'Gallagher, A.: Effects of noise on Lamb-mode acoustic-emission arrival times determined by wavelet transform. J. Acoust. Emiss. **23**, 1–24 (2005)
166. Bancroft, S.: An algebraic solution of the GPS equations. IEEE Trans. Aerosp. Electron. Syst. **21**, 56–59 (1985)
167. Pullin, R., Baxter, M., Eaton, M.M.J., Holford, K.M., Evans, S.L.: Novel acoustic emission source location. J. Acoust. Emiss. **25**, 215–223 (2007)
168. Eaton, M.J., Pullin, R., Holford, K.M., Featherston, C.A.: AE wave propagation and novel source location in composite plates. In: 28th European Conference on Acoustic Emission Testing, Berlin, Germany (2008)
169. Baxter, M.G., Pullin, R., Holford, K.M., Evans, S.L.: Delta T source location for acoustic emission. Mech. Syst. Signal Process. **21**, 1512–1520 (2007)

170. Eaton, M.J., Pullin, R., Holford, K.M.: Acoustic emission source location in composite materials using Delta T Mapping. Compos. Part A Appl. Sci. Manuf. **43**, 856–863 (2012)
171. Chlada, M., Prevorovsky, Z., Blahacek, M.: Neural network AE source location apart from structure size and material. J. Acoust. Emiss. **28**, 99–108 (2010)
172. Blahacek, M., Chlada, M., Prevorovsky, Z.: Acoustic emission source location based on signal features. Adv. Mater. Res. **13–14**, 77–82 (2006)
173. Kalafat, S., Sause, M.G.R.: Lokalisierung von Schallemissionsquellen in Faserverbundwerkstoffen mit künstlichen neuronalen Netzwerken. In: 19. Kolloquium Schallemission, pp. 1–8. Augsburg, Germany (2013)
174. Kalafat, S., Sause, M.G.: Acoustic emission source localization by artificial neural networks. Struct. Heal. Monit. 1–15 (2015).
175. Balageas, D., Fritzen, C.-P., Gemes, A.: Structural Health Monitoring. ISTE, London (2006)
176. Ciang, C.C., Lee, J.-R., Bang, H.-J.: Structural health monitoring for a wind turbine system: a review of damage detection methods. Meas. Sci. Technol. **19**, 122001 (2008)
177. Sause, M.G.R., Horn, S.R.: Influence of specimen geometry on acoustic emission signals in fiber reinforced composites: FEM-simulations and experiments. In: 29th European Conference on Acoustic Emission Testing, pp. 1–8. Vienna, Austria (2010)
178. Sause, M.G.R., Scharringhausen, J., Horn, S.R.: Identification of failure mechanisms in thermoplastic composites by acoustic emission measurements. In: 19th International Conference on Composite Materials, Montreal, Canada (2013)
179. Burks, B., Kumosa, M.: A modal acoustic emission signal classification scheme derived from finite element simulation. Int. J. Damage Mech. **23**, 43–62 (2013)
180. Pahr, D.H., Rammerstorfer, F.G., Rosenkranz, P., Humer, K., Weber, H.W.: A study of short-beam-shear and double-lap-shear specimens of glass fabric/epoxy composites. Compos. Part B Eng. **33**, 125–132 (2002)
181. Sause, M.G.R., Plöckl, M., Horn, S.R., Forberich, B., Scharringhausen, J.: Bestimmung der GIc und GIIc Kennwerte von thermoplastischen Faserverbundwerkstoffen mittels Schallemissionsanalyse. In: Verbundwerkstoffe und Werkstoffverbunde 2013, Bayreuth, Germany (2013)
182. Spruiell, J.E.: A Review of the Measurement and Development of Crystallinity and Its Relation to Properties in Neat Poly(Phenylene Sulfide) and Its Fiber Reinforced Composites. Oak Ridge, Tennesse, USA (2005)
183. Hahn, H.T., Lagace, P.A., O'Brien, T.K.: Composite Materials: Fatigue and Fracture. ASTM International, West Conshohocken (1991)
184. Wisnom, M.R.: On the increase in fracture energy with thickness in delamination of unidirectional glass fibre-epoxy with cut central plies. J. Reinf. Plast. Compos. **11**, 897–909 (1992)
185. Cui, W., Wisnom, M.R., Jones, M.: An experimental and analytical study of delamination of unidirectional specimens with cut central plies. J. Reinf. Plast. Compos. **13**, 722–739 (1994)
186. Sause, M.G.R., Monden, A.: Comparison of predicted onset of failure mechanisms by nonlinear failure theory and by acoustic emission measurements. In: 16th European Conference on Composite Materials. Sevilla, Spain (2014)
187. Scott, A.E., Mavrogordato, M., Wright, P., Sinclair, I., Spearing, S.M.: In-situ fibre fracture measurement in carbon-epoxy laminates using high resolution computed tomography Compos. Sci. Technol. **71**, 1471–1477 (2011)
188. Zhou, Y., Jiang, D., Xia, Y.: Tensile mechanical behavior of T300 and M40J fiber bundles at different strain rate. J. Mater. Sci. **36**, 919–922 (2001)
189. Durand, L.P.: Composite Materials Research Progress. Nova Science, New York (2008)
190. Lomov, S., Karahan, M., Bogdanovich, A., Verpoest, I.: Monitoring of acoustic emission damage during tensile loading of 3D woven carbon/epoxy composites. Text. Res. J. **84**, 1373–1384 (2014)
191. Nairn, J.A., Hu, S.: The formation and effect of outer-ply microcracks in cross-ply laminates: a variational approach. Eng. Fract. Mech. **41**, 203–221 (1992)

192. Nairn, J., Hu, S.: The initiation and growth of delaminations induced by matrix microcracks in laminated composites. Int. J. Fract. **57**, 1–24 (1992)
193. Matthias Deuschle, H., Kroplin, B.-H.: Finite element implementation of Puck's failure theory for fibre-reinforced composites under three-dimensional stress. J. Compos. Mater. **46**, 2485–2513 (2012)
194. Deuschle, H.M.: 3D Failure Analysis of UD Fibre Reinforced Composites: Puck's Theory Within FEA. PhD-thesis, University of Stuttgart (2010)
195. Puck, A., Mannigel, M.: Physically based stress-strain relations for the inter-fibre-fracture analysis of FRP laminates. Compos. Sci. Technol. **67**, 1955–1964 (2007)
196. Puck, A., Mannigel, M.: Physically based non-linear stress-strain relations for the inter-fibre fracture analysis of FRP laminates. Compos. Sci. Technol. **67**, 1955–1964 (2007)
197. Wisnom, M.R.: Size effects in the testing of fibre-composite materials. Compos. Sci. Technol. **59**, 1937–1957 (1999)
198. Moosburger-Will, J., Sause, M.G.R., Horny, R., Horn, S., Scholler, J., Llopard Prieto, L.: Joining of carbon fiber reinforced polymer laminates by a novel partial cross-linking process. J. Appl. Polym. Sci. **132**, 42159 (2015)
199. Bohse, J., Chen, J., Brunner, A.: Acoustic emission analysis and micro-mechanical interpretation of mode I fracture toughness tests on composite materials. Fract. Polym. Compos. Adhes. **27**, 15–26 (2000)
200. Carlsson, L.A., Gillespie, J.W., Trethewey, B.R.: Mode II interlaminar fracture of graphite/epoxy and graphite/PEEK. J. Reinf. Plast. Compos. **5**, 170–187 (1986)
201. Kaiser, J.: Untersuchungen über das Auftreten von Geräuschen beim Zugversuch. Dissertation, Technische Hochschule München (1950)
202. Fowler, T.J.: Acoustic emission testing of fiber reinforced plastics. In: Preprint 3092. American Society of Civil Engineers, New York (1977)
203. Fowler, T.J.: Acoustic Emission Testing of Fiber Reinforced Plastics. Proc. Pap. J. Tech. Counc. ASCE. **105(TC2)**, 281–289 (1979)
204. Rodriguez, G.: Development and implementation of a testing concept for the bearing of a carbon fiber shaft. Thesis, Technical University Munich (2013)
205. Tonatto, M.L.P., Faria, H., Marques, A.T., Amico, S.C.: Effect of environmental conditioning on burst pressure of carbon/epoxy filament wound composite. In: 16th European Conference on Composite Materials, pp. 22–26. Sevilla, Spain (2014)
206. Hill, E.K., Dion, S.T., Karl, J.O., Spivey, N.S., Ii, J.L.W.: Neural network burst pressure prediction in composite overwrapped pressure vessels. Neural Netw. **25**, 187–193 (2007)
207. Walker, J.L., Workman, G.L., Russell, S.S., Hill, E.V.K.: Neural network/acoustic emission burst pressure prediction for impact damaged composite pressure vessels. Mater. Eval. **55** (1997)
208. Gorman, M.R.: Burst prediction by acoustic emission in filament-wound pressure vessels. J. Acoust. Emiss. **9**, 131–139 (1990)
209. Hamstad, M.A., Patterson, R.G.: Considerations for acoustic emission monitoring of spherical kevlar/epoxy composite pressure vessels. In: ASME Energy Technology Conference on Composites in Pressure Vessels and Piping, Houston, TX, USA (1977)
210. Bunsell, A.R., Thionnet, A.: Life prediction for carbon fibre filament wound composite structures. Philos. Mag. **90**, 4129–4146 (2010)
211. Dong, L., Mistry, J.: Acoustic emission monitoring of composite cylinders. Compos. Struct. **40**, 43–53 (1997)
212. Höck, B., Regnet, M., Bickelmaier, S., Henne, F., Sause, M.G.R., Schmidt, T., Geiss, G.: Innovative and efficient manufacturing technologies for highly advanced composite pressure vessels. In: Proceedings of 13th European Conference on Spacecraft Structures, Materials + Environmental Testing, Braunschweig, Germany (2014)
213. Hamstad, M.A., Sause, M.G.R.: Acoustic emission signals versus propagation direction for hybrid composite layup with large stiffness differences versus direction. In: 31st Conference of the European Working Group on Acoustic Emission, pp. 1–8. Dresden, Germany (2014)

214. Burks, B., Hamstad, M.A: On the anisotropic attenuation behavior of the flexure mode of carbon fiber composites. In: 19th International Conference on Composite Materials, pp. 1–9. Montreal, Canada (2013)
215. Burks, B., Hamstad, M.A.: The impact of solid–fluid interaction on transient stress wave propagation due to Acoustic Emissions in multi-layer plate structures. Compos. Struct. **117**, 411–422 (2014)
216. Hamstad, M.A.: A waveform-based study of AE wave propagation by use of eight wide-band sensors on a composite pressure vessel. In: 30th European Conference on Acoustic Emission, pp. 12–15. Granada, Spain (2012)
217. Henne, F., Ehard, S., Kollmannsberger, A., Hoeck, B., Sause, M., Drechsler, K.: Thermoplastic in-situ fiber placement for future solid rocket motor casing manufacturing. In: SAMPE Europe SETEC 14—Efficient Composite Solutions to Foster Economic Growth, Tampere, Finland (2014)
218. Fowler, T.J.: The origin of CARP and the term "Felicity Effect". In: 31st Conference of the European Working Group on Acoustic Emission, pp. 1–8. Dresden, Germany (2014)

Chapter 5
Electromagnetic Emission

Originating from the field of geophysics science, electromagnetic emission (EME) analysis is a nondestructive measurement technique to monitor crack formation and propagation. Similar to AE analysis, the EME method is capable of providing real-time information on microscopic failure mechanisms on a qualitative basis. In both cases the reliability of quantitative information still hinges on the lack of a detailed understanding of the correlation between the source mechanism and the measured signal. Therefore, EME analysis is currently considered a method under development and is mostly limited to applications in materials research and has not yet been used for structural health monitoring applications.

In this chapter the principle of operation of electromagnetic emission is presented first. Subsequent to that, aspects of the source mechanism, the detection systems, and some applications of the method used as in situ technique are introduced. Due to the novelty of the method, some applications are demonstrated to monitor failure of reference materials such as polymers and carbon fibers to elucidate basic relationships between failure mechanisms and the EME phenomenon. At the end of the chapter, the established approaches are then extended to applications monitoring failure of fiber reinforced composites.

5.1 Principle of Operation

EME signal generation is due to a temporary electrification of newly generated crack surfaces caused by charge separation [1]. Based on physical considerations, spatial movement of these charges is detectable as electromagnetic fluctuation in time, which is referred to as electromagnetic emission. However, there is no generally established theory, which adequately describes the mechanical and electrical dynamics of this charge movement. A comprehensive review on this lack of consistency of current models and their discrepancy with recent experiments is

M.G.R. Sause, *In Situ Monitoring of Fiber-Reinforced Composites*,
Springer Series in Materials Science 242, DOI 10.1007/978-3-319-30954-5_5

given by Frid et al. [2]. Some authors attributed the dynamic to the mechanical vibration of the crack surface [1, 3]. To overcome these discrepancies, new suggestions were made to describe EME in the form of charge surface vibrational waves [2, 4, 5]. However, there are still inconsistencies of the proposed models and experimental results, e.g., regarding the occurrence of EME under shear crack propagation [2, 5–8].

In comparison to AE analysis, the major advantage of EME analysis is the negligible influence of dispersion and attenuation on the EME signals. But EME signals are affected by the conductive properties of the cracking material [2, 4, 9–12]. This determines the time constant of charge equalization. Most investigations of the EME phenomena were carried out in minerals (granite, marble, concrete) [1–5, 10, 11]. In recent years various authors report on EME in reinforced polymer materials, like carbon nanotube reinforced polyether ether ketone [13], glass fiber reinforced epoxy resins [9], and carbon fiber reinforced epoxy resins [14, 15]. Other authors also report on EME in highly conductive materials like metals [16–18], which suggest that, in principle, EME is observable in all types of materials.

Another advantage of EME analysis is the possibility to fabricate sensor systems with ideally flat response within the frequency range of the signals. Among these are capacitive sensor systems and coil sensors operating far away from their resonance frequencies. But currently no commercial sensor systems are available for EME detection, although transient recorder cards and preamplifiers as being used for AE acquisition are well suited for digitization of the detected EME signals.

Due to their different source mechanism and their weak interaction with the nearby medium, the detected EME signals are a valuable alternative to investigate the formation and propagation of cracks in solid materials. Some of the possibilities of EME analysis are schematically visualized in Fig. 5.1. The occurrence of EME

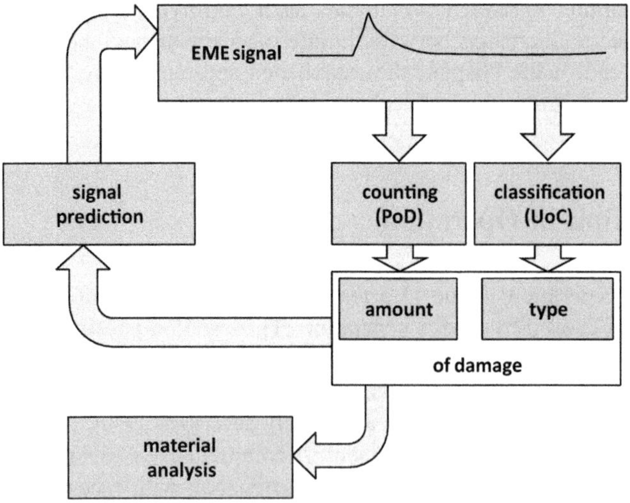

Fig. 5.1 Analysis methods used with EME signals and EME signal prediction as forward approach

signals can be used to determine the time of failure and their accumulation. In addition their activity can be interpreted in terms of the accumulated damage and the activity of failure in the material. As with every detection system, the successful acquisition of an EME signal still is subject to a probability of detection (PoD) governed by the sensitivity of the detection system and several other factors that will be pointed out in Sect. 5.4.

For materials like fiber reinforced composites, the occurrence of different types of failure mechanisms is also associated with different types of EME signals being released. Given a known relationship between the source mechanism and the signatures of the EME signals, the classification of the EME signals allows to differentiate between different source types. This provides unique access to the dynamics of a crack source as will be demonstrated in Sect. 5.2. Similar to AE analysis, the classification procedure is subject to an uncertainty of classification (UoC), which is dominated by the classification strategy and the uniqueness of relation between EME signatures and their underlying source types.

Other than for AE, a source localization procedure based on arrival time differences is hardly practical, since the propagation speed of the EME waves is close to the speed of light. For the usual distances between sensor systems, the arrival time differences are in the order of fractions of nanoseconds, which is still below the typical temporal resolution of ultrasonic transient recorder cards. However, in combination with simultaneous AE acquisition, the time of occurrence of EME signals can precisely determine the time of emission t_0 and therefore supplement the AE localization procedure [3].

Based on a suitable EME source model, it is convenient to perform a forward prediction of the EME signal being detected at certain positions in the vicinity of the source [12]. This signal prediction aspect is vital to establish physically based EME source models and to allow for an improved understanding of the EME source process and the influence of the signal detection chain.

5.2 Source Mechanism

In order to understand the EME signals of material failure, it is of vital importance to understand their source mechanism. This involves some particular challenges regarding the physical origin of the electrical charges and their relation to the formation and propagation of cracks in materials. As a common denominator of most theories, the type of EME source discussed in the following is restricted to the active formation and propagation of a crack in a material. On the atomistic scale, this process of crack formation is directly related to breakage of bonds at the position of the crack tip. This somehow links the phenomenon of fracture and the generation of charges as release of chemical bonds, but is not sufficient to understand the nature of electromagnetic emission sources. Instead there are numerous published theories trying to explain the details of the relationship between the occurrence of fracture and the finally detected EME signal. Since there is no finally accepted theory explaining the EME phenomenon, the aim of this section is to provide an overview on the existing model descriptions and to present some recent advances toward forward prediction of EME signals by means of finite element modeling.

For discussion of effects involving EME, the basis for all further discussion are Maxwell's equations. In combination with Lorentz force law, these yield the foundation of classical electrodynamics. As a convenient way to write Maxwell's equation, the differential form is used in the following:

$$\nabla \cdot \boldsymbol{E} = \frac{\tilde{\rho}}{\xi_0} \tag{5.1}$$

$$\nabla \cdot \boldsymbol{B} = 0 \tag{5.2}$$

$$\nabla \times \boldsymbol{E} = -\frac{\partial \boldsymbol{B}}{\partial t} \tag{5.3}$$

$$\nabla \times \boldsymbol{B} = \mu_0 \left(\boldsymbol{J} + \xi_0 \frac{\partial \boldsymbol{E}}{\partial t} \right) \tag{5.4}$$

Here the conventional abbreviations of the electric field $\boldsymbol{E}$ and magnetic field $\boldsymbol{B}$ are applied, including the permittivity of free space ξ_0, permeability of free space μ_0, the electric charge density (charge per unit volume) $\tilde{\rho}$, and the electric current density (current per unit area) $\boldsymbol{J}$.

To derive the nature of electromagnetic emission sources, it is easy to infer from the first Maxwell's equation (Gauss' law) that electric charges showing a temporal or spatial oscillation in time produce a corresponding oscillating electric field. Introducing homogeneous, isotropic, nondispersive, linear materials, the electric field is directly linked to the displacement-field $\boldsymbol{D}$ by the electric permittivity ξ_0 and the polarization field $\boldsymbol{P}$:

$$\boldsymbol{D} = \xi_0 \boldsymbol{E} + \boldsymbol{P} \tag{5.5}$$

Similarly, the magnetizing field $\boldsymbol{H}$ is defined by the permeability μ_0 and the magnetization field $\boldsymbol{M}$:

$$\boldsymbol{H} = \frac{1}{\mu_0} \boldsymbol{B} - \boldsymbol{M} \tag{5.6}$$

Therefore it is straightforward to conclude that oscillations of electric charge density on the microscale are potentially a source to generate a macroscopically detectable electric field oscillation.

However, there are two key questions which need to be addressed to establish a meaningful description of the full EME phenomenon:

(a) What is the source of the variation of electrical charge density?
(b) What is the link between the electrical charge density variation and the mechanical (fracture) dynamics?

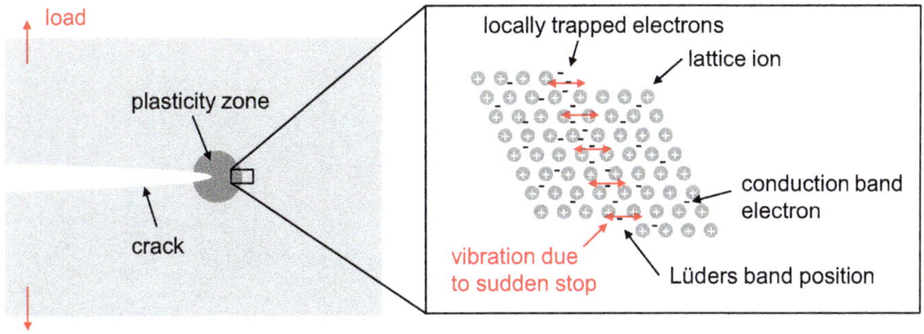

Fig. 5.2 Dislocation model as possible source description of EME

Despite the variety of proposed EME source models, the individual approaches can be roughly categorized as dislocation models, discharge models, capacitor models, and the more recently proposed electrical surface wave model.

Being one of the first proposals to describe the EME phenomenon, Misra et al. suggested in 1978 the dislocation model [19, 20]. Their source description is based on the elastic-plastic transition region ahead of the crack tip. As visualized in Fig. 5.2, after a certain crack increment, the zone of plastic deformation arrests in a mechanically stable configuration. This is associated with a sudden arrest of the conduction electrons being trapped at the position of the dislocation band. This sudden stop of the Lüders band dislocations is assumed to induce a type of electromagnetic radiation similar to the Bremsstrahlung process used to generate electromagnetic waves in the X-ray frequency range. During propagation of the dislocations, Misra et al. further assumed a discrete redistribution of the conduction electrons yielding a Hertzian dipole formed by the conduction electrons moving relative to the positive ions of the crystal lattice [19]. Although parts of this EME source mechanism are plausible, there is some experimental evidence which is in conflict to the proposed description. One contrasting argument was provided by Molotskii et al., who calculated the expected maximum EME frequency based on the propagation velocity of the dislocation bands [21]. For some materials this calculated frequency was found to be one order of magnitude less than the measured EME frequencies. Other experimental evidence is the occurrence of EME in a broad range of solid materials ranging from rather ductile materials like metals and polymers to comparatively brittle types like ceramics, glasses, and minerals. Especially for brittle materials, it is well known that the movement of plasticity bands is negligible, yet the EME has even been found to be stronger in such materials than in ductile materials [8, 22–30]. Also for all materials other than metals, there is no such trap for the conduction electrons, rendering this attempt impossible to describe EME for solids in general [12, 18, 26].

Another early attempt to describe the occurrence of EME was given by Finkel et al. [31]. They demonstrated that a rapid crack in an alkali halide crystal can produce a stochastic arrangement of positive and negative electric charges on both sides of the crack walls causing a strong electrostatic field of the potential order of

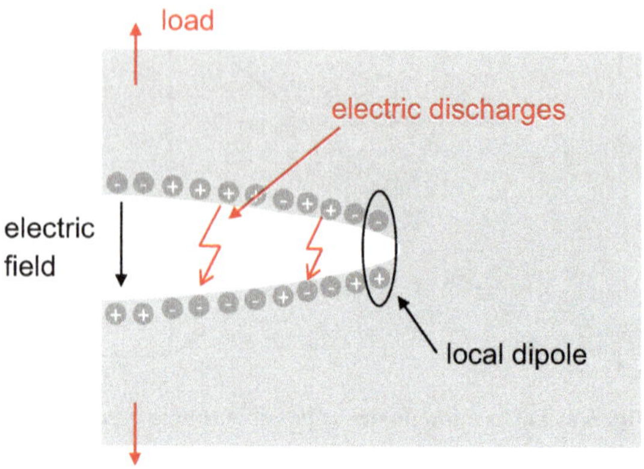

Fig. 5.3 Discharge model proposed as source of EME, generating electrical discharges due to stochastically arrangement of local dipoles

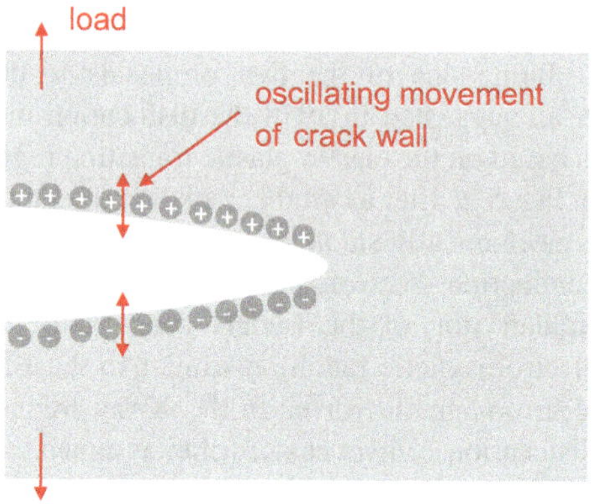

Fig. 5.4 Capacitor model using opposite charges on the crack surfaces to generate EME

10^7 V/cm. As a consequence of such high field strengths and the close distance of the crack faces, Finkel et al. speculated whether this may cause electrical discharges and these may form the origin of EME (cf. Fig. 5.3). However, the EME spectrum of such electric discharges are well known to be of the white noise type [6] rather than the spectra of EME pulses exhibiting some preferred frequencies as observed in most of the materials [8, 12, 14, 23, 24, 26, 27, 32, 33]. Therefore this type of source mechanism is not able to sufficiently describe the experimentally observed EME signals.

A more recently proposed type of EME source is the capacitor model, which goes back to the considerations of Finkel et al. and early experiments of Miroshnichenko et al. [6, 31]. The actual EME source type was proposed by O'Keefe and Thiel assuming the presence of electrical charges on the crack walls moving apart due to a crack opening process [7]. This is schematically shown in Fig. 5.4. As has been demonstrated by Miroshnichenko et al., the movement of such charged capacitance plates is indeed linked to the generation of an electric field [6]. Similarly, following the consideration of Finkel et al. [31], one could assume that the crack process itself is able to generate a reasonable amount of charges on the crack walls. As a first model, the capacitor model is able to capture the temporal decay of an EME signal due to the separation of crack walls being drawn apart. This increases the capacitance causing an associated voltage drop being measured as loss in electric field strength. However, there are several arguments which are not adequately addressed by just the capacitor model. One of the key arguments against this type of EME model is the symmetry-break argument. In order to operate in the proposed way, there needs to be a charge imbalance during formation of the crack. Consequently, a break of electrical symmetry needs to occur at some point. Several ideas have been proposed in order to justify the occurrence of the symmetry break. Some of these arguments are a possible pre-polarization of the material, applied physical gradients due to piezoelectricity or pseudo-piezoelectricity, temperature gradients, deformation gradients, or an impurity concentration gradient [7]. Other authors claimed that the electrification of the crack walls could originate from inhomogeneous elastic strains in the vicinity of the crack [34] or that crack side electrification stems from piezoelectrification and contact (or separating) electrification [35]. Even for the assumption that the two crack walls would be fully charged in a stochastically pattern as discussed by Finkel et al. [31] and thus retaining an overall charge neutrality, the induced EME of the local dipoles would be very weak due to the cancelation of the macroscopic field. Also for such a scenario, frequent changes of the electric field polarity would occur. Besides the symmetry-break argument, Frid et al. pointed out the lack of applicability of the capacitor model to the description of EME signals for shear cracks [2], since a relative movement of the charges in parallel to the crack extension would not sufficiently generate EME signals of similar strength as for the movement in parallel to the applied tensile load. This was found to be in contrast to the experimental findings in [27, 28]. Also, the absence of EME signals under simultaneous occurrence of AE signals and vice versa was raised as an argument against the capacitor model [1, 22, 27].

Based on the lack of consistency of the previous EME source descriptions, Frid et al. proposed an alternative EME source model as seen in Fig. 5.5 [2]. Given the typical shape of an experimental EME pulse, they concluded the overall signal to originate from three contributions:

1. Generation of an electrical charge source with strength proportional to area of crack growth
2. Charge oscillations on both sides of the crack due to electrical surface waves
3. Decay of electrical surface waves as function of time

For the first contribution, there have been several indications being reported in literature, which indicate the proportionality between crack length and the generated EME signal strength [12, 23, 36]. Also, the crack arrest has been clearly correlated to the occurrence of amplitude decrease [23, 36]. Therefore it is concluded that progression of the crack tip (and likewise generation of new crack surface) is causing a generation of new charges at the crack tip since this also involves a progressive fracture of bonds.

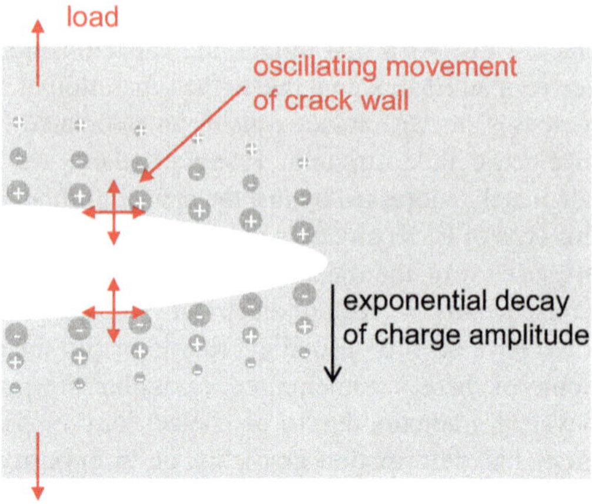

Fig. 5.5 Electrical surface wave model as source description of EME

For the second contribution, Frid et al. propose the existence of an electrical surface wave [2]. During the rupture of atomic bonds at the crack tip, the atoms on both sides of the bonds are moved to nonequilibrium positions relative to their steady state. If each atom vibrates individually, the situation would resemble the Einstein model of lattice vibrations, and the frequency of these oscillations would approach 10^{15} Hz. However, in a solid the movement of one atom is coupled to the surrounding atoms resulting in a Debye bulk model rather than an Einstein model causing slightly lower effective frequencies, but still in similar orders of magnitude [37]. During rupture of one atomic bond, the positive charges on these surfaces move away together in one direction from the equilibrium position (one crack wall), while the negative charges move in unison in the other direction from the equilibrium position (the same crack wall). Therefore this retains overall charge neutrality, while there is an initiation of an electrical charge wave propagating along the crack wall. Their speed and mode of propagation is assumed to be similar to other types of surface waves like Rayleigh waves [38] or surface vibrational optical waves [39]; thus, their amplitude decays exponentially into the bulk material.

The model also readily explains the decay of amplitude as typically observed in EME signals. Given the presence of charges at the newly formed fracture surface, there is relaxation of charges over time dominated by the dispersive propagation of the electrical charge wave and the flow of charges from the position of the crack tip into the bulk material, which is governed by the electrical conductivity of the material.

In the publication by Frid et al., plenty of experimental evidence has been given that the electrical surface wave model is a valid description of the EME source [2]. However, the origin of the oscillating crack wall movement is subject to further discussion. In recent experiments it was demonstrated that the presence of the oscillating part in the detected EME signal is subject to the experimental setup, namely, the bandwidth of the recording setup [12]. Also, recent advances in AE source modeling point out the existence of further contributions to the mechanical movements occurring during crack propagation which are likely to act as source for the oscillating part observed in EME signals.

Therefore getting back to the question "What is the link between the electrical charge density variation and the mechanical (fracture) dynamics?" one has to further consider the dynamics involved in the generation and propagation of cracks.

For the case of a mode I crack, Freund et al. have comprehensively reviewed the crack propagation using a linearly elastic isotropic continuum [40–44]. The relationship between the instantaneous dynamic energy release rate per unit area of crack plane G_d (J/m^{-2}) and the instantaneous dynamic stress intensity K_{Id} (Pa m$^{1/2}$) of the moving crack in (5.7) is given by [44] as

$$G_d = \frac{(1-\nu)^2}{E}(K_{\mathrm{Id}})^2 = \frac{(1-\nu)^2}{E} g_c (K_{\mathrm{Ic}})^2 \tag{5.7}$$

where K_{Ic} (Pa m$^{1/2}$) is the equivalent static stress intensity factor of a stationary crack having the same instantaneous length as the dynamic crack. The function g_c is the complex universal function of crack speed which may be approximated by the simple relation in (5.8) as defined by [40, 41, 45]:

$$g_c \approx 1 - \frac{c_{\mathrm{cr}}}{c_{\mathrm{R}}} = \frac{c_{\mathrm{R}} - c_{\mathrm{cr}}}{c_{\mathrm{R}}} \tag{5.8}$$

where c_{cr} is the crack tip speed and c_{R} is the Rayleigh wave velocity. This classical linear elastic fracture mechanics result points out that the ultimate limit of crack propagation speed is the Rayleigh wave velocity. However, in reality most materials do not tend to approach the Rayleigh wave velocity. Some reasons for this behavior were reported by Tromans [46]. Based on ab initio calculations of the crystal lattice structure and the interatomic potentials in polycrystalline matter, he was able to demonstrate the decrease in critical crack tip speed c_{cr} due to the effects of tip bifurcation, branching, and the coalescence of cracks.

Naturally, at the moment of initiation, the propagation speed starts at zero and may approach the speed limit for crack propagation after a certain acceleration phase. Similarly at the end of crack propagation, another change of crack propagation speed may be expected. This is either due to a lack of strain energy concentration at the crack tip, causing a crack arrest or due to the changes in the boundary conditions as induced by the presence of a free edge. Thus, the temporal stages of crack growth may be subdivided in three phases as schematically shown in Fig. 5.6a. First the crack starts to accelerate (phase 1) and may reach the critical propagation velocity (phase 2) if the boundary conditions allow such. Subsequently the crack decelerates (phase 3) to adapt to the changing boundary conditions (crack arrest or approaching free surface). Clearly, the relative share of the three phases may look characteristically different, possibly hiding the contributions of phase 1 and phase 3 and, hence, leaving an almost straight line. In the context of acoustic emission source functions, such different possibilities of crack activity have been also described by C. Scruby in [47] for the case of a linear elastic material.

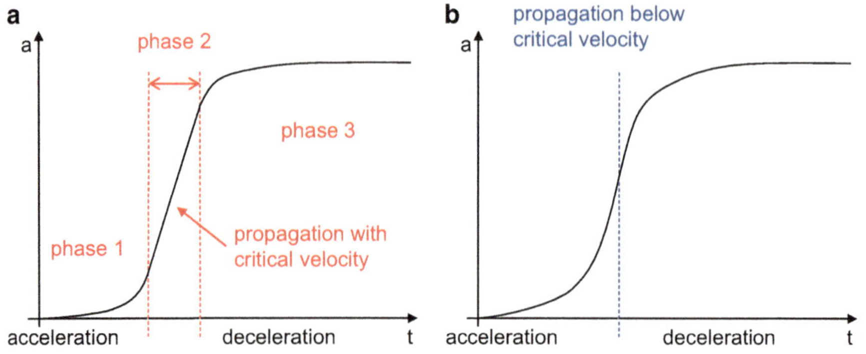

Fig. 5.6 Schematically representations of temporal stages of crack growth for case reaching critical propagation velocity (**a**) and case staying below critical propagation velocity (**b**)

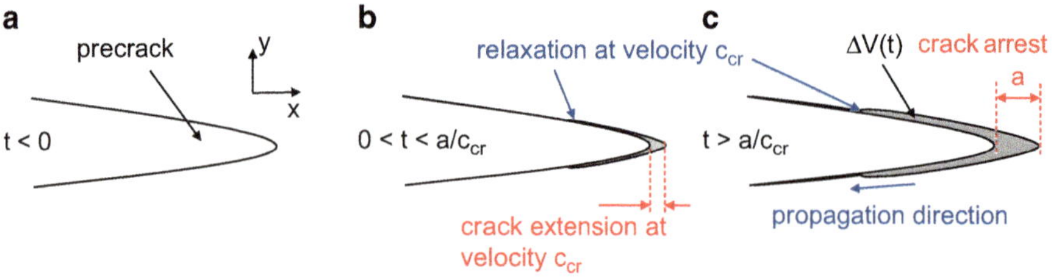

Fig. 5.7 Visualization of crack growth in a simplified 2D model for a linear elastic material (based on [48])

In the same context, Wadley and Scruby proposed a simplified 2D model to explain the dynamics of crack growth in a linear elastic material for a mode I load situation [48]. As shown in Fig. 5.7, the parabolic precrack is subject to a tensile load in the y-direction causing a crack extension a along the x-axis within a certain duration a/c_{cr} and within a certain velocity c_{cr}. Accordingly, the crack front passes a certain volume ΔV driving the excitation of an acoustic emission signal as discussed in Sect. 4.2. However, this simplified model already predicts the existence of a backward motion propagating along the crack walls exhibiting Rayleigh speed velocity c_{R}. This additional "vibration" of the crack surface is another important contribution to the overall crack dynamics, which likely causes another spatial movement of the charges being generated by the progression of the crack tip.

While the simplified 2D model in Fig. 5.7 and the corresponding 1D representation of the crack tip position in Fig. 5.6 are valuable to understand the individual contributions to the generation of the dynamics of the crack movement, it is still challenging to transfer these observations to a real experimental situation. As evident from fracture surface analysis, the occurrence of fast crack growth in brittle materials typically does not come with a plane crack front. Except for some specifically designed single crystal structures [49–52], the usual engineering

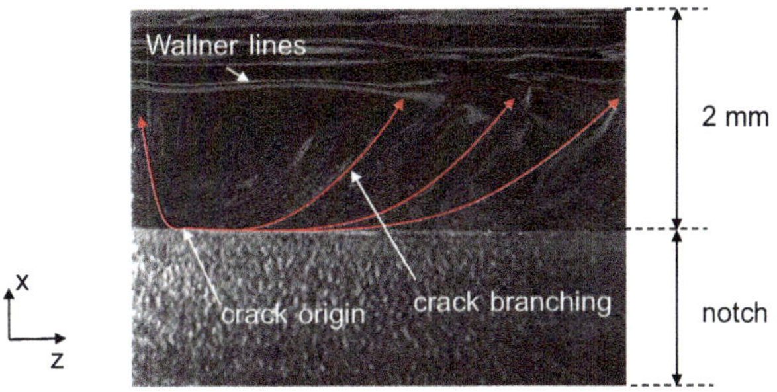

Fig. 5.8 Microscopy image of a crack surface of a RTM6 resin. The *red arrows* indicate the direction of crack propagation starting at a distinct point on the notch of the specimen

materials exhibit a distinct microstructure, grain boundaries, include defects or impurities, and simply do not have a perfect geometry. As a consequence, the failure of these materials is usually initiated at flaws, and the direction of crack growth is governed by the nonsymmetric and inhomogeneous stress field caused by the imperfect material and specimen geometry. As seen from the exemplary microscopy image of a crack surface of a RTM6 resin in Fig. 5.8, the crack initiates at a distinct point at the edge of the specimen. Subsequently, the crack grows in two different stages: a rough area with almost radial marks extruding from the position of the crack origin and a smooth area with rib marks at the upper half of the crack surface. The radial marks are parallel to the direction of crack propagation, and the sliver-like patterns indicate regions where crack branching takes place and therefore indicates an accelerating, unstable crack. The upper half of the fracture surface is mirror smooth with straight, horizontal rib marks (Wallner lines). This indicates smaller crack propagation velocities and a propagation direction perpendicular to these lines, which is induced by the presence of tensile and compressive stress states in the lower and upper part of the bending specimen cross-section shown in Fig. 5.8. Hence, the overall assumptions of acceleration and deceleration can be related to morphological investigations of the fracture surface, but the position of the crack is hardly be captured by a 1D description along the x-axis. Instead, the 3D orientation of real crack surfaces is likely to cause further differences to the simplified model geometries discussed so far.

5.2.1 Modeling of Electromagnetic Emission Sources

In order to overcome some of the limitations of established analytical descriptions for the investigation of arbitrary 3D geometries, the development of a source model based on finite element modeling is a suitable approach. Recently, Gade et al. proposed a phenomenological EME source model using a finite element modeling approach [12]. The general idea of this modeling strategy is to generate charges at the position of the crack tip, to allow for charge relaxation by incorporating the electrical conductivity of the material, and to couple the electromagnetic calculation with structural mechanics to perform modeling of crack growth following the implementation of Sect. 4.2.

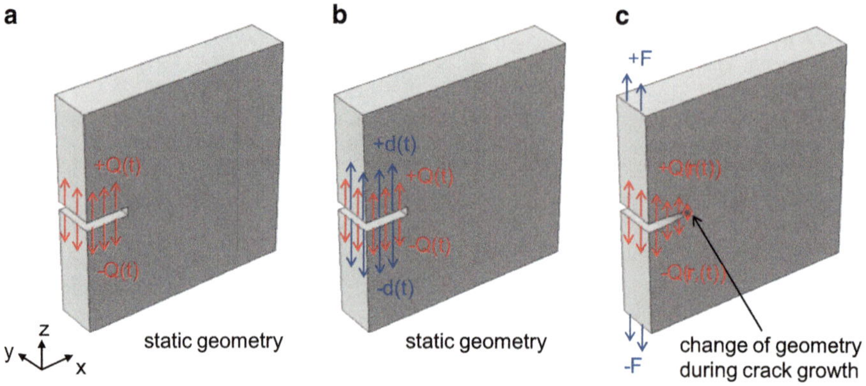

Fig. 5.9 Different types of electromagnetic emission source model descriptions employing charging of fracture surface as function of time (**a**), charging of fracture surface in combination with mechanical vibration (**b**), and a full source model description coupling the dynamic changes of the source geometry based on fracture mechanics with the generation of charges in space and time (**c**)

In analogy to the approach to categorize acoustic emission source models in Sect. 4.2, the different modeling strategies to describe electromagnetic emission sources of crack propagation may be distinguished as shown in Fig. 5.9.

The first type of electromagnetic emission source models considers a static crack geometry and explicitly defines the position and intensity of charges present on the crack surface (cf. Fig. 5.9a). This type of model may be understood as a direct implementation of analytical source functions, since it requires defining all dynamics of the charges as function of time $Q(t)$.

The second type of electromagnetic emission source model is the type of model of Fig. 5.9b, which now includes a stationary crack geometry with superimposed crack wall vibration prescribed by an analytical function $\boldsymbol{d}(t)$. The generation of charges $Q(t)$ is based on the progression of the crack tip and needs to be derived from a known crack growth function. The latter could be inversely derived from experimental EME measurements or could be computed by an AE source model such as presented in Sect. 4.2.

Finally, the third type of electromagnetic emission source model is a model capable to calculate the crack dynamics based on experimental loading conditions and fracture mechanics laws (cf. Fig. 5.9c). At the same time charges are generated on the fracture surface according to the position of the crack tip, i.e., $Q(\boldsymbol{r}, t)$.

5.2.1.1 Implementation and Validation

In order to implement a source model of the first type (Fig. 5.9a), fracture of a single-edge-notched beam (SENB) specimen was modeled using the “AC/DC module” and the “structural mechanics module” of the software COMSOL

Multiphysics. To reduce the computational intensity of the model, several simplification steps may be introduced. A transient computation of the electric field components using the reduced version of the Maxwell's equations (5.1) to (5.6) above was chosen, since the magnetic field components are not detected by the sensor systems and are thus not relevant to the interpretation of the experimental findings.

Further considerations to simplify the modeling approach can be made regarding the nature of charge relaxation.

As experimentally observed and described by [36], the EME signals are not affected by charge relaxation as long as the underlying dynamics are much faster than the relaxation process. The latter involves Ohm's law using the material's electric conductivity $\tilde{\sigma}$

$$\boldsymbol{J} = \tilde{\sigma}\boldsymbol{E} \tag{5.9}$$

the equation of continuity

$$\frac{\partial \tilde{\rho}}{\partial t} + \nabla \cdot \boldsymbol{J} = 0 \tag{5.10}$$

and Gauss' law

$$\nabla \cdot \boldsymbol{D} = \tilde{\rho} \tag{5.11}$$

The combination of (5.9), (5.10), and (5.11) results in the differential equation for electric charge density in a homogeneous medium:

$$\frac{\partial \tilde{\rho}}{\partial t} + \frac{\tilde{\sigma}}{\xi_r}\tilde{\rho} = 0 \tag{5.12}$$

This equation has a solution of the form

$$\tilde{\rho}(t) = \tilde{\rho}_0 \mathrm{e}^{-t/\tilde{\tau}} \tag{5.13}$$

introducing the charge relaxation time $\tilde{\tau} = \xi_r/\tilde{\sigma}$. For highly conductive materials, $\tilde{\tau}$ may approach 10^{-19} s, whereas poor conductors reach $\tilde{\tau}$-values in the order of 10^2 s and approach infinity for ideal insulators. For modeling of charge relaxation, it is thus important to consider the observation scale of the computation relative to the expected $\tilde{\tau}$-values.

In [12] an implementation assuming $\tilde{\tau} \cong t$ is presented. In this case, the computation requires to solve (5.10), which may be written in terms of the material properties ξ_r and $\tilde{\sigma}$ as

$$-\nabla \cdot \frac{\partial}{\partial t}(\xi_r \nabla \mathrm{U} + \mathbf{P}) - \nabla \cdot (\tilde{\sigma}\nabla \mathrm{U} - \mathbf{J}) = 0 \tag{5.14}$$

As long as the induced electric fields are free of curls, this is sufficient for a near-field approximation, i.e., the expected wavelengths need to be much larger than the geometrical dimensions considered. As will be discussed in Sect. 5.3, the wavelengths of the electric fields in the frequency range of interest are indeed much larger than the typical dimensions of the experimental setup. Hence, the computation is carried out for the reactive near field around the source and this assumption is applicable. Moreover, the skin effect and wave propagation effects need to be negligible as well. This approach was found to be suitable to describe the dynamics intrinsic to the EME source as presented in [12].

However, for materials such as polymers, carbon fibers, or carbon fiber reinforced polymers, it may readily be estimated from the conductivity values and the materials permittivity given in Table B.3 of Appendix B that $\tilde{\tau} \gg t$ (cf. Sect. 5.2.1.4). For the model implementation, this observation allows to reduce the computational intensity further by switching to an electrostatic formulation of the problem, yielding a reduced version of (5.14):

$$-\nabla \cdot (\xi_r \nabla \mathrm{U} - \mathbf{P}) = \tilde{\rho} \tag{5.15}$$

In this context, the charge distribution can be considered as static input to (5.15) resulting in a corresponding electric potential U. However, this does not necessarily imply that the electric charge density is constant. Instead, the use of Gauss' law in this reduced variant does allow definition of an electric charge density as function of time $\tilde{\rho}(t)$ as used for the EME source model outlined in Fig. 5.9 and reduces the computational intensity of the coupled approaches.

Figure 5.10 shows the model geometry, consisting of the SENB specimen made from RTM6 epoxy resin, the platelike EME sensors, and the surrounding air volume. To avoid interaction with nearby materials and to eliminate influences of the size of surrounding conductive media, a spherical infinite element domain of 5 mm thickness was used to stretch the electric field to infinity. Such an infinite element domain applies a rational coordinate scaling to a layer of domains

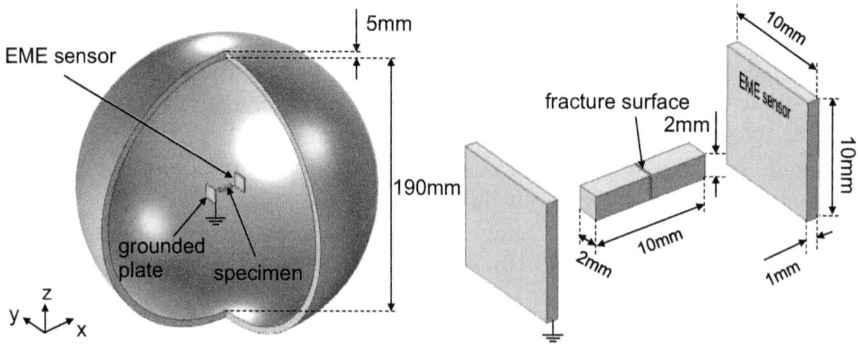

Fig. 5.10 Overall dimensions of the model configuration (*left*) including details of the modeled specimen and EME detector system (*right*)

surrounding the region of interest. When the dependent variables vary slowly with radial distance from the center of the region of interest, the finite elements coordinate can be stretched in the radial direction such that boundary conditions on the outside of the infinite element layer effectively operate at a very large distance from the region of interest. The coordinate scaling along the radial direction was implemented as nearly singular $1/r$ stretching.

For the specimen and the EME sensors, the corresponding material properties of RTM6 epoxy resin and copper as reported in Tables B.1 and B.3 in Appendix B were used. Capacitance plate sensors of 10 mm $\times$ 10 mm width and height and 12 mm distance to the fracture plane were modeled and implemented as will be described in detail in Sect. 5.4. As has been previously shown in [12], the explicit modeling of the EME sensors and incorporation of the system transfer function allow to compute EME signals of very close similarity to the experimentally measured signals. For the present configuration, the attached circuitry was modeled as combination of a 10 MΩ resistor and a 70 pF capacitor to account for the first stage of the preamplifier circuit used in the experiments (cf. Sect. 5.4). This is sufficient to emulate the high-pass behavior of the EME sensor system and enables a direct comparison of the modeled signals with experimental signals.

For an EME source model of the first type, as source function for the EME, a time-dependent surface charge $Q(t)$ is used. Within the model this was implemented as surface charge density $\tilde{\rho}(t)$ located at the hypothetical fracture surface position (cf. Fig. 5.10). Based on the EME source operation hypothesis formulated above, it is assumed that one contribution of $Q(t)$ is due to charge separation following the path of the crack tip causing an increasing accumulation of charges at the fracture surface during crack growth as seen in Fig. 5.11a. The second contribution is attributed to the movement of charges owing to mechanical movement (such as vibration of the crack surface) and is modeled as rise and decay of a sinusoidal oscillation with a frequency of 80 kHz (cf. Fig. 5.11b). The superposition of both contributions defines the assumed analytical change of charges $Q(t)$ existing on the fracture surface as function of time.

Electrically insulated boundaries are simulated by the zero-charge condition $\boldsymbol{n} \cdot \boldsymbol{D} = 0$, while at grounded boundaries the electric potential is set equal to zero. As initial conditions for the model, a zero potential is chosen everywhere.

To achieve convergent numerical solutions, quadratic Lagrange elements were used for the spatial discretization, and a maximum mesh element scale of 0.5 mm was selected for the RTM6 specimen. For the two small copper plates of the EME sensors, a resolution with maximum mesh element size of 1 mm proved sufficient. The air domain was meshed with a maximum element size of 11 mm and with a maximum element growth rate of 1.4 per element. This ensures an adequate resolution in the area between specimen and EME sensor and reduces the computationally required degrees of freedom. The transient calculation is done via a generalized-α algorithm with a time step size of 0.5 µs.

The computation results are shown in Fig. 5.12a as voltage signal detected at the EME sensor compared to the surface charge density on the fracture surface. The major differences between the two curves are due to the high-pass behavior of the

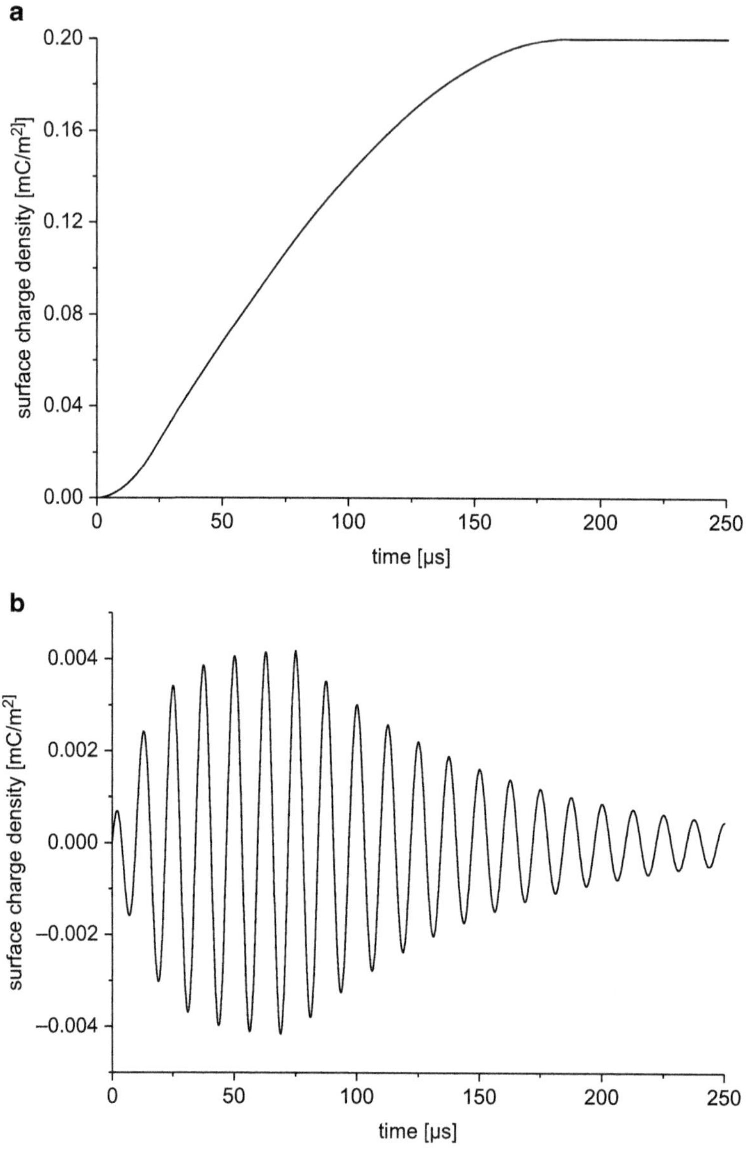

Fig. 5.11 Implementation of analytical source function in EME source model comprising (**a**) contribution of charge accumulation due to crack growth and (**b**) contribution due to crack surface oscillation

modeled EME sensors causing a noticeable decay of signal voltage compared to the buildup of charge density. However, the step-function type of charge generation is well preserved in the EME sensor signal and constitutes the dominating part of the detected signal. A comparison of both signals after application of a 50 kHz

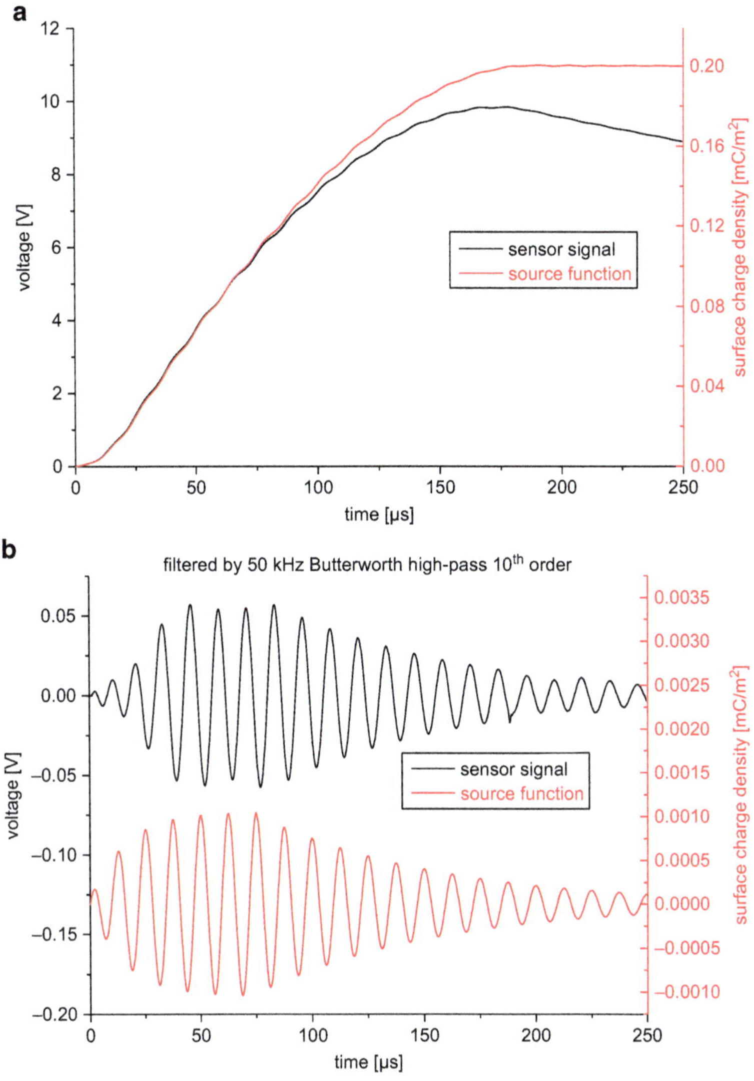

Fig. 5.12 Resulting signal voltage at EME sensor compared to generation of surface charges as function of time (**a**) and same comparison for filtered sensor signal compared to oscillating part of source function (**b**)

Butterworth high-pass filter of tenth order reveals close similarity between the fluctuation of charges assumed due to crack surface oscillation and the detected signal at the sensor position (cf. Fig. 5.12b). The overall shape of the vibrating part of the surface charges is well captured in the detected EME signal, and the minor differences as well as the observed phase shift between both signals are solely caused by the high-pass filter. Thus, as previously demonstrated in [12], such generation of surface charges and variation of their strength in time will cause an EME signal of close similarity to those seen in the experiment.

As extension of the previous EME source model description, the origin of the oscillating part of the source function is now modified. Thus, the EME source model of second type (cf. Fig. 5.9b) is implemented using a generation of charges as function of time as caused by the progression of the crack tip. The increase of surface charges due to crack growth is chosen identical to the previous model as seen in Fig. 5.11a. As second contribution, the crack walls are now forced to vibrate in an oscillatory motion using a prescribed displacement condition $d_x(t)$ applied to the hypothetical fracture surface as given in Fig. 5.11b with a maximum amplitude of 0.5 μm. All other computation settings were kept identical to the previous model. As seen from the resulting sensor voltage in Fig. 5.13a, the overall rise and decay as expected due to the growth of charges and the high-pass character of the sensing system is found in close analogy to the findings of the previous model. However, the superimposed oscillations seen in the sensor signal do not match the expected shape of the source function of Fig. 5.11b. Instead, close resemblance is observed for the source displacement component evaluated along the y-direction. This is shown in the filtered sensor signal of Fig. 5.13b. In addition, the strength of the EME signal rises until 150 μs, which is owed to the increased number of charges applied on the fracture plane. This oscillating contribution is already easy to see in the sensor signal of Fig. 5.13a giving rise to the question of correct choice of the displacement magnitude.

Consequently, the EME source model of Fig. 5.9c combines the previously presented acoustic emission source model and the generation of charges as function of time. To this end, the same geometry as used in the two previous models is chosen. The specimen is loaded up to 23 N static force in three-point bending until the fracture condition is fulfilled (cf. implementation in Sect. 4.2). In a subsequent modeling step, charges are applied to the fracture surface according to the evaluation of the crack tip position, i.e., following the degradation function. All dynamic displacements of the fracture plane are directly computed and are not prescribed by an analytical function. The duration of crack growth in this case was found to be approximately 150 μs. The evaluated surface charge density follows the previous choice and was selected to provide the same amount of charges at the end of fracture to comply with the previous cases.

Therefore, the sensor signal of Fig. 5.14a yields close resemblance to the two previous cases. Systematic differences are observed for the oscillating part of the signal as seen in the filtered version of Fig. 5.14b. For the EME signal, a first oscillation occurs at the first instance of crack growth followed by a damped oscillation until the crack reaches the upper part of the specimen. At the moment of rupture, a second oscillation is observed in the EME signal. This part of the EME signal is described best by the mean source displacement along the x- and y-axis. The first vibration of the formed fracture surface is in good correspondence with the respective signature in the EME signal. However, the oscillation was found to be damped much faster. The most likely reason for this discrepancy is the contributions of other displacement components, i.e., along the z-direction.

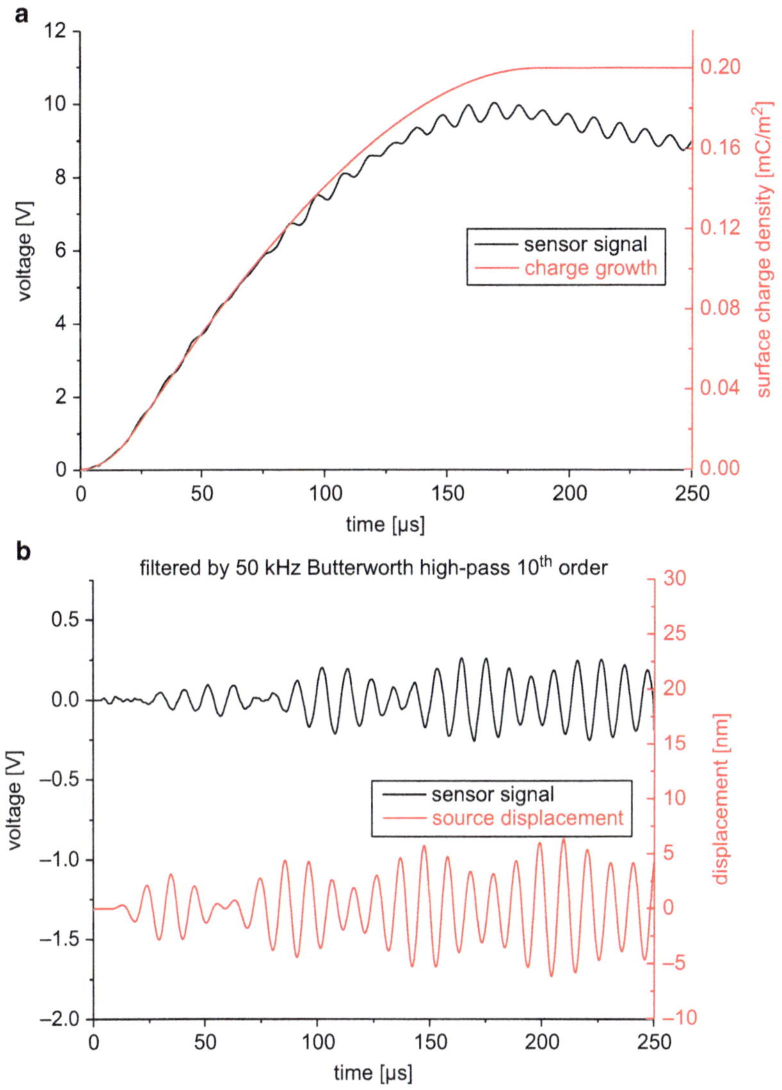

Fig. 5.13 Resulting signal voltage at EME sensors compared to generation of surface charges as function of time (**a**) and same for filtered sensor signal compared to source oscillation (**b**)

Although similar signatures were observed in these displacement signals, it was not suitable to combine them into one single "surface vibration" being directly comparable to the EME signal. At the moment of rupture of the specimen, another high displacement oscillation occurs along the y-direction. Due to the averaging process, the propagation time of the acoustic wave is slightly delayed when compared to the EME signal.

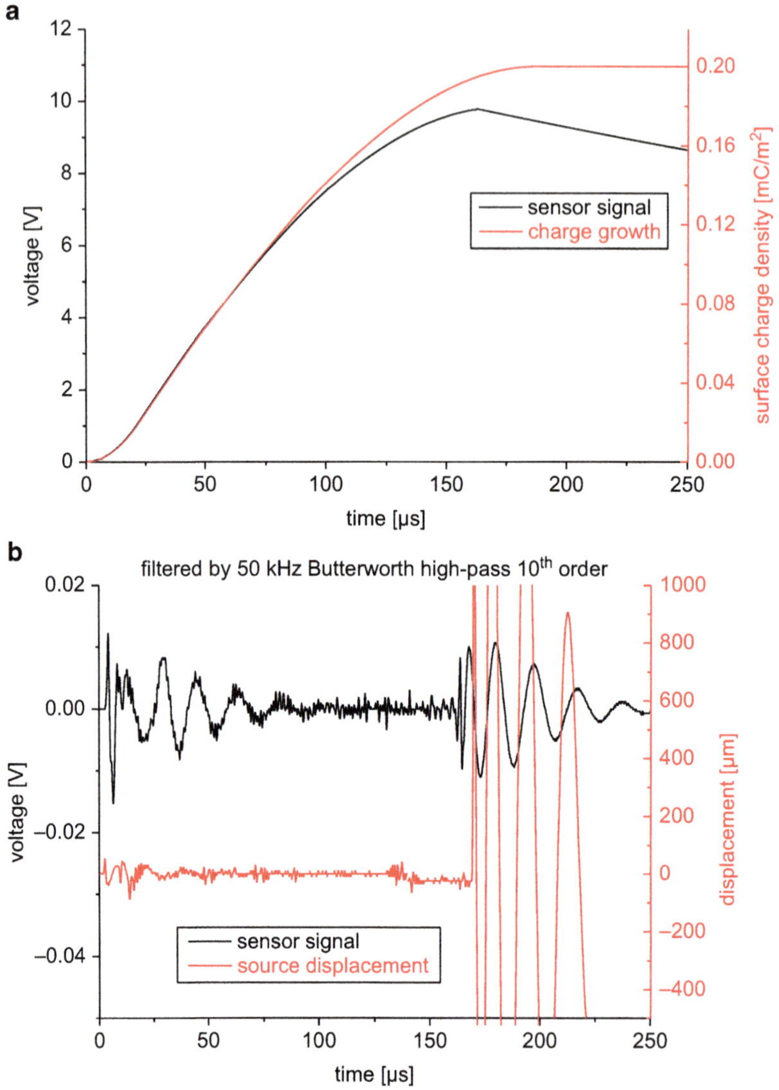

Fig. 5.14 Resulting signal voltage at EME sensors compared to generation of surface charges as function of time (**a**) and same for filtered sensor signal compared to source oscillation (**b**)

Overall, the proposed EME source models allow to compare and discuss the origin of the individual contributions of the EME signal. In combination with validated sensor modeling (see Sect. 5.3), this provides a solid basis to investigate some more details of EME source operation.

5.2.1.2 Source Radiation Pattern

As first aspect to study, the effect of source orientation relative to the sensor shall be investigated. Based on the distinct orientation of the fracture surface within the specimen, it may be expected that the emitted electromagnetic waves exhibit some sort of radiation pattern. If the EME source has a preferential orientation relative to the crack surface, a noticeable change in the intensity should be detectable as function of detection angle. In order to investigate this effect, the single-edge-notched beam (SENB) configuration as shown in Fig. 5.15 with a specimen size of 25 mm × 5 mm × 5 mm (length × height × width) is used. The notch depth was prepared with 2 mm. All tests were carried out using a three-point bending fixture made from PMMA and supports made from PVC in combination with copper capacitance plates of 6 mm × 8 mm (height × width) as EME sensors. Acquisition settings were chosen in analogy to acoustic emission measurements and are summarized in Table 5.1.

Keeping the sensor orientation constant, the bending fixture was rotated by an angle ϕ around the z-direction. This allows a systematic variation of the angle between the crack surface normal and the sensor plate normal between 0° (parallel) and 90° (perpendicular) in steps of 10°. The distance d of the sensor to the rotation axis (see Fig. 5.15) was kept constant at 14 mm. From the detected EME signals, the absolute energies were quantified using the definition of Appendix C. In addition, a tenth order Butterworth band-pass filter ranging from 20 to 100 kHz was applied to

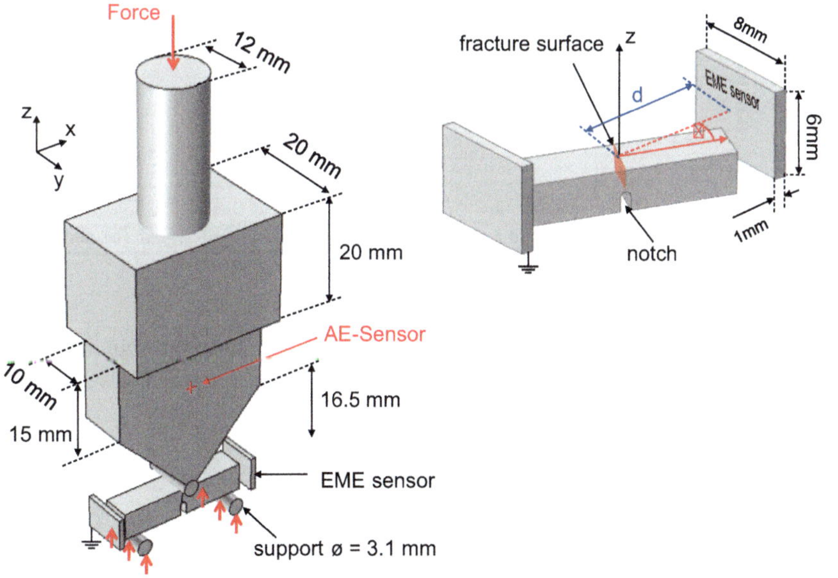

Fig. 5.15 Experimental setup for flexural testing of SENB specimens with EME acquisition (*left*) and definition of rotation angle and source-sensor distance (*right*)

Table 5.1 EME acquisition settings used for SENB tests

Acquisition settings	Value
Preamplification	40 dB
Threshold	35 dB
Triggering	Independent
Acquisition rate	10 MS/s
Bandpass range	1 kHz to 3 MHz
Mounting system	Floating arm
Sensor type	Capacitance plate
Number of sensors	1 pair

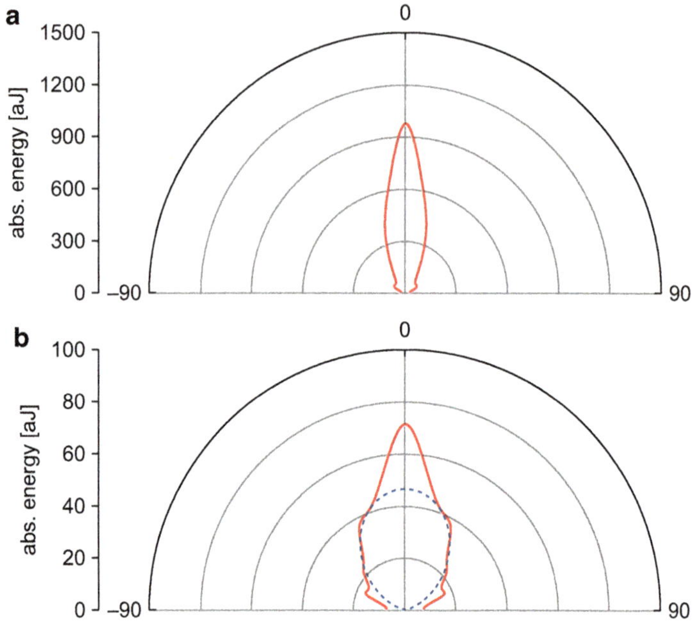

Fig. 5.16 EME signal energies as function of detection angle for full signal (**a**) and filtered signal (**b**)

the signals to independently investigate the oscillating part of the signals. Subsequently, the absolute energies of the filtered signals were quantified as well.

As shown in Fig. 5.16 for the filtered and unfiltered signal, a substantial difference of the signal energy was measured as function of the detection angle. While the strongest signals are detected for a parallel orientation of crack surface and sensor plate, the measured signals decrease significantly in energy with increasing angle. This indicates a directional field distribution. The angular directivity differs for the different parts of the signals. Figure 5.16a, b shows the energies for the detected signals quantified for the base part and for the oscillating part. The total signal energy is dominated by the energy of the base part. As discussed in the previous sections, the base signal is attributed to the separation of charges during crack growth. This part exhibits some kind of dipole characteristic, since two crack surfaces with opposite charges form a dipole moment with a direction parallel to the crack surface normal. The detected energies show a stronger angular dependence than the energies of the oscillating part. Only for angles up to 40° energies significantly exceeding the level of the noise were detected with a maximum at 0°. At $\phi = 10°$ the energy of the signals has dropped to 53 % when compared to the energy at $\phi = 0°$. The oscillating part of the signals is generated by the vibration of the crack surface as has been demonstrated by the source models discussed previously. When this vibration is assumed to be perpendicular to the crack surface, i.e., has a strong directional orientation, a similar angular dependence of the signal intensity is expected. The measured behavior seems to be more complex than one would expect for a simple dipole characteristic. For a point dipole, the potential scales with $\cos(\phi)$, so one could expect the energy scaling with $\cos^2(\phi)$ (as indicated by the dashed line in Fig. 5.16b). The detected signal energies only partially confirm this simple dipole characteristic. However, since the sensor is in the reactive near-field zone of the source and may not be approximated by a point-like position, it is likely to always detect some field components of other angles. This could account for the stronger detection of energies for small angles than one would expect for a simple dipole characteristic. Additionally the spatial characteristics of an electrical field depend on the type of source and on the surrounding matter. For example, the presence of conductors near the EME source affects the formation of the electric field as shown in Fig. 5.28 in Sect. 5.3.

5.2.1.3 Influence of Distance Between Source and Sensor

Similar to AE signals, it may be expected that due to geometric spreading, signal intensity will decay with increasing distance to the sensing system. To evaluate the detectable signal energy as function of distance, the EME signal energy was measured in the SENB configuration shown in Fig. 5.15 as function of source-sensor distance along the detection angles 0°, 45°, and 90°. All acquisition settings were chosen following Table 5.1. For the different angles, the distance of the EME sensor to the source was varied ranging from the closest distance possible up to the distance where the signal-to-noise ratio inhibits detection of the signal. The recorded signals and the filtered signals (band pass setting as in Sect. 5.2.1.2) were then analyzed in terms of the detected energy. In Fig. 5.17a, the results for the calculated energies are displayed for the full signals and for the oscillating part of the signals in Fig. 5.17b, respectively.

As expected, the measured energy decreases with increasing sensor distance. This was observed for all three angles. As discussed in the previous subsection, the different parts of the signals show a different angular dependence. The dependence on the distance appears to be almost the same for the base signal and the oscillating part. For both parts the signals are only detectable within a distance of a few millimeters to the source. The characteristics of the plotted data show no intuitive

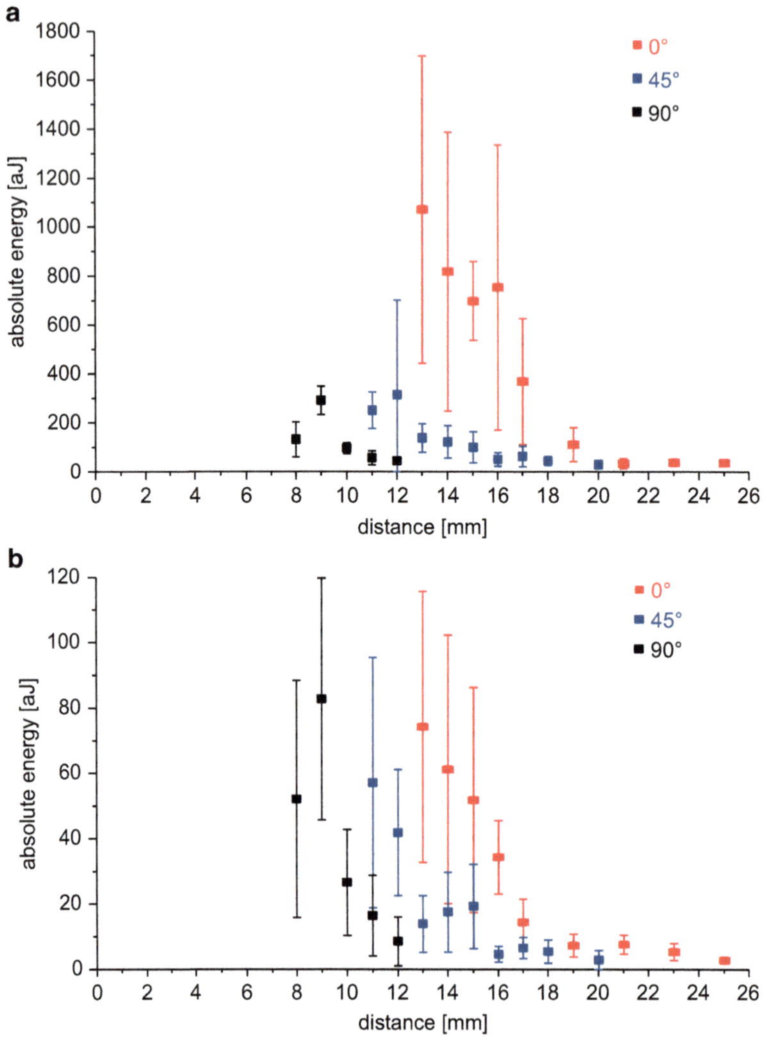

Fig. 5.17 EME signal energies as function of source-sensor distance for full signal (**a**) and filtered signal (**b**) (based on [12])

dependence of distance d. This may be due to multiple influences that all depend on the position of the sensor. The main effect is assumed to arise from the spatial characteristic of the electric field. The potential generated by a dipole decreases with a $1/d^2$ dependence. This would result in a decrease with $1/d^4$ for the measured energies. However, such dependence as function of distance was not observed. Since the real charge distribution is unclear, multipole moments of different order may also appear during the fracture process exhibiting different kinds of distance dependencies. Furthermore, with an increase of d, the distance between the capacitor plates also increases and thus the capacitance decreases. For constant field strength, the smaller the capacitance gets, the higher the voltage between the plates would be.

Another effect on the signal intensity, which might even be of larger relevance than the increase of capacitance with distance, is the influence of other conducting parts of the experimental setup. Although the fixture was built from nonconducting materials, some other elements inevitably consist of conducting materials. The most important one is an acoustic emission sensor which is positioned 25 mm above the specimen and was used for simultaneous detection of acoustic emission signals. Since all conductors near the source influence the voltage on the sensor plate, this influence becomes more important at larger sensor distances.

5.2.1.4 Influence of Electrical Properties

As next fundamental investigation, the influence of the electrical properties of the cracking medium is investigated. Based on the modeling approach considered in [12], the properties of the modeled materials are now systematically varied. As first parameter, the surface charge density is investigated. Figure 5.18a shows computation results of the reference case of an RTM-6 specimen using the source functions plotted in Fig. 5.11.

Compared to the reference case with a final surface charge density of 0.02 C/m^2, the two other results were computed for a surface charge density reduced by a factor of two and four, respectively. As seen from Fig. 5.18a, the resulting EME signal amplitude decreases directly proportional to the amount of charges used as source. Consequently, the level of charge imbalance will contribute significantly to the detection of EME signals.

As second numerical investigation, the material of the bending geometry of Fig. 5.10 was changed to properties of a typical carbon fiber reinforced epoxy taken from [53] and to those of steel as noted in Table B.3 of Appendix B. As seen from the comparison of EME signals with equal surface charge density, there is only little influence of the change in electrical conductivity and electric permittivity. For the computation results of RTM6 and CFRP, almost no difference is observed (black and red lines superimpose in Fig. 5.18b). Only the substantial change in conductivity of steel causes a noticeable change in the EME signal as found in the inset of Fig. 5.18b. This allows to conclude for the cases of materials studied herein that the influence of charge relaxation is expected to be negligible within the typical durations of a few 100 μs.

5.2.1.5 Sources in Fiber Reinforced Composites

For application of EME measurements to fiber reinforced composites, several other factors regarding the EME source need to be considered. For these materials, the formation of a crack is due to fracture of one or more of the constituent materials or

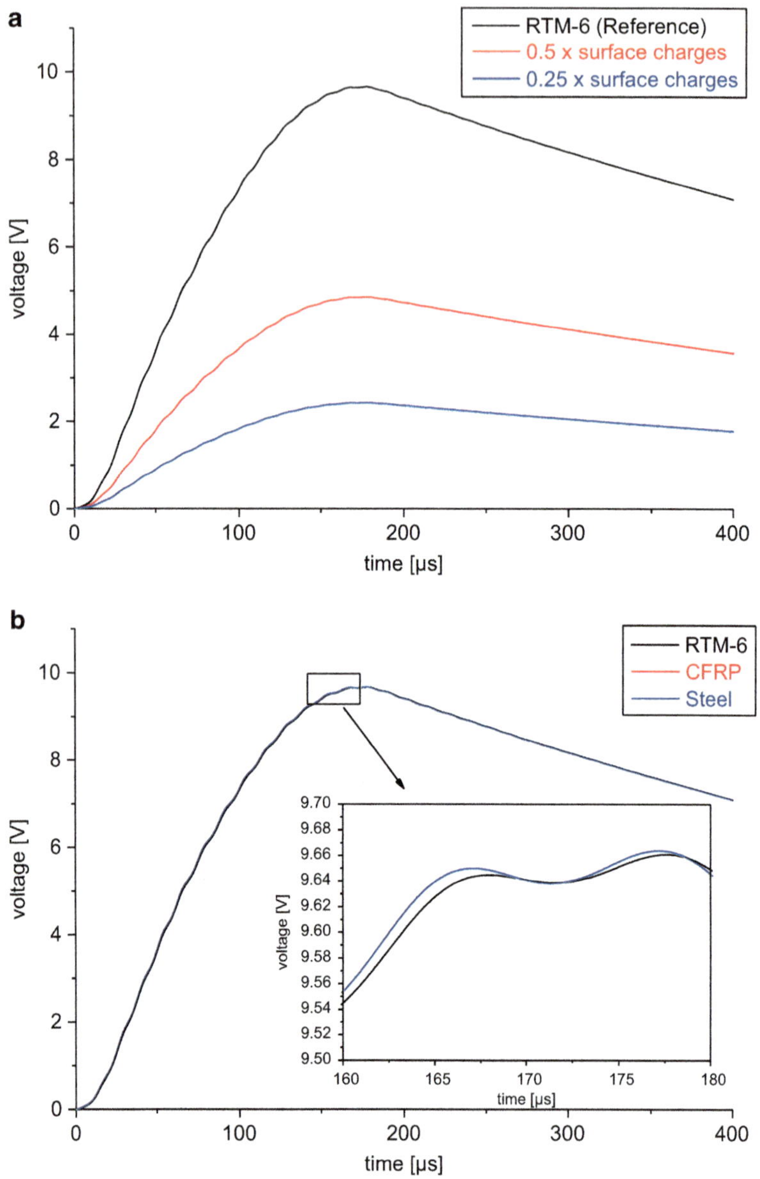

Fig. 5.18 Variation of charge density applied to the fracture surface (**a**) and variation of materials (**b**)

due to crack formation at their interfaces. In this case, the difference in electrical properties of the fibers and the matrix will contribute to the generation of charges along the fracture plane, and the anisotropic electrical conductivity of the composite will govern charge relaxation effects. Based on previous studies [12] and other findings in literature [9, 13–15], it may readily be expected that failure mechanisms in composites cause measurable electromagnetic emission (see also Sect. 5.5). This underlines that charge relaxation effects in fiber reinforced composites will not completely suppress the EME signatures arising from crack growth and the oscillation of the fracture surface as described in Sect. 5.2.1.1.

Following the validated modeling approach, three basic types of failure modes are investigated in order to point out some basic possibilities of EME measurements in composite materials. To this end, inter-fiber failure, inter-ply delamination, and fiber breakage are studied. As model configuration, a typical tensile test specimen with 195 mm × 20 mm × 1.9 mm (length × width × thickness) according to DIN-EN-ISO 527-4 is investigated. The electrical material properties used for the electrostatic simulations within the "AC/DC module" of COMSOL Multiphysics are those of Sigrafil CE1250-230-39 as given in Table B.3 of Appendix B. No explicit modeling of the stacking sequence was carried out in the following, since the electric conductivity tensor is not required in electrostatic modeling. Instead, only the orientation of the fracture surface was varied to account for directivity effects. As for the configuration of Fig. 5.10, the tensile specimen was located within a sphere of 300 mm diameter and adjacent infinite element domain to avoid interaction with any nearby boundaries. In all cases investigated, the charge density on the fracture surface was chosen to be 0.02 C/m^2. Since this value is arbitrarily chosen, it is important to note that the appearance of the field lines and the comparability of the iso-voltage surfaces throughout Figs. 5.19, 5.20, and 5.21 are not affected by this choice. By keeping this value constant throughout the

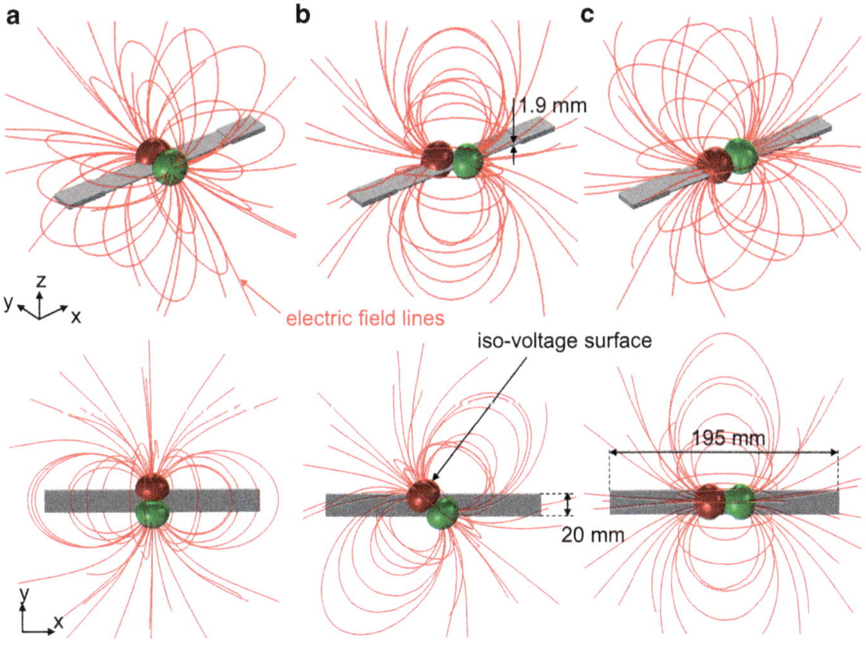

Fig. 5.19 Electric field lines surrounding charged fracture surface and iso-voltage surfaces as 3D view and projection to xy-plane for inter-fiber failure at 0° angle (**a**), 45° angle (**b**), and 90° angle (**c**)

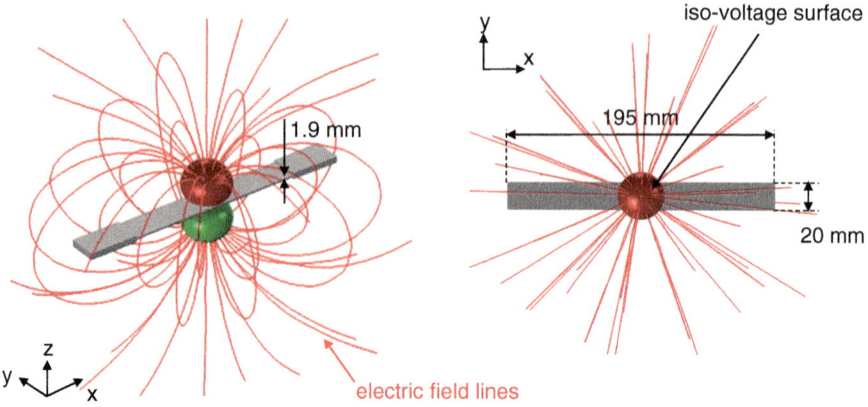

Fig. 5.20 Electric field lines surrounding charged fracture surface and iso-voltage surfaces as 3D view and projection to xy-plane for inter-ply delamination

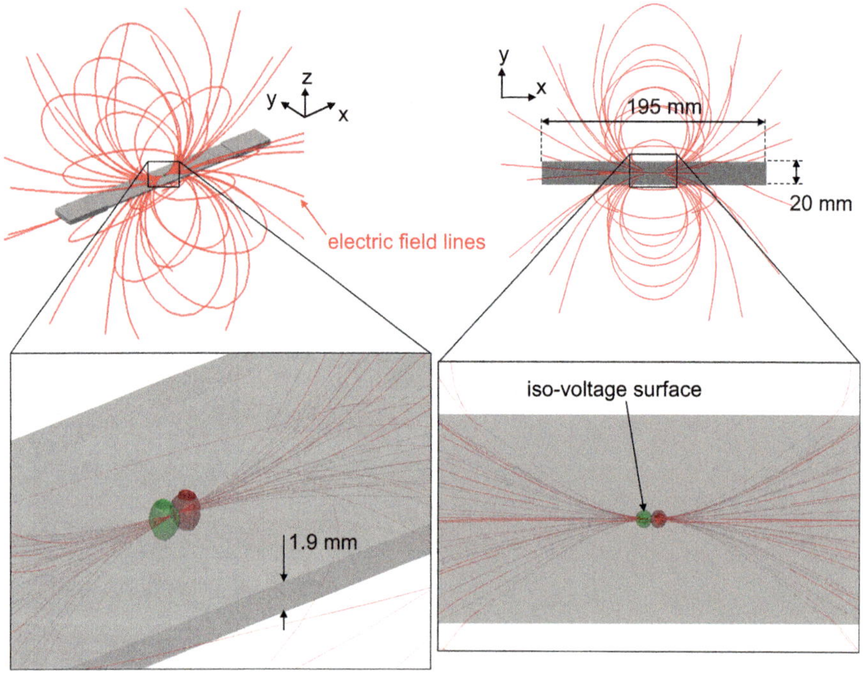

Fig. 5.21 Electric field lines surrounding charged fracture surface and iso-voltage surfaces as 3D view and projection to xy-plane for fiber breakage including details close to fracture surface

simulations, it is instead assumed that a similar amount of charges per crack area is generated regardless of the fracture type. This might only be partially justified for the fiber breakage case as compared to the other two cases. All mesh settings confirm with the choices reported in Sect. 5.2.1.1.

As a first example, a study of the electric field lines due to inter-fiber failure is presented. The size of the fracture surface was implemented as 2.5 mm × 0.5 mm (width × height). The result of the computation is shown in Fig. 5.19a–c for three different orientations of fracture surfaces within the *xy*-plane. In addition, the iso-voltage surfaces surrounding the modeled fracture surface are shown in red (positive) and green (negative) color. The iso-voltage surfaces may readily be understood as radiation pattern as discussed in Sect. 5.2.1.2. As clearly seen from the orientation of the iso-voltage surfaces as well as the formation of electric field lines, the orientation of the inter-fiber failure will directly translate into the radiation pattern of the respective EME source. Given the findings of Sect. 5.2.1.2, this is not surprising, but points out the possibility to measure the orientation of fracture surfaces by EME as discussed in Sect. 5.5.2.

Choosing another orientation of the fracture surface as typical for inter-ply delamination results in the electric field lines and iso-voltage surface shown in Fig. 5.20. Here the size of the delamination fracture surface was implemented as 5 mm × 5 mm (length × width). The orientation of the delamination was chosen to be within the *xy*-plane at the center of the laminate with respect to the *z*-axis. Due to the change in orientation of the charged fracture surface, the radiation pattern is now found to be dominant in the out-of-plane direction of the laminate. Despite of the small increase in the fracture surface area, the iso-voltage planes are of almost similar size as for the inter-fiber failure cases.

For the configuration of a charged fracture surface as typical for fiber breakage, some obvious differences arise as seen in Fig. 5.21. The size of the charged fracture surface was chosen to be 240 μm × 240 μm (width × height) being oriented within the *yz*-plane. This corresponds to a typical fracture surface of more than 1000 filaments. Overall, this is much smaller than the fracture surfaces investigated before. Given the assumption of constant charges per surface area, this translates into a much smaller signature, having iso-voltage planes of the same level being much closer to the source position. Nevertheless, the overall radiation pattern seen from the magnification of details in Fig. 5.21 also resembles a dipole characteristic. Due to the difference in rise-times of the sources and the resulting oscillations as presented in Sect. 4.2, it may readily be expected that the signature of the corresponding EME signals will be characteristically different to those of inter-fiber failure of the same orientation.

5.2.2 Test Sources for Electromagnetic Emission

In order to validate measurement setups and to qualify sensor systems, it is quite useful to have an artificial type of EME source. However, there is no national or international standard covering this topic, and there are not even fully established routines found in literature. Consequently, various designs have been proposed, which exhibit different strengths and weaknesses. In this section some simple test source concepts are presented, which are easy to fabricate and easy to use for generation of artificial electromagnetic emission signals. The natural choice to emit an electromagnetic wave is a metallic conductor with a driving voltage applied. This combination is further referred to as emitter[1] for simplicity. To act as a test source for electromagnetic emission, three criteria may be raised to distinguish different artificial source types:

1. The emitter should operate sufficiently far outside its electric resonance frequency to ensure a reasonably flat frequency characteristic.
2. The dimensions of the emitter should be chosen such that it can be located at the same position as the real electromagnetic emission source to be detected.
3. The emitter should exhibit a source radiation characteristic similar to a real EME source.

In general, all sorts of metallic objects are used as emitter, whereas different electric systems and excitation functions are used as signal drivers. In many cases, simple plates or open-ended wires were found to be most useful as emitter component.

In order to fulfill the first criterion, the emitter and its circuit need to be selected appropriately. Theoretically it is possible to estimate the resonance frequency of any electrical system based on the geometry and the electrical properties of the antenna material and concise knowledge of the emitter circuit. While this is straightforward to derive for some emitter configurations, in reality the near-field aspect [54–56] may cause additional resonances, which are difficult to predict by the analytical calculations. Thus it is advisable to directly measure the resonance frequency of a test source emitter based on the electric impedance of the system. Such measurement is readily be achieved by use of a network analyzer or by means of monitoring the detected signal at a sensor system, given the latter is reasonably flat with frequency.

In order to comply with the second criterion, the dimensions of the emitter system should be reasonably small. Since this is a general matter of the experimental setup, a typical emitter size should not be much larger than a few millimeters to be able to resemble a point-like EME source originating from material failure. If the EME source to be investigated cannot be approximated as point source, this criterion becomes almost obsolete. In cases of an EME source size ranging from some millimeters to few centimeters, the emitter is certainly more appropriate when it is of similar dimensions.

Compared to the other two criteria, the third one is the most challenging one. Based on the discussion of source radiation characteristics of EME sources in Sect. 5.2.1.2, it is obvious that this effect needs to be incorporated to form a representative artificial EME source. As schematically shown in Fig. 5.22, it is possible to fabricate such EME test sources by using two aligned metallic parts.

[1] The term emitter should not be misunderstood as system operating with far-field characteristics, since the EME configuration is usually in the reactive near-field (cf. Sect. 5.3).

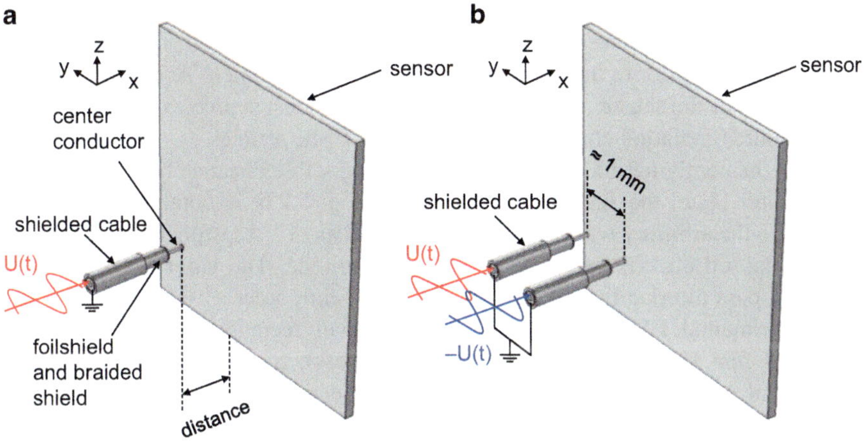

Fig. 5.22 Emitter configuration used for generation of artificial EME using single shielded cable (**a**) and similar configuration using two shielded cables yielding artificial EME source with characteristic source radiation pattern (**b**)

For this purpose, cables as typically used for Bayonet Neill-Concelman (BNC) cables may be used. At the end of the cable, parts of the shielding are removed to reveal the center conductor to a length of few millimeters. It is important to retain the shielding up to the position, where the center conductor is revealed to open space to avoid symmetry breaks between the two center conductors. Using only one such cable, an easy EME test source is fabricated, which can be driven by a signal voltage $U(t)$. Keeping a constant distance to the capacitance sensor position (a plate in Fig. 5.22a, b), this allows to test the acquisition settings and allows to evaluate the system transfer function (cf. Sect. 5.4). To fabricate an EME source with a characteristic source radiation, two shielded cables may be placed next to each other. Applying a signal voltage $U(t)$ of opposite polarity and identical phase to the two cables as seen in Fig. 5.22b, this yields an EME source with characteristic radiation pattern.

Since the relative movement between source and detector will also cause noticeable EME signals, it is recommended to generally use nonconductive holders or supporting arms to position the test source at the designated spot and to keep it at this constant position during the measurements.

For all geometrical configurations, the question arises, which type of source function $U(t)$ to use for generating artificial EME. This certainly depends on the intended purpose. For calibration of sensor systems as carried out in Sect. 5.4, Gaussian enveloped sinusoidal signals with varying center frequency may be used to sweep a certain frequency bandwidth. For generation of artificial EME signals, i.e., signals representative for a real EME source, different source functions are necessary. As demonstrated in Sect. 5.2.1, the most likely type of source function seen for an EME source are step-function types with superimposed oscillations. Such source signals are straightforward to implement in arbitrary waveform generators and may then be applied as driving voltage to the emitter configurations shown in Fig. 5.22. As one of the intents of the test source, this allows checking the detector systems prior to the measurement of the test specimen. And it also allows optimization of the sensor positions, given the chosen test source exhibits the same type of source radiation characteristic as the real EME source.

Using the configuration of Fig. 5.22a and a negligible distance between emitter and detector plate, the EME signal as seen in Fig. 5.23a is obtained. The input function to the arbitrary waveform generator consists of a step-function signal with a superimposed 80 kHz oscillation of small amplitude. The latter is only visible after high-pass filtering by a Butterworth filter of tenth order as seen in Fig. 5.23b. The experimental EME sensor consists of a small rectangular copper plate of 5 mm × 8 mm × 1 mm (height × width × thickness). Signals are detected by a commercial acoustic emission system using a preamplification factor of 37 dB. Comparing the measured EME signal before (Fig. 5.23a) and after filtering (Fig. 5.23b) readily reveals both contributions of the input signal. However, the overall shape of the detected EME signal and phase shifts occur due to the nonlinear transfer function and the band-pass characteristic of the measurement system (cf. Sect. 5.4).

Using a test source configuration as schematically show in Fig. 5.24a and the same procedure as for the single BNC cable as described above, it is possible to evaluate the detected signal intensities as function of detection angle. Experimentally, two capacitance plate sensors were mounted in orthogonal arrangement, i.e., one sensor plate normal in parallel to the y-axis and one sensor plate normal in parallel to the z-axis. Following the sensor modeling approach of Sect. 5.4, this configuration was implemented within the "AC/DC module" of COMSOL Multiphysics and modeled using the geometry of the experimental configuration accordingly. The distance between the center of the artificial source and the sensor plate was 22 mm, and as test signal a sinusoidal voltage of 1 kHz frequency was chosen. Measurements and simulations were performed for angle increments of 10° ranging from 0 to 90° rotation of the test source. A comparison of the obtained experimental and numerical peak-to-peak voltages is given in Fig. 5.24b. Within the margin of error, the artificial test source shows the expected cosine shape (dashed line). Also, the detected signal voltages at sensor 1 and sensor 2 exhibit the expected phase shift of 90° due to their orthonormal orientation relative to the test source.

One of the drawbacks of arbitrary waveform generators is their low operational voltage (typically 20 V or less). In typical experimental configurations involving a distance between test source and detector of a few centimeters, this might already be at the limit of detectability. Hence, these comparatively low voltages do not sufficiently resemble strong EME sources as sometimes seen in the experiment. However, it is not advisable to change the distance between detector and source to apparently increase the test source strength, since this will have an impact on the relative position of EME sensors and their position relative to other metallic objects close by. Thus such modifications might falsify the measurements of the test sources. To overcome this problem, either more powerful arbitrary waveform

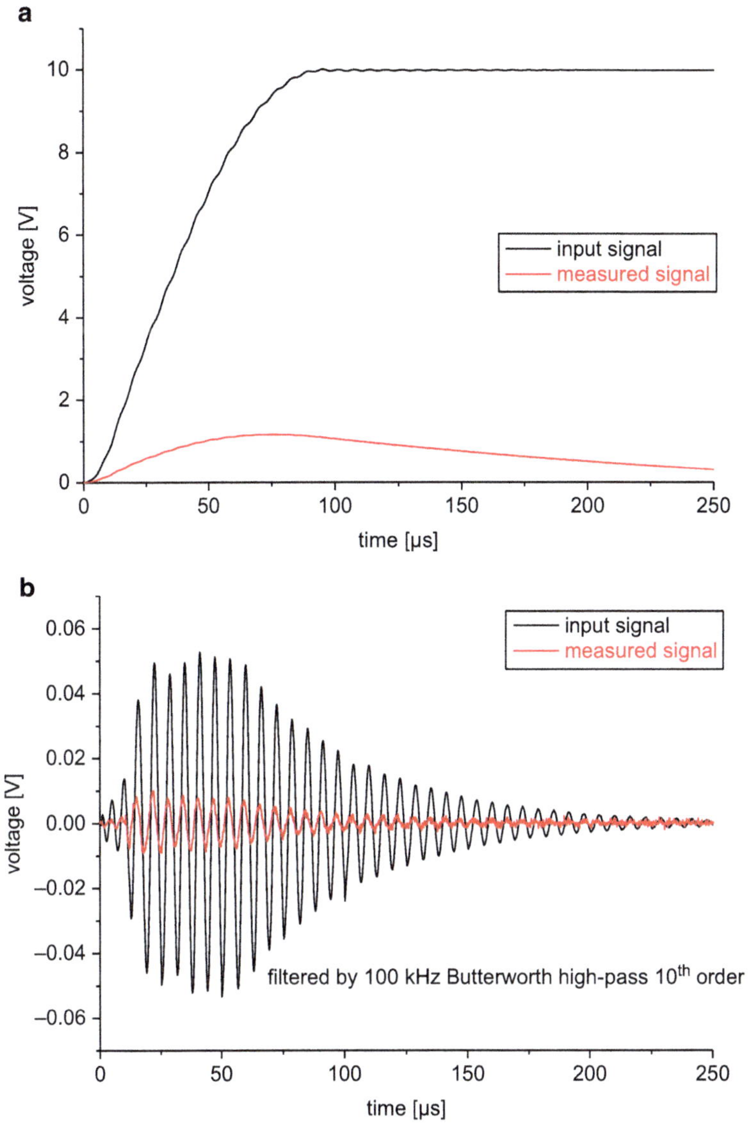

Fig. 5.23 Comparison of test source signals detected by capacitance sensor in negligible distance to test source using BNC wire configuration (**a**) and same signals after application of 10 kHz high-pass filter (**b**)

generators may be applied to generate stronger EME signals or an alternative test source may be applied. Experiments using a configuration as shown in Fig. 5.25a involving a piezoelectric lighter as voltage source have demonstrated the possibility to yield highly reproducible and strong EME signals.

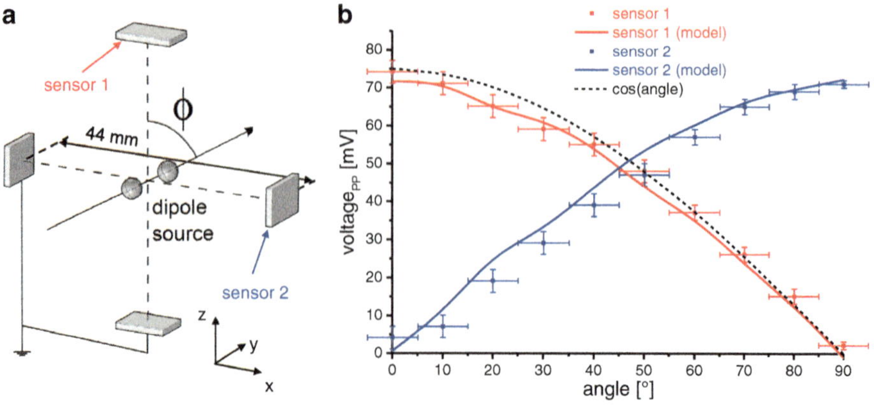

Fig. 5.24 Artificial test source arrangement (**a**) and evaluation of peak-to-peak voltages for experiment and model in comparison to cosine function (**b**)

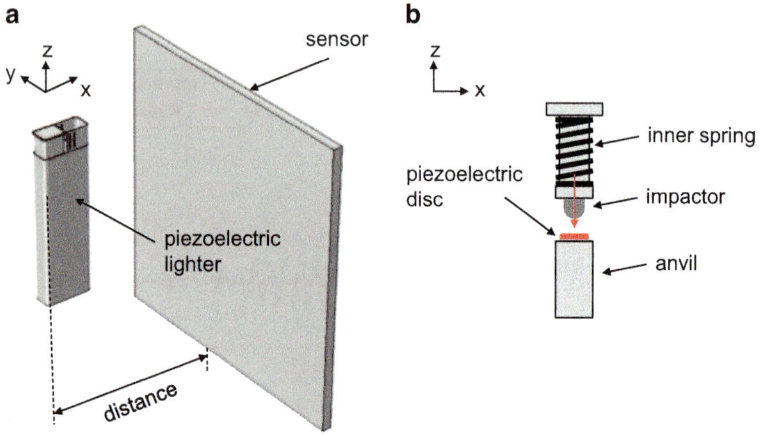

Fig. 5.25 EME test source configuration using piezoelectric lighter in close distance to the EME sensor (**a**) and part of the spring mechanism on the interior of ignition device (**b**)

In a piezoelectric lighter, a spring system is used to impact a piezoelectric crystal, which produces a strong electric field in the kilovolt range. This spring mechanism is schematically shown in Fig. 5.25b, but may slightly vary in different commercial lighters. In its original application, this spring-driven impact on the piezoelectric material generates a spark that ignites the inflammable gaseous medium. After removal of the gas of a commercial lighter, the ignition device is still operating and may be used as EME test source. At a constant distance of 30 mm and constant orientation relative to the sensor, this allows generation of reproducible and strong EME signals as seen in Fig. 5.26a detected by wire loop sensors (see Sect. 5.4) with a total preamplification factor of 25 dB. In contrast to the approach

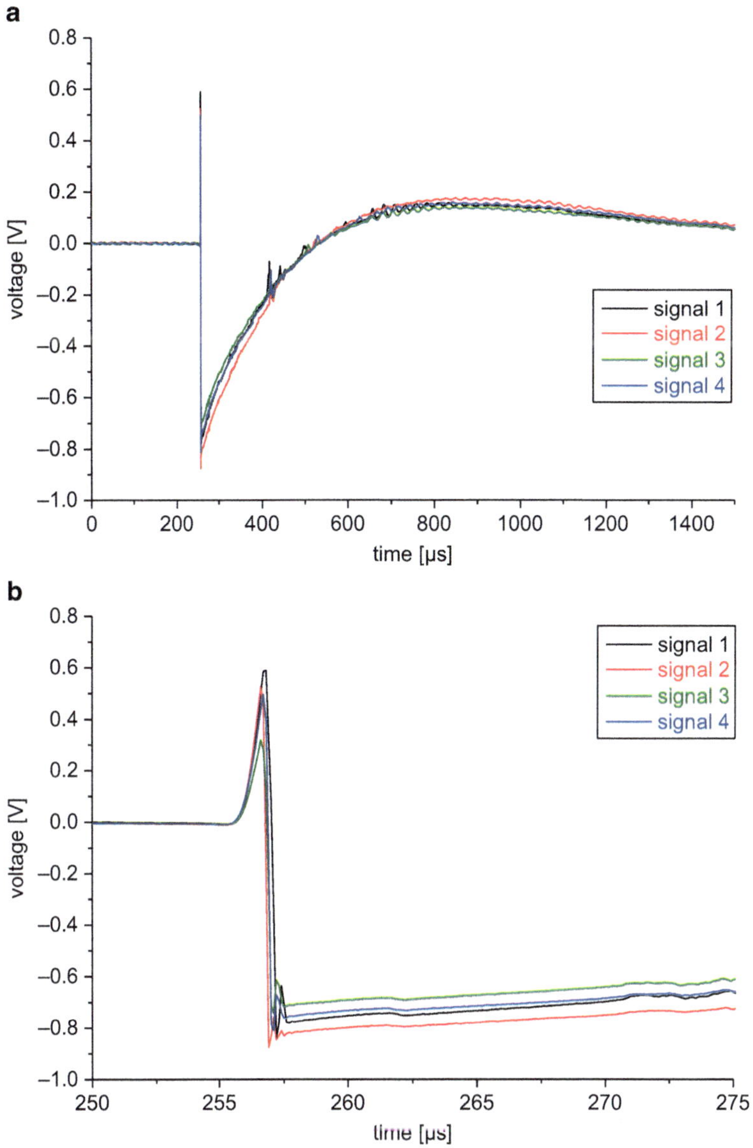

Fig. 5.26 Typical EME signals of piezoelectric lighter source detected by wire loop sensors in 30 mm distance (**a**) and same signals zoomed to reveal first steplike signature in the test signal (**b**)

described above, this EME source does not show a distinct source radiation characteristic. This is due to the metallic parts surrounding the piezoelectric material, which tend to partially shield and distribute the generated electromagnetic field. Experiments using various orientations to the detector did not reveal a characteristic radiation pattern.

5.3 Signal Propagation

After the charges are present at the crack tip, according to Maxwell's equations, the spatial movements of such charges will implicitly cause a fluctuation of the electric and magnetic field strength and therefore form the source for an electromagnetic wave [57]. The basic physical principles are well established, and for a comprehensive reading on this topic, the reader is referred to some recent literature in this field [54–56]. To understand the propagation of electromagnetic waves in the context of electromagnetic emission, it is first of all necessary to distinguish the relevant aspects unique to this particular physical phenomenon from well-known terms and phenomena in classical physics. A first distinction shall be made between the term electromagnetic radiation and the term electromagnetic emission.

Electromagnetic radiation is associated with electromagnetic fields free to propagate without the continuing influence of the moving charges that produced them, because they have achieved sufficient distance to those charges. The electromagnetic wave travels through space carrying radiant energy and thus are decoupled from their source. This description is only valid in the far field of the source and, therefore, can make use of several simplifications regarding the source and the structure of the electromagnetic wave. For these electromagnetic waves in free space, Maxwell's equations directly result in two main classes of solutions, namely, plane waves and spherical waves propagating at the speed of light in vacuum conditions. These are convenient to describe using the angular frequency ω and the wave number k. The propagation velocity of an electromagnetic wave of particular frequency is given by the phase velocity c_P:

$$c_P = \frac{\omega}{k} \tag{5.16}$$

Frequency and wave number are directly related to each other by the dispersion relationship:

$$\omega = f(k) \tag{5.17}$$

In the case of electromagnetic wave propagation in solid materials, the frequency is not directly proportional to the wave number, causing dispersion upon propagation. However, the impact of this dispersion for the interpretation of electromagnetic emission is not nearly as severe as discussed for the guided acoustic emission waves in Sect. 4.3. For a wave package of a narrow range of frequencies, the group velocity is defined by the gradient of the dispersion relation:

$$c_G = \frac{\partial \omega}{\partial k}. \tag{5.18}$$

In an identical meaning as for acoustic waves, the group velocity is the velocity at which the energy propagates through the medium.

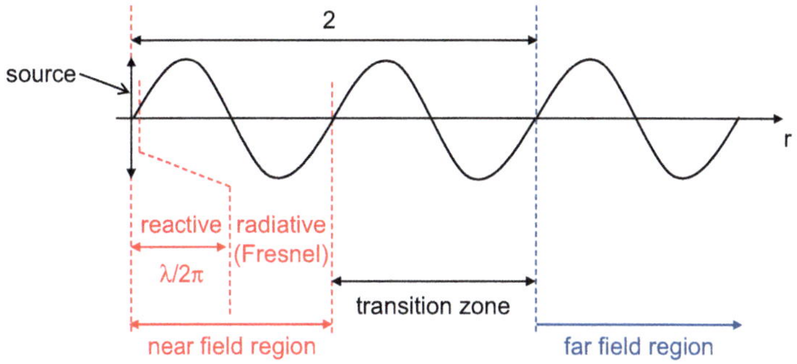

Fig. 5.27 Distinction between near-field and far-field regions based on wavelength and distance to source

For the electromagnetic emission phenomenon discussed herein, the frequency range from 10 Hz to 10 MHz results in typical wavelengths ranging from 30,000 km to 30 m. This covers the full range from the ELF (extremely low frequencies) to the HF (high frequencies) band. Still, the typical spatial dimensions of the source are much smaller than the emitted wavelength.

In analogy to electromagnetically short antennas, this gives rise to the distinction of four characteristic distances to the source position as shown in Fig. 5.27. Here the near-field and far-field regions are defined by ratios of the distance r of the source to the wavelength λ of the emitted wave. The near field in this case is defined as distance $r \ll \lambda$. Within the near field, further distinction can be made between the reactive and radiative region. The outer boundary of the reactive near-field region is commonly considered to be at a distance $r = \lambda/2\pi$. The radiative near field (also known as the "Fresnel region") covers the rest of the near-field region. The latter is the boundary to the far field of the source, which is defined for any distance $r \gg 2\lambda$. The transition zone is understood as any distance $\lambda \leq r \leq 2\lambda$ and forms an intermediate region in which both near-field and far-field effects are important. In this region, near-field behavior decreases gradually, leaving far-field effects as dominant behavior.

For electromagnetic emission, the used frequencies and typical source sizes imply that the position of the sensing systems are located in the reactive near field of the source, which e.g., for frequencies below 10 MHz is any distance $r < 4.77$ m. Due to the limited sensitivity of the detection systems, the typical source-sensor distance can even be expected in the centimeter range, and thus the reactive near field would still be applicable for frequencies up to 100 MHz. Hence, a particular focus is given in the following to explain the effects of wave propagation in this reactive near-field region.

In the region very close to the source, the prediction of the electric and magnetic field strength is comparatively complex, because geometric details of the source and its vicinity as well as the electrical boundary conditions can have a strong impact on the field lines. For a simple and ideal configuration like a dipole antenna, the near field may still appear predictable, but is already hard to achieve by use of analytical tools since multipole expansion is no longer valid and the full solution of

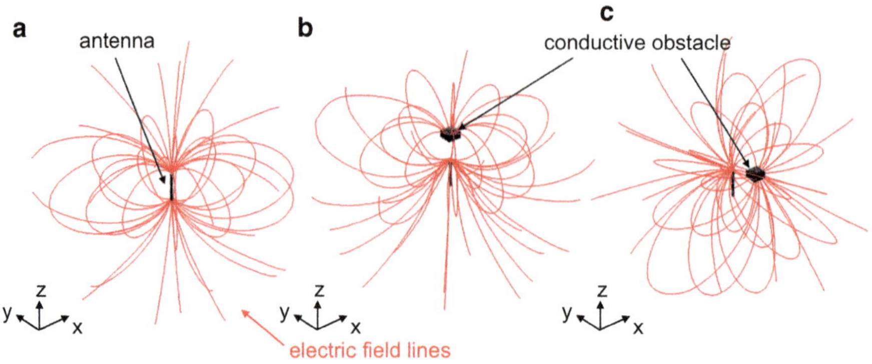

Fig. 5.28 Near field surrounding dipole antenna visualized using the electric field lines (**a**) and same antenna with added conductive obstacle directly in parallel to dipole axis (**b**) and perpendicular to dipole axis (**c**)

Maxwell's equations is generally required. Using numerical methods as demonstrated by the respective finite element solution shown in Fig. 5.28a, the electric field lines are straightforward to obtain.[2]

Moreover, the addition of a metallic obstacle in the vicinity of the antenna causes a strong distortion of the near-field appearance as visualized in Fig. 5.28b, which is readily obtained using Maxwell's equation in conjunction with numerical methods, but is hard to predict solely by using analytical methods due to the boundary conditions faced in a geometrically complex situation.

The term reactive near field indicates the nature of interaction within this particular distance to the source. Here the detection process of electromagnetic waves includes a substantial interaction with the source. This is in contrast to the detection process in the far field, where the electromagnetic wave is fully decoupled from the source, and hence, no interaction takes place. It is well known from other fields of science that measurements in the reactive near field may give ambiguous results. Hence it is of vital importance to concisely understand the interaction between the source and the detection system.

For the case of an antenna as source representation, this interaction can readily be understood by electrostatic and magnetostatic effects. For an alternating current flowing in the antenna, a purely magnetic component is created in the near field, which starts to collapse as the antenna current begins to reverse. This causes a transfer of the magnetic field energy back to electrons in the antenna due to self-induction. Consequently, this alternating process returns energy from the magnetic field back to the antenna in a regenerative way. A similar consideration can be made for the electric field as electric charges are generated. If charges are forced to locate in one section of the antenna by the excitation voltage, a local electric field forms

[2] It is worth noting that in the near-field region, the vectors of electric and magnetic field are not always perpendicular, but may exhibit a phase shift relative to each other.

around that section of antenna due to the self-capacitance of the antenna. When the excitation signal reverses, charges can dislocate from their forced regions, and the previously generated electric field assists in pushing electrons back in the new direction of their flow in a similar fashion as the discharge of a unipolar capacitor. Due to this process, energy returns from the electric field back to the antenna current.

Because of this energy storage and return, it is straightforward to conclude that any additional inductive or electrostatic interaction in the range of the reactive near field will cause a transfer of field energy from the antenna to another conductor in close distance. Some prominent technical examples making use of this effect are voltage transformers or transponder systems.

The amplitude of the electromagnetic field components close to the source can be quite high, but decay with $1/r$ behavior if there is no disturbance of the electromagnetic field. Hence, the field energy is strongest close to the region near the source and may fall below detection limits very fast if the source strength is weak. However, for an antenna as source, the usual understanding is that there is no loss of field energy unless there is another object, such as a sensor in the area close to the antenna. Thus, the near fields only transfer energy to very close conductors and cause a measurable drop of energy at the antenna, when interacting with the conductors. In the context of electromagnetic emission, these additional conductors could be the detecting sensors, but also all other nearby objects such as shielding or load frame components. To demonstrate this effect, the full configuration is enclosed in a typical rectangular shielding chamber of 100 mm × 100 mm × 100 mm (length × width × height) as seen in Fig. 5.29a (two surfaces are chosen transparent to visualize interior). For experimental measurements, the shielding chamber typically comprises the largest piece of conductive material in the reactive near field of the EME source. In addition, the source configuration of Sect. 5.2.1 is modified to include two EME capacitance sensors with orthogonal orientation relative to each

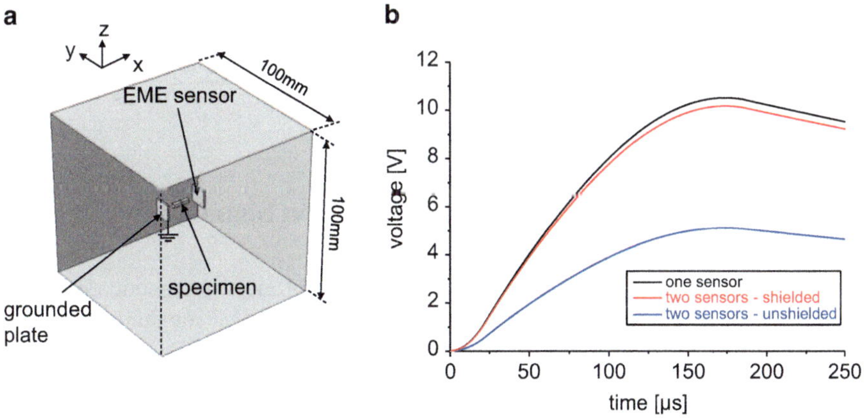

Fig. 5.29 Model geometry including shielding chamber (**a**) and comparison of signal voltages at sensor 1 with and without the presence of sensor 2 in shielded and unshielded configuration (**b**)

other. Simulations of the detected EME signal at sensor position 1 are performed with and without the presence of the second EME sensor system. All other model settings confirm to the approach of Sect. 5.2.1.

As consequence of the addition of a second EME sensing system in the reactive near-field range of the EME source, the measurable voltage drops slightly from a maximum of 10.5–10.1 V. This effect demonstrates the impact of the reactive near field and the direct interaction with the EME sensor system. However, given the presence of the large conductive shielding chamber, the addition of the small amount of conductive material of the EME sensor is almost negligible for practical measurements. Nevertheless, other metallic pieces in close vicinity to the EME source will have similar impact and may reduce the applicability of the presented possibilities to detect EME signals as such or to measure the orientation of the fracture surface. In direct comparison, a model using a spherical infinite element domain surrounding the specimen as in Fig. 5.10 would mimic an unshielded case. If an additional EME sensor is added to this unshielded case, a significant voltage drop by a factor of 0.5 occurs. In this case, the amount of conductive material is doubled by the presence of the second sensor system, and the electric field strength redistributes accordingly.

For a real EME source, the behavior in the reactive near field is generally expected to be more complex than for an antenna. Another important distinction between an EME source and an antenna is given by the resulting near field of an EME source. This is of transient nature since it is linked to the generation and decay of charges of the fracture surface. So the description of alternating currents or charges is only correct as far as the spatial movement is concerned. The driving energy for the EME signal is the crack process, and the resulting field energy is therefore increased as long as charges are generated at the crack tip. Hence, the flow of energy back and forth from the field to the source as described for the antenna above is of limited relevance in this context. Instead, the electromagnetic emission process is better described by a single cycle unipolar generation of field energy. In principle, the generated electric fields may couple back to the charges via self-capacitance. But a coupling of field energy back to the charges is not expected during the duration of source activity, since the respective frequencies and wavelengths do not allow back-coupling within the durations of a few 100 μs.

5.4 Detection of Electromagnetic Emission Signals

Unlike the other testing methods reviewed, the type of electromagnetic emission measurements addressed herein are not yet covered by technical standards and have not been turned into a commercial measurement technique so far. Consequently, no common standard for EME detection is existent today. The standardization of a detector concept is also challenging due to the variety of reasons already raised in the previous sections. First, the detector establishes a coupled interaction with the near field of the EME source being influenced in sensitivity and response by the presence of nearby conductors. Hence, no unique transfer function can be expected for one sensing system, but rather a unique transfer function for each particular configuration. Second, no general guidelines exist for fabricating an EME sensing system. To this end, various proposals for EME detection systems have been made in literature [1–8, 12], all of them resulting in unique sensing systems.

The physical principle for EME detection can be fully established from Maxwell's equations. Thus, the mode of operation can be readily obtained for every individual configuration by use of analytical or numerical methods. Therefore, the intent of this section is to provide an overview of sensing systems that have been proposed in literature so far and to elucidate their strength and weakness in terms of EME detection and EME signal interpretation. To facilitate an objective comparison, most of the effects are discussed based on modeling results using the validated approach of EME sensor modeling established in [12] applying a test source configuration as presented in Sect. 5.2.

The second goal of this section is reviewing the important aspect of electromagnetic shielding. Since this comprises a standard approach in the design of modern electronic equipment, only the major aspects as relevant to EME measurements are discussed. Along this line, the reduction of shielding effectiveness for different challenges faced in practical EME testing is outlined, and results of experimental shielding measurements are presented. Focus is given to the bandwidth of the frequency spectrum relevant to the detection of EME and to provide some guidelines for design of shielding chambers to carry out mechanical testing in combination with EME detection.

5.4.1 EME Detector Concepts

As discussed in Sect. 5.3, EME detection constitutes a measurement in the near-field range of the source. The relevant quantities to detect are the emitted electric or magnetic field components. Whether the one or the other will be detected is subject to the individual sensing system, since every sensor may be built to detect either electrical field components or magnetic field components or a mixture of both. For detection of the electric field component, most detector systems apply the generation of surface charges on a conductive sensing material to convert the incident electric field to a measurable electric potential. For this field component detector geometries like plates or other conductive parts acting as capacitance plates are typically used. For magnetic fields the detector principle often relies on inductance using such geometrical configurations like coils. Other detection systems like Hall sensors, magnetoresistive devices, or superconducting quantum interference devices (SQUID) magnetometers could be applied to the same end. However, based on the considerations of the previous sections and the detector systems used in the presented experiments, the following discussion of sensing systems will focus only on the detection of the electrical field components.

Gade et al. previously presented a modeling approach, which allows computing EME signals of almost identical shape to experimental EME signals [12].

To perform a validation of the modeling approach, all required details of the experimental configuration were implemented as 3D geometry. Using a system transfer function accounting for the influence of the preamplifier and acquisition system, the modeled signals at the EME sensor in Fig. 5.30a may be converted to an EME signal for comparison with a respective experimental signal as plotted in Fig. 5.30b.

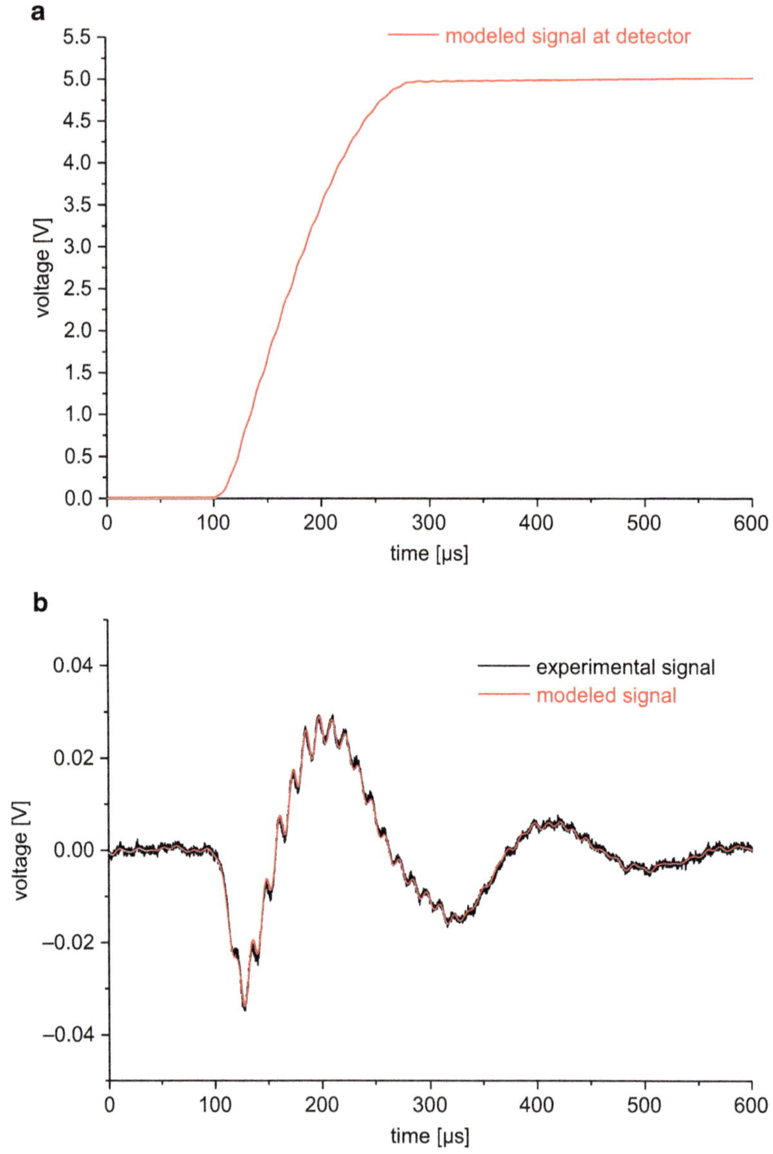

Fig. 5.30 Validation of EME sensor modeling using modeled signal voltage at detector (**a**) subject to system transfer function and comparison to experimental signal (**b**) (based on [12])

An input surface charge density with a maximum of 3.3 nC/m^2 applied as steplike source function with superimposed oscillations resulted in a calculated electric potential at the surface of a capacitance plate sensor as seen in Fig. 5.30a. After convolution by the system transfer function, the modeled signal matches the amplitude and shape of the experimental signals as compared in Fig. 5.30b. This demonstrates that the proposed EME sensor modeling approach is suitable to describe the experimentally obtained signals.

To compare different sensing systems and to point out some basic principles, all detector geometries are evaluated within the same validated modeling environment in the following. As modeled source type, the SENB test specimen geometry as presented in the source modeling of Sect. 5.2 is used. This comprises a bar of solid material (10 mm $\times$ 2 mm $\times$ 2 mm) with a constant charge density of 0.020 C/m^2 on the fracture surface at the center of the specimen. A sphere of 100 mm diameter surrounds the test object and the sensing systems to allow visualization of the electric field lines in a spherical volume of free space. The modeled sphere is terminated by infinite element domains of 5 mm thickness, causing a virtual scaling of the spatial coordinates to infinity as seen in Fig. 5.10 in Sect. 5.2. Thus the modeled setup for the source and detector system has no nearby boundaries to interact with. The floating potential at the conductive surfaces of the detector system is input to a circuit simulation consisting of a parallel series of a 70 pF capacitor and a 10 MΩ resistor to implicitly account for the measurement systems band-pass characteristic. The resulting EME signal is then evaluated as voltage across the 10 MΩ resistor. All computations are carried out using the "AC/DC module" of COMSOL Multiphysics.

In the following, the results of different EME detector concepts are presented. First, the geometrical configuration of the different concepts is outlined, and the interaction of the detector geometry with the electric field lines is shown. Second, the resultant transfer functions are discussed in terms of their absolute values and their frequency characteristics. To facilitate the reading, these two aspects are presented for three basic categories of sensor types. These are capacitance plate sensors, wire sensors, and coil sensors. Finally, some basic factors of influence in sensor configuration and selection are analyzed, and their relative sensitivity to detect EME is discussed.

5.4.1.1 Capacitance Plate Sensors

As first basic category of EME detector systems, the results from capacitance plate sensors are demonstrated. The geometrical configuration of the capacitance plate sensor is shown in Fig. 5.31a including the modeled lateral dimensions of the plate of a thickness of 1 mm. The configuration in Fig. 5.31a uses two solid plates modeled using electrical properties of copper of Table B.3 in Appendix B. While the surfaces of one of the plates is set as electrical ground (i.e., U = 0 V), the generated charges on the surface of the other plate are used as input to the electrical circuit simulation. Modeling of a charged surface at the center of the test specimen

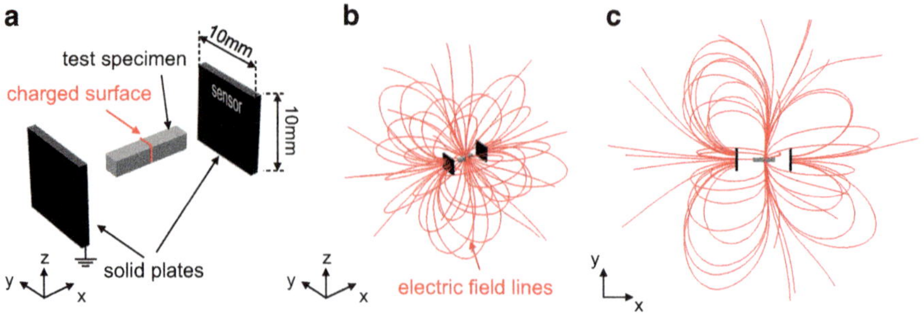

Fig. 5.31 Geometrical configuration of capacitance plate sensor (**a**), electric field lines in 3D view (**b**), and electric field lines projected to *xy*-plane (**c**)

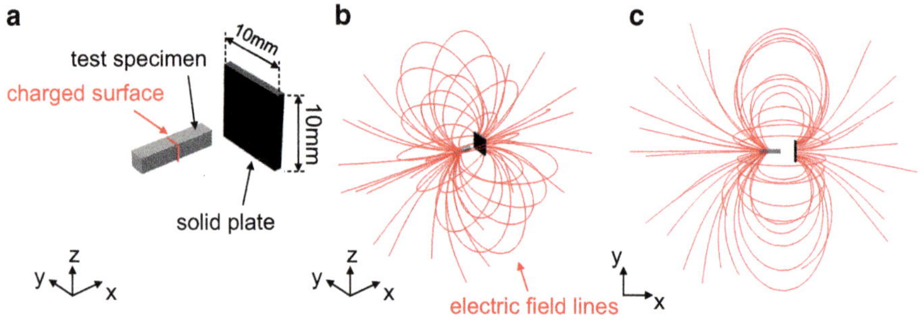

Fig. 5.32 Geometrical configuration of floating capacitance plate sensor (**a**), electric field lines in 3D view (**b**), and electric field lines projected to *xy*-plane (**c**)

causes the formation of electric field lines as shown in Fig. 5.31b, c. As seen from the 3D view in Fig. 5.31b, the field lines emerge at the charged surface and terminate at the conductive capacitance plate sensors, either on the grounded plate or on the detector plate. The resultant pattern of electric field lines is even better visualized in the projection of the field lines to the *xy*-plane shown in Fig. 5.31c. The field lines "cutoff" at some distance to the detector plates is due to the presence of the infinite element domain at this position. As mentioned above this boundary domain allows a scaling of the spatial coordinates toward infinity, so the field lines are truncated at the entry to the infinite element domain.

As small modification to the configuration of Fig. 5.31a is the removal of the grounded capacitance plate as seen in Fig. 5.32a. For such configurations, the ground wire is left unconnected, so the signal wire is operating on a floating ground. This small modification causes some distinct differences in the shape of the field lines. In contrast to the field lines in Fig. 5.32b, c, the floating capacitance plate sensor forms a dipole-like pattern together with the charged surface. All field lines emerge at the charged surface and terminate in the capacitance plate thus transmitting all field energy into the sensing system. However, such ideal situations as given by the infinite element boundary domain are hardly realized in practice. Instead,

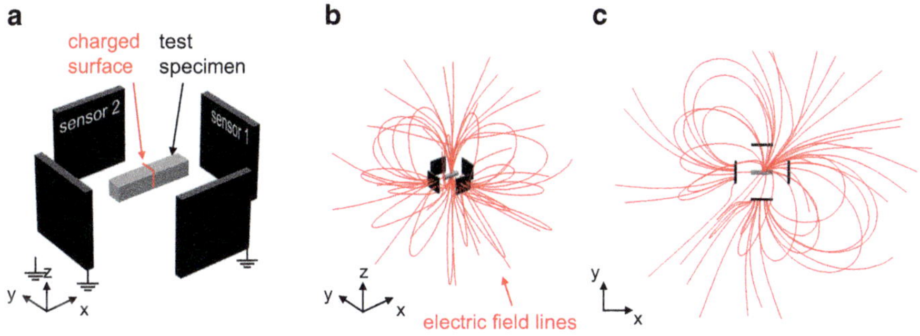

Fig. 5.33 Geometrical configuration of two orthogonal capacitance plate sensors (**a**), electric field lines in 3D view (**b**), and electric field lines projected to xy-plane (**c**)

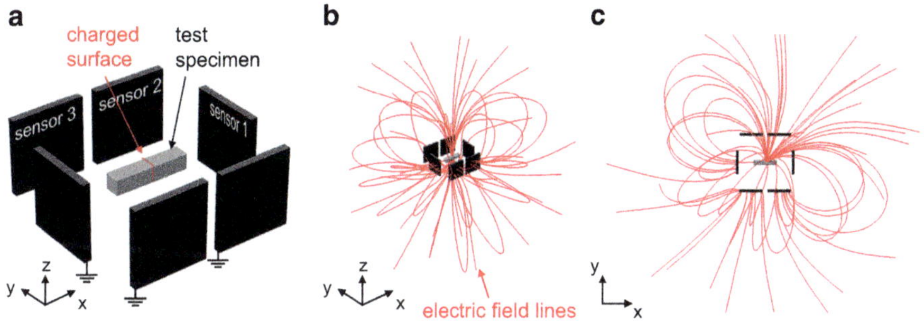

Fig. 5.34 Geometrical configuration of three orthogonal capacitance plate sensors (**a**), electric field lines in 3D view (**b**), and electric field lines projected to xy-plane (**c**)

field lines will terminate into other nearby conductive media (cf. Fig. 5.28 in Sect. 5.3) and therefore reduce the effectiveness of this sensor configuration.

In order to detect EME signals with more than one distinct angle relative to the source, the geometrical arrangements of Figs. 5.33a and 5.34a are also interesting. Using an assembly of either two capacitance plate pairs with perpendicular orientation relative to each other or three capacitance plate pairs with different angles relative to the source allows measuring the orientation of the fracture surface as will be demonstrated in Sect. 5.5.2. However, as pointed out in Sect. 5.3, the increased amount of conductive materials in the surrounding area of the EME sources reduces the detection sensitivity of the individual sensor. This may also be readily derived from the reduced density of electric field lines terminating in the detector plates in Figs. 5.33b, c and 5.34b, c. Moreover, due to the orientation of the grounded capacitance plates, the electric field lines are also oriented along the principal axis of the geometrical configuration (i.e., 45° orientation in the xy-plane for configuration in Fig. 5.33).

5.4.1.2 Wire Sensors

The second basic group of EME detector systems is identified as all configurations using single-ended wires. In contrast to the capacitance plate sensors, the basic geometry of the detector system is thus build from a conductive wire instead of a conductive platelike material. As simplest of all configurations, the open wire sensor is shown in Fig. 5.35a. Here a small piece of a conductive wire is placed close to the EME source or is formed into a wire loop. In practice, various orientations of the wire loop may be used following the considerations of the source radiation patterns established in Sect. 5.2.1.2. For the arrangement shown in Fig. 5.35a, a 100 μm diameter wire made of copper was modeled as open loop enclosing the charged surface. As seen from the resultant electric field lines in Fig. 5.35b, c, this yields a similar dipole-like pattern as observed for the floating capacitance sensor of Fig. 5.32.

Another geometrical arrangement using wire sensors is shown in Fig. 5.36a. Here the wire is formed into a spiral loop of 10 mm diameter with open end. This flat structure may then be placed next to an EME source in similar configurations as presented for the capacitance plates in the previous section. Based on the dipole-like appearance of the electric field lines, it may be concluded that

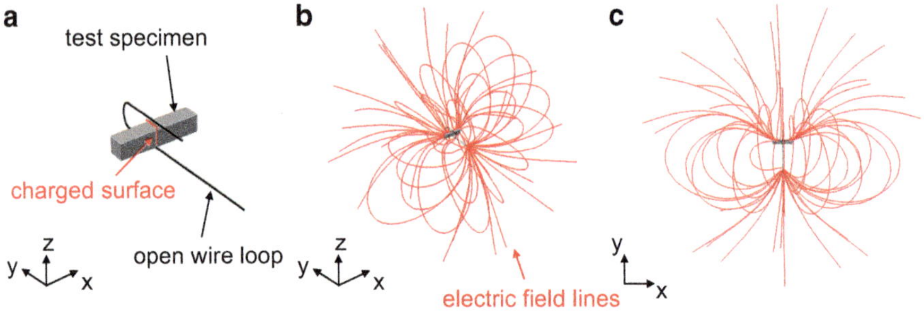

Fig. 5.35 Geometrical configuration of open wire sensor (**a**), electric field lines in 3D view (**b**), and electric field lines projected to *xy*-plane (**c**)

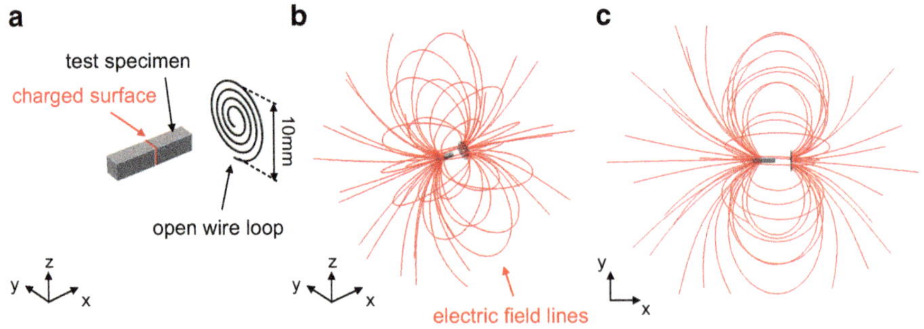

Fig. 5.36 Geometrical configuration of open wire plate sensor (**a**), electric field lines in 3D view (**b**), and electric field lines projected to *xy*-plane (**c**)

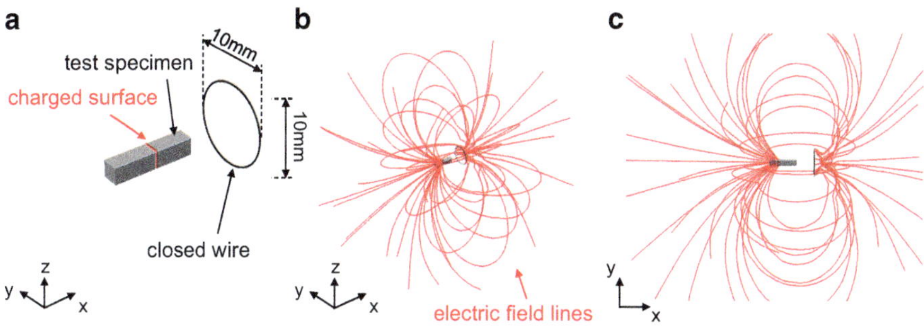

Fig. 5.37 Geometrical configuration of closed wire plate sensor (**a**), electric field lines in 3D view (**b**), and electric field lines projected to *xy*-plane (**c**)

this configuration does effectively operate almost identical to a floating capacitance plate. Since the overall dimensions of the modeled sensors are approximately identical, the good match of the respective transfer functions shown later in Fig. 5.41 is not surprising. Hence, this configuration is termed "open wire plate" sensor.

A further possibility of wire-based sensor configurations is the use of closed wire loops shown in Fig. 5.37a. Similar to the previous configuration, this resembles the overall shape of a floating capacitance plate and is thus referred to as "closed wire plate" sensor in the following. Here the signal wire is attached at an arbitrary position along the circumference and the ground wire is left unconnected. As seen from the Fig. 5.37b, c, the electric field lines interact with the sensing system in a similar fashion as for the floating capacitance plate. As for the configuration shown in Fig. 5.36a, this is due to the large wavelength of the electric field in the frequency range up to 10 MHz, which does not allow to resolve the geometrical difference between the configurations in Figs. 5.32a, 5.36a, and 5.37a.

5.4.1.3 Coil Sensors

The third basic group of EME detector systems comprises all sorts of coil configurations. Traditionally, coils are associated with inductance as primary principle of interaction with electromagnetic fields. However, since the electric field component is the quantity of interest, different approaches are required.

Together with the attached circuit, all coils form an electrical resonator. But as long as the resonance frequency is sufficiently far away from the frequency range of interest, such systems may still be used to detect EME signals with comparatively flat transfer characteristics. And likewise, one can design resonator circuits with resonance frequencies within the frequency range of interest, which are sensitive, but narrow-band EME detector systems.

For the two modeled coil configurations shown in Figs. 5.38a and 5.39a, a wire of 200 μm diameter was wound into coils of ten turns in circular cross-section yielding an inductance of 250 nH. Together with the attached circuit, this yields a

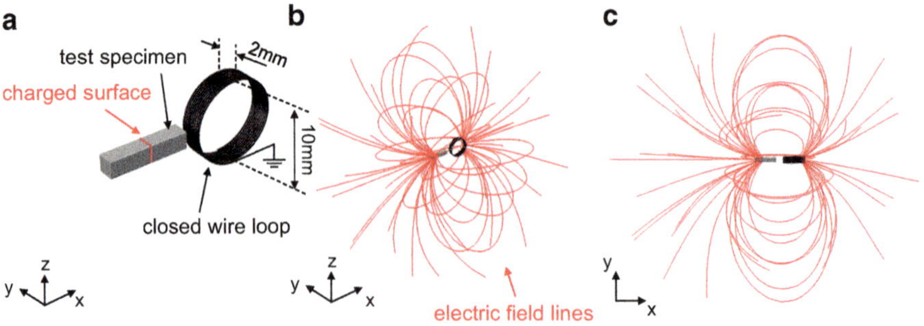

Fig. 5.38 Geometrical configuration of coil sensor with main axis in x-direction (**a**), electric field lines in 3D view (**b**), and electric field lines projected to xy-plane (**c**)

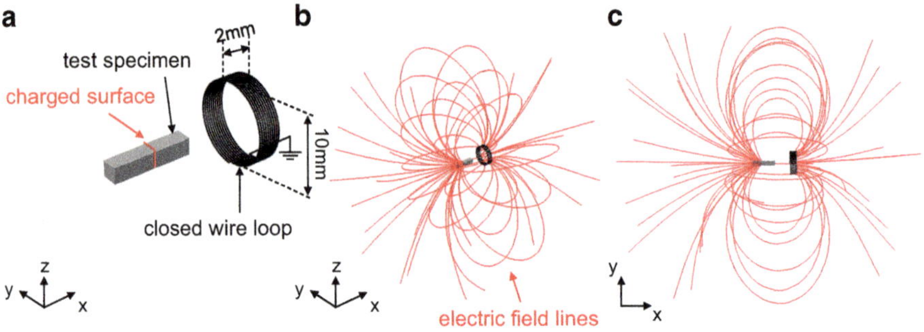

Fig. 5.39 Geometrical configuration of coil sensor with main axis in y-direction (**a**), electric field lines in 3D view (**b**), and electric field lines projected to xy-plane (**c**)

resonance frequency of 38 MHz and a bandwidth of 227 Hz with reasonable sensitivity, which is sufficiently far away from the investigated frequency range up to 5 MHz. To resemble a real experimental configuration, one cross-section of the coil is grounded and the other cross-section is input the modeled electrical circuit.

In such conditions, the coil operates as simple metallic conductor, which causes the formation of field lines as seen in Figs. 5.38b, c and 5.39b, c for the two different coil configurations, respectively. Hence, the configuration of electric field lines is again best described by a dipole-like shape with emerging lines at the charged surface and terminating lines at the surface of the coil.

5.4.1.4 Comparison and Common Aspects

To compare the sensitivity of the different EME sensor systems investigated in the previous section, it is useful to calculate their transfer functions. To this end, the modeled test specimen is replaced by a small copper antenna, which is driven by a sinusoidal voltage with 5 V peak amplitude. A frequency sweep is carried out for each configuration ranging from 10 Hz to 5 MHz with nonlinear frequency steps adapted to the logarithmic frequency scale. The circuit is kept identical to the previous section. The transfer function is then derived from the voltage drop at the 10 MΩ resistor divided by the source voltage. To validate this procedure, experimental measurement results using the described configuration for the floating capacitance plate and the wire sensor are shown in Fig. 5.40a, b as reference.

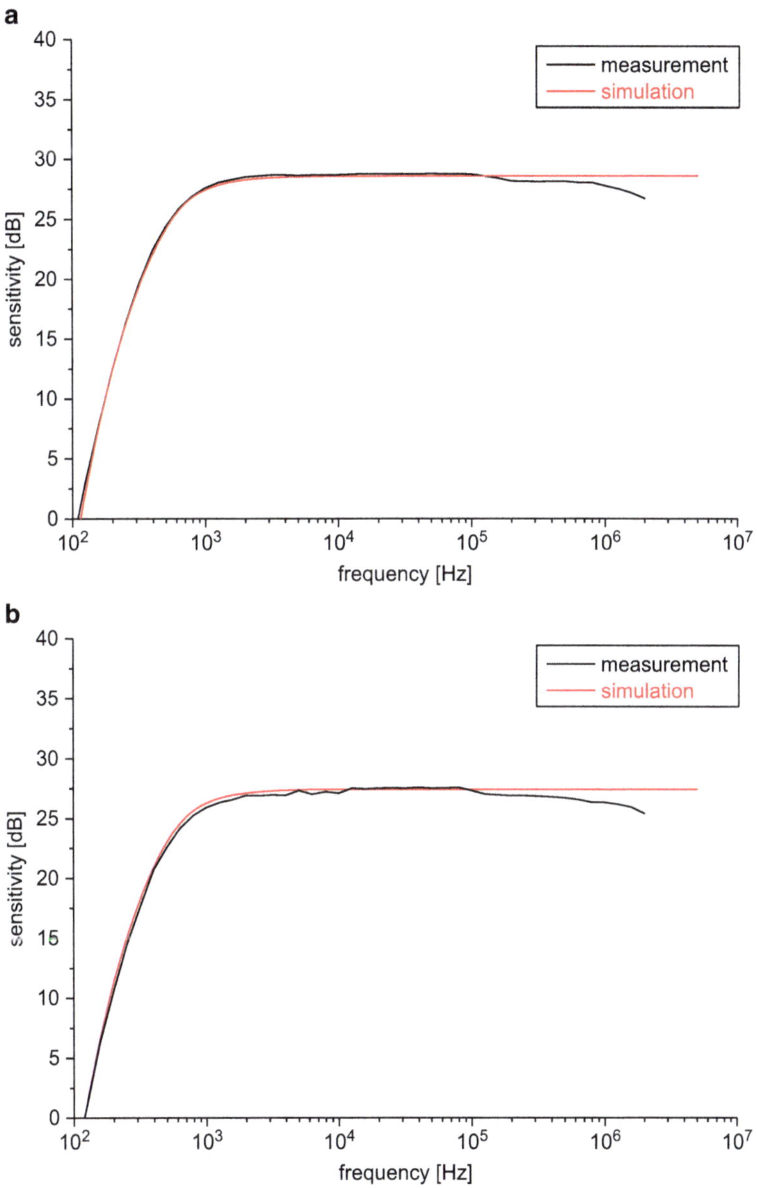

Fig. 5.40 Frequency characteristic of EME sensor using floating capacitance plate (**a**) and open wire sensor (**b**)

The corresponding modeling results were amplified by 26 dB to conform with the experimentally used preamplifier gain. Quantitatively, the 1.8 dB difference in sensitivity of the floating capacitance plate and the open wire sensor are well described by the model configuration. While the high-pass characteristic of both sensor systems is captured by the modeled high-pass given by the resistor and capacitor in the attached circuit, the falloff at higher frequencies is not reflected well in the present model configuration. This low-pass characteristic is owed to the 3 MHz cutoff frequency of the acquisition system used to record the experimental data and therefore is not accounted for in the EME sensor model. However, since the frequencies found for the experimental EME signals so far do not exceed 1 MHz, the present implementation of EME sensor modeling is assumed to be sufficiently validated within the frequency range of interest.

Based on this experimental validation, it may be assumed that the evaluated transfer functions of the remaining EME sensor systems are sufficiently accurate to allow quantitative comparison of the different systems. A comparison of the modeling results for the different sensor systems described in the previous sections is shown in Fig. 5.41.

The floating capacitance plate was found to show the highest sensitivity among the cases studied. Also, except for the high-pass characteristic given by the modeled circuit, there is no deviation from a flat frequency response for this detector system. All other floating sensor systems yield almost similar sensitivity, their differences being attributed to the size of the detector surface and their relative orientation to the source. The latter is seen best for the coil sensor investigated. For this sensor type, there is no indication of resonances in the plotted frequency range as expected from the calculated resonance frequency of 38 MHz.

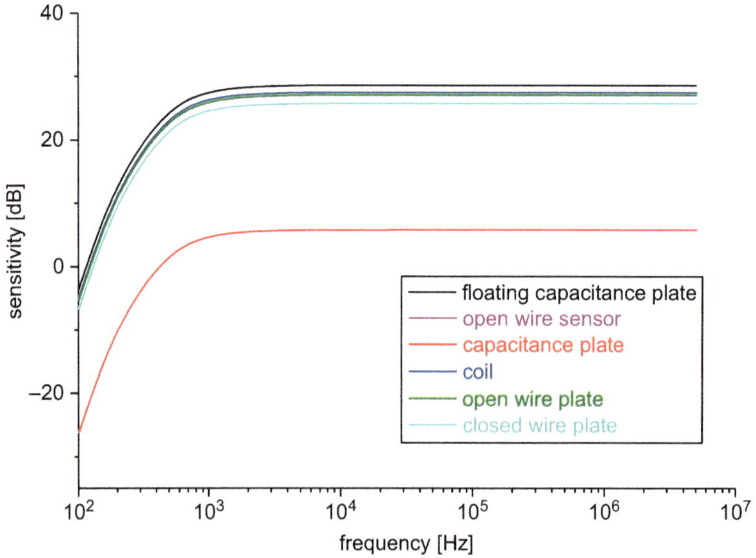

Fig. 5.41 Comparison of transfer functions of different EME sensor systems

One obvious result is the reduced sensitivity of the capacitance plate sensor with grounded plate as compared to all other EME sensor types in Fig. 5.41. This is due to the chosen configuration involving an explicitly grounded metallic plate close to the EME source. As discussed in Sect. 5.4.1.1, this causes a partial termination of electric field lines at the grounded plate and therefore reduces the sensitivity of the sensing plate by 23 dB compared to the floating capacitance plate (an EME detector of otherwise identical geometry).

However, these results need to be interpreted carefully. In reality all sensor systems will need a respective electrical ground to allow a voltage measurement. Surrounding conductive materials (i.e., fixtures, shielding chamber) which are electrically grounded will thus reduce the sensitivity of the sensor system. In addition, an increase in distance to the EME source will cause a significant reduction of the measured signal strength as demonstrated in Sect. 5.2.1.3. But based on the modeling results of Sect. 5.2, it may be concluded that for a fixed source-sensor distance, the potential at the position of the sensor is a linear function of the electric charges present at the crack surface. Consequently, the EME sensor transfer function could also be expressed in terms of measured voltage per surface charges, yielding the same curves as shown in Fig. 5.41 but resulting in different dB values.

Many other EME sensor configurations may be derived from the basic categories presented in the previous sections, all of which will yield unique properties if the presented aspect ratios are changed. To judge on their sensitivity and bandwidth, a direct measurement of the sensor (and system) transfer function using the artificial EME sources as described in Sect. 5.2.2 is recommended for every new design.

Apart from this, all EME sensor configurations share some common aspects, which shall be discussed next. Among these aspects is the choice of sensor material, the positioning of the sensor system, the preamplifier choice, and the acquisition system.

Material

The basic choice for fabrication of the EME detectors is a conductive material. Trials with different electrically conductive materials did not reveal substantial differences in the efficiency of the sensor system as long as the material is a reasonable electrical conductor. This naturally includes all types of metals and their alloys. Due to the good availability and high electrical conductivity, copper alloys are the first choice for an EME detector material. Moreover, copper is easy to machine and comes with good surface compatibility to allow soldering of electrical wires for connecting the EME detector to cables.

Positioning

Similar to the considerations made for the artificial EME sources in Sect. 5.2.2, there are some common guidelines to the positioning of EME sensor systems. It is advantageous to fabricate all types of clamps or fixtures from nonconductive materials. Otherwise the presence of these sensor holders will adversely affect the sensor performance. Furthermore, the EME sensor needs to be kept at a constant position during the experiment, since any relative movement to the EME source will induce an additional signal voltage. Hence, EME sensor fixtures may either rest at a fixed (global) position or may be attached to the test specimen to retain a fixed position relative to the specimen. The decision between either of these configurations is subject to the respective experiment. Some practical examples for each configuration are found in Sect. 5.5 to assist in an appropriate selection.

Preamplifier

As indicated in the previous sections, the system response of an EME sensor is directly influenced by the attached circuit. For all practical considerations, this means a cable ranging from the EME detector to the preamplifier and a second cable ranging from the preamplifier to the acquisition system. The importance of these settings may readily be understood from the change in the detected EME signals shown in Fig. 5.42b for a variation of input capacitance ranging from 70 pF to 140 pF and 210 pF. The value of the connected capacitance is dominated by the input capacitance of the selected preamplifier and the length of the cable between sensor system and preamplifier.

In analogy to the developments in acoustic emission measurement technology, it was further found to be suitable to use a cascade of preamplifiers in order to optimize the signal-to-noise ratio of the measurement chain. Thus, one J-FET preamplifier is preferentially located directly attached or physically close to the EME detector system and provides a first amplification of the detected signal voltage. In a subsequent stage, the amplified signal is fed into another preamplifier with broad operational bandwidth to retain the large bandwidth provided by the EME sensor systems. A typical circuit for the first preamplifier stage is shown in Fig. 5.42a and is based on the related work used for acoustic emission sensors by Hamstad et al. and Shiwa et al. [58–60].

Acquisition

As final piece of the measurement chain, the amplified signal needs to be recorded by an acquisition system. To this end, digital storage oscilloscopes or transient recorders may be used. Similarly, typical commercial acoustic emission systems capable of transient signal acquisition can be applied for this purpose. Compared to the typical acoustic emission signals, one difference of the acquired signals is the characteristic shape of EME signals, which may require some modifications to classically used trigger settings. Also, EME signals of larger (macroscopically) fracture events were found to have durations ranging into the millisecond ranges. If these are meant to be stored as single event, the acquisition depth needs to be adequate to store such long signals at the required digitization rate.

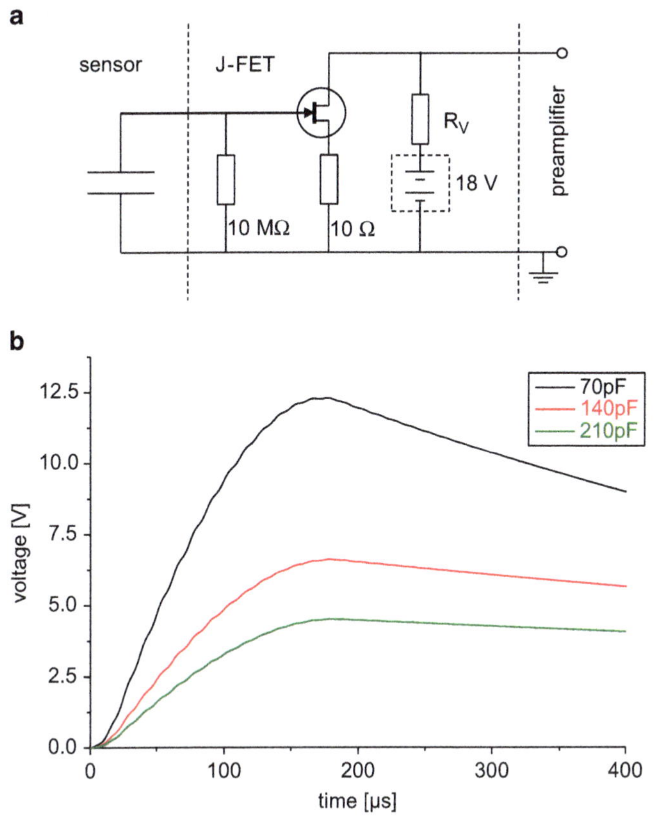

Fig. 5.42 Typical preamplifier circuit for first stage (**a**) and variation of attached circuit parameters (**b**)

5.4.2 Electromagnetic Shielding

Although the detection of EME signals may seem experimentally straightforward based on the descriptions of the previous section, there is one considerable experimental drawback involved in their measurement. Since electromagnetic fields obey the superposition principle, any other electromagnetic field present during the detection process may interfere with the useful EME signal of the source of interest. Other external electromagnetic fields within the frequency range of the measurement may also significantly contribute to the detected signal level and shape.

In most locations, there is a tremendous amount of sources for electromagnetic waves in the relevant frequency spectrum. The sum of the contribution of all of the sources is usually considered as electromagnetic noise, consisting of several characteristic peaks and a comparatively continuous background noise floor. Hence, electromagnetic shielding is the practice of reducing this electromagnetic noise by blocking external fields with barriers made of conductive or magnetic materials.

Thus, EME measurements are carried out within a shielding chamber suitably reducing the surrounding electromagnetic noise level and thus allowing detection of the electromagnetic field of the EME source. Generally, shielding effectiveness needs to be distinguished for electric fields and magnetic fields, because different physical mechanisms are relevant to reach the shielding of the one or the other.

As general measure for the respective shielding effectiveness (SE), the following definition is conventionally used in electronics engineering for electric fields and magnetic fields [61–63]:

$$\mathrm{SE} = 20\log\frac{|\boldsymbol{E}_{\text{unshielded}}|}{|\boldsymbol{E}_{\text{shielded}}|} \tag{5.19a}$$

$$\mathrm{SE} = 20\log\frac{|\boldsymbol{H}_{\text{unshielded}}|}{|\boldsymbol{H}_{\text{shielded}}|} \tag{5.19b}$$

The amount of reduction depends very much on the material properties, the chamber thickness, the size of the shielded volume, and the frequency of the electromagnetic fields. Moreover, the size, shape, and orientation of apertures (e.g., penetrations, holes) in a shielding chamber may reduce the shielding effectiveness and therefore need to be considered explicitly.

Especially in the reactive near-field distance, the electric and magnetic field requires different methods of shielding, since the electric waves are typically being reflected, whereas magnetic fields are absorbed instead.

Hence it is convenient to distinguish between three categories of electromagnetic shielding:

1. Electric field shielding
2. Magnetic field shielding
3. Shielding enclosures

This distinction is quite useful, since electromagnetic fields first of all consist of coupled electric and magnetic fields with individual requirements to shield the one or the other field component.

The electric field results in forces acting on the charge carriers in the shielding material. If this material is an ideal conductor, the electric field component induces a current that causes internal charge displacement generating an opposed electric field canceling the externally applied electric field. In contrast, oscillating magnetic fields generate eddy currents inside a conductive material that drive an opposing magnetic field canceling the externally applied magnetic field. For static magnetic fields, there is no shielding effect unless the shielding material moves relative to the static magnetic field. In both cases, the result is that electromagnetic fields do not penetrate the shielding material. As a distinctly different category, shielding enclosures are also used to protect a setup from unwanted external electromagnetic fields. With increasing frequencies, it may happen that a simple shield starts to emit an electromagnetic wave, which sometimes is even more intense than the original source. This effect starts to dominate, when the dimensions of the shield approach

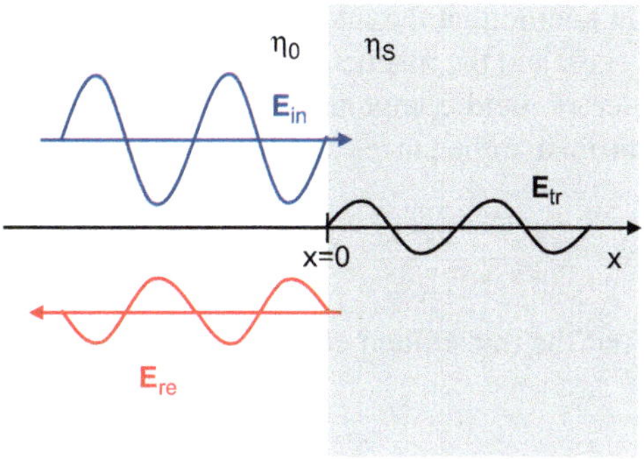

Fig. 5.43 Plane wave incident on a conductive material

fractions of the wavelength and thus act as antenna more than a shield. For such cases and to shield electromagnetic fields in a broad frequency range, it is useful to encapsulate a setup completely inside a perfectly conductive material.

Several factors limit the shielding effectiveness in practice. The finite electrical resistance of the shielding material causes the excited internal field not to completely cancel the incident external field. Also, for the case of magnetic shielding, many conductive materials exhibit a ferromagnetic response at low frequencies including hysteretic losses during generation of the eddy currents, and thus the external field is not completely attenuated. As technical implication, any geometric imperfections of the shield itself, such as apertures, holes, or electrical connectors, reduce the overall shielding effectiveness.

In order to understand the shielding of electromagnetic fields, one can first of all consider the incidence of a plane wave propagating from one material into another. As with other laws of refraction, the incident electromagnetic wave is partially reflected and partially transmitted when incident to an interface between two materials of different electrical properties. For the case of orthogonal incidence of a plane wave from one material with impedance η_0 to another material with impedance η_S, the incident electric field component $\boldsymbol{E}_{\text{in}}$ is partially transmitted $\boldsymbol{E}_{\text{tr}}$ and partially reflected $\boldsymbol{E}_{\text{re}}$ (Fig. 5.43).

The magnetic field orientation in a plane wave is perpendicular to the electric field, and the magnetic field amplitude in free space with intrinsic impedance η_0 is related to the electric field amplitude by

$$|\boldsymbol{H}_{\text{in}}| = \sqrt{\frac{\xi_0}{\mu_0}}\,|\boldsymbol{E}_{\text{in}}| = \frac{1}{\eta_0}\,|\boldsymbol{E}_{\text{in}}| \tag{5.20}$$

Similarly, the magnetic field amplitude in the shielding material is directly linked to the electric field amplitude by

$$|\boldsymbol{H}_{\text{tr}}| = \frac{1}{\eta_S}\,|\boldsymbol{E}_{\text{tr}}| \tag{5.21}$$

In addition, at the interface between both materials, continuity of the electric field vector and the magnetic field vector is required. Thus the amplitude of the reflected electric field component may be written in terms of a characteristic ratio of both intrinsic impedances and the incident amplitude:

$$|\boldsymbol{E}_{\text{re}}| = |\boldsymbol{E}_{\text{in}}| \frac{\eta_S - \eta_0}{\eta_S + \eta_0} \tag{5.22}$$

And the transmitted electric field amplitude in a similar way as

$$|\boldsymbol{E}_{\text{tr}}| = |\boldsymbol{E}_{\text{in}}| \frac{2\eta_S}{\eta_S + \eta_0} \tag{5.23}$$

Equations (5.22) and (5.23) imply that as η_S approaches η_0, the transmission increases and the reflection decreases. For matching impedances $\eta_S = \eta_0$, the incident field is fully transmitted.

As additional effect, a real material has finite conductivity (i.e., $\tilde{\sigma} > 0$), and the transmitted electric field amplitude will decrease with distance to the surface due to the skin effect. The electric field amplitude as function of depth below surface may thus be written as function of the skin depth δ of the material:

$$|\boldsymbol{E}_{\text{tr}}(x)| = |\boldsymbol{E}_{\text{tr}}(x = 0)| \mathrm{e}^{-\frac{x}{\delta}} \tag{5.24}$$

The skin depth δ is usually well approximated assuming

$$\delta \cong \sqrt{\frac{2}{\omega \tilde{\sigma} \mu_r \mu_0}} \tag{5.25}$$

using the conductivity of the material $\tilde{\sigma}$, the angular frequency ω, the relative magnetic permeability of the material μ_r, and the permeability of free space μ_0.

When the electromagnetic wave is incident to a material with finite thickness h, the transmission through the shielding material is readily obtained by combination of the above effects. The incident wave is partially reflected upon penetration of the material, then attenuated within the material, and finally partially transmitted into free space as schematically shown in Fig. 5.44. The law of reflection for transmission at the interface from inside the shielding into free space at $x = h$ is given as:

$$|\boldsymbol{E}_{\text{tr}}| = |\boldsymbol{E}_{\text{tr}}(x = h)| \frac{2\eta_0}{\eta_0 + \eta_S} \tag{5.26}$$

By combination of the equations above, the finally transmitted electric field amplitude may be written for all materials being substantially thicker than one skin depth δ as:

$$|\boldsymbol{E}_{\text{tr}}| = |\boldsymbol{E}_{\text{in}}| \frac{2\eta_S}{\eta_S + \eta_0} \left(\frac{2\eta_0}{\eta_0 + \eta_S} \right) \mathrm{e}^{-\frac{h}{\delta}} \tag{5.27}$$

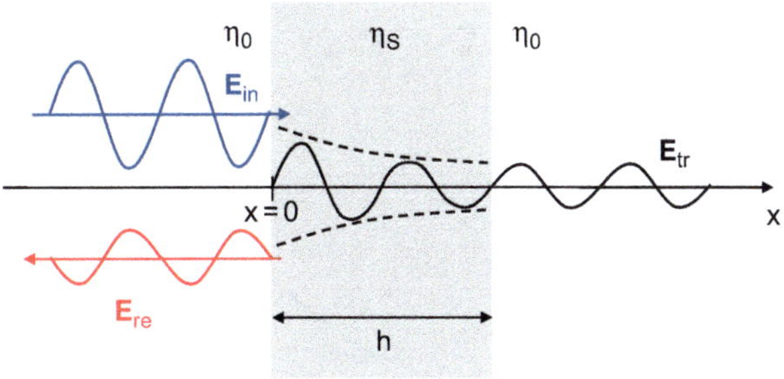

Fig. 5.44 Plane wave incident on a shielding material of finite thickness

It can be derived from (5.27) that good conductors ($\tilde{\sigma} \gg \omega\xi$) will yield the best shielding effectiveness, since this implies $\eta_S \ll \eta_0$.

Based on the definition of the shielding effectiveness in (5.19) and the components of (5.27) the overall contribution to the shielding effectiveness may be split into two parts accounting for the reflection loss R(dB) and the absorption loss A(dB):

$$\mathrm{SE} = 20 \ \log\left(\frac{4\eta_S\eta_0}{(\eta_0 + \eta_S)^2}\right) + 20 \ \log\left(\mathrm{e}^{-\frac{h}{\delta}}\right) = R(\mathrm{dB}) + A(\mathrm{dB}) \qquad (5.28)$$

The magnitude of R(dB) is thus independent of the thickness and comprises a constant contribution to the shielding effectiveness, while the contribution of A (dB) is directly influenced by the ratio between thickness and the skin depth of the material.

Since the reduction of the electrical field component is of primary interest for EME measurements, some shielding effectiveness curves are shown in Fig. 5.45 for a variation of material thickness. Using aluminum with electrical properties as listed in Table B.3 in Appendix B, the calculated shielding effectiveness following (5.28) as function of thickness is plotted in Fig. 5.45. For the frequency range of interest for EME measurements, the thinnest configuration of 100 μm thickness (i.e., aluminum foil) already provides a shielding effectiveness above 100 dB. For practical applications a shielding effectiveness >40 dB was found to be sufficient to reduce the surrounding electromagnetic noise beyond the detectable limit. However, handling of such thin shields in the framework of mechanical testing is quite difficult, so it is more realistic to use shielding chambers within the thickness range of some millimeters. For a thickness of 1 mm, the contribution of A(dB) starts to dominate and is found even more pronounced for a thickness of 10 mm both of them yielding even higher shielding effectiveness.

As materials being used for electromagnetic shielding, metals are obviously the first choice. In order to construct shielding chambers of a decent size and with a thickness providing sufficient mechanical stability, this limits the type of metallic materials to typical engineering alloys based on aluminum, steel, or copper. Other metals, such as silver, gold, or platinum would also be interesting for shielding applications but are economically not reasonable to apply. Using (5.28) the

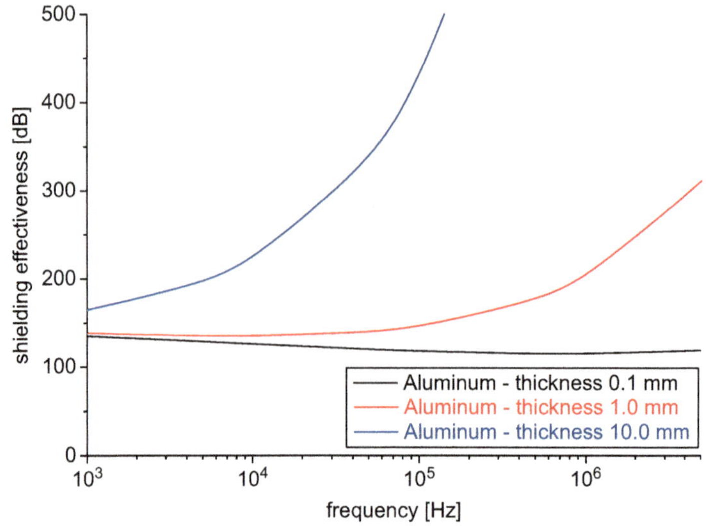

Fig. 5.45 Shielding effectiveness of aluminum shield as function of thickness and frequency

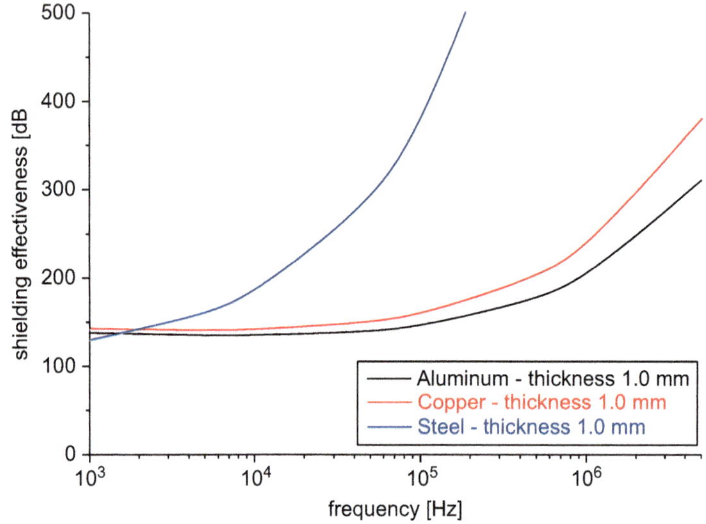

Fig. 5.46 Shielding effectiveness of shield with 1.0 mm thickness as function of material and frequency

shielding effectiveness for a chamber made from 1.0 mm thick, aluminum, steel, and copper applying the electrical properties of Table B.3 in Appendix B are plotted in Fig. 5.46. Based on their different properties, it can be observed that steel would yield the best shielding solution in the relevant frequency range. However, based on handling considerations, a shielding chamber made of steel is less desirable than one made of an aluminum alloy. Since the gain in shielding effectiveness at frequencies below 10 kHz is also fairly low, this makes aluminum the best choice for a solid shielding chamber for an EME application.

As alternative to metallic materials, attempts have been made to use carbon fiber reinforced materials to shield electromagnetic fields [64]. Another commonly used material modification is to coat the inside of the enclosure with a metallic ink. The ink itself consists of a carrier material loaded by suitable metallic particles and thus provides some shielding capability.

Although the sensing systems discussed in Sect. 5.4.1 are not intended to detect the magnetic field component, it may occur that the detector system or the preamplifier electronics also need protective shielding against external magnetic fields. For frequencies above 100 kHz, the dominating effect of shielding is the generation of eddy currents. However, for parts of the relevant frequency range (below about 100 kHz), static or slowly varying magnetic fields are not well shielded by the generation of eddy currents. This is also readily explained by the skin effect. Similar to the electric field, also the magnetic field penetrates the shielding material inducing a voltage. For an incident magnetic field causing a charge flux

$$Q(t) = Q_0 \sin(\omega t) \tag{5.29}$$

this results in a voltage

$$U(t) = \frac{dQ}{dt} = Q_0\, \omega \cos(\omega t) \tag{5.30}$$

Therefore, the induced voltage increases with increasing ω, causing increasing eddy currents to cancel the incident magnetic field. In a similar fashion as the electric field described above, the amplitude of the magnetic field also decreases exponentially with distance to the surface. This skin effect is minimal for low frequencies, allowing most of the magnetic field to be transmitted through the shielding material. But for sufficiently high frequencies, the shielding maximizes and will reach a sufficient level.

To improve the shielding effectiveness at lower frequencies, a modification of the shielding material toward high magnetic permeability metal alloys or ferromagnetic metal coatings is suitable. Since the shielding material is acting as path for the magnetic field lines, the best design in this context is a fully closed container surrounding the setup (i.e., a shielding enclosure).

As general drawback with shielding enclosures of finite size, the geometrical arrangement itself may be subject to resonances, thus reducing the shielding effectiveness at these frequencies. For a simple rectangular box of length a, height h, and width b, the resonance frequencies are given for an arbitrary material by

$$f_{ijk} = \frac{c}{2\pi\sqrt{\mu_r \xi_r}} \sqrt{\left(\frac{i\pi}{a}\right)^2 + \left(\frac{j\pi}{h}\right)^2 + \left(\frac{k\pi}{b}\right)^2} \tag{5.31}$$

This equation can be simplified for the case of an empty box (vacuum conditions) to

$$f_{ijk} = 150\sqrt{\left(\frac{i}{a}\right)^2 + \left(\frac{j}{h}\right)^2 + \left(\frac{k}{b}\right)^2} \tag{5.32}$$

The fundamental resonance case $i = j = k = 0$ is forbidden by definition of (5.32). Thus for a realistic size of a shielding enclosure of dimensions 200 mm × 300 mm × 100 mm (length × width × height), the first resonances are expected for 901, 1250, and 1580 MHz. Since these are well outside the frequency range of EME signals, the primary resonances and their higher orders are not relevant to consider for construction of shielding applications.

So far all considerations have been made for shields or enclosures of sufficient geometrical size to prevent anything within the shielded volume from electrically coupling to anything outside. Such an ideal shielded enclosure with sufficient conductivity and no electrical discontinuities would perfectly shield whatever is inside the enclosure from whatever is outside. In reality all shielding enclosures exhibit deficiencies which reduce the shielding effectiveness and thus require separate discussion in the following subsections. In most practical applications, there are apertures, cable penetrations, or seams which need to be considered in their effect on the shielding effectiveness.

5.4.2.1 Apertures

For practical use of EME in combination with mechanical testing, the aim is to shield the specimen and the EME detector system. Since the specimen itself is attached to load fixtures, there are only two obvious possibilities:

1. Shielding of the specimen, detectors, fixtures, and testing machine within a shielding chamber
2. Fabrication of shielding chamber for specimen and detectors with apertures to connect to fixtures

While the first one would not only require a sufficiently large shielding chamber to enclose the full test equipment, it even comes with the disadvantage of enclosing other potential noise sources, such as the electromotor or computer control systems with the specimen and the detector.[3] Thus it may not necessarily constitute the best solution for this kind of application. Instead, the fabrication of shielding chambers with apertures was found to be the best economical and technical solution. As basic rule of thumb, any apertures must be significantly smaller than the incident wavelength of the electromagnetic waves to be shielded. Else parts of the incident waves

[3] Noise measurements with the same EME sensor system did not reveal a substantial difference of the noise floor when comparing a commercial walk-in shielding chamber and a self-made shielding chamber directly attached around the specimen (as used in [12]).

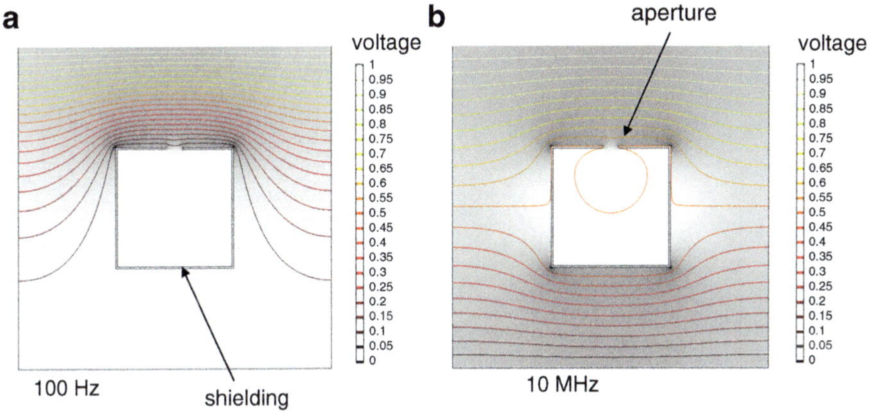

Fig. 5.47 Leakage of electric field component due to 20 mm aperture at 100 Hz (**a**) and 10 MHz (**b**). Contour lines show electric potential and *gray* scale represents magnitude of electric field strength (*white* being zero)

might be transmitted into the shielded volume, and the total shielding effectiveness might be reduced, because of the discontinuity of electric conductivity of the shielding enclosure. This "leakage" is shown as a result from a numerical study using an electromagnetic wave incident to a metallic enclosure as calculated within the AC/DC module of COMSOL Multiphysics. The enclosure includes an aperture of 20 mm at the top edge. At 100 Hz, the aperture is still smaller than the wavelength and the shielding effectiveness of the chamber is not affected (cf. Fig. 5.47a). As seen from the same calculation at 10 MHz frequency in Fig. 5.47b, the electric field component is able to penetrate into the shielded volume, causing a reduction of the overall shielding effectiveness.

The wavelengths of electromagnetic waves in the frequency range up to 10 MHz are larger than 30 m. Hence, all practically relevant aperture sizes for laboratory size specimens (order of few millimeters) may be assumed to be much smaller than the incident wavelengths.

Some measurements of the shielding effectiveness using an aluminum chamber of 320 mm × 320 mm × 120 mm (length × width × height) and 5 mm thickness are shown in Fig. 5.48a. First, the EME sensor is mounted unshielded in free space to collect the locally present noise floor, which is the black FFT spectrum in Fig. 5.48a. Mounting the sensor inside the box, but leaving one side surface open, causes a substantial reduction of the collected noise background as seen by the blue line. If the cover plate is mounted on the box, the shield chamber is fully closed and the incident background noise almost eliminated within the frequency range up to 5 MHz. The remaining falloff in the detected intensity as function of frequency is dominated by the characteristics of the EME sensor and the acquisition system. In the fully enclosed shield chamber, a circular aperture of 10 mm diameter is drilled and then the measurement is repeated. There is no significant increase of the noise floor. Even after drilling of five apertures of 10 mm diameter close by, no change in the detected noise floor is observed.

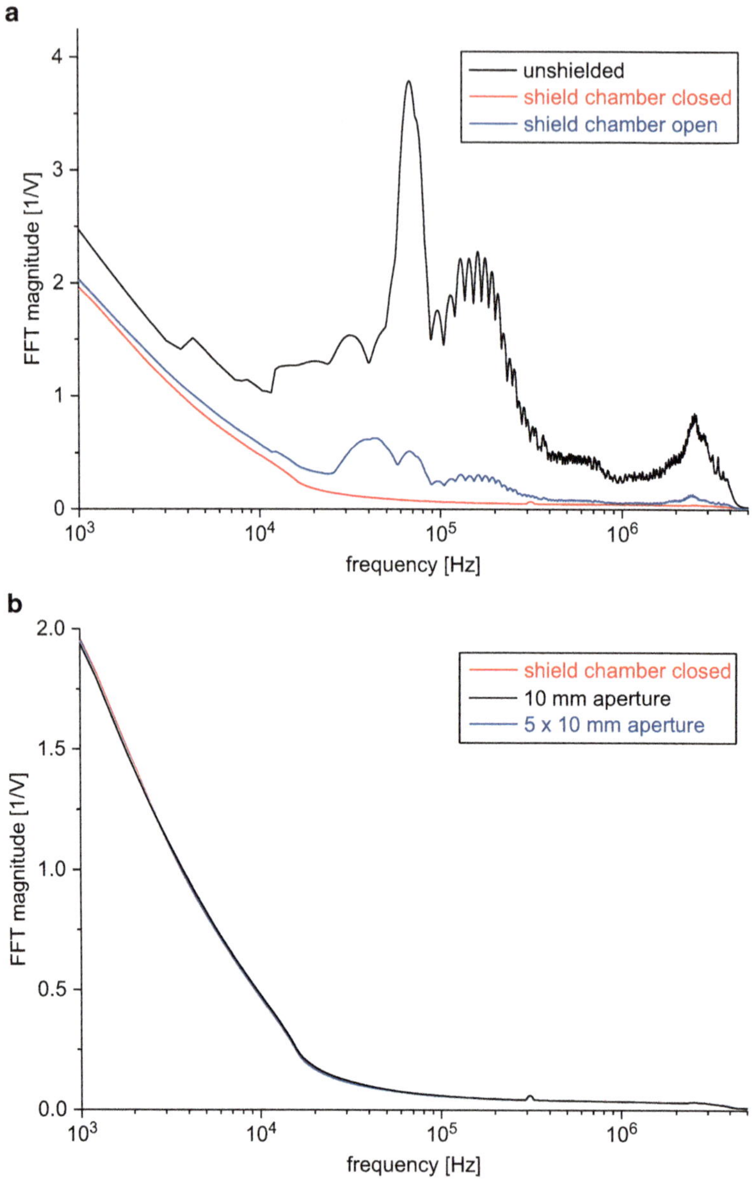

Fig. 5.48 Measurements of the electric noise floor using shield chamber (**a**) and repetitive measurements after application of apertures (**b**)

Such apertures are obviously required as lead through for mechanical fixtures or cables, but are also useful to look into the shielding chamber to monitor specimen behavior. If some leakage occurs due to the presence of apertures, it is possible to increase the depth of an aperture by adding a metallic skirt, which acts as a waveguide below cutoff frequency.

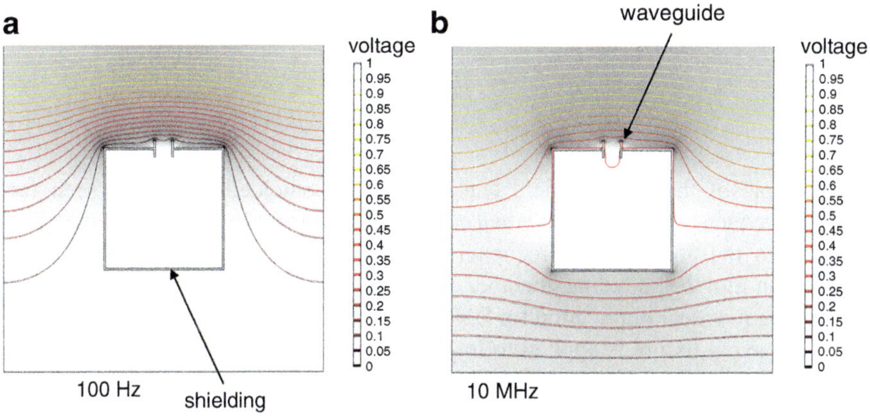

Fig. 5.49 Reduction of leakage of electric field component using waveguide shown at 100 Hz (**a**) and 10 MHz (**b**). Contour lines show electric potential and *gray* scale represents magnitude of electric field strength (*white* being zero)

5.4.2.2 Waveguides Below Cutoff

Using the geometry and computation setup of the previous section, the aperture is modified by adding a metallic skirt perpendicular to the aperture diameter as seen in Fig. 5.49. This modification is motivated by the idea that energy will not propagate within an electromagnetic waveguide at frequencies below its cutoff frequency. Based on the cross-section of the obtained waveguide, the attenuation as function of waveguide length d may be estimated as:

$$\text{attenuation [dB]} \cong 27\frac{d}{a}\sqrt{1-\left(\frac{f}{f_c}\right)^2} \tag{5.33}$$

for a rectangular waveguide of cross-sectional lengths a and b with $b < a$. The cutoff frequency of the lowest propagation mode (the TE_{10} mode) for a hollow waveguide is given by using the propagation velocity $c = 3 \times 10^{-8}$ m/s of the wave as

$$f_c = \frac{c}{\lambda_c} = \frac{c}{2a} \tag{5.34}$$

Similarly, for a circular waveguide of diameter a, one may estimate the attenuation as:

$$\text{attenuation [dB]} \cong 32\frac{d}{a}\sqrt{1-\left(\frac{f}{f_c}\right)^2} \tag{5.35}$$

Equation (5.35) uses the cutoff frequency of the lowest frequency propagation mode (the TE_{11} mode in this case):

$$f_c = \frac{0.586\,c}{a} \tag{5.36}$$

An example of the reduced field leakage into the shielded volume is seen for 10 MHz frequency in Fig. 5.49b when compared to Fig. 5.47b without a waveguide attached to the aperture. At frequencies below the cutoff, such as 100 Hz, no penetration into the shielding chamber occurs.

Another practical implication for EME testing is the inverse use of a waveguide. If a mechanical fixture is applied to load the specimen inside the shield chamber, the typical material choices are metallic fixtures due their high stiffness, strength, and wear resistance. However, metals come with high electric conductivity and will constitute a preferential path of leakage into the shielding chamber. This effect is shown in Fig. 5.50a using the same modeling concept as described above. Clearly, the metallic rod acts as waveguide allowing the electric field to penetrate the shielding chamber. In the EME experiments, it was found that a highly conductive material will not work as fixture system, since it compromises the effect of the shielding chamber almost completely. The usage of low conductive materials, such as a polymer rod seen in Fig. 5.50b, does not affect the shielding effectiveness of the chamber. Hence, fixtures for EME measurements need either to be fabricated out of materials with poor electric conductivity or metallic fixtures need to be electrically isolated so that no conductive path from the outside to the inside of the shielding chamber exists. In this context, PMMA was found to be a primary material for fabrication of fixtures at lower loads, while GFRP was found to be a reasonable alternative for construction of fixtures subject to higher mechanical loads.

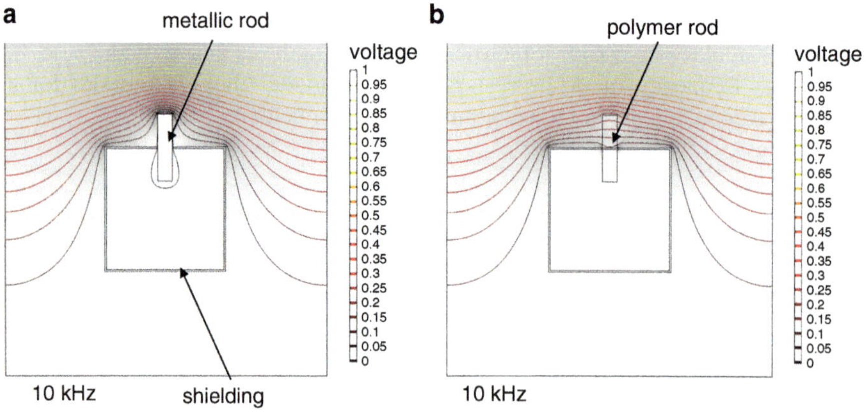

Fig. 5.50 Penetration of electric field components in shield chamber due to metallic rod penetrating the shield (**a**) and reduced penetration using polymer rod (**b**). Contour lines show electric potential and gray scale represents magnitude of electric field strength (*white* being zero)

5.4.2.3 Seams

Similar to the leakage effects due to apertures, other effects come into play when fabricating shielding chambers in practice. Obviously, no metallic box exists, which is fully closed, since this is hardly useful in practical applications. Instead, shield chambers either have some larger apertures or are closed by lids or covers during operation. In the context of EME measurements, these are used to mount the specimen and the sensor system as well as to wire the cables inside the shield chamber. Potentially a leakage of the electric field may occur at the positions of bad electrical contact, which typically are the interfaces between the shielding chamber and the covers and lids. As seen in Fig. 5.51, there is a variety of connector mechanisms available to ensure a good electrical connection between two metallic parts.

Generally, seams between the cover and the shield chamber may form a more significant source of leakage than apertures, since they typically are of much larger dimension. Seams in the order of half-wavelengths may act as dipole-like radiation source causing a substantial increase of the field strength within the shielded volume. Generally, it is not advisable to have simple touching seams as seen in Fig. 5.51 for constructing shield chambers. The disadvantage of these configurations is that they may have bad electrical contact over long distances, while they physically still appear to have contact. Similar challenges arise with rivets or screws, since they provide a good local contact, but may cause waviness of the casing and therefore inhibit good electrical contact in the area between their spots. Seams of shielding chambers are thus typically constructed using overlapping seams, finger stocks, or conductive gaskets to assure a good electrical contact along the full length of the seam.

The leakage of such seams is shown in Fig. 5.52a, b applying the same modeling setup as in the previous sections for two geometries. Instead of a perfect shield chamber, the upper edge is modified by incorporating the geometric dimensions of a seam as seen in the magnified images. At frequencies of 50 MHz and above, the two seams allow a significant portion of the incident field strength to penetrate the shield chamber. However, this may easily be compensated by use of a longer overlapping distance applying the concept of using a waveguide below cutoff frequency.

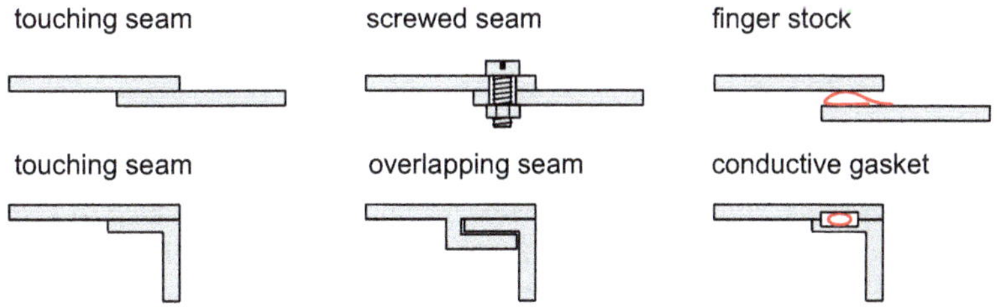

Fig. 5.51 Scheme of different types of connector mechanisms used to combine electrically conductive materials for shielding chambers

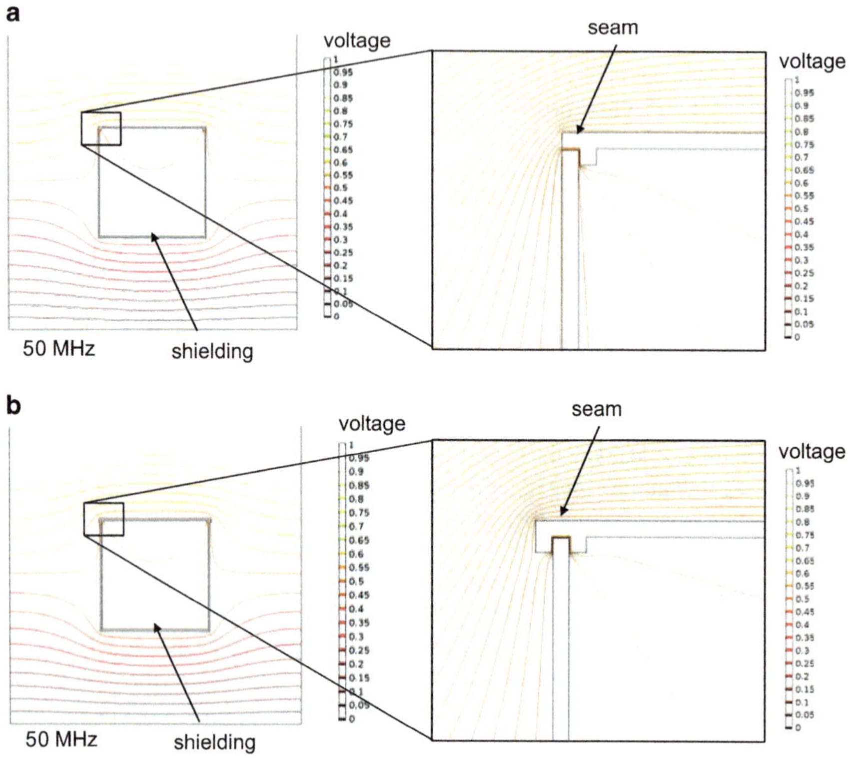

Fig. 5.52 Leakage of electric field component due to two different seams at 50 MHz. Contour lines show electric potential and *gray* scale represents magnitude of electric field strength (*white* being zero)

5.4.2.4 Cable Penetrations

As final component required to build an EME shield chamber for mechanical testing, the EME detectors need to be connected with acquisition equipment on the outside of the chamber. Thus it is necessary to consider cable penetrations. In the first instance, these consist of an aperture to allow physical penetration of the shield chamber by the cables. However, the conductive materials in the cable will cause a similar reduction of the shielding effectiveness of the chamber as discussed for the metallic rod used as fixture system above. Here, a single unshielded, unfiltered wire penetrating a shielded enclosure can completely eliminate any shielding benefit that the enclosure will provide.

For practical EME applications, it has been found a useful practice to keep the shield of the incident wires at the same electric potential as the shielding chamber at all relevant frequencies. For this reason, the wire shielding needs to make a low-inductance connection to the enclosure as seen in Fig. 5.53a. This is typically achieved by a conductive shell enclosing the signal and ground wire penetrating the

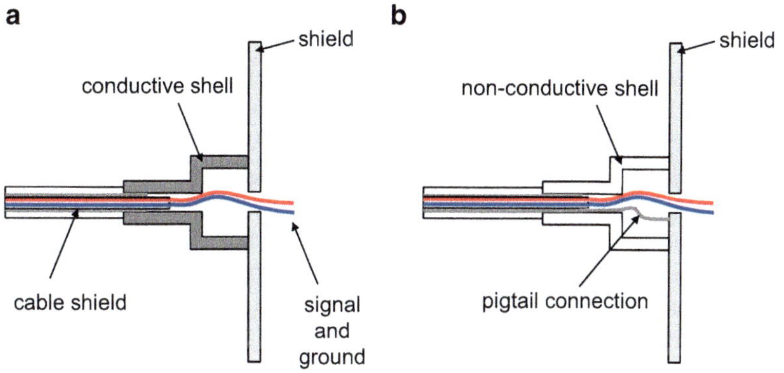

Fig. 5.53 Shielding of cable connections when penetrating shielding chambers using conductive (**a**) and nonconductive (**b**) shell

shield. Such conductive shell connectors are readily available for BNC cables as female-to-female connectors allowing direct connection of a BNC cable on both sides of the shield. Using nonconductive shells in combination with a pigtail connection as in Fig. 5.53b is not recommended, since this will result in a connection with significant inductance.

5.5 Application to Composites

Based on the previously discussed possibilities of the EME method, the focus of this section is the presentation of some experimental studies carried out under laboratory conditions, which have been conducted to assess the capabilities of EME measurements of material failure. The section starts with presentation of some exemplary studies and measurement setups used to detect EME signals due to crack formation. Subsequently some first applications with respect to measurement of fracture surface orientation are provided. The section closes with a discussion of the method sensitivity to detect material failure.

5.5.1 Measurement of EME Due to Crack Formation

Within this section examples of various experimental configurations are given, which have been used to study EME due to crack formation in materials. Since the direct application of a barely understood measurement method to the complexity of failure modes found in composites is quite challenging, some results in application to study polymer failure and failure of carbon fibers are presented first. This introduces some basic test concepts and allows understanding the generation of EME signals in general.

5.5.1.1 Polymer Failure

In order to induce crack growth in pure polymeric material, several test concepts are established in fracture mechanics. In general, distinction is made between the suitability of an experimental setup for particular fracture modes, such as a pure mode I loading. Due to the broad range of fracture behavior of polymers ranging from brittle to ductile, a common test standard for a particular fracture mode is challenging.

An experimental setup used for mode I testing has to fulfill several requirements. From the mechanical point of view, the load introduction has to ensure that a crack opening mode is induced in a notched geometry without significant contributions of plastic deformation at other position than at the notch tip (e.g., at the positions of load introduction). In order to allow a direct comparability of the detected EME signals, identical test geometries are beneficial. To this end simple compact-tension (CT) specimens, double-edge-notched tension (DENT) specimens, and tapered double-cantilever beam (TDCB) specimens as shown in Fig. 5.54 were evaluated for their applicability with a broad range of technically relevant polymers such as epoxy resins (RTM6), polypropylene (PP), polyether ether ketone (PEEK), and polytetrafluoroethylene (PTFE).

For the RTM6 specimen, fully unstable crack growth was observed for the CT and DENT geometry once a crack has initiated. Therefore, only one EME signal was detected during these measurements. For the more ductile thermoplastic polymer types, no crack initiation was found in the CT and DENT geometry. Instead, only large plastic deformation occurred at the positions of the pinholes used for load introduction. Thus no EME signals were recorded. In contrast, the TDCB geometry shown in Fig. 5.54 provided many EME signals for the RTM6

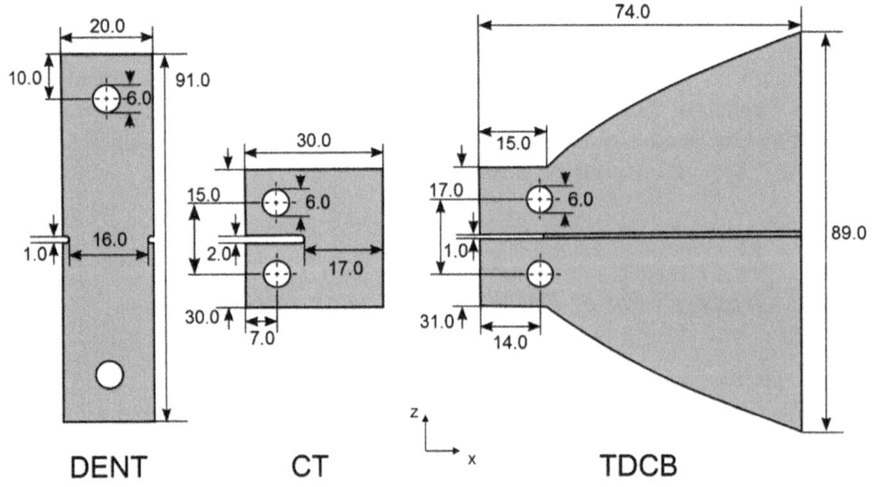

Fig. 5.54 Comparison of different specimen geometries as investigated for mode I testing of polymers (all dimensions in mm)

specimens investigated. This is due to the self-inhibiting crack growth caused by the tapered design of the cantilever beams. For the more ductile materials, large plastic deformation was observed at the pinholes used for load introduction. Consequently the tapered design of the cantilever beams was retained, but the contour was optimized to avoid excessive plastic deformations at the pinholes. An increase of specimen thickness (along y-axis) was not feasible, since curing of RTM6 specimens causes exothermic reactions which lead to specimen degradation for thicknesses values above 5 mm. As a result of a numerical geometry optimization procedure, the most promising geometry that was found to allow crack growth for brittle and ductile specimens is shown in Fig. 5.55. For this configuration, the EME sensor was realized by two horizontally oriented wire loops as shown in red. Those wires form two effective capacitance plates with plate normal directed along the z-axis.

As seen in Fig. 5.55, besides the EME sensor, two acoustic emission sensors of type KRNBB-PC are also attached to the TDCB specimens, which are used to detect the respective acoustic emission signal. All acquisition settings are listed in Table 5.2. All measurements are carried out in an aluminum shielding chamber of 320 mm × 120 mm × 320 mm (length × width × height) and 5 mm wall thickness.

All TDCB specimens were loaded by the fixture shown in Fig. 5.56a. The crosshead velocity was chosen as 5 mm/min with force measurement performed by a 5 kN load cell. The fixture was machined out of stainless steel components with a pull-rod out of polyoxymethylene. This is to avoid the presence of a conductive path to the volume outside the shielding chamber. The pull-rod was designed to yield negligible compliance for the load capacity of 5 kN. Representative load-displacement curves for each of the four polymers are plotted in Fig. 5.56b. While RTM6 and PEEK specimens exhibit almost brittle failure, the two types PP and

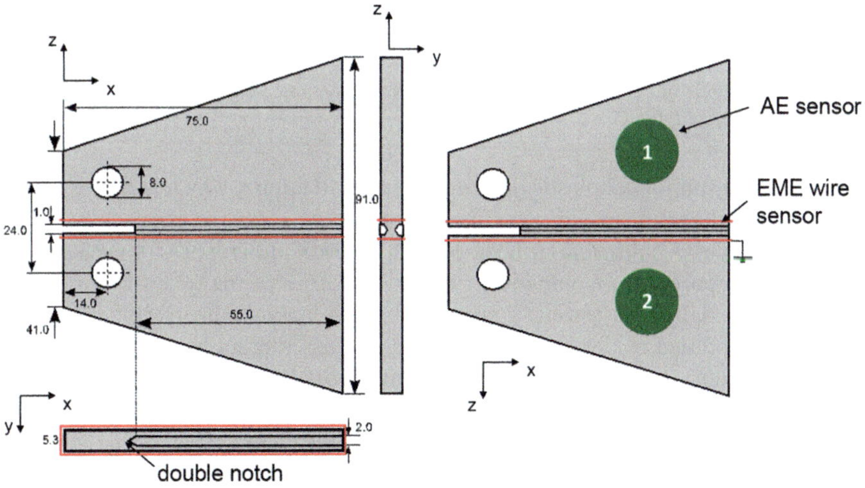

Fig. 5.55 Optimized TDCB specimen geometry used for mode I testing of polymers including AE and EME sensor positions (all dimensions in mm)

Table 5.2 EME and AE acquisition settings used for TDCB tests

Acquisition settings	Value EME	Value AE
Preamplification	23 dB	20 dB
Threshold	35 dB	35 dB
Triggering	Individual channel	Individual channel
Acquisition rate	5 MS/s	10 MS/s
Bandpass range	1 kHz to 3 MHz	1 kHz to 3 MHz
Mounting system	Fixed on specimen	PMMA nut and clamp
Sensor type	Closed wire loop	KRNBB-PC
Number of sensors	1 pair	2

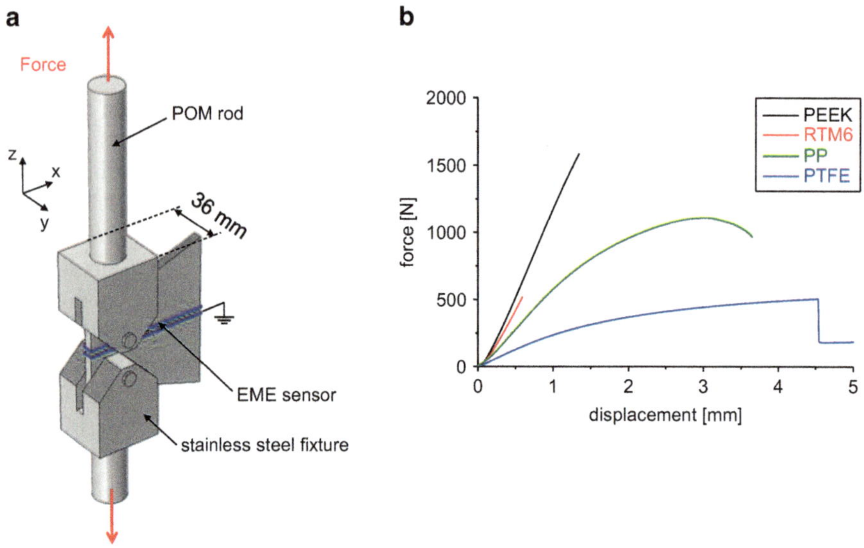

Fig. 5.56 Schematic of measurement setup (**a**) and respective force-displacement curves for the investigated polymer types (**b**)

PTFE undergo substantial plastic deformation prior to failure, which is expressed in the pronounced nonlinearity of the curves seen in Fig. 5.56b.

A representative collection of EME signals is given in Fig. 5.57 for each of the four different polymers. Comparison is made relative to the acoustic emission signals detected by AE sensor 2 shown as black curve. The EME signals in Fig. 5.57 are plotted as detected (red lines) and after filtering by a seventh order Butterworth high-pass filter of 20 kHz to reveal the oscillating part of the signal. As seen from the EME signals of Fig. 5.57, characteristic signals are obtained for each polymer type. First of all this is expressed by the different shapes of the signals, comprising different rise-times and decay times as may be perceived in the unfiltered signals. Based on the theory presented in Sect. 5.2, this corresponds to different durations of crack propagation. The latter may arise due to the differences in the overall propagation length or due to differences in the crack velocity. Since the crack length may readily be derived from microscopic measurements of the fracture surfaces after failure, this allows evaluation of the maximum crack velocity for each of the cases shown in Fig. 5.57. These evaluate as 1475 m/s for RTM6,

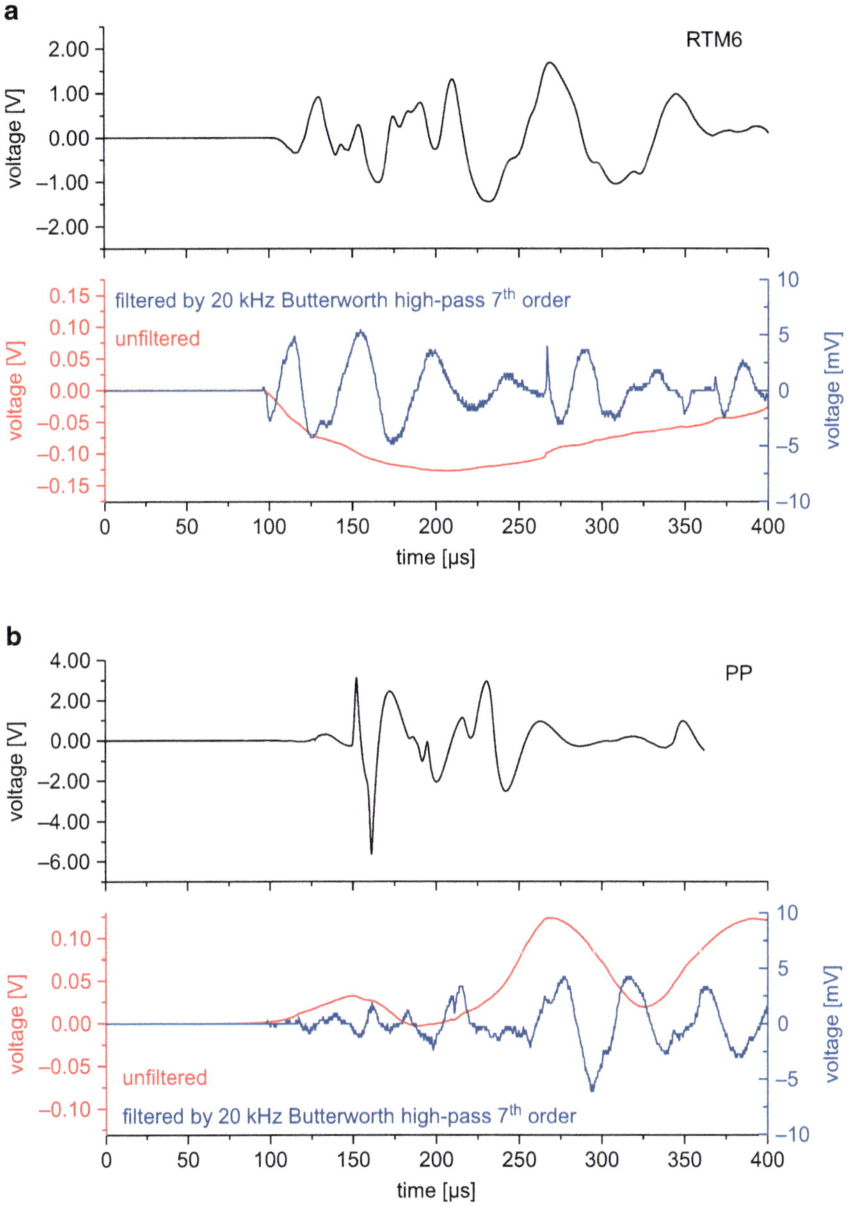

Fig. 5.57 Set of representative AE signals (*black*) and EME signals (*red* and *blue*) as acquired during tests of TDCB specimen geometries in RTM6 (**a**), PP (**b**), PTFE (**c**), and PEEK (**d**)

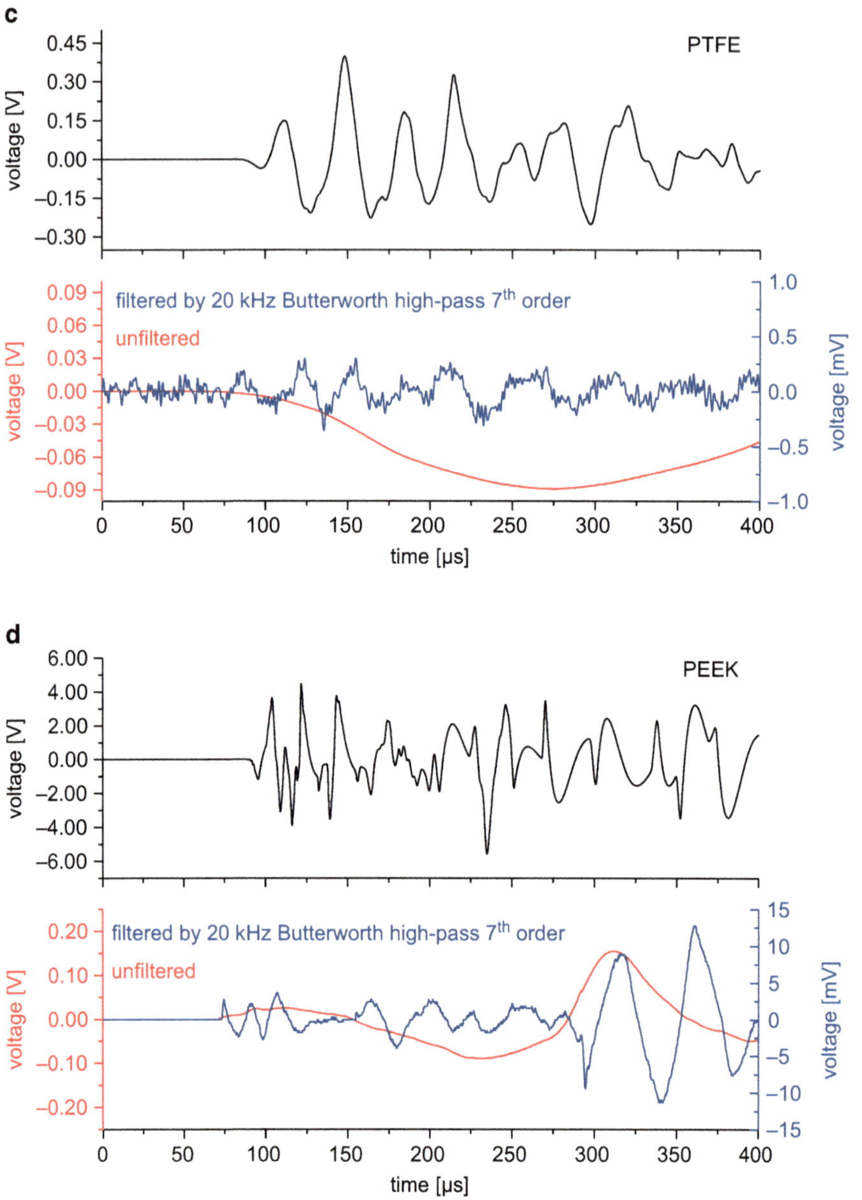

Fig. 5.57 (continued)

1596 m/s for PP, 294 m/s for PTFE, and 2130 m/s for PEEK and therefore are well below the expected Rayleigh wave velocities of the four materials ($c_R = 2586$ m/s for RTM6, $c_R = 2625$ m/s for PP, $c_R = 1321$ m/s for PTFE, and $c_R = 2493$ m/s for PEEK).

In addition, characteristically different amplitude levels of the EME signals are observed. These can be attributed to systematic differences in the number of generated charges per fracture surface as well as to the differences in absolute crack propagation lengths.

After application of the 20 kHz high-pass, the EME signals tend to appear similar to the respective AE signals. Since the "oscillating" part of the EME signal should be directly proportional to the movement of the crack walls, which act as AE source, this similarity is not unexpected. However, only the first few cycles of the AE signal are expected to show close resemblance, since the later parts of the AE signal are additionally superimposed by edge reflections incident at the sensor position. Still, for the present investigations, the similarities of the AE and EME signals were found to be significant enough to attribute them to the same source mechanism.

5.5.1.2 Fiber Filament Failure

For the investigation of fiber filament failure, an experimental setup as shown in Fig. 5.58 was developed. To this end, the fiber filament is first bonded to a PEEK rod mounted in a testing machine and then moved into a drop of two component epoxy adhesive. The adhesive is cured in situ, and after sufficient curing time, the fiber filament is loaded to failure by a high-precision universal testing machine using 20 μm/min test speed. The drop of adhesive rests on a replaceable PMMA plate, which is attached to an aluminum frame by two PA6 screws. Within the

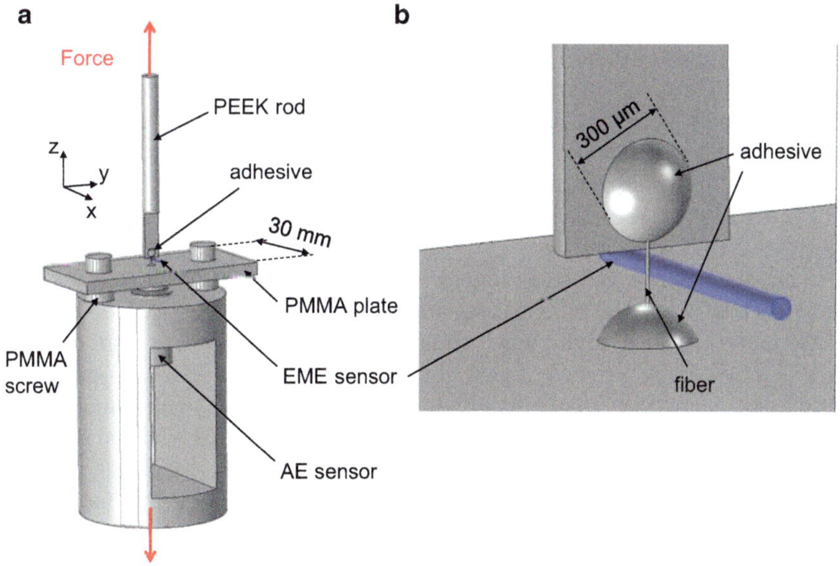

Fig. 5.58 Schematic of measurement setup (**a**) and magnified section close to fiber filament (**b**)

Table 5.3 EME and AE acquisition settings used for single fiber filament tensile tests

Acquisition settings	Value EME	Value AE
Preamplification	40 dB	40 dB
Threshold	30 dB	32 dB
Triggering	Individual channel	Individual channel
Acquisition rate	40 MS/s	40 MS/s
Bandpass range	1 kHz to 3 MHz	1 kHz to 3 MHz
Mounting system	Floating arm	Screw thread
Sensor type	Open wire loop	KRN-BB PC
Number of sensors	1	1

aluminum frame, a KRNBB-PC type AE sensor is mounted to detect the generated AE signal. The acoustic propagation path thus comprises the adhesive drop and the PMMA plate along the thickness direction, since the tip of the AE sensor is positioned directly below the fiber position. As EME sensor system, an open wire configuration is used as schematically seen in Fig. 5.58b.

Acquisition settings for AE and EME are reported in Table 5.3. The advantage of this sensor system is the flat transfer function and the ease of handling to adjust the sensor position relative to the fiber position. Moreover, due to the small dimensions of the sensor system, a very short source-sensor distance can be realized. Since this was found to be a major parameter to improve the detectable signal energy, it is a key to detect potentially weak EME signals of carbon fiber fracture. All measurements are carried out in an aluminum shielding chamber of 240 mm $\times$ 90 mm $\times$ 240 mm (length $\times$ width $\times$ height) and 2.5 mm wall thickness.

Some exemplary EME signals for testing of Sigrafil C030 type carbon fiber filaments with a mean diameter of (6.5 $\pm$ 1.2) μm are shown in Fig. 5.59. For single fiber filament tensile tests, failure typically occurs after exceeding the ultimate strength of the filament, so only one EME signal can be detected. It was found that the EME signals of carbon fiber breakage are fundamentally different to the EME signals found for polymers. Two representative EME signals obtained from a tensile test of one fiber are plotted in Fig. 5.59a, b. As first noticeable difference, the polarity of the EME signals of Fig. 5.59 is found to be different for both cases, while the AE signals have the same polarity. Since no unique polarity was observed in the present experimental setup, this indicates that the assumed charge disbalance does not occur in a preferential orientation.

The unfiltered EME signal has strong similarity to the system response of a sharp step function; thus, these signals are dominated by the 1 kHz high-pass filter used in the acquisition card. On the high- frequency side, the signal bandwidth is limited by the 3 MHz low-pass filter. Therefore, it is feasible to conclude that the rise-time of the signal is less or equal to the cutoff frequency and thus is in the order of $\leq$333 ns. Compared to the calculated crack duration of 1.2 ns for the free fiber filament in [65], this result is still plausible, but without faster digitization rates than 40 MS/s and an increased cutoff frequency no direct measurement can be made.

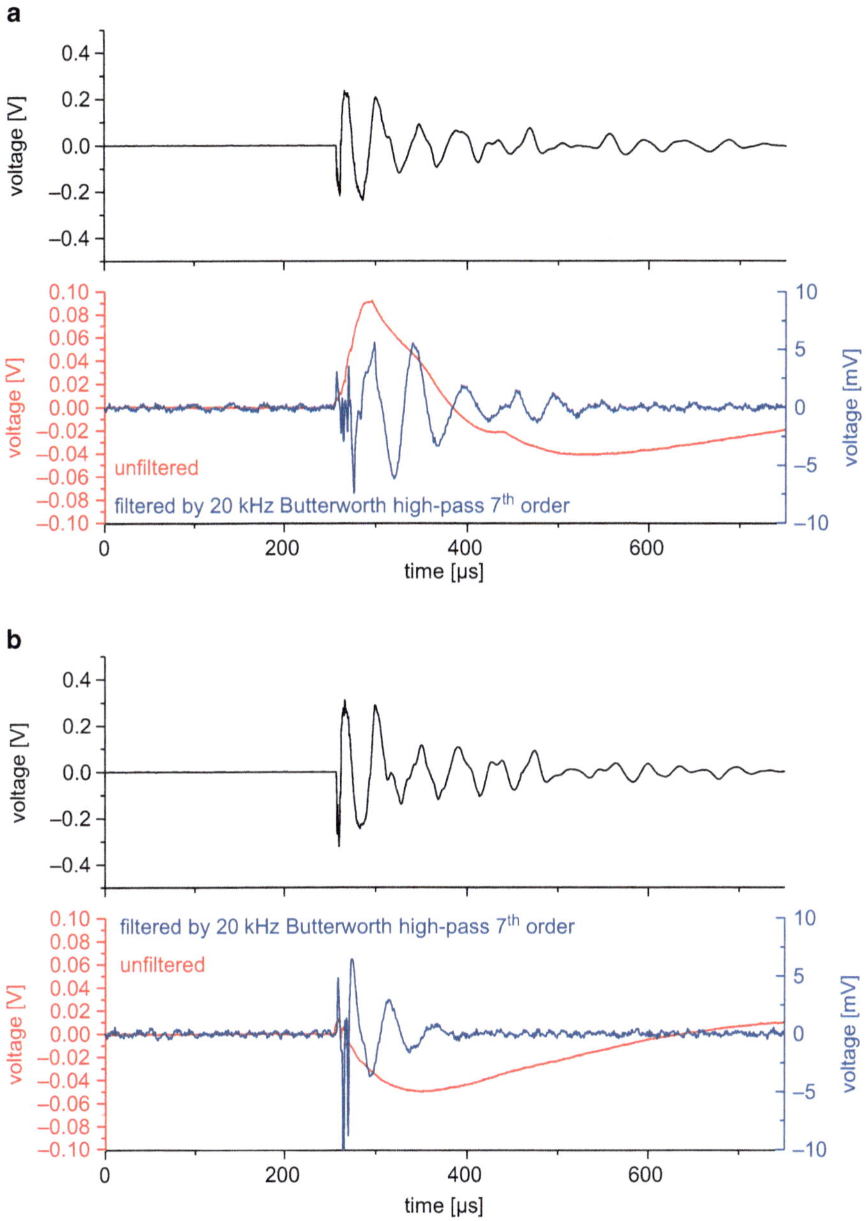

Fig. 5.59 Set of representative AE signals (*black*) and EME signals (*red* and *blue*) as acquired during tensile testing of single carbon fiber filaments

After application of a 20 kHz high-pass filter, a close similarity to the detected AE signals is observed. This is again attributed to the same origin of the oscillating part of the EME signal and the source of the acoustic emission.

5.5.1.3 Composites

Beyond the detection of EME signals in pure materials, the use of these signals for analysis of composite failure is also a promising approach. As presented in several publications before [9, 13–15], fiber reinforced composites are known to generate electromagnetic emission signals. The present section demonstrates the detection of such EME signals in configurations as typically used to measure the mode I and mode II fracture toughness values in double-cantilever-beam (DCB) configuration and end-notched flexure (ENF) configuration.

As pointed out in Sect. 5.3, the presence of conductive materials in the proximity of the EME source may lead to adverse effects with respect to the alignment of the electric field lines. Due to these restrictions, all specimen fixtures are completely machined from nonconductive materials, and the full setup is completely included in a shielding chamber. For the present experimental setups, the only metallic components are the EME sensor itself, the AE sensor casing, the attached cables, and the shielding chamber.

The first configuration is used to measure the interlaminar crack growth in mode I loading as seen in Fig. 5.60. The bulk of the fixture is machined from PMMA parts, which are connected by epoxy adhesive and PA6 screws. The load is introduced by a pultruded GFRP rod, and the precracked specimen is connected by PVC bolts to the load blocks fabricated from PMMA. The load blocks in turn are adhesively bonded to the test specimen. In addition, a wire is glued to the surface of the specimen to act as EME sensor, and a KRNBB-PC AE sensor is mounted on the specimen using a clamp system. All acquisition settings for AE and EME are listed in Table 5.4. The cross-head displacement was chosen as 2 mm/min in accordance with the speed of typical test standards. Since the measurement of the mode I fracture toughness value was not envisaged in this test scenario, no optical tracking of the crack tip or localization of the AE sources by means of a second sensor (cf. Sect. 4.7.4.1) was carried out. All specimens used are fabricated as unidirectional Sigrafil CE1250-230-39 prepreg laminate cured according to the material supplier's specification. All specimens were cut to dimensions of 155 mm $\times$ 25 mm $\times$ 3 mm (length $\times$ width $\times$ thickness) with an embedded ETFE foil of 60 mm length acting as crack initiation side.

During the test, numerous AE and EME signals can be detected, thus requiring a different approach than for the single events discussed and presented in the previous two sections. Among the multitude of EME signals, four characteristic groups can be identified, which are shown in Fig. 5.61 as unfiltered (red) and filtered signals (blue). Since these signal types occurred in all tests on fiber reinforced composites, they may already indicate some sort of distinctly different EME source groups.

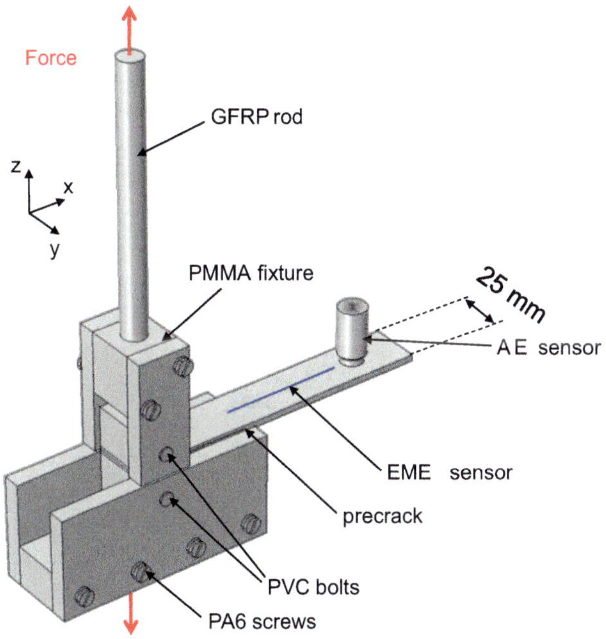

Fig. 5.60 Schematic of measurement setup showing details of specimen fixture for DCB test and AE and EME sensor arrangement

Table 5.4 EME and AE acquisition settings used for mode I (DCB) testing

Acquisition settings	Value EME	Value AE
Preamplification	40 dB	20 dB_{AE}
Threshold	28 dB	40 dB_{AE}
Triggering	Individual channel	Individual channel
Acquisition rate	5 MS/s	10 MS/s
Bandpass range	1 kHz to 3 MHz	1 kHz to 3 MHz
Mounting system	Epoxy	PMMA nut and clamp
Sensor type	Open wire	KRN-BB PC
Number of sensors	1	1

The first signal type shown in Fig. 5.61a exhibits a very quick rise of the unfiltered EME signal being superimposed by the 1 kHz response of the acquisition system filter. The corresponding filtered signals reveal some lower-frequency oscillations. However, the fast rise in the beginning of the signal indicates a fast crack procedure with similar or faster rise-times than those observed for fiber failure in the single filament tensile tests.

Signals of similar characteristic, but lower rise-time, were found as seen in Fig. 5.61b. Their rise-time was in the range between 10 μs and 200 μs indicating either growth of rather large cracks or growth of smaller cracks with low propagation velocity.

As distinctly different type of EME signals, the oscillating type seen in Fig. 5.61c was recorded. For the unfiltered case, these EME signals do not show a signal, which is dominated by the 1 kHz high-pass filter. Instead their bandwidth seems to be systematically above 1 kHz yielding almost similar shape after application of the 20 kHz high-pass. These signals could arise from a variety of mechanisms. As first possibility, a sequence of fracture events could cause a

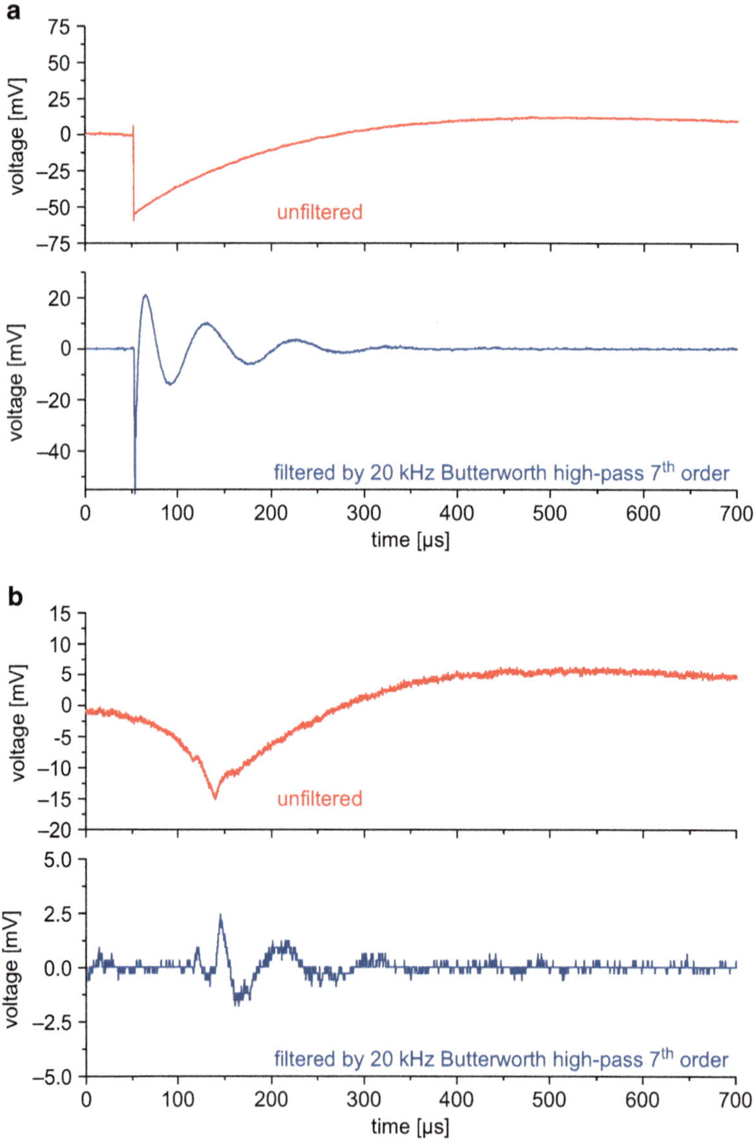

Fig. 5.61 Set of typical EME signals recorded during DCB test in unfiltered (*red*) and filtered state (*blue*)

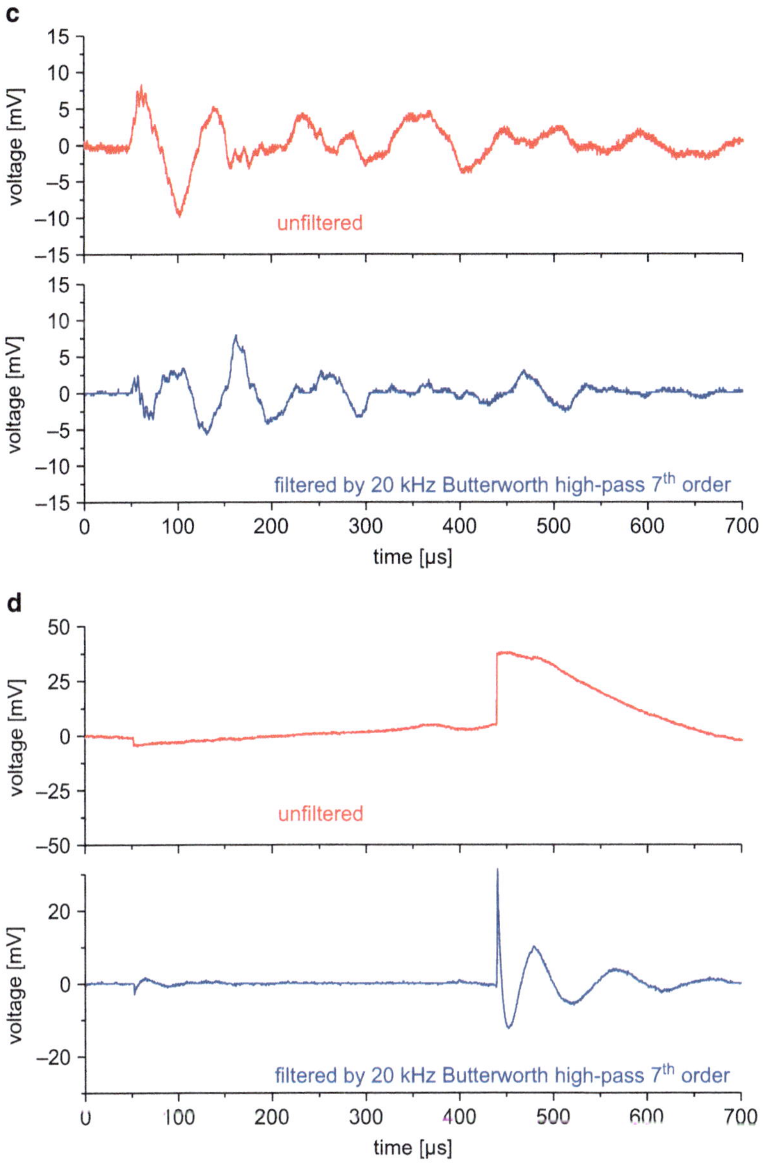

Fig. 5.61 (continued)

periodic appearance of an EME signal like this. Second, the EME signal could be the result of a wave motion of a previously charged surface. Such could happen subsequently to a larger fracture event, when a substantial fracture surface is still charged and acoustic waves are passing these areas. Finally, these EME signals could also result from cross talk between AE and EME sensors. The latter cases can be excluded in scenarios where no AE sensors are used or no AE signal is recorded synchronous in time.

The last type of EME signal found to be characteristic in failure of fiber reinforced composites is shown in Fig. 5.61d. This signal is an obvious example of a cascade of fracture events recorded in one signal. At around 50 μs time, a first small EME signature is seen. After approximately 400 μs a second, larger EME signature is seen, which likely constitutes a new fracture event. Although this effect is obvious from Fig. 5.61d, there are many EME signals with merged signatures, which cannot be clearly separated. Since the trigger mode for the present setup operates on the unfiltered signals, this is not surprising, given the rising slope of the unfiltered signal of Fig. 5.61d. Since this does not fall below the threshold level, no termination criterion is reached to stop the acquisition of the first event. If only filtered signals are used, this might limit the number of merged EME signals, but would also limit the approaches to estimate the rise-time of the underlying source.

Beyond the many AE and EME signals recorded after crack initiation in DCB tests, some specific signal pairs are shown exemplary in Fig. 5.62. These are the AE and EME signals corresponding to the crack initiation at the ETFE insert in Fig. 5.62a and the initiation of interlaminar crack growth starting from a precrack in Fig. 5.62b. For comparison to the AE signal, only the filtered EME signals are shown. As for the previous investigations on polymer fracture and fiber failure, the filtered EME signals show close similarity to the AE signals especially in the first 100 μs. Also, the EME signals show a systematic onset which precedes the AE signals by 50–100 μs. This is due to the difference in propagation velocities of the acoustic waves and the electromagnetic waves. Despite of the close amplitude values of the AE signals, the EME signals of interlaminar crack growth starting at the precrack were generally found to be of higher amplitude. This could indicate a systematic difference in the nature of the EME source induced by the presence of the ETFE insert. Alternatively, this could also be due to a systematic difference in the interlaminar crack growth, which is known to behave differently when starting from precracks or inserts.

A quantification of the signal amplitudes and their absolute energy is shown in Fig. 5.63 for the detected AE and the detected EME signals during crack propagation. Since both sensor systems use different amplitude threshold levels, a distinct difference on the low end of the distribution is found. However, for the strongest signals, a difference of 26 dB is observed for the corresponding unfiltered EME and AE signals. Hence, it is likely that the majority of the EME signals corresponding to the weaker AE signals are not detected with the present system, since they likely fall below the detection threshold. Given the presently used measurement chain of a J-FET preamplifier mounted close to the EME sensor and the subsequent UBBV-NF-35 preamplifier, it is not possible to increase the preamplification factor without saturating stronger signals. Thus, refinement of the signal acquisition electronics is required to account for the broad dynamic range of signal amplitudes seen in fiber reinforced composites.

As second configuration used to demonstrate the use of EME measurements for composite materials, the mode II test shown in Fig. 5.64 was selected. Like the

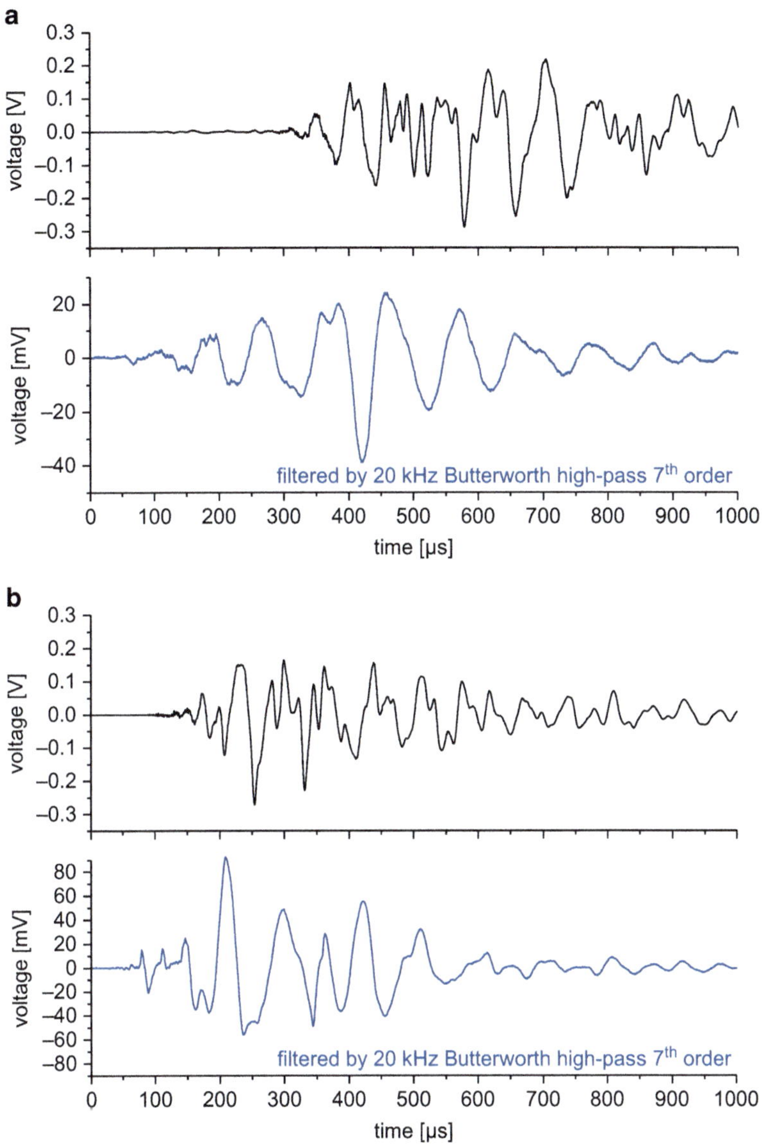

Fig. 5.62 Two exemplary pairs of AE signals (*black*) and filtered EME signals (*blue*) detected at crack initiation starting at ETFE insert (**a**) and starting from precrack (**b**)

previous experimental setups, all components are fabricated from nonconductive materials. Namely, the bulk fixtures are made from PMMA with supports out of PVC and a pultruded GFRP rod for load transfer to the outside of the shielding chamber. Same as above, the specimens used in the tests are fabricated as unidirectional Sigrafil CE1250-230-39 prepreg laminate cured according to the material

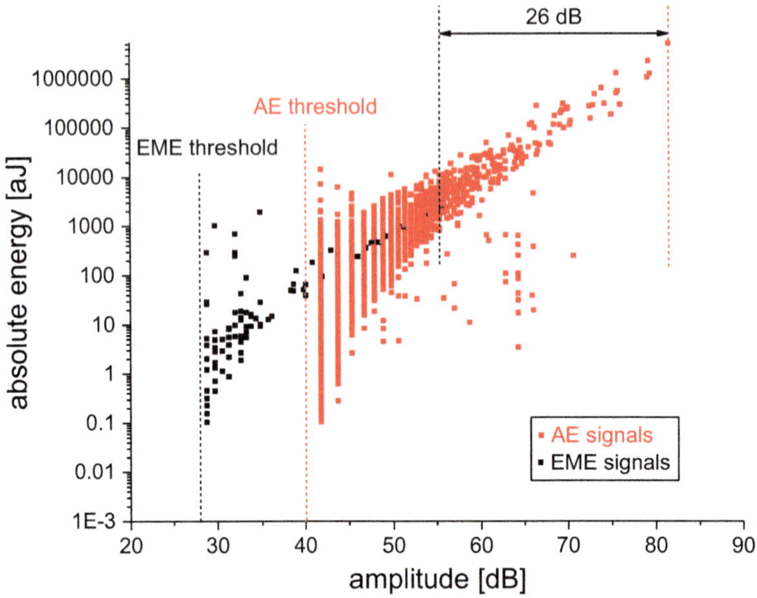

Fig. 5.63 Comparison of AE signals energy and EME signals energy as function of amplitudes

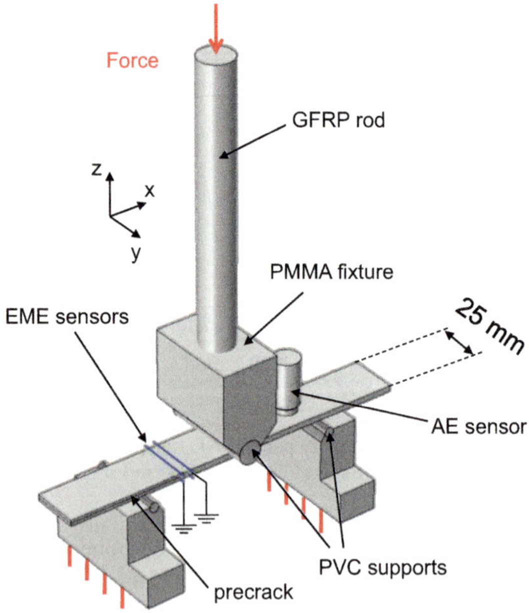

Fig. 5.64 Schematic of measurement setup showing details of specimen fixture for ENF test and AE and EME sensor arrangement

Table 5.5 EME and AE acquisition settings used for mode II (ENF) testing

Acquisition settings	Value EME	Value AE
Preamplification	40 dB	20 dB_{AE}
Threshold	27 dB	40 dB_{AE}
Triggering	Individual channel	Individual channel
Acquisition rate	2 MS/s	10 MS/s
Bandpass range	1 kHz to 3 MHz	1 kHz to 3 MHz
Mounting system	Duct tape	PMMA nut and clamp
Sensor type	Open wire	KRN-BB PC
Number of sensors	2	1

supplier's specification. Specimens were cut to dimensions 180 mm × 25 mm × 3.5 mm (length × width × thickness) with an embedded ETFE foil of 45 mm length acting as crack initiation side.

The occurrence of acoustic emission during the test is monitored by a KRNBB-PC sensor attached to the specimen by means of a clamp system at the position seen in Fig. 5.64. Electromagnetic emission is monitored by two EME sensors realized by open-ended wires arranged in an array seen in Fig. 5.64 and more detailed in Fig. 5.66b in the xz-cross-section. The acquisition settings were chosen as reported in Table 5.5.

All tests are carried out in displacement-controlled mode using a universal testing machine with a 5 kN load cell. Specimens are first loaded to generate a precrack in the material starting from the ETFE foil using a cross-head velocity of 2 mm/min. Subsequently, specimens are shifted within the three-point bending setup to move the position of the crack tip to the center of the EME sensor array. Then the specimen is loaded again by a cross-head velocity of 2 mm/min until crack propagation is observed.

Two exemplary pairs of AE and EME signals are shown in Fig. 5.65 as recorded during the crack initiation starting at the ETFE foil and as starting from the precrack. For this comparison only the signal of one of the two EME sensors is shown and compared to the AE sensor signal after filter application of a seventh order Butterworth high-pass of 20 kHz. As for the previous findings, close similarity of the EME signal appearance to those of the AE signals is found. Similar to the EME signals of the mode I investigation, there is a systematic difference in the measured EME amplitude of both cases. Here the EME signals starting from the precrack appear to be more intense than the first crack initiation starting at the embedded ETFE foil. Therefore, the specific load condition (mode I or mode II) does not seem to hinder the detectability or occurrence of noticeable EME signals. Given the 90° rotation in the present case when compared to the DCB arrangement of Fig. 5.60, it even seems likely to be able to detect EME signals in various orientations. For the ENF experiments, AE and EME signals were even found to be of similar amplitudes, so almost no difference in signal amplitude was measured. This is likely due to the close distance of the EME sensors relative to the position of

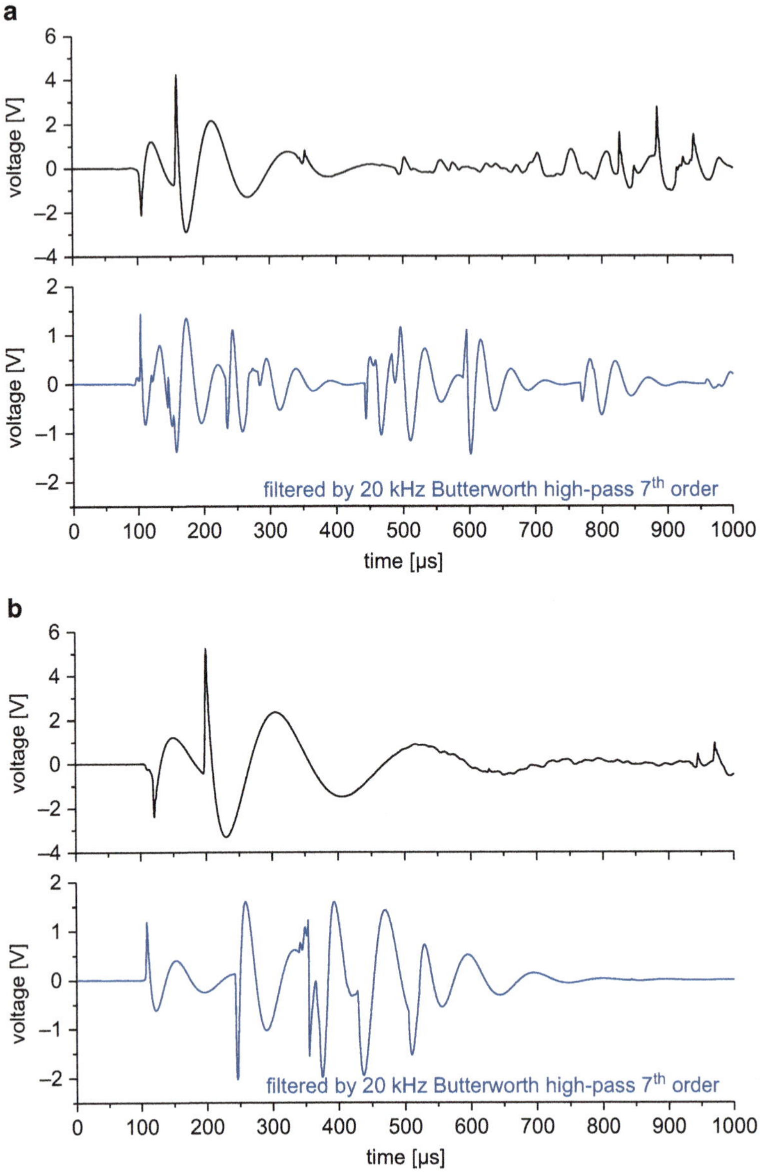

Fig. 5.65 Two exemplary pairs of AE signals (*black*) and filtered EME signals (*blue*) detected at crack initiation starting at ETFE insert (**a**) and starting from precrack (**b**)

the EME source. This was confirmed by the lack of detection of EME signatures, when the crack tip has propagated sufficiently far away. Thus no EME signals were detected by the present configuration after the crack tip has propagated beyond 20 mm distance to the sensor system.

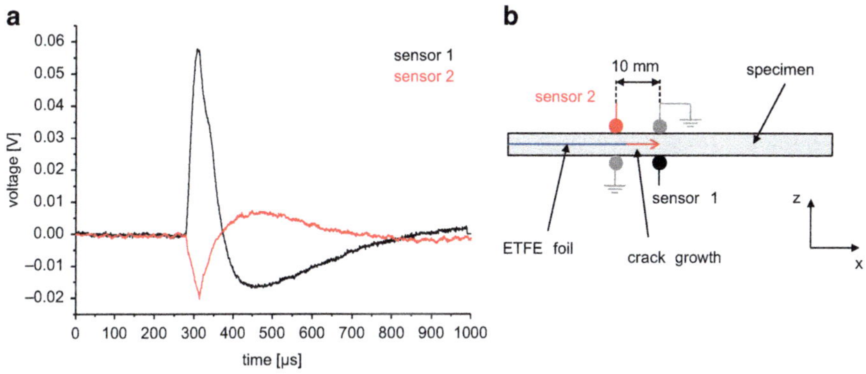

Fig. 5.66 Pairs of EME signals detected at sensor 1 and sensor 2 at crack initiation (**a**) and respective schematic of sensor arrangement (**b**)

Given the arrangement of the two EME sensors, signal pairs were detected in most of the cases. One example of such an EME signal pair is seen in Fig. 5.66a. The specific sensor array used in this experiment is seen in Fig. 5.66b. For this experiment, two of the wires were chosen at ground potential, while one wire on top and bottom was selected as sensing system. Due to the intersected arrangement, this allows to analyze the nature of an EME source in various means. Due to the position of the ETFE foil at the center of the laminate, the interlaminar crack is initiated at this ply level and thus propagates at the center between the two EME sensors. As seen from the EME signals detected at the top and bottom of the plate, the two signals are detected with opposite polarity. This indicates a directed electric field orientation as expected from a crack surface charged with opposing polarity on each side. Since the crack propagates toward the sensor 1, this likely explains the stronger signal amplitude when compared to the sensor 2. Moreover, this finding underlines the possibility to use the characteristic radiation pattern of the EME source to perform orientation measurements of the fracture surface.

5.5.2 *Measurement of Fracture Surface Orientation*

To demonstrate the use of EME measurements for measurement of the fracture surface orientation, other test configurations than DCB and ENF test are required, since the fracture surface is at an approximately constant angle in these setups.

Thus, cross-ply specimens with $[0/90/90/0/90]_{sym}$ were fabricated from the unidirectional Sigrafil CE1250-230-39 prepreg laminate cured according to the material supplier's specification. All specimens were cut to dimensions of 95 mm × 15 mm × 2.2 mm (length × width × thickness) according to the recommendations of DIN-EN-ISO 14125. As test configuration a three-point bending setup as seen in Fig. 5.67 was chosen. This ensures predominant failure at the position of the load nose, i.e., at the middle of the specimen.

Fig. 5.67 Schematic of measurement setup showing details of specimen fixture for three-point bending test and AE and EME sensor arrangement

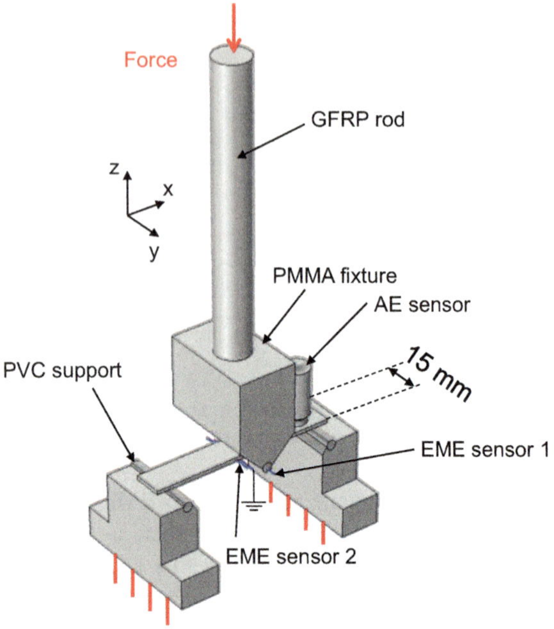

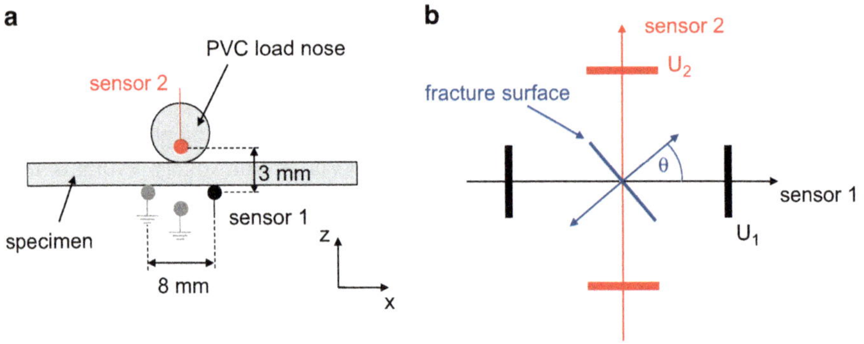

Fig. 5.68 Arrangement of EME sensors used in three-point bending test (**a**) and concept to derive orientation of fracture surface by measurement with two sensors (**b**)

In analogy to the previous setups, the test was carried out inside a shielding chamber using only nonconductive materials for the specimen fixtures (cf. Fig. 5.67). For all tests a universal test machine with a slow cross-head velocity of 0.02 mm/s was applied to allow good separation of single failure events. The occurrence of acoustic emission during the test is monitored by a KRNBB-PC sensor attached to the specimen by means of a clamp system at the position seen in Fig. 5.67. The concept for the arrangement of EME sensors for the measurement is shown in the *xz*-cross-section of Fig. 5.68a with acquisition parameters as reported in Table 5.6. Similar to the principle of the ENF measurements above, four

Table 5.6 EME and AE acquisition settings used for flexural testing

Acquisition settings	Value EME	Value AE
Preamplification	54 dB	20 dB_{AE}
Threshold	30 dB	40 dB_{AE}
Triggering	On acoustic emission channel	
Acquisition rate	10 MS/s	10 MS/s
Bandpass range	1 kHz to 3 MHz	1 kHz to 3 MHz
Mounting system	Floating arm	PMMA nut and clamp
Sensor type	Open wire	KRN-BB PC
Number of sensors	2	1

open-ended wires are used as EME sensors. Two wires are grounded as marked in Fig. 5.68a. The wire acting as sensor 1 and the corresponding ground are glued to the bottom of the specimen. The wire of sensor 2 is embedded in an open hole of the PVC load nose, and the corresponding ground is floating sufficiently low below the bottom of the specimen to avoid collision during the test. Due to this alignment, this allows to evaluate two components of the electric field at approximately orthogonal angle.

The concept used to determine the inclination angle θ from the measurement is shown in Fig. 5.68b. Given the principle of operation of the EME source in Sect. 5.2, the two EME sensors should be able to detect orthogonal components of the electric field. As seen from the measurement using the test source in Sect. 5.2.2, the orientation of the source is directly obtained from a sinusoidal relationship of the two signal amplitudes. The inclination angle θ of Fig. 5.68b is obtained using the EME sensor voltages U_1 and U_2 of the oscillating signal part as:

$$\theta = \tan^{-1}\left(\frac{U_1}{U_2}\right) \tag{5.37}$$

The use of the oscillating part is recommended, since the exact amplitude of the low-frequency components are typically difficult to obtain as quantitative values due to saturation of system components as well as due to the 1 kHz high-pass characteristic.

In the following, failure of one exemplary measurement is evaluated using this concept and is discussed relative to microscopy observations after failure based on the considerations made in [66]. Typically, the cross-ply specimens investigated herein fail in compressive mode at the position of the load nose with subsequent inter-ply delamination and failure of multiple plies. Due to the failure within the 0° plies, the orientation of inter-ply delamination and the orientation of failure within the 90° plies, fracture surfaces of macroscopically visible extent are generated with distinctly different orientation.

During the first significant load drop of one specimen, a sequence of two strong EME signatures was detected. The first event is shown in Fig. 5.69a as unfiltered signal and in Fig. 5.69b as signal filtered by a 20 kHz Butterworth high-pass of

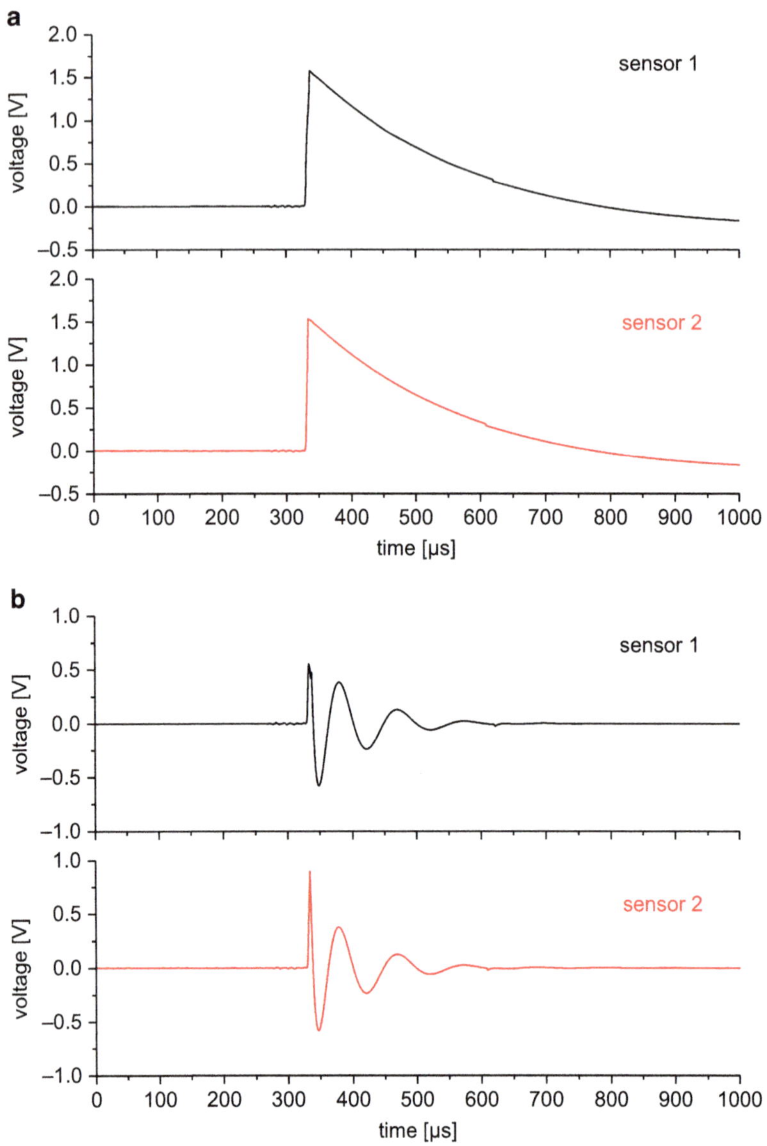

Fig. 5.69 First EME event detected during substantial load drop recorded by two EME sensors in unfiltered view (**a**) and as filtered signals (**b**)

seventh order. Since EME signals of both sensors are shown in identical scales, it is obvious that there is almost no difference in signal amplitude seen before and after filtering, when comparing signals of both sensors. An application of (5.37) to the measured EME signal amplitudes thus results in an inclination angle of $\theta = 44.5°$. This corresponds well to the orientation of the failure in the 0° ply seen in Fig. 5.71, when taking into account the superimposed deflection and rotation during loading.

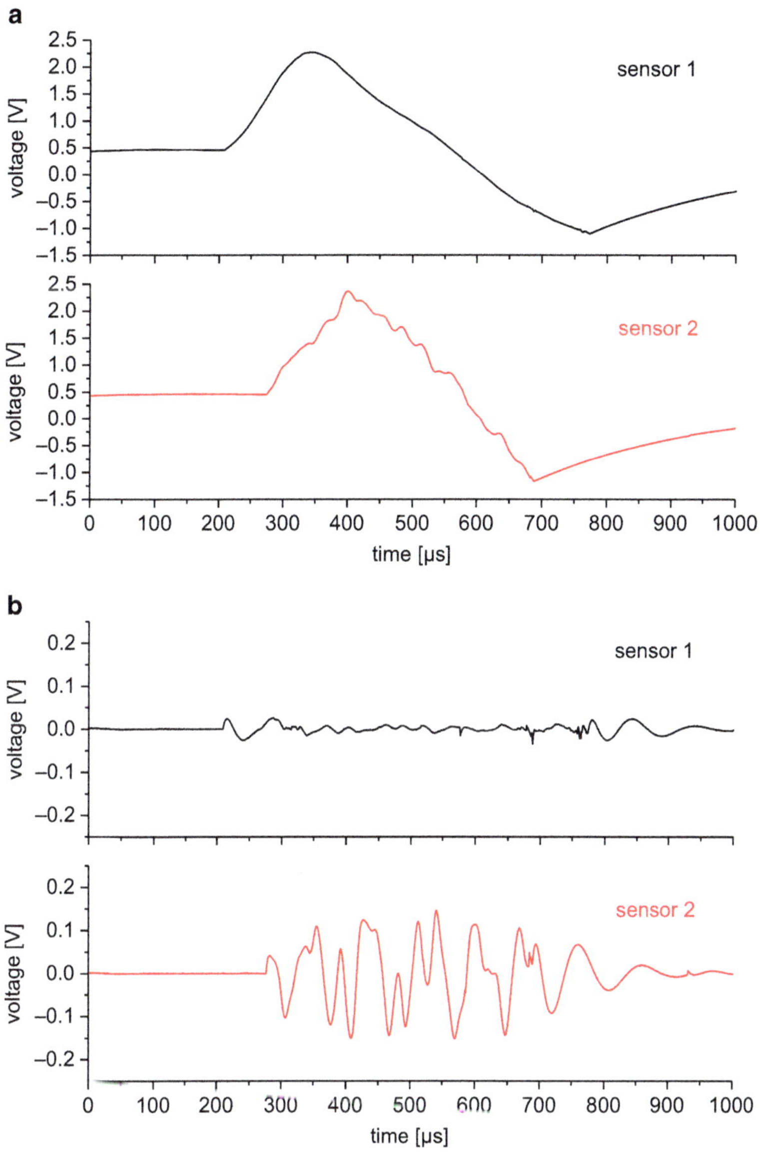

Fig. 5.70 Second EME event detected during substantial load drop recorded by two EME sensors in unfiltered view (**a**) and as filtered signals (**b**)

Only few milliseconds later, a second strong event was observed as seen in Fig. 5.70a as unfiltered and in Fig. 5.70b as filtered signal in analogy to Fig. 5.69. While the unfiltered signals reveal almost no difference in signal amplitude, the oscillating part of Fig. 5.70b exhibits a distinct difference in signal intensity. Applying equation (5.37) to these amplitudes, a fracture angle of $\theta = 11.0°$ is

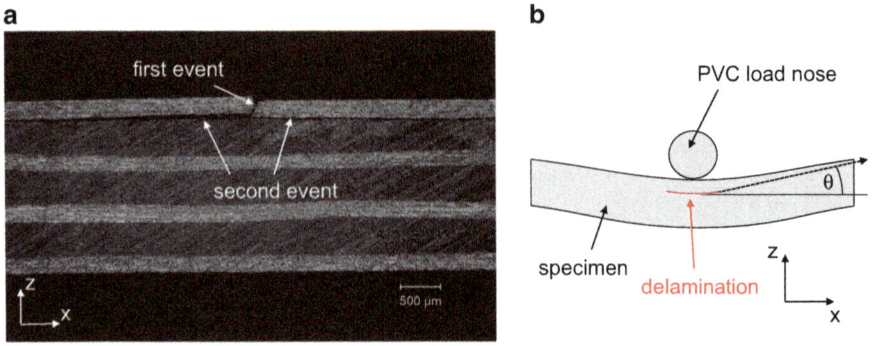

Fig. 5.71 Microscopy image of fracture mode as corresponding to the EME events shown in Figs. 5.69 and 5.70 and corresponding scheme of situation under load

obtained. This narrow angle can directly be related to the orientation of the inter-ply delamination of Fig. 5.71 when taking into account the curvature of the specimen under load as seen from the scheme on the right side of Fig. 5.71.

5.5.3 Detectability of Failure Mechanisms

As has been demonstrated in the previous sections and throughout various literature sources [9, 13–15], EME is generally detectable for the failure mechanisms encountered in fiber reinforced composites. Since measurement of EME signals of single filament failure was found to be possible, it may readily be assumed that the method is generally suitable to aid in interpretation of composite failure.

However, given the novelty of the method and the lack of established technical standards, it is difficult to assess the detectability in a similar fashion as for digital image correlation in Sect. 3.3.3 or for acoustic emission in Sect. 4.2.4. Thus, more intense research and standardization efforts will be required to establish the probability of EME detection in a similar level of detail.

Instead, this section will provide a guideline as first estimate for a couple of experimental factors to consider when conducting EME measurements. Similar to the other methods investigated, the first focus is given to the absolute limit of the measurement system.

5.5.3.1 Absolute System Limits

Based on the experimental investigations presented in this chapter and the efforts to establish a thorough implementation of the measurement chain in Sect. 5.4, it is possible to allow for some first conclusions regarding the absolute system limits. Other than for AE sensors, there is no technically relevant lower detection limit of the EME sensor system itself. Since this EME measurement is in principle only a measurement of an AC voltage, the sensitivity of this measurement can practically be realized for all technically relevant orders of magnitude of voltages. Instead, the dominating factor of influence with respect to the detection limits of the measurement system is currently owed to shielding aspects, since the electromagnetic noise detected by the sensor system constitutes the real practical limit of applicability of the measurement method. Because this will be discussed in the next step explicitly, this is neglected for the first discussion. Apart from the EME sensor itself, the only remaining part in the measurement chain are the system electronics. Since EME signals were observed to span similar orders of magnitude in induced signal voltage as for AE sensor signals, this constitutes the primary challenge of the signal acquisition stage. For an AE sensor, the generated voltage may be assumed to be proportional to the surface motion present at the sensor's contact area to the solid [67–69]. For an EME sensor, the generated signal voltage will depend primarily on the distance and orientation relative to the source. Thus the generated sensor signal voltage may not be directly related to the charges on the fracture surface, but possibly to the incident electric field strength.

Consequently, the requirement to the acquisition chain is to assure a sufficient digitization, without saturating the preamplifier or the digitization equipment. Given the assumption of negligible saturation effects, this causes the lower detection limit to be dominated by the electronic noise floor of the system electronics, which is below 1 mV for a typical commercial system operating in the frequency range from 1 kHz to 1 MHz.

5.5.3.2 Signal-to-Noise Ratio

As indicated in the previous section, the relevant detection limit for EME measurements was found to be given by the suppression of electromagnetic noise. For a specific sensor sensitivity and generated electric field strength of the EME source, the ability to detect the corresponding EME signal will substantially depend on the background noise level. To this end, the noise floor of the system electronics was already introduced in the previous section. However, there are numerous other sources for electromagnetic noise included within the measurement chain. As a first step to gain a high signal-to-noise ratio, it is thus required to avoid detection of electromagnetic noise.

The EME sensor itself is susceptible to all incident electromagnetic radiation, not only from the EME source. Therefore, the use of a shielding chamber as presented in Sect. 5.4 is a recommended step to resolve the weak signatures of EME sources as originating from material failure. The level of shielding effectiveness will determine the overall quality of the measurement, but may certainly not be reduced beyond a certain level. As additional sources to detect electromagnetic noise, all cable connections and preamplifier circuits may be considered. In order to reduce the incident electromagnetic noise to a minimum, one step consists of mounting a low-noise J-FET preamplifier directly to the EME sensor. In the spirit of [58–60], this was implemented for the presented experiments as seen in Fig. 5.42 and was found to increase the signal-to-noise ratio by 18 %. While this approach helps to improve the signal quality, it is further required to keep a high shielding effectiveness for all further cables and remaining electrical components (filters, preamplifiers). In practice it was found to be suitable to include the cables to the main preamplifier stage within the shielding chamber. This effectively avoids detection of external electromagnetic noise sources and keeps the signal-to-noise ratio at a high level.

As a second step to increase the overall signal-to-noise ratio, it is generally possible to increase the sensitivity of the sensor system. However, for the present EME sensor system discussed in Sect. 5.4, there was no significant difference observed as function of sensor geometry. Thus, the simple detection by using a metallic conductor seems to be limited in this aspect. Therefore, other sensor concepts, such as Hall effect sensors, magnetometers, or superconducting quantum interference devices (SQUIDs) could be explored as alternatives to detect electromagnetic waves.

Another obvious optimization routine consists in considering the source-sensor distance and the radiation pattern of the expected source types. Based on the decay of electric field strength as function of distance to the charged fracture surface, it is advisable to position the EME sensor as close as possible to the fracture surface to obtain high EME signal amplitudes. Similarly, the expected EME source radiation patterns may allow selecting preferred EME sensor positions to optimize the incident electric field strength.

5.5.3.3 Acquisition Mode

Another factor of practical relevance to the detectability of EME signals in experiments deals with the chosen acquisition mode. Like AE signals, EME signals are of transient nature. Thus the acquisition mode of system electronics may either operate in continuous storage mode causing a large data amount for reasonable acquisition rates or may operate in triggered mode to store only parts of the stream after exceeding a trigger event.

The continuous storage mode comes with the distinct advantage of extensive capabilities for offline analysis, but requires respective hardware and software systems. For the triggered mode, other challenges arise, which shall be discussed briefly.

For typical commercial transient recorders, the triggered acquisition mode is based on exceedance of a certain voltage threshold. Depending on system quality, further parameters are used to determine if the trigger event is a valid transient signal to be stored or not. For acquisition systems optimized for AE signal detection, the shape of EME signals comprises a first challenge to overcome by modified hardware and software settings. Since this was hard to be implemented in practice, another approach was followed throughout some of the experiments presented in this chapter. Typically the trigger condition was chosen to simultaneously record the EME signals, once an AE signal is detected. This provided reliable detection conditions, but also indicated an important factor only partially solved so far.

As previously discussed in literature [1, 3, 5, 10, 14], there is a general discrepancy of the number of AE signals versus the number of EME signals. In general, less EME signals than AE signals are acquired during an experiment. While it might be intuitive to attribute this discrepancy to the lack of sensitivity of EME sensors, it is likely that this is not the only reason. Apart from the aspects of the source radiation pattern, it was demonstrated that EME measurements are sensitive enough to detect even the weakest failure modes (i.e., fiber breakage signals) also detectable by AE measurements. Instead, one essential difference could be given by the origin of the source mechanism for EME signals and AE signals. Thus it can be speculated, if every AE source is also an EME source. Since the generation of an EME signal requires the generation of surface charges due to breakage of bonds, this might not be the case for some AE sources. For stick-slip friction or internal friction, this might apply. Since these are most relevant AE sources, the absence of EME for such sources could be used to securely identify this AE source type. Another possible scenario could be the formation of cracks not exhibiting a charge imbalance as required for the proposed EME source model description in Sect. 5.2. In this context it cannot be excluded that the EME phenomenon could potentially be suppressed in some exotic material configurations. If crack formation happens without charge imbalance in these materials or the process is of stochastically nature in its strength, the detection of EME signals might not be a reliable process to detect crack formation in such materials. However, the latter seems not applicable for the experiments and materials presented herein, since these showed outstanding repeatability in their EME occurrence.

References

1. Yamada, I., Masuda, K., Mizutani, H.: Electromagnetic and acoustic emission associated with rock fracture. Phys. Earth Planet. Inter. **57**, 157–168 (1989)
2. Frid, V., Rabinovitch, A., Bahat, D.: Fracture induced electromagnetic radiation. J. Phys. D. Appl. Phys. **36**, 1620–1628 (2003)
3. Sedlak, P., Sikula, J., Lokajicek, T., Mori, Y.: Acoustic and electromagnetic emission as a tool for crack localization. Meas. Sci. Technol. **19**, 045701 (2008)
4. Rabinovitch, A., Frid, V., Bahat, D.: Surface oscillations—a possible source of fracture induced electromagnetic radiation. Tectonophysics **431**, 15–21 (2007)
5. Lacidogna, G., Carpinteri, A., Manuello, A., Durin, G., Schiavi, A., Niccolini, G., Agosto, A.: Acoustic and electromagnetic emissions as precursor phenomena in failure processes. Strain **47**, 144–152 (2011)
6. Miroshnichenko, M., Kuksenko, V.: Study of electromagnetic pulses in initiation of cracks in solid dielectrics. Sov. Phys. Solid State **22**, 1531–1533 (1980)
7. O'Keefe, S.G., Thiel, D.V.: A mechanism for the production of electromagnetic radiation during fracture of brittle materials. Phys. Earth Planet. Inter. **89**, 127–135 (1995)
8. Rabinovitch, A., Frid, V., Bahat, D., Goldbaum, J.: Fracture area calculation from electromagnetic radiation and its use in chalk failure analysis. Int. J. Rock Mech. Min. Sci. **37**, 1149–1154 (2000)

9. Koktavy, P.: Experimental study of electromagnetic emission signals generated by crack generation in composite materials. Meas. Sci. Technol. **20**, 015704 (2008)
10. Koktavy, P., Pavelka, J., Sikula, J.: Characterization of acoustic and electromagnetic emission sources. Meas. Sci. Technol. **15**, 973–977 (2004)
11. Sikula, J., Majzner, J., Sedlak, P., Mori, Y.: Electromagnetic and acoustic emission fine spectra. Adv. Mater. Res. **13–14**, 169–174 (2006)
12. Gade, S.O., Weiss, U., Peter, M.A., Sause, M.G.R.: Relation of electromagnetic emission and crack dynamics in epoxy resin materials. J. Nondestruct. Eval. **33**, 711–723 (2014)
13. Sedlak, P., Enoki, M., Ogasawara, T., Sikula, J.: Electromagnetic and acoustic emission in PEEK/carbon nanotube composites. In: 29th European Conference on Acoustic Emission Testing, Vienna (2010)
14. Sklarczyk, C., Winkler, S., Thielicke, B.: Die elektrische Emission beim Versagen von Faserverbundwerkstoffen und ihren Komponenten. Materwiss. Werkstofftech. **27**, 559–566 (1996)
15. Aman, S., Aman, A., Morgner, W.: Monitoring of carbon fibre breakage in composites based on microwave emission. Compos. Sci. Technol. **84**, 58–64 (2013)
16. Misra, A., Prasad, R.C., Chauhan, V.S., Srilakshmi, B.: A theoretical model for the electromagnetic radiation emission during plastic deformation and crack propagation in metallic materials. Int. J. Fract. **145**, 99–121 (2007)
17. Winkler, S.R.: Field emissions caused by fracture and yielding. Int. J. Fract. **136**, 221–235 (2005)
18. Jagasivamani, V., Iyer, K.J.L.: Electromagnetic emission during the fracture of heat-treated spring steel. Mater. Lett. **6**, 418–422 (1988)
19. Misra, A.: A physical model for the stress-induced electromagnetic effect in metals. Appl. Phys. **16**, 195–199 (1978)
20. Misra, A., Ghosh, S.: Electromagnetic radiation characteristics during fatigue crack propagation and failure. Appl. Phys. **23**, 387–390 (1980)
21. Molotskii, M.I.: Dislocation mechanism for the misra effect. Sov. Tech. Phys. Lett. **6**, 22–23 (1980)
22. Rabinovitch, A., Bahat, D., Frid, V.: Comparison of electromagnetic radiation and acoustic emission in granite fracturing. Int. J. Fract. **71**, R33–R41 (1995)
23. Rabinovitch, A., Frid, V., Bahat, D.: Parametrization of electromagnetic radiation pulses obtained by triaxial fracture of granite samples. Philos. Mag. Lett. **77**, 289–293 (1998)
24. Rabinovitch, A.: A note on the amplitude-frequency relation of electromagnetic radiation pulses induced by material failure. Philos. Mag. Lett. **79**, 195–200 (1999)
25. Rabinovitch, A., Frid, V., Bahat, D.: Gutenberg-Richter-type relation for laboratory fracture-induced electromagnetic radiation. Phys. Rev. E. **65**, 011401 (2001)
26. Frid, V.: Electromagnetic radiation associated with induced triaxial fracture in granite. Philos. Mag. Lett. **79**, 79–86 (1999)
27. Frid, V.: Electromagnetic radiation method water-infusion control in rockburst-prone strata. J. Appl. Geophys. **43**, 5–13 (2000)
28. Bahat, D., Rabinovitch, A., Frid, V.: Fracture characterization of chalk in uniaxial and triaxial tests by rock mechanics, fractographic and electromagnetic radiation methods. J. Struct. Geol. **23**, 1531–1547 (2001)
29. Bahat, D., Frid, V., Rabinovitch, A., Palchik, V.: Exploration via electromagnetic radiation and fractographic methods of fracture properties induced by compression in glass-ceramic. Int. J. Fract. **116**, 179–194 (2002)
30. Goldbaum, J., Frid, V., Rabinovitch, A., Bahat, D.: Electromagnetic radiation induced by percussion drilling. Int. J. Fract. **111**, 15–20 (2001)
31. Finkel, V.M., Golovin, Y.I., Sereda, V.E.: Electric effects at fracture of LiF crystals in connection with the crack control problem. Sov. Phys. Solid State **17**, 492–495 (1975)
32. Rabinovitch, A., Bahat, D., Frid, V.: Similarity and dissimilarity of electromagnetic radiation from carbonate rocks under compression, drilling and blasting. Int. J. Rock Mech. Min. Sci. **39**, 125–129 (2002)

33. Sklarczyk, C., Altpeter, I.: The electric emission from mortar and concrete subjected to mechanical impact. Scr. Mater. **44**, 2537–2541 (2001)
34. Petrenko, V.F.: On the nature of electrical polarization of materials caused by cracks. Application to ice electromagnetic emission. Philos. Mag. Part B **67**, 301–315 (1993)
35. Ogawa, T., Oike, K., Miura, T.: Electromagnetic radiations from rocks. J. Geophys. Res. **90**, 6245 (1985)
36. Rabinovitch, A., Frid, V., Bahat, D., Goldbaum, J.: Decay mechanism of fracture induced electromagnetic pulses. J. Appl. Phys. **93**, 5085 (2003)
37. Hill, T.L.: An Introduction to Statistical Thermodynamics. Dover, New York (1986)
38. Rayleigh, L.: On waves propagated along the plane surface of an elastic solid. Proc. London Math. Soc. **s1–s17**, 4–11 (1885)
39. Srivastava, G.P.: The Physics of Phonons. Taylor & Francis, New York (1990)
40. Freund, L.B.: Crack propagation in an elastic solid subjected to general loading—I. Constant rate of extension. J. Mech. Phys. Solids **20**, 129–140 (1972)
41. Freund, L.B.: Crack propagation in an elastic solid subjected to general loading—II. Non-uniform rate of extension. J. Mech. Phys. Solids **20**, 141–152 (1972)
42. Freund, L.B.: Crack propagation in an elastic solid subjected to general loading—III. Stress wave loading. J. Mech. Phys. Solids. **21**, 47–61 (1973)
43. Freund, L.B.: Crack propagation in an elastic solid subjected to general loading—IV. Obliquely incident stress pulse. J. Mech. Phys. Solids **22**, 137–146 (1974)
44. Freund, L.B.: Dynamic Fracture Mechanics. Cambridge University Press, Cambridge (1990)
45. Rose, L.R.F.: Recent theoretical and experimental results on fast brittle fracture. Int. J. Fract. **12**, 799–813 (1976)
46. Tromans, D.: Crack propagation in brittle materials: relevance to minerals comminution. Int. J. Res. Rev. Appl. Sci. **13**, 406–427 (2012)
47. Scruby, C.B.: Quantitative acoustic emission techniques. Nondestruct. Test. **8**, 141–208 (1985)
48. Wadley, H.N.G., Scruby, C.B.: Elastic wave radiation from cleavage crack extension. Int. J. Fract. **23**, 111–128 (1983)
49. Guozden, T.M., Jagla, E.A.: Supersonic crack propagation in a class of lattice models of mode III brittle fracture. Phys. Rev. Lett. **95**, 224302 (2005)
50. Guo, G., Yang, W., Huang, Y.: Supersonic crack growth in a solid of upturn stress–strain relation under anti-plane shear. J. Mech. Phys. Solids **51**, 1971–1985 (2003)
51. Sanders, W.T.: On the possibility of a supersonic crack in a crystal lattice. Eng. Fract. Mech. **4**, 145–153 (1972)
52. Rosakis, A.J.: Cracks faster than the shear wave speed. Science **284**, 1337–1340 (1999)
53. Todoroki, A., Tanaka, M., Shimamura, Y.: Measurement of orthotropic electric conductance of CFRP laminates and analysis of the effect on delamination monitoring with an electric resistance change method. Compos. Sci. Technol. **62**, 619–628 (2002)
54. Girard, C., Joachim, C., Gauthier, S.: The physics of the near-field. Reports Prog. Phys. **63**, 893–938 (2000)
55. Mikki, S.M., Antar, Y.M.M.: A theory of antenna electromagnetic near field—Part I. IEEE Trans. Antennas Propag. **59**, 4691–4705 (2011)
56. Schmitt, R.: Electromagnetics Explained: A Handbook for Wireless/RF, EMC, and High-Speed Electronics. Newnes, New York (2002)
57. Maxwell, J.C.: A dynamical theory of the electromagnetic field. Philos. Trans. R. Soc. London **155**, 459–512 (1865)
58. Hamstad, M.A.: Improved signal-to-noise wideband acoustic/ultrasonic contact displacement sensors for wood and polymers. Wood Fiber Sci. **29**, 239–248 (1997)
59. Hamstad, M.A., Fortunko, C.M.: Development of practical wideband high-fidelity acoustic emission sensors. In: Chase, S.B. (ed.) Nondestructive Evaluation of Aging Infrastructure, pp. 281–288. International Society for Optics and Photonics, San Diego (1995)
60. Shiwa, M., Inaba, H., Carpenter, S.H., Kishi, T.: Development of high-sensitivity and low-noise integrated acoustic emission sensor. Mater. Eval. **50**, 868–874 (1992)

61. Smuga, J.R.: Conductively filled poly(methyl methacrylate) composites: manufacture and testing processes for EMI shielding effectiveness. PhD-thesis, Edinburgh Napier University (2012)
62. Ott, H.: Electromagnetic Compatibility Engineering. Wiley, New York (2009)
63. Paul, C.R.: Introduction to Electromagnetic Compatibility. Wiley Series in Microwave and Optical Engineering. Wiley, New York (2006)
64. Préault, V., Corcolle, R., Daniel, L., Pichon, L.: Influence of skin effect on the effective shielding effectiveness of composite materials. J. Appl. Phys. **115**, 154904 (2014)
65. Sause, M.G.R., Richler, S.: Finite element modelling of cracks as acoustic emission sources. J. Nondestruct. Eval. **34**, 1–13 (2015)
66. Alaca, B.B., Gade, S.O., Sause, M.G.R.: Untersuchung von elektromagnetischen und akustischen Emissionen während Bruchvorgängen in faserverstärkten Kunststoffen. In: 20. Kolloquium Schallemission. pp. 1–8, Garmisch-Partenkirchen, Germany (2015).
67. McLaskey, G.C., Glaser, S.D.: Acoustic emission sensor calibration for absolute source measurements. J. Nondestruct. Eval. **31**, 157–168 (2012)
68. Sause, M.G.R., Hamstad, M.A., Horn, S.: Finite element modeling of conical acoustic emission sensors and corresponding experiments. Sensors Actuators A Phys. **184**, 64–71 (2012)
69. Ono, K., Cho, H., Matsuo, T.: Transfer functions of acoustic emission sensors. J. Acoust. Emiss. **26**, 72–90 (2008)

Chapter 6
Computed Tomography

Many of the early developments of X-ray-based computed tomography (CT) go back to the pioneering work of Cosslett and Nixon [1–4]. The method of computed tomography is understood as any computer-aided tomographic process (usually involving X-rays) to produce 3D representations of the scanned object both externally and internally. This inspection method has been substantially improved since its technological beginnings in the 1970s and allows inspecting fiber reinforced composite materials nowadays with remarkable precision. Some of the key uses for industrial CT scanning involve flaw detection, failure analysis, metrology, assembly analysis, and reverse engineering applications [5]. X-ray-based computed tomography is an established technology, which is readily available from commercial companies. Consequently, the focus of this chapter is not to cover all fundamentals of the measurement method but its application and limitation in the context of defect sensing in fiber reinforced materials. For a more detailed review on the method of computed tomography, the reader is instead referred to some recent books on this topic [6–8].

To start with the chapter, first an overview on the principle of operation of CT is provided, and some basic configurations of CT are presented. The subsequent section has its focus on the present limitations in terms of CT resolution, including artifacts and typical failure modes as encountered in fiber reinforced materials. One section is used to review the important concepts involving combinations of mechanical loading of materials with CT imaging. As further development of DIC methods, the important aspect of digital volume correlation (DVC) and its application to fiber reinforced materials are discussed. Finally some applications of CT to assess the failure progression in composite materials along with possibilities to use CT information in modeling approaches are presented.

M.G.R. Sause, *In Situ Monitoring of Fiber-Reinforced Composites*,
Springer Series in Materials Science 242, DOI 10.1007/978-3-319-30954-5_6

6.1 Principle of Operation

A conventional X-ray inspection is based on a simple projection of the test object onto a detector using a point-like X-ray source. This simple geometrical arrangement as seen in Fig. 6.1 causes a projection of the object volume to the detector and has been termed “shadow microscopy.” The magnification factor of the test object is primarily determined by the distance between source and object as well as the distance between the object and the detector.

In order to yield a difference in detected X-ray intensity, two main effects can be distinguished:

1. The detected intensity may be decreased due to a material with higher X-ray absorption.
2. The detected intensity may be decreased due to a test object with higher thickness.

Hence, one projection direction may not allow to distinguish between both effects, since a reduction in detected X-ray intensity could be due to either effect or a combination of both.

In contrast to the conventional 2D projection, the process of CT uses projections of the test object acquired at different angles and combines their information to yield a volumetric structure of the test object. This is only possible by usage of a reconstruction algorithm operating on discretized versions of the projection images.

The typical CT configuration basically requires three components, an X-ray source, a moveable table carrying the test object, and a detector system. Although there are many types of CT configurations available, only the most common type of cone-beam CT will be explained in the following, since it was used for the experimental results shown in this chapter.

For a CT scan, the test object is placed on a rotary table. Similar to the principle of “shadow microscopy,” in CT the test object can be moved closer to the source to yield higher magnification or closer to the detector to yield lower magnification. One very important factor for cone-beam CT is that the part must fit in the field of view of the detector to allow proper reconstruction (Fig. 6.2). After the initial alignment, the test object is rotated 360° and 2D X-ray projections are taken at

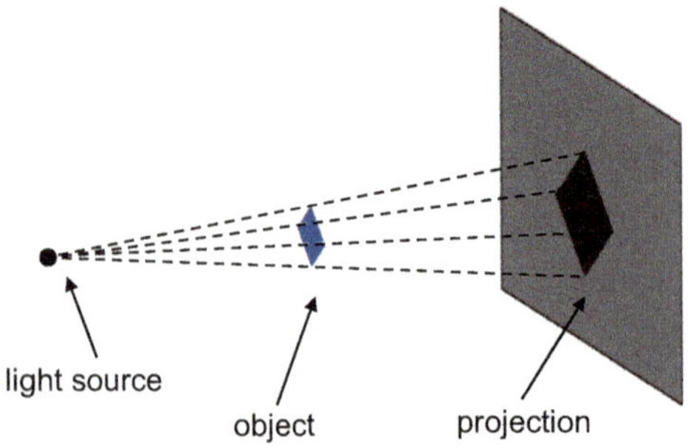

Fig. 6.1 Arrangement of source, test object, and detector in “shadow microscopy”

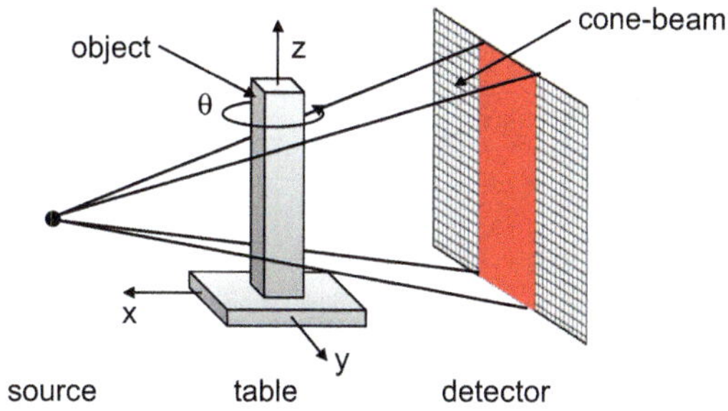

Fig. 6.2 Typical arrangement of cone-beam CT

distinct angles. For typical CT acquisitions, the number of X-ray images ranges from several hundreds to several thousands in a single rotation of the test object. The individual images are then reconstructed to create a 3D volume representation of the test object. If the test object does not allow a full 360° rotation, special reconstruction algorithms are used to allow scans based on projected images acquired at fractions of the full 360° rotation.

In the following, some key aspects of CT are discussed in more detail to allow a better understanding on how the X-ray source, the detector, and the reconstruction algorithm influence the imaging process. In the context of this book, the main application is to visualize the interior volume of fiber reinforced materials.

In order to improve the resolution of X-ray microscopes, one of the key challenges was to achieve a smaller source size, while keeping or increasing the brightness of the source to allow for reasonable exposure times. This lead to the geometrical arrangement of X-ray tubes nowadays referred to as micro-focus or nano-focus tubes as seen in Fig. 6.3 which constitute the most frequently used X-ray source.[1]

As with any conventional X-ray tube, the first step is a generation of a free electron gas using thermionic emission by a heating voltage U_{H} applied to a coil located inside a vacuum tube. The free electrons are focused to leave a filament grid and form an electron beam using a voltage U_{G}. By using an acceleration voltage U_{acc} between the emission coil and an anode, the electron beam is further accelerated and directed towards a target material. The target material (typically tungsten or molybdenum) is hit by the electron beam, and X-rays are released in form of deceleration radiation and characteristic radiation. To achieve a high resolution, a small focal spot size and thus a small interaction volume between electron beam and target material are necessary. In order to minimize the focal spot size, the electron beam is typically focused using magnetic lens systems minimizing its cross-section at the position of target impact. In addition, high-resolution tube systems use target materials, which are comparatively thin and are arranged perpendicular to the incident beam (transmission tubes). This is in contrast to the usual

[1] In principle, radioisotopes could also be used to generate 2D projections of objects as frequently done in field applications. However, the handling aspects and the intensity decay of these sources practically inhibit their use for computed tomography applications.

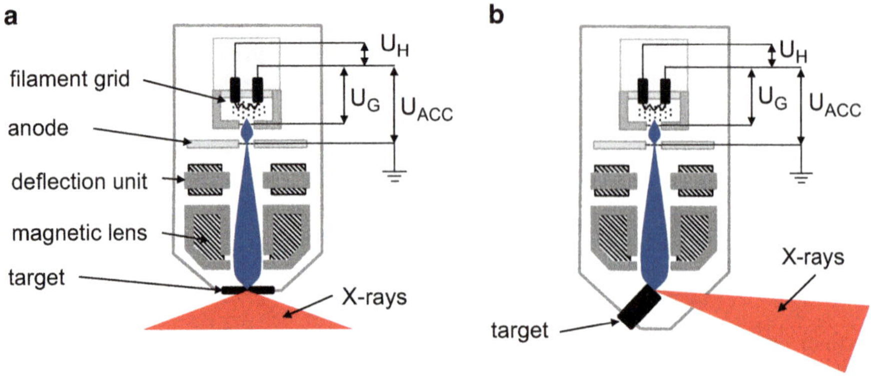

Fig. 6.3 Configurations of X-ray transmission tubes (**a**) and tubes using directional design (**b**)

configuration with oblique incidence of the electron beam onto a thick target material, a configuration which is known as directional tubes. In any case the X-rays are released within the target material and radiate outward from their position. This causes a typical cone-shaped X-ray beam spreading starting at the source spot.

A superior technological alternative as an X-ray source for CT is synchrotron radiation. When electrons traveling at ultrarelativistic speeds are forced by a magnetic field to travel in a curved path, synchrotron radiation is produced. In theory, the ultrarelativistic speed will change the observed radiation frequency due to the Doppler effect. The relativistic length contraction changes the radiated frequencies further, thus effectively yielding electromagnetic radiation in the X-ray spectrum. The radiation pattern can be distorted from an isotropic dipole pattern into an extremely forward-pointing cone of radiation depending on the velocity of the electrons. Synchrotron radiation is the brightest artificial source of X-rays and may be achieved artificially in synchrotrons or storage rings. The electromagnetic radiation produced in this way has a characteristic polarization, and the high intensity photon beam allows fast exposure times for CT scanning. In addition, the high brilliance of the synchrotron beam generates a small divergence and enables a small source spot size.

Another important component to achieve high-resolution volumetric representation of objects is the detector system. For computed tomography this is a device that captures X-rays and usually converts them to light in the visible spectrum. Among the most relevant detector concepts used, there are scintillator systems and flat-panel detectors. In practice this includes line scanners and array scanners to produce 1D or 2D projections of the test object.

For the scintillator systems, the incident X-rays are first converted by a scintillator crystal into visible light. A photo cathode is then used to convert the visible light into electrons, which are accelerated and focused onto a fluorescent screen. The projected image on the fluorescent screen is then detected by a CCD camera system and constitutes the final 2D digital image. As technological alternative, the digital flat-panel detectors use a thin scintillation foil directly applied on a matrix array of photodiodes operating in the range of emitted visible light. The detected photon intensity is then turned into a 2D digital image. Some of the recent developments consider the overall pixel resolution of the detector arrays, which has a direct impact on the detail visibility. In addition, digital detectors come in a variety of pixel pitches which is the space between pixels. Typically a detector with a smaller pixel pitch relates to longer image acquisition time and higher costs and requires a higher photon flux to produce an image. A larger pixel pitch requires less energy to light up the pixels and is cheaper to manufacture but reduces resolution.

As fundamental principle of computed tomography, the 2D projections by itself are meaningless without further processing by a reconstruction algorithm. The task of the latter is the generation of 2D slices transverse to the axis of rotation and the subsequent stacking of these individual slices to produce a 3D volume. The usual algorithm, which is applied to this end, is the filtered back projection, which is based on the Radon transformation developed in 1917 by Radon [9]. Since the ongoing developments in this particular field expand beyond the level of detail intended for this chapter, the interested reader is referred to some recent literature in this area for more precise information [6–8].

6.2 Detail Visibility

Within this section the important aspect of detail visibility is reviewed. The term detail visibility accounts for our perception of the scanned objects. Since this is not equivalent to the system resolution, it needs some further specification. In most of the imaging technologies, detail visibility is used for the (human) ability to identify a certain object as what it is. Thus it can be understood as the necessary resolution of the image in terms of pixels or voxels sufficient to allow spotting a detail, which is different to its surrounding. However, in a general situation, an exact definition of detail visibility is quite challenging, since our perception is substantially influenced on whether the relevant structure is isolated against a background or whether the structure is periodic with a certain grid length. In the context of this chapter, the term will be used in conjunction with the aspect of visualization of defects in fiber reinforced materials. Some examples of CT images are presented at the end of this section to provide a subjective feeling of detail visibility as achieved by the state-of-the art computed tomography devices.

Before, the two main contributions influencing detail visibility are discussed. First, the overall object resolution as well as induced uncertainties of the experimental configuration of the 2D projections is derived. Second, the aspects of image artifacts arising from material combinations, reconstruction algorithms, and other effects are presented.

6.2.1 Object Resolution

Based on the concept of shadow microscopy, the geometrical magnification M is readily derived by the length of the focus-object-distance (FOD) and the length of the focus-detector-distance (FDD) as

$$M = \frac{\mathrm{FDD}}{\mathrm{FOD}} \tag{6.1}$$

This simple geometrical relationship is shown in Fig. 6.4a schematically and as measurement result of a composite material in Fig. 6.5.

In principle, (6.1) would allow arbitrary large magnification factors, given the size of the detector is scaled accordingly. However, the actual object resolution is not just determined by the geometrical magnification factor. To assess the final resolution of the measurement system, there are three contributions from the experimental equipment to take into account:

1. Pixel size $\widetilde{P}$
2. Focal spot size $\widetilde{F}$
3. Mechanical stability

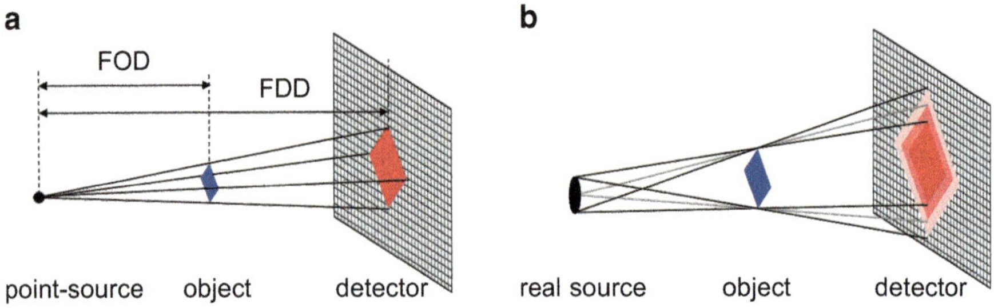

Fig. 6.4 Definition of geometrical magnification factor (**a**) and relation between penumbra effect and focal spot size (**b**)

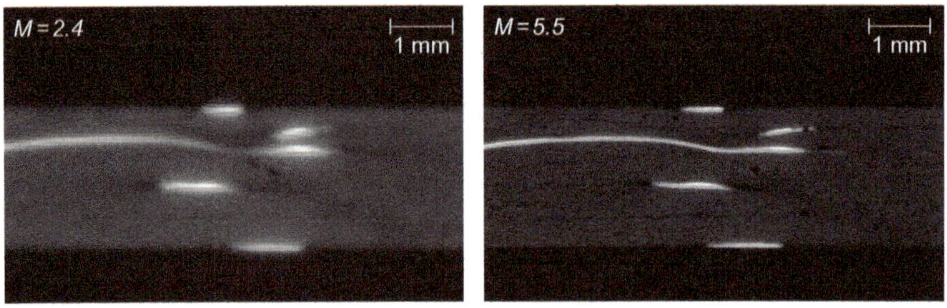

Fig. 6.5 Comparison of CT scan slice of same object with different magnification factors

The acquired digital images are spatially discretized by the pixel array of individual size $\widetilde{P}$. This constitutes the fundamental length scale of spatial resolution at the position of the detector. The absolute resolution of the volume produced from such 2D images is generally referred to as voxel size $\tilde{V}$, which is defined by the pixel size $\widetilde{P}$ and the geometric magnification factor M:

$$\widetilde{V} = \frac{\widetilde{P}}{M} \tag{6.2}$$

The overall resolution of the system is further limited by the effect of unsharpness resulting in blurred images. The geometric unsharpness d_g is mostly due to the focal spot size $\widetilde{F}$. The latter causes a penumbra effect as seen in Fig. 6.4b and may be written in terms of the geometric magnification as

$$d_g = \widetilde{F}(M - 1) \tag{6.3}$$

Thus, the higher M, the larger the impact on d_g as seen by the schematically comparison in Fig. 6.6.

Finally, the mechanical stability of the configuration may result in unintended movement of the object and thus may introduce additional movement unsharpness d_{mov}. The latter is linked to the rotation of the test object and may be induced by tilting of the object during scanning. However, given the test object is tightly fixed and has sufficient stiffness, this effect is usually negligible. As another factor of mechanical movement during image acquisition, temperature gradients may cause a slow change in the dimensions of the test object or the geometric arrangement as such. Under usual operating conditions, the CT systems are designed in a way that the thermal expansion of the system is minimal and therefore the effect is usually also negligible. For the dimensional change of the test object, the thermal conductivity and the thermal expansion coefficients are relevant quantities. These strictly depend on the scanned material and thus may induce according thermal expansion of the test object. However, within the usual change of temperature, this effect is negligible for most materials. The sum of all of these movement effects will lead to

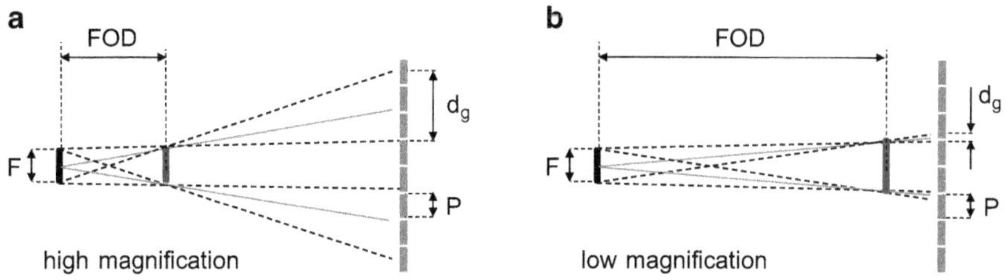

Fig. 6.6 Relation between magnification factor and geometrical unsharpness for high (**a**) and low (**b**) magnification factor

an additional unsharpness proportional to M in the acquired image and, hence, will directly influence the resolution of the volume.

The total unsharpness d_{tot} of the system may then be derived by combination of the geometric unsharpness, the movement unsharpness, and the pixel size, yielding

$$d_{tot} = \frac{1}{M}\sqrt{d_{mov}^2 + d_g^2 + \widetilde{P}^2} = \sqrt{\left(\frac{d_{mov}}{M}\right)^2 + \left(\frac{\widetilde{F}(M-1)}{M}\right)^2 + \widetilde{V}^2} \tag{6.4}$$

Thus for a detector pixel size $\widetilde{P} \ll d_{tot}$ and $d_{mov} \to 0$, the focal spot size is the limiting factor for image resolution. In contrast, for pixel size $\widetilde{P} \gg d_{tot}$ and $d_{mov} \to 0$, the pixel size will be the limiting factor.

In high-resolution imaging, the ultimate resolution is usually limited by the focal spot size $\widetilde{F}$. For the case of microstructure as found in fiber reinforced materials, an empirical value of the maximum detail detectability may be approximately 1/3 of the focal spot size $\widetilde{F}$.

As an example of this relationship, a series of cross-sectional images of the same object are shown as function of the focal spot size $\widetilde{F}$ in Fig. 6.7 for a constant magnification factor $M = 5.5$. Clearly, the smallest spot size comes with best detail visibility. When changing $\widetilde{F}$ to larger sizes, this induces noise in the images, therefore reducing the possibility to resolve details of the composite microstructure.

6.2.2 *Artifacts*

In the context of computed tomography, the term artifact is applied to any systematic discrepancy between the attenuation values in the reconstructed volume and the true attenuation coefficients of the test object [10, 11]. CT images are easily affected by various artifacts since the final volumetric image is based on a multitude of independent measurements. The reconstruction algorithm assumes that all these measurements are consistent, so any inconsistency during the measurement will directly transfer into an error in the reconstructed volume. As general terminology used for artifacts in computed tomography, Barrett and Keat suggest the following four artifact types [10]:

(a) Streaking
(b) Shading
(c) Rings
(d) Image distortion

Further it is possible to group the artifacts according to their origins into four categories:

Fig. 6.7 Comparison of CT scan slice of same object with different spot sizes $\widetilde{F}$ at constant magnification factor $M = 5.5$

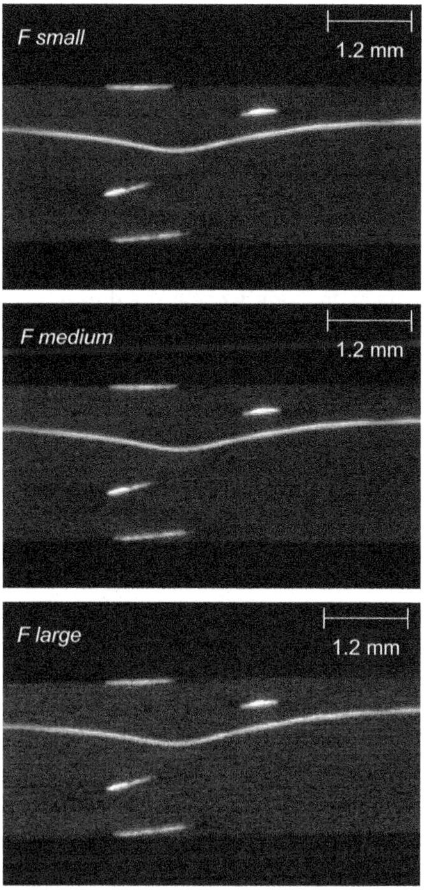

1. Physics-based artifacts, which result from the physical processes involved in the acquisition of the projection images
2. Hardware-based artifacts, which result from imperfections in the acquisition system
3. Reconstruction artifacts, which are caused by the deficiencies in the image reconstruction process
4. Motion artifacts, which are caused by movement of the test object during acquisition

In the following, typical artifacts of these four categories are discussed following the considerations made by [10] for medical applications, but are generalized in their meaning to the field of material analysis, and distinct examples in application to fiber reinforced materials are presented.

6.2.2.1 Physics-Based Artifacts

Beam Hardening

As a polychromatic X-ray beam passes through matter, low-energy photons are preferentially absorbed, and the attenuation is no longer a linear function of absorber thickness. Hence, the beam becomes "harder" because the lower-energy photons are absorbed more rapidly than the higher-energy photons, and therefore the mean energy of the beam increases [10, 12]. This may induce several types of artifacts frequently occurring during scans of fiber reinforced materials. Among the types of beam hardening artifacts, there are the so-called cupping artifacts, the appearance of dark bands or streaks between objects with high density, and the glooming of high-density inclusions. Several algorithms are available in modern reconstruction programs to minimize the effect of beam hardening to some extent [13–15].

Cupping Artifact

The occurrence of cupping artifacts is readily explained following [12] by considering the attenuation profile observed for a uniform cylinder as seen in Fig. 6.8a. Ideally, the attenuation profile would resemble a certain shape allowing a reconstruction of the cylinder in its expected shape. However, the beams passing through the center of the cylinder suffer more from beam hardening than those passing the edges. Since higher-energy X-rays are attenuated less during propagation, the projection areas with more beam hardening suffer less from attenuation than theoretically expected. The reconstruction of these falsified projections thus causes a characteristic cupped shape at the edges of the cylinder.

This type of artifact is frequently observed at the edge of the material, i.e., the interface between the surrounding air and the solid. This causes the white shade at the edge of the cross-section of the fiber reinforced material seen in Fig. 6.8b and should not be misinterpreted as material with higher density.

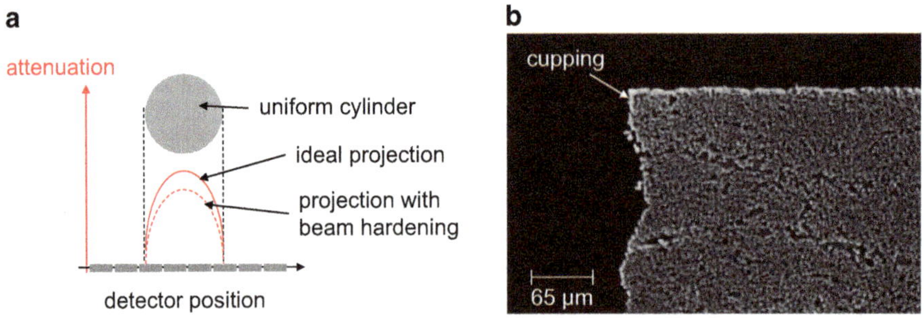

Fig. 6.8 Attenuation profiles obtained with and without beam hardening for an X-ray beam passing through a uniform cylinder (**a**) and example of cupping artifact in CT scan (**b**)

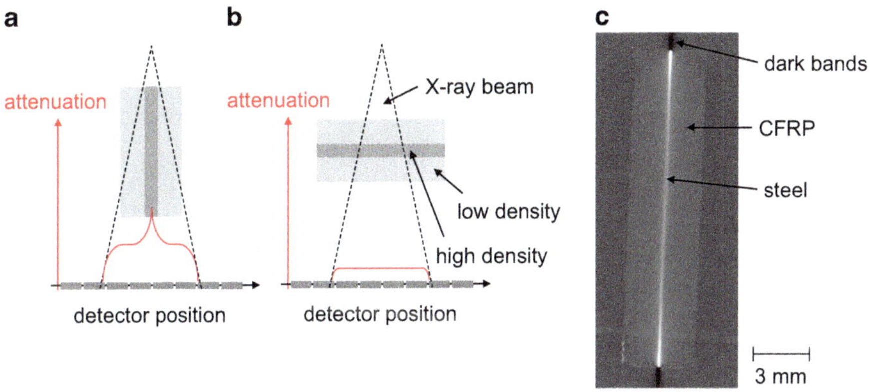

Fig. 6.9 Origin of dark bands due to beam hardening effects (**a**, **b**) and example of dark bands due to high-density inclusion (steel layer) in a carbon fiber reinforced polymer (**c**)

Dark Bands

Another type of artifact, which may occur due to beam hardening, are dark bands caused by spatially close high-density inclusions or oriented high-density materials. In such a situation, the incident beam is affected more by beam hardening for the path between source and detector where the high-density object is aligned with the beam direction (Fig. 6.9a) and suffers less for the path perpendicular to this arrangement (Fig. 6.9b). As consequence, the regions behind projection directions with strong beam hardening appear with higher density than in reality. This effect is visible as dark bands extruding from these projection directions. As an example for this effect, a reconstructed slice of a CT scan of a fiber-metal laminate is shown in Fig. 6.9c. In this case, the thin steel layer has much higher density than the surrounding fiber reinforced material and therefore acts as high-density inclusion. Along the long axis of the metal sheet, strong dark bands extrude from the position of the specimen edge.

High Density Foreign Material Artifact

Similar to the previous two effects, the presence of a material with comparatively high density to the surrounding material can generally lead to visible streaking artifacts. These typically occur because the density of the material may exceed the measurement range of the detector system and thus results in an incorrect attenuation measurement. In the context of fiber reinforced materials, any metallic inclusions may cause this type of artifact. Since metallic materials attenuate the X-ray beam much more than the polymeric matrix system or the typical reinforcement fibers, the attenuation values of objects behind the metallic inclusion are measured too high. Due to the application of the reconstruction algorithm to this

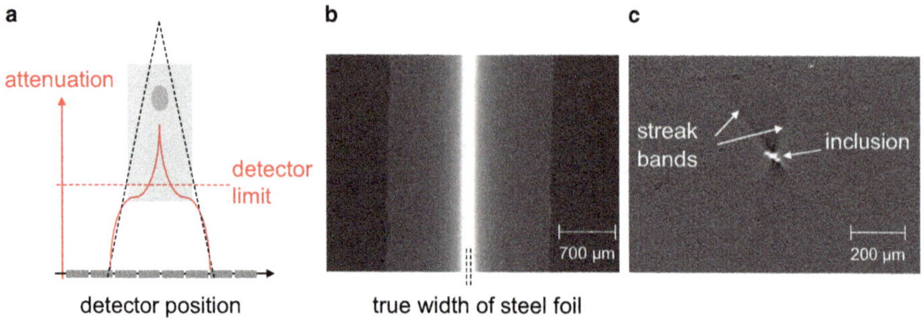

Fig. 6.10 Wrong measurement of attenuation profile due to detector limit (**a**) and example of gloom due to high-density inclusion (steel layer) in a carbon fiber reinforced polymer (**b**) and example of streak bands due to high-density particle inclusion in a carbon fiber reinforced polymer (**c**)

incorrect measurement of the attenuation profile, bright and dark streaks in transverse directions to the rotation axis may appear in the CT image.

The effect of this incorrect measurement of the density is visible as bright glow around the high-density object. For the case of the fiber-metal laminate shown in Fig. 6.10b, this causes a characteristic glow at the interface between the thin metallic sheet and the fiber reinforced material reducing visibility of details at this interface. As another example, streak bands due to a particle of high density included in a fiber reinforced polymer are shown in Fig. 6.10c.

Partial Volume Averaging

The algorithms used in CT data reconstruction assume that the object is completely covered by the detector at all view angles and that the attenuation is caused by the object only. In cases where this situation is not fulfilled, the CT reconstruction starts to show truncated sections. Due to the divergence of the X-ray beam, this is usually visible at the end of the reconstructed volume if the scanned region is smaller than the object (see Fig. 6.11a). This occurs because the finite size of the detector does not allow measuring the full attenuation profile of the cone beam, since the red areas are not equally scanned at all angles. This effect is exemplarily shown for a reconstruction of a piece of fiber reinforced material in Fig. 6.11b.

Another artifact related to the partial volume effect is mostly relevant to the specific scanning configuration of helix tomography. Here line scanners instead of array scanners are used to produce each reconstructed slice. The feed of the test object is done mechanically, so the thickness of each slice is determined by the feeding speed. Considering the scheme in Fig. 6.12, the challenge is to provide an equal lateral resolution along all three axes for reconstructing the volume image. If the slice thickness is chosen large compared to the lateral size of the pixel, its aspect ratio gets distorted and the contained information is smeared out (cf. Fig. 6.12a). Here the length of the voxel along the feed axis is much larger than in the directions

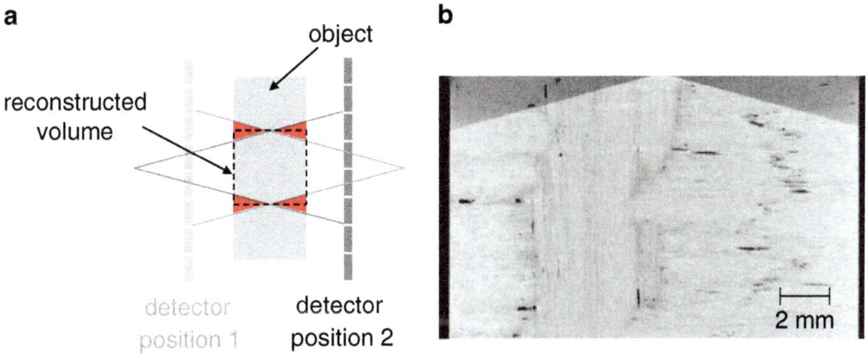

Fig. 6.11 Scan configuration (**a**) and example of partial volume effect upon reconstruction (**b**)

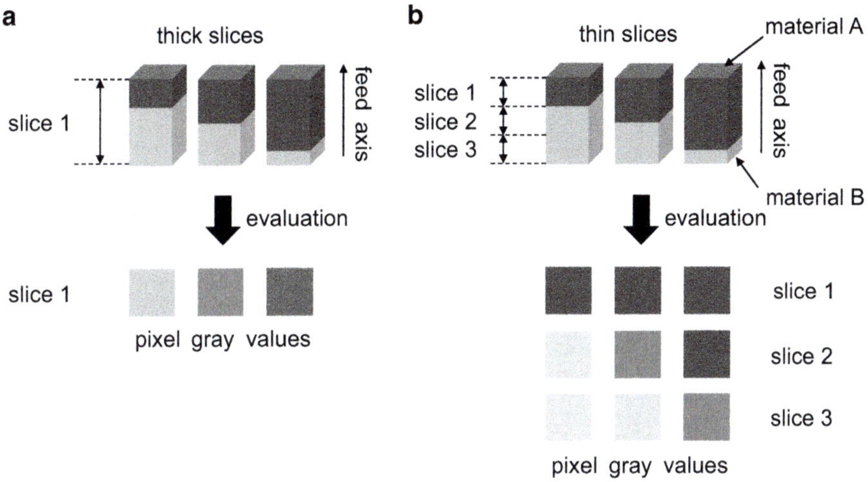

Fig. 6.12 Origin of spillover effect and distortion of voxel aspect ratio causing blurred information in the feed direction due to thick slices (**a**) compared to thin slices (**b**)

perpendicular. This averaging effect is called spillover and manifests itself in a blurred image information along the direction of object feed. Selecting thin slices instead allows keeping a decent aspect ratio of the voxel and retains sharp gradients of material density (cf. Fig. 6.12b).

Quantum Mottle (Noise)

Another trivial type of artifact is due to quantum mottle. The origin of this effect is the incident background radiation seen at the detector, which superimposes to the scan beam. This results in a grain noise structure seen on the image and thus

Fig. 6.13 Example of grain noise in CT image

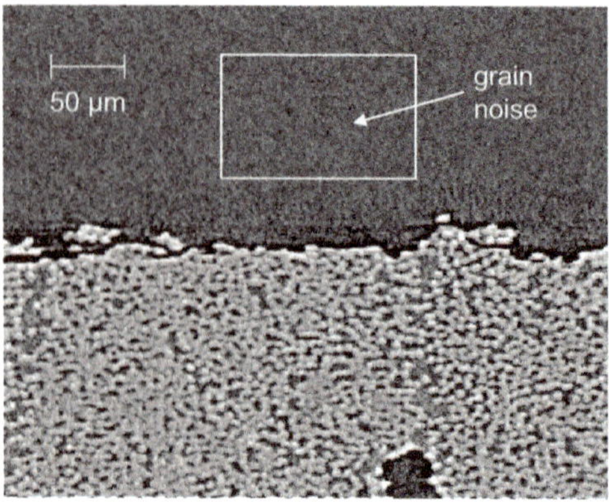

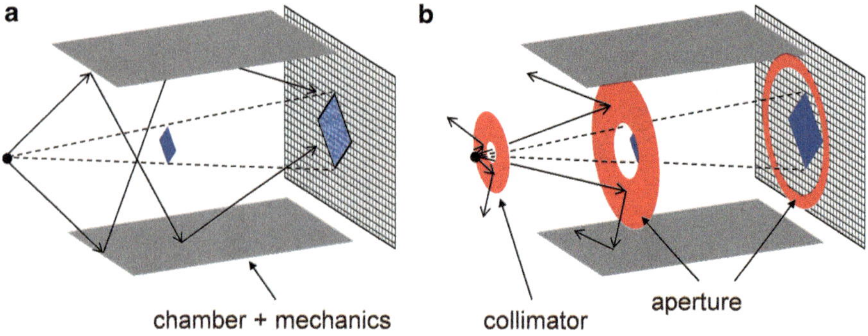

Fig. 6.14 Scheme of measurement configuration without (**a**) and with (**b**) additional collimator and apertures to reduce the intensity of secondary X-rays incident to the detector

resembles an artificial scatter of density. This noise is composed of the general cosmic background noise and the secondary reflections of the X-ray beam, which are incident to the detector at other angles than through the test object. As seen in Fig. 6.13, this is typically pronounced for scans with a low signal-to-noise ratio. The latter may occur for setups with X-rays barely able to penetrate the test object due to its thickness or absorption, but also when attempting to distinguish materials with close-by density such as carbon fibers and polymer matrix systems (see Fig. 6.13).

Some X-ray CT devices use additional shielding measures to reduce the amount of secondary X-ray photons to the detector. In this context "secondary" refers to any backscatter produced by interaction of the primary X-ray beam with other pieces of the equipment, such as the shielding cabinet or the included mechanics. In the configuration shown in Fig. 6.14b, the presence of apertures collimates the primary beam and shields the cabinet from the incident X-rays so the amount of secondary X-rays gets substantially reduced.

Photon Starvation

Another potential source of streaking artifacts is photon starvation. This may occur for objects with highly attenuating areas in some of the projection directions. Similar to the dark bands, these are caused by the relative alignment of the test object and the source-detector axis. Other than for beam hardening the present effect is due to a lack of photons arriving at the detector in general, rather than a falsified attenuation profile. Therefore some of the orientations suffer from noisy projection images. The consequences are streaks in the reconstruction plane, i.e., perpendicular to the rotation axis. For scanning of materials, this effect may be avoided by previously selecting the orientation with highest attenuation and suitably adjusting the X-ray intensity to avoid photon starvation at these angles. However, for geometries with strongly distorted *x*/*y* aspect ratio, this may come with the cost of saturating the X-ray detector at the other projection angles, so suitable trade-off might be required.

Undersampling

As a fundamental principle to computed tomography, the number of projections should be sufficient to allow a reasonable reconstruction of the specimen volume. If the number of projections gets too low (undersampling), the first artifact to observe is ray aliasing at sharp edges and at narrow details. This is seen for the object in Fig. 6.15 being visible as discrete stripes at some distance to the scanned object.

If the overall number of projections gets even lower, severe imperfections of the scanned object can be expected. This is seen in Fig. 6.16 for a decreasing number of projection images as basis for the same reconstruction algorithm. For the full set of 736 images, the rectangular cross-section is well reconstructed, but also the occurrence of dark bands is seen as extension to the longer axis edges of the material. With reduced number of images, the rectangular cross-section is hard to be reconstructed accurately, and an increasing amount of ray aliases occur.

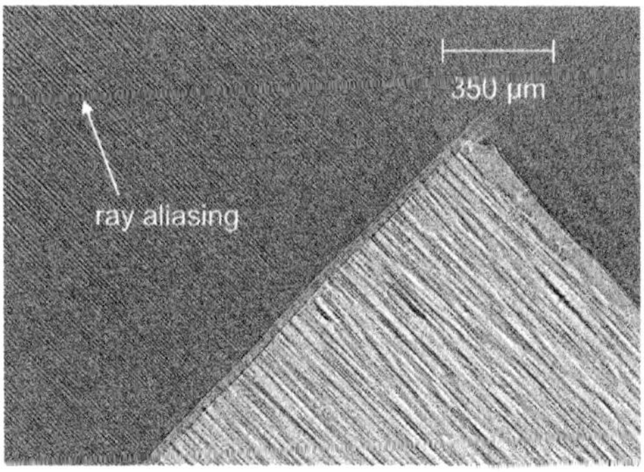

Fig. 6.15 Example of ray aliasing due to undersampling

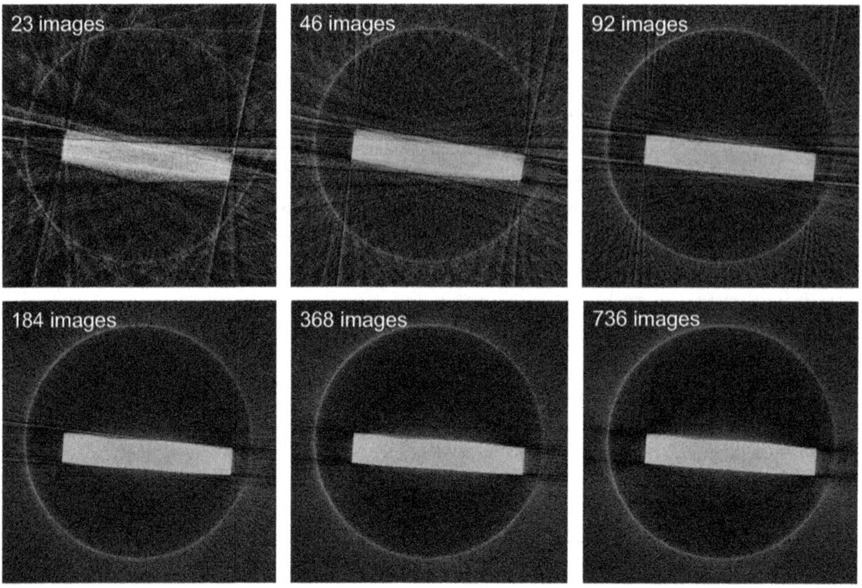

Fig. 6.16 Example of image reconstructed with different numbers of projections

6.2.2.2 Hardware-Based Artifacts

Ring Artifact

The ring artifact is probably the most common hardware-based artifact. It appears as image of one or more "rings" within the reconstruction plane. If one of the pixels of the detector is corrupted or badly calibrated, the detector will give a consistently false reading at each angular position. During the reconstruction process, this turns into a circular artifact with the circle extension being perpendicular to the rotation axis as seen in Fig. 6.17. The appearance of such ring artifacts naturally lowers the image quality, since it tends to overlap with parts of the image reducing the visibility of details in these areas. Modern CT systems allow recalibrating the detector system to reduce the occurrence of ring artifacts by simply removing the information of dead pixels in the detector unit. Similarly, some reconstruction algorithms explicitly focus on suppression of ring artifacts by removing the dead pixel content during the reconstruction process.

Tube Arcing

During operation of X-ray tubes, the occurrence of short circuits within the tube may cause image artifacts due to tube arcing. The electrical shortcut typically emanates from a spark between the cathode and tube cover, which causes a spontaneous loss of X-ray intensity. The occurrence of this effect is linked to the tube design and the particular reliability of the individual tubes.

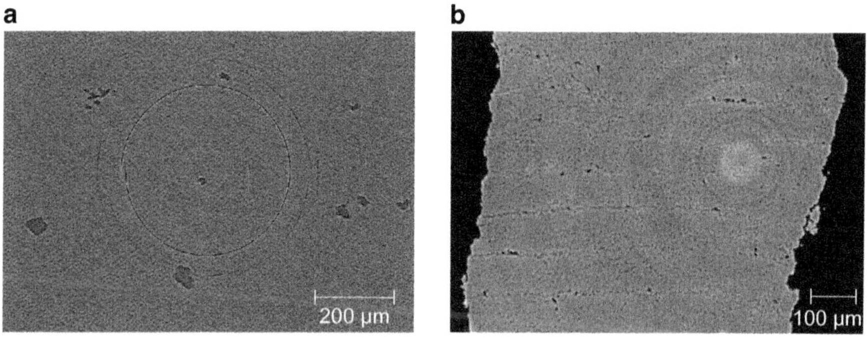

Fig. 6.17 Examples of ring artifacts due to dead pixel (**a**) and due to bad calibration range (**b**)

6.2.2.3 Reconstruction Artifacts

The next class of artifacts arises from the (physical) imperfection of the incident X-ray beams and the according assumptions made in the reconstruction process. Since this is different for the configurations of cone-beam CT and helical CT, the resulting types of artifacts are distinguished based on these two arrangements in the following.

Cone-Beam Effect

For the cone-beam CT configurations, further distinction is made between line-scan detectors resulting in fan-shaped beam geometries and the array detectors resulting in a cone-beam shape of the X-ray beam (cf. Fig. 6.18). Due to the physical size of the beam geometries, the collimators can be closed more for the line scanners but need to stay open more for the array detectors. As a consequence of the cone-beam geometry, the photons collected at the detector position do not only originate from the ideal transmission path along an ideally flat plane configuration achieved in the fan-shape geometry but are also incident at other angles. This effect becomes more pronounced the farther away the scan object is from the ideal rotation axis. Thus to minimize the effect, it is most suitable to position the test object at the center of the rotation axis. However, modern CT systems are readily equipped with cone-beam reconstruction algorithms, so this effect does not play an important role in such systems anymore.

Windmill Artifacts

The windmill artifacts are linked to the use of helical CT device for the scan. The two most well-known artifacts for these systems are the zebra artifact and the stair-step artifact. As zebra artifact, a periodic modulation of light stripes parallel to the

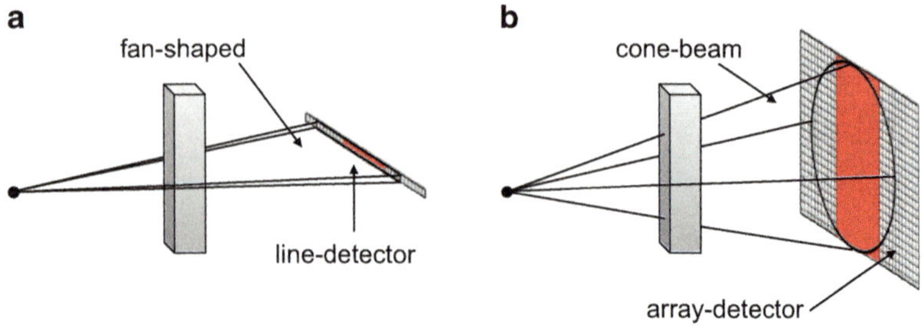

Fig. 6.18 Difference between line-scan arrangement (**a**) and array detectors (**b**)

axis of rotation may appear in the images, which is due to the helical interpolation process causing a modulation of the noise level. This effect starts to be more pronounced at some distance to the axis of rotation. As a stair-step artifact, the misalignment of the individual slices during stacking along the axis of rotation is known. This causes steplike contours responsible for the name of this artifact.

6.2.2.4 Motion Artifacts

Motion artifacts may result from sudden movement or relaxation of a test specimen being scanned. This results in unequal projections of the angles before and after movement and therefore needs to be avoided for proper reconstruction of the volume. For most of the CT scans involving fiber reinforced materials, motion artifacts may not seem to be of huge relevance. If the specimen is tightly fixed to the rotary table, there is no strong likelihood of sudden movement of the specimen. From medical imaging, the consequence of such sudden movement in CT imaging is well known, and motion artifacts usually appear as shading or streaks in the volumetric image or may even render the whole scan useless if the movement was too high.

However, for the in situ load stages discussed herein, there is a certain likelihood of motion artifacts during scanning. For the test object being scanned still subject to mechanical load, the occurrence of partial failure during the acquisition process may cause small movements of the scan object and thus cause the aforementioned artifacts. This is exemplarily seen in the example of a fiber reinforced material recorded within a tensile load stage as seen in Fig. 6.19. While the bottom part of the CT scan is free of motion artifacts, the upper half of the fractured material was subject to sudden secondary failure causing an additional motion during the acquisition procedure turning into the blurred representation seen best at the edges.

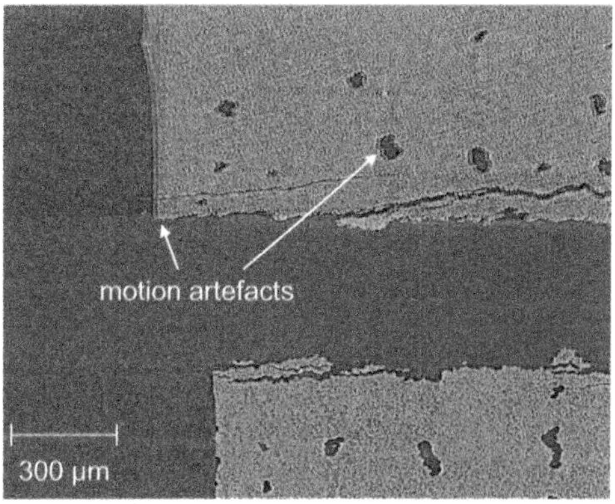

Fig. 6.19 Motion artifact in upper half of fractured material as observed in in situ load stage

6.2.3 Detectability of Defects in Fiber Reinforced Materials

High-resolution computed tomography can be used to identify microstructures in composite materials, such as fiber distributions [16–18], void volume fractions [19–24], and damage mechanisms such as cracks and fiber breaks [25–34] occurring within the volume of the material. A high level of confidence can be placed in the results as the features are viewed directly at the relevant length scale. In the following some representative examples of computed tomography images of fiber reinforced materials are presented to demonstrate the current capabilities of high-resolution computed tomography devices in application to fiber reinforced materials. Such high-resolution X-ray CT has gained considerable interest as a powerful tool for material studies directed to understand interaction of composite failure mechanisms [31, 35, 36]. Advances in the last decade now provide established 3D imaging routines in the micrometer and sub-micrometer range. This method is gaining importance in engineering and material science to determine the internal characteristics of materials and their behavior [37]. In the case of fiber reinforced materials, CT has been used to identify material characteristics like fiber orientation and volume fraction at the sub-ply level, and failure mechanisms, such as ply cracks [31, 32, 34, 38] and fiber failure [39, 40]. Among the multitude of possibilities, a particular focus is given to the detectability of damage mechanisms in the following, since this is the main topic within this book.

Figure 6.20a–d shows CT results of a fiber reinforced laminate with $[(0/90)_2]_{sym}$ stacking sequence made from a Sigrafil CE1250-230-39 prepreg. The 17 mm × 150 mm × 2.2 mm specimen was first subject to tensile load according to DIN-EN-ISO 527-4 to induce some failure and was unloaded prior to ultimate failure, unmounted and scanned without further preparation. The according acquisition parameters of the scan are summarized in Table 6.1.

As seen in Fig. 6.20, the individual layers can clearly be distinguished based on their different fiber orientations. Also, a high level of detail can be observed for the inter-fiber cracks and inter ply delamination present in the laminate. The details in Fig. 6.20b, c allow to spot inter-fiber failure in the 90° plies, inter-ply delamination,

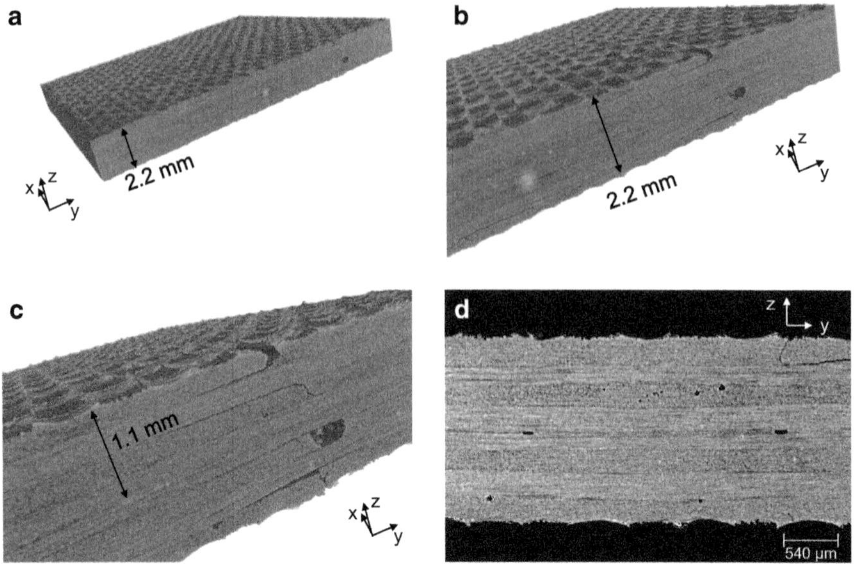

Fig. 6.20 3D representation of $[(0/90)_2]_{sym}$ laminate (**a**) and magnifications allowing to spot inter-fiber failure and inter-ply delamination in 3D volume according (**b**, **c**) and respective virtual cross-section (**d**)

Table 6.1 Acquisition parameters for scan of tensile specimens

Measurement parameter	Value
Detector type	Amorphous Silicon CsI detector $(3072 \times 2400)\text{px}^2$
Filter	None
X-ray tube voltage	90 kV
X-ray tube current	150 μA
Focal spot size	1.5 μm
Voxel size	7.8 μm
Acquisition angles	2000
Image exposure time	1.0 s
3D reconstruction software	Phoenix datos|x2 reconstruction

and inclusions like voids. Clearly one key advantage beyond classical microscopy investigations is the possibility to produce virtual cross-sections at arbitrary positions of the scanned volume. In contrast, optical and electron microscopy may offer even higher spatial resolution, but can only be carried out at surfaces, requiring either a preparation step after loading (losing any relation to the applied load or damage evolution) or via in situ analysis under a nonrepresentative stress state (i.e., monitored at free surface as opposed to the state within the bulk material). Polishing and cutting techniques commonly required for cross-sectional imaging can also induce artifacts which may be mistaken for features of interest.

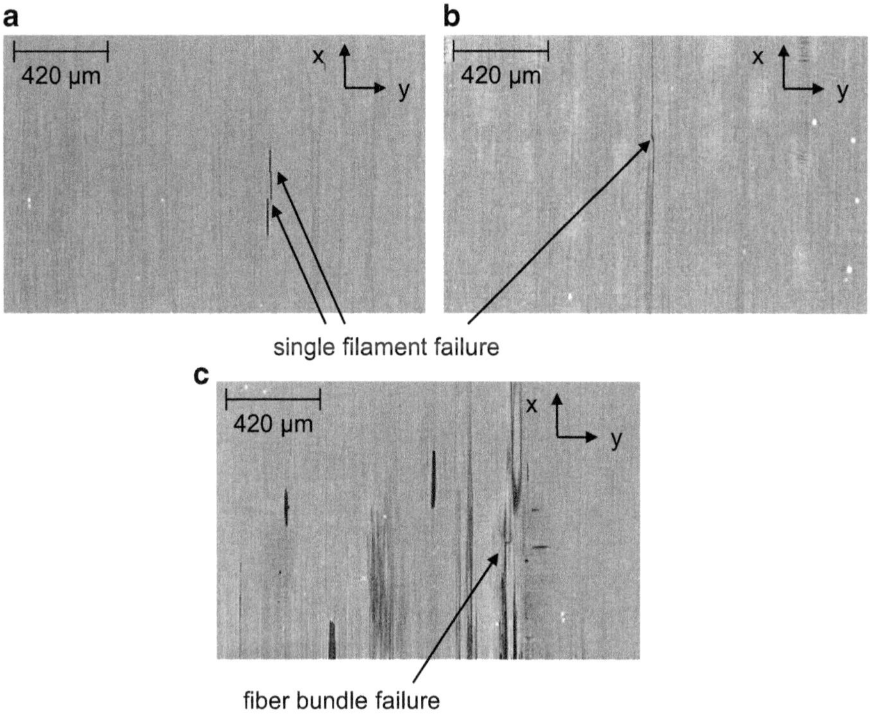

Fig. 6.21 Visibility of single carbon fiber filament failure (**a**, **b**) and fiber bundle failure (**c**) in 2D cross-sections of $[(0/90)_2]_{sym}$ laminate

Using a cross-section in the *xy*-plane, this allows inspecting the 0° layers for the occurrence of fiber breakage. As seen from Fig. 6.21a, b, the occurrence of single carbon filament failure (typical diameter 7 μm) is clearly visible from the cross-sectional images. Moreover, the initiation of matrix cracks at these positions causing splitting of the laminate in parallel to the fiber axis is also readily observed. In addition, Fig. 6.21c shows an example of a fiber bundle failure.

Another advantage of CT is to segment the scanned volume to retrieve only a particular part of the specimen. As seen in the 3D visualization in Fig. 6.22, this allows to artificially remove one part of a fractured body to reveal the internal fracture surface in a unidirectional T800/913 prepreg laminate. The high-resolution scan settings reported in Table 6.2 allow spotting a high level of detail at the fracture surface, such as the morphology and broken fiber filaments.

Using similar high-resolution settings reported in Table 6.3, CT scans also allow removal of the matrix material based on its density and to spot the remaining single carbon fiber filaments (cf. Fig. 6.23). This can be used to evaluate their orientation, to quantify their distribution, or to spot fiber filament breaks as seen in the lower left corner of the present HTA fiber bundle in a polyurethane resin. However, these

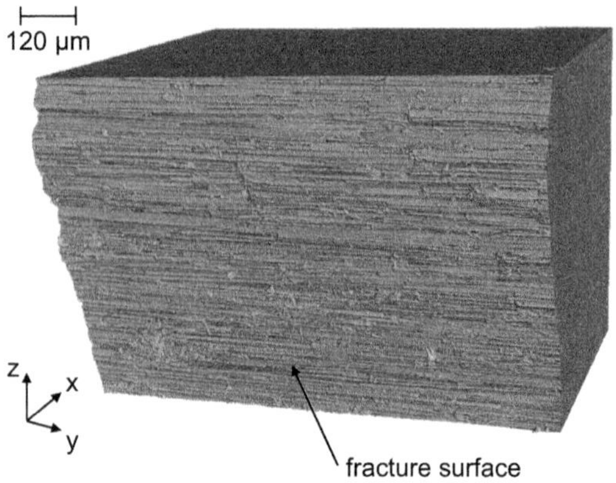

Fig. 6.22 Visibility of details on fracture surface of fiber reinforced polymer in high-resolution scan

Table 6.2 Acquisition parameters for scan of fracture surface

Measurement parameter	Value
Detector type	Amorphous Silicon CsI detector $(3072 \times 2400)\text{px}^2$
Filter	None
X-ray tube voltage	50 kV
X-ray tube current	170 µA
Focal spot size	1.1 µm
Voxel size	3.4 µm
Acquisition angles	1000
Image exposure time	1.5 s
3D reconstruction software	Phoenix datos|x2 reconstruction

Table 6.3 Acquisition parameters for high-resolution scan

Measurement parameter	Value
Detector type	Amorphous Silicon CsI detector $(3072 \times 2400)\text{px}^2$
Filter	None
X-ray tube voltage	70 kV
X-ray tube current	220 µA
Focal spot size	0.6 µm
Voxel size	0.7 µm
Acquisition angles	1800
Image exposure time	2.0 s
3D reconstruction software	Phoenix datos|x2 reconstruction

high-resolution images are only possible for specimen sizes small enough to permit short focus-object distances.

In particular for fiber reinforced materials, the damage zone is sometimes geometrically complex, and severe fragmentation during fracture is happening frequently. Thus conventional cross-sectioning techniques using resin embedding

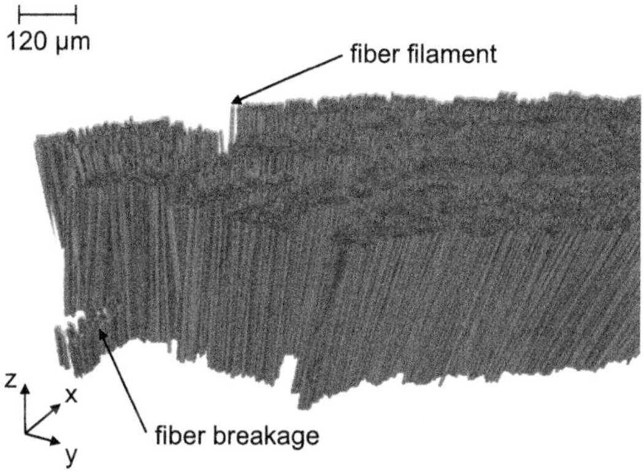

Fig. 6.23 Visibility of single carbon fiber filaments and fiber failure in 3D volume

Table 6.4 Acquisition parameters for scan of impact damage

Measurement parameter	Value
Detector type	Amorphous Silicon CsI detector $(3072 \times 2400)\text{px}^2$
Filter	None
X-ray tube voltage	50 kV
X-ray tube current	200 μA
Focal spot size	1.5 μm
Voxel size	27.5 μm
Acquisition angles	1000
Image exposure time	1.5 s
3D reconstruction software	Phoenix datos|x2 reconstruction

and mechanical grinding and polishing are somewhat limited as many of the small fragments are not conserved by this approach. Using a CT scan for such damage zones, these small details can readily be visualized. As seen for the example of impact damage in a carbon fiber/epoxy laminate with glass fiber rovings in Fig. 6.24, all small fiber filaments residues and epoxy resin fragments can be spotted, and the geometrical complexity is much better represented in a 3D image than in a simple 2D cross-section.

The analysis of porosity inside a laminate has achieved high attention within the last decade [19–24]. Due to the direct visibility of gaseous inclusions within the laminate, the direct quantification of their volumetric fraction is tempting, but standardized procedures are still subject to recent research efforts. As an example of the visibility of voids on the micro- and mesoscale, two examples are shown in Fig. 6.25a, b for a CT scan with the parameter settings of Table 6.1. Extensive studies have been carried out to investigate the visibility and accuracy of void dimensions as function of CT scan parameters [21, 23, 24]. Since the quantification of such inclusions is not within the scope of this book, this aspect will not be

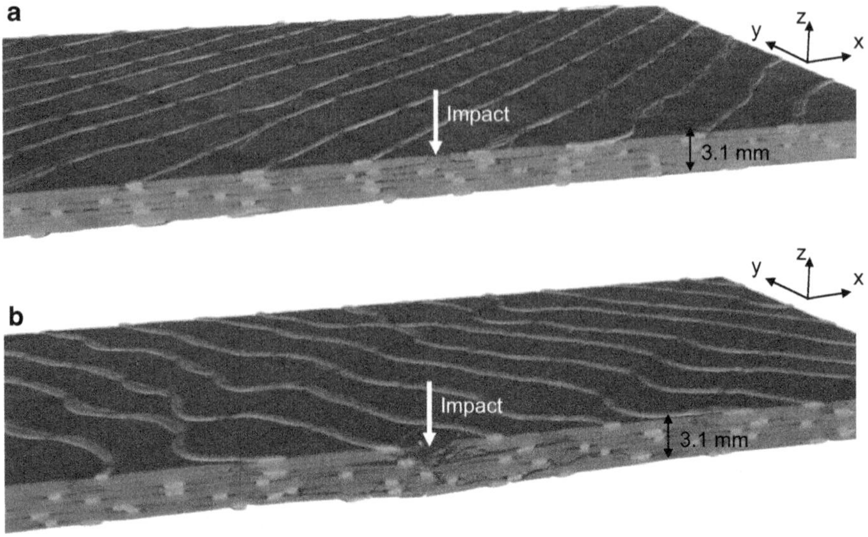

Fig. 6.24 Scan of damage zone as due to impact damage in 3D view for low- (**a**) and high-energy impact (**b**)

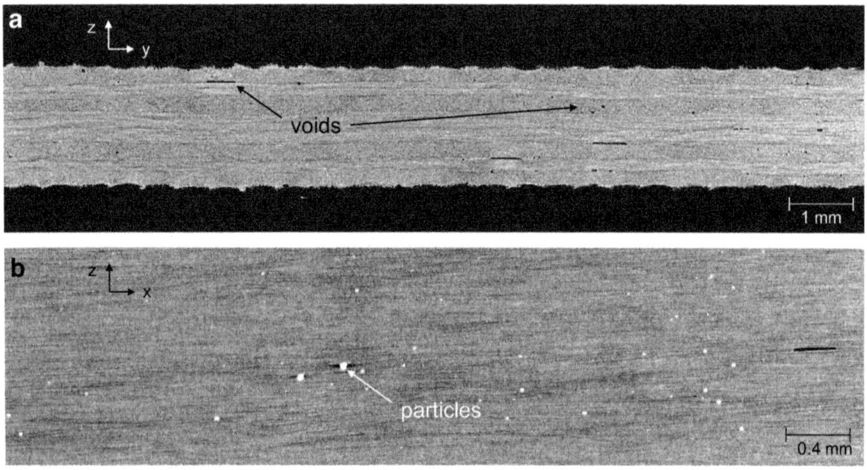

Fig. 6.25 Detectability of voids and foreign material inclusion in $[0/90/0/90]_{sym}$ laminate

covered in the following. Also visible in Fig. 6.25b are examples of foreign material inclusions identified as bright particles. Such inclusions may result from impurities during the production process, but may also be part of the resin systems, i.e., as flame retardants.

6.3 Volumetric Inspection of Materials

As demonstrated within the previous sections of this chapter, computed tomography provides a microscopy method suitable to visualize the interior structure of a fiber reinforced composite. Still the approaches presented so far are merely a substitute to other microscopy techniques. In the following, distinction is thus made between concepts used for ex situ loading, removal of the specimen, and subsequent scanning in contrast to approaches using in situ loading of materials and subsequent scanning.

6.3.1 Concepts for Ex Situ Loading

The basic concept for ex situ loading is a straightforward combination of a test rig as conventionally used for mechanical loading of a material and a CT scanner. As schematically shown in Fig. 6.26, the specimen is first loaded within the test rig until a certain damage state is reached. The relevance of the load level for inspection may either be based on signatures in stress-strain curves or based on secondary methods such as acoustic emission. The next step consists of dismounting the specimen and taking to the CT device. Subsequently a CT scan is made and the specimen is potentially remounted in the mechanical load rig to increase the mechanical load level. To this end a sequence of damage states may be recorded

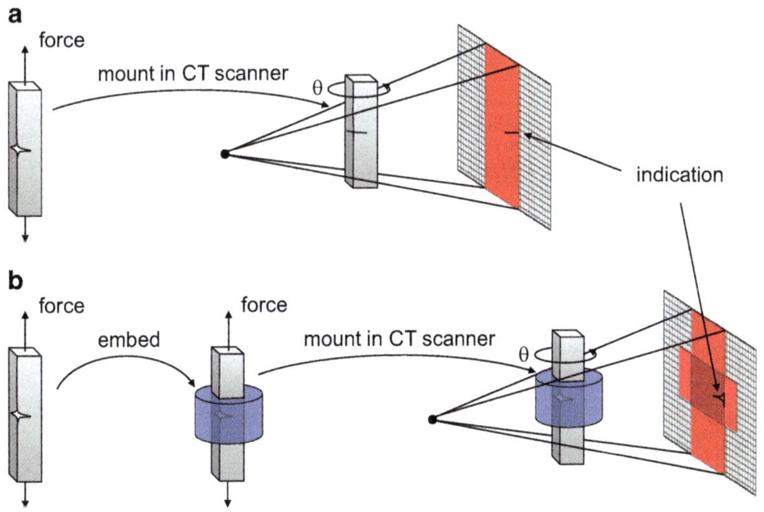

Fig. 6.26 Scheme of ex situ loading in combination with CT scan (**a**) and by using potting material to embed damage zone (**b**)

as discrete volumetric images to allow assessment of the present damage status of the specimen [29, 41–45].

However, the release of the applied mechanical load during dismounting may cause a closure of small cracks and, therefore, render them invisible during the CT inspection. Hence it might be necessary to preserve the stress-strain state within the material during scanning. To this end an approach using epoxy resin as potting material to embed the specimen under load can be applied [44]. Typically fast-curing epoxy resins without high-density fillers are used to avoid generation of artifacts within the epoxy resin. Optical transparence is not of vital importance since the X-rays will penetrate the material, but it helps to later select the correct portion of the specimen for a precise cutting process. For the purpose of segmentation and to increase visibility of the specimen, a noticeable difference in material density between the potting material and the embedded fiber reinforced polymer is recommended. This may be subject to trial and error, since typically the difference in density values is very low, when comparing values of epoxy resins used as potting material and values of the epoxy resins used as matrix materials. Ultimately, this procedure will assure that a certain portion of the original stress-strain state is preserved within the specimen, and small cracks will be better visible. As disadvantage, the specimen cannot be remounted to the test rig for further loading, and there is some justified doubt, whether this procedure fully inhibits specimen relaxation. Therefore any quantitative information like separation of crack walls or other geometric measurements need to be taken with necessary care. Also, from a practical point of view, some initial experience is required to perform the potting process within a mechanical test rig since the relevant portion of the specimen usually is in the middle of free air or barely accessible due to the load fixtures. In practice this means that suitable potting compounds need to be fabricated for each specimen type. Moreover, with the specimen being under load during this procedure, according safety measures are required to avoid unintended failure during a hold cycle.

6.3.2 Concepts for In Situ Loading

In order to perform in situ observations of material failure, further modifications to the established test setups are required. As major difference to the ex situ approach, the configurations discussed in this section use an arrangement as shown in Fig. 6.27. Here, the load rig is mounted in situ within the CT scanning device [25, 26, 31, 34, 35, 40, 46–61]. For the implementation of these concepts, three experimental factors may be distinguished and shall be discussed in the following. These are the type of X-ray source, the type of X-ray detector, and the type of load rig.

As general remark to all of the techniques mentioned within this section, it is important to understand that none of these CT techniques really operates in a real-time sense. Although comparatively fast scans are possible nowadays and the

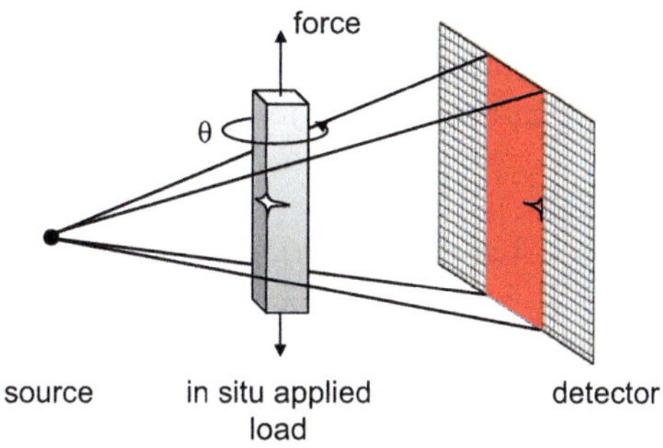

Fig. 6.27 Experimental configuration for in situ loading during CT scan

reconstruction time may be of similar order, the specimen still is required to be kept under a certain load for durations of minutes to hours during scan operation. This will allow stress relaxation within the material and therefore will alter the "true" geometry of the specimen. Hence all practical implementations make use of a short duration to allow for stress relaxation of the specimen before scan operation in order to avoid motion artifacts. Therefore, the name in situ is only considered to be correct in the sense that the specimen does not need to be dismounted from an external load rig to be inspected by CT. Nevertheless, this approach enables several new possibilities for analysis of failure behavior of fiber reinforced materials, so it is worthwhile to consider some of the technical details of this approach more precisely.

6.3.2.1 Type of X-ray Source

The majority of the published approaches were carried out using synchrotron-based CT scanners in conjunction with specifically developed load rigs [25–28, 31, 32, 34–36, 38, 40, 42, 45, 46, 49, 57, 61–64]. The main reason for this trend is the high brilliance and high intensity of the synchrotron beams in combination with a small spot size assuring high-resolution imaging as discussed in Sect. 6.1. Since commercial devices hardly could compete with any of these three aspects, only few attempts have been made to use commercial CT scanners for in situ inspection with a level of detail as discussed in Sect. 6.2. Nowadays there are some first CT scanners which provide voxel resolutions in the sub-micrometer range and are thus suitable to resolve the microstructure of a fiber reinforced polymer. One of the key arguments in favor for commercial X-ray CT scanners is without doubt the price and the availability. However, the intensity of commercial X-ray tubes is hardly comparable with the photon flux reached at typical synchrotron beam lines, so the contrast resolution is still not within the same range, and scan durations are typically much longer. Since longer scan times increase the chance of motion artifacts and may also induce motion unsharpness, these should be kept at a minimum. Therefore, a suitable trade-off between the necessary detail visibility[2] and the scan duration is required.

6.3.2.2 Type of X-ray Detector

Similar to the X-ray source, the type of X-ray detector is of relevance when considering in situ inspection of materials. Because both devices are usually coupled to form one scanning device, it is not straightforward to discuss them independently, but since a variety of configurations is encountered in practice, the relevant parameters of the detector system shall also be briefly discussed. Regardless of the X-ray source, one can distinguish between line detector and array detector systems as previously discussed in Sect. 6.2. The key advantage of line scanners in the present context is their fast scan times as compared to array detectors. If the in situ inspection is restricted to a particular cross-section within the material (cf. Fig. 6.28a), this enables very fast scan times even with typical commercial X-ray sources. Here typical scan durations were found to be in the order of few minutes and thus reducing the likelihood of motion artifacts. Obviously, the drawback of this detector configuration is the lack of 3D information of the material interior. So unless damage is expected to occur at a distinct position (i.e., at a notch) or is expected to be randomly distributed, it is very likely to miss damage formation in some parts of the specimen under load. Hence, to retrieve information out of a certain volume of the material, either subsequent line scanning or an inspection using an array detector system is the only choice (see Fig. 6.28b). For all other technical aspects of the detector system, the same considerations as for conventional X-ray CT scans apply.

6.3.2.3 Type of Load Rig

As key component of the in situ loading concept, the load rig is discussed next. From the mechanical engineering perspective, one challenge is to apply a certain stress on the test specimen, while being able to rotate the whole configuration ideally by 360° without having any additional obstacles within the projection area. Since any applied load also needs to be compensated within the rig, the typical scheme of a load test rig is of composed of the components shown in Fig. 6.29. These comprise the specimen fixtures, a stationary (fixed) part and a moving part driven by an actuator imposing a displacement to the specimen. In typical laboratory size machines, the vertical glides are simple columns, and the stationary and

[2] As the first approximation, an increase of the number of projections and an increase of the exposure time can be assumed to be beneficial for detail visibility if low photon intensity is given. Both parameters add to the overall scan duration.

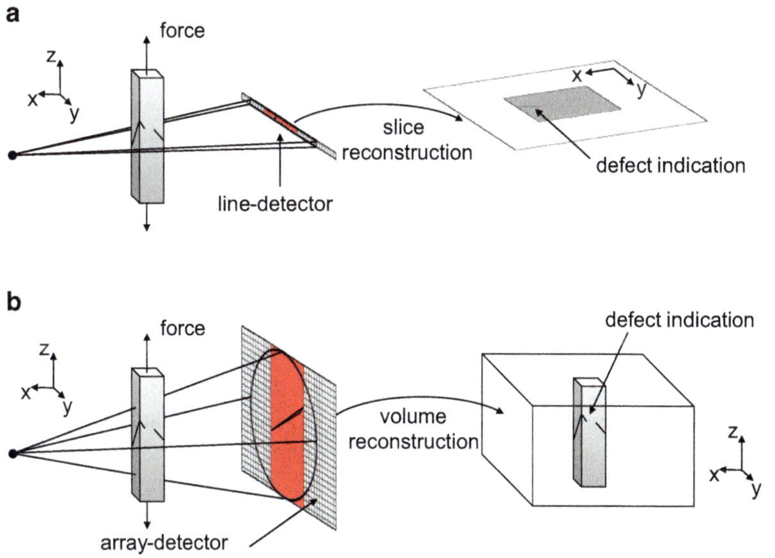

Fig. 6.28 Defect indication as retrieved from in situ measurement using line detector (**a**) and using array detector (**b**)

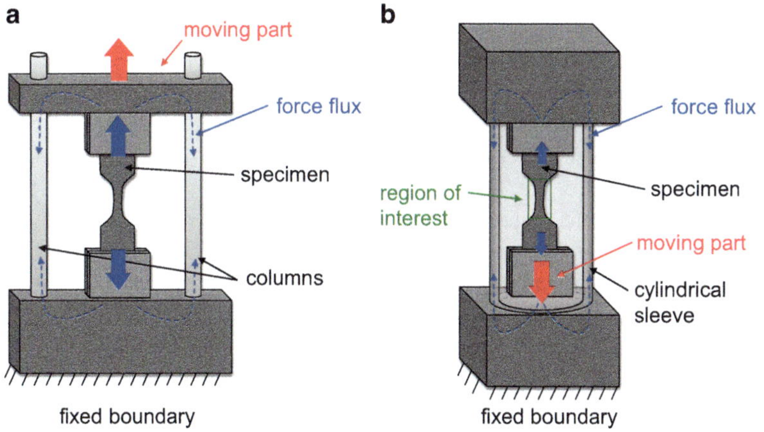

Fig. 6.29 Scheme of conventional load test rig using columns (**a**) and load test rig using cylindrical sleeve for load deflection (**b**)

movable parts of the machine are simple beams. If the whole configuration now is rotated around its center axis, it will be subject to substantial artifacts caused by the presence of the columns when passing the projection area. Hence, a configuration using a radially symmetric geometry as shown in Fig. 6.29b is favored from this perspective. Moreover, having the bottom fixture as moving part also enables a simple mechanical design for the upper part of the test rig. Thus the upper part is chosen as stationary fixture, and the lower part is fully instrumented by an actuator and a load cell. The force flux for both cases is indicated by blue dashed arrows.

In order to ensure high detail visibility, further considerations with respect to the material used for the load rig are required. Since metallic parts exhibit comparatively high X-ray attenuation, a presence of these parts within the projection area should be avoided as much as possible. To scan the region of interest shown in Fig. 6.29b, materials with low X-ray attenuation should be applied at all positions within the projection path. Nevertheless, these parts are meant to carry the full mechanical load as applied to the specimen, so they need to be of a sufficient cross-section to avoid buckling or failure of the load rig. Here polymer materials are among the first choices when designing this part of the load rig, which is typically done in the shape of a tube. Transparent polymers are technically not required since the X-rays will penetrate the polymer material, but they allow for visual inspection of the specimen alignment and also were found to improve the handling during mounting of specimen. To select a transparent polymer, further aspects may also be considered. First, the material should be able to withstand X-ray intensity for an extended period without optical or mechanical degradation. Second, the microstructure of the material should be homogeneous and free of textures, which could cause artifacts reducing the detail visibility of the scanned specimen.

In order to achieve highest possible resolution, the limiting factor in terms of the described configuration is the distance between the center of the test specimen and the outer dimension of the test rig. Hence, this value should be minimized to achieve high-resolution images. Due to the combination of the aforementioned aspects, the load deflection concept using a tube (cylindrical sleeve) instead of columns is generally favored, since this geometry is fairly stable to withstand buckling and comes with a larger effective cross-section than two or more columns retaining the same outer dimensions of the test rig.

In order to measure the applied force, a load cell is included in the test configuration. Since this requires some electronic readout, the preferred spot to mount the load cell is at the lower part of the test rig. That way, the cables can easily be attached to the rotary table and do not need to hang in free space avoiding possible interference with the X-ray source and the projection directions.

Finally, the actuator may consist of either a manually operated mechanical system or an electrically driven motor. In the following some exemplary configurations as used for fiber reinforced materials are presented to explain some established concepts for in situ CT scanning. In order to classify the configurations, distinction is made between the different load cases.

Tensile or Compressive Load Test

As pointed out in the previous section, the basis for design of load rigs is a cylindrical tube made from a material with low X-ray attenuation able to carry the applied load. Therefore only cross-sectional drawings of the load rigs will be shown in the following. One characteristic difference between the concepts in use is

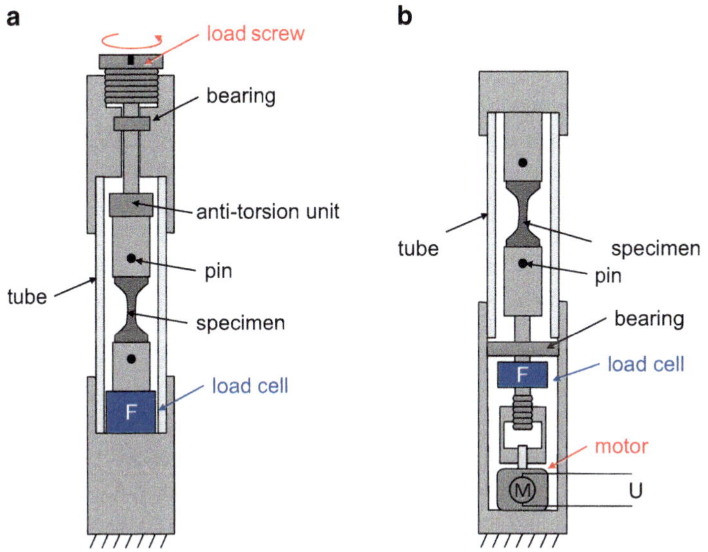

Fig. 6.30 Comparison between manually driven load rig for tensile tests (**a**) and computer-controlled load rig driven by electronic stepper motor (**b**)

the type of load control. While manual load drivers are mechanically more straightforward to implement and more economical, the advantage of a fully load controlled systems certainly is the possibility to automatize the full procedure. In Fig. 6.30 a comparison of two systems as commercially available is shown. Figure 6.30a shows a simple load screw in combination with an anti-torsion unit to stress a specimen pinned to the load rig. This concept is well suitable for isotropic materials causing no failure in the pin section when designing the dimensions of the tapered region and the thickness of the specimen accordingly. The anti-torsion unit is vital in this configuration to avoid any torque being applied to the specimen during the loading procedure. Otherwise superimposed tensile-torsion stress states could be the result, or the mounted specimen could even fail due to excessive torque during load application.

For the configuration of Fig. 6.30b, the specimen is also attached using pins. The basic difference to Fig. 6.30a is the use of an electric stepper motor in combination with a mechanical driver system to stress the specimen in close relation to the operation principle of typical macroscopic load test rigs. Here, the moving part is typically supported by additional bearings and also protected to avoid the occurrence of excessive torque applied to the test specimen. In both configurations, the load applied to the specimen is measured by a load cell positioned at the bottom of the load path.

In contrast to the concepts of Fig. 6.30, for fiber reinforced composites, a different fixture concept was found to be necessary. Similar to macroscopic test specimens, a connection to the load test rig by pins is not reasonable due to shear-out failure for many laminate configurations. Following the established test standards for tensile or compression testing, the typical approach is to produce tapered test specimens by bonding reinforcements at the end of the specimen. The specimen is then gripped using mechanical or hydraulic jaws to apply load.

In very small load test rigs, the fabrication of jaws comes with some additional challenges, so a more practical solution was found to directly bond the specimen within the load rig using standard two-component epoxy adhesives. Since the duration of one experiment with CT scanning (e.g., hours) largely exceed the duration for bonding and curing (e.g., minutes), this seems to be a reasonable approach. For macroscopic specimens this is quite different because the total test duration is usually within few minutes, and therefore a quick replacement of specimens is appreciated. Using a suitable embedding length of the material within the test rig, this allows failure in the free section and avoids failure of the adhesive bond. A major disadvantage of this concept is the necessity to have replaceable parts of the load rig tensile or compression bars, since the removal of specimen and adhesive from the bars degrades the geometry and therefore requires replacement after few un-mounting and re-mounting cycles. The schematic arrangement of the load rig as used for tensile tests and for compression test of fiber reinforced composites is shown in Fig. 6.31a, b, respectively. Similar to the macroscopic test concepts, one key difference is to keep the free length of the specimen (i.e., distance between edges of lower and upper bar) smaller for compression tests to avoid buckling.

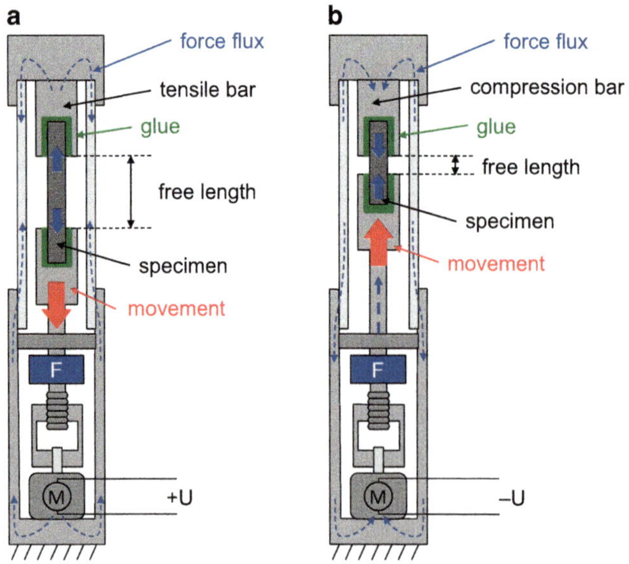

Fig. 6.31 Load rig used for tensile tests (**a**) and compression tests (**b**) of fiber reinforced materials

Mode I Test

As distinct difference to tensile or compression load, the generation of other relevant load states in fiber reinforced composites requires some further modification to the test rigs of Fig. 6.31. As for the macroscopic investigations of laminates, the damage progression under mode I load is of key interest to the design and material optimization of composites. To observe such damage progression with volumetric imaging as offered by CT scanning, the modifications shown in Fig. 6.32 are introduced. In this load rig, the specimen is attached to the lower load bar and moved in the vertical direction. This relative movement may either be due to an actuator at the bottom as seen in Fig. 6.32 or by a load screw at the top as seen in Fig. 6.30a. The material is prepared including a precrack as in the case of macroscopic specimens. At the top load bar a sharp wedge is mounted sliding into the precrack. Due to the steepness of the wedge, the laminate splits and a damage zone develops around the precrack tip.

If the curvature of the wedge is sufficiently large and the material is brittle enough, the crack tip position is sufficiently far away from the wedge tip. Thus the crack progression occurs under a mode I load situation. If the wedge curvature is too sharp, the laminate is instead "cut" and the real load type causing crack progression will likely be a superimposed load state. The force in this configuration is measured at the bottom of the load path and needs to be carefully interpreted since friction effects may add a substantial amount to the force value and thus may not be directly used in the same way as in macroscopic load configurations (i.e., double-cantilever beam configurations). Instead, closer relationship of the mechanical properties is expected to macroscopic wedge-peel tests as carried out for adhesives or thermoplastic tapes [65–69].

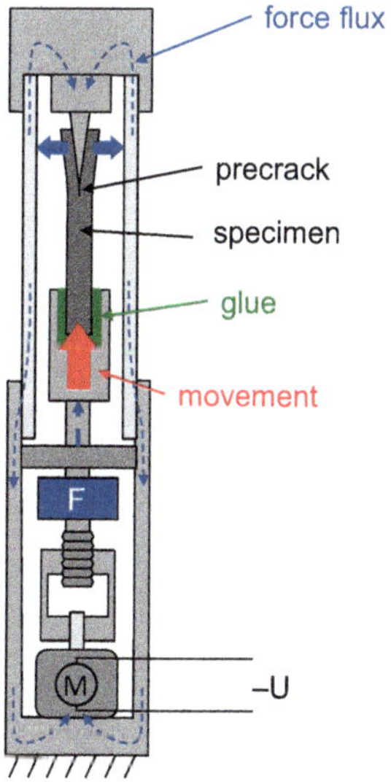

Fig. 6.32 Test rig configuration for wedge-loaded mode I investigation of fiber reinforced materials

Flexural Test and Mode II Test

Another load configuration being distinctly different to the previous cases, the flexural load case is shown in Fig. 6.33. Such loads can either be used to generate flexural stresses or to induce mode II-driven cracks as in the macroscopically applied end-notched flexure specimens. Distinction is made between a load rig applying manual load introduction (cf. Fig. 6.33a) and a computer-controlled system using a motor (cf. Fig. 6.33b). In both cases one challenge is to have supports being present at different positions to allow a deflection of the mounted specimen. Since the support at the center of the tube will always be present within the projected images, the same considerations as for the tube apply. If possible, the material of the support should be affected less by X-rays to assure mechanical stability during exposure, and it should cause a minimum of artifacts. For the case in Fig. 6.33a, the center support is tightly mounted to the tube and therefore remains at a constant position. To introduce a flexural load, two load screws transfer their displacement to the specimen using a slide bearing on which the supports are located. Here one disadvantage is the lack of a proper load cell measurement and the challenge to synchronize the upper and lower load screw in order to avoid skew load situations.

An alternative option is shown in Fig. 6.33b. Here the upper and lower supports remain at a constant position, while the center support is the moving part. By an elongated drill hole within the tube, the support is guided to allow only movement in the designated direction (i.e., the horizontal axis in Fig. 6.33b). Using an actuator

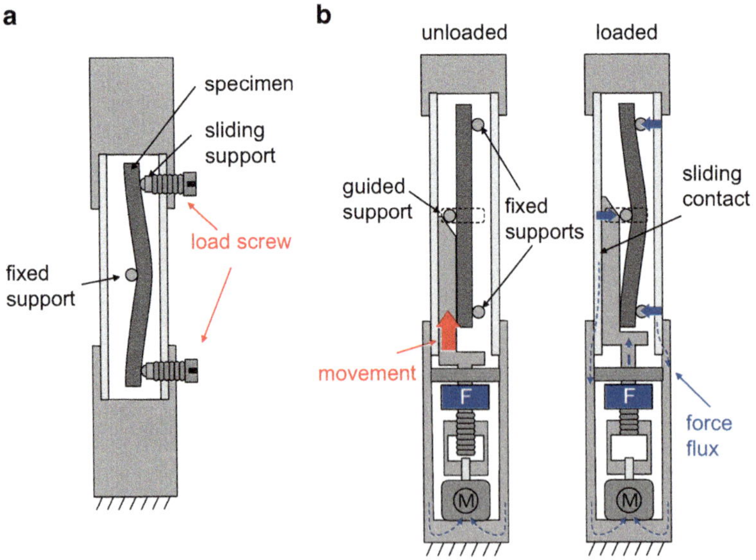

Fig. 6.33 Test rig configuration for flexural testing of fiber reinforced materials using load screws (**a**) and computer-controlled load rig (**b**)

system at the bottom of the test rig, a pull bar is moved in the vertical direction to stress the specimen. A wedge at the tip of the bar allows pressing the support firmly to the specimen and causes deflection. During this operation, the load bar is sliding against the tube, and thus the acting loads are also transferred to the tube in order to avoid significant deflection of the bar itself. The load cell is mounted at the bottom of the configuration to measure the required load. However, this load is composed of contributions due to friction and due to specimen deflection and may not be directly used. Hence using a material with known stiffness for calibration of the system, the measured load may be corrected to obtain true mechanical properties of the measured material.

Thermal Loading

As a fundamentally different type of load source relevant to fiber reinforced materials, a change in temperature may also generate stresses in the material. These stresses are solely caused by the thermal expansion of the material, which may induce some strong gradients in the case of anisotropic materials as in fiber reinforced composite laminates. Moreover, joints between metallic parts and fiber reinforced polymers are subject to a strong gradient in the thermal expansion properties which may induce severe stress at the interface between both materials. Here the basic concept is to generate such stresses in situ to allow accompanied CT scanning. As seen in the schematically drawing in Fig. 6.34, the concept of the test stage is relatively similar to the designs explained in the previous paragraphs. The specimen is mounted at the bottom part and is enclosed in a sealed cylindrical tube made from X-ray stable material. To avoid the occurrence of artifacts, the cylindrical design is kept. To monitor the temperature within the chamber, a temperature element is positioned close to the specimen. Either electric heat elements or Peltiér elements are applied to cause a change in temperature reaching either above or below room temperature.

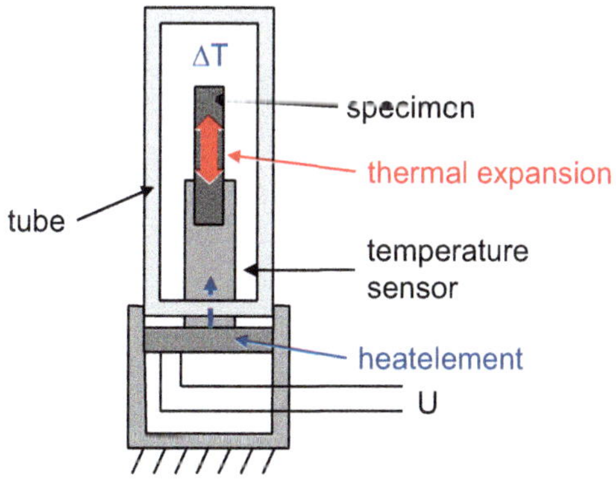

Fig. 6.34 Test rig to generate thermal stress in material specimen

Imaging Requirements

Whatever type of load is being applied to the specimen, one requirement in CT scanning is to have the specimen immobile during the scan duration. Otherwise motion artifacts will appear which may render the measurement useless or cause less detail visibility.

In the case of fiber reinforced polymers, the stress relaxation effect of the matrix system needs to be considered in this context. Consider the load scheme shown in Fig. 6.35 for two subsequent steps of in situ CT scans. The aim is to load a specimen up until a distinct load level F_1 and perform a CT scan at this load level to finish step one. Without material creep, the specimen would simply keep its internal stress state at load F and scanning could be performed immediately. In reality all materials will show some stress relaxation after reaching the programmed load. Hence after the quasi-static load phase to reach F_1 (blue arrows in Fig. 6.35), a wait period needs to be present to allow the material to reach a relaxation load F_1'. At this relaxation load, the CT scan can take place to avoid motion artifacts. Subsequently, the load is further increased to F_2 and the next CT scan is performed in step 2 after allowing stress relaxation of the material to the load level F_2'. This sequence is carried on until the specimen finally fails.

This procedure immediately allows concluding that the mechanical experiments need to be performed in displacement-controlled mode. In force-controlled mode, the material would undergo a more or less constant creep and therefore

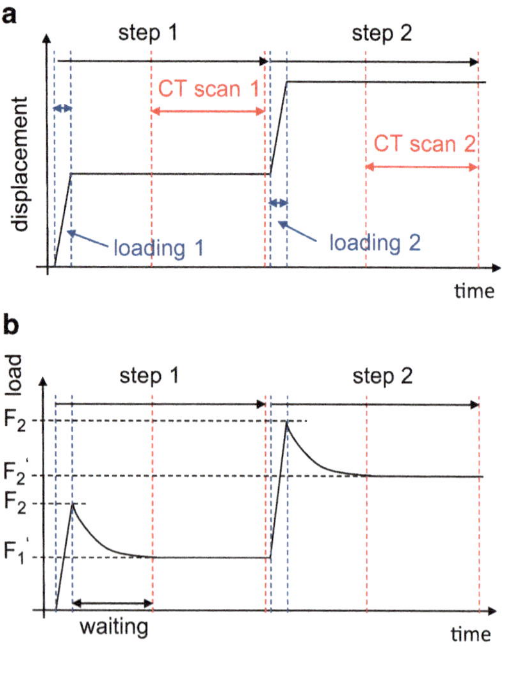

Fig. 6.35 Typical displacement profile (**a**) and load profile (**b**) during in situ CT scanning including effects of relaxation and sequence of actions

continuously deform causing no stable conditions to perform CT scanning. It should be noted that the ratio of durations for loading and scanning is not to scale from Fig. 6.35, since the CT scan operation may easily fall in the order of hours, whereas the load duration typically still is within a few minutes. Also the level of stress relaxation is exaggerated for better visibility. However, careful considerations are required when interpreting the imaged damage state. Firstly, the quantified dimensions and corresponding stress-strain states might be of reduced amount, since the specimen undergoes relaxation prior to scanning. Secondly, upon subsequent loading, the material starts from a different stationary state, so the material behavior concluded from such interrupted load profiles may be altered compared to the conclusions drawn from continuous load profiles until failure.

6.4 Digital Volume Correlation

The approach to use the correlation of a sequence of digital images is already well established and is being applied more and more in material testing applications. Volumetric imaging-based X-ray computed tomography [25, 26, 31, 34, 35, 40, 46–61], magnetic resonance imaging [70], or, for partially transparent or opaque materials, laser confocal scanning microscopy [71–75] are also used frequently. Based on the volumetric image information, a consequent extension of the digital image correlation method presented in Chap. 3 was suggested by Bay et al. [76]. They were among the first to consider the correlation of two subsequent volume images [76]. This method of DVC allows to measure quantitatively the 3D internal displacement and deformation of a material. The main distinction between approaches of 2D-DIC, 3D-DIC, and DVC is shown in Fig. 6.36. The primary difference between DIC and DVC is the hierarchy of information derived from the measurement. In simple 2D-DIC, a plane surface measurement is performed, not being able to detect out-of-plane components w and thus resulting in simple 2D

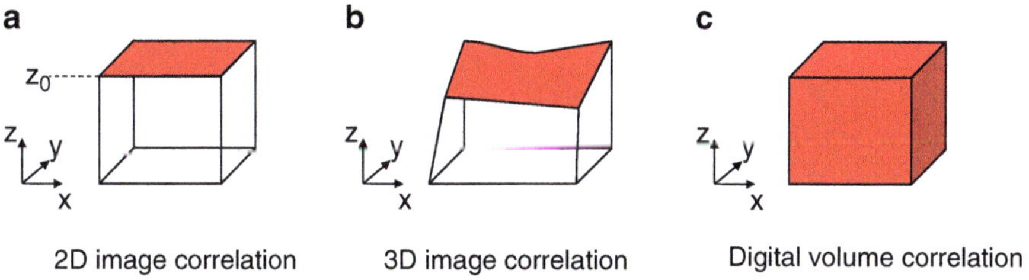

$$u = \begin{pmatrix} u(x,y,z_0) \\ v(x,y,z_0) \\ 0 \end{pmatrix} \quad u = \begin{pmatrix} u(x,y,z(x,y)) \\ v(x,y,z(x,y)) \\ w(x,y,z(x,y)) \end{pmatrix} \quad u = \begin{pmatrix} u(x,y,z) \\ v(x,y,z) \\ w(x,y,z) \end{pmatrix}$$

Fig. 6.36 Comparison of displacement vectors $\boldsymbol{u}$ as obtained from 2D-DIC (**a**), 3D-DIC (**b**), and DVC (**c**) including schematically representation of information hierarchy displacement information $\boldsymbol{u}$ of the outer surface. The extension to 3D-DIC allows measuring the out-of-plane component of the displacement and results in a 3D displacement measurement $\boldsymbol{u}$ of the surface. For the DVC approach, the displacement is not just measured at the surface of the material but originates from the full volume and therefore constitutes a true 3D measurement of the displacement vector $\boldsymbol{u}$ in every point of the volume.

Among the applications of the DVC technique, there are obvious candidates like the ability to validate the result of 3D-FEM computations. Successful applications include materials such as wood [41, 42, 49] and composite materials [58].

In full analogy to the procedure used in DIC, the basic steps for DVC are the following:

1. First recording a volumetric image of the material
2. Deforming the volume due to applied loads
3. Second recording of a volumetric image of the material
4. Correlation of both volumetric images using subvolumes and computation of displacement vector for each subvolume

The acquisition of a volumetric image and the in situ deformation of the material were presented throughout Sects. 6.1, 6.2, and 6.3, and the only remaining step now is the correlation of two volumetric images. The basic concept is shown in Fig. 6.37. Since a computation of the displacement vector of a single voxel is not suitable due to the ambiguity faced in the image correlation process (cf. Sect. 3.1), a local neighborhood around one voxel is used. As seen in Fig. 6.37, in DVC this is a local subvolume subset rather than a 2D subset as introduced for DIC. Apart from that, the subvolume is subject to a general displacement, rotation, and deformation as introduced in DIC before.

In order to perform the correlation process in DIC, speckle-type patterns are applied to the inspected surface to provide locally unique subsets. Since such application of speckle patterns is not possible in the interior of the material, other image signatures are required to perform the correlation process. To this end, the material microstructure itself can be used to provide a unique correspondence of

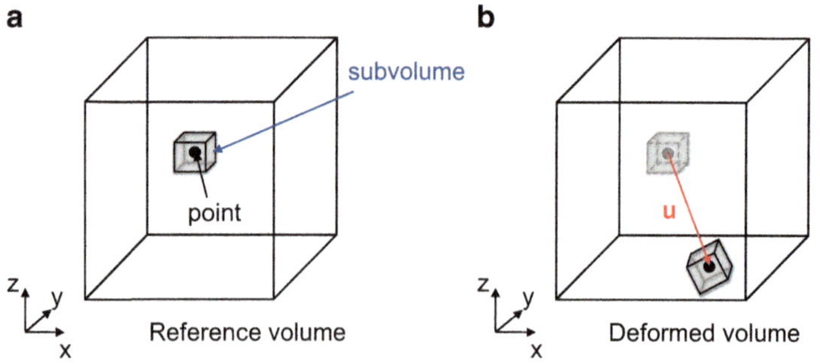

Fig. 6.37 Tracking deformation states between reference volumetric image (**a**) including local volumetric neighborhood (subvolume) and deformed volume (**b**) as being used for DVC

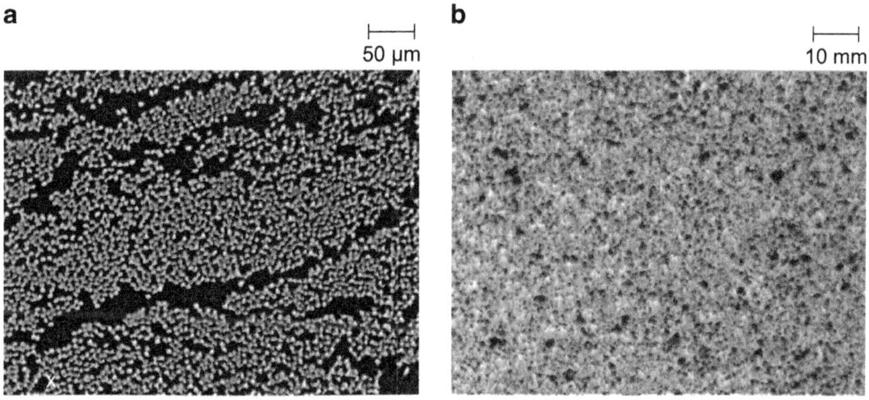

Fig. 6.38 Examples of CT cross-sections of fiber reinforced material (**a**) and typical speckle pattern used in DIC (**b**)

subvolumes. For some materials a stochastically distribution of particles or voids is found on the microscale. Accordingly, subvolumes of such materials yield unique information to perform a correlation process. For fiber reinforced materials, the arrangement of the fiber filament cross-section is of such a length scale to allow successful DVC [58]. A comparison of two cross-sectional images of a fiber reinforced material in Fig. 6.38 to a speckle-type pattern reveals the close similarities.

Although the principle of DVC constitutes a straightforward extension of the DIC approach, one of the key differences is the computational intensity faced. To this end, further distinction is required between two typical implementations of DVC. The first algorithms developed were based on a correlation procedure using only discrete voxel sizes. This comes with lower computational intensity and is readily be implemented by search algorithms in the spatial [76] or frequency domain [74]. However, this allows a calculation of the displacement vector only by increments of more than half the voxel size, which is typically not feasible for many materials being subject to small deformations only. Thus the strain resolution of this simple implementation would be 16.7 % strain for a 3-voxel cube, 10.0 % strain for a 5-voxel cube, 7.1 % strain for a 7-voxel cube, and so on. For the typical strain values in the linear elastic range of many materials with fractions of 1 %, a sub-voxel accuracy of the correlation procedure is required. Therefore, the present focus of the method development is on the improvement of the algorithms used for such sub-voxel correlation [76–81]. Since an extensive discussion of the image correlation algorithms is beyond the scope of this book, the reader is referred to [82] on the basics of image correlation and to Pan et al. [77] for a recent review and modification of existing correlation algorithms.

To demonstrate the use of this method, Figs. 6.39 and 6.40 show some exemplary results of volume cross-sections as collected from an in situ CT scan of a carbon fiber reinforced epoxy material. Following the procedure outlined in

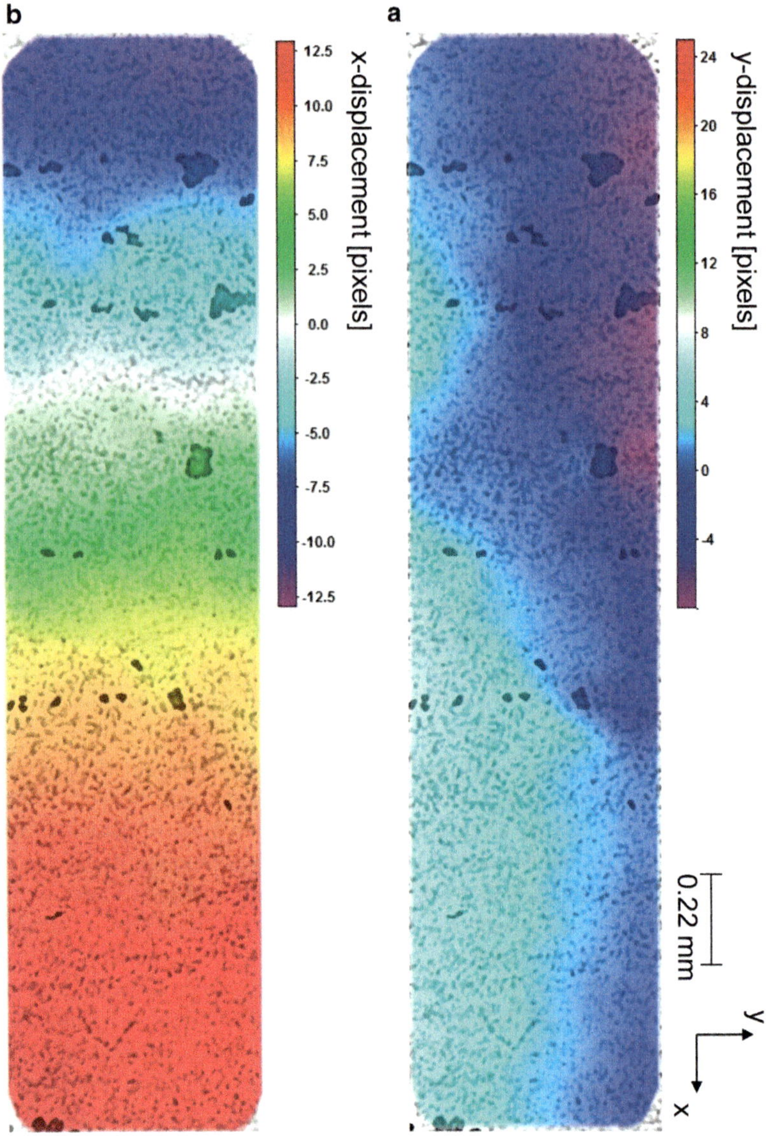

Fig. 6.39 Arbitrary *xy*-cross-section out of volumetric image of in situ loaded fiber reinforced polymer. Computed displacement fields in *y*-direction (**a**) and *x*-direction (**b**) are superimposed to the CT cross-sectional image

Sect. 6.5.2.1, miniaturized specimens were prepared and loaded by a compressive force in situ in parallel to the *y*-axis. CT scans were performed in the initial state and at a load level prior to ultimate failure. For the DVC process, the software package DaVis was applied, and the resultant displacement field is visualized as

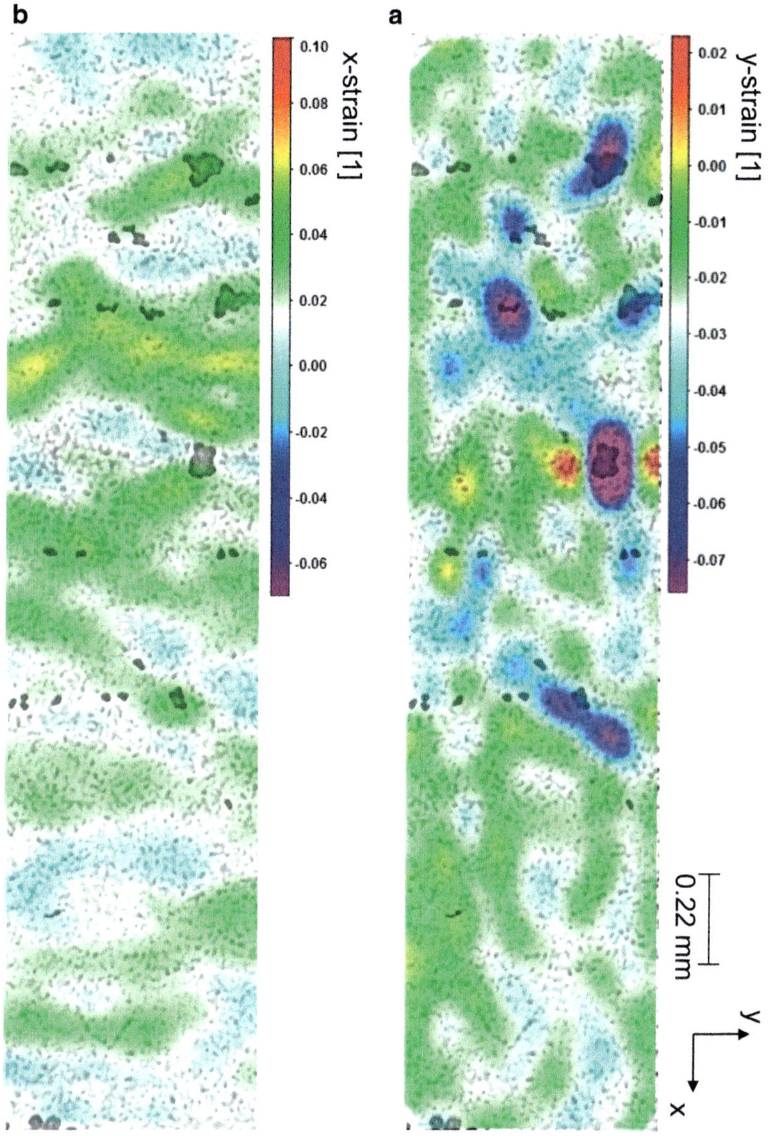

Fig. 6.40 Arbitrary *xy*-cross-section out of volumetric image of in situ loaded fiber reinforced polymer. Computed strain fields in *y*-direction (**a**) and *x*-direction (**b**) are superimposed to the CT cross-sectional image

x-displacement and *y*-displacement in Fig. 6.39 superimposed to the selected cross-section. For technical reasons, the displacement scale is given in pixels with one pixel corresponding to 1.36 μm. As seen from the black regions in the *xy*-slice, the material includes several voids of diameters up to 100 μm. In the computed displacement field, the material predominantly shows the compression due to the applied compressive force in y-direction and the effect of Poisson's contraction in the x-direction. For the corresponding local strains shown in Fig. 6.40, the situation is different. Here the larger voids show a significant strain concentration indicating regions of higher compliance. As will be seen from the analysis of the failure modes in Sect. 6.5.2, these regions form a weak link in the material contributing significantly to the final failure mode. Similar to the analysis of DIC images presented in Chap. 3, this analysis can thus be used to identify strain concentrations in the volume of a material to perform assessment of possible failure locations prior to failure or to analyze the evolution of failure initiating at such spots.

6.5 Application to Composites

The use of computed tomography to inspect the microstructure of fiber reinforced materials has been outlined in the previous sections. Based on the scope of the book, the last section of this chapter has its focus on capabilities of the method to detect and quantify damage initiation and damage progression. Thus only exemplary applications related to in situ testing and ex situ testing approaches are presented in the following. Other fields of application of computed tomography as relevant to composites such as porosity quantification or measurement of fiber orientation are instead covered by recent literature [16–24].

Before proceeding to the first subsection on ex situ testing approaches, a brief introduction to the use of segmentation tools to visualize the internal microstructure is given. This step is necessary to provide the basic analysis routines used to evaluate the damage progression as presented in Sects. 6.5.1 and 6.5.2.

As discussed in Sect. 6.1 and 6.2, the gray scale images obtained in computed tomography basically represent the material density. In the absence of image artifacts and for reasonable resolution, this may thus be used to segment different materials based on their difference in density. A trivial example for this procedure is given in Fig. 6.41. The selected xz-slices of a computed tomography scan result from an inspection of a Sigrafil CE1250-230-39 prepreg fabricated with consolidation pressures below the manufacturer recommendations. Due to the missing pressure, this results in the formation of pores upon curing and is readily visible in Fig. 6.41. Both laminates were scanned at a time, so the histogram of gray values on the left side of Fig. 6.41 is representative for both laminates. Within the histogram two clear peaks are found, which correspond to the density values of air and the composite material. Based on the large distances of the peak maximums, it may seem trivial to segment this image. The threshold of Fig. 6.41a is thus selected at the half distance between both peak maxima. The corresponding segmentation is indicated as yellow line in the xz-slice in Fig. 6.41a on the right side. Obviously, some of the needle pores in the lower laminate are not segmented correct. This is due to the presence of image artifacts as discussed in Sect. 6.2, causing slightly different gray values for the case of air enclosed by the composite material. By moving the segmentation threshold towards the material peak, more

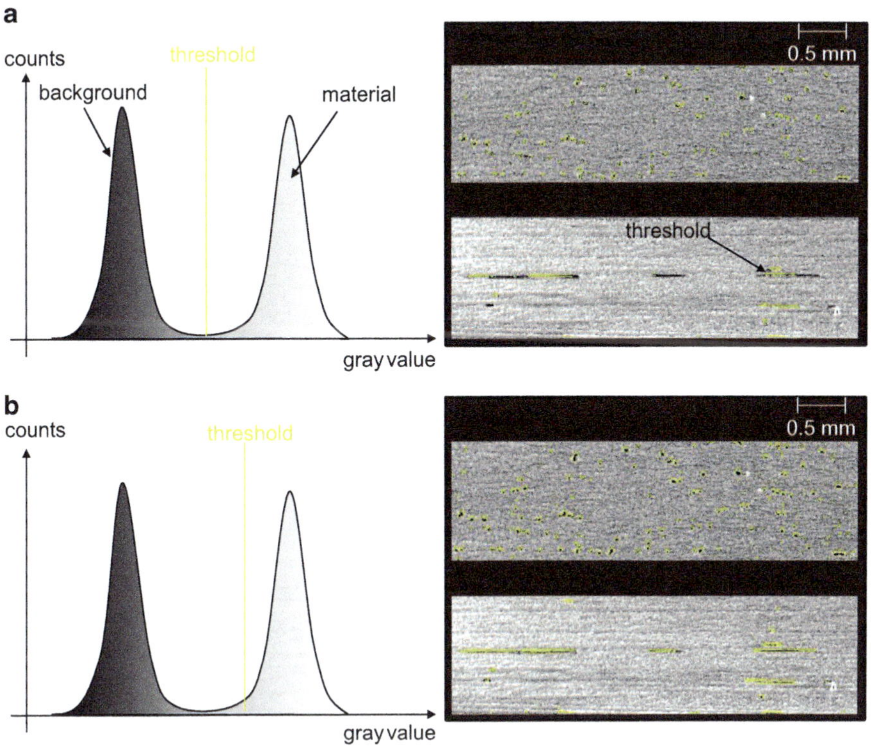

Fig. 6.41 Selection of threshold for image segmentation at half distance between peak maxima (**a**) and close to material peak (**b**)

realistic segmentation is reached as seen in Fig. 6.41b. Since this procedure is driven by visual decision criteria, its suitability for extraction of quantitative information is generally questionable. Consequently, various strategies to obtain valid decision boundaries have been proposed in literature. Recently, this has seen a strong focus on the application to quantify the degree of porosity as well as to segment matrix and fibers [16–24]. Despite of these strong research efforts, there is still no standardized procedure to perform image segmentation in computed tomography, and consequently, extraction of quantitative volumetric information is still challenging.

After image segmentation, there is a broad range of possibilities to analyze and interpret the data, ranging from simple volumetric ratios to topological analysis. As an example of both, the two laminates of Fig. 6.41 alongside with a reference laminate are subject to a topological analysis with their results shown in Fig. 6.42. On the top of Fig. 6.42a–c, the 3D visualization of the inspected laminates is shown. All three laminates were scanned at the same time and are segmented using the threshold of Fig. 6.41b. At the bottom of each figure, a topological analysis is shown for the detected porosity. The false color range indicates the volumetric size

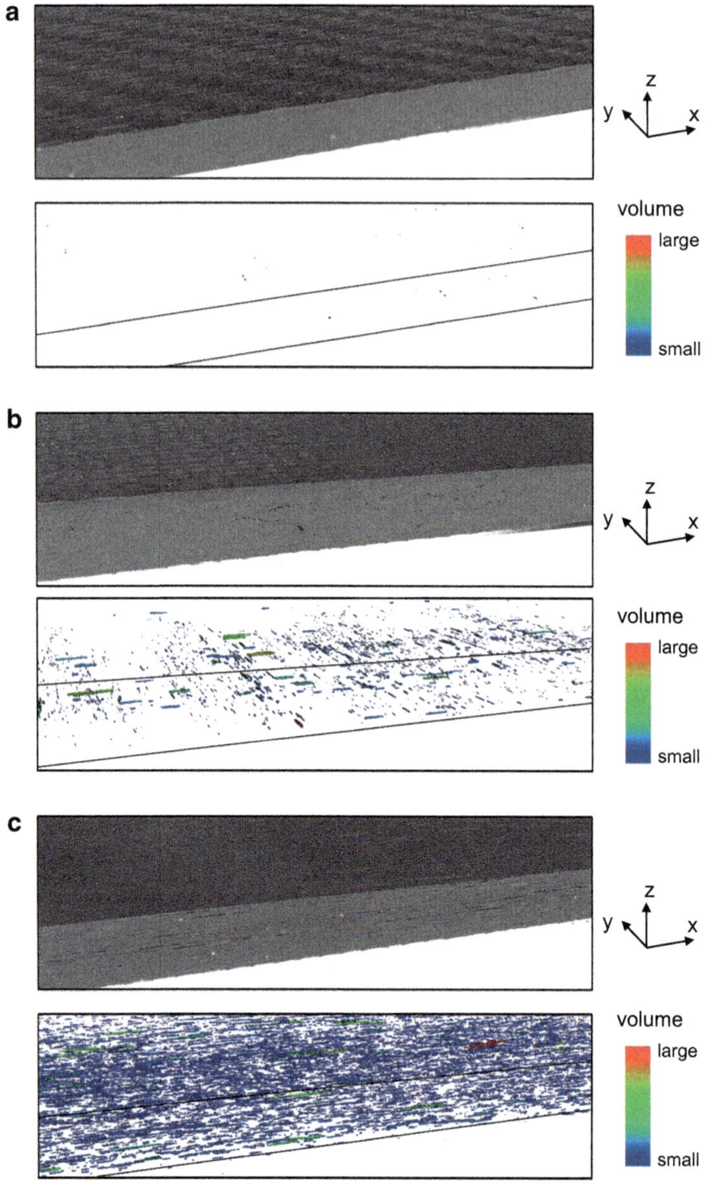

Fig. 6.42 Visualization of porosity for well-consolidated laminate (**a**), laminate with some needle pores (**b**), and laminate with extraneous porosity (**c**)

of the single pores (or the pore network if existent). As additional guide to the eye, two black lines are added to the figures, representing the edges of the laminates. A reduced consolidation pressure results in the formation of a substantial amount of needle-shaped pores (cf. Fig. 6.42b) in combination with extraneous porosity (cf. Fig. 6.42c). Despite of the difficulty to extract quantitative information from these measurements, the visualization of these voids prior to testing certainly helps to assure that the specimen is free of porosity as for the laminate of Fig. 6.42a. However, it is important to mention that the scan resolution will obviously influence the porosity assessment as well. As demonstrated by Kastner et al. [21], the voxel resolution as well as the choice of the threshold will have tremendous impact on the quantified porosity. Thus, the choice of scan parameters as well as the size of the test specimen will need to be considered for such analysis routines.

In the context of damage assessment, it is worth noting that it is generally hard to distinguish between the indications of porosity and those given by the formation of cracks. This is owed to the fact that cracks in the composite material are basically air inclusions of a certain shape and orientation. Since the visualization of these indications is based on their density, it is thus impossible to separate both indications without further knowledge. As practical example for this challenge, a 3D visualization of a tapered section of a laminate is shown in Fig. 6.43. This scan visualizes the tapered section of a Sigrafil CE1250-230-39 prepreg test specimen of $[(0/90)_2]_{sym}$ stacking sequence fabricated with interspersed plies as described in Sect. 4.7.3.2.

The specimen was subject to a tensile load until failure within the reduced section (not shown in Fig. 6.43). For Fig. 6.43b, the gray value range of the composite material is chosen transparent, so that only the volumes with density close to air remain. For improved visibility, the air volume surrounding the specimen was also chosen transparent, so only enclosed air volumes remain visible in

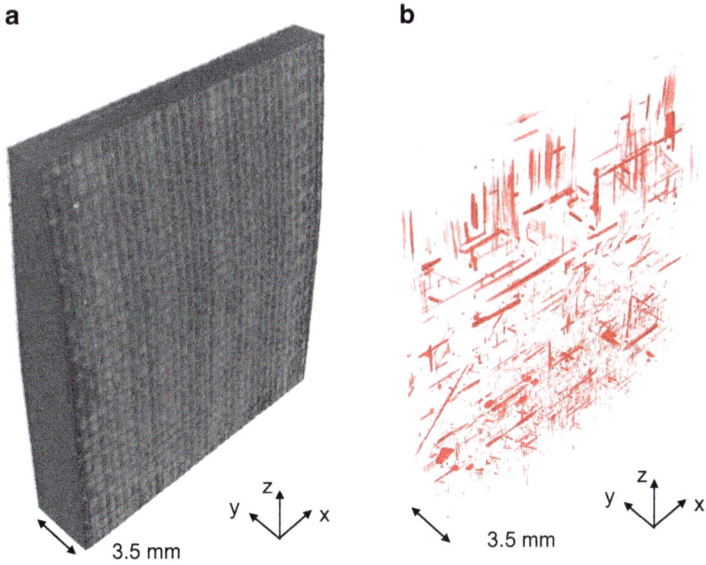

Fig. 6.43 3D CT scan of tapered laminate section (**a**) and visualization of cracks and porosity by setting composite material and surrounding air transparent (**b**)

Fig. 6.43b. This kind of visualization allows evaluating the orientation and position of pores and cracks. However, there is no way to distinguish between the occurrences of either kind. The laminate has certainly seen some damage in the scanned region before final failure of the specimen, so parts of the indications of Fig. 6.43b are due to the presence of cracks. But as evident from computed tomography scans prior to testing, there is always a certain amount of pores in the reinforcement section, which is difficult to avoid by the chosen processing conditions. Hence, a considerable amount of the indications of Fig. 6.43b are due to needle pores. Because of their shape and orientation in parallel to the fibers of the individual plies, it is not possible to distinguish them from the occurrence of inter-fiber cracks. Also, it is likely, that the presence of a pore acts as initiation site for an inter-fiber crack. Consequently, mixtures of both indications are also likely to occur.

Thus for damage assessment by computed tomography, the following procedure is recommended:

1. Inspection of the relevant part of the test specimen prior to mechanical testing to assure there are no significant pores or other inclusions
2. Comparative inspection after testing of same part of the test specimen

This procedure allows evaluating the difference before and after testing and thus may be used to resolve the ambiguity of indications. In cases of almost negligible indications prior to testing (such as in Fig. 6.42a), the indications after testing may be directly used to perform damage assessment.

6.5.1 Ex Situ Testing

Based on the consideration presented in Sect. 6.3, it is suitable to distinguish between inspection approaches performed ex situ and in situ. For the ex situ approaches, the specimen is loaded in an external universal test machine until a certain damage state is reached. Subsequently, the specimen is dismounted and inspected by a CT scanner. Depending on the type of analysis, the specimen is then mounted again in the universal test machine and loaded until the next damage state is reached. In contrast, in situ approaches combine the capabilities of a universal testing machine with a CT scanning device and are described in their applications in Sect. 6.5.2.

6.5.1.1 Visualization of Damage Progress

In this subsection exemplary applications for tracking of damage progress using ex situ CT approaches are shown. Since there are numerous applications found in literature that could be added to this section [25, 26, 31, 34, 35, 40, 46–61], focus will be given on two representative examples. As first example, the growth of a mode II interlaminar crack is investigated. This kind of example is representative

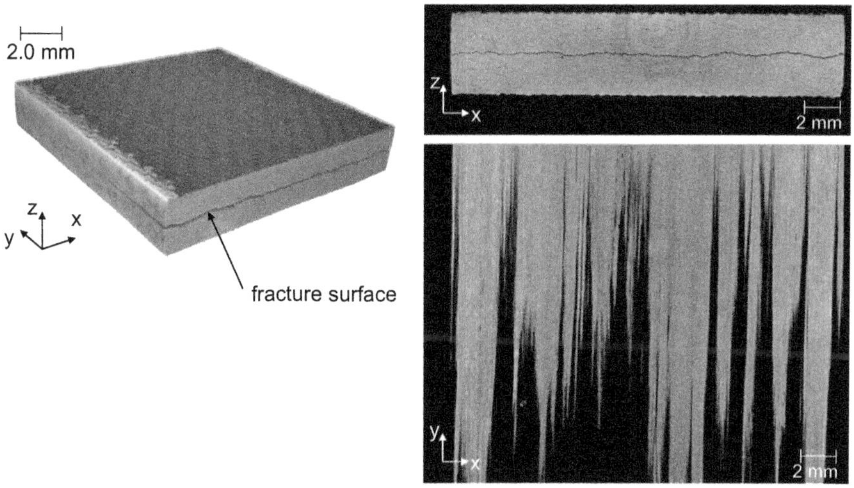

Fig. 6.44 3D visualization of mode II fracture surface in unidirectional laminate and according cross-sections in *xz*-plane and *xy*-plane

for a macroscopic indication and is used to demonstrate some basic concepts in tracking of damage progression. As second example, the growth of inter-fiber cracks in cross-ply laminates is presented. For this example, the major challenge stems from the visualization of relevant indications and the compromise between high image resolution and large inspection volume.

As first example, a typical specimen as used in end-notched flexure test to determine the mode II fracture toughness is analyzed (cf. Sect. 4.7.2.2). In this configuration the crack initiates at the precrack position and starts to grow interlaminar due to the applied bending load. The typical appearance of the fracture surface after testing is shown as 3D volume in Fig. 6.44. The *xz*-cross-section and the *xy*-cross-section of Fig. 6.44 are used in addition to visualize the complexity of the fracture surface achieved in this test condition. For this unidirectional prepreg laminate fabricated out of the Sigrafil CE1250-230-39 material, the crack growth in mode II condition causes splitting of the laminate in parallel to the fiber axis. Thus, the real crack front is hardly described as an ideal rectangle, but has certain waviness along the *z*-axis. In addition, the crack extends to different distances along the *y*-axis depending on the *z*-position.

In order to evaluate the damage progression during the mode II test, the specimen was first scanned directly after precracking and a second time after the first load drop according to the evaluation procedure of ASTM D7905. A comparison of both scans is shown in Fig. 6.45. For this visualization, the composite material and the surrounding air volume were chosen transparent, so that only indications of the enclosed volume remain visible. To improve the visibility of the crack front, a small wedge was inserted at the precrack position to open the interlaminar crack during the CT inspection. For the evaluation the 3D result is projected to the *xy*-plane rendering all indications semitransparent.

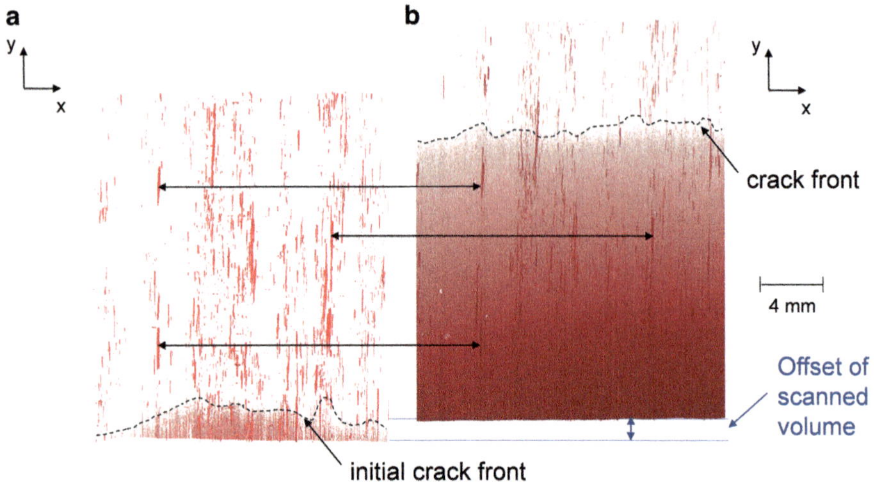

Fig. 6.45 Comparison of mode II fracture surface in unidirectional laminate after precrack procedure (**a**) and after mode II ENF test (**b**) using visualization of cracks and porosity by setting material as transparent

The position of the crack front after precracking is marked as dashed line in Fig. 6.45a. There is not a plane crack front, but a distinct profile with different crack extension along the *y*-axis as function of the *x*-position. Also, there is a certain level of pores readily visible inside the laminate. In this application they were found to be quite useful, since their shapes can be used as characteristic fingerprint to determine the offset between the first scan and the second scan as indicated by the black arrows of Fig. 6.45. During the mode II test procedure, the crack front advances by 17.4 mm in average as seen from Fig. 6.45b. The crack front retains a distinct profile. A 3D evaluation of the newly generated crack surface of Fig. 6.45 allows to quantify the total crack surface, which evaluates as 1074 mm^2 for the case shown. Unlike other attempts to measure crack lengths using X-ray projection methods, the use of the 3D volume considers the full topology of the crack surface and therefore especially considers the waviness along the *z*-axis. For metallic materials such approaches have already demonstrated their capabilities for quantitative estimation of mixed-mode load conditions [83, 84]. The quantification based on visual observation of the crack front (averaged at both edges) according to the ASTM D7905 results in 348 mm^2. Thus the real fracture surface is underestimated by a factor of 3. However, the data reduction routine in ASTM D7905 does not directly apply the value of the fracture surface and, hence, does not benefit from such improved evaluation methods. Instead, the volumetric measurement of the fracture surface may constitute an alternative approach of fracture toughness determination in the spirit of Griffith's energy relation.

As further example, the growth of an inter-fiber network in cross-ply laminates is presented. The damage sequence recorded for a $[0/90_3/0]_{sym}$ laminate made from

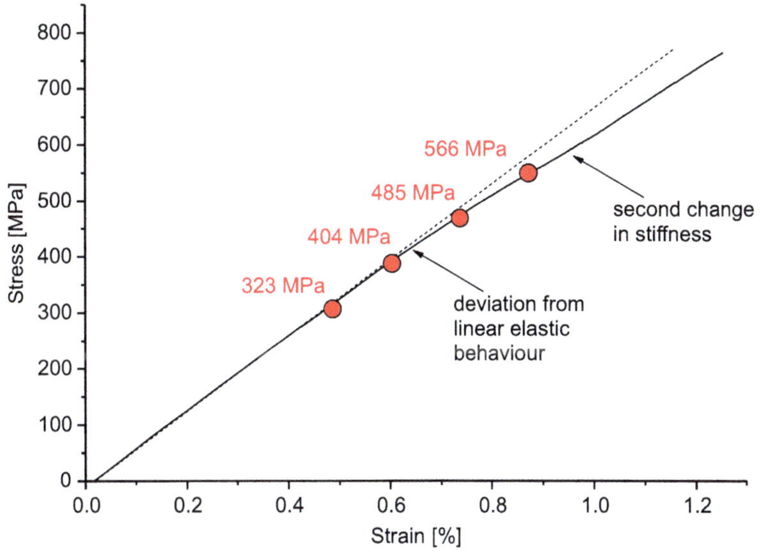

Fig. 6.46 Stress-strain curve of exemplary $[0/90_3/0]_{sym}$ specimen with levels of CT inspection marked in *red*

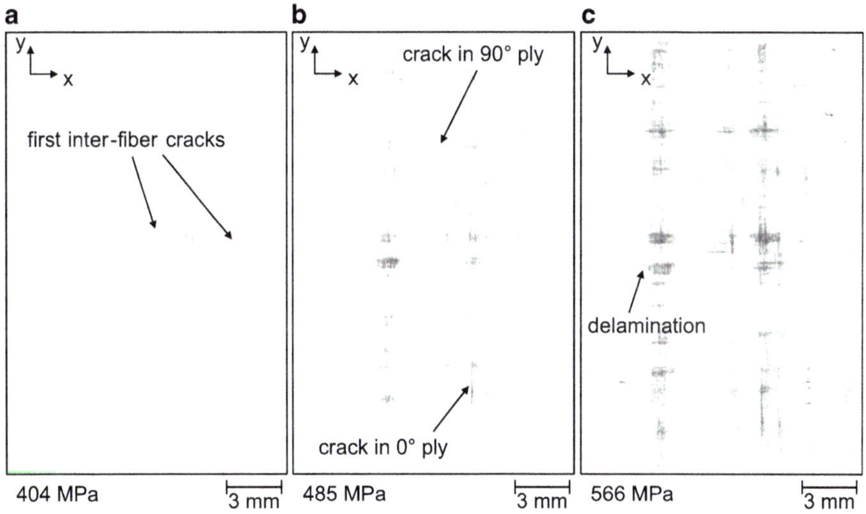

Fig. 6.47 Sequence of images recorded at distinct stress levels shown as projection of the 3D volume to the *xy*-plane with material set transparent

Sigrafil CE1250-230-39 prepreg subject to the load profile shown in Fig. 6.46 is given in Fig. 6.47. The specimen was loaded along the *y*-axis to the first level of 323 MPa, unmounted from the test machine and then scanned using the computed tomography parameters given in Table 6.1. Subsequent to the scan, the specimen was remounted and loaded until the second stress level of 404 MPa is reached. These steps are repeated until reaching the stress level of 566 MPa. As seen from the stress-strain curve of a specimen from the same plate loaded until failure, there are several changes in specimen stiffness, which can readily be associated with microstructural changes within the materials, such as crack formation.

In Fig. 6.47, the scans are shown as projection of the scanned volume to the *xy*-plane following the visualization described above with the composite material set as transparent. Since almost no porosity was observed in the scan at the lowest stress level of 323 MPa, all indications visible in Fig. 6.47 may readily be understood as formation of damage in the material.

In Fig. 6.47a, recorded at the region of first nonlinearity of the stress-strain curve, the formation of a microscopic crack network occurs. This is indicated by the light black lines in the *xy*-projection, corresponding to the formation of first inter-fiber cracks in the 90° and 0° plies. At increased stress levels, the black indications densify at the previous positions and appear at new locations (cf. Fig. 6.47b). Depending on the orientation of the lines, these may readily be correlated to the occurrence of further inter-fiber cracks, which can be related to their specific depth position by the inspection of the 3D volume. Some areas show larger black clouds extending from the position of previously grown inter-fiber cracks. These areas are located at the depth positions between the individual plies and may thus easily be understood as formation of delamination. At highest stress levels, a significant amount of microscopic damage has already been accumulated and has led to the formation of the crack network seen in Fig. 6.47c. In addition to numerous inter-fiber cracks along the 0° and 90° directions, there is a substantial area showing inter-ply delamination. Beyond this stress level, the damage is expected to accumulate further and finally cause the rupture of full plies. Since this is typically not compensated by the remaining plies, this will ultimately cause the rupture of the full laminate.

Thus for stress-strain curves with significant signatures as seen in Fig. 6.46, the direct transfer of those procedures to other load scenarios, specimen types, and material combinations is straightforward. However, depending on the selected CT scan parameters, the presence of image artifacts, and the volume size, it may result in less visible details. Also, for several applications no significant deviations of the stress-strain behavior from linearity occur prior to ultimate failure. To this end, Sect. 7.2.4.1 presents a suitable combination between acoustic emission analysis and CT to deal with such conditions.

6.5.1.2 Extraction of Geometries from Volume Data

Apart from visualization of damage progression itself, the extraction of existing damage from volumetric images can be used as input for modeling approaches.

The application of computed tomography to perform dimensional measurements is already standard in quality control and has seen a long history of developments. Using the image segmentation methods presented above, it is feasible to extract 3D representations of the segmented entities. This process is known as reverse engineering and is referring to the creation of a digital dataset based on the physical representation of the object. Compared to alternative methods, like tactile or optical measurement systems, computed tomography comes with the large benefit to also show internal structures and thus allows virtual disassembly of parts. The outer contour of these parts can be directly turned into point clouds or into tessellated surfaces (e.g., as triangulated STL files). These files can then be compared relative to a CAD drawing or may be used to inspect the dimensional accuracy of a part [5].

In the context of damage tracking followed herein, these possibilities are consequently extended by the extraction of crack geometries as digital objects for damage progression modeling or for validation of measurement methods. To this end, two examples are presented in the following to demonstrate the possibilities of this approach for assessment of material failure.

In a first example, the extraction of a fracture surface of a mode II crack is presented. As schematically shown in Fig. 6.48, the full procedure basically consists of two subsequent steps. In the first step, the crack surface is extracted from a CT scan and digitized to yield a digital representation of the 3D geometry. Within the second step, this digital object is imported to a FEM program and embedded to a CAD object for further processing. Although these steps are conceptually easy to follow, the practical handling of the surface extraction, digitization, and embedding comes with several challenges and may result in inaccuracies if carried out inappropriately. Thus, the sequence is described in more detail in the following using

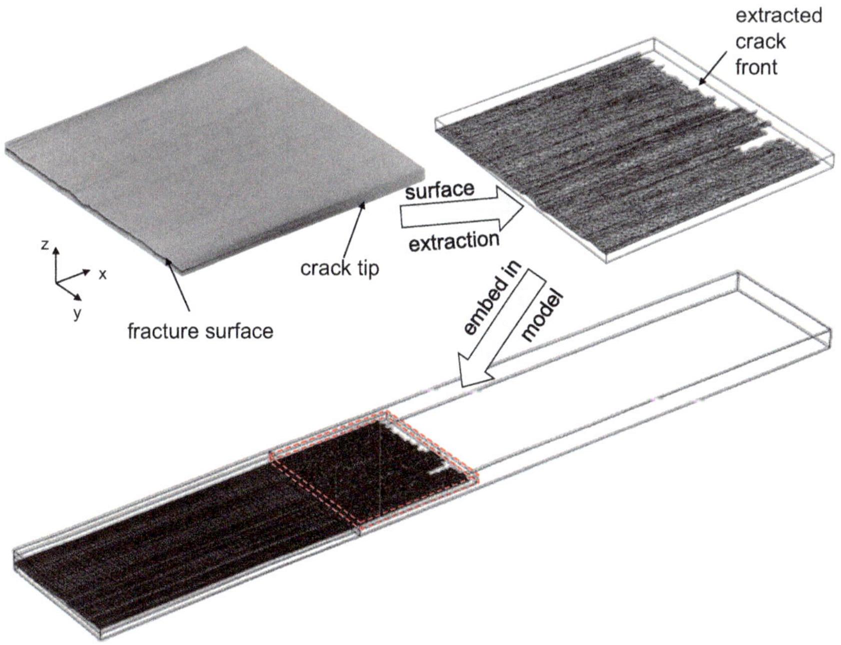

Fig. 6.48 Scheme of procedure used to transfer CT scan of fracture surface into FEM-model

the software packages VGStudio MAX, MeshLab, and COMSOL Multiphysics as representative tools to perform surface extraction, mesh processing and geometry embedding, and FEM computation.

To extract the fracture surface, a suitable CT scan of the region surrounding the damage zone is performed. In this context, the volume may already be reduced to yield only the region of interest close to the fracture surface for further processing as seen in Fig. 6.48.

The CT scan should be of sufficient resolution and contrast, to allow proper image segmentation using the threshold techniques discussed in the beginning of Sect. 6.5. For proper extraction of the crack tip geometry, the selection of the segmentation threshold is of particular relevance and should be considered in detail when performing the segmentation step (cf. Fig. 6.49). In this context it is worth noting that most of the postprocessing steps cause a reduction of the volume of the extracted geometry. Thus shrinkage of the extracted object is technically possible, while artificial growth of the geometry is generally not deemed a feasible approach. As a consequence, the threshold selection should be carried out in a way to slightly overestimate the crack front position as seen in the example of Fig. 6.49c. Within VGStudio MAX it is possible to perform a surface extraction step using thresholding techniques. The obtained segmentation boundary can then be exported as STL file format to yield a tessellated surface for further processing. In this context, a "normal" level of detail was found to be appropriate for the surface extraction algorithm in VGStudio MAX. Nevertheless, a high level of accuracy in

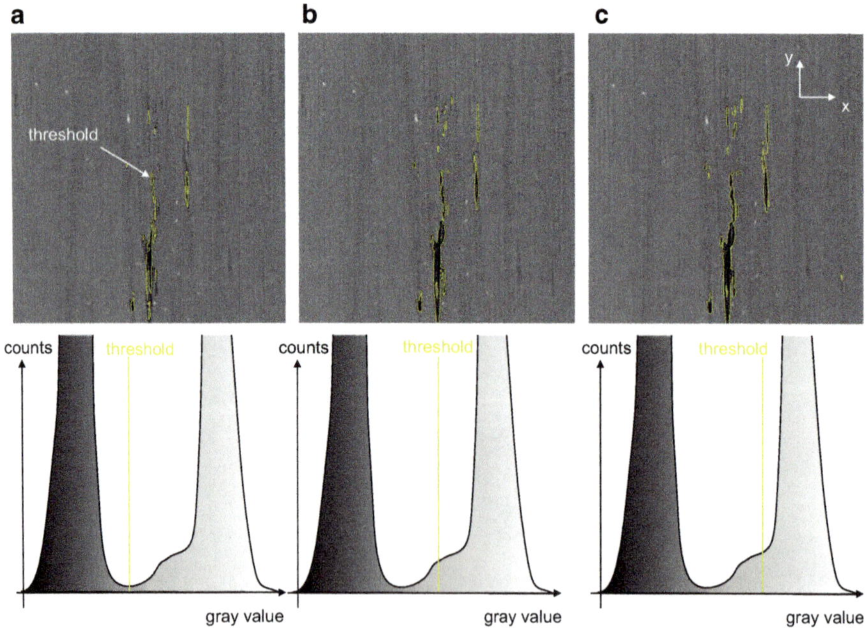

Fig. 6.49 Influence of selected threshold on geometry of crack tip

this step may result in large files for further processing and may need adjustment subject to the extracted volume, scan resolution, and computational capabilities for processing.

As next step, the obtained STL file typically needs further processing before used in a FEM environment. This is mostly due to several issues arising from the tessellation algorithm applied to the extracted surface. First, there is a large likelihood that the obtained fracture surface is not fully closed, which is caused by imperfections of the CT scan as well as the selected segmentation threshold. Such holes need to be closed, to yield an appropriate representation of the fracture surface. Second, the tessellation algorithm is likely to generate topologically non-manifold faces and vertices in regions of geometric singularities (i.e., at the crack tip). Since these cause difficulties for generation of computational meshes, these need to be removed as well. And finally, the presence of pores or crack bifurcations may cause isolated entities, which may be considered in FEM approaches, but may also lead to difficulties in mesh generations. Thus it was found to be useful to remove such isolated surfaces before further processing.

Since all these operations are hardly implemented by manual mesh processing strategies, they require an appropriate environment to perform such STL file modifications. To this end, the software platform MeshLab was used for processing and simplification of the extracted surface.

After import of the STL file, the first step consists of a removal of isolated entities by the command "remove isolated pieces" applied to the imported mesh. Based on the selected option, this is used to remove all isolated objects depending on their diameter or number of faces as seen in the transition between Fig. 6.50a, b.

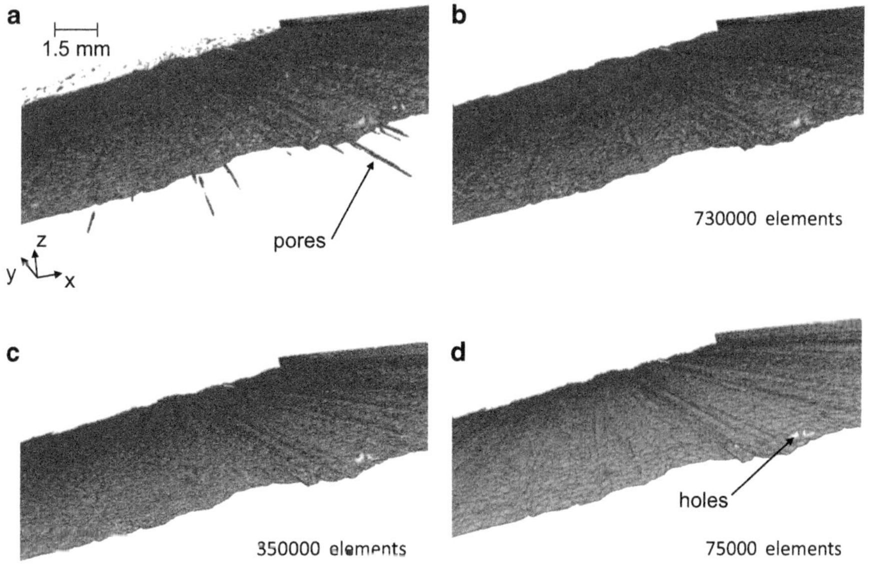

Fig. 6.50 Use of "remove isolated pieces" command to remove isolated entities (**a**) and "quadratic based edge collapse strategy" to simplify the mesh to different number of elements (**b–d**)

For further processing within a FEM program, the topology of the obtained mesh may need some processing. Some of the mesh refinement strategies present in the next steps will require a mesh with two-manifold edges. This means that each edge should be connected to exactly two neighboring faces. Non-closed objects are typically one-manifold (or even zero-manifold for plane edges) and objects containing artifact faces are three- or more manifold. Before further processing, the vertices, edges, or faces causing this lack of topological integrity of the surface need to be removed. Examples for such non-two-manifold edges are enclosed faces or overlapping faces. These are difficult to spot by the bare eye, but are easy to select and delete by using the "select non-manifold edge" filter. Careful removal without compromising the integrity of the geometry is used to yield a suitable mesh representation for further processing.

Subsequently, the mesh is subject to a simplification scheme. Among the different choices, the "quadratic based edge collapse strategy" option following the scheme of Hoppe was found to be particularly useful [85]. Two stages of mesh simplification are seen in Fig. 6.50c, d as compared to the exported mesh from VGStudio MAX in Fig. 6.50b. For such mesh refinements, it is key to preserve the topology of the extracted surface and to avoid removal of geometric details, which are required for the computation result. The latter aspect is hard to determine at this stage, so the simplification routines should be evaluated in their impact on the computation results by comparing results of different simplification levels. Clearly, this step is used to increase the numerical efficiency of the modeling procedure and should thus be considered carefully to reach a satisfying level of detail, while keeping the numerical intensity at a minimum. Newly generated or existing holes are filled using the "fill holes" or "close holes" command as shown in Fig. 6.51. Since a fracture surface is by definition free of holes, the latter is a requirement for further computation tasks. As final step, the simplified and repaired surface is exported to another STL file.

Within the FEM program COMSOL Multiphysics, STL files may be imported during geometry creation. For the exemplary fracture surface of a mode II crack, only the crack tip was scanned by CT and is imported into the FEM program. The remaining parts of the specimen may either be imported from other CAD formats or

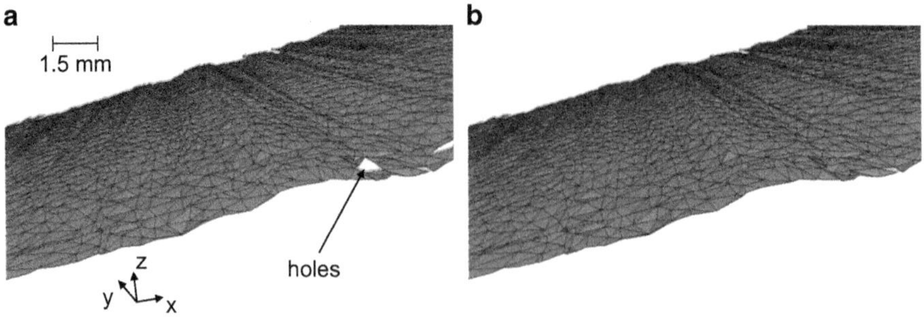

Fig. 6.51 Use of "fill holes" command to artificially close the generated holes

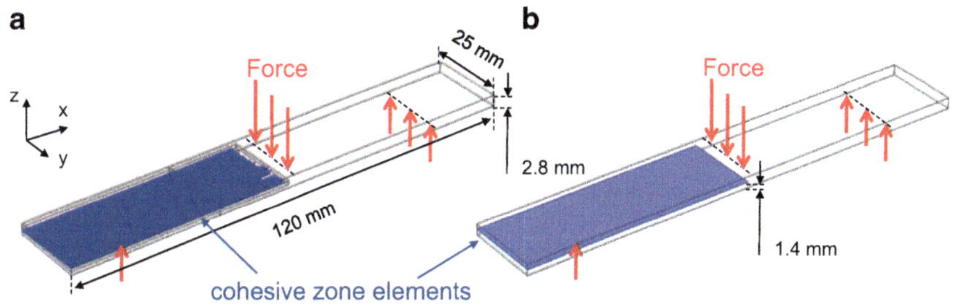

Fig. 6.52 FEM-model configuration used to calculate strain fields due to static load using extracted fracture surface (**a**) and reference case with plane fracture surface (**b**)

are being generated directly within the FEM program. From the imported crack surface, only the lower half was kept, and the remaining crack surface was joined with a surrounding volume using a sequence of intersection and union commands to yield the volume shown in Fig. 6.48. The so-obtained volume was then embedded within the full 3D geometry of a mode II specimen as shown in Fig. 6.52a. The full fracture surface of the specimen is reached by extrusion of the crack contour at the left edge of the embedded volume and finally yields the blue fracture surface seen in Fig. 6.52a. For comparison a second model with identical outer geometry is generated. As a major difference, the second model uses a simple rectangular crack representation at the half-thickness position of the laminate as seen in Fig. 6.52b.

For further computations, a line load acting in z-direction is applied at the center of the specimen, while at the position of the lower supports, displacement constraints are chosen, inhibiting movement in x-direction and z-direction. Material properties are those of Sigrafil CE1250-230-39 as listed in Table B.1 in Appendix B. The full fracture surface is modeled by cohesive zone elements following the implementation described in Sect. 4.2.

Within the "Structural Mechanics module" of COMSOL Multiphysics, the configurations of Fig. 6.52 are subject to a stationary analysis using an applied force equal to the residual load after crack growth as obtained in the experiment. The computation results are visualized as false-color plots of the shear angle in Fig. 6.53. This representation of the calculated deformation is chosen in analogy to the evaluation of mode II crack growth using DIC techniques presented in Sect. 3.4. The typical 2D image seen by a DIC system is shown as an additional inset for both cases.

It is evident from Fig. 6.53 that the different implementations of crack geometry result in noticeable differences in the computed shear angle values around the crack position and along the delaminated area. Both cases allow identifying the position of the crack tip by the steplike change in shear angle from 0° to -0.5° or less. However, the remaining signatures of the strain field around the position of the crack tip are affected by the real shape of the crack tip, causing less distinct areas of positive shear angle above and below the crack tip for the case shown in Fig. 6.53b.

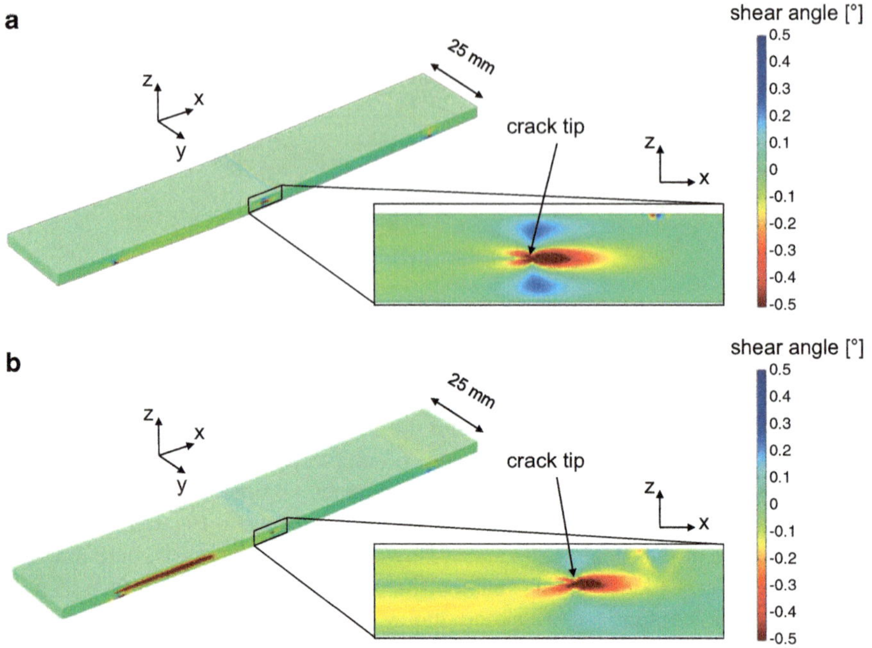

Fig. 6.53 Comparison of calculated shear angle in ENF test for laminate with perfect center crack (**a**) and shear angle due to real fracture surface as obtained from CT scan (**b**)

Moreover, the present finding is in contrast to the typical observations of crack growth by DIC methods as seen in Figs. 3.48, 3.50, and 3.51 in Sect. 3.4. Thus, it may be speculated if the shear strain signatures seen by DIC methods are really due to the deformation of the speckle pattern or are more a result of the local chipping of the speckle pattern due to interlaminar crack growth.

As second example for numerical computation utilizing an extracted fracture surface, the configuration shown in Fig. 6.54 is presented. Applying the same software environments as for the previous example, the fracture surface of a RTM6 specimen as used for flexural testing in Sect. 5.2 is extracted from a CT scan and is joined with a 3D volume of the full specimen. This test specimen is then embedded within a model of the full test setup including a 3D representation of the lower supports and the geometry of the pushrod introducing the mechanical load. The dimensions of the model as well as the selected boundary constraints are given in Fig. 6.55. To induce a mechanical load, the upper surface of the pushrod was subject to a force of 78.8 N acting in z-direction, while movement of the edges at the bottom of the support was constrained along the x-direction and the z-direction. As further boundary constraints, the extracted fracture surface was modeled by cohesive zone elements following the approach presented in Sect. 4.2 with a von Mises failure criterion and the material properties of RTM6 as given in Tables B.1 and B.4 of Appendix B.

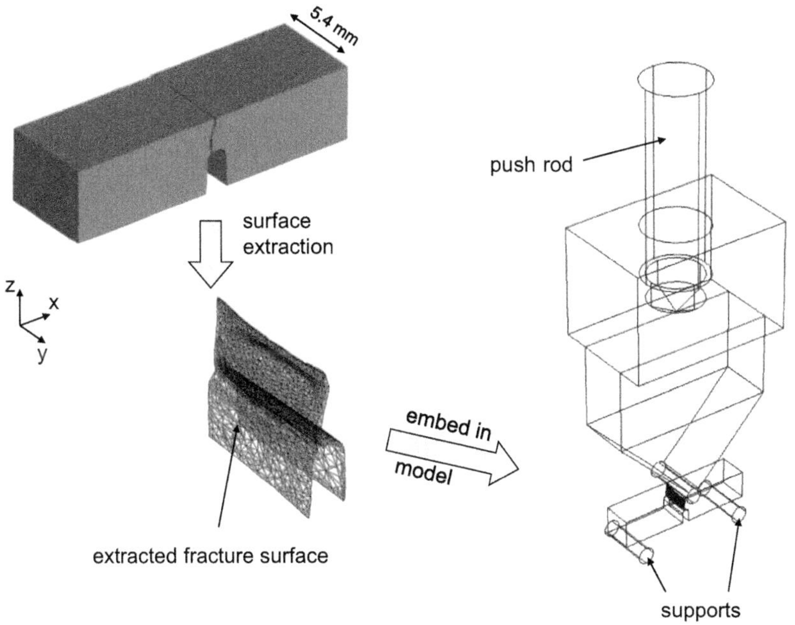

Fig. 6.54 Scheme of procedure used to transfer CT scan of fracture surface into FEM-model

Within the "Structural Mechanics module" of COMSOL Multiphysics, the configuration of Fig. 6.55 is subject to a transient analysis. The acting force was identical to the failure load of the specimen. All dimensions including the fracture plane are chosen identical to the experimentally used specimen. Based on the material properties and the von Mises failure criteria, this causes crack initiation and crack propagation at this load level.

As seen from the false-color visualization of the degradation function $\check{C}(\boldsymbol{r})$ in Fig. 6.56, the applied external load results in crack initiation at the notch position with crack growth from the bottom to the top of the specimen. The position of the crack front is determined by the evaluation of $\check{C}(\boldsymbol{r})$ and does not obey a spatially and temporally homogeneous growth. Instead, the topology of the fracture surface causes an inhomogeneous growth of the crack front along the y-axis. Due to the initial acceleration, the crack velocity increases first (cf. $t = 12.5$ μs to $t = 25.0$ μs) and decelerates when approaching the top of the specimen (cf. $t = 50.0$ μs to $t = 100.0$ μs). The dynamics of this crack growth confirm well with results of a fractographic analysis of the resultant fracture plane (cf. Fig. 5.8). Here a crack initiation zone with sliver-like pattern is observed followed by pronounced Wallner lines when reaching the deceleration zone. In order to perform a physically correct description of the dynamics of the crack growth, it is thus required to consider the topology of the fracture plane. A simple approach defining a straight rectangular plane was found to yield inaccurate results even for such relatively simple failure modes [86]. Especially the complex fracture planes of composite materials are

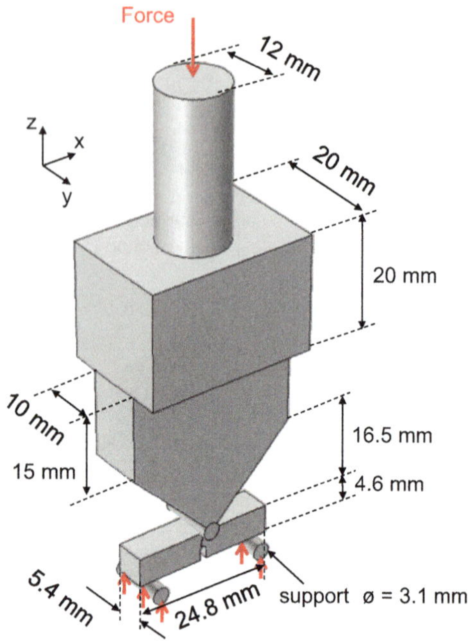

Fig. 6.55 FEM-model configuration used to calculate crack growth in flexural test configuration (based on [86])

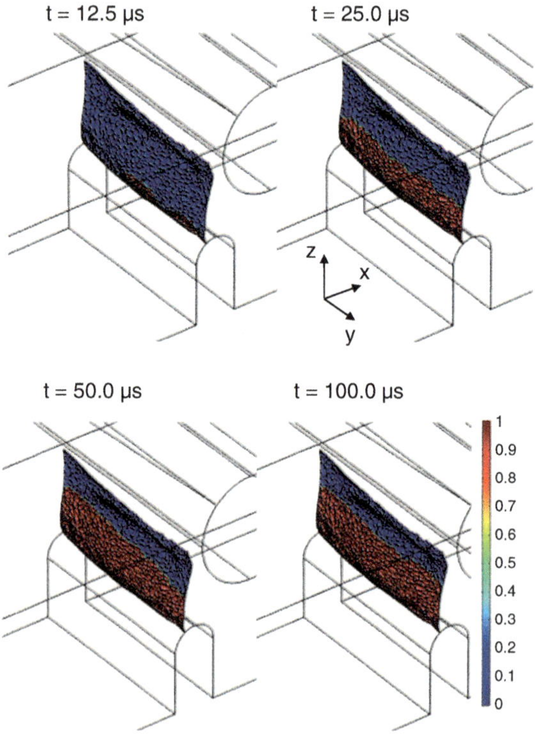

Fig. 6.56 Crack growth in RTM6 specimen as function of time visualized as false-color diagram of the degradation function $\check{C}(\boldsymbol{r})$ (based on [86])

expected to benefit from approaches to directly obtain the geometric dimensions from computed tomography measurements (see also Sect. 7.2.4.2). However, other approaches using meshless methods [87–93] or refined FEM approaches [84, 94] do not need to define an explicit fracture plane and therefore should be able to capture the dynamics intrinsically. For such advanced modeling methods, the extraction of fracture planes may be then directly be used to compare model predictions to experimental results.

6.5.2 *In Situ Testing*

As primary difference to the previous section, the approaches considered as in situ testing using computed tomography typically combine a mechanical test rig with a CT device as presented in Sect. 6.4. Many applications to visualize and detect damage progression in composite materials by in situ approaches have already been published [25, 26, 31, 34, 35, 40, 46–61]. Thus, the aim of this section is not to repeat their findings but to demonstrate the capabilities of such approaches using some representative examples. To this end, results of a study are presented, which investigate a subset of the fundamental microscopic failure modes of fiber reinforced composites.

In order to induce a particular failure mechanism, the load frame concept introduced in Sect. 6.3 is applied. This comprises an aluminum load frame component in combination with a PMMA tube used as load deflection. The load stage itself is a fully computer-controlled commercial device of type "Microtest" applying a 500 N load cell for precise force measurements. For later combination with AE measurements (see Sect. 7.2.4.2), the load frame was modified to form a waveguide to transfer the AE signal from the source position to the sensor position [60]. A 3D CT scan of a mounted specimen showing the details of the transfer components is found in Fig. 6.57 next to a photography image of the system. As explained in Sect. 6.3, the system applies a load deflection concept introducing tensile or compressive load by movement of the lower bar. The upper bar is mounted to the surrounding PMMA tube and acts as fixed counterpart. To improve the acoustic propagation path between the specimen and the sensor, a waveguide concept is applied, which is explained in more detail in Sect. 7.2.4.2. The AE sensor itself is mounted in a screw thread machined into the upper part of the stage.

The specimens investigated are fabricated as prepreg laminate using the carbon/epoxy system Sigrafil CE1250-230-39 following the curing cycle recommended by the manufacturer. Miniaturized samples of nominally 1.7 mm × 2.0 mm × 22.0 mm (depth × width × height) are cut from the laminates utilizing a water-cooled low-speed saw with precision diamond blade. Samples are attached to the aluminum load bars by an UHU Plus endfest 300 epoxy adhesive. Mechanical testing is carried out in displacement-controlled mode with 0.2 mm/min displacement rate.

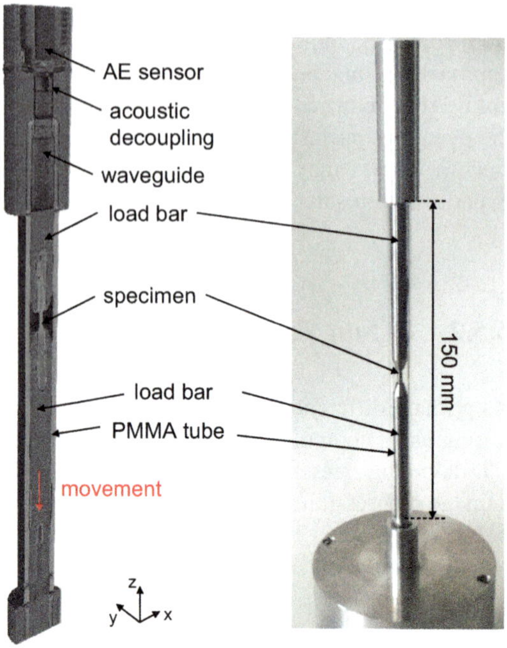

Fig. 6.57 3D scan result of load rig showing cross-section of the mounted specimen, the tensile bars, and the mounted acoustic emission sensor (*left*) and according photography image (*right*)

The generated load-displacement curves are recorded by the software program "Deben Microtest."

The different types of load bars are shown in Fig. 6.58. Depending on the intended type of load, several load bar types are used. Tension and compression tests were carried out by using the load bars of Fig. 6.58a. Here the specimen is directly bonded to the load bars by an epoxy adhesive. The large embedding length seen in Fig. 6.57 offers a comparatively large contact surface to the adhesive and is sufficient to transfer tensile or compressive loads into the specimen until failure in the free section. For compressive loading, the adhesive bonding has the further advantage to support the specimen at the edges, thus avoiding end crushing by a shear load introduction. As further load concept, a mode I configuration was realized by a wedge cracking method. This is shown in Fig. 6.58b holding the sharp wedge on the bottom load bar and a respective slit for the specimen machined into the upper load bar.

In mechanical testing, a comparatively stiff load frame should allow a direct measurement of the specimen strain by evaluation of the cross-head displacement, but the strain rate intrinsically changes during increased loading [95]. For the given case, the stiffness of the full load frame was evaluated as 635 N/mm for tensile and compressive load and was not found to be sufficiently high to evaluate the specimen strain directly from the movement of the stepper motor. A typical measurement curve for the stress-strain relationship during tensile testing transverse to the fiber

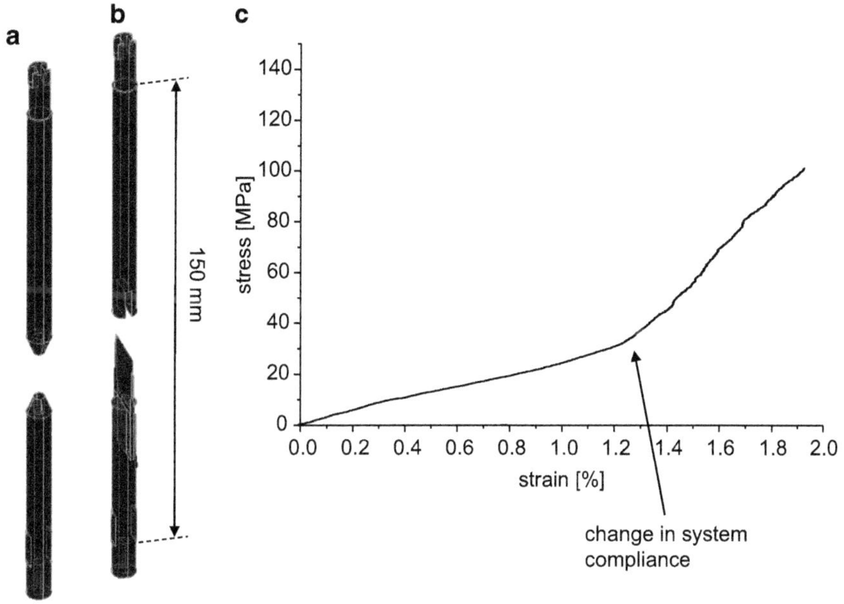

Fig. 6.58 Several load bars as used for tension and compression (**a**) and mode I (**b**) loading. Typical stress-strain curve from tensile test exhibiting change of system compliance (**c**)

direction is shown in Fig. 6.58c. As seen from the figure, the curve exhibits a characteristic change of compliance which unlikely originates from the material. Instead such changes in system compliance were found in various load configurations, being partially at characteristic load levels and partially being unique for each measurement. Consequently, a direct compensation of the load frame compliance by a linear correction factor is not possible. For this case a nonlinear compensation would be required to account for changes in system compliance, i.e., due to screw threads, bolt connections, and other connectors. Since this was found to be partially at unique load levels, no generally applicable compliance correction could be made. Therefore, further discussion will be limited to the observed failure mode and the stress level during failure.

In the following, results are shown for the occurrence of inter-fiber failure in a unidirectional composite due to tensile and compressive load as well as fiber failure due to tensile and compressive load. Further results are presented on tracking of interlaminar crack growth resulting from mode I loading. All scans were performed using a Nanotom 180 CT scanner applying the scan parameters of Table 6.5 and the "phoenix datos|*x*2 reconstruction" software for image acquisition and VGStudio MAX for postprocessing. Given the high voxel resolution, details of the fracture plane and inclusions (e.g., pores) in the bulk composite are readily visible.

Table 6.5 Acquisition parameters for CT scan

Measurement parameter	Value
Detector type	Amorphous Silicon CsI detector (3072 × 2400)px^2
Filter	None
X-ray tube voltage	50 kV
X-ray tube current	170 μA
Focal spot size	0.6 μm
Voxel size	1.7–3.8 μm
Acquisition angles	1000
Image exposure time	1 s
3D reconstruction software	Phoenix datos|x2 reconstruction

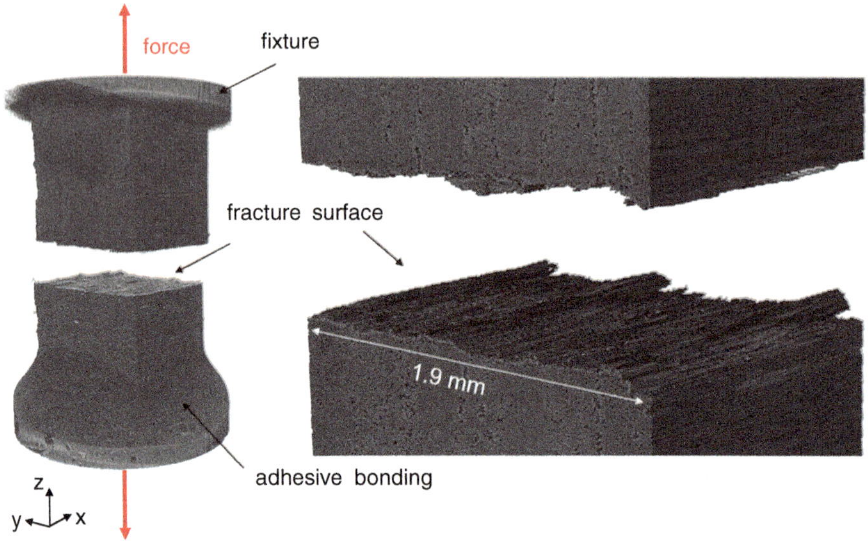

Fig. 6.59 3D scan of the failure mode due to tensile load transverse to fiber axis including details of the fracture surface (reprinted from [60])

6.5.2.1 Inter-fiber Failure

For tensile loading transverse to the fiber direction, the expected failure mode is a comparatively smooth fracture surface normal to the applied load. An exemplary CT scan after final failure of the specimen as shown in Fig. 6.59 confirms this assumption. Although the details of the fracture plane reveal a certain roughness of the fracture surface, the overall orientation is found to be normal to the load axis.

Beyond the possibility to extract the fracture surface as discussed in Sect. 6.5.1.2, the 3D capabilities of the scan allow further inspections to be made. Since failure of unidirectional composite transverse to the fiber axis is dominated by the presence of laminate imperfections (e.g., inclusions, dry spots, and resin rich

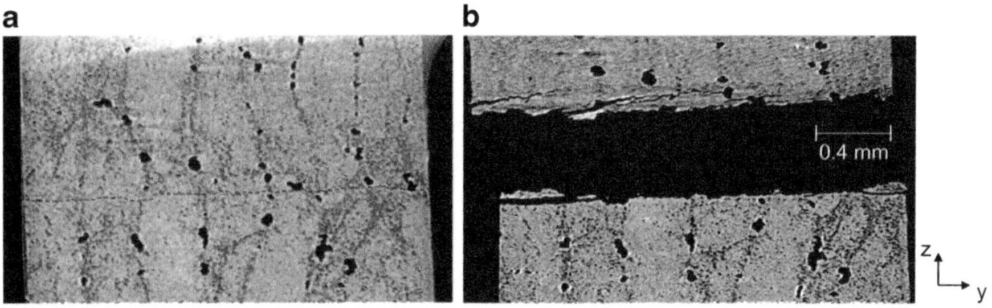

Fig. 6.60 Exemplary *yz*-cross-section before (**a**) and after failure (**b**) due to tensile load transverse to fiber axis

areas), the 3D analysis may readily be used to spot such regions and to investigate their relation to formation of the fracture plane. In the spirit of [28], an example of such inspection is shown in Fig. 6.60.

Here the scan of Fig. 6.60a reveals a substantial amount of porosity in the laminate, which has been quantified to be in the order of 5 %. From the position of the pores, it is obvious that they link and agglomerate at the positions between the individual plies, which can readily be observed in Fig. 6.60a. A direct comparison to the contour of the fracture plane shown in Fig. 6.60b can thus be made. As seen from the red line in Fig. 6.60a, the later position of the fracture surface links several pores on the right side of the specimen. However, based on the cross-section shown in Fig. 6.60, a deviation of the crack from the *y*-axis would be expected including some of the bigger pores seen in Fig. 6.60a. Instead, the crack stays almost in parallel to the *y*-axis. There are several reasons for this behavior. First, for the given load scenario, the presence of a crack tip will immediately result in transverse failure of the material due to the superimposed normal stress. Energetically it is not favorable to deviate from this angle (cf. Sect. 2.2). Unless there are strong stress concentrators, the crack will keep its propagation direction as seen in Fig. 6.59. Moreover, the specific cross-section shown in Fig. 6.60 is only a part of the full volume. Since some of the larger pores seen in Fig. 6.60a do not extend along the full dimension of the specimen along the *x*-axis, they do not necessarily constitute weak paths to dominate crack propagation.

To investigate the relation between porosity and formation of the fracture plane further, two examples of compression transverse to the fiber axis are discussed next. A direct comparison of the specimen shape before and after failure is shown in Fig. 6.61.

Based on Puck's failure theory, an angle of approximately 53° would be expected for an otherwise perfect and homogenous laminate [96, 97]. As seen from the 3D scan result in Fig. 6.61, the shape and orientation of the fracture surface in this load case does coincide well with the prediction. However, parts of the specimen were certainly subject to bending loads due to the partial or full buckling of the specimen after crack initiation. Accordingly, smaller fragments of the material have separated from the remaining specimen and are thus not included

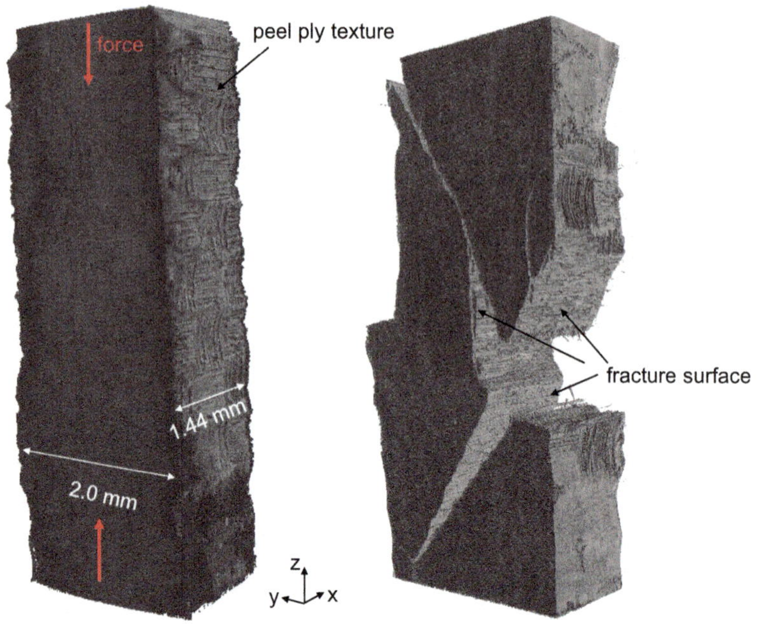

Fig. 6.61 Comparison of specimen loaded by compressive force transverse to fiber axis before and after failure

in the scan. Using an *yz*-cross-section, the resultant fracture angle can readily be measured as shown in the example of Fig. 6.62. As average of ten measurements, this evaluates as $(53.6 \pm 1.5)^\circ$ and therefore agrees well to the prediction of Puck's inter-fiber failure criterion.

To study the influence of porosity in such load cases, a relevant *yz*-cross-section of a laminate before and after fracture is shown in Fig. 6.63. The dashed line in Fig. 6.63a resembles the contour of the final fracture surface seen in Fig. 6.63b. As a result of the acting forces, a distinct fracture surface develops. This is partially due to the collapse of pores and due to multiple fractures along the paths dictated by the position of the pores. The red line of Fig. 6.63a clearly indicates that the connection of individual pores finally results in the formation of the fracture plane and thus acts as preferential path for crack growth.

6.5.2.2 Fiber Failure

In order to investigate fiber failure due to tensile and compressive load, different specimen sizes were prepared to conform with the maximum force capacity of 500 N of the used load cell. Specimens are fabricated from the Sigrafil CE1250-230-39 carbon/epoxy prepreg laminate using miniaturized samples of

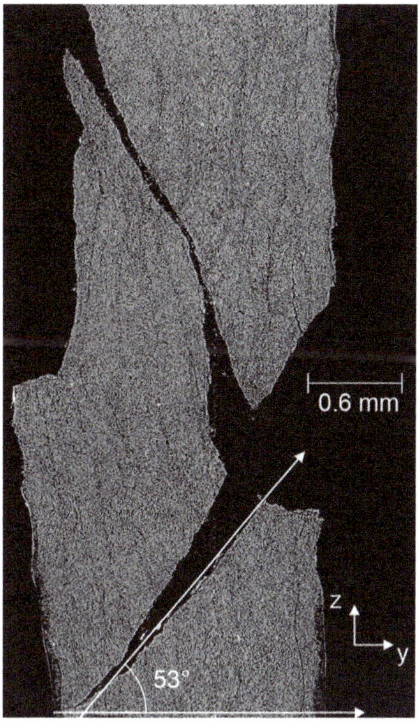

Fig. 6.62 Evaluation of fracture angle due to compressive force transverse to fiber axis

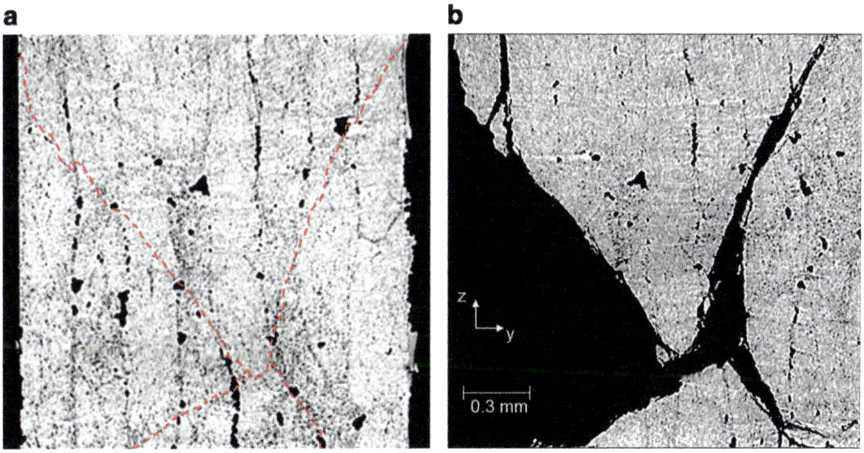

Fig. 6.63 Exemplary *yz*-cross-section before (**a**) and after failure (**b**) due to compressive load transverse to fiber axis

0.14 mm × 1.35 mm × 22.0 mm (depth × width × height). All other settings were chosen as for the previous cases.

As seen from the comparison of the specimen before and after failure in Fig. 6.64, the failure mode is distinctly different to the inter-fiber failure cases

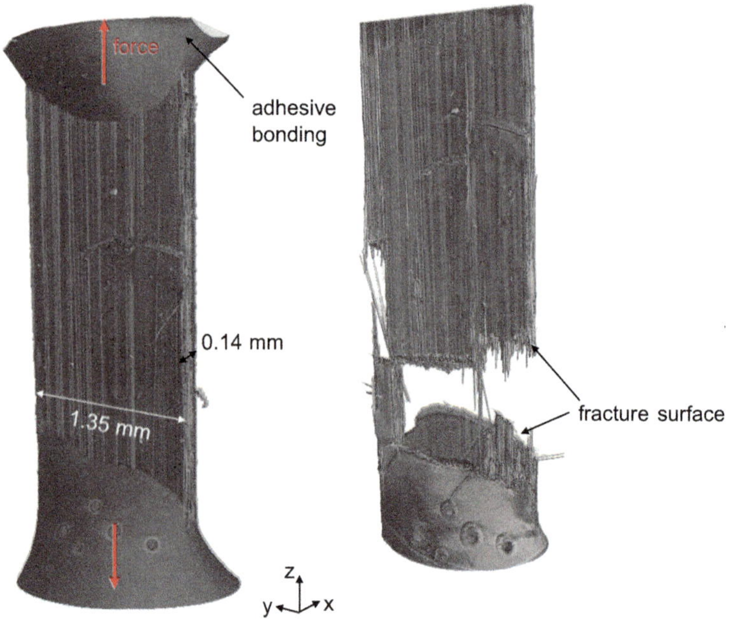

Fig. 6.64 Comparison of specimen loaded with tensile force parallel to fiber axis before and after failure

since this load configuration requires the rupture of the reinforcement fibers. Inspired by macroscopic test configurations for tensile tests parallel to the fiber axis, a tapered region as shown in the *xz*-cross-section of Fig. 6.65a was prepared using the epoxy adhesive. This ensures failure of the material in the reduced part of the specimen as seen in Fig. 6.64. In addition to the rupture of the fiber filaments, splitting along the *z*-axis occurs during failure. The latter is seen well in the comparison of the specimen cross-section in Fig. 6.65b, c. Here several inter-fiber cracks occur in parallel to the fiber axis.

The influence of pores on the occurrence of fiber filament failure has been studied in [28]. Therein it was concluded that the presence of voids increases the likelihood of a single filament fiber break in the immediate vicinity. Other experimental studies with a focus on the accumulation of single fiber filament breaks and their relevance to global failure of the laminate have provided substantial insight on this failure mode [25, 26, 56]. Moreover, substantial refinement of the theoretical framework to describe the agglomeration of fiber failure has been proposed by Swolfs et al. [98–100].

To induce failure of the specimens with compressive load applied parallel to the fiber axis, the same experimental configuration as for the tensile test is used. The taper provided by the epoxy was found to be quite useful to stabilize the specimen and to avoid buckling failure modes. As visualized the comparison of 3D scans before and after failure of the composite in Fig. 6.66, the predominant failure mode

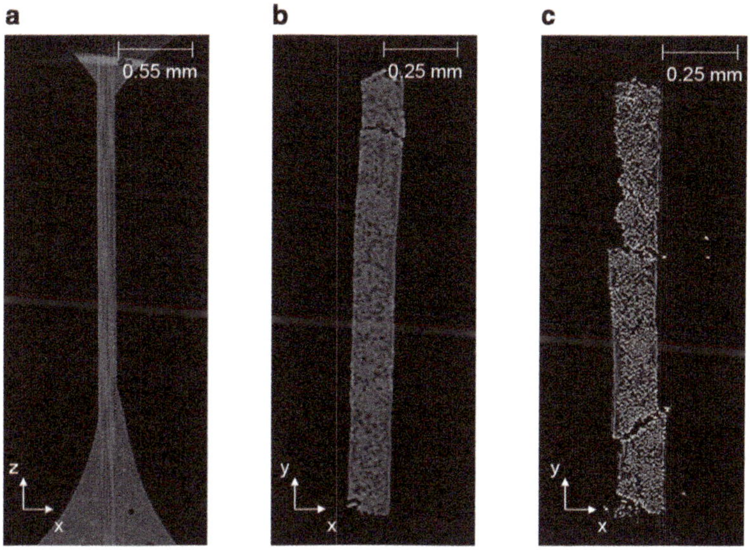

Fig. 6.65 Comparison of tensile specimen cross-section in *xz*-plane (**a**) and *xy*-plane (**b**) before loading and in *xy*-plane after loading (**c**)

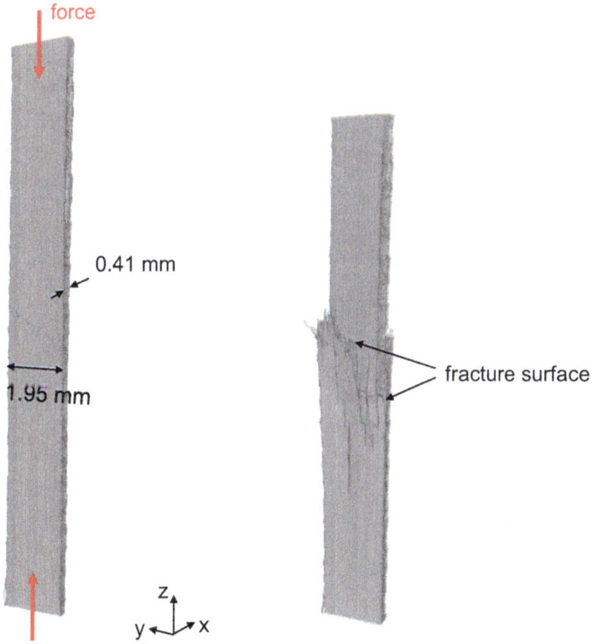

Fig. 6.66 Comparison of specimen loaded with compressive force parallel to fiber axis before and after failure

observed for the carbon/epoxy material used in this load configuration, is a brooming failure mode at the center of the specimen. This is likely induced by band kinking of the fibers prior to the final failure. As seen from the cross-sectional images in Fig. 6.67, the upper half of the specimen impacts the bottom half upon failure as typically observed in macroscopic testing. This impact causes much of the substantial damage visible in Figs. 6.66 and 6.67. The used scan resolution allows to resolve many details on the fracture mode, such as fiber fragments, the extent of delamination, splitting of the beams, and other geometrical changes. Note also the rotated alignment of the upper part of the beam compared to the bottom part. This is seen best by comparing the cross-sectional images in the *xy*-plane of Fig. 6.67 taken at the dashed red lines as indicated in the *xz*-plane cross-section.

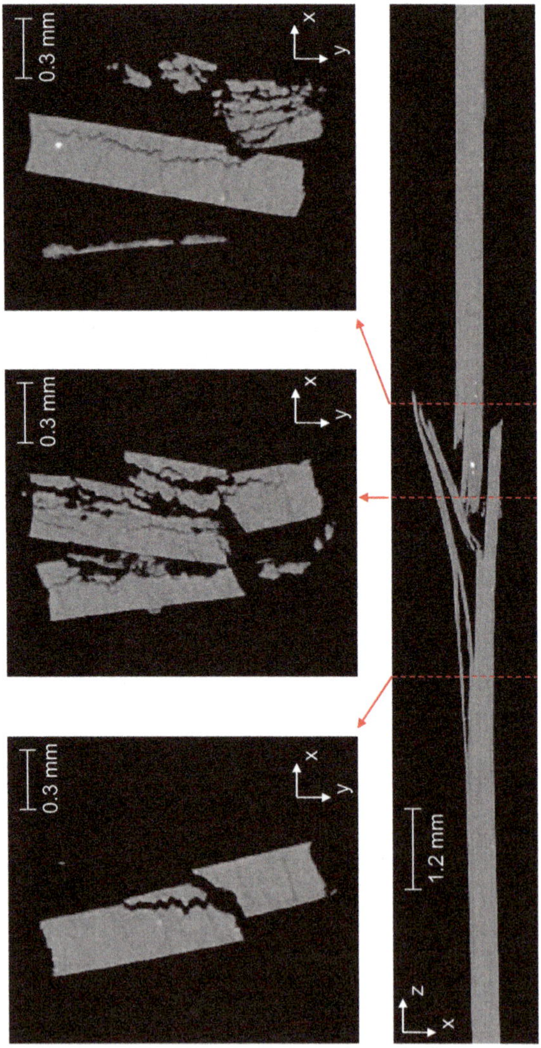

Fig. 6.67 Cross-sectional view of specimen after failure in directions perpendicular to applied load (*xy*-plane) and as cross-section aligned with applied load (*xz*-plane)

6.5.2.3 Interlaminar Failure

For initiation of an interlaminar mode I crack, a different load configuration than for tensile or compressive testing is required. As discussed in Sect. 6.3, the specimen is hold in an upright position using a fixture at the bottom part. At the top position, a sharp knife is used to induce a mode I crack due to wedge loading of the two beams. To facilitate crack initiation at the center part of the prepared specimen, a precrack is introduced prior to testing. For this test procedure, specimens are fabricated from the Sigrafil CE1250-230-39 carbon/epoxy prepreg laminate using miniaturized samples of 1.7 mm × 2.2 mm × 32.0 mm (length × width × height). The displacement rate of the indentation knife was chosen as 1 mm/min.

The resultant crack propagation after exceeding the 121 N for initiation of the precrack is shown in Fig. 6.68 for two subsequent states of knife indentation. As clearly revealed by these images, the knife indentation causes a splitting of the beam resulting in a typical mode I load condition at the position of the crack tip.

The in situ scan of this mode I crack allows examining details of the crack front, such as bifurcations or branching of the crack tip, surface roughness or interaction with particles, and voids or other defects. As an example for this kind of analysis, Fig. 6.69 presents cross-sections in the *xy*-plane (i.e., plane perpendicular to the direction of crack growth) taken at different positions along the *z*-axis. Figure 6.69a is taken at a position close to the present position of the crack tip. As noticeable from the absence of the crack on the left side of the cross-sectional image, the crack does not grow with a planar crack front along the *y*-axis (cf. also Fig. 6.44). When

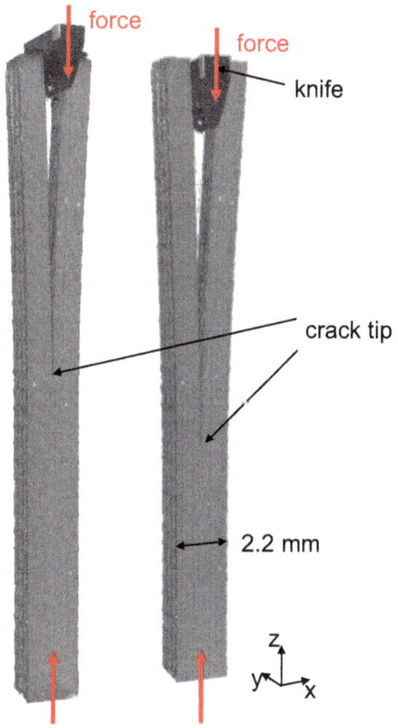

Fig. 6.68 3D visualization of interlaminar crack growth at two subsequent stages of knife indentation

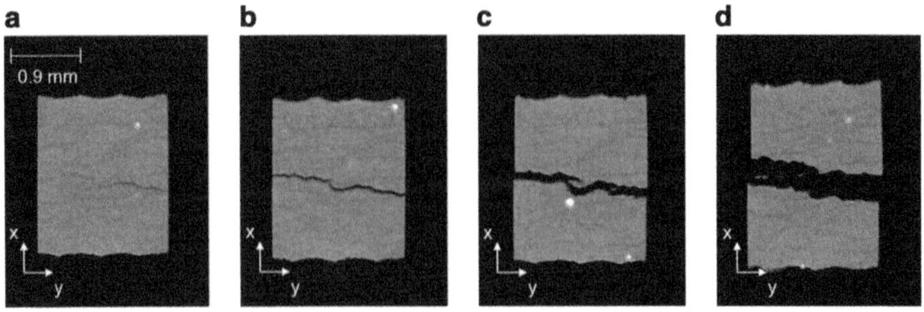

Fig. 6.69 Cross-sections in the xy-plane at z-positions of approximately 2 mm distance to the precrack

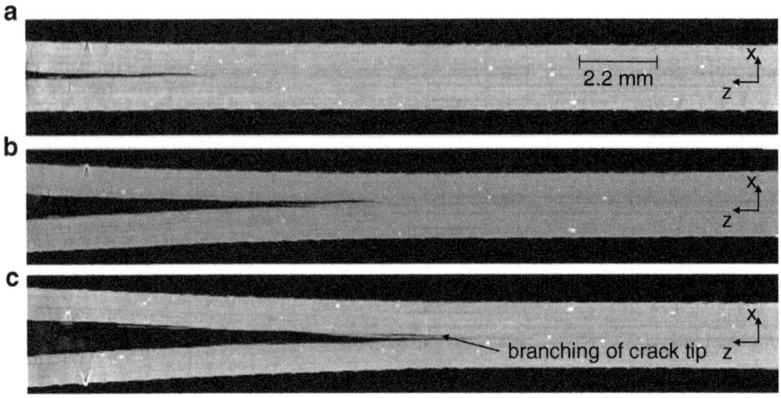

Fig. 6.70 Sequence of images in xz-cross-section recorded in subsequent load stages to visualize interlaminar crack propagation and branching of crack tip

moving towards the position of the indentation knife, Fig. 6.69b–d reveals the roughness of the fracture surface along the y-axis as well as some crack bifurcations and splitting of the laminate upon crack growth. The latter is particularly well seen for the fiber filament residues visible as small gray particles between the two fracture surfaces.

The progression of the crack tip may also be visualized for a sequence of images taken at different indentation depths of the knife. As visible in cross-sectional images in the xz-plane in Fig. 6.70 at identical y-position, the crack tip starts to propagate starting at the precrack position (Fig. 6.70a). Due to the loading of the indentation knife, the mode I condition at the crack tip causes interlaminar crack propagation at a distance of more than 3 mm ahead of the knife position. Thus, the crack is solely driven by a mode I load and is not simply cut by the indentation knife. As seen from the cross-sectional images in Fig. 6.70b, c, the crack splits the laminate and propagates with occasional branching of the crack tip. This sequential imaging mode is therefore well suited to detect influences of enclosed particles or to study deviations in the crack front due to the presence of defects, the textile architecture, or other likely crack deflectors.

References

1. Cosslett, V.E., Nixon, W.C.: X-ray shadow microscope. Nature **168**, 24–25 (1951)
2. Cosslett, V.E., Nixon, W.C.: X-ray shadow microscopy. Nature **170**, 436–438 (1952)
3. Cosslett, V.E., Nixon, W.C.: X-Ray Microscopy. Cambridge University Press, Cambridge (1960)
4. Nixon, W.C.: Improved resolution with the x-ray projection microscope. Nature **175**, 1078–1079 (1955)
5. Flisch, A., Wirth, J., Zanini, R., Breitenstein, M.: Industrial computed tomography in reverse engineering applications. In: Computerized Tomography for Industrial Applications and Image Processing in Radiology, Berlin, Germany (1999)
6. Buzug, T.M.: Computed Tomography: From Photon Statistics to Modern Cone-Beam CT. Springer, Berlin (2008)
7. Kalender, W.A.: Computed Tomography: Fundamentals, System Technology, Image Quality, Applications. Wiley, New York (2011)
8. Shull, P.J.: Nondestructive Evaluation: Theory, Techniques, and Applications. Marcel Dekker, Inc., New York (2002)
9. Radon, J.: Über die Bestimmung von Funktionen durch ihre Integralwerte längs gewisser Mannigfaltigkeiten. Berichte Sächsische Akad. der Wissenschaften. **69**, 262–277 (1917)
10. Barrett, J.F., Keat, N.: Artifacts in CT: recognition and avoidance. Radiographics **24**, 1679–1691 (2004)
11. Herman, G.T.: Fundamentals of Computerized Tomography: Image Reconstruction from Projections. Springer, Berlin (2009)
12. Brooks, R.A., Di Chiro, G.: Beam hardening in X-ray reconstructive tomography. Phys. Med. Biol. **21**, 390–398 (1976)
13. Jin, P., Bouman, C.A., Sauer, K.D.: A method for simultaneous image reconstruction and beam hardening correction. In: 2013 I.E. Nuclear Science Symposium and Medical Imaging Conference (2013 NSS/MIC), pp. 1–5. IEEE (2013)
14. Van de Casteele, E., Van Dyck, D., Sijbers, J., Raman, E.: A model-based correction method for beam hardening artefacts in X-ray microtomography. J. Xray. Sci. Technol. **12**, 43–57 (2004)
15. Van Gompel, G., Van Slambrouck, K., Defrise, M., Batenburg, K.J., de Mey, J., Sijbers, J., Nuyts, J.: Iterative correction of beam hardening artifacts in CT. Med. Phys. **38**, S36 (2011)
16. Kastner, J., Plank, B., Salaberger, D.: High resolution X-ray computed tomography of fibre- and particle filled polymers. In: 18th World Conference on Nondestructive Testing, 16-20 April 2012, Durban, South Africa, pp. 16–20 (2012)
17. Salaberger, D., Arikan, M., Paier, T., Kastner, J.: Characterization of damage mechanisms in glass fibre reinforced polymers using X-ray computed tomography. In: 11th European Conference on Non-destructive Testing (ECNDT 2014), Prague, Czech Republic, pp. 1–9 (2014)
18. Krause, M., Hausherr, J.M., Burgeth, B., Herrmann, C., Krenkel, W.: Determination of the fibre orientation in composites using the structure tensor and local X-ray transform. J. Mater. Sci. **45**, 888–896 (2009)
19. Sugimori, M., Lam, F.: Macro-void distribution analysis in strand-based wood composites using an X-ray computer tomography technique. J. Wood Sci. **45**, 254–257 (1999)
20. Little, J.E., Yuan, X., Jones, M.I.: Characterisation of voids in fibre reinforced composite materials. NDT E Int. **46**, 122–127 (2012)
21. Kastner, J., Plank, B., Salaberger, D., Sekelja, J.: Defect and porosity determination of fibre reinforced polymers by X-ray computed tomography. In: 2nd International Symposium on NDT in Aerospace, Hamburg, Germany, pp. 1–12 (2010)
22. Stoessel, R., Kiefel, D., Oster, R., Diewel, B., Llopard Prieto, L.: μ-computed tomography for 3d porosity evaluation in Carbon Fibre Reinforced Plastics (CFRP). In: International Symposium on Digital Radiology and Computed Tomography, Berlin, Germany (2011)

23. Plank, B., Mayr, G., Reh, A., Kiefel, D., Stoessel, R., Kastner, J.: Evaluation and visualisation of shape factors in dependence of the void content within CFRP by means of X-ray computed tomography. In: 11th European Conference on Non-destructive Testing (ECNDT 2014), Prague, Czech Republic (2014)
24. Kiefel, D., Stoessel, R., Plank, B., Heinzl, C., Kastner, J.: CFRP porosity characterisation using μ-computed tomography with optimized test parameters supported by XCT-simulation. In: Proceedings of Conference on Industrial Computed Tomography (iCT2014), Wels, Austria, pp. 35–43 (2014)
25. Scott, A.E., Mavrogordato, M., Wright, P., Sinclair, I., Spearing, S.M.: In-situ fibre fracture measurement in carbon-epoxy laminates using high resolution computed tomography. Compos. Sci. Technol. **71**, 1471–1477 (2011)
26. Scott, A.E., Sinclair, I., Spearing, S.M., Mavrogordato, M., Bunsell, A.R., Thionnet, A.: Comparison of the accumulation of fibre breaks occurring in a unidirectional carbon/epoxy composite identified in a multi-scale micro-mechanical model with that of experimental observations using high resolution computed tomography. In: Matériaux 2010, Nantes, France, pp. 1–9 (2010)
27. Scott, A.E., Hepples, W., Kalantzis, N., Wright, P., Mavrogordato, M.N., Sinclair, I., Spearing, S.M.: High resolution damage detection of loaded carbon/epoxy laminates using synchrotron radiation computed tomography. In: ICCM-18 18th International Conference on Composite Materials, pp. 1–6. ICC Jeju, Korea (2011)
28. Scott, A.E., Sinclair, I., Spearing, S.M., Mavrogordato, M.N., Hepples, W.: Influence of voids on damage mechanisms in carbon/epoxy composites determined via high resolution computed tomography. Compos. Sci. Technol. **90**, 147–153 (2014)
29. Allen, T., Scott, A., Hepples, W., Spearing, S.M., Reed, P.A., Sinclair, I., Testing, N., Resistance, D.: Investigating damage resistance of hybrid composite—metallic structures using multi-scale computed In: ECCM16—16th European Conference on Composite Materials, Sevilla, Spain, pp. 22–26 (2014)
30. Schilling, P.J., Karedla, B.R., Tatiparthi, A.K., Verges, M.A., Herrington, P.D.: X-ray computed microtomography of internal damage in fiber reinforced polymer matrix composites. Compos. Sci. Technol. **65**, 2071–2078 (2005)
31. Moffat, A.J., Wright, P., Buffière, J.Y., Sinclair, I., Spearing, S.M.: Micromechanisms of damage in 0° splits in a [90/0]s composite material using synchrotron radiation computed tomography. Scr. Mater. **59**, 1043–1046 (2008)
32. Moffat, A.J., Wright, P., Helfen, L., Baumbach, T., Johnson, G., Spearing, S.M., Sinclair, I.: In-situ synchrotron computed laminography of damage in carbon fibre-epoxy [90/0]s laminates. Scr. Mater. **62**, 97–100 (2010)
33. Moffat, A.J., Wright, P., Renault, A., Sinclair, I., Spearing, S.M.: Analysis of transverse ply cracks using computed tomography. In: 17th International Conference on Composite Materials, Edinburgh, UK (2009)
34. Wright, P., Moffat, A., Sinclair, I., Spearing, S.M.: High resolution tomographic imaging and modelling of notch tip damage in a laminated composite. Compos. Sci. Technol. **70**, 1444–1452 (2010)
35. Wright, P., Moffat, A., Renault, A., Sinclair, I., Spearing, S.M.: High resolution computed tomography for modelling laminate damage. In: 17th International Conference on Composite Materials, Edinburgh, UK (2009)
36. Garcea, S.C., Sinclair, I., Spearing, S.M.: Characterisation of fatigue micromechanisms in toughened carbon fiber-polymer composites using synchrotron radiation computed tomography. In: 16th European Conference on Composite Materials, Sevilla, Spain (2014)
37. Busignies, V., Leclerc, B., Porion, P., Evesque, P., Couarraze, G., Tchoreloff, P.: Quantitative measurements of localized density variations in cylindrical tablets using X-ray microtomography. Eur. J. Pharm. Biopharm. **64**, 38–50 (2006)

38. Wright, P., Fu, X., Sinclair, I., Spearing, S.M.: Ultra high resolution computed tomography of damage in notched carbon fiber—epoxy composites. J. Compos. Mater. **42**, 1993–2002 (2008)
39. Sinclair, R., Preuss, M., Maire, E., Buffiere, J.Y., Bowen, P., Withers, P.J.: The effect of fibre fractures in the bridging zone of fatigue cracked Ti-6Al-4V/SiC fibre composites. Acta Mater. **52**, 1423–1438 (2004)
40. Aroush, D.R.-B., Maire, E., Gauthier, C., Youssef, S., Cloetens, P., Wagner, H.D.: A study of fracture of unidirectional composites using in-situ high-resolution synchrotron X-ray microtomography. Compos. Sci. Technol. **66**, 1348–1353 (2006)
41. Forsberg, F., Sjödahl, M., Mooser, R., Hack, E., Wyss, P.: Full three-dimensional strain measurements on wood exposed to three-point bending: analysis by use of digital volume correlation applied to synchrotron radiation micro-computed tomography image data. Strain **46**, 47–60 (2010)
42. Forsberg, F., Mooser, R., Arnold, M., Hack, E., Wyss, P.: 3D micro-scale deformations of wood in bending: synchrotron radiation muCT data analyzed with digital volume correlation. J. Struct. Biol. **164**, 255–62 (2008)
43. Badel, P., Vidal-Sallé, E., Maire, E., Boisse, P.: Simulation and tomography analysis of textile composite reinforcement deformation at the mesoscopic scale. Compos. Sci. Technol. **68**, 2433–2440 (2008)
44. Scott, A.E., Clinch, M., Hepples, W., Kalantzis, N., Sinclair, I., Spearing, S.M.: Advanced micro-mechanical analysis of highly loaded hybrid composite structures. In: ICCM 17—17th International Conference on Composite Materials, Edinburgh, UK (2009)
45. Zauner, M., Keunecke, D., Mokso, R., Stampanoni, M., Niemz, P.: Synchrotron-based tomographic microscopy (SbTM) of wood: development of a testing device and observation of plastic deformation of uniaxially compressed Norway spruce samples. Holzforschung. **66**, (2012)
46. Stampanoni, M., Groso, A., Isenegger, A., Mikuljan, G., Chen, Q., Bertrand, A., Henein, S., Betemps, R., Frommherz, U., Böhler, P., Meister, D., Lange, M., Abela, R., Boehler, P.: Trends in synchrotron-based tomographic imaging: the SLS experience. In: Developments in X-ray Tomography V, pp. 63180M–63180M–14 (2006)
47. Groso, A., Abela, R., Stampanoni, M.: Implementation of a fast method for high resolution phase contrast tomography. Opt. Express. **14**, 8103 (2006)
48. Maire, E., Carmona, V., Courbon, J., Ludwig, W.: Fast X-ray tomography and acoustic emission study of damage in metals during continuous tensile tests. Acta Mater. **55**, 6806–6815 (2007)
49. Baensch, F., Zauner, M., Sanabria, S.J., Sause, M.G.R., Pinzer, B.R., Brunner, A.J., Stampanoni, M., Niemz, P.: Damage evolution in wood: synchrotron radiation micro-computed tomography (SRμCT) as a complementary tool for interpreting acoustic emission (AE) behavior. Holzforschung **69**(8), 1015–1025 (2015)
50. Cazaux, J., Erre, D., Mouze, D., Patat, J.M., Rondot, S., Sasov, A., Trebbia, P., Zolfaghari, A.: Recent developments in X-ray projection microscopy and X-ray microtomography applied to materials science. J. Phys. IV **03**, 2099–2104 (1993)
51. Chotard, T.J., Boncoeur-Martel, M.P., Smith, A., Dupuy, J.P., Gault, C.: Application of X-ray computed tomography to characterise the early hydration of calcium aluminate cement. Cem. Concr. Compos. **25**, 145–152 (2003)
52. Steppe, K., Cnudde, V., Girard, C., Lemeur, R., Cnudde, J.-P., Jacobs, P.: Use of X-ray computed microtomography for non-invasive determination of wood anatomical characteristics. J. Struct. Biol. **148**, 11–21 (2004)
53. Trtik, P., Dual, J., Keunecke, D., Mannes, D., Niemz, P., Stähli, P., Kaestner, A., Groso, A., Stampanoni, M.: 3D imaging of microstructure of spruce wood. J. Struct. Biol. **159**, 46–55 (2007)

54. Van den Bulcke, J., Masschaele, B., Dierick, M., Van Acker, J., Stevens, M., Van Hoorebeke, L.: Three-dimensional imaging and analysis of infested coated wood with X-ray submicron CT. Int. Biodeterior. Biodegrad. **61**, 278–286 (2008)
55. Helliwell, J.R., Sturrock, C.J., Grayling, K.M., Tracy, S.R., Flavel, R.J., Young, I.M., Whalley, W.R., Mooney, S.J.: Applications of X-ray computed tomography for examining biophysical interactions and structural development in soil systems: a review. Eur. J. Soil Sci. **64**, 279–297 (2013)
56. Scott, A.E., Sinclair, I., Spearing, S.M., Thionnet, A., Bunsell, A.R.: Damage accumulation in a carbon/epoxy composite: comparison between a multiscale model and computed tomography experimental results. Compos. Part A Appl. Sci. Manuf. **43**, 1514–1522 (2012)
57. Hu, X., Wang, L., Xu, F., Xiao, T., Zhang, Z.: In-situ observations of fractures in short carbon fiber/epoxy composites. Carbon **67**, 368–376 (2014)
58. Brault, R., Germaneau, A., Dupré, J.C., Doumalin, P., Mistou, S., Fazzini, M.: In-situ analysis of laminated composite materials by X-ray micro-computed tomography and digital volume correlation. Exp. Mech. **53**, 1143–1151 (2013)
59. Rolland, H., Saintier, N., Robert, G.: Damage mechanisms into short glass fibre reinforced thermoplastic during in-situ microtomographic tensile tests. In: ECCM16—16th European Conference on Composite Materials, Sevilla, Spain (2014).
60. Kalafat, S., Zelenyak, A.-M., Sause, M.G.R.: In-situ monitoring of composite failure by computing tomography and acoustic emission. In: 20th International Conference on Composite Materials, Copenhagen, Denmark, pp. 1–8 (2015)
61. Borstnar, G., Mavrogordato, M.N., Sinclair, I., Spearing, S.M.: Micro-mechanistic analysis of in-situ crack growth in toughened carbon/epoxy laminates to develop micro-mechanical fracture models. In: ECCM16—16th European Conference on Composite Materials, Sevilla, Spain (2014)
62. Silva, F.A., Williams, J.J., Chawla, N.: 3D Microstructure Visualization of SiC Particle Reinforced Al Matrix Composites by X-ray Synchrotron Tomography. In: ICCM 17—17th International Conference on Composite Materials, Edinburgh, UK (2009)
63. Borstnar, G., Mavrogordato, M.N., Helfen, L., Sinclair, I., Spearing, S.M.: Interlaminar fracture micro-mechanisms in toughened carbon fibre reinforced plastics investigated via synchrotron radiation computed tomography and laminography. Compos. Part A Appl. Sci. Manuf. **71**, 176–183 (2015)
64. Rodríguez Hortalá, M., Requena, G., Seiser, B., Degischer, P., Di Michiel, M., Buslaps, T.: 3D-characterization of continuous fibre reinforced composites. In: ICCM 17—17th International Conference on Composite Materials, Edinburgh, UK
65. Henne, F., Ehard, S., Kollmannsberger, A., Hoeck, B., Sause, M., Drechsler, K.: Thermoplastic in-situ fiber placment for future solid rocket motor casing manufacturing. In: SAMPE Europe SETEC 14—Efficient Composite Solutions to Foster Economic Growth, Tampere, Finland (2014)
66. Sargent, J.P.: Durability studies for aerospace applications using peel and wedge tests. Int. J. Adhes. Adhes. **25**, 247–256 (2005)
67. Martiny, P., Lani, F., Kinloch, A.J., Pardoen, T.: Numerical analysis of the energy contributions in peel tests: a steady-state multilevel finite element approach. Int. J. Adhes. Adhes. **28**, 222–236 (2008)
68. Ferracin, T., Landis, C.M., Delannay, F., Pardoen, T.: On the determination of the cohesive zone properties of an adhesive layer from the analysis of the wedge-peel test. Int. J. Solids Struct. **40**, 2889–2904 (2003)
69. Khan, M.A., Mitschang, P., Schledjewski, R.: Identification of some optimal parameters to achieve higher laminate quality through tape placement process. Adv. Polym. Technol. **29**, 98–111 (2010)
70. Reiser, M.F., Semmler, W., Hricak, H.: Magnetic Resonance Tomography. Springer, Berlin (2007)

71. Kalinina, O., Kumacheva, E.: A "core-shell" approach to producing 3D polymer nanocomposites. Macromolecules **32**, 4122–4129 (1999)
72. Zhang, Y., Ouyang, H., Lim, C.T., Ramakrishna, S., Huang, Z.-M.: Electrospinning of gelatin fibers and gelatin/PCL composite fibrous scaffolds. J. Biomed. Mater. Res. B. Appl. Biomater. **72**, 156–165 (2005)
73. Chestnut, M.H.: Confocal microscopy of colloids. Curr. Opin. Colloid Interface Sci. **2**, 158–161 (1997)
74. Franck, C., Hong, S., Maskarinec, S.A., Tirrell, D.A., Ravichandran, G.: Three-dimensional full-field measurements of large deformations in soft materials using confocal microscopy and digital volume correlation. Exp. Mech. **47**, 427–438 (2007)
75. Germaneau, A., Doumalin, P., Dupré, J.-C.: Comparison between X-ray micro-computed tomography and optical scanning tomography for full 3D strain measurement by digital volume correlation. NDT E Int. **41**, 407–415 (2008)
76. Bay, B.K., Smith, T.S., Fyhrie, D.P., Saad, M.: Digital volume correlation: three-dimensional strain mapping using X-ray tomography. Exp. Mech. **39**, 217–226 (1999)
77. Pan, B., Wu, D., Wang, Z.: Internal displacement and strain measurement using digital volume correlation: a least-squares framework. Meas. Sci. Technol. **23**, 045002 (2012)
78. Leclerc, H., Périé, J.N., Roux, S., Hild, F.: Voxel-scale digital volume correlation. Exp. Mech. **51**, 479–490 (2011)
79. Smith, T.S., Bay, B.K., Rashid, M.M.: Digital volume correlation including rotational degrees of freedom during minimization. Exp. Mech. **42**, 272–278 (2002)
80. Verhulp, E., van Rietbergen, B., Huiskes, R.: A three-dimensional digital image correlation technique for strain measurements in microstructures. J. Biomech. **37**, 1313–1320 (2004)
81. Lee, S.Y., Kim, G.W., Han, B.H., Cho, M.H.: Strain measurement from 3D micro-CT images of a breast-mimicking phantom. Comput. Biol. Med. **41**, 123–130 (2011)
82. Sutton, M.A., Orteu, J.J., Schreier, H.: Image Correlation for Shape, Motion and Deformation Measurements: Basic Concepts,Theory and Applications. Springer, Berlin (2009)
83. Toda, H.: A 3D measurement procedure for internal local crack driving forces via synchrotron X-ray microtomography. Acta Mater. **52**, 1305–1317 (2004)
84. Ferrié, E., Buffière, J.-Y., Ludwig, W., Gravouil, A., Edwards, L.: Fatigue crack propagation: in-situ visualization using X-ray microtomography and 3D simulation using the extended finite element method. Acta Mater. **54**, 1111–1122 (2006)
85. Hoppe, H.: New quadric metric for simplifying meshes with appearance attributes. In: VISUALIZATION'99 Proceedings of the 10th IEEE Visualization 1999 Conference (VIS '99), Washingto, DC, USA (1999)
86. Sause, M.G.R., Zelenyak, A.-M.: Modellierung von Schallemissionsquellen auf Basis von volumetrischen Bildinformationen. In: 20. Kolloquium Schallemission, Garmisch-Partenkirchen, Germany, pp. 1–9 (2015)
87. Guo, Y., Nairn, J.: Calculation of J-integral and stress intensity factors using the material point method. Comput. Model. Eng. Sci. **6**, 295–308 (2004)
88. Guo, Y., Nairn, J.: Three-dimensional dynamic fracture analysis using the material point method. Comput. Model. Eng. Sci. **16**, 11–25 (2006)
89. Nairn, J.: Material point method calculations with explicit cracks. Comput. Model. Eng. Sci. **4**, 649–663 (2003)
90. Guilkey, J.E., Weiss, J.A.: Implicit time integration for the material point method: quantitative and algorithmic comparisons with the finite element method. Int. J. Numer. Methods Eng. **57**, 1323–1338 (2003)
91. Więckowski, Z.: The material point method in large strain engineering problems. Comput. Methods Appl. Mech. Eng. **193**, 4417–4438 (2004)
92. Wang, B., Karuppiah, V., Lu, H., Komanduri, R., Roy, S.: Two-dimensional mixed mode crack simulation using the material point method. Mech. Adv. Mater. Struct. **12**, 471–484 (2005)

93. Tan, H., Nairn, J.A.: Hierarchical, adaptive, material point method for dynamic energy release rate calculations. Comput. Methods Appl. Mech. Eng. **191**, 2123–2137 (2002)
94. Moës, N., Belytschko, T.: Extended finite element method for cohesive crack growth. Eng. Fract. Mech. **69**, 813–833 (2002)
95. Hamstad, M.A., Gillis, P.P.: Effective strain rates in low-speed uniaxial tension tests. Mater. Res. Stand. **6**, 569–573 (1966)
96. Schürmann, H.: Konstruieren mit Faser-Kunststoff-Verbunden. Springer, Berlin (2005)
97. Puck, A., Schürmann, H.: Failure analysis of FRP laminates by means of physically based phenomenological models. Compos. Sci. Technol. **62**, 1633–1662 (2002)
98. Swolfs, Y., Verpoest, I., Gorbatikh, L.: Issues in strength models for unidirectional fibre-reinforced composites related to Weibull distributions, fibre packings and boundary effects. Compos. Sci. Technol. **114**, 42–49 (2015)
99. Swolfs, Y., Gorbatikh, L., Verpoest, I.: Stress concentrations in hybrid unidirectional fibre-reinforced composites with random fibre packings. Compos. Sci. Technol. **85**, 10–16 (2013)
100. Swolfs, Y., Gorbatikh, L., Romanov, V., Orlova, S., Lomov, S.V., Verpoest, I.: Stress concentrations in an impregnated fibre bundle with random fibre packing. Compos. Sci. Technol. **74**, 113–120 (2013)

Chapter 7
Combination of Methods

For the previous chapters on the methods of DIC, AE, EME, and CT the intent was to discuss the scientific and technological basis and to provide factors of influence as relevant in application to composite materials. However, attention was given to the capabilities of the individual methods only. The intent of the present chapter is to demonstrate the synergetic effects that arise by suitable combinations of the methods reviewed intensively in Chaps. 3–6 as well as some additional methods not extensively discussed in the previous chapters. Still the main focus is on the application to in situ detection of failure mechanisms occurring in fiber-reinforced composites under mechanical load.

In this context, some method combinations are well-established and are presented with reference to relevant literature. For other combinations, recent experimental developments have been made, which require a more detailed presentation.

This chapter starts with a comparison of the established methods. Focus is given to the possibilities to distinguish between different failure types as well as to the sensitivity of the individual methods with respect to the size of the damage zone. The important aspect of portability from laboratory environments to field applications is discussed and practical technological scaling aspects as well as physical limitations of some of the methods are elucidated to provide an overview of possible field applications. Subsequently several method combinations and their applications are presented. Since a comprehensive review on this topic would be beyond the scope of this book only representative examples are used to demonstrate the strength of some combinations. Finally this book closes by providing an outlook on currently ongoing developments of the experimental methods that are likely having an impact on further applicability of the methods within the next decades.

M.G.R. Sause, *In Situ Monitoring of Fiber-Reinforced Composites*,
Springer Series in Materials Science 242, DOI 10.1007/978-3-319-30954-5_7

7.1 Comparison of In Situ Methods

In order to select a particular method for in situ testing of a composite material a comparison of the individual capabilities is useful, but quite challenging. Due to the different principles of operation involved in the detection process, the sheer amount of different failure types occurring in a composite material under load and the partially diffuse limits of application inhibit an entirely equivalent judgement of the individual methods. Hence an attempt of comparison is made, which is based on three different aspects used to provide a first overview to the reader.

As first aspect, the suitability of a method to detect a particular failure mechanism as occurring in fiber-reinforced materials is discussed. To this end some generalization of failure mechanisms as extensively discussed in Sect. 2.1 is made.

As second aspect the detection sensitivity of the individual methods for these defects is assessed. This includes some considerations of the absolute detection limits of the methods but also empirical limits that have been obtained from experimental work.

As third aspect, the applicability of the individual methods to field testing is reviewed. In this context "field testing" considers two basic implications. The first implication is the change in size from a material specimen in the range of few centimeters to a range of some meters as realistic for some composite material applications. Second, the implications of outdoor situations such as temperature, noise level, humidity, or portability are covered as opposed to classical indoor situations.

One important aspect covered within this book is the general suitability of the method to detect failure mechanisms in composite materials in situ. In this context the broader meaning of "in situ" is being able to resolve the failure process as a function of loading either online or inline assuring a certain discretization of the load stages until final failure. Moreover, for a downselection of methods, only the aspect of volumetric imaging is considered. Since for the evolution of failure in composite materials, internal damage is as relevant as surface damage, only methods are discussed in the following that have subsurface-sensing capabilities. This basically excludes all sorts of in situ microscopy, since these attempts (despite their high value) are limited to the detection of surface damage.

In order to facilitate the reading of the subsequent sections, the prerequisites of the individual in situ method configurations are first of all defined in order to avoid misunderstandings.

Digital image correlation (DIC): Speckle patterns are applied to relevant surfaces of the investigated part. Images of the speckle pattern are recorded as a function of loading with reasonable high speed (e.g., in the order of Hz for quasi-static load conditions) and sufficient spatial resolution to be able to spot the relevant damage type as discussed in Chap. 3.

Acoustic emission analysis (AE): Acoustic emission sensors with reasonable broad bandwidth of sensitivity are mounted on the investigated part. The spacing and position is chosen adequately so that a sufficient probability of detection for interpretation is reached and source localization is possible with an acceptable localization error. Acquisition settings are chosen appropriately to also allow for source discrimination as described in Chap. 4.

Electromagnetic emission analysis (EME): Electromagnetic emission sensors are placed close to the investigated part with sufficient number to be able to detect the release of EME. Shielding of the test configuration is appropriate to reach adequately low background noise. Acquisition settings are chosen to allow for detection of EME signals following the recommendations of Chap. 5.

Passive Thermography (PTT): The camera system is positioned to monitor the relevant surface of the investigated part. Temperature images of the surface are recorded as a function of loading with reasonable high speed (e.g., in the order of Hz for quasi-static load conditions) and sufficient spatial resolution to spot the heat signatures of the relevant damage type.

Shearography (SH): The shearography probehead is positioned to monitor the relevant surface of the investigated part. Shearography images of the surface are recorded as a function of load at reasonable intervals to track damage evolution. System resolution is assumed to be sufficient to spot the strain signatures of the relevant damage type.

In situ *computed tomography (CT)*: The investigated part is inspected in situ using a load stage concept as described in Chap. 6. Acquisition settings of the CT-scan are chosen adequately to reach sufficient spatial resolution to spot the intended failure.

X-Ray diffraction (XRD): The orientation of the laminate is chosen appropriately to allow detection of the relevant failure types. The part is mounted inside a load stage following considerations of Chap. 6 and imaging parameters are chosen adequately to achieve sufficient spatial resolution to spot the relevant failure type.

Acousto-Ultrasonics (AU): Pulsable acoustic emission sensors with reasonable broad bandwidth are mounted on the investigated part. The spacing and position is chosen adequately so that a sufficient probability of detection for interpretation is reached and source localization is possible with an acceptable localization error and acquisition settings are appropriate to also allow for source discrimination as described in Chap. 4. Further, the density and position of sensors is chosen to allow for detection of the mutual pulses by a sufficient number of other sensors.

As a general rule-of-thumb for the discussions in the following subsections, all methods are assumed to operate under technically optimal conditions. All deviations to this ideal state (i.e., change of position affecting resolution) will be specifically discussed in terms of the detectability, the loss of sensitivity, or the limitations caused by field measurements.

7.1.1 In Situ Capabilities

A first distinction among the methods can be made with respect to their in situ capabilities. The basic difference is the moment in time when the individual methods provide some information on the failure progress inside the inspected part. Distinction can be made between online, time resolved and inline. In this context, online capabilities are understood in a way, that the method is able to provide first results during the duration of the experiment. In contrast, time-resolved means the method is able to track failure evolution in a timewise discrete way, but more or less extensive post-processing of the recorded data is required before information is accessible. As further distinction, the experiment might be carried out in situ, but the method may need some interruption of the load process to record data. The latter approach enables applicability of many other nondestructive testing methods. Within the scope of this book restriction is made to in situ computed tomography as typical example of this class of nondestructive testing methods. A summary of the capabilities of the individual methods is given in Table 7.1.

For DIC, the images of the speckle pattern are recorded during the measurement, which does not interrupt the loading process. But the displacement field information is only readily interpretable after a post-processing step, thus the method cannot be fully considered as online method. This is not true for the application of DIC-techniques used as optical extensometers. These can readily be applied for strain sensing of the specimen, but make use of lower resolution and only regional calculation of the strain components, which renders them useless in the context of defect detection since this requires full-field measurements. For inline purposes the DIC method is readily applicable, given a reference image at zero load is available.

AE is detected during the measurement with high temporal discretization. It can be considered as an online method, since display of signal activity or source localization are standards in commercial software packages providing an almost instant view on the damage progression. Some approaches still need post-processing steps, which reduces the speed of information to a time-resolved

Table 7.1 Categorization of in situ capabilities of methods discussed in the following

Method	Online	Time resolved	Inline
DIC	(x)	x	(x)
AE	x	(x)	–
EME	x	(x)	–
PTT	x	(x)	(x)
SH	–	x	x
CT	–	–	x
XRD	x	(x)	x
AU	x	(x)	x

Note: x: applicable, (x): partially applicable, – : not applicable method in these cases. However, the method requires an active damage progression or active friction in damaged areas to record data, so it cannot be applied to parts subject to constant load without any damage progression or even to dismounted parts.

Similar to AE, the detection of EME requires an active damage progression during the measurement. Thus the method is also considered as an online method, since commercial AE equipment is capable of detecting and processing EME signals in a similar fashion as AE signals. For the time being, the information is limited to activity curves and data acquisition, since no further software packages are currently available to make use of the remaining information of the EME signals. The further categorization follows the considerations of AE signals above.

For PTT the images are recorded during the measurement and can be displayed fully online during the measurement. But the application of filters or background subtractions usually involves offline steps, leaving the method as time-resolved image series. As further advantage, an existing passive thermography setup can easily be modified to turn into an active thermography setup by addition of an external heat source. That way it is possible to detect the presence of damage also inline, e.g., without applied load or during load-hold steps.

A measurement using SH is typically not performed online, since the data processing step requires some computational intensity to turn the recorded data into an image allowing spotting defect positions. As additional benefit, SH is also applicable to inspect parts inline. To this end, SH operates as classical NDT method using an external heat source to generate a small deformation or by applying a small amount of mechanical load directly to the part.

For CT inspection in the aforementioned sense, the only applicable category is inline inspection, since the post-processing time and scan duration is so large, the specimen is required to be kept at a load-hold step. Of course, as classical NDT method, CT is capable of performing scans also for parts without any applied loads.

In contrast to CT-scans, XRD inspection is carried out during the continuous measurement of a part and allows viewing the projection images online. Similar to the other imaging techniques it is still hard to evaluate the image content online, so generally post-processing will yield more detailed information to examine. Because this information may still be linked to the previously applied load this method is fully time resolved. Finally, XRD is also capable of inline measurements, since the X-ray diffraction constitutes an absolute quantity, which can also be evaluated during load holds.

AU generally comes with the same capabilities as AE since it fully includes this method. As further aspects, the active pulsing and detection of signals contribute to the information value and is available online, but also after post-processing. As major distinction to AE, the possibility to actively scan the part under investigation enables the possibility to perform a damage assessment also during load-hold steps or even for a part completely free of external load.

7.1.2 Detectability of Failure Mechanisms

A key aspect of an inspection technology is their ability to sense a particular defect in a material. As pointed out in Sect. 2.1, the range of failure mechanisms observed in fiber-reinforced materials is quite complex, and it is an intuitive conclusion that not all methods will be able to detect all sorts of failure mechanisms. In the present section, focus is given to the aspect whether the method is by concept capable to sense the different failure mechanisms. The detection sensitivity to each mechanism will be discussed in the subsequent section.

Before starting a discussion of the individual method's capabilities, the failure mechanisms reported in Table 7.2 are briefly defined with reference to Sect. 2.1 for a more detailed explanation of each type. All types of inter-fiber failure involving fracture of the matrix material or at the interface between fiber and matrix will be termed matrix cracking. This is regardless of the size of the crack, which can cover a broad range from the order of few micrometers to several centimeters. Although the failure type delamination is technically also a matrix crack, it is still distinguished as a different failure type in Table 7.2, due to its characteristically different orientation. Further distinction is made between the occurrence of single filament failure and fiber bundle failure. While the first considers the occurrence of fiber breakage within few filaments in close spatial distance (i.e., $>100\ \mu m$), the second type considers fiber failure in the form of substantial dimensions like complete rovings or even layerwise failure. Fiber pullout is explicitly understood as debonding between the fiber filaments and the matrix material and is thus expected to reside on the microscale. The phenomenon of fiber bridging ranges in size from the microscale to the mesoscale, therefore spanning several orders of magnitude. Main distinction to matrix cracking is due to the orientation of the crack front being perpendicular to the fiber direction.

Table 7.2 Applicability of individual in situ methods to detect failure mechanisms and anomalies in fiber-reinforced composites

Method	Failure mechanisms						Anomalies	
	Matrix cracking	Delamination	Single fiber filament failure	Fiber bundle failure	Fiber pullout	Fiber bridging	Voids	Undulations
DIC	x	x	no	x	no	?	?	?
AE	x	x	x	x	x	x	no	no
EME	x	x	x	x	?	x	no	no
PTT	x	x	x	x	no	?	(x)	(x)
SH	x	x	no	x	no	x	?	x
CT	x	x	x	x	x	x	x	x
XRD	x	(x)	no	no	no	(x)	x	(x)
AU	x	x	x	x	x	x	?	?

Note: x: applicable, (x): partially applicable, no: not applicable, ?: not validated but theoretically possible

Similarly, specific types of anomalies are of interest to be detected by in situ methods, since they often act as stress concentrator, therefore may induce damage at preferential spots. Among these anomalies, Table 7.2 distinguishes between voids as gaseous inclusions of various size and undulations of the fiber filaments. The latter is understood as waviness of many fiber filaments simultaneously rather than the specific orientation of one particular fiber.

For DIC, the Chap. 3 has demonstrated the broad range of applicability to sense the failure mechanisms marked in Table 7.2. General deficits exist in sensing of very small defect types causing weak strain signatures. Among these are single filament failure and fiber pullout. Hence it is not possible to detect these failure mechanisms when recordings of macroscopic specimens are used. The aspect of fiber bridging has not been explicitly investigated in Chap. 3, so it remains unclear if the method is applicable to this end. However, if the crack is of a certain size, there is no general reason, why the resulting strain signature should not be detectable. For voids and undulations, the applicability has also not been investigated in Chap. 3. In this case it is questionable, whether the presence of these anomalies will cause significant distortion of the strain field so they can be detected by DIC.

As presented throughout Chap. 4, AE is generally able to sense all sorts of active failure types in fiber-reinforced materials. Due to operational principle being based on active damage progression, this restricts detection of voids and undulations.

Similar considerations apply to EME detection, since this also requires an active failure type. However, the occurrence of fiber pullout has not been explicitly investigated by EME so it remains unclear, whether the method is suitable to this end. Given the detectability of the other failure mechanisms, there is no general physical limitation why this should not be detectable. But the current experimental setups still suffer from several drawbacks with respect to the sensitivity when compared to AE (cf. Sect. 5.5).

For PTT imaging, the image resolution and the strength of the heat signature are the relevant quantities to discuss regarding the detectability. As demonstrated in Sect. 7.2, PTT is capable to detect the heat signatures of single filament failure as well as several sorts of matrix cracking, including delamination and possibly fiber bridging. However, the occurrence of failure types with minimal heat release such as fiber pullout is quite unlikely and therefore assumed to be generally not applicable. Given the modification to active thermography as described in the previous subsection, the method is also in principle capable of detecting anomalies like voids or undulations.

For the suitability of SH, similar considerations as for DIC apply. Although the strain sensing capabilities of SH are superior to DIC, there is currently no likely solution to detect single filament failure or fiber pull out by means of SH. For the anomaly types, SH has been reported to be applicable for detection of undulations [1], but there are no indications in literature, that SH testing may be used for the detection of voids.

Using appropriate settings as high-resolution configuration, there is no technical limitation for CT to spot the failure types and anomalies listed in Table 7.2. The general limitation for CT is the detail visibility defined by the chosen measurement configuration and the specimen size as discussed in detail in Chap. 6 and will be emphasized in more detail in the following subsection.

For the measurement setup using XRD, a particular focus is given to the detection of matrix cracks in the form of inter-fiber cracks. Hence this method is specifically developed for that purpose making it insensitive for other failure mechanisms. There is limited applicability to delamination and fiber bridging, in case the orientation of the incident beam is chosen accordingly. In addition, the XRD method provides sensitivity by means of absorption measurements to quantify voids and some limited applications to quantify undulations by means of diffraction pattern analysis.

As a further method considered herein, AU again includes all sensing capabilities of AE, since it fully includes this method. As extension to the aspects mentioned before AU is in principle applicable to detect undulations or voids, since both anomalies will cause a local change of the acoustic properties. However this has not become a standard application, yet.

7.1.3 Detection Sensitivity

Besides the general applicability of an in situ method to detect a particular type of defect, the question of detection sensitivity arises. Certainly some of the presented methods might be able to spot a particular type of damage given some boundary conditions are fulfilled, but might easily fail to detect the same type of damage with smaller dimensions. Thus for practical implications it is worthwhile to compare the methods with respect to their sensitivity to detect a given type of failure mechanism or anomaly. In order to allow a fair comparison, three conditions are assumed in the following:

1. The method generally operates under in situ conditions as discussed in Sect. 7.1.1.
2. The method is subject to ideal operating conditions as outlined in the beginning of Sect. 7.1.
3. The detectable size of the anomaly or failure mechanism is used as a benchmark to compare methods.

Likewise, if the method requires any modifications to the above conditions (e.g., to move from online to time-resolved in order to perform the operation) the sensitivity of the method is generally ranked lower than without any modifications.

For DIC, the possibility to detect failure mechanisms has been discussed intensively in Chap. 3. As concluded from the experimental investigations and the modelling, there is a certain limitation in detection sensitivity as soon as the damage is located deep enough below the surface. Similarly, small defect sizes inhibit detection of particular failure types, since they may easily fall below the detection limit. For some the anomalies listed in Table 7.3 the method might be applicable, but has not been tested so far.

Table 7.3 Detection sensitivity of individual in situ methods for failure mechanisms and anomalies in fiber-reinforced composites

Method	Failure mechanisms						Anomalies	
	Matrix cracking	Delamination	Single filament failure	Fiber bundle failure	Fiber pullout	Fiber bridging	Voids	Undulations
DIC	x	x	no	x	no	(x)	(x)	(x)
AE	xxx	xxx	xx	xxx	xx	xxx	no	no
EME	xx	xx	x	xx	(x)	xx	no	no
PTT	x	x	x	x	no	(x)	x	x
SH	xx	xx	no	xx	no	xx	(x)	x
CT	xx	xx	xx	xx	x	xx	xxx	xx
XRD	xx	x	no	no	no	x	xx	(x)
AU	xxx	xxx	xx	xxx	xx	xxx	(x)	(x)

Note: x: working, xx: sensitive, xxx: very sensitive, (x): not validated but theoretically possible, no: not applicable

The method of AE testing has been investigated for many decades and the current technical limitations have been extensively discussed in Chap. 4. In general, AE has extraordinary benefits in terms of the detection sensitivity, because the sensing systems are able to detect active failure mechanisms covering several orders of magnitude in size (cf. Sect. 4.2). Some limitations occur for failure mechanisms being very small in size (e.g., single filament failure) because of attenuation effects, the respective AE signal may fall below the detection threshold after short distances of wave propagation. For passive defects, such as voids or undulations, AE is not applicable.

Being similar to AE testing, EME is generally able to detect the same types of failure mechanisms as AE. But the distance between sensing system and source position has a strong impact on the detection system, which currently turns the detectability into a unique assessment for each setup. Also, the present detection systems are not standardized or readily available from commercial companies turning their use into a very individual configuration for every experiment. Thus, the method is ranked in a similar fashion as AE testing but with generally lower detection sensitivity. However, as presented in Chap. 5, for a particular configuration the EME systems may show similar detection sensitivity as an AE system.

In PTT imaging, the detection sensitivity is dominated by the resolution of the images and the thermal resolution of the camera and the heat signature intensity as compared to the thermal background. For the camera resolution, the present technology of thermography cameras does not show as high of a resolution as conventional systems. Accordingly, detail visibility is generally less than for imaging methods operating in the visual spectrum. The thermal resolution of current camera systems was found to be suitable to detect transient heat signatures as occurring during loading of fiber-reinforced materials. However, the main challenge arises from the discrimination of these heat signatures relative to the thermal background. First, the material generally heats up during mechanical loading (much more pronounced for cyclic loading). With increasingly strong heat signatures at already damaged zones (i.e., due to internal friction) it becomes impossible to detect superimposed transient heat signatures indicating further failure. Second, the heat signature itself needs to be strong enough to be discriminated against the noisy thermal background that exists for fiber-reinforced materials. For the cases studied in Sect. 7.2.3 it was found, that only very strong acoustic emission sources could be related to such heat signatures. Thus the method may generally be assumed to be less sensitive to a majority of the more microscopic failure mechanisms occurring in the material. For the branch of active thermography however, the method has some sensitivity to detect the occurrence of voids or the local change in heat flux due to undulations, but again is not the most sensitive method among the list in Table 7.3.

The detection of failure mechanisms using SH has seen substantial applications in conventional nondestructive testing. Since the method is readily transferable to in situ applications, some of the general conclusions on detection sensitivity of SH testing are still valid. The method is well suited to detect these failure mechanisms which are indicated by strong out-of-plane deformation signatures. In contrast to DIC, the method generally benefits from a higher detection sensitivity in these cases. But for failure mechanism being small in size or located below the surface, the same limitations arise. Here the method in its present technological state is not assumed to be suitable to detect the occurrence of single filament failure or likewise dimensions of the other failure mechanisms. For the anomalies, SH testing has seen some applications to visualize undulations with certain sensitivity [1] but it is questionable, whether the systems are ready to detect the presence of voids.

Volumetric imaging using CT generally benefits from a high level of detail visibility. Thus, given suitable scan duration and sufficiently small volumes of the material, the method is able to detect all failure mechanisms and anomalies with very high detection sensitivity (cf. Sect. 6.2). In the context of in situ testing, general drawbacks originate from the lack of full online capabilities and the limitations in scanned volume. Whenever the limits of the ideal operating conditions are approached, the method loses detection sensitivity to small failure mechanisms and anomalies very quickly. Thus the overall ranking in Table 7.3 is lower than it would be for a CT system operating in standard scan operations, i.e., without the specific implications of the in situ mode with superimposed mechanical or thermal loads.

For XRD only a subset of the failure mechanisms listed in Table 7.3 are detectable. The method has been specifically adapted to detect the occurrence of matrix cracking or delamination (and also fiber bridging if large enough) and shows reasonable sensitivity for these cases. Due to the nature of diffraction techniques, very precise alignment of the apparatus and the specimen is required with strong reduction in the detection capabilities when the limits of the ideal operating conditions are approached. For detection of voids, XRD generally has a high sensitivity because of the change in X-ray attenuation. Similarly the change in fiber orientation due to undulation causes noticeable changes in the diffraction spectrum and could be used to detect the presence of undulations to a certain extent.

AU testing extends the detection sensitivity of the AE method to detect the presence of anomalies. Since the presence of voids or undulations generally may influence the guided wave propagation, the method may show some sensitivity to detect the presence of either of these defects. Although the interaction of guided waves with internal damage as discussed in Sect. 4.3 may result in measurable quantities it is certainly not most sensitive to the occurrence of these two types of anomalies. Hence, it currently remains an open question, whether the AU method is able to sense these anomalies with reasonable accuracy.

7.1.4 Extension of Methods to Large Scale and Field Testing

One further challenge for all test methods is the harsh change in operating conditions when moving from a laboratory environment to an application in the field. In the context of the in situ testing approach followed within this book, the term "field testing" needs some further specification prior to the discussion.

In the following field, testing is understood to apply to all kinds of testing applications that leave the well-defined operating conditions as outlined in the beginning of Sect. 7.1. This obviously includes all sorts of outside inspection of fiber-reinforced material parts and thus includes all aspects of harsh climates or remote inspection sites. But for many methods, strong limitations arise long before reaching these rather severe inspection conditions. Many methods suffer from strong drawbacks, when the size of the part to be inspected is already larger than a few centimeters, but the part being tested still resides within a laboratory with standardized climate. Other limitations arise due to typical noise sources as being present for large-scale testing or outside conditions. Further, some specific limitations, which may or may not exist within a laboratory environment or in outside conditions will be discussed for each method. Consequently, the aim of the present section is to discuss the key limitations of each of the methods in their ability to inspect larger parts made from fiber-reinforced materials, potentially in an outside inspection scenario as summarized in Table 7.4.

The method of DIC is generally applicable for field-testing applications, since the experimental setup basically consists of a camera system recording a speckle-type pattern. Challenges in terms of failure detection arise from the loss of sensitivity, when investigating larger objects. As pointed out in Sect. 3.2, the resolution of DIC depends on the field of view which in turn defines the image resolution. Thus monitoring of a larger object with only one camera system causes a strong loss of detection sensitivity when compared to monitoring of a smaller object. As a consequence, one could use multiple camera systems, or if attempts are made to solely measure the macroscopic displacement and deformation of a structural part, a full monitoring approach by one camera system may still be feasible. But many of the other analysis types proposed in Chap. 3 will lose their significance beyond a certain spatial resolution. Therefore, a scale-up to monitor failure types as outlined in Sect. 3.3 is not assumed to be successful beyond an imaged object size

Table 7.4 Applicability of individual in situ methods in the field

Method	Applicability	Sensitivity	External noise sources	Other relevant limitations
DIC	x	Scales with field of view	Lighting conditions, structure movement	Visibility of specific defects not feasible beyond a certain resolution, large amount of data for long-term monitoring
AE	x	Density of network	Extraneous noise, vibrations and EMI signals	Acoustic signal spreading, attenuation, signal interpretation
EME	?	Density of network	EMI signals, vibrations	Housing in shielding chamber
PTT	x	Scales with field of view	Transient or constant heat sources, turbulences	Visibility of specific defects not feasible beyond a certain resolution, large amount of data for long-term monitoring
SH	x	Constant for one system	Vibrations, heat sources	Laser handling
CT	(x)	Scales with field of view	Vibrations, heat sources	Closed system, X-Ray handling, large amount of data for long-term monitoring
XRD	?	Scales with field of view	Vibrations	Geometric alignment, X-Ray handling
AU	x	Density of network	Extraneous noise, vibrations and EMI	Acoustic signal spreading, attenuation, change of material properties and dimension, signal interpretation

Note: x: applicable, (x): partially applicable, ?: not validated but theoretically possible

of few cm^2. If critical spots in structural parts are known, one attempt is to directly observe these areas instead of monitoring the whole part.

Among the external noise sources, a sudden or periodic change in lighting conditions may be named first. This happens frequently in outside conditions, but may also happen in laboratory environments if no constant lighting conditions (e.g., closing sun-flaps) can be assured during the test. The consequence may be useless sequences of images or falsified quantitative displacement and strain information. As second important noise source, an absolute movement or vibration of the structure may occur. Since the cameras field of view is usually fixed, a large displacement or rotation of a structure might move the relevant part to be imaged outside the cameras field of view. In particular for large-scale testing, some parts may undergo strong deformation before the first occurrence of detectable damage. Thus the camera position may need to be adjusted previously, so the relevant part of the structure is within the field of view after reaching this critical deformation stage. Finally, one general challenge in long-term testing is the amount of data collected. At constant image size, the major impact on the amount of data collected is the acquisition rate. For a long-term monitoring operation, the acquisition rates may easily fall in the ranges below 0.1 Hz causing less chance to track the details of damage evolution within the monitored structure.

AE testing has a long tradition as a method being applied in the field to monitor fiber-reinforced parts [2]. This is due to several beneficial aspects as has been reviewed so far. The loss of sensitivity of one AE sensor may be compensated by adding more sensors to the same structure until a certain density of the AE sensor network is reached. In modern equipment it is possible to record signals from several channels simultaneously, thus the sensitivity could in principle be kept at a similar level as in laboratory testing, given the operator is willing to mount the required number of sensors to the structure.

In large-scale testing and in outside monitoring applications a variety of external noise sources may occur which will reduce the effective sensitivity, since the acquisition settings may need to be modified. In this context, the occurrence of extraneous acoustic noise is the most relevant source. In some test configurations, the acoustic noise floor may easily render any attempts by AE monitoring useless. Potential reasons are manifold, but among the typical sources are all sorts of friction (e.g., between inspected part and supports or fixtures), valves or other parts of hydraulic equipment as well as rain drops, snow, or hail. If technically possible, the potential noise sources may be removed from the structural part (e.g., mounting valves offset to the structure) or lubricants may be used to minimize stick–slip friction. However some of the noise sources may not be eliminated leaving the operator with the only chance to decrease the sensitivity of the equipment, so to acquire only those AE signals being stronger than the background noise. As a further type of external noise source, the equipment may detect electromagnetically induced (EMI) signals. These may originate from any electrical devices operating nearby and are typically detected because of insufficient shielding of the AE measurement chain. An improvement of the electrical shielding of the measurement equipment or spatial decoupling of the potential noise source may be used to avoid detection of EMI signals. Beyond the external noise sources, some further limitations of the AE method arise when moving from small-scale testing to large-scale testing. First, the aspects of signal propagation in larger structures need to be carefully considered. Thus either the same source-sensor distance as in a small-scale test is required, or wave propagation and attenuation will have an adverse effect on the information included in the detected AE signals. Also, the wave propagation in structural parts may generally be more complicated as shown in the simple cases of Sect. 4.3. Therefore a careful revision of the approaches to perform source localization and source classification are required when being applied to a test object of larger size.

The method of EME testing has so far not seen any application to testing of larger parts or outside applications. Instead, the method is still under development and is hence discussed with an optimistic view on potential application in such scenarios. Similar to AE testing, the sensitivity of the method is a function of the number of sensors applied during the test. Therefore the same limitations in terms of the network density arise as for AE systems. However, since the overall sensitivity of EME sensors is much less than for commercial AE sensors and the maximum source-sensor distance is still very small (very few centimeters) a comparatively dense sensor network would be required for monitoring of a larger part. Further, a multitude of external noise sources exist. As demonstrated in Sect. 5.4, one of the key requirements to perform EME measurements is to operate within a shielding chamber. This is to reduce the electromagnetic background to a level, where EME signatures are not fully hidden and are thus detectable. The scaling of such shielding chambers as presented in Chap. 5 to test scenarios of larger parts is generally challenging, but technically possible. Commercially available walk-in shielding chambers have been tested with present EME equipment and were found to show sufficient shielding capabilities from external electromagnetic background. Another potential challenge in such testing scenarios is seen by structural vibrations of the test rig or the structure. Since the EME signals are generated by movement of charges, there is no distinction possible between charge movements due to active defects (e.g., vibration of the crack wall) or due to the vibration of the material itself. It may even be expected that the macroscopic movement or vibration of the structure will form the dominant EME signature and therefore may cover the EME signatures resulting from damage formation.

Testing with PTT is done in a variety of field applications, so there is general applicability of the method despite of some general limitations of the equipment concerning outside applications. As being an imaging technique, for the sensitivity of the method the same considerations as outlined for DIC apply. Thus, the sensitivity to detect details will suffer from an increasing field of view. In addition to these limitations, the temperature information may also be subject to changes when moving away from the test object. Any turbulence in the air may cause visible artifacts and may cause difficulties to spot tiny heat signatures as relevant to damage detection. However, the overall loss of sensitivity due to scaling is not as severe as for DIC, so this imaging technique may allow monitoring of larger parts, while keeping reasonable detection sensitivity.

Among the external noise sources the occurrence of additional heat sources is the most severe. If the part to be tested is exposed to sunshine or if the mechanical actuator dissipates its heat to the tested part then additional heat backgrounds are superimposed and may mask transient heat signatures. Also artificial lights may cause reflections visible in the thermal images and may cause zones of the part not able to be inspected. The latter is even more relevant if the inspected parts are subject to larger displacements or rotations, since this might cause additional movement of reflection spots on the structure. Especially for outside applications, movement of the air between the test object and the camera system as well as the local cooling due to the air flow may cause additional artifacts in the camera images. Some of these effects may be avoided by shielding the test object against any external heat source and moving artificial light sources to avoid reflections being visible within the camera images. For monitoring applications the image acquisition rate needs to be considered further, since this is the key quantity to define the overall amount of data being recorded.

As a further method to be considered for field applications, SH testing is among the methods being frequently applied in such conditions. The systems are typically used as handheld devices able to scan several cm^2 at once or as specific large-scale systems to allow inspection of larger structures at once. The general benefit of these systems is that their detection sensitivity is constant for one system, since this is more or less fixed in its distance to the test object. Among the various external noise sources, the occurrence of any superimposed displacement is one of the limitations of the systems. Thus additional vibrations or transient heat sources causing deformation due to thermal expansion may need specific considerations of the operator to distinguish image artifacts from defect indications. An additional aspect to be considered for the operation of shearography devices is the protective measures required to allow laser operation. Since the interference of laser-based speckle patterns forms the basis of this method, potentially hazardous reflections of the projected area need to be shielded to avoid any health risks.

Considering CT for a field application comes with some conceptual barriers that need to be overcome. For currently existing systems, the size of the scanned object is limited by the size of the CT-scanner. While nowadays some large-scale scanners are developed which allow scanning of objects in the meters range, the general principle of shadow microscopy is not overcome by those attempts. Thus one always needs to consider the trade-off between the scanned volume and the level of detail to be detected. Also, the application of in situ stages is an emerging field and so far no test rigs have been reported in literature, which are located beyond the range of laboratory test rigs (i.e., 250 kN machines). However, there is no existing limitation in extending the in situ test concept to even larger scales. Among the limitations of the method as far as external noise sources are concerned, a majority of the typical image artifacts have been shown in Sect. 6.2. Specifically for a large-scale test, the lack of penetration of the test object as well as geometry-related image artifacts are likely to occur. Further, motion of the test object may originate from structural vibrations or thermal expansion due to localized heat sources. Most of these effects (including stress-relaxation) will cause visible artifacts and therefore reduce the visibility of details in the volumetric images. Since CT using X-rays is the current technological standard, suitable protective measures are required. X-ray CT-scanners usually operate in sealed rooms or chambers to avoid any hazardous X-ray dose for the operator. Any testing equipment and the test object itself needs to be contained inside such a space. This obviously limits any attempts to perform such operation in remote site inspection scenarios. Another relevant limitation is the amount of data being recorded for one volumetric scan. This amount of data is much more challenging than for the 2D imaging methods discussed so far. Since one volume scan typically consists of several hundred to thousand projection images, the data amount increases accordingly for just one volumetric image. On the positive side the amount of volumetric images taken during one experiment is still comparatively low due to the long scan durations.

Although the principle of XRD has been known for decades, there have been no substantial attempts to bring the method into a large-scale or field-testing application. In contrast to conventional X-ray inspection, which is used in remote inspection, the major drawback is the precise alignment of XRD device and specimen. Whilst this is also a requirement for X-ray-based CT application, the applications using XRD in combination with inspection of fiber-reinforced materials may seem to be too limited to enforce further development of the technique beyond the laboratory scale. This is certainly to a large portion also caused by the required protective measures to avoid exposure to X-rays and thus causes similar limitations as discussed for CT-scanning.

The method of AU has already found its way into field applications as it complements the AE technology by an interesting further aspect. As such, the inspection of fiber-reinforced materials by guided waves alone is also carried out frequently and comes with a couple of further challenges in addition to those already mentioned in the AE description above. For the aspect of guided wave inspection alone, the density of the sensor network will also define the overall inspection sensitivity of the method. If the distance between sensors is chosen too large, the guided waves excited by one actuator may not be detected by another sensor. However, the average distance from actuator to sensor is allowed to be much larger than for AE operation alone, since the excited guided waves are typically of much larger amplitude and hence may propagate much longer distances before falling below the detection threshold.

Because the mode of propagation in fiber-reinforced materials in AE and AU is typically a guided wave, the same limitations in terms of wave propagation apply as discussed previously. Hence, the changes in signatures of the guided waves may need careful revision when brought from geometrically simple structures (such as material test specimens) to more complex structural parts. For operation in outside conditions, additional factors, such as changes in material properties due to temperature changes as well as dimensional changes due to thermal expansion or mechanical loads may need to be considered.

7.2 Established Method Combinations

While the previous section had a strong focus on the strength and weakness of each individual method, one clear strategy is to combine methods in order to reduce some of the weaknesses of the single method. To this end, some particular method combinations are discussed in the following subsections to provide an overview of established combinations and to demonstrate their application to fiber-reinforced composites. In general, the fusion of information obtained by individual methods has become an interesting strategy to yield improved insight into the processes occurring in materials during failure [3–7].

7.2.1 Imaging Methods and Acoustic Emission

One conceptual drawback of acoustic emission analysis is the lack of imaging capabilities with respect to defects not being active. Thus acoustic emission signals only appear as a transient type of information at the moment, when the source is active. Using sensor arrays and localization algorithms as presented in Sect. 4.6 the detected acoustic emission signals can be used to calculate the spatial location of the source. Depending on the number of sensors, this may result in 1D, 2D, or 3D source coordinates, which visualize the damage progression. However, the localization result itself is often difficult to interpret without an additional visual observation of the specimen.

Imaging methods benefit from their ability to capture damage progression in a visual sense, but often lack sensitivity to detect the occurrence of damage and may not cover several orders of length scale in observation. Therefore skilled decision needs to be made, when selecting the type of resolution and the field of view. But the combination of both methods is able to compensate some of the drawbacks associated with the individual methods.

Consider the example of a tensile test of a unidirectional T700/PPS material seen in Fig. 7.1. Here, the localized source position in x-direction between the two acoustic emission sensors is shown as false-color density scale in addition to the simultaneously recorded images of the tensile test. Due to the sensor arrangement, no 2D source localization is possible in this test configuration. As consequence, the source positions may not be attributed to particular (x, y) coordinates using solely acoustic emission measurement. The corresponding images reveal the occurrence of fiber filament and fiber bundle failure at distinct load levels. Simultaneously recorded acoustic emission signals localized at the respective x-coordinate of the observed failure can therefore be attributed to these particular locations.

Also, the ability to have the full image of the specimen in its present damaged state as reference provides better understanding of the formation and densification of acoustic emission source positions at specific spots. Hence, weak spots in the material may be spotted by acoustic emission analysis very early and may be

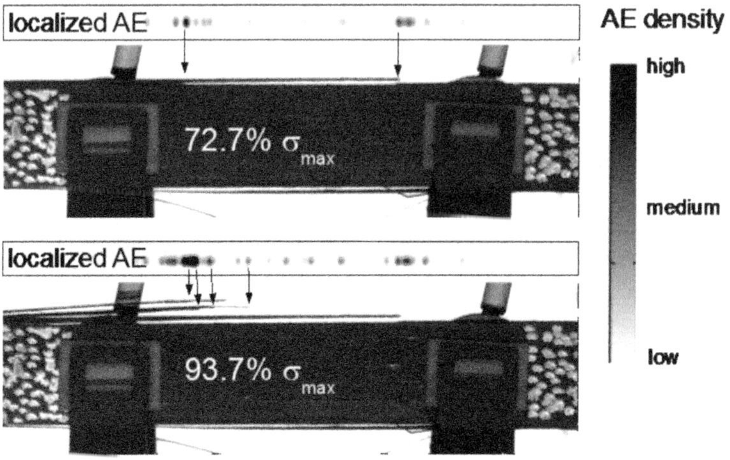

Fig. 7.1 Localized AE source positions during tensile test of unidirectional T700/PPS specimen synchronized with camera images at two distinct load levels

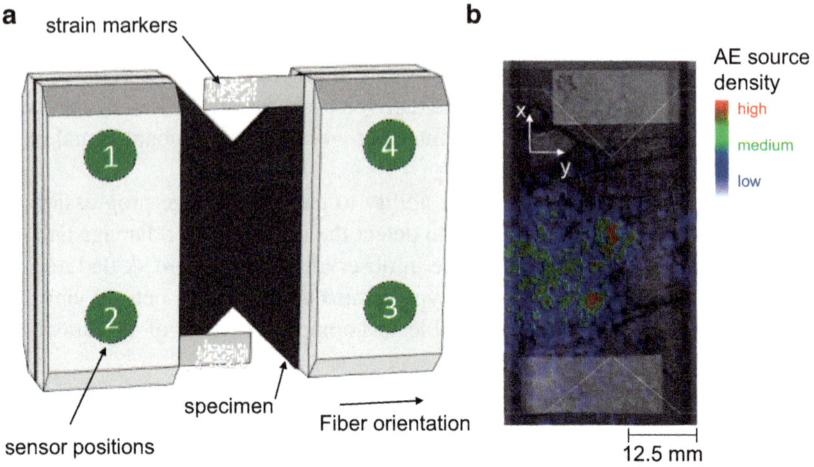

Fig. 7.2 Localized AE source positions during V-notched rail shear test of unidirectional T700/PPS specimen superimposed to camera images

interpreted as formation of the according damage visible in the recorded images at later stages of the experiment.

Another example of combination between acoustic emission measurements and imaging is shown in Fig. 7.2. Here V-notched rail shear testing of an unidirectional T700/PPS laminate was monitored by four acoustic emission sensors to allow source localization in 2D as seen in Fig. 7.2a. The experiment was carried out inside a temperature chamber at 160 °C and images of the specimen were taken at discrete time intervals to allow visual monitoring of the damage progression. As seen by the superposition of localized (x, y) source positions to the deformed specimen, this allows visualizing the damage progression within the material. As seen from the image in Fig. 7.2b, the majority of the acoustic emission sources start to develop on one side of the specimen and shift in position in accordance with the specimen deformation, the latter being caused by severe inter-ply failure and several further failure mechanisms occurring at the notch positions.

Such superposition of the acoustic emission source localization on recorded images of the test structure was found to be of particular use for larger test objects. Here structurally weak spots are easy to identify by early onsets of acoustic emission signals and are much easier to correlate to the relevant part of the geometry when put into the same coordinate system and merged together to yield one sequence of images to interpret as a function of load.

Another natural application of the combination between imaging and acoustic emission is the field of micromechanics. Specifically developed micromechanical experiments allow a direct microscopic observation of the acoustic emission source. As recently presented in [8], this allows to generate particular source types and allows to benchmark modelling approaches in their ability to predict acoustic emission signals in a quantitative way. As seen in Fig. 7.3, the core of the

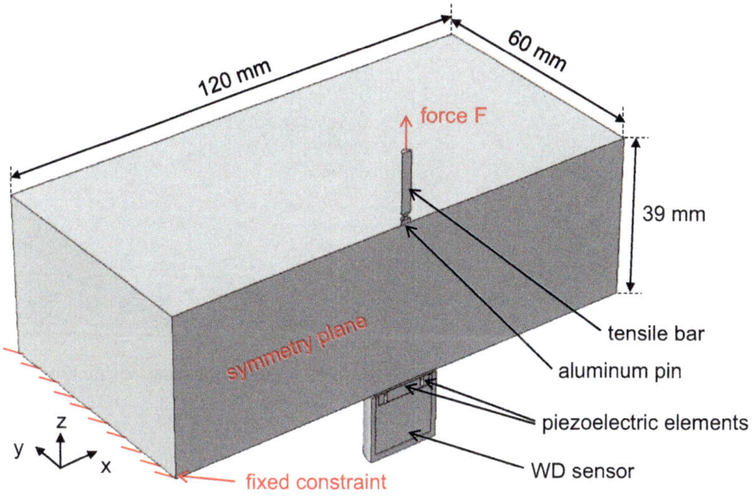

Fig. 7.3 3D-model of the experimental setup including cross-sectional view of acoustic emission sensor used (reprinted from [8])

micromechanics test stage consists of an aluminum block with dimensions of 120 mm × 120 mm × 39 mm (length × width × height). At the center of the top side of the aluminum block, a pin with 2 mm diameter and 1 mm height was machined out of the block. At the opposite side of the block, an acoustic emission sensor is positioned at the center. In order to ensure reproducible mounting conditions, a spring system to firmly press the sensor to the aluminum block is used (not shown in Fig. 7.3).

For the presented cases, the first part of the preparation is to apply a small droplet of epoxy resin on top of the aluminum pin. To facilitate the positioning of the epoxy resin droplet a small depression was machined into the end of the aluminum pin. To prepare the micromechanical stage for generation of a matrix crack a tensile bar made from polyether ether ketone (PEEK) of 2 mm diameter and 80 mm length is used. The tensile bar is first moved into the liquid epoxy resin and then retracted to yield a tapered contour of 380–800 μm (see Fig. 7.4). Subsequently, curing of the liquid resin is carried out using heating foils attached to the aluminum block and an additional heating sleeve wrapped around a small cylinder covering the aluminum pin and parts of the tensile bar. A curing cycle comprising a heating rate of 2 °C/min up to the curing temperature of 180 °C is chosen. The curing temperature is kept constant for 150 min with subsequent cooling to room temperature at a rate between 0.5 and 2 °C/min. Because of its low thermal conductivity, the tensile bar made from PEEK minimizes the dissipative heat flux and therefore assures a constant temperature of the resin during curing. The tensile bar and the aluminum block are mounted in a universal test machine so that thermal expansion of the components and chemical shrinkage of the resin can be compensated by a closed loop force control. The test machine control adjusts the tensile bar position to guarantee zero

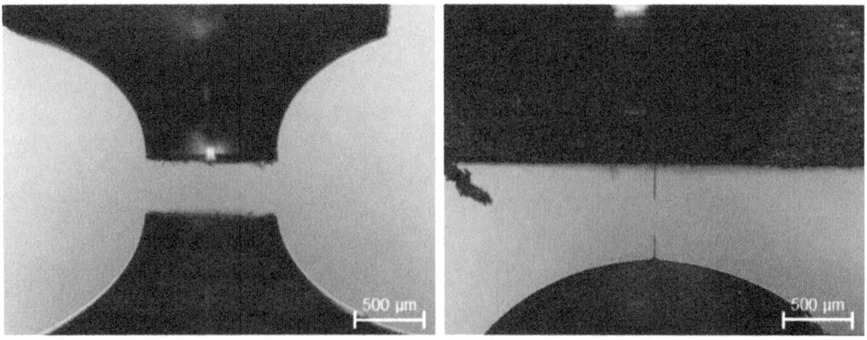

Fig. 7.4 Microscopy images of the failure mechanisms investigated: matrix cracking (*left*) and fiber breakage (*right*) (reprinted from [8])

Table 7.5 Acquisition settings used for micromechanics experiments

Acquisition settings	Value
Preamplification	20 dB_{AE}/40 dB_{AE}
Threshold	45 dB_{AE}
Triggering	Individual channel
Acquisition rate	40 MS/s
Band-pass range	20 kHz to 1 MHz
Couplant	Apiezon-L grease
Mounting system	Spring loaded
Sensor type	WD
Number of sensors	1

force acting during curing, which is necessary to avoid excessive forces acting on the test specimen, potentially causing preliminary failure due to thermal expansion and cure shrinkage of the resin.

To prepare fiber breakage, a two-component epoxy is used to bond a HTA carbon fiber to the end of a flat-topped tensile bar made from PEEK. The fiber is then moved into the resin droplet utilizing a micrometer stage and an optical microscope. The free fiber length was between 350 and 450 μm. The embed length was chosen larger than 100 μm to reach fiber breakage before fiber pullout occurs. After embedding, the resin droplet is cured in situ.

After preparation of the test geometry, the universal test machine applies a tensile force using a displacement controlled mode with velocities dependent on the selected failure mechanism. A velocity of 20 μm/min for fiber breakage and 50 μm/min for matrix cracking was found to be suitable. The occurrence of failure was monitored by an optical microscope with a magnification factor of 100. The images obtained after failure are shown in Fig. 7.4 for the two failure types. The respective acoustic emission acquisition settings are reported in Table 7.5.

Having such direct microscopic observation of the failure mechanism allows performing numerical simulation following the approach reported in [8], which has been outlined in Sect. 4.2. The modelled signals are the result from a full 3D computation including piezoelectric conversion within the sensor model along with application of a P-Spice circuit simulation and subsequent band-pass filtering. After amplification of the modelled signals by 20 or 40 dB_{AE} in accordance with the experimental settings, this allows for direct comparison of the acoustic emission signal amplitudes in the voltage scale.

For the case of fiber breakage, the comparison is found in Fig. 7.5a, b. As seen from the voltage scale and the time-frequency signature given in the Choi-Williams distribution there is very good agreement between the modelled and the

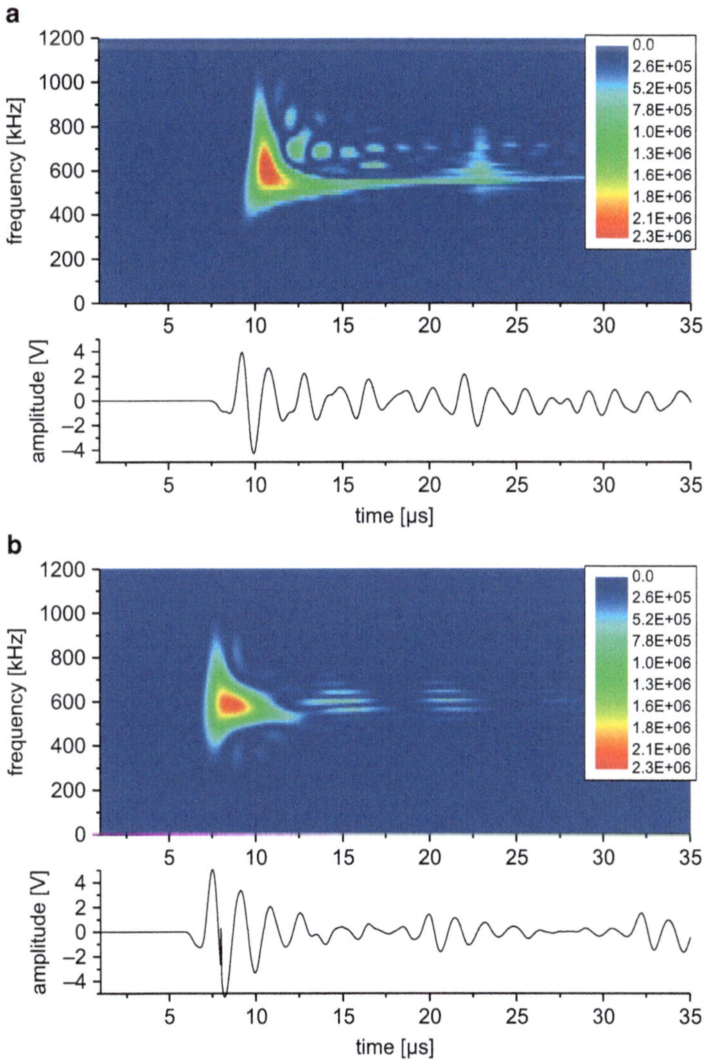

Fig. 7.5 Comparison between simulated (**a**), (**c**), and experimental (**b**), (**d**) results of fiber breakage and matrix cracking, respectively (reprinted from [8])

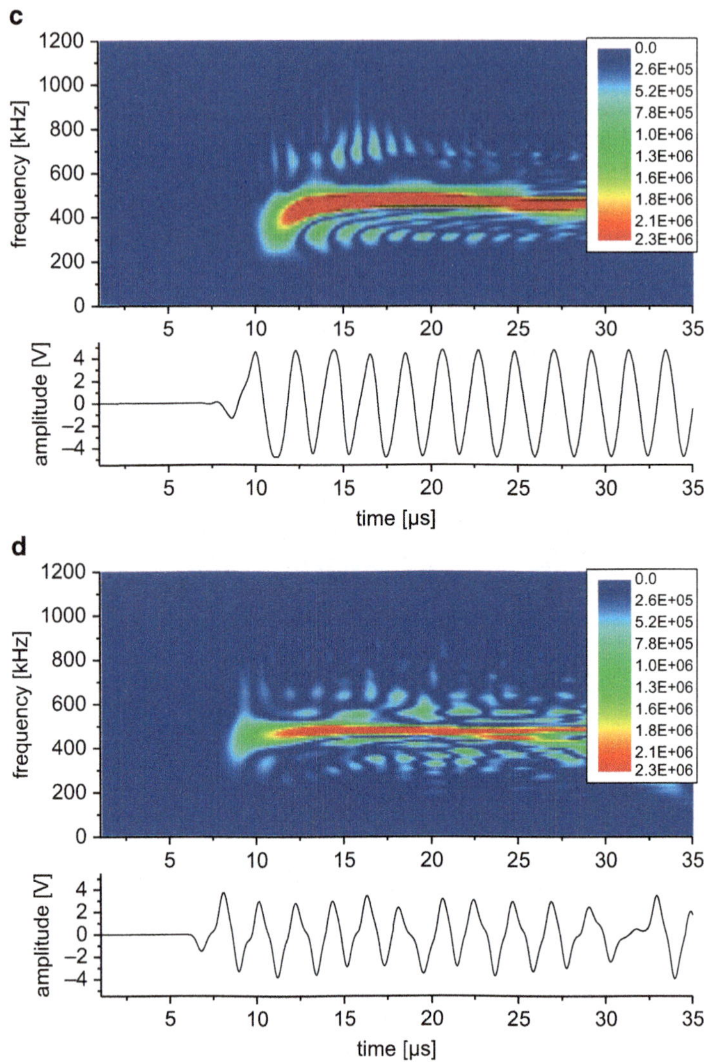

Fig. 7.5 (continued)

experimental signal. In particular, the signal amplitudes show almost identical peak values and the echoes of the initial pulse are adequately captured.

Also for the matrix cracking case shown in Fig. 7.5c, d there is very good agreement in the voltage scales and the time-frequency signature seen in the Choi-Williams distribution. Slight differences arise in the modelled signal after $t = 10$ μs. This is due to the repetitive approach-retract cycles of the newly formed fracture surface. In the modelling part, those fracture planes are partially restricted in their relative motion because of the selected symmetry plane. Therefore their dominant movement direction is in the z-direction. In the experimental part, the fracture surface might experience additional sliding and torsional motions as well as additional interlocking of rough surface parts, which may account for the smoother appearance of the spikes in the modelled signals.

The combination of imaging methods and acoustic emission analysis allows to visually examine particular source types and to directly correlate the respective AE signals. Since imaging methods in this sense are restricted to surface observations, their results may be error-prone when compared to a volumetric analysis method such as acoustic emission. But for many applications, the correspondence is unique and allows safe concluding on the occurrence of particular failure modes in fiber-reinforced materials.

7.2.2 Digital Image Correlation and Acoustic Emission

As an extension of the combination of information outlined in the previous section, the results of digital image correlation may also be merged with results from acoustic emission measurements. Similar to simple imaging of the test object, this concept has recently emerged and has already been applied for various configurations and materials [3, 4, 9–13]. When compared to simple imaging of the specimen, DIC keeps the beneficial aspect of being an imaging technique, but certainly facilitates the detection of damage locations due to visible strain concentrations as described in Sects. 3.3 and 3.4. Beyond the initial comparison made in Sect. 4.7.2.3, the following example is used to elucidate the link between both methods.

To this end, tensile tests of the Sigrafil CE1250-230-39 cross-ply specimens of Sect. 4.7.3.2 were simultaneously monitored by acoustic emission and DIC using the acquisition parameters of Tables 4.2 and 3.7, respectively. Here the focus is on the formation and growth of inter-fiber cracks in the 90° plies of the $[0/90]_{sym}$, $[(0/90)_2]_{sym}$ and $[(0/90)_3]_{sym}$ stacking sequences. Based on the considerations in Chap. 3 such transverse cracks should be detectable to a certain extent below the specimen surface. Hence a comparison between the localized AE sources as function of macroscopic stress level and the DIC results is made in Figs. 7.6, 7.7, and 7.8.

The left side of Fig. 7.6 presents the localized AE sources identified as matrix cracking as function of the specimen stress. The right side shows the evaluation results of identified positions of transverse cracks from DIC measurements shown as horizontal black lines. These are reported integrated over time, i.e., all identified transverse crack positions until final failure of the specimens are marked. The DIC image shows the field of the first principal strain in the last image prior to ultimate failure of the specimen. Clearly some of the transverse cracks are still visible by the strain concentrations found in the DIC image. However, some of the signatures are hard to spot, since the occurrence of further transverse cracks in close neighborhood smears out the signatures or starts to dominate the strain scales.

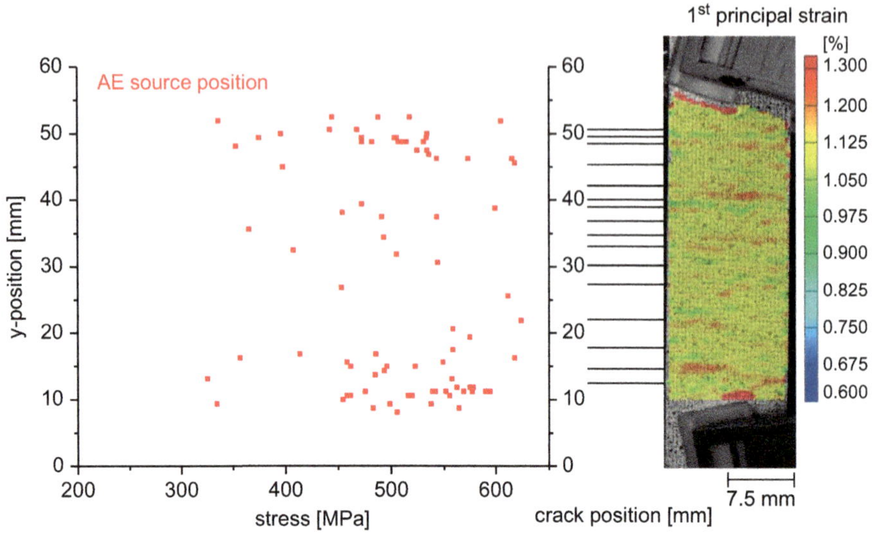

Fig. 7.6 Localized AE source positions of matrix cracking in $[0/90]_{sym}$ specimen as function of the specimen stress level (*left*) and DIC results of localized strain concentrations (*black lines* indicate *y*-position) with snapshot of strain field of first principal strain at last image prior to failure (*right*)

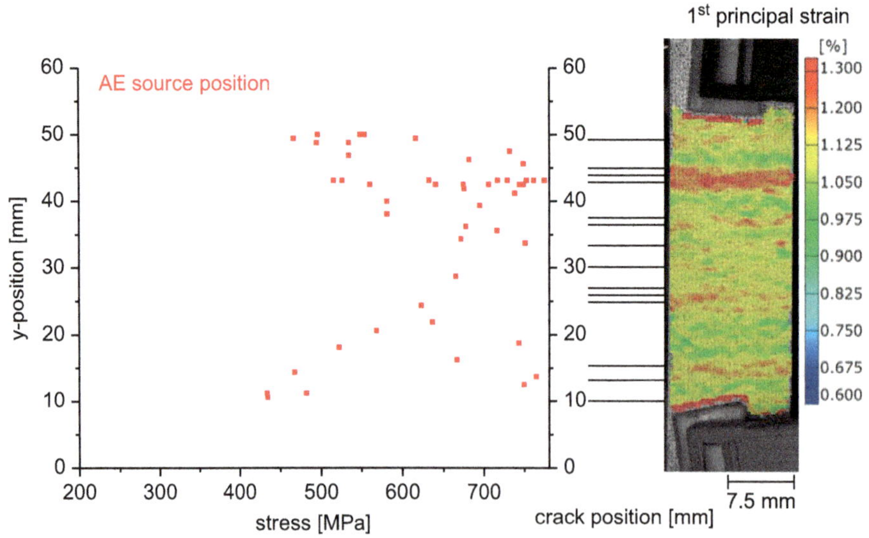

Fig. 7.7 Localized AE source positions of matrix cracking in $[(0/90)_2]_{sym}$ specimen as function of the specimen stress level (*left*) and DIC results of localized strain concentrations (*black lines* indicate *y*-position) with snapshot of strain field of first principal strain at last image prior to failure (*right*)

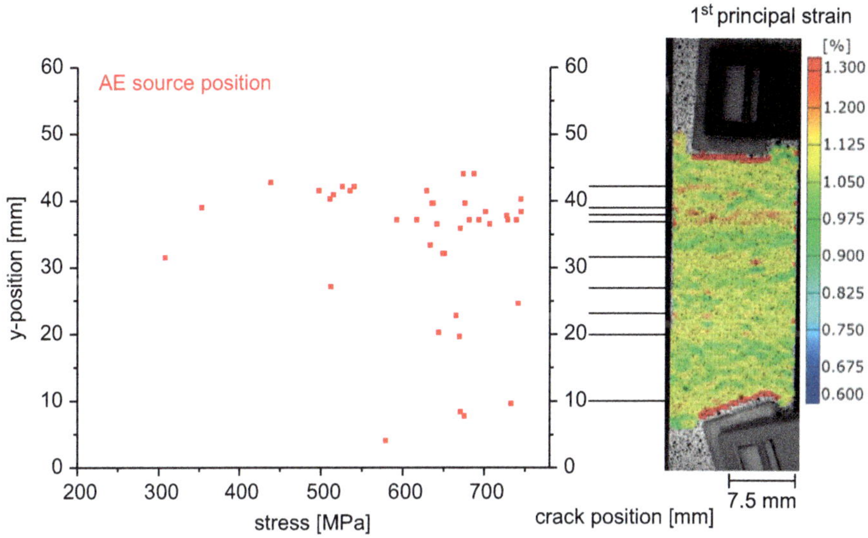

Fig. 7.8 Localized AE source positions of matrix cracking in $[(0/90)_3]_{sym}$ specimen as function of the specimen stress level (*left*) and DIC results of localized strain concentrations (*black lines* indicate *y*-position) with snapshot of strain field of first principal strain at last image prior to failure (*right*)

For the three different stacking sequences shown in Figs. 7.6, 7.7, and 7.8, good agreement was found in the observed positions of transverse crack growth in AE and DIC. Considering that the source localization accuracy of the AE signals suffers from the growing number of cracks acting as an internal obstacle (cf. also Sect. 4.7), the occurrence of the signatures in DIC indicating the occurrence of transverse crack growth coincides well with the occurrence of the respective AE source locations. This allows tracking of the characteristic concentration of transverse cracks as seen in the examples of the $[(0/90)_2]_{sym}$ and the $[(0/90)_3]_{sym}$ stacking sequence.

Based on the conclusions in Sects. 3.3 and 4.2, acoustic emission measurements show superior sensitivity to detect failure mechanisms. However, for the present case only one-dimensional localization was possible using acoustic emission, whereas the DIC results provide full-field 2D information of the size of the defects. For a reasonable field of view, the DIC measurements provide significantly better spatial representations of failure locations and allow easy tracking of their growth.

In summary, the combination of information of both methods substantially increases the reliability of the drawn conclusions. While AE measurements still serve as benchmark in terms of detection sensitivity, the DIC signatures add full-field information and are easier to interpret in terms of their significance, once conforming AE source positions are found at the same spot.

7.2.3 Thermography and Acoustic Emission

In order to complement acoustic emission measurements by a visual method, still other options exist. One approach which has been reported in literature recently is the use of thermography in combination with acoustic emission monitoring [3, 7]. To this end, two approaches exist, that can be employed in practice:

1. Load-hold experiments in combination with active thermography to perform classical NDT inspection.
2. Passive monitoring of the experiment using thermography camera.

The first option is highly valuable in terms of the derived information, such as visual representation of damaged zones and the possibility to track and locate damage as an independent method. However, using the state-of-the-art equipment, the second option comes with much more benefits as to merge information of AE measurements and thermography since it constitutes a similar in situ approach with high temporal resolution.

To demonstrate the high synergy between the two methods and to provide an estimate to which level thermography and acoustic emission may supplement each other, two representative examples are provided in the following. As first example, results of an open-hole tensile test using a Sigrafil CE1250-230-39 cross-ply specimens as described in Sect. 4.7.3.2 with $[(0/90)_3]_{sym}$ stacking sequence are presented. As second example, an edge-notched tensile test result applied to a $[(0/90)_2]_{sym}$ laminate made from the same material is shown. Acquisition settings for acoustic emission measurements confirm with Table 4.2, while the recording parameters of the thermography equipment are reported in Tables 7.6 and 7.8.

During quasi-static loading of an open-hole specimen, several acoustic emission signals were detected and have been classified into the occurrence of matrix cracking, interfacial failure, and fiber breakage as outlined in Sect. 4.5. After reaching a static load of 21 kN, a cyclic load with 0.4 Hz and ±2 kN was applied to allow damage progression due to periodic loading and unloading. Figure 7.9 shows the localization results as a function of the load time with superimposed load-time curve and a schematic representation of the specimen dimension. As expected from the stress concentration effect of the hole inside the laminate, most of the AE source positions fall into the expected area of the damage zone close to the position of the hole.

Table 7.6 Thermography acquisition parameters

Measurement setup	Value
Camera	FLIR SC7000 with InSb-Detektor of 640 × 512 pixels
Lenses	50 mm, spectral range 3–5 μm
Thermal resolution	20 mK
Acquisition rate	100 Hz
Averaging	×4
Integration time	1250 μs

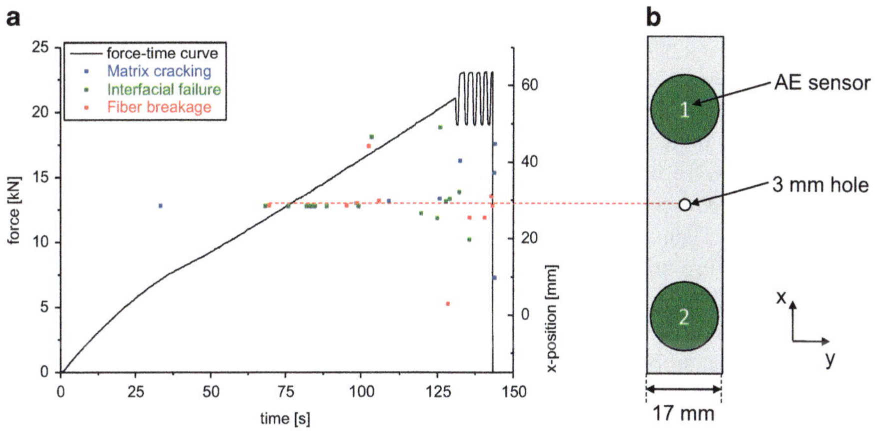

Fig. 7.9 Localized AE source positions during open-hole tensile test as function of the load time with superimposed load-time curve (**a**) and schematic representation of the specimen (**b**)

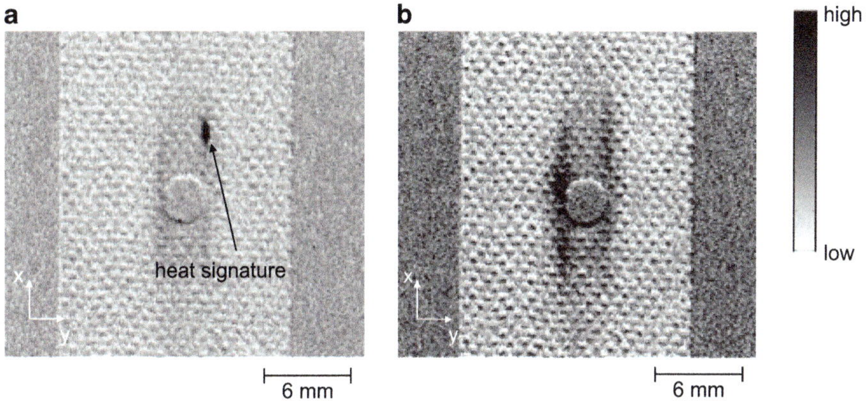

Fig. 7.10 Thermography images recorded during loading and unloading stage at $t = 132$ s (**a**) and at $t = 142$ s (**b**). Temperature scale is adapted for each image to provide highest visibility

Within the simultaneously recorded thermography images, first heat signatures appear during the first cycles of the loading and unloading stage. As seen in Fig. 7.10a, a first strong transient heat signature occurs at the onset of the first cyclic loading period (around 132 s in Fig. 7.9) at a designated position. This is followed by generation of more heat in the vicinity of the hole exhibiting a stripe-like appearance before ultimate failure occurs. The stripe-like heat signatures are likely caused by failure initiated at the hole in between the fibers oriented in parallel to the load axis. Due to the high image acquisition rate and the known position of the heat signatures this allows comparing the occurrence of these heat signatures with the acoustic emission source positions. Therefore the first heat signature can be directly related to the occurrence of matrix cracking and interfacial failure at around 132 s. The subsequent indication of the thermography images (cf. Fig. 7.10b) are more difficult to be correlated to a particular signal. This is because of the close sequence of the acoustic emission signals resulting from multiple failure types including fiber breakage. Since each failure type will cause a heat signature, the challenge arises from the heat flux in the material, which inhibits spotting individual heat spots. Moreover, the cyclic loading will generate heat from friction at the damaged areas. Thus the heat signatures of the individual failure types all merge into one signature and form a good visual representation of the overall affected damage zone. The extent of this zone along the x-direction is in good accordance with the calculated acoustic emission source positions.

The second example considers a tensile test of a double-edge-notched specimen. As seen from the load profile in Fig. 7.11, this type of specimen was subject to a sequence of static load increases followed by a short period of loading and unloading cycles of 0.4 Hz with an amplitude of ±2 kN. In the cyclic loading stages, the experiment was monitored by a thermography camera with acquisition settings of Table 7.7. During the experiment, many acoustic emission signals were

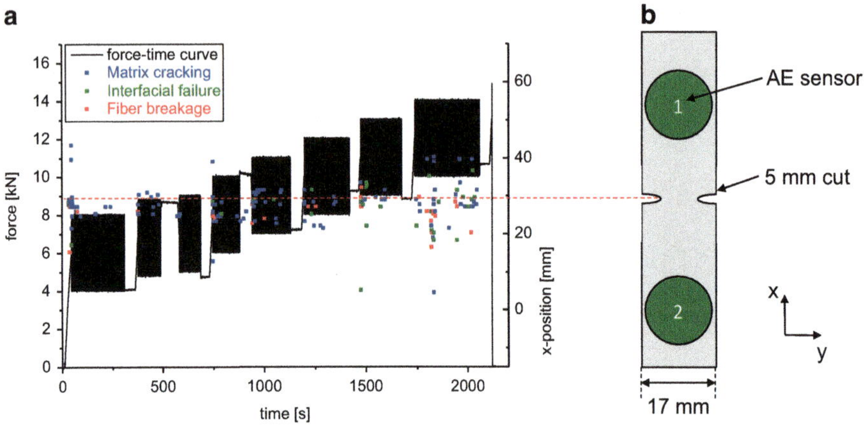

Fig. 7.11 Localized AE source positions during double-edge-notched tensile test as function of the load time with superimposed load-time curve (**a**) and schematically representation of the specimen (**b**)

Table 7.7 Thermography acquisition parameters

Measurement setup	Value
Camera	FLIR SC7000 with InSb-Detektor of 640 × 512 pixels
Lenses	50 mm, spectral range 3–5 μm
Thermal resolution	20 mK
Acquisition rate	100 Hz
Averaging	×2
Integration time	1250 μs

recorded and were post-processed analogous to the previous example. The resulting localized source positions are plotted as function of load time in Fig. 7.11 with color-coded information of the source type. Since the presence of the notch causes stress concentration at the center of the specimen it is expected to see most of the AE source locations being present at this position. However, there is a reasonable distribution of source positions extending well above and below the position of the notches.

In Fig. 7.12 a selection of thermography images exhibiting heat signatures at distinct positions is shown for the loading and unloading periods with center load levels of 6, 8, 9, and 10 kN. For all cases, the moment of occurrence and the extent of the heat signatures allow direct comparison to the detected acoustic emission signals. As seen from the comparison of damage zone sizes evaluated based on the heat signatures and as based on acoustic emission measurements there is good agreement between both methods (cf. Table 7.8). However, the 7 kN load step lacks a clear heat signature, while still some acoustic emission signals are detected and localized. Also, for the broad heat signatures, it is difficult to correlate the occurrence of a particular signal to the heat signature. For small heat signatures such as the 8 kN load case, most of the corresponding acoustic emission signals are all classified as matrix cracking, so the occurrence of inter-fiber failure at this position is likely the case for these heat signatures. For the heat signatures of Fig. 7.12 at

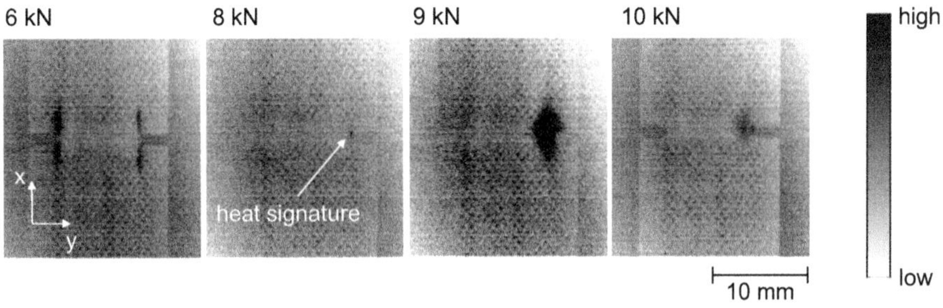

Fig. 7.12 Set of thermography images exhibiting heat signatures recorded during loading and unloading stage at 6, 8, 9, and 10 kN. Temperature scale is adapted for each image to provide highest visibility

Table 7.8 Comparison of damage zone size evaluated based on heat signatures and based on localized acoustic emission sources

Load (kN)	Damage zone size thermography (mm)	Damage zone size AE (mm)
6	±7	±15
7	–	±3
8	±1	±1
9	±5	±5
10	±7	±5.5
11	±10	±14.2
12	±15	±17.5

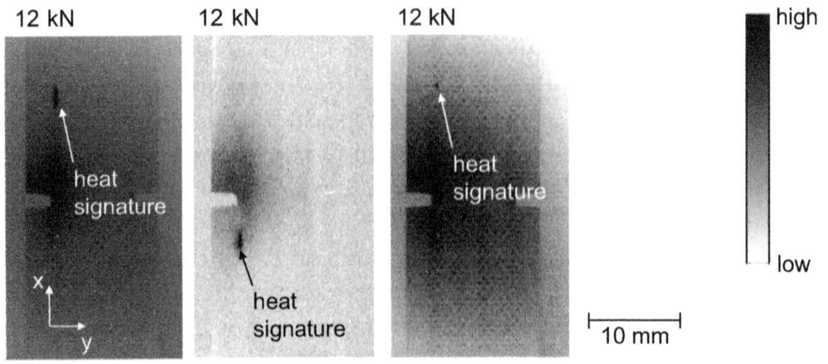

Fig. 7.13 Set of thermography images exhibiting heat signatures recorded during loading and unloading stage at (12 ± 2) kN. Temperature scale is adapted for each image to provide highest visibility

6, 9, and 10 kN, the corresponding acoustic emission signals originate from different source types, so the signature is likely caused by an extended damage zone covering more than one particular failure mechanism.

A further set of spatially localized heat signatures is shown in Fig. 7.13 all recorded during the 12 kN loading and unloading stage. AE signals at these source positions in space and time all correspond to the occurrence of matrix cracking. Thus it can be concluded, that these signatures also originate from the occurrence of inter-fiber failure in the laminate.

At the first cycle of (14 ± 2) kN load, the specimen finally fails. This rupture process is shown in the thermography images of Fig. 7.14. As seen by the generation of heat at the left edge a first crack starts to appear, quickly propagating in parallel to the applied load. The heat zone starts to broaden, possibly also indicating damage growth in perpendicular to the applied load, which could be due to inter-ply delamination. Subsequent to this first signature, a second heat signature appears at the opposite edge and starts to propagate in parallel and perpendicular to the applied load. Finally both zones merge at the notch position and cause severe fragmentation with heat signatures of particles being visible. After rupture, the amount of heat generated by the macroscopic failure spreads out in the remaining specimen.

The exemplary results shown above demonstrate the high synergistic potential of thermography imaging in combination with acoustic emission measurements. Still the detection sensitivity of acoustic emission measurements is better than the detection of heat signatures. However, the appearance of transient heat signatures allows spotting damage in a similar fashion as acoustic emission measurements, yet the signatures are difficult to be attributed to a particular failure mechanism. But the usage of thermography visualizes damage progression with reasonable spatial and temporal resolution [14], while the classification and localization of acoustic emission sources may be used to add quantitative information in terms of the underlying failure mechanism.

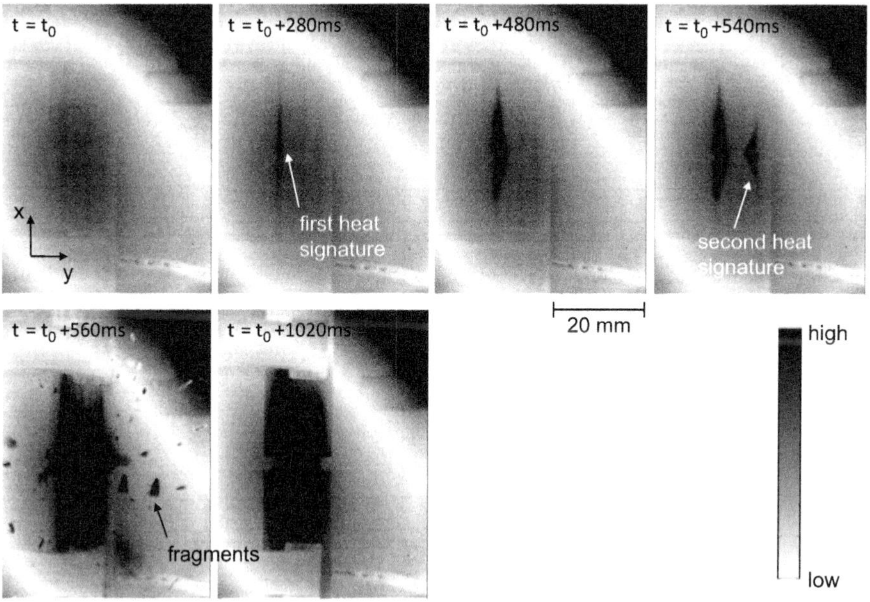

Fig. 7.14 Sequence of thermography images recorded during rupture of the specimen. Temperature scale is adapted for each image to provide highest visibility

7.2.4 *Computed Tomography and Acoustic Emission*

Computed tomography constitutes another approach to combine an imaging technique with the results of acoustic emission measurements. In contrast to the previous imaging techniques, computed tomography is used as an example for a volumetric imaging technique and thus forms a method combination able to interpret and understand damage progression within the full volume of the material. As discussed in Sect. 6.3 it is useful to distinguish between ex situ approaches and in situ approaches in this context.

7.2.4.1 Ex Situ Computed Tomography

In addition to the previous benefits of acoustic emission measurements, two additional possibilities arise for investigating fiber-reinforced materials. First, the whole concept of ex situ inspection is based on the idea to image a certain damage state at a given damage progression. But without any secondary method being applied during loading of the test specimen, this is rather difficult to accomplish. As a representative example, consider the test results of three different Sigrafil prepreg laminate types subject to a tensile test shown in Fig. 7.15. The experimental setup confirms with the description given in Sect. 4.7.3.2. Based on the stress–strain curve

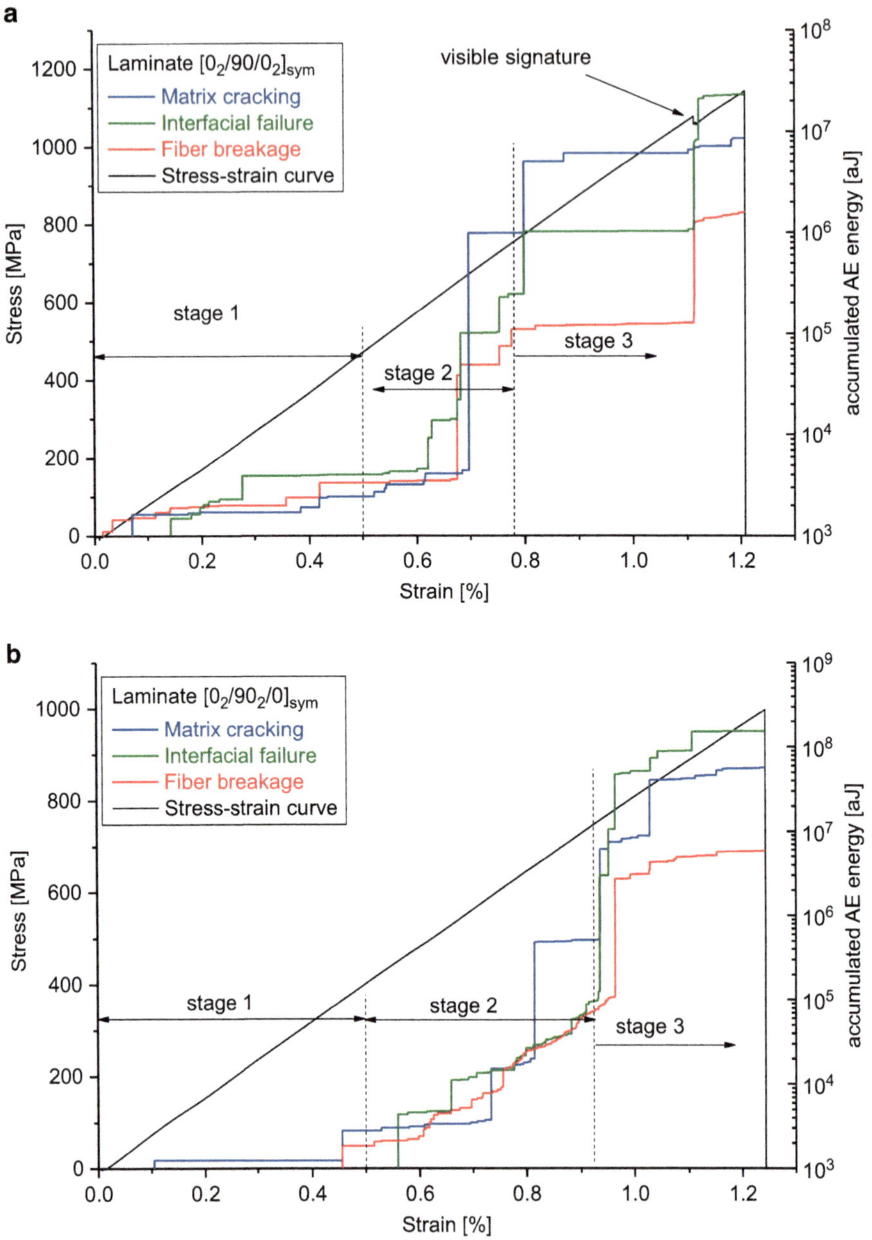

Fig. 7.15 Example of three tensile tests with different laminate stacking including accumulated acoustic emission energy in logarithmic scale

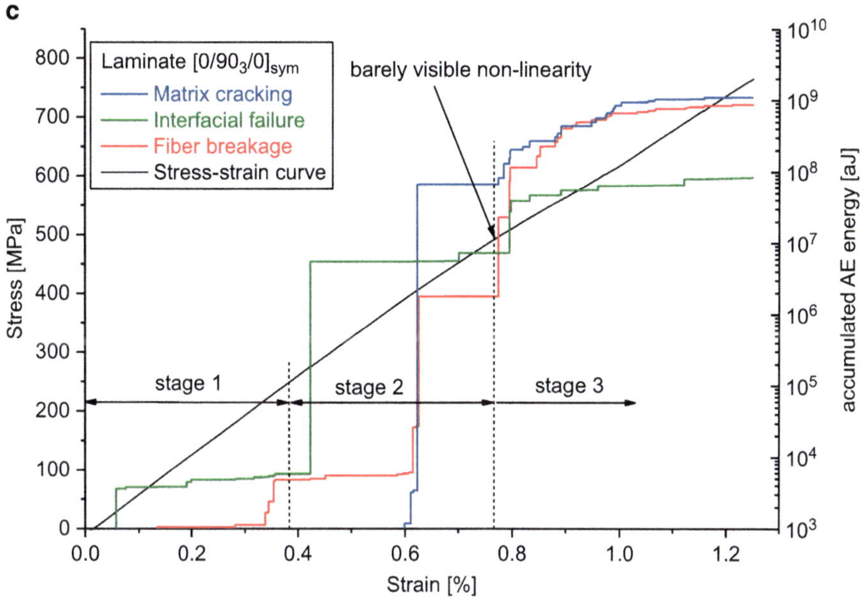

Fig. 7.15 (continued)

of the materials it is often not possible to determine the damage onset or to evaluate the level of damage progression. Thus without any secondary method, such as digital image correlation, thermography or acoustic emission there is no intuitive way to define distinct load levels relevant to be inspected by CT-scanning. Hence, the specimen could only be scanned at arbitrary load levels defined relative to the average ultimate tensile strength. This could potentially result in redundant scans without damage being present or may select load levels of unstable damage states and cause false conclusions. As alternative approach, the onset or levelling of failure mechanisms based on acoustic emission measurements as shown in Fig. 7.15 may be used to define characteristic stages of damage progression and, consequently, characteristic load levels to perform CT inspection. For these examples, distinction can be made for three significant stages of damage progression. As the first damage occurs almost immediately after load onset, this stage is considered as load level of only microscopic damage being present within the material. In this context note the logarithmic acoustic emission energy scale in Fig. 7.15, so the total amount of acoustic emission energy during stage 1 is almost negligible compared to the remaining signals. In some cases (e.g., Fig. 7.15a) the energy level settles which indicates a state of crack arrest after first occurrence of microscopic damage. Such load levels are well suited for a first inspection to track the potential damage initiation sites like matrix cracks in off-axis plies or first delamination (cf. examples in Fig. 7.16).

Within stage 2, energetically relevant acoustic emission release is recorded. More than the spurious signals detected within the first stage, this is indicative of

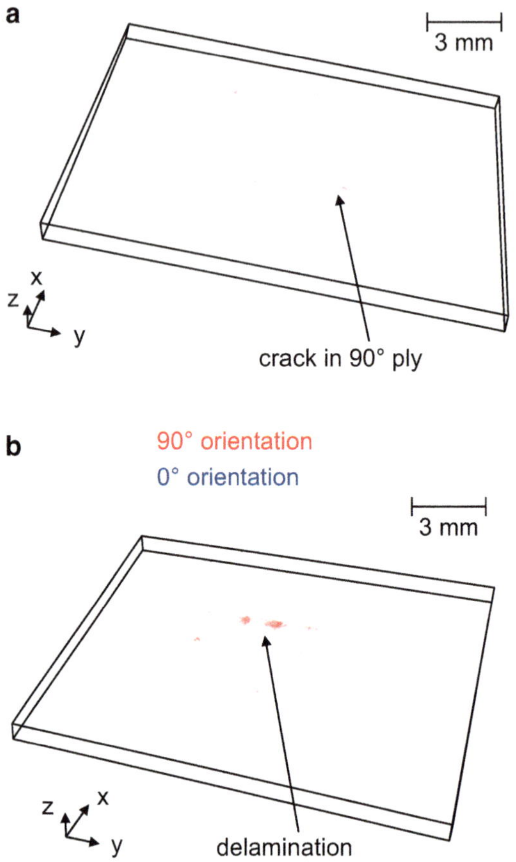

Fig. 7.16 Exemplary set of CT-scan images recorded at the end of stage 1 in $[0_2/90/0_2]_{sym}$ laminate (**a**) and $[0_2/90_2/0]_{sym}$ laminate (**b**)

a damage onset which affects the structural integrity of the material. Whilst this acoustic emission still originates from cracks on the microscopic scale, their accumulation will reach an extent which will leave the structure in a pre-damaged state after unloading from this stage. For the cross-ply stackings investigated here, this damage consists mostly of inter-fiber cracking in the 0 and 90° plies and interfacial delamination in between the plies. Also the first occurrence of fiber breakage is noted and is due to the fact of weak filaments failing long prior to ultimate failure. As seen from the corresponding CT-scans in Fig. 7.17 there are numerous indications of inter-fiber cracks and delamination, which can readily be observed after unloading a specimen from stage 2.

Subsequently, stage 3 ranges from this pre-damaged state to ultimate failure of the specimen. Depending on the type of stacking, this region might also show several distinct signatures, which might be useful to identify further levels for inspection. For some cases, within stage 3, also significant signatures in the stress–strain curve can be identified. This may range from substantial load drops (cf. Fig. 7.15a) to weak nonlinearity of the stress–strain response originating from

Fig. 7.17 Exemplary set of CT-scan images recorded at the end of stage 2 in $[0_2/90/0_2]_{sym}$ laminate (**a**) and $[0_2/90_2/0]_{sym}$ laminate (**b**)

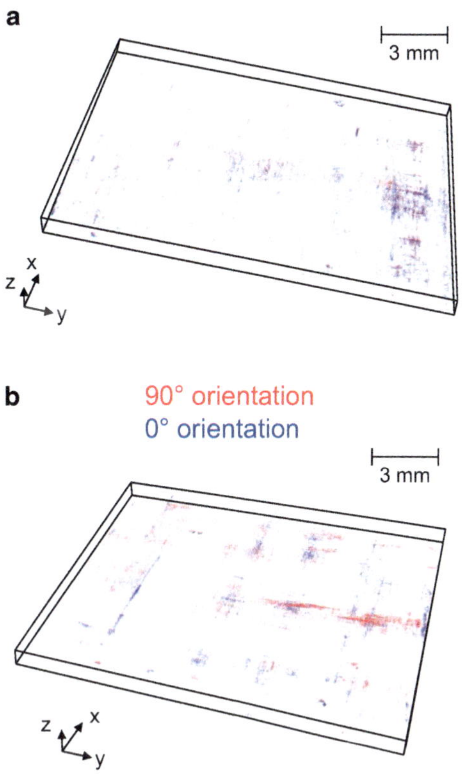

the reduction of specimen stiffness (cf. Fig. 7.15c). Within stage 3, the detected acoustic emission energy tends to increase substantially, sometimes reaching the exponential growth behavior. This is because of the immense amount of microscopic damage that accumulates, interacts, and finally causes failure of individual layers or the full cross-section. In stage 3, all sorts of failure types can be found and it is almost impossible to directly relate the occurrence of one acoustic emission signal with the CT-images due to the multitude of options for correlation, given the amount of signals within a time interval and the accuracy of source localization procedures.

However, for lower load levels, an obvious strength of the combination between CT and AE is the possibility to validate the source mechanism of the detected AE signals. To this end, sufficiently high resolution may be achieved by state-of-the-art CT-scanners as discussed in Sect. 6.2. In this context the detail visibility and accuracy of spatial localization in CT is superior to AE source localization and thus comprises a major challenge for a direct comparison. Since the typical plate-like structure also inhibits source localization along the thickness direction of the laminate, the AE source position is always a projection of all failure mechanisms occurring within the laminate to the 3D-surface. Despite of these difficulties, it is possible to perform a direct correlation for some cases. The first approach, which was already successfully used for coating materials in [15] is to rely on the characteristic source positions of a partition of signals, rather than the individual signals. To this end, a Sigrafil prepreg laminate with $[0/90]_{sym}$ stacking was investigated. The experimental setup confirms with the description given in Sect. 4.7.3.2. The specimen was loaded to stage 3 (strain level 1.2 %) as discussed previously, was removed from the test machine and was scanned in this damaged state. Subsequently, the specimen was loaded until ultimate failure. For this specimen quite different source positions and signal densities were found for each classified failure mechanism, which makes it particularly interesting to correlate with CT-scan observations.

In Fig. 7.18 a comparison of the signals attributed to matrix cracking and a scan of the center part of the specimen is shown. The AE signals' source positions along the x-axis are plotted as the function of strain and are used to demonstrate the evolution of the occurrence of matrix cracking. The information on source location is thus only one-dimensional, i.e., the source coordinate is only known for the x-axis. The CT-scan volume was chosen to provide high spatial resolution inhibiting the scan of a larger volume than presented in Fig. 7.18. For visibility of the cracks, the material was rendered transparent, so only air inclusions remain to reveal the pattern of inter-fiber cracks, inter-ply delamination and pores.

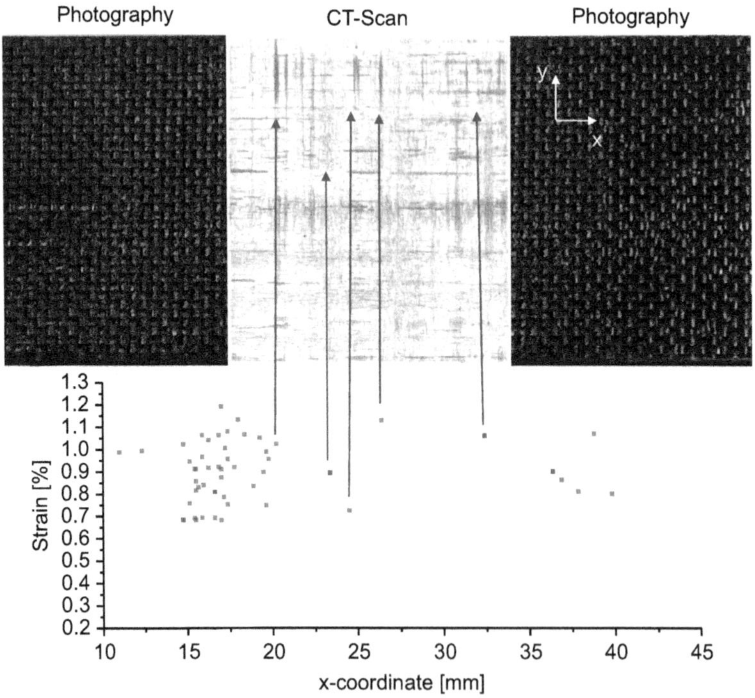

Fig. 7.18 Comparison between AE source localization pattern for matrix cracking and CT-scan after reaching 1.2 % strain (only signals of highest AE energy are related by *arrows*)

Remarkably, the signals of matrix cracking falling within the range of $x = 20$ mm to $x = 35$ mm show very high acoustic emission energies. Therefore they should correspond to the occurrence of large inter-fiber cracks, which was found to be valid for the CT-scan shown in Fig. 7.18.

However, some inter-fiber cracks readily visible in Fig. 7.18 are not accompanied by an according AE signal source position. The reasons can be manifold. First, it could be that the occurrence of the crack has fallen into the dead-time of the recording equipment (see Appendix C). Second, it could be, that the signal was detected, but no valid source position could be calculated due to overlapping signals. Third, the signal could be attributed to interfacial failure instead. The latter is in fact quite likely, since some of the inter-fiber cracks seen in the CT-scan also cause substantial amount of inter-ply delamination. Apart from that, the localization pattern was found to be in reasonable agreement and demonstrates how AE and CT can be combined to increase the reliability of the AE method.

For the group of signals identified as interfacial failure, the corresponding localization as function of strain is shown in Fig. 7.19. For comparison, the same CT-scan as in Fig. 7.18 is shown. Compared to the matrix cracking signals, the overall number of signals of interfacial failure is much higher. Hence it is difficult to correlate individual signals to particular signatures in the CT-scan. Nevertheless,

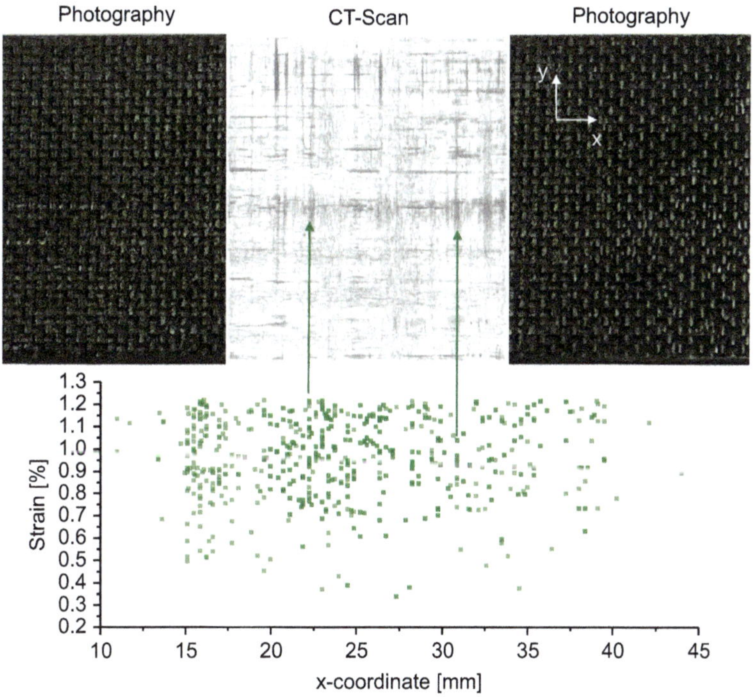

Fig. 7.19 Comparison between AE source localization pattern for interfacial failure and CT-scan after reaching 1.2 % strain

two positions of repetitive identical source locations are marked in Fig. 7.19 and indicate two positions of large inter-fiber cracks, where no associated matrix cracking signal has been found. These cracks also are accompanied by a zone of inter-ply delamination (seen as gray band at half the specimen width). The growth of these inter-fiber cracks may have initiated inter-ply delamination simultaneously and thus may have caused a signal corresponding rather to this type of mechanism. The cracks in parallel to the load axis (x-direction) and the extent of the inter-ply delamination are within the same range as the position of signals belonging to the group of interfacial failure. Therefore, this cluster may be understood and credited to these phenomena in this specimen.

An obviously different localization pattern is observed for the fiber breakage AE signals. Their source position is shown as the function of strain in Fig. 7.20. A large amount of signals localizes at two distinct positions at $x = 15$ mm and $x = 38$ mm. As seen from the photography image after ultimate failure of the specimen, the left agglomeration of fiber breakage signals corresponds to the position of specimen failure. Since this is due to the rupture of the plies with fibers in parallel to the load axis, this finding is not surprising. But the distinctly different source location positions may readily be understood as further evidence for the difference in source

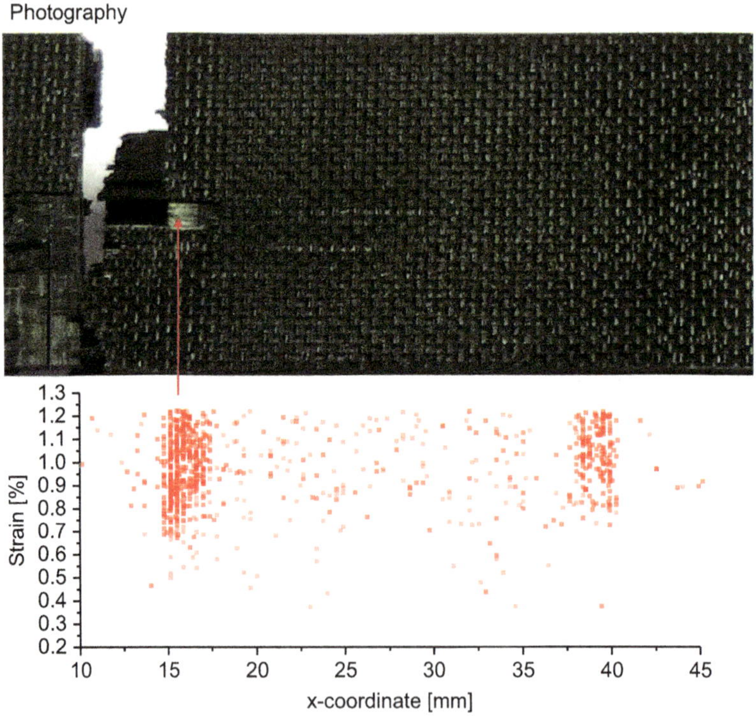

Fig. 7.20 Comparison between AE source localization pattern for fiber breakage and photography after final failure

mechanism, which is well supported by the CT-scans and the failure mode of the specimen. Likewise, other specimens without preferential densification of fiber breakage source positions were observed to fail either at multiple positions or by vast fragmentation of the specimen.

Further possibilities to combine ex situ CT observations with AE measurements are attempts to perform a direct one-to-one correlation of an acoustic emission signal with signatures in a CT-scan. To this end, investigations using a double-edge-notched tensile specimen were performed. As seen from the 3D CT-scan of the specimen after failure, the dominant position of failure is at the position of the reduced cross-section (cf. Fig. 7.21). However, based on the crack-pattern seen in Fig. 7.22, there is also a substantial amount of damage located at distances far away from the position of the notches. Since only few acoustic emission signals were

Fig. 7.21 3D image of CT-scan of double-edge-notched tensile specimen after ultimate failure

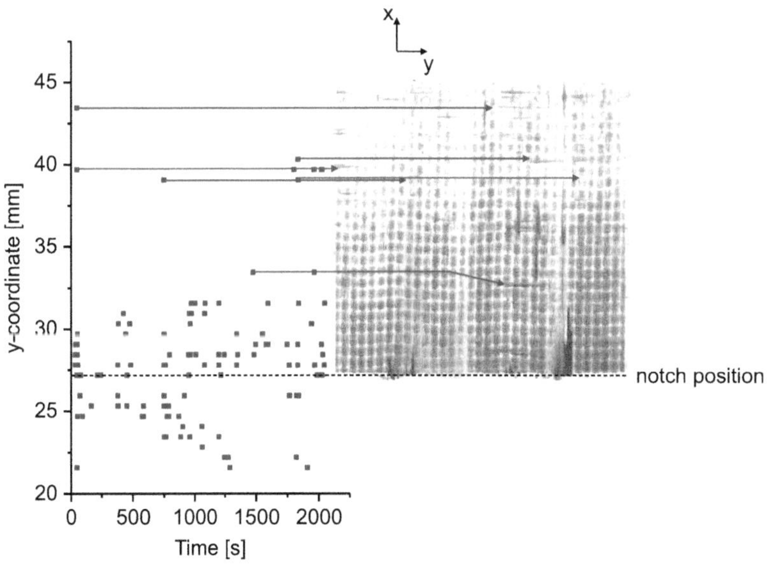

Fig. 7.22 Comparison between AE source localization pattern of matrix cracking and CT-scan after failure of specimen

localized outside the position of the notches, this allows a direct correlation of the isolated signals with respective signatures in the CT-scan.

In Fig. 7.22 a comparison of the signals attributed to matrix cracking and a projection of all inter-fiber cracks, inter-ply delamination and pores is shown. The AE signals' source positions along the *y*-axis are plotted as the function of time of the experiment. The CT-scan volume was chosen to provide high spatial resolution, thus allowing spotting the occurrence of single fiber filament failure. Similarly as for the previous example, it is possible to identify an inter-fiber crack signature for all AE signals in the corresponding CT-scan. For all source positions above $y = 32$ mm, the correspondence was found to be unambiguous. For the remaining source positions, the situation is more difficult, since the area close to the notches suffers from the final rupture of the specimen, and it is not straightforward to correlate the multiple AE source positions with the details of the CT-scan.

For the case of interfacial failure shown in comparison to the CT-scan in Fig. 7.23, again good and unambiguous assignment of the source positions can be made with signatures corresponding to the occurrence of inter-ply delamination. The latter was mostly observed in areas close to cracks in parallel to the *y*-axis.

Finally, for the AE signals identified as fiber breakage, only three distinct signals were found outside the notch positions, which fall within the volume of the CT-scan. As seen in Fig. 7.24, these do not occur at positions of the inter-fiber cracks discussed before. Accordingly, virtual *xy*-cross-sections were made within the CT-volume at different *z*-positions in the region of localized fiber breakages. To account for the uncertainty of the AE source coordinate, several cross-sections were

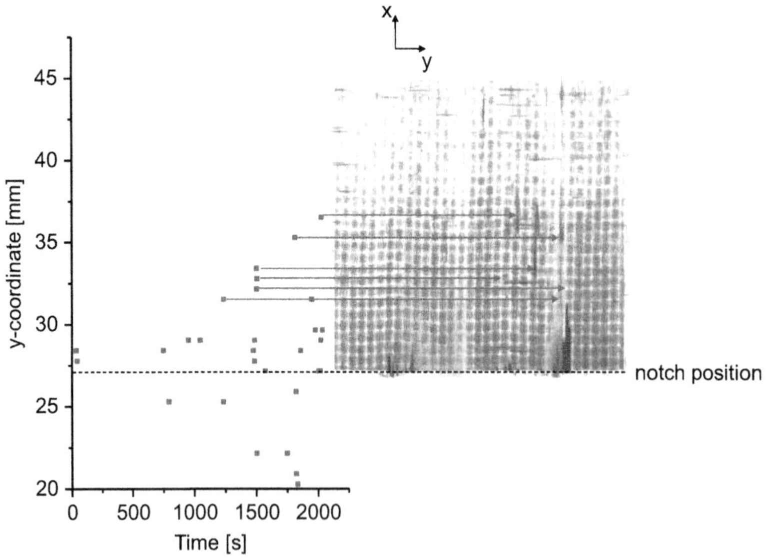

Fig. 7.23 Comparison between AE source localization pattern of interfacial failure and CT-scan after failure of specimen

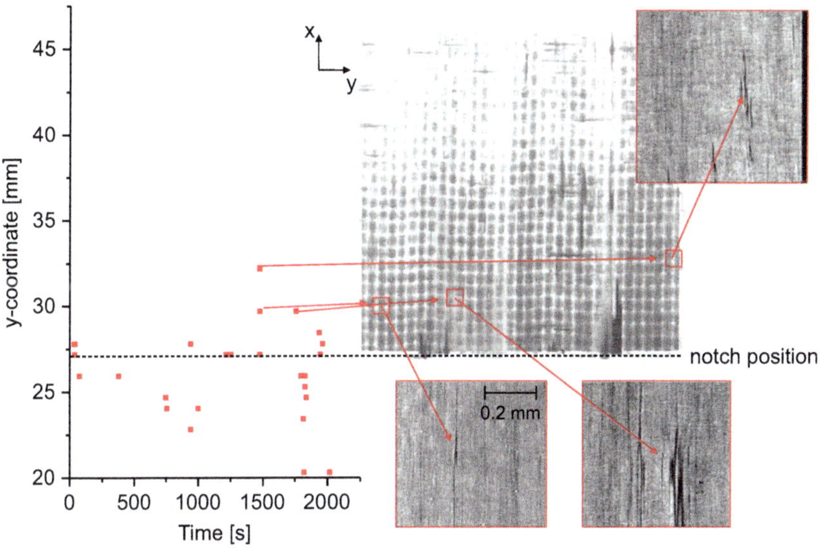

Fig. 7.24 Comparison between AE source localization pattern of fiber breakage and CT-scan after failure of specimen with details of particular cross-sections revealing the occurrence of single filament failure

evaluated along the *y*-axis, considering an uncertainty of localization of ±2 mm. Based on the acoustic emission energy of the signals, it is expected that the signals correspond to the occurrence of fiber breakage likely with a cross-section of less than three filaments.

Only three distinct positions with signatures of fiber breakage were located in the designated regions of the CT-scan as indicated in the magnified *xy*-cross-section planes shown in Fig. 7.24. Two of them are part of an inter-fiber crack running along the y-axis. The latter might either be the consequence of the fiber breakage or the cause for the filament failure. Given the resolution of the CT-scan it is clear, that all three cross-sections are of a typical dimension as expected for one single filament.

7.2.4.2 In Situ Computed Tomography

In contrast to the ex situ approach, the combination of in situ CT-scanning with AE measurements requires some modifications to the approaches described in Sect. 6.3. In addition to the challenges faced by the load introduction and the geometrical stability, the generated acoustic emission signals also need to be detected by attached sensor systems. For reasonable specimen sizes and miniaturized AE sensors these may be attached to the specimen in a conventional way as reported in literature before [16–18].

But in order to achieve the highest resolution, i.e., to resolve single fiber filaments, the specimen cross-section will typically approach dimensions of

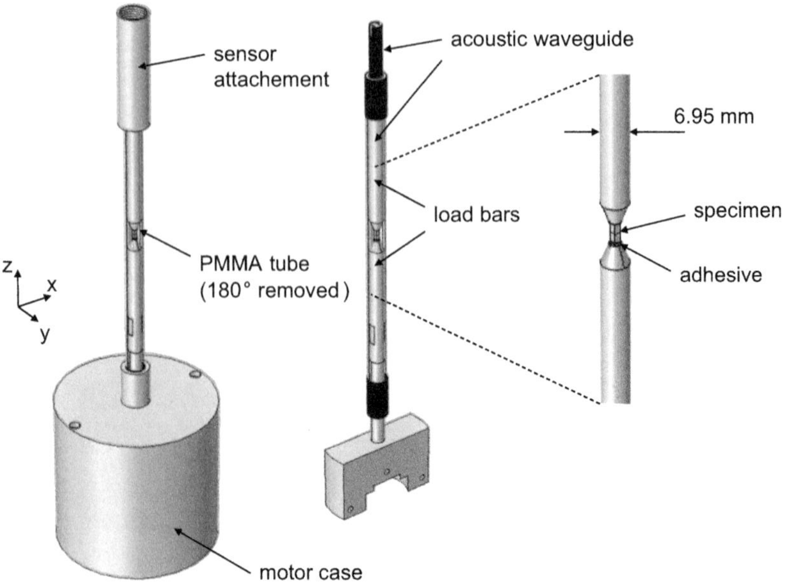

Fig. 7.25 3D drawing of in situ load rig optimized for AE acquisition showing full device (*left*) and interior design with some parts removed (*center*) as well as details of specimen attachment (*right*) (reprinted from [19])

few mm^2, which does not allow a direct mounting of the substantially larger AE sensors. Hence, in such cases another approach is necessary to allow AE monitoring of specimen failure. To this end the load rigs presented in Sect. 6.3 can be modified to include a waveguide to transfer the AE signal to an attached AE sensor. Figure 7.25 shows a 3D drawing of such a load rig as used for tensile/compression testing. The top of the load rig has been adapted to allow acoustic emission monitoring by using a KRNBB-PC-type sensor. The primary modification involves a slender waveguide-like design of the specimen fixture and an acoustic decoupling to the surrounding parts before reaching the sensor position. Using these modifications allow a reasonable interpretation of the first arrival of the recorded AE signal without significant interference with signal reflections due to the surrounding components.

Considering the waveguide design used in the following experiments, it is helpful to calculate the dispersion curves of the design. The L-modes, F-modes, and T-modes of an Aluminum rod of 6.95 mm diameter made from AlMg3 as found in Table B.1 of Appendix B are shown in Figs. 7.26, 7.27, and 7.28. The indexing of guided wave families follows the software used for the computation, which is described in [20]. For the frequency range up to 1 MHz there are numerous guided wave modes found for this rod diameter. Thus, it may happen that the detected AE signal may suffer from the transmission due to the generation of multiple guided wave modes.

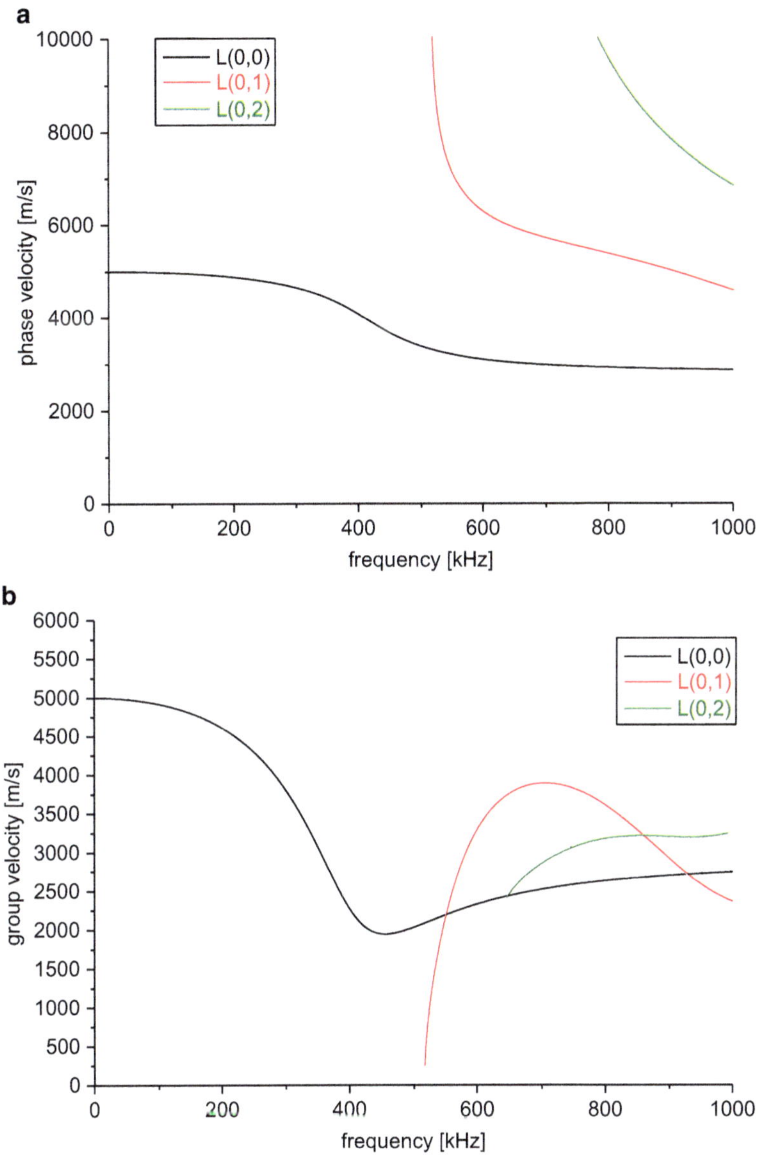

Fig. 7.26 Calculated phase velocities (**a**) and group velocities (**b**) of L-modes for an Aluminum rod of 6.95 mm diameter

To evaluate the severity of this effect, the transfer function of the waveguide was assessed using numerical methods following [21, 22]. Several failure mechanisms were modelled at the position of the mounted specimen and their source function was compared to the detected AE signal. As a simple source function, a cosine-bell step-function with 1 μs rise-time and 3 N force was used. The resulting wave

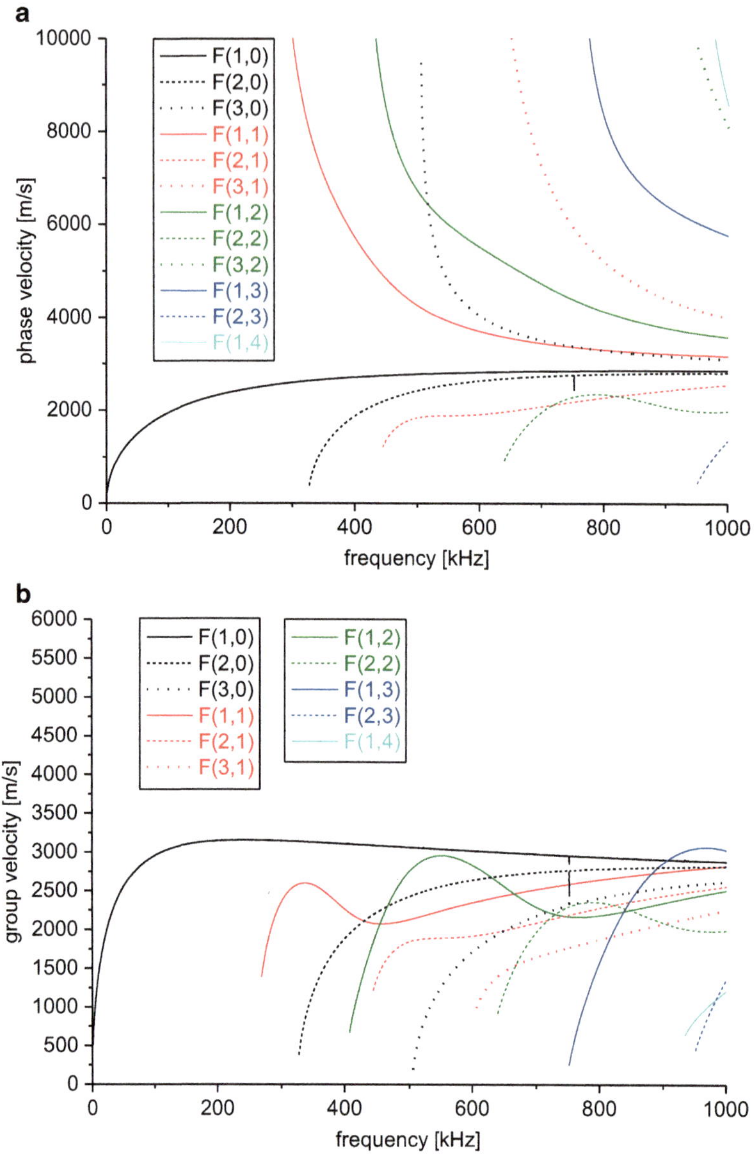

Fig. 7.27 Calculated phase velocities (**a**) and group velocities (**b**) of F-modes for an Aluminum rod of 6.95 mm diameter

propagation within the rod is shown in Fig. 7.29 at three distinct time steps. As seen from the false-color images, the dominant wave propagation mode for this case is a longitudinal pressure wave moving along the rod, corresponding to the $L(0,0)$ mode.

For the source position in this case being at the axis of the rod (i.e., at the center) it is not likely to excite any other mode than the longitudinal mode, since no torque

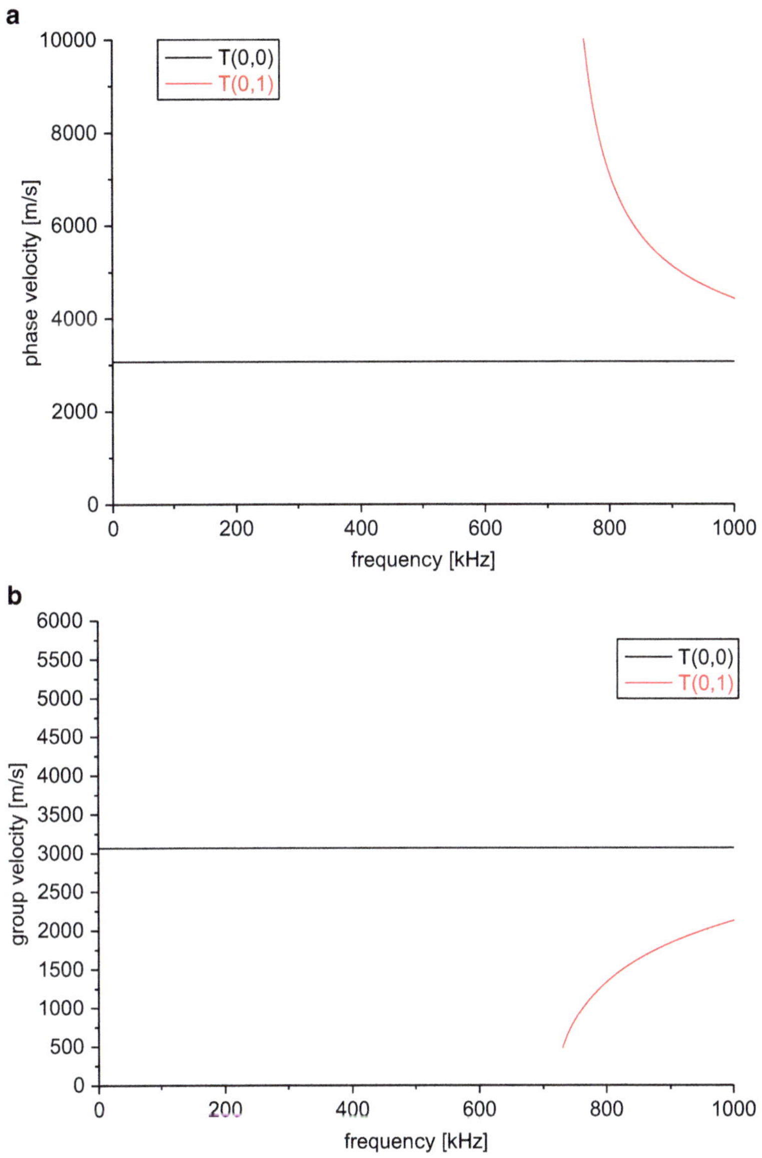

Fig. 7.28 Calculated phase velocities (**a**) and group velocities (**b**) of T-modes for an Aluminum rod of 6.95 mm diameter

or bending moment is induced by the source. To check this influence, a second study was conducted using a source position offset to the center by half a radius. Despite of a noticeable excitation of the *F*(1,0) mode, the overall appearance of the wave propagation and the calculated transfer function was found to be quite similar.

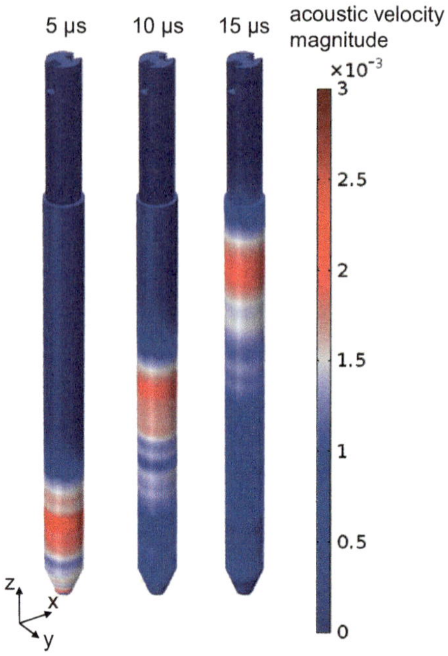

Fig. 7.29 Sequence of images showing wave propagation in the waveguide

Thus, the transmission for the present case seems to be comparatively flat for the *L* (0,0) mode and the *F*(1,0) mode propagation within the rod.

An evaluation of the source function and the detected out-of-plane displacement at the detector position is given in Fig. 7.30a. Although the two signals seem to be substantially different at first sight, this observation is no longer valid after transformation to the frequency domain as evident from the FFT of both signals in Fig. 7.30b. In fact, both signals are found to be very similar, resulting in a transfer function, which is almost unity as seen in Fig. 7.30b. The reason for the visual difference in Fig. 7.30a is the drift of the rod superimposed to the evaluation at the detector position. This also causes the noticeable deviation to unity of the transfer function below 100 kHz. However, for the present applications, the design of the waveguide is assumed to be sufficient to detect the acoustic emission signal of material failure.

Using this experimental setup, four different fundamental AE source configurations were studied. These comprise the occurrence of inter-fiber failure in unidirectional composites due to tensile and compressive load perpendicular to the fiber axis and fiber failure due to tensile and compressive load parallel to the fiber axis.

Inter-Fiber Failure

To induce inter-fiber failure, the mechanical testing procedure follows the techniques described in Sect. 6.5.2. For the following tests, the specimens are fabricated as prepreg laminate from the carbon/epoxy system Sigrafil

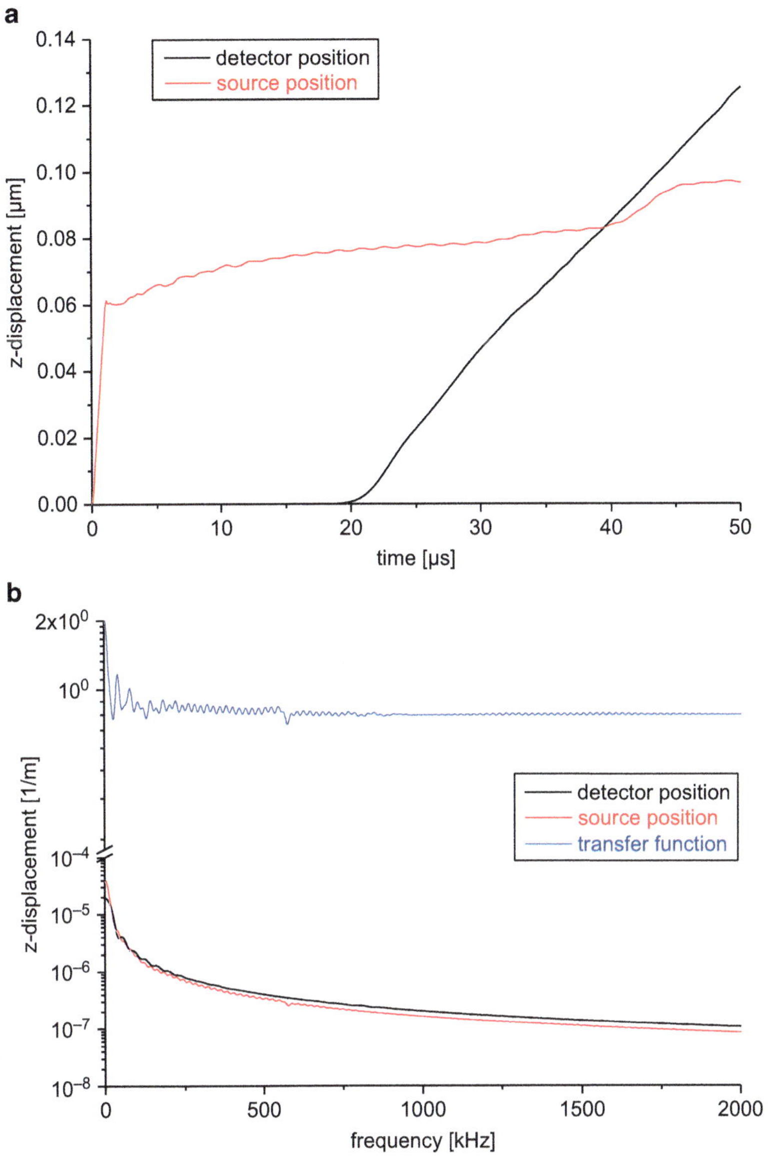

Fig. 7.30 Comparison of source function displacement and out-of-plane displacement at detector position (**a**) and respective signals in frequency domain including calculated transfer function of waveguide

CE1250-230-39 following the curing cycle recommended by the manufacturer. Miniaturized samples of nominally 1.2 mm × 1.9 mm × 22.0 mm (depth × width × height) are cut from the laminates using a water-cooled low-speed saw with precision diamond blade. Samples are attached to the aluminum load bars by an

Table 7.9 Acquisition settings used for in situ experiments

Acquisition settings	Value
Preamplification	20 dB_{AE}
Threshold	38 dB_{AE}
Triggering	Individual channel
Acquisition rate	20 MS/s
Band-pass range	1 kHz to 3 MHz
Couplant	None
Mounting system	Screw thread
Sensor type	KRNBB-PC
Number of sensors	1

UHU Plus endfest 300 epoxy adhesive. Mechanical testing is carried out in displacement controlled mode applying 0.2 mm/min displacement rate. During mechanical loading and CT-scan operation, AE measurements are performed by a KRNBB-PC-type AE sensor with flat frequency response up to 1 MHz and a PCI-2 data acquisition card with acquisition settings reported in Table 7.9. For evaluation of the failure mode, a CT-scan was made after recording an AE signal.

For comparison to the experimental signal, a respective AE signal was modelled for the presented case. All computation results shown in the following were obtained by the "Structural mechanics module" in the software program Comsol Multiphysics following established routines for modelling of acoustic emission sources [8, 23], signal propagation [24] and detection [25]. Explicit modelling of the KRNBB-PC sensor is not necessary, since the voltage signal for this type is directly proportional to the out-of-plane displacement. Hence, the modelled AE signals are evaluated as displacement normal to the top surface of the load bar, i.e., the z-direction in Fig. 7.29. All mesh settings were sufficient to provide >99 % coherence as established in [24], with a maximal mesh size of 0.5 mm and respective refinement in regions of narrow details. The modelling procedure consists of two subsequent analysis steps following the way of implementation as described in Sect. 4.2 and for the present example specifically described in [19].

As first step, the model is subject to a static load applied along the z-axis direction with a value identical to the fracture load in the experiment (i.e., 200 N). In the second step, a transient analysis is performed allowing crack growth and AE release. As time step 10 ns were chosen to provide sufficient resolution during crack growth and signal propagation. The cohesive zone element approach requires a definition of an internal fracture plane. Assuming a planar fracture plane for this purpose might not take into account roughness or curvatures as existent in reality. Therefore, the present implementation uses the procedure described in Sect. 6.5.1.2, to extract the resultant fracture surface. The example of Fig. 7.31 shows the fracture surface formed due to tensile load perpendicular to the fiber axis, which was extracted from the CT-scan. The extracted and simplified fracture surface was then embedded within a FEM-model to calculate the respective acoustic emission signal. This procedure is schematically shown in Fig. 7.31, including the specific dimensions of the fracture surface. The extraction of the fracture surface was performed using "VGStudio MAX" with subsequent simplification in "MeshLab".

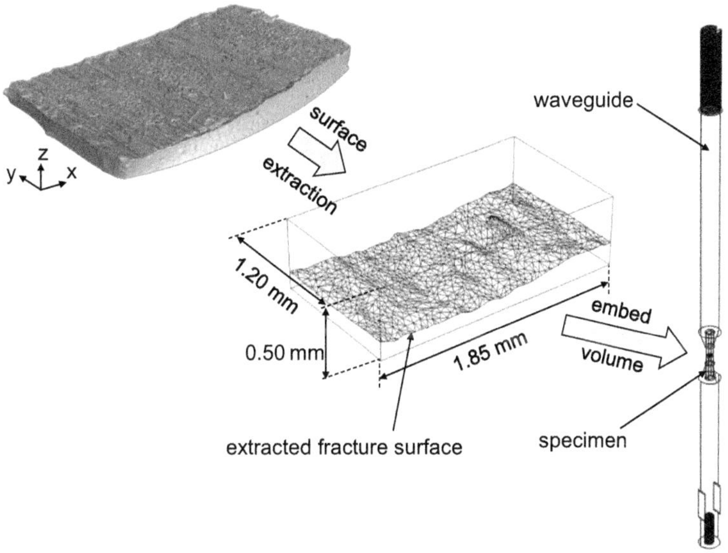

Fig. 7.31 Scheme of procedure used to transfer CT-scan of inter-fiber fracture surface into FEM-model (reprinted from [19])

Due to the presence of the static load in z-direction, a static displacement field exists within the specimen. As pointed out in Sect. 4.2, this static load and displacement field is used as an initial boundary condition for a second computation step. This initiates the transient degradation of the stiffness vector of the thin elastic layer based on the evaluation of Puck's inter-fiber failure criterion. A sequence of images during the crack propagation is shown in Fig. 7.32.

The false-color range of the degradation function $\check{C}(\boldsymbol{r})$ indicates the extension of the crack by red areas, while blue colors resemble areas, where the material is still in contact. The images visualize the state of crack growth between $t = 0.2$ μs and $t = 1.2$ μs. The initiation point is located at the lower left corner of the embedded fracture plane. The initiation is solely owed to the stress concentration at this edge based on the 3D topology of the fracture plane and the applied load level. After initiation, the crack grows along the fracture plane due to the stress concentration at the crack front. The extension of the fracture plane is driven by the roughness of the fracture surface and the velocity of crack growth approaches the speed of a Rayleigh wave, which is the general limit of crack propagation velocity.

Due to the orientation of the local stress components, the newly formed fracture surface is subject to a dominant mode-I-type load situation. This causes a crack opening during propagation of the crack and the resulting movement of the fracture surfaces constitutes the actual source mechanism for AE release. The initiation of the dynamic motion of the fracture planes are conveniently visualized using diagrams of the acoustic velocity magnitude to visualize the birth of the AE signal around the crack. As seen from Fig. 7.33, the near-field of this crack type consists of a dominant radiation perpendicular to the crack surfaces, which extends along the

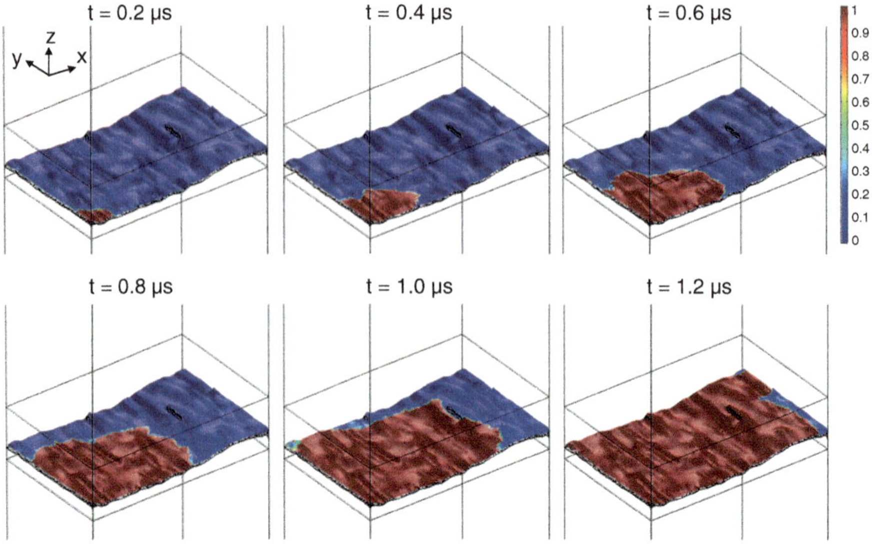

Fig. 7.32 Calculated crack growth in in situ tensile test with fiber orientation perpendicular to load axis at distinct time intervals with false-color range to indicate the value of the degradation function $\check{C}(\boldsymbol{r})$ (reprinted from [19])

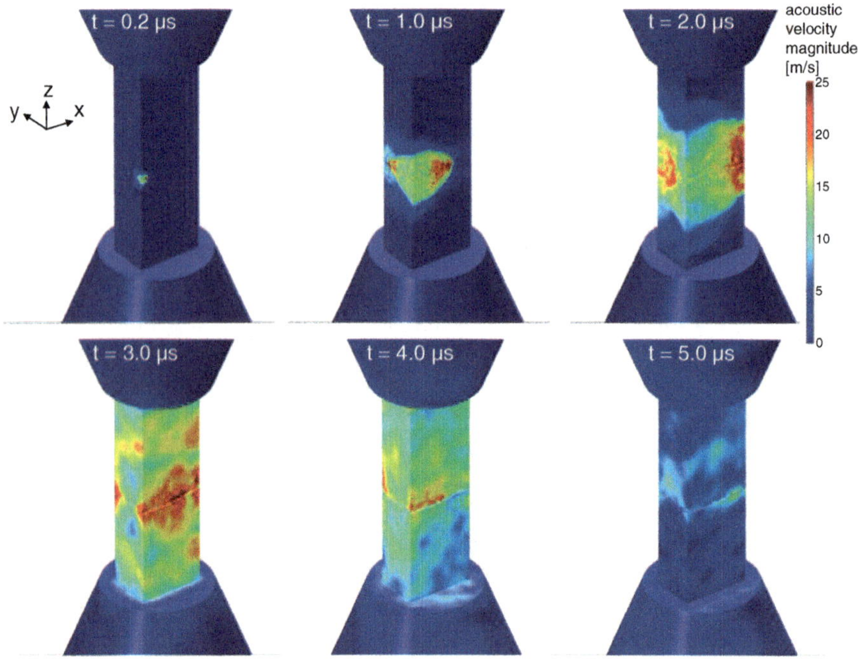

Fig. 7.33 Plot of acoustic velocity magnitude during and after crack growth (based on [19])

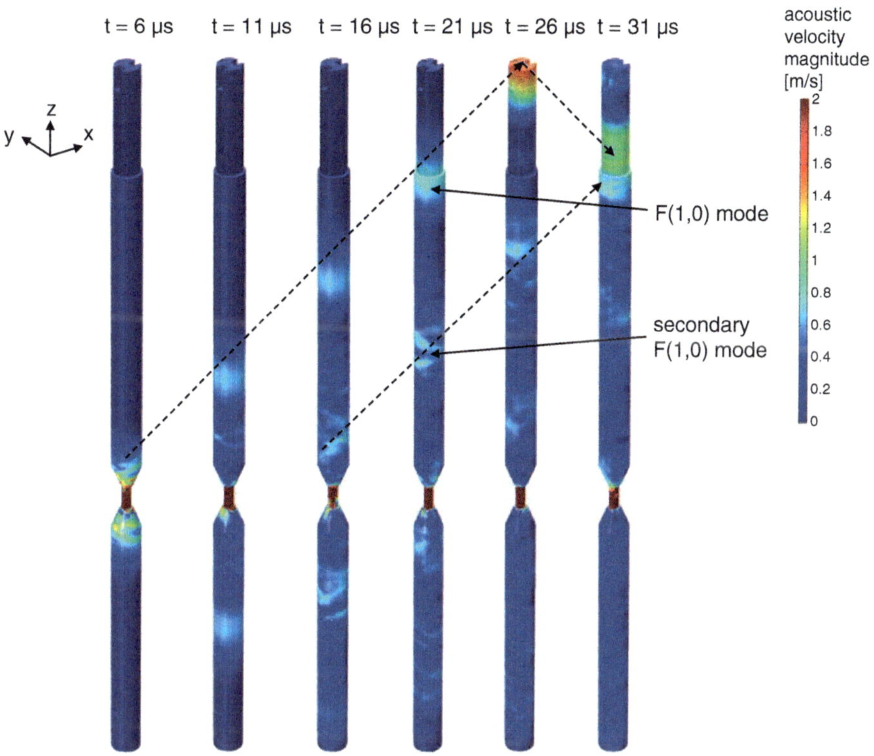

Fig. 7.34 Plot of acoustic velocity magnitude during signal spreading in waveguide geometry (based on [19])

x- and y-axis during the progression of the crack. After $t = 3.0$ μs the first reflections at the edges of the small specimen superimpose to the emitted waves of the source. Thus it is more convenient to adapt the false-color scale and observe the signal spreading in the attached waveguide.

Within the waveguide (i.e., tensile bar) the incident waves are converted into guided waves, which propagate towards the AE sensor position. As seen from the acoustic velocity fields in Fig. 7.34 at six distinct time steps, there are several guided wave modes visible within the waveguide. The two wave packages with highest intensity were identified as primary and secondary $F(1,0)$ modes by means of time-frequency analysis using Choi-Williams diagrams and calculated dispersion curves of the 6.95 mm aluminum rod using the software provided by [20] following the routines established in [26, 27].

The displacement normal to the top surface of the waveguide is evaluated as AE signal and is plotted in Fig. 7.35 for the calculated duration of $t = 60$ μs. For comparison, the respective experimental AE signal obtained for the investigated fracture surface is shown in Fig. 7.35 as well. The behavior of the experimentally used KRNBB-PC sensor can readily be estimated as flat frequency response with

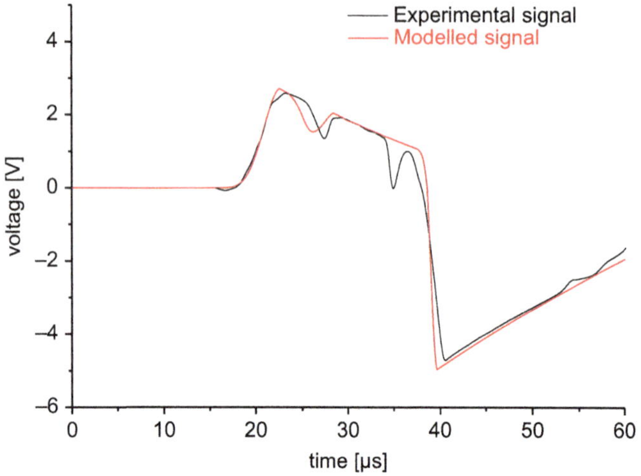

Fig. 7.35 Comparison of experimental and simulated acoustic emission signal at detector position (reprinted from [19])

constant conversion factor of 6 kV/m also accounting for the preamplifier gain. Thus the modelled sensor voltage may directly be compared to the experimental sensor voltage. In addition, the modelled signal was filtered by a sixth-order Butterworth high-pass of 1 kHz to confirm with the bandwidth limitations of the experimental setup. As seen from the good quantitative agreement between both signals, the proposed model setup is able to describe the generated AE signal very well. Some differences arise with respect to the number and position of dips after 25 µs. Based on visual observations of the signal propagation path, this is likely owed to interactions with parts not included within the model, which cause additional reflections arriving at the sensor position after the first wave mode has passed beyond the sensor position (cf. Fig. 7.34). Since the experimental signal will also move into the sensor and will interact with the surrounding metallic sensor fixture this will contribute to the shape of the signal to a certain extent. However, except for the first rise in the signal, the shape is dominated by the experimentally used band-pass filter, which causes the sharp falling slope between 20 and 40 µs and the rising slope from 40 µs on.

On an extended time scale, the existence of the guided wave modes and their reflections become more apparent when superimposing the calculated dispersion curves to a Choi-Williams distribution of the experimental signal as shown in Fig. 7.36. All relevant modes exhibit a first arrival within a time window of about 10 µs (between 52 and 62 µs in Fig. 7.36) for the given distance between the source and sensor. For the detected frequency components only the $L(0,0)$ as first signal arrival and later the dominant $F(1,0)$ are present within the signal. The discrepancy in the dominant frequency components of the $F(1,0)$ dispersion curve and the Choi-Williams distribution is dominated by the high-pass filter and can still be attributed to this wave mode, since no other guided wave mode exists at the observed

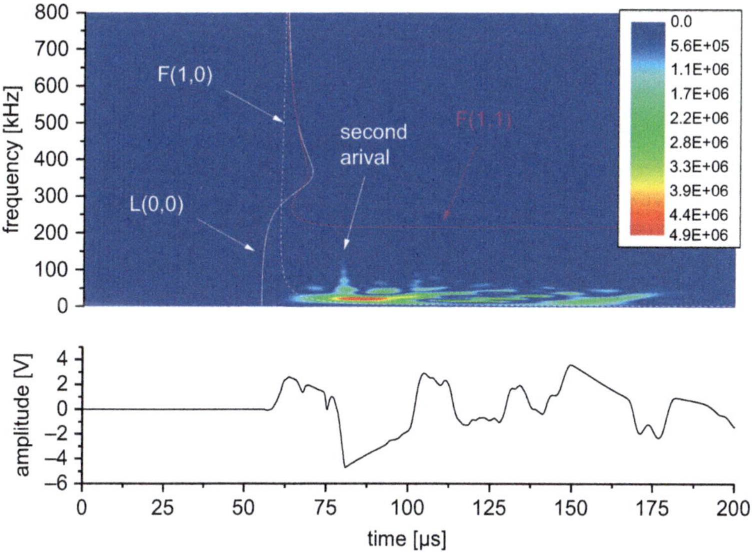

Fig. 7.36 Choi-Williams distribution (*top*) for acoustic emission signal (*bottom*) recorded during failure in tensile load transverse to fiber axis

frequency. The second wave package arrival seen in Fig. 7.36 also matches the signature of a *F*(1,0) wave mode. This is emitted by the source as second event, which was evaluated from the model to be due to the first vibration cycle of the two fracture planes. Such a vibration cycle is given by the retracting fracture planes for about 8 μs, which are reflected at the load bars and start to approach each other at about 15 μs. This impact causes the second signature and in superposition to the first *F*(1,0) signal the highest amplitude in the signal.

A distinctly different failure mode is induced by compressive load applied transverse to the fiber axis. As seen from the Choi-Williams distribution of Fig. 7.37, the acoustic emission signals are dominated by the occurrence of the *F* (1,0) guided wave mode. At the beginning of the signal, a small signature due to the first arrival of the *L*(0,0) mode is visible.

Compared to the tensile load case, the signal appears to be different in several aspects. First, the overall signal intensity is less by almost a factor of five.

The primary reason for this difference in amplitude is the orientation of the fracture surfaces and their vibration. Unlike the dominant mode-I failure of the tensile case, the compressive load is likely to induce a sequence of individual failures under mixed-mode conditions. Such a sequence may correspond to the formation of the fracture planes A, B, C, and D seen in Fig. 7.38 either individually or several at a time. This is confirmed by the sequence of *F*(1,0) mode arrivals in the Choi-Williams distribution of Fig. 7.37, which are unlikely reflections, since these are not present in other signals detected in this configuration (cf. e.g., Fig. 7.36). Instead, they might be the individual release of one fracture plane after another

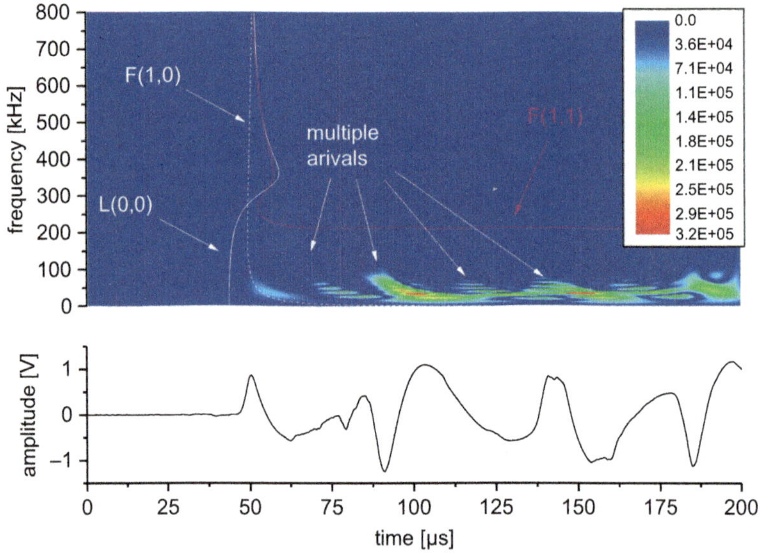

Fig. 7.37 Choi-Williams distribution (*top*) for acoustic emission signal (*bottom*) recorded during failure in compressive load transverse to fiber axis

Fig. 7.38 Fracture mode for failure due to compressive load transverse to fiber axis

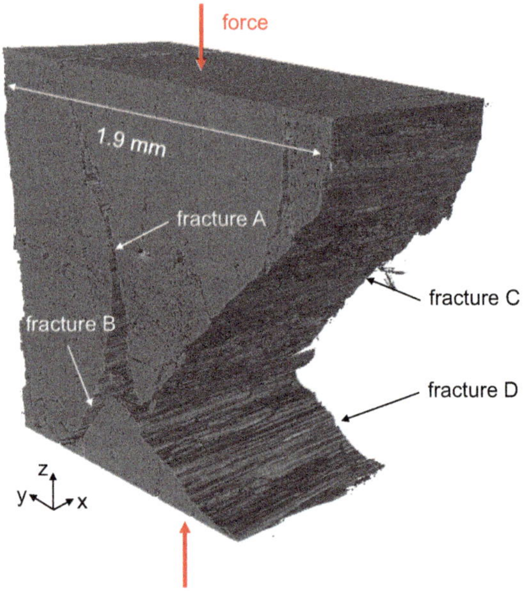

causing the final appearance of the specimen as seen in Fig. 7.38. From the viewpoint of failure theories, the full sequence is still considered as single failure mode, but given the temporal resolution of the acoustic emission measurement, such sequential damage can readily be resolved.

Fiber Failure

In order to investigate fiber failure due to tensile and compressive load, different specimen sizes were prepared to comply with the maximum force capacity of 500 N of the in situ test frame and load cell. Specimens are fabricated from the Sigrafil CE1250-230-39 carbon/epoxy prepreg laminate using miniaturized samples of 0.14 mm × 1.35 mm × 22.0 mm (depth × width × height). All other settings were chosen as for the previous cases.

As seen from the Choi-Williams distribution in Fig. 7.39, the acoustic emission signal is significantly different compared to the recorded signals of inter-fiber failure. This is caused by the failure mode of the signals as shown in Fig. 7.40. The dominant movement of the fracture surface is expected to be along the z-direction giving rise to a strong excitation of the $L(0,0)$ mode as identified by the respective dispersion curve in Fig. 7.39. In addition, the failure mode is not symmetric along the y-axis likely causing a superimposed movement turning into a simultaneous excitation of a flexural mode. Since the dominant frequency content of the source mechanism seems to be located around 250 kHz, the resultant guided wave mode is the $F(1,1)$ instead of the previously encountered $F(1,0)$ wave mode. A possible secondary arrival of a $L(0,0)$ or $F(1,0)$ is found around 110 μs and may be caused by a sequence of damage as described previously. However, given the nature of tensile failure it is not very likely to obtain a sequence of damage with an intermediate settling of 30 μs.

For failure of the test specimens subject to compressive loading parallel to the fiber axis, the acoustic emission signal shown in Fig. 7.41 was found to be

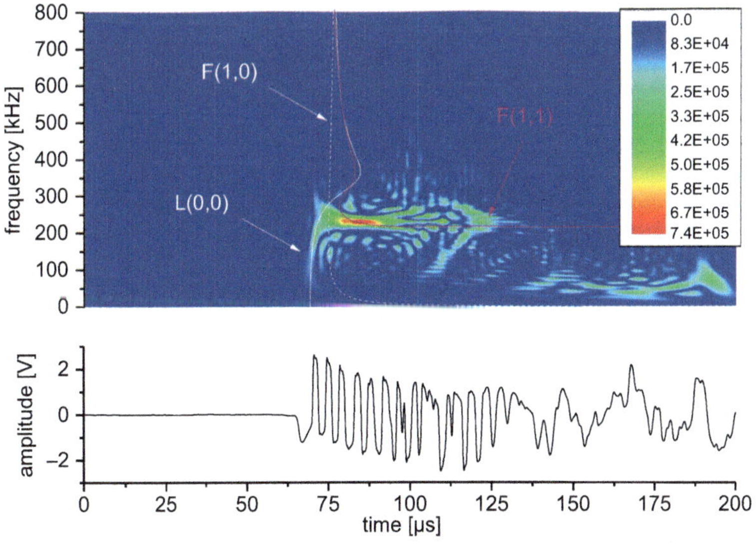

Fig. 7.39 Choi-Williams distribution (*top*) for acoustic emission signal (*bottom*) recorded during failure in tensile load parallel to fiber axis

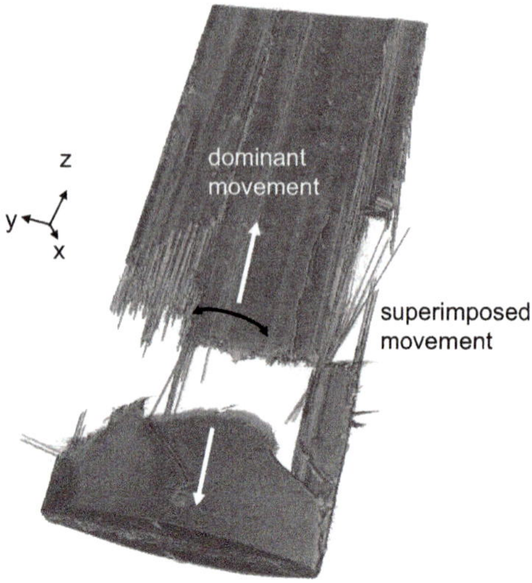

Fig. 7.40 Fracture mode for failure due to tensile load parallel to fiber axis

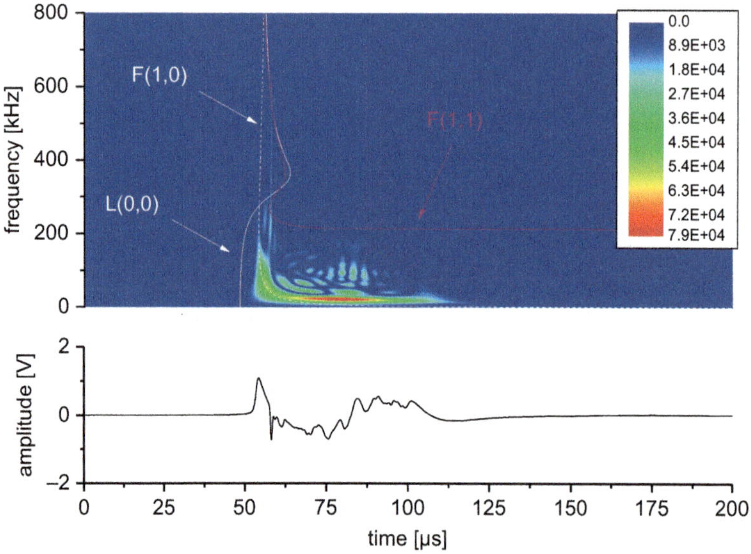

Fig. 7.41 Choi-Williams distribution (*top*) for acoustic emission signal (*bottom*) recorded during failure in compressive load parallel to fiber axis

representative. Unlike the tensile load, the detected acoustic emission signals are dominated by an almost pure $F(1,0)$ wave mode. To analyze the resultant fracture plane, the specimen fragments were moved apart subsequent to fracture to reveal the scan result shown in Fig. 7.42. As seen in the resultant fracture surface, the expected displacement components during fracture are in parallel to the z-axis, but

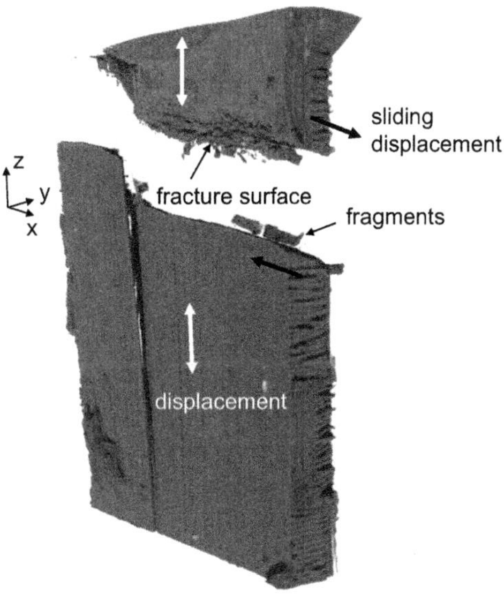

Fig. 7.42 Band-kinking fracture mode for failure due to compressive load parallel to fiber axis

also along a sliding displacement in the *yz*-plane. The latter component is expected to cause a preferential stimulation of a flexural wave mode incident to the waveguide. It can also be observed from Fig. 7.41, that there are no noticeable secondary arrivals of other wave modes. Based on the conclusions for the other failure modes above, this indicates that the observed damage of Fig. 7.42 occurs not as pronounced sequence of individual failure events but merges all into one acoustic emission signal.

The present findings demonstrate the valuable possibilities to investigate composite failure mechanisms offered by an in situ combination of computed tomography and acoustic emission detection. Among other aspects, the experiments demonstrate the sensitivity of the acoustic emission method to detect and analyze sequences of damage macroscopically understood as a single failure event (such as in the case of compressive inter-fiber failure). Moreover, it could be demonstrated, that all failure mechanisms investigated cause distinctly different signal types. Most of all, these could be distinguished by the presence of specific guided wave modes. The use of a flat-with-frequency acoustic emission sensor in combination with a waveguide of very flat transfer function provides a good basis to investigate the frequency content of the detected signals. Here it was found, that the tensile fiber failure stimulates guided wave modes at higher frequencies. For the fiber failure due to compressive load, this was not the case, but it is likely, that the dominant contribution to generate the acoustic emission signal is not the band-kinking of the fibers, but the longitudinal splitting of the laminate. For the inter-fiber failure modes, dominant stimulation at lower frequencies was observed.

However, the detected frequencies in these experiments might not be directly related to experiments in macroscopic specimens. This is due to the used waveguide, which allows only distinct wave modes to propagate at distinct frequencies (cf. also Sect. 4.3). Also, the overall size of the fracture zone will add to the expected rise-time, thus dominating the excited bandwidth (cf. Sect. 4.2.1). Since failure of fiber filaments will consist of much smaller fracture surfaces than investigated here a substantial shift of bandwidth towards higher frequencies is likely. Instead, the present in situ studies might act as reference for validation of computational methods as demonstrated for the case of inter-fiber failure due to tensile load. Such validated models of failure mechanisms may then be embedded in macroscopic models as seen in Sect. 4.2 to compute quantitative acoustic emission signals for larger structures.

7.2.5 Computed Tomography and Digital Image Correlation

As further suitable combination of the methods presented herein, the application of digital image correlation techniques to computed tomography images may be named. This has been established in the form of "digital volume correlation" and has already turned into an active field of research and is ready to be used for composite materials. The principles of this approach have already been introduced in Sect. 6.4 and are thus not repeated here.

7.2.6 Electromagnetic Emission and Acoustic Emission

Acoustic emission analysis and electromagnetic emission analysis are established measurement techniques for monitoring of crack formation and propagation as discussed throughout Chaps. 4 and 5. While acoustic emission is already used for monitoring of the durability of engineering structures and is properly established in technical terms, electromagnetic emission analysis still is subject to basic research efforts and has not yet turned into a technologically relevant measurement technique. However, both methods share an important common link, which is the source mechanism driving the excitation of an acoustic wave as well as the electromagnetic wave.

Both techniques are capable of providing qualitative and quantitative statements, but the reliability of quantitative information relies on the detailed understanding of the correlation between the source mechanism and the measured signal. In this context, the temporal evolution of the displacement field around the crack tip is the source of acoustic emission and also constitutes a relevant contribution to the generation of the electromagnetic emission signal. The combination of both measurements is thus expected to provide valuable information to understand crack formation and propagation on the microscale. However, during propagation of the acoustic wave, the characteristics of the signal (e.g., frequency content) suffer from attenuation, dispersion, and propagation in guiding media. In addition, the characteristics of the signals detected at the surface of the solid are further altered by the detection process using piezoelectric sensors. Consequently, to derive a fundamental correlation between the (visco-) elastic properties of the cracking medium and the crack surface motion, an experimental investigation by a method which is not influenced by these mechanisms is desirable.

In comparison to acoustic emission analysis, the major advantage of electromagnetic analysis is the negligible influence of the propagation medium on the detected signals, as well as the possibility to ensure flat response of the detecting sensor within the frequency range of the signals. As demonstrated in Chap. 5, electromagnetic emission signals provide a direct observation to investigate the crack wall movement during and after crack formation. As major difficulties of this method, the intensity of electromagnetic emission signals is strictly dependent on the orientation between crack surface and the detecting sensors and the distance to the sensor. In comparison to these effects, the electromagnetic emission signals show only negligible influence on the conductive properties of the material investigated. Given the observation of electromagnetic emission in a broad range of materials [28–37] it can be assumed, that measurement of electromagnetic emission of fiber-reinforced polymers should generally be possible for a broad range of fiber/matrix combinations.

Consequently, acoustic emission and electromagnetic emission measurements may be combined to provide a comprehensive measurement approach for each mechanism, which generates both signals.

The simultaneous detection of acoustic emission and electromagnetic emission can then been used to aid in a variety of aspects:

1. Measure the source excitation time t_0 for source localization procedures
2. Measurement of crack duration
3. Measurement of crack tip velocity
4. Measurement of source dynamics for acoustic emission source modelling
5. Measurement of crack orientation

Since Chap. 5 already presented some examples to demonstrate the combined use of electromagnetic emission and acoustic emission, only some specific aspects mentioned above are discussed in the following.

The use of the source excitation time t_0 has been previously proposed in [37] and for a detailed discussion, the interested reader is referred to this original publication. As has been pointed out in Sect. 4.6, the application of some source localization procedures requires to solve for the unknown source coordinates and the unknown source excitation time t_0. Typically this is derived by solving an overdetermined equation system. An alternative strategy has been proposed by [37], since the EME measurement allows directly obtaining t_0. Because electromagnetic waves propagate at the speed of light, their signal onset directly corresponds to the absolute value of t_0. As seen in the exemplary graphs of Fig. 7.43, the EME signals recorded during failure of single fiber filaments systematically precede the AE signals by some microseconds. For the given experimental setup of Fig. 5.58 in Sect. 5.5, this is a well-expected value, since the distance between the source and sensor is about (4 ± 1) mm (depending on the exact location of filament failure) resulting in approximately (3 ± 1) μs time of propagation to reach the sensor position. Similar behavior was found for all pairs of EME and AE signals investigated in the experiments throughout Chap. 5. Therefore, the measured onset of the EME signal may directly be used for source localization procedures, i.e., to obtain

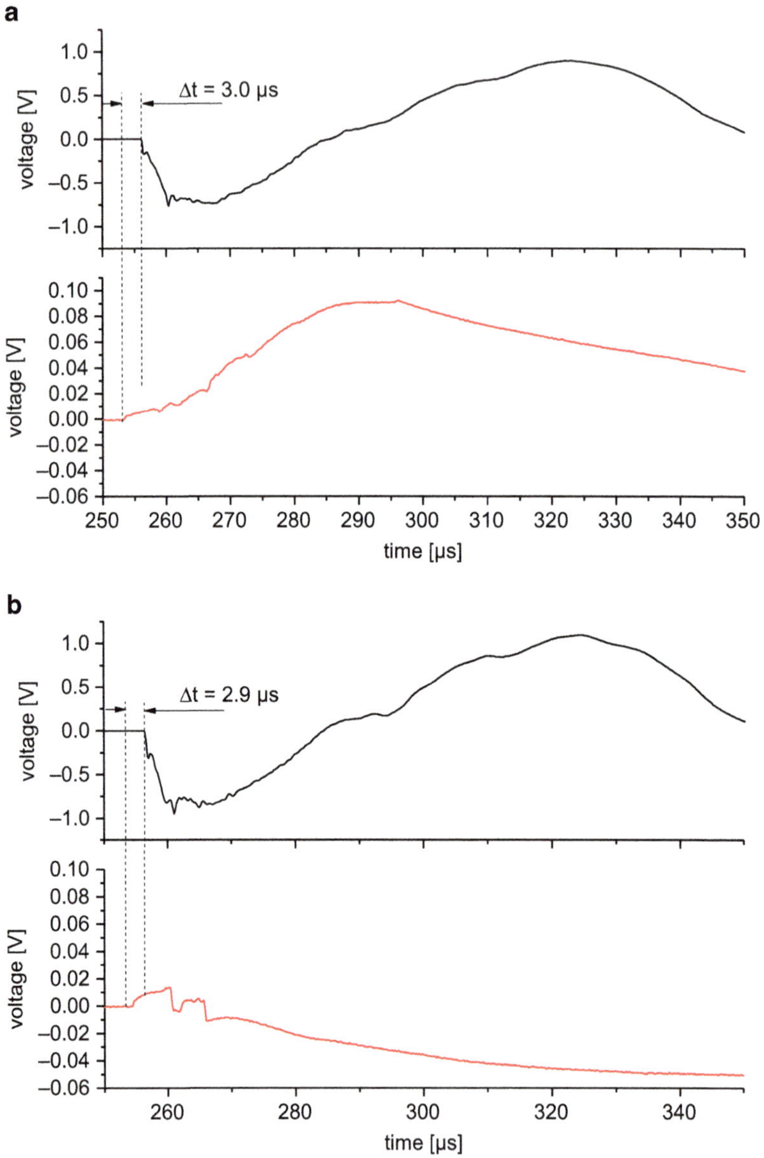

Fig. 7.43 Representative set of EME signals (*red*) and AE signals (*black*) for failure of single filament failure as presented in Sect. 5.5

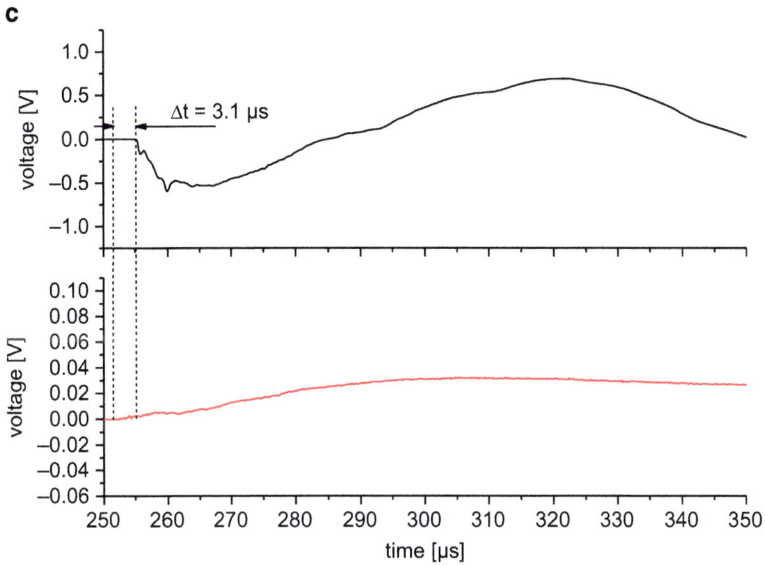

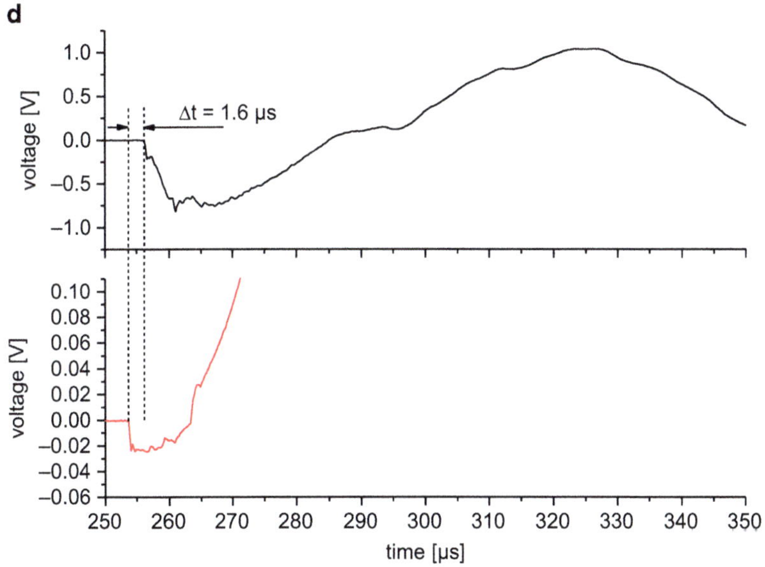

Fig. 7.43 (continued)

Δt-values not between sensors, but as absolute measure relative to the time of excitation t_0 as seen in Fig. 7.43. However, this procedure is certainly restricted to the simultaneous occurrence of EME and AE signals, which is not always the case as discussed in Chap. 5.

The extraction of quantitative data to describe the crack dynamics is exemplified in Fig. 7.44. As demonstrated in Sect. 5.2, the EME source provides a direct

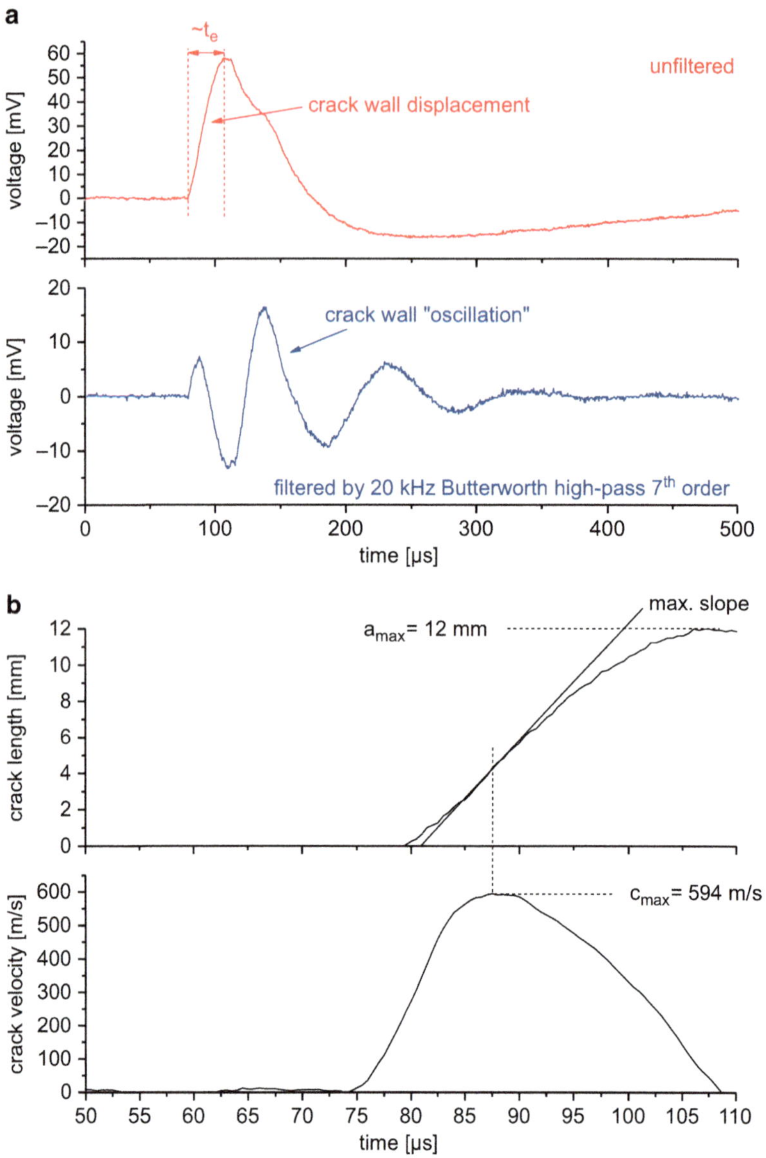

Fig. 7.44 Exemplary signal of an EME signal in composite failure to determine crack duration and source oscillation (**a**) and respective results to obtain crack tip position and crack velocity (**b**)

window to observe the movement of the charged crack walls as the function of time. In the "unfiltered" signals the dominating contribution is due to the generation of charges as a consequence of the newly formed fracture surface. For reasonably flat frequency sensors the first rise of the signal can thus be related to the duration of crack growth t_e. The subsequent decay is partially owed to the charge relaxation, but was found to be dominated by the 1 kHz high-pass used in the presented experimental setup of Fig. 5.64 in Sect. 5.5. If the respective fracture surface is known (e.g., by macroscopic or microscopic measurements) the voltage scale may directly be converted into a length scale of crack growth, since the initial and final boundary condition is known. For the given case in Fig. 7.44b, this was evaluated as 12 mm and thus allows measuring the crack tip position as the function of time. Likewise, the derivative of this signal may be used to measure the crack velocity as the function of time. In the present example this evaluates 594 m/s as the maximum velocity, which is reasonably less than the Rayleigh wave speed expected for fracture of a carbon fiber-reinforced polymer failing parallel to the fiber axis.

Using a high-pass filter setting as shown in Fig. 7.44a, the "unfiltered" signal may also be reduced to reveal only the oscillating part of the signal. As presented throughout Sect. 5.5, this oscillation is in good agreement to the detected AE signals of flat-with-frequency sensors. Based on the considerations on EME source operation in Sect. 5.2, it may be readily understood as the oscillation of the newly formed crack walls. Thus, the information on its frequency, amplitude rise, and decay are valuable measures to compare to respective modelling attempts. As has been demonstrated in Sect. 5.4, the EME signals of a fracture event can be quantitatively predicted by respective modelling means. Likewise, the results of AE models presented in Sect. 5.2 may directly be validated in the near-field of the source by using such EME measurements.

Finally, Sect. 5.5.2 has presented the possibility to perform orientation measurements of the fracture surface by means of EME. Based on the distinct EME radiation pattern, this allows to inversely determine the internal orientation of the fracture plane. Since such information may not be obtained by acoustic emission analysis using guided waves,[1] the use of an array of EME sensors complements the possibilities of AE in this respect. Therefore, the recording of multiple EME and AE signals of one fracture event in a composite material can be used to obtain a multitude of information:

1. 2D position of fracture (AE and EME source localization)
2. Type of fracture (AE source classification procedures)
3. Orientation of fracture plane (EME radiation pattern analysis)

The availability of this information provides a substantial increase in the knowledge on failure evolution in composite materials provided in situ. However, one of the general difficulties is given by the absence of EME signals for many AE signals. As discussed in Sect. 5.5, this may likely occur due to two reasons.

First, the overall sensitivity of the EME equipment may not be at the same level as for comparable AE equipment. There is an ongoing development to increase the sensitivity of the EME sensors, but so far it is not absolutely clear, if all EME

[1] For volumetric propagation media, moment tensor inversion procedure constitutes a possibility to analyze the orientation of the fracture plane.

signals can be detected by the used sensor systems, or what fraction of signals is hidden in the background noise.

As second reason, it may be speculated if every AE source is also an EME source. For all types of friction-based AE sources, there is no breaking of bonds on the microscopic scale and, consequently, no EME should be expected. Also, the amount of charges present on the surface may change as the function of material. If this drops below a certain value, no measurable EME will be generated. Similarly, if no charge imbalance occurs, no EME signal is to be expected. For the latter effect, it is required to clarify if the charging of the fracture surface discussed in Sect. 5.2 is a statistical phenomenon or if the amount of generated charges obeys a strict relationship to the material properties and the spatial dimension of the fracture zone.

7.2.7 Acousto-Ultrasonics

As discussed in Sects. 7.1 and 2.4.6, the method of acousto-ultrasonics is a combination of guided wave testing with acoustic emission analysis. In this approach the capabilities of acoustic emission measurements are fully incorporated. The major improvement consists in the ability to additionally sense defects not currently active in the material by means of guided wave analysis. As has been demonstrated in a variety of publications in the past [38–42] this allows continuous assessment of the damage progression and allows visualization of damaged areas in a broad range of materials and applications. Since this has been covered by many scientific publications it is not intended to repeat these findings in detail within this book. Instead the reader may refer to specific publications dealing with the method, i.e., a special issue in the Journal of Acoustic Emission [43].

In the context of this book, the acousto-ultrasonics method is used in the following to provide additional insight on the limiting factors of the acoustic emission method in terms of source classification and source localization. As pointed out in Sects. 4.5 and 4.6 the uncertainty of localization and uncertainty of classification are directly affected by changes of the propagation path between source and sensor. Thus the question arises, whether the formation of damage and the dimensional change of the material will affect the extracted feature values, such as Δt-values used for source localization or frequency parameters used for source identification. To assess the severity of the effect, tensile tests of six different stacking sequences are presented as representative data basis for some first conclusions.

The basic experimental configuration is identical to the setup presented in Sect. 4.7.3.2, but uses an additionally mounted piezoelectric pulser directly below the bottom AE sensor as seen in Fig. 7.45. By an arbitrary waveform generator, a 10 V pulse with duration of 10 μs and 20 ns rise-time is generated to excite an acoustic emission signal having a characteristic bandwidth between 1 and 800 kHz (cf. Fig. 7.46a, b). The use of a signal bandwidth with reasonably high frequencies

Fig. 7.45 Experimental setup used for acousto-ultrasonics and tensile specimen geometry

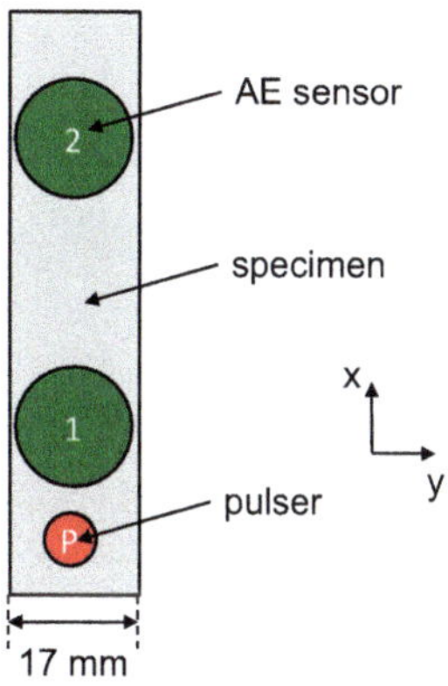

is based on the findings of Sect. 4.3, which indicate higher sensitivity to interact with damage for shorter wavelengths.

Pulses are repeated every 2 s and are detected in addition to the regular AE signals as obtained from specimen failure during tensile loading. All recorded AE signals are post-processed in the same way to obtain the typical AE features. Due to the periodicity of the signals of the piezoelectric pulser, these are easy to distinguish from the rest of the AE signals and are hence marked as pulser signals in the following. For further interpretation of these signals, only signals of detector 2 were evaluated, since these signals have passed through 45–50 mm of potentially damaged specimen length.

Figure 7.47 shows an example of the calculated Δt-values of the pulse travelling from one sensor to the other. As seen by the gradual decrease of the Δt-value, the increasing amount of damage and the gradual change of specimen dimensions cause a measurable shift of almost 1 µs during the experiment. Using the group velocity of the S_0 mode in this configuration, this translates into a resultant shift in source position of 5.8 mm. This finding constitutes the extreme case of evaluation results found for the investigated laminates. Compared to the other influences on the uncertainty of localization as described in Sect. 4.6, this value is within the same order as other effects that affect the source localization accuracy. Since the shift in Δt-values is expected to be proportional to the propagation length this effect is likely even more severe for larger structures. Hence, substantial contributions to source localization errors may occur. For the other laminate configurations the Δt-value changes were typically less, but also faced nonmonotonic trends. For the quasi-isotropic laminate, first an average decrease of around 0.40 µs was found and with the onset of substantial damage, the trend changed, resulting in an average increase of the Δt-values by 0.97 µs. The findings of the other laminates including their corresponding deviation in source positions are given in Table 7.10.

For the other extracted feature values, the plot of Partial Power 2 (150–300 kHz) versus weighted Peak-Frequency is used as a representative diagram. Superimposed to the partition of signals due to specimen failure, a small cluster corresponding to the detected pulser signals is readily visible for each case shown in Fig. 7.48.

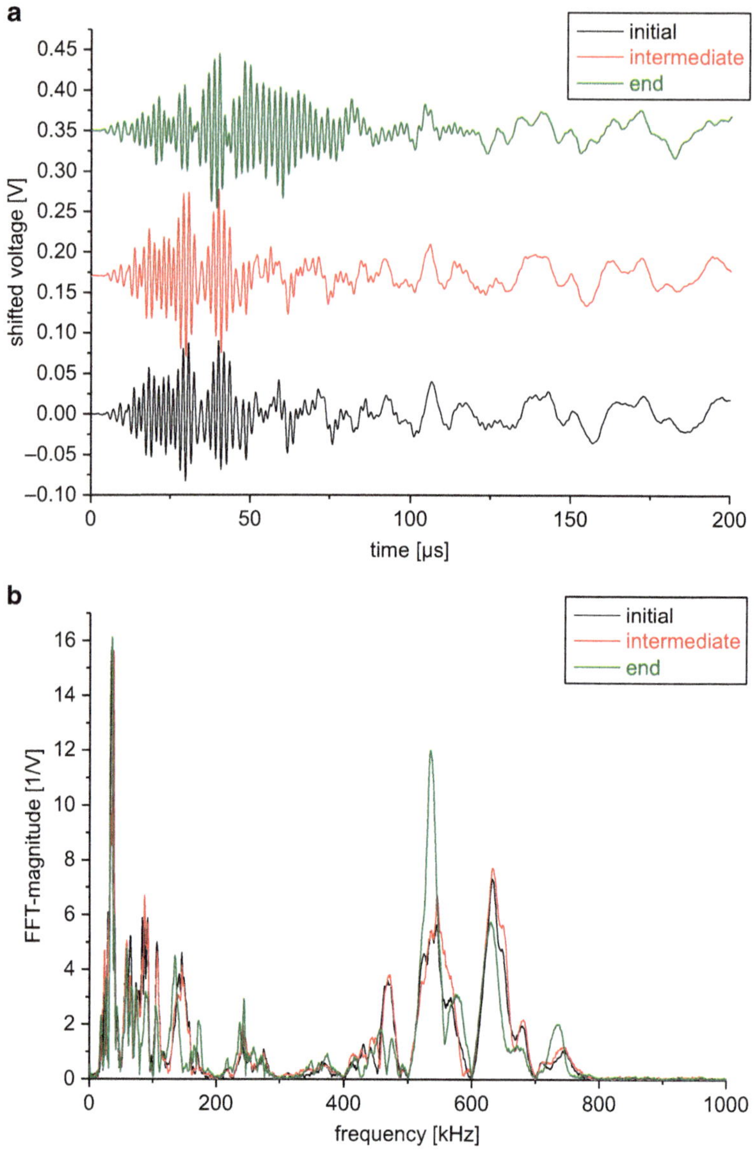

Fig. 7.46 Comparison of initial, intermediate, and end signal detected at sensor 2 (**a**) and corresponding frequency spectra (**b**) for a $[0/+45/-45/90]_{sym}$ laminate

The absolute number of AE signals is subject to the respective laminate type. For better visibility, the extent of the cluster corresponding to the pulser signals is marked as black ellipsoid. During the progression of the experiment (and hence damage) the pulser signals frequency content does not change substantially. This is already seen quite well in the three representative signals shown in Fig. 7.46. Since

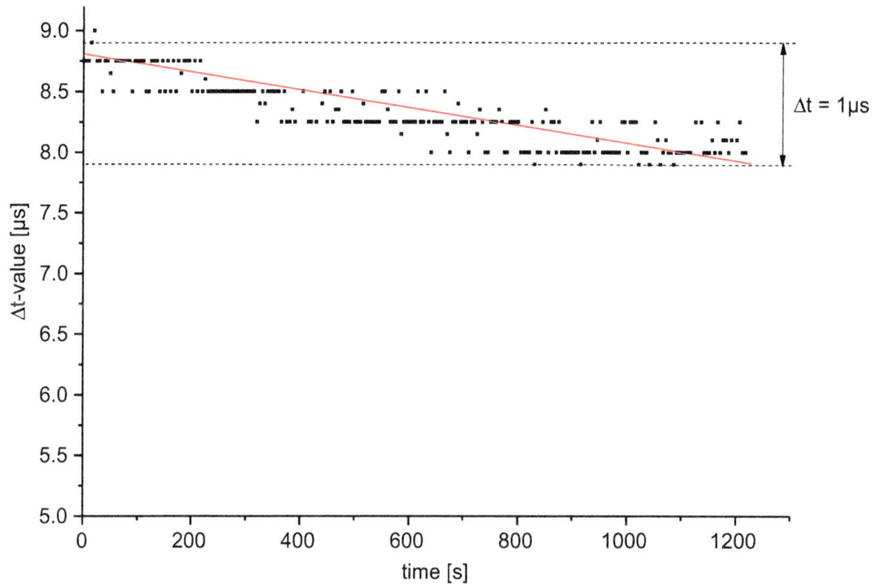

Fig. 7.47 Calculated Δt-values for $[0/90]_{sym}$ laminate as the function of experiment duration

Table 7.10 Evaluation results of average Δt-values of all laminate configurations

Laminate type	Δt-value (μs)	Deviation of source location (mm)
$[0_4]_{sym}$	0.27	1.9
$[0/+45/-45/90]_{sym}$	First drops 0.40 then increases 0.97	First drops 2.2 then increases 5.4
$[0_2/90/0_2]_{sym}$	0.20	1.3
$[0_2/90_2/0]_{sym}$	0.20	1.0
$[0/90_3/0]_{sym}$	0.35	1.9
$[0/90]_{sym}$	1.00	5.8

highest frequencies are expected to be more sensitive to the interaction with damage within the laminate, the findings below may be considered as extreme case. For all laminate configurations, the position of the pulser signals cluster falls at the range of weighted Peak-Frequency between 500 and 600 kHz at lowest Partial Power 2 levels. Despite the damage growth in the specimen, the resulting spread of the signals does not move outside the corresponding fiber breakage cluster or switches to another cluster. Compared to the other factors discussed in Sect. 4.5 the minor shift in frequency positions observed in the partitions of Fig. 7.48 would not substantially interfere with attempts to perform source classification.

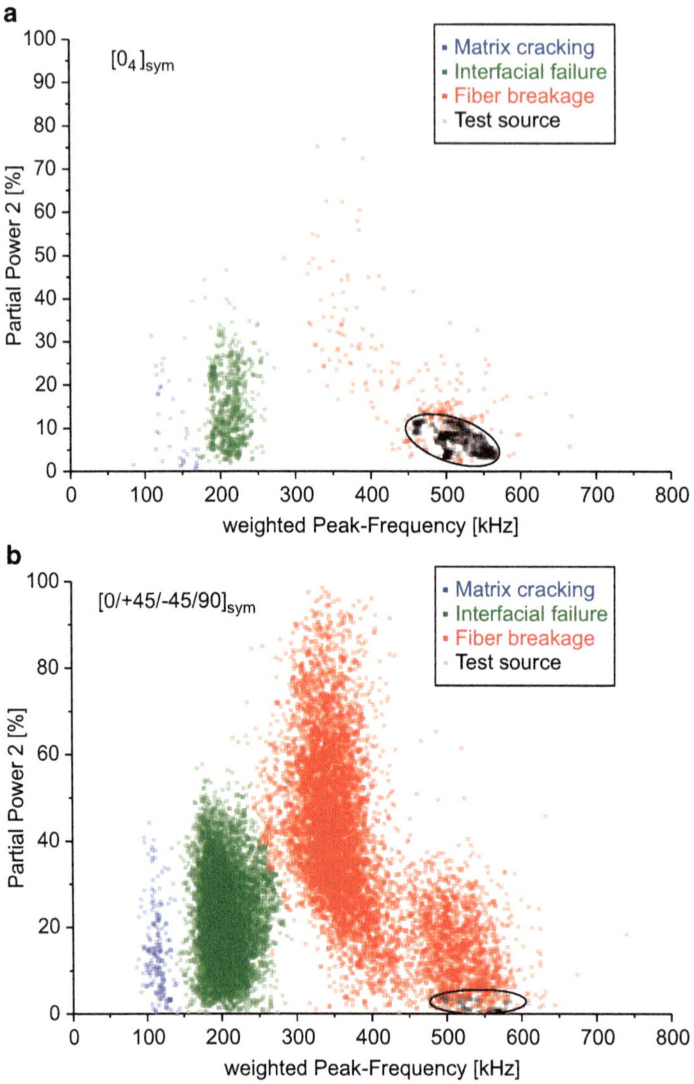

Fig. 7.48 Representative partitions for each laminate configuration with superimposed cluster of pulser signals marked with black ellipsoid

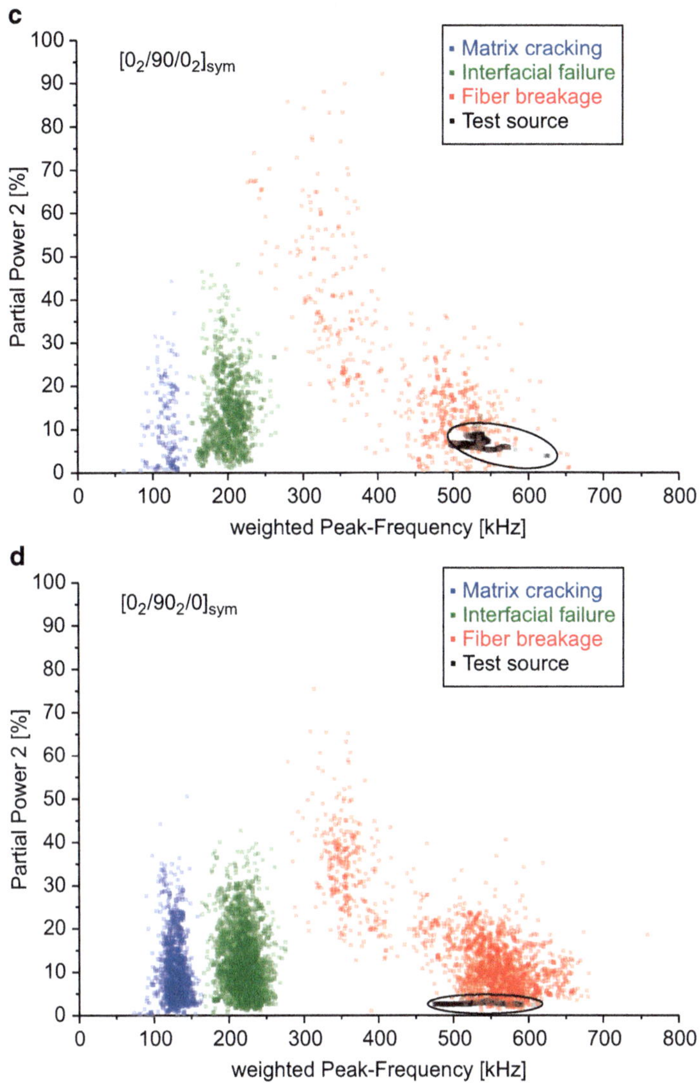

Fig. 7.48 (continued)

7.3 Future Developments

The intent of this book is to present the current state-of-the-art of in situ test methods as available for composites. Still some of the challenges and drawbacks associated with each method are at the edge of being overcome by technological improvements. For every method discussed herein, there are ongoing developments and a constant flow of new ideas may push the capabilities of the various techniques beyond currently existing limits. Therefore, the aim of this final section is to provide a short outlook on emerging developments of each method and to discuss the future prospects of method combinations as presented in Sect. 7.2.

7.3.1 Digital Image Correlation

Digital image correlation has already made its way to a mature measurement technology. One urgent issue to overcome is the missing standardization in the optical measurement technologies. This is currently discussed intensively in various standardization organizations and deals with the general applicability of DIC as strain measurement tool as well as the explicit allowance to perform strain measurements by an optical technique instead of a strain gauge. A further item achieving much attention these days is the standardization of the equipment as such and the need to calibrate systems having a standardized measurement reference, i.e., objects with optimal speckle patterns undergoing known displacements and deformations. Being an imaging technique DIC benefits from the technological evolution in camera systems. From an economical perspective, the use of high-quality and high-resolution camera systems becomes more and more feasible. Also, a further boost in resolution of camera systems extending the range beyond 20 megapixels will substantially impact the detail visibility of DIC systems.

One further challenge stems from the fusion of information of different DIC systems. As a technical alternative to monitor one test object by one system, it is also possible to use several systems to perform such monitoring. In this case, the primary task is to combine the recorded information in one coordinate system.

At the heart of the DIC method, the image correlation algorithm itself is still an ongoing development. Pushing the limits in correlation with sub-pixel accuracy is the key to get to even better quantitative strain measures and to allow spotting of strain concentrations in fiber-reinforced materials. For this material another challenge still originates from the distinction between natural anomalies and defect indications. This is a general problem of fiber-reinforced materials, which due to their inhomogeneity and local fluctuation in reinforcement fibers start to produce strain signatures which are not necessarily linked to the formation of defects. This part of DIC will certainly benefit from the development of damage tracking algorithms in machine vision, since discrimination between indications and material structure is a current field of research common to many imaging-based inspection methods applied to fiber-reinforced materials.

7.3.2 Acoustic Emission

As an established measurement technology, acoustic emission is already applied to many composite applications and has already seen some standardization attempts.

Still deficiencies exist in the primary calibration of the sensing equipment and more specific standards for composite applications. Also, new analysis routines are hardly incorporated in existing standards. Since the quality of interpretation might benefit by incorporation of source classification procedures or improved localization routines it would be appropriate to perform some standardization in their context for future applications. For the field of AE source identification, the methods proposed herein are generally of high value and were found to be well suited for the presented purposes. In order to extend the proposed classification strategies to larger, thicker, or geometrically more complex structures made from fiber-reinforced materials, still more research efforts are required. The same applies for attempts to perform valid source classification in the presence of external noise. In particular the ability to update source classification strategies as damage starts to develop or as initiation of further AE sources happens [44, 45] is of vital importance to get to a generally accepted source classification strategy. Similarly, the process of source localization may need even closer attention than seen so far. Based on the change of propagation medium and the growth of damage, simple localization algorithms come with the price of increased uncertainty at higher load or damage levels. Accordingly an implementation using an automatic update of the source localization algorithm during operation could be used to overcome such loss of accuracy.

A particular type of AE source, which has received only little attention in the past, is the AE from stick–slip friction. For a fiber-reinforced materials, the rough fracture surfaces that form within the material is naturally causing such AE sources to a certain extent. It may be speculated that friction AE is of a different characteristic than the AE of fracture processes, but there is currently no proof by experiments or modelling work. Overall, the reliable implementation of fracture mechanics-based modelling strategies to yield quantitative AE signals for friction sources and material fracture could be understood as key to increase the overall confidence put into the AE method.

Using such validated AE methods, they could certainly serve to indicate the initiation and growth of damage as discussed herein, but also to perform damage prediction and lifetime prediction of structures in service. Currently existing standards are partially making use of this concept already, but their applicability is always limited to the generic type of structures the standard covers (e.g., lift devices [46, 47] or pressure vessels [48, 49]).

Another limitation to overcome in the future is the design of AE sensors. The currently used commercial devices are comparatively bulky and are not well suited for a permanent monitoring of lightweight materials. To this end various groups have already recommended embedded sensor systems for AE monitoring [50, 51], most of them being still not as sensitive as existing solutions and are not appropriate for frequency analysis due to system resonances. However, the general trend to fabricate integrated sensing systems is directly compatible with fabrication of fiber-reinforced materials and may be of stronger relevance in future structural designs. Alongside with these attempts, the ongoing development in MEMS is generally of high relevance to the field of AE. Following the idea to have commercially available sensor types of low cost and low weight, the integration of AE as permanent monitoring technology is much more economical and technically competitive. In the same context, the wireless operation of AE sensors is receiving much attention. The removal of wires is a weight-saving factor by itself, but also limits the need for access to the structure and especially in the field of remote inspection sites might be superior to currently existing solutions.

7.3.3 Electromagnetic Emission

Electromagnetic emission is a phenomenon, which has already demonstrated its value in basic research applications. It nicely complements acoustic emission technology and adds more information to interpret material failure. To become a standalone measurement technology, there are several aspects that need to be addressed by future research. First, the technology is still emerging, so substantial improvement with respect to the sensing technologies is necessary. Here the key parameters to optimize are the signal-to-noise ratio within the experimentally used bandwidth and likewise the sensitivity of the sensing system relative to background noise. If any of these aspects can be improved by several orders of magnitude, the necessity of shielding chambers might become obsolete. A closely related topic is the uniqueness of the transfer function of each sensing system, which makes comparison of the measurement results very challenging. To this end, first guidelines have been established by means of this book, but these are not generally applicable and might fail to be applicable for other experimental configurations than those presented herein. Likewise, the development of miniaturized (MEMS type) sensing systems or other sensor concepts as presented in Chap. 5 could boost the technology if the current drawbacks can be overcome by such sensor systems.

From the basic research point of view, a scientifically sound description of the source mechanism is required. Beyond the initial assessments presented within this book, there is need for validation of the presented theories in other material systems and material combinations. Based on the electrical properties of different materials, the present description of the EME source could turn out to be valid or could lack in the level of detail to describe the EME source operation.

Having such a validated EME method, it may act as an ideal tool to complement AE or may act as an independent tool for inspection of fiber-reinforced materials. The insight gained is likely to be used for material development and improvement of failure theories, because it provides essentially new information on the formation and propagation of cracks. Particular attention should be paid to the difference in the absolute number of AE and EME signals. Despite the generally less sensitivity of the EME system it might be speculated if every AE source also operates as an EME source. The phenomenon of stick–slip friction is well-known to cause noticeable AE but since there is no real crack formation it can be hypothesized if such mechanisms will also cause EME. If this is not the case, such simultaneous recording might well be suited for advanced source discrimination. This would be a valuable approach to improve the understanding of a well-known phenomenon in acoustic emission. In cyclic loading of composites, the Felicity-ratio becomes smaller than one when approaching the ultimate limit of the material. Some researchers have speculated if the AE signals detected before the previous load limit are mostly due to friction sources [52, 53]. If EME is able to safely discriminate between AE due to crack growth and AE due to friction this would help to clarify a long-living ambiguity in interpretation of AE results in such load scenarios.

7.3.4 In Situ Computed Tomography

For the applications involving in situ computed tomography, many technological developments have already moved the frontiers of the method in the last 10 years. Naturally, the method benefits from all advances in the design of the computed tomography devices. There is still a trend to achieve smaller focal spot sizes since this has still not reached its conceptual limits. Also, the detector systems still profit from ongoing developments. In the context of in situ CT the most important factor would be the considerable reduction of the scan duration. This could be achieved by more sensitive detector systems or by substantial reduction of the signal-to-noise ratio of the detector. Alongside with the attempts to achieve higher resolution, there are various efforts to scale the whole CT concept to be able to inspect much larger parts. Similar benefits in terms of detail visibility are expected from the reconstruction algorithms, which are still subject to numerical improvements. Furthermore, new concepts to reduce scan durations or to reduce the required X-ray intensities have been proposed by means of dual-source-CT and dual-energy CT [54–57]. Also other imaging modes like bright-field, dark-field, and phase-contrast imaging have demonstrated their potential for improving the visibility of details [58, 59].

In the specific field of in situ CT, the amount of data being processed for damage tracking or digital volume correlation is one of the key challenges for algorithms. Hence, many efforts are currently put into development of smarter algorithms to address existing fields of application like damage mark-up or damage tracking as well as digital volume correlation. The extension of the latter to a technological level currently seen for DIC will enable a much wider field of application of digital volume correlation to perform research on the mechanical behavior of materials.

Finally, X-rays are just one type of electromagnetic waves, which are well suited for volumetric imaging of fiber-reinforced materials. Other volumetric imaging sources, like microwave tomography or Terahertz-tomography are currently emerging and might be used for future inspection of fiber-reinforced materials.

7.3.5 Method Combinations

The combination of various measurement systems has become a field of research on its own. Instead of relying on the information provided by one measurement system, there is a strong trend to fuse information of different equipment. One challenge directly related to these approaches is the lack of compatibility between different measurement systems. Frequently the equipment is provided by different manufacturers and there is no technical solution to directly merge the acquired information of different systems. In principle there are common data exchange formats, which are capable to store and distribute the acquired information. However, there is a general lack of implementation of these formats in the commercial equipment due to the missing standardization. But data fusion of measurement systems has been identified as key requirement for future technologies. With the advances in information technology, it is likely that such data exchange might become a standard at some point. Until then, specific software needs to be developed, which allows an easy fusion of discrete data pools originating from many different measurement systems. This software needs to be ready for 3D visualization and needs to show a high level of integration in the existing software packages of commercial devices as the currently used portable document file (pdf) standard according to ISO 32000-1.

References

1. Lee, J.-R., Molimard, J., Vautrin, A., Surrel, Y.: Application of grating shearography and speckle shearography to mechanical analysis of composite material. Compos. Part A Appl. Sci. Manuf. **35**, 965–976 (2004)
2. Ono, K.: Structural integrity evaluation using acoustic emission techniques. J Acoust. Emiss. **25**, 1–20 (2007)
3. Cuadra, J., Vanniamparambil, P.A., Hazeli, K., Bartoli, I., Kontsos, A.: Damage quantification in polymer composites using a hybrid NDT approach. Compos. Sci. Technol. **83**, 11–21 (2013)
4. Iliopoulos, S., Tsangouri, E., Aggelis, D.G., Pyl, L., Vantomme, J., Van Marcke, P., Areias, L.: Digital image correlation, acoustic emission and ultrasonic pulse velocity for the detection of cracks in the concrete buffer of the Belgian nuclear supercontainer. In: 31st Conference of the European Working Group on Acoustic Emission. pp. 1–7. Dresden (2014)
5. Chotard, T.J., Smith, A., Boncoeur, M.P., Fargeot, D., Gault, C.: Characterisation of early stage calcium aluminate cement hydration by combination of non-destructive techniques: acoustic emission and X-ray tomography. J. Eur. Ceram. Soc. **23**, 2211–2223 (2003)
6. Scott, A.E., Clinch, M., Hepples, W., Kalantzis, N., Sinclair, I., Spearing, S.M.: Advanced micro-mechanical analysis of highly loaded hybrid composite structures. In: ICCM 17—17th International Conference on Composite Materials, Edinburgh (2009)
7. Munoz, V., Vales, B., Perrin, M., Pastor, M.L., Welemane, H., Cantarel, A., Karama, M.: Coupling infrared thermography and acoustic emission for damage study in CFRP composites. In: The 12th International Conference on Quantitative InfraRed Thermography—QIRT 2014, Bordeaux (2014)
8. Sause, M.G.R., Richler, S.: Finite element modelling of cracks as acoustic emission sources. J. Nondestruct. Eval. **34**, 1–13 (2015)

9. Vanniamparambil, P.A., Bolhassani, M., Carmi, R., Khan, F., Bartoli, I., Moon, F.L., Hamid, A., Kontsos, A.: A data fusion approach for progressive damage quantification in reinforced concrete masonry walls. Smart Mater. Struct. **23**, 015007 (2014)
10. Fischer, G., Bohse, J.: Observation and analysis of fracture processes in concrete with acoustic emission (AE) and digital image correlation (DIC). In: 31st Conference of the European Working Group on Acoustic Emission. pp. 1–8. Dresden (2014)
11. Mao, W.G., Wu, D.J., Yao, W.B., Zhou, M., Lu, C.: Multiscale monitoring of interface failure of brittle coating/ductile substrate systems: a non-destructive evaluation method combined digital image correlation with acoustic emission. J. Appl. Phys. **110**, 084903 (2011)
12. Rouchier, S., Foray, G., Godin, N., Woloszyn, M., Roux, J.-J.: Damage monitoring in fibre reinforced mortar by combined digital image correlation and acoustic emission. Constr. Build. Mater. **38**, 371–380 (2013)
13. Aggelis, D.G., Verbruggen, S., Tsangouri, E., Tysmans, T., Van Hemelrijck, D.: Characterization of mechanical performance of concrete beams with external reinforcement by acoustic emission and digital image correlation. Constr. Build. Mater. **47**, 1037–1045 (2013)
14. Yekani Fard, M., Sadat, S.M., Raji, B.B., Chattopadhyay, A.: Damage characterization of surface and sub-surface defects in stitch-bonded biaxial carbon/epoxy composites. Compos. Part B Eng. **56**, 821–829 (2014)
15. Sause, M.G.R., Haider, F., Horn, S.: Quantification of metallic coating failure on carbon fiber reinforced plastics using acoustic emission. Surf. Coatings Technol. **204**, 300–308 (2009)
16. Baensch, F., Zauner, M., Sanabria, S.J., Sause, M.G.R., Pinzer, B.R., Brunner, A.J., Stampanoni, M., Niemz, P.: Damage evolution in wood: synchrotron radiation micro-computed tomography (SRμCT) as a complementary tool for interpreting acoustic emission (AE) behavior. Holzforschung **69**, 1015–1025 (2015)
17. Maire, E., Carmona, V., Courbon, J., Ludwig, W.: Fast X-ray tomography and acoustic emission study of damage in metals during continuous tensile tests. Acta Mater. **55**, 6806–6815 (2007)
18. Katsaga, T., Young, R.P.: Acoustic emission and x-ray tomography imaging of shear fracture formation in large reinforced concrete beam. J. Acoust. Emiss. **25**, 294–307 (2007)
19. Kalafat, S., Zelenyak, A.-M., Sause, M.G.R.: In-situ monitoring of composite failure by computing tomography and acoustic emission. In: 20th International Conference on Composite Materials. pp. 1–8. Copenhagen (2015)
20. Seco, F., Jiménez, A.R.: Modelling the generation and propagation of ultrasonic signals in cylindrical waveguides. In: Santos, A. (ed.) Ultrasonic Waves, pp. 1–28. Intech Open Access, Rijeka (2012)
21. Zelenyak, A.-M., Hamstad, M., Sause, M.: Modeling of acoustic emission signal propagation in waveguides. Sensors **15**, 11805–11822 (2015)
22. Zelenyak, A.-M., Hamstad, M.A., Sause, M.G.R.: Finite element modeling of acoustic emission signal propagation with various shaped waveguides. In: 31st Conference of the European Working Group on Acoustic Emission. pp. 1–8. Dresden (2014)
23. Sause, M.G.R., Horn, S.: Simulation of acoustic emission in planar carbon fiber reinforced plastic specimens. J. Nondestruct. Eval. **29**, 123–142 (2010)
24. Sause, M.G.R., Hamstad, M.A., Horn, S.: Finite element modeling of lamb wave propagation in anisotropic hybrid materials. Compos. Part B Eng. **53**, 249–257 (2013)
25. Sause, M.G.R., Hamstad, M.A., Horn, S.: Finite element modeling of conical acoustic emission sensors and corresponding experiments. Sensors Actuators A Phys. **184**, 64–71 (2012)
26. Hamstad, M.A.: Comparison of wavelet transform and Choi-Williams distribution to determine group velocities for different acoustic emission sensors. J. Acoust. Emiss. **26**, 40–59 (2008)
27. Choi, H.-I., Williams, W.: Improved time-frequency representation of multicomponent signals using exponential kernels. IEEE Trans. Acoust. Speech Signal Process. **37**, 862–872 (1989)
28. Rabinovitch, A.: A note on the amplitude-frequency relation of electromagnetic radiation pulses induced by material failure. Philos. Mag. Lett. **79**, 195–200 (1999)

29. Rabinovitch, A., Frid, V., Bahat, D., Goldbaum, J.: Fracture area calculation from electromagnetic radiation and its use in chalk failure analysis. Int. J. Rock Mech. Min. Sci. **37**, 1149–1154 (2000)
30. Rabinovitch, A., Bahat, D., Frid, V.: Similarity and dissimilarity of electromagnetic radiation from carbonate rocks under compression, drilling and blasting. Int. J. Rock Mech. Min. Sci. **39**, 125–129 (2002)
31. Frid, V.: Electromagnetic radiation method water-infusion control in rockburst-prone strata. J. Appl. Geophys. **43**, 5–13 (2000)
32. Rabinovitch, A., Frid, V., Bahat, D.: Parameterization of electromagnetic radiation pulses obtained by triaxial fracture of granite samples. Philos. Mag. Lett. **77**, 289–293 (1998)
33. Frid, V.: Electromagnetic radiation associated with induced triaxial fracture in granite. Philos. Mag. Lett. **79**, 79–86 (1999)
34. Gade, S.O., Weiss, U., Peter, M.A., Sause, M.G.R.: Relation of electromagnetic emission and crack dynamics in epoxy resin materials. J. Nondestruct. Eval. **33**, 711–723 (2014)
35. Sklarczyk, C., Altpeter, I.: The electric emission from mortar and concrete subjected to mechanical impact. Scr. Mater. **44**, 2537–2541 (2001)
36. Sklarczyk, C., Winkler, S., Thielicke, B.: Die elektrische Emission beim Versagen von Faserverbundwerkstoffen und ihren Komponenten. Materwiss. Werkstofftech. **27**, 559–566 (1996)
37. Sedlak, P., Sikula, J., Lokajicek, T., Mori, Y.: Acoustic and electromagnetic emission as a tool for crack localization. Meas. Sci. Technol. **19**, 045701 (2008)
38. Gallego, A., Ono, K.: An improved acousto-ultrasonic scheme with lamb wave mode separation and damping factor in CFRP plates. J Acoust. Emiss. **30**, 109–123 (2012)
39. Vary, A.: The acousto-ultrasonic approach. In: Duke, J.C. (ed.) Acousto-Ultrasonics, pp. 1–21. Springer, Boston (1988)
40. Guo, N., Cawley, P.: Lamb wave propagation in composite laminates and its relationship with acousto-ultrasonics. NDT E Int. **26**, 75–84 (1993)
41. Kwon, O.-Y., Lee, S.-H.: Acousto-ultrasonic evaluation of adhesively bonded CFRP-aluminum joints. NDT E Int. **32**, 153–160 (1999)
42. Philippidis, T.P., Aggelis, D.G.: An acousto-ultrasonic approach for the determination of water-to-cement ratio in concrete. Cem. Concr. Res. **33**, 525–538 (2003)
43. Ono, K.: Special issue: acousto-ultrasonics. J. Acoust. Emiss. **12**, 1–102 (1994)
44. Serir, L., Ramasso, E., Zerhouni, N.: Evidential evolving Gustafson–Kessel algorithm for online data streams partitioning using belief function theory. Int. J. Approx. Reason. **53**, 747–768 (2012)
45. Doan, D.D., Ramasso, E., Placet, V., Boubakar, L., Zerhouni, N.: Application of an unsupervised pattern recognition approach for AE data originating from fatigue tests on CFRP. In: 31st Conference of the European Working Group on Acoustic Emission. pp. 1–8. Dresden (2014)
46. Anastassopoulos, A., Tsimogiannis, A., Kouroussis, D.: Unsupervised classification of acoustic emission sources from aerial man lift devices. In: 15th World Conference on NDT, Roma (2000)
47. Bingham, A.H., Ek, C.W.: Acoustic emission testing of aerial devices and associated equipment used in the utility industries, Ausgabe 1139. ASTM International (1992)
48. Downs, K.S., Hamstad, M.A.: Acoustic emission from depressurization to detect/evaluate significance of impact damage to graphite/epoxy pressure vessels. J. Compos. Mater. **32**, 258–307 (1998)
49. Hill, E.K., Dion, S.T., Karl, J.O., Spivey, N.S., Ii, J.L.W.: Neural network burst pressure prediction in composite overwrapped pressure vessels. Neural Netw. **25**, 187–193 (2007)
50. Marin-Franch, P., Martin, T., Tunnicliffe, D.L., Das-Gupta, D.K.: PTCa/PEKK piezo-composites for acoustic emission detection. Sensors Actuators A Phys. **99**, 236–243 (2002)
51. Barbezat, M., Brunner, A.J., Flüeler, P., Huber, C., Kornmann, X.: Acoustic emission sensor properties of active fibre composite elements compared with commercial acoustic emission sensors. Sensors Actuators A Phys. **114**, 13–20 (2004)

52. Hamstad, M.A.: A discussion of the basic understanding of the felicity effect in fiber composites. J. Acoust. Emiss. **5**, 95–102 (1986)
53. Awerbuch, J., Gorman, M.R., Madhukar, M.: Monitoring acoustic emission during quasi-static loading-unloading cycles of filament-wound graphic-epoxy laminate coupons. Mater. Eval. **43**, 754–764 (1985)
54. Robert, C., Dinten, J.M., Rizo, P.: Dual-energy computed tomography for ceramics and composite materials. In: Review of Progress in Quantitative Nondestructive Evaluation. pp. 481–488. Springer, Boston (1996)
55. Kalender, W.A.: Computed Tomography: Fundamentals, System Technology, Image Quality, Applications. Wiley, New York (2011)
56. Petersilka, M., Bruder, H., Krauss, B., Stierstorfer, K., Flohr, T.G.: Technical principles of dual source CT. Eur. J. Radiol. **68**, 362–368 (2008)
57. Flohr, T.G., McCollough, C.H., Bruder, H., Petersilka, M., Gruber, K., Süss, C., Grasruck, M., Stierstorfer, K., Krauss, B., Raupach, R., Primak, A.N., Küttner, A., Achenbach, S., Becker, C., Kopp, A., Ohnesorge, B.M.: First performance evaluation of a dual-source CT (DSCT) system. Eur. Radiol. **16**, 256–268 (2006)
58. Momose, A., Takeda, T., Itai, Y., Hirano, K.: Phase-contrast X-ray computed tomography for observing biological soft tissues. Nat. Med. **2**, 473–475 (1996)
59. Takeda, T., Momose, A., Hirano, K., Haraoka, S., Watanabe, T., Itai, Y.: Human carcinoma: early experience with phase-contrast X-ray CT with synchrotron radiation—comparative specimen study with optical microscopy. Radiology **214**, 298–301 (2000)

Appendix A: Acoustic Emission—Parameters of Influence

For detection of acoustic emission signals, a variety of parameters need to be considered during the experiment. The aim of the following list is to provide an overview on the potential influence of some of these settings. Moreover, it is also meant to act as a guide for selecting suitable parameters for detection of acoustic emission signals in composites.

A.1 Acquisition System Electronics

- Threshold setting should be 3–4 dB above noise floor. If threshold is set to high a potential loss of signals below the threshold level could occur.
- Choice of trigger settings (e.g., PDT/HDT/HLT). Erroneous setup may cause triggering on echoes or system may not trigger on relevant signals.
- High-pass or low-pass filtered signals may be used for triggering another set of sensors, whereas triggered channels record without filtering but higher reliability.
- Independent triggering of channels may possibly not yield a set of signals to allow source localization because of missing signals.
- Synchronous triggering of channels may record some signals, which are hidden in the noise floor.
- Dynamic range of waveform recording relative to the range of amplitudes present in the experiment. Possible saturation of signals if chosen to low or loss of signals if they fall below threshold.
- Suitable selection of gain. Possible saturation of signals if chosen to low or loss of signals if they fall below threshold.
- Bit resolution of acquisition card. If not sufficient, wrong frequency content for digitized signals may result and nonlinear digitization effects may occur.
- Poor digitization rate may cause wrong recording of signal frequency content and may increase the uncertainty for localization.

M.G.R. Sause, *In Situ Monitoring of Fiber-Reinforced Composites*,
Springer Series in Materials Science 242, DOI 10.1007/978-3-319-30954-5

- High acquisition rates of signals may lead to downtimes of system due to data storage.
- Signal termination due to missing signal depth (amount of memory of recording system for individual signal) may cause lack of information for signals with long durations.
- Memory handling concepts of either onboard storage or hard disk storage. Both concepts may lose signals during storage if interfering with other high-priority tasks running on the system.
- Usage of waveform streaming during experiment may possibly limit the available digitization rate.
- Bad signal-to-noise ratios due to the quality of equipment or induced electronic noise may cause wrong computation of signal features.
- High number of channels processed by one computer may saturate the electronic processing capabilities.
- Band pass settings (either analog in preamplifier/card or digital band pass in acquisition settings) may cut-off frequency ranges of interest.
- Dead-time after triggered signal acquisition may lose further signals arriving within this time window.
- Application of energy or amplitude filters may skip low amplitudes or energy signals.
- Arrival-time filters allowing only recording of signals within certain arrival time at different sensors may lose signals not detected by more than one sensor.
- Use of load filter in fatigue testing to record only signals above a certain load level.

A.2 Sensing Technology

- Too low sensor density may provide detectability of AE sources only within certain range around the sensors
- Signal-to-noise ratio can be influenced by quality of the sensor and preamplifier and the cables used
- Sensor types (e.g., resonant, multi-resonant, near flat with frequency) can have significant impact on detected frequency spectra and signal amplitudes
- Sensor frequency characteristics (e.g., flat, resonant) can have significant impact on detected frequency spectra and signal amplitudes
- Sensor response type (e.g., displacement sensitive, velocity sensitive, combination) can have significant impact on detected frequency spectra and signal amplitudes
- Couplants may produce intrinsic AE during temperature exposure, may cause additional attenuation, and may reduce transmission of shear (in-plane) components

- Use of waveguides may cause frequency filtering, additional attenuation effects, resonances and may show different sensitivity to the in-plane and out-of-plane components than for sensor system mounted directly on specimen
- The position of the sensor relative to radiation pattern of a oriented sources may cause substantial differences in the detected energy or frequency of the signal
- The propagation direction between the source and the sensors in anisotropic materials may cause substantial changes to the detected signals
- The aperture of the sensor may reduce the ability to see certain (higher) frequencies and the presence of sensors with larger apertures will also disturb the wave propagation
- The mismatch in acoustic impedance between the sensors' wear plate and the material under test may cause different sensitivity
- The sensor system needs to match the input capacitance and impedance of the preamplifier system, since they form a coupled system
- Usage of integrated preamplifier may increase signal-to-noise ratio, but comes with less flexibility for selection of preamplifier gain
- Use of integrated preamplifiers may also simplify mounting procedure of measurement chain, e.g., in industrial environments operating on cranes
- The detector position (surface side) of an AE sensor on hybrid materials may show different sensitivity due to guided waves in individual layers

A.3 Material and Geometry

- Material may have anisotropic propagation velocities to consider for sensor placement
- As a function of frequency (anisotropic) propagation velocities may change as function fo frequency due to dispersion effects
- The attenuation of the inspected material needs to be considered to select appropriate sensor positions
- Attenuation due to geometric spreading or other effects (e.g., thermoelastic effect) may cause fast decay above a certain frequency level and may not be detectable by narrow band acquisition systems after some distance
- The effective rise-time of an AE source is linked to the generated frequency bandwidth and may not allow detection of certain source types with acquisition systems of narrow bandwidth
- The physical size of the source will partially determine the amount of energy release (e.g., bundle of filaments vs. single filament) and may cause substantial differences in detectability of the AE signals
- For materials with high AE activity, one continuous source or multiple sources may exist at the same spot
- The orientation of the source inside material may cause characteristic radiation patterns and may limit detectability for certain sensor positions

- The depth-position of the AE source (level below surface) will govern the relative excitation of different Lamb-wave modes
- For materials with pronounced ductility (high fracture toughness) the source rise-times may become too long to produce AE in the ultrasonic or audible range
- Fracture of inner parts of the material (e.g., inclusions, fibers, etc.) may produce AE due to different fracture toughness on the microscale with sufficiently short rise-times
- For materials with very low modulus large rise-times are expected which may become too long to produce AE in the ultrasonic or audible range
- The sample thickness will govern the number of Lamb-wave modes in the typical AE bandwidth
- The specific specimen layup will have substantial impact on the degree of anisotropic signal spreading and anisotropic attenuation
- The volume fraction of the constituents of the composite will define the elastic properties and therefore also the acoustic wave velocities and acoustic attenuation
- The diameter of reinforcement fibers may have an impact on the AE signal amplitude, since larger diameter fibers cause a larger fracture surface and at identical fracture toughness will produce a higher amplitude signal
- A change in temperature may induce change to the material properties or may induce a brittle to ductile transition, which will impact the AE source rise-times and the size of fracture surfaces, which in turn will influence the detectability, and possible also the overall AE activity
- Environmental changes such as moisture uptake may change the material properties or the overall AE activity
- Prior or initial damage may cause increased attenuation or changes in the signal path and potentially higher initial AE activity
- Presence of flaws near a source may affect radiation pattern in the near-field, while larger defects may cause changes in the propagation path, e.g., reflections
- Void content may cause higher attenuation levels, or initiation of damage due to stress concentration at pores
- Residual stresses are superimposed to external loads, which may cause an early initial AE onset
- Autofrettage cycles may cause continuous AE signals, which cannot be detected as transient signals
- Thermal expansion may also cause residual stresses which may cause AE signals
- Signals with multiple paths arriving at sensor location (e.g., water path, reflections at edges) may have a wrong signal arrival and may not allow identification of guided wave modes
- Signals superimposed by echoes or reflections may not be useful for frequency analysis or pattern recognition techniques based on frequency analysis
- Close position of the sensor relative to source may detect waves, which have not fully developed into guided wave modes
- Distinction is necessary between the generation of volume waves, guided waves, and surface waves

- The possibility of leakage of waves into fluids, attached solids, or air may need to be considered during sensor placement and signal interpretation
- Repeatability of AE of a specimen geometry also depends on the accuracy of geometry and the occurrence of typical manufacturing defects
- Strength distributions of the individual materials may cause scatter and characteristic ranges of signal amplitudes
- Fiber orientation relative to load "axis" may cause preliminary failure of specific (possibly disoriented) fiber bundles, while small misalignment of fibers may also cause early matrix cracking
- Some fibers are tightly aligned and carry load before other (loose) fibers are starting to carry the load, which may cause shifts of expected signal onsets corresponding to fiber breakage to lower load levels
- Stress concentration effects caused by geometry intended (e.g., fracture toughness testing or open-hole-tensile) or unintended (e.g., tapered areas, broken filaments) may cause AE onsets at unexpected (low) load levels
- Generation of relative magnitude of extensional versus flexural modes may affect source localization accuracy when based on a particular mode, since wave modes are subject to different attenuation levels

A.4 Experimental Configuration

- Fast loading rates may cause superimposed signals or continuous AE, which will not allow distinguishability of signals and, hence, possibilities to localize and classify AE signals
- Attenuation effects due to propagation into adjacent media may cause a substantial reduction of detection sensitivity at some spots
- Bad signal-to-noise ratio might be caused by general background noise, presence of continuous AE, presence of friction sources, bad cable shielding or other effects
- Is the testing scheme suitable to generate AE or induce damage in general?
- Sufficiency of load or test level to record AE data relative to the average strength of the item (Is the analysis based only on precursor signals, or even absence of signals?)
- Fixtures and grips may cause additional AE (i.e., damage is induced in these sections)
- Fatigue testing may cause extraneous noise due to friction at fixtures
- Hydraulic test equipment may cause extra noise signals due to servo valve operation
- Pneumatic pressurized grips or actuators may cause even more noise signals
- Uniformity of loading or pressurization may cause less repeatable loading conditions and therefore may induce non-repeatable AE onsets

- Uniaxial and biaxial loading conditions may cause different evolution of failure mechanisms from specimen to specimen and therefore different generation of AE sources, resulting in different energy release of AE sources
- For material with high AE activity, the method may also be seen as statistical approach (lots of AE data per experiment, hence loss of some signals in dead-time and related settings is not as relevant)
- For material with low AE activity each AE signal is important to capture since it contributes a substantial amount to the total signature
- For the detection of AE onsets a high gain to capture low amplitude signals may result in different interpretation than low gain measurements
- For monitoring of structural integrity, the gain needs to be chosen adequate to capture signals representative of the damage state (these may be low or high amplitude signals)
- For detection of failure evolution (e.g., materials testing) the gain needs to be set to avoid saturation of high amplitude signals, while keeping it high enough to detect low amplitude signals

Appendix B: Material Properties

The following tables list the material properties used in the analytical and numerical calculations. For better readability, distinction is made between elastic and piezo-electric properties as well as values used for failure criteria (Tables B.1, B.2, B.3 and B.4).

Table B.1 Elastic properties of materials used in modeling

	Density [kg/m^3]	Poisson-Ratio	Stiffness [GPa]
T800/913 (unidirectional, 60 % fiber volume fraction)	1550	–	$C_{11} = 154.0$
			$C_{12} = C_{13} = 3.7$
			$C_{22} = C_{33} = 9.5$
			$C_{23} = 5.2$
			$C_{44} = 2.5$
			$C_{66} = C_{55} = 4.2$
Sigrafil CE 1250-230-39 (unidirectional, 60 % fiber volume fraction)	1500	–	$C_{11} = 131.0$
			$C_{12} = C_{13} = 3.7$
			$C_{22} = C_{33} = 8.1$
			$C_{23} = 3.1$
			$C_{44} = 2.1$
			$C_{55} = C_{66} = 6.1$
T700/PPS (unidirectional, 60 % fiber volume fraction)	1600	–	$C_{11} = 152.7$
			$C_{12} = C_{13} = 4.7$
			$C_{22} = C_{33} = 11.7$
			$C_{23} = 4.4$
			$C_{44} = 3.6$
			$C_{55} = C_{66} = 4.5$

(continued)

M.G.R. Sause, *In Situ Monitoring of Fiber-Reinforced Composites*,
Springer Series in Materials Science 242, DOI 10.1007/978-3-319-30954-5

Table B.1 (continued)

	Density [kg/m^3]	Poisson-Ratio	Stiffness [GPa]
AS4/3502 (unidirectional, 60 % fiber volume fraction)	1550	–	$C_{11} = 147.1 + i3.0$
			$C_{12} = C_{13} = 4.1 + i0.9$
			$C_{22} = C_{33} = 10.6 + i0.3$
			$C_{23} = 3.1 + i0.2$
			$C_{44} = 3.8 + i0.1$
			$C_{55} = 6.0 + i0.2$
			$C_{66} = 6.0 + i0.3$
High strength alloy steel	7850	0.33	200.0
Stainless steel	7970	0.29	219.0
Aluminum AlMg3	2660	0.33	70.0
Aluminum 6063-T83	2700	0.33	69.0
Titanium Ti-6Al-4V	4430	0.40	110.2
Brass	8530	0.33	113.4
Al-filled epoxy	2700	0.40	30.0
Ag-filled epoxy	1700	0.45	2.7
HDPE	959	0.38	1.1
PTFE	2200	0.33	0.4
ETFE	1750	0.33	1.5
PPS	1350	0.36	3.8
PEI	1290	0.40	2.9
PI	1430	0.40	3.3
PVC	1760	0.40	2.9
PMMA	1190	0.32	6.2
RTM6	1140	0.38	2.8
HTA40	1770	0.20	240.0
T700S	1800	0.20	230.0
Al_2O_3	3965	0.22	400.0

Table B.2 Piezoelectric properties of PZT-5A as used in modeling

	PZT-5A
Density [kg/m^3]	7750
Stiffness [GPa]	$C_{11} = C_{22} = 120.3$
	$C_{12} = 75.2$
	$C_{13} = C_{23} = 75.1$
	$C_{33} = 110.9$
	$C_{44} = C_{55} = 21.1$
	$C_{66} = 22.6$
Piezoelectric constants [C/m^2]	$S_{31} = S_{32} = -5.4$
	$S_{33} = 15.8$
	$S_{24} = S_{15} = 12.3$
Electrical permittivity	$\xi_{11} = 919.1$
	$\xi_{22} = 919.1$
	$\xi_{33} = 826.6$

Subscript indices follow the conventions for piezoelectric materials

Table B.3 Electric properties of materials used in modeling

	Electrical conductivity [Sm^{-1}]	Relative permittivity
RTM 6 (@10 kHz)	6.6×10^{-9}	4.26
Copper	5.998×10^{7}	1
Aluminum	3.030×10^{7}	1
High strength alloy steel	4.032×10^{6}	1
Sigratex CE 1250-230-39 (unidirectional-parallel)	5.50×10^{6}	4.26
Sigratex CE 1250-230-39 (unidirectional-perpendicular)	0.204×10^{6}	4.26
Carbon fiber	1.0×10^{3}	1
PEEK (@10 kHz)	7.5×10^{-11}	3.21
PTFE (@10 kHz)	$<10^{-18}$	2.10
PP (@10 kHz)	3.5×10^{-11}	1.59
PE (@10 kHz)	8.7×10^{-11}	2.32
PA6 (@10 kHz)	4.0×10^{-9}	4.97

For non-literature values, measurement frequencies are denoted in brackets

Table B.4 Failure strength values of material used for failure criteria

	Sigratex CE 1250-230-39 (unidirectional)	T700/PPS (unidirectional)	RTM6	PPS	T700S	HTA40
Parallel tensile strength $R_{\parallel}^{+}$ [MPa]	1660	2400	–	–	–	–
Transverse tensile strength $R_{\perp}^{+}$ [MPa]	54	20	–	–	–	–
Parallel compressive strength $R_{\parallel}^{-}$ [MPa]	703	800	–	–	–	–
Transverse compressive strength $R_{\perp}^{-}$ [MPa]-	88	70	–	–	–	–
In-plane shear strength $R_{\perp\parallel}$ [MPa]	80	63	–	–	–	–
Tensile strength R [MPa]	–	–	76	20	4061	3006

Subscript notation refers to conventions used in Puck's failure criterion

Appendix C: Acoustic Emission Signal Parameters

This appendix is used to define the common acoustic emission parameters as used in this work and in technical standards. The definition of a characteristic parameter or "feature" like an energy derived from a transient signal is related to the concept known as feature extraction. In acoustic emission these signal features have long been used to interpret the type of failure which caused the signal. Basically, a feature is used to reduce the amount of information carried by the signal to a single value. As shown in Fig. C.1 for a typical voltage signal in the time domain $U(t)$ and frequency domain $\tilde{U}(f)$ large numbers of features are available to describe the signal. Although many features are already part of national and international standards the exact feature definitions of different providers for acoustic emission equipment slightly differ. Further, for the interpretation of acoustic emission signals with pattern recognition techniques it is useful to define new features like the "weighted Peak Frequency." Since the definition and calculation of signal features is an important step preceding the analysis routines of Chap. 4, all features used in this book are briefly summarized. While the basic signal features in Table C.1 are almost self-explanatory and covered by the above mentioned standards, this is not necessarily the case for the derived features presented in Table C.2. These show discrepancies between the different providers of acoustic emission software and are hence defined here in the way used within this book.

Most of the features used to describe an acoustic emission signal are composed of extreme values, derived from the signal in time- or frequency domain as shown in Fig. C.1. The "RMS"- and "ASL"-values are used in a similar manner like in electrical engineering, where these are used to describe the effective voltage within a characteristic time interval. The three different count-based frequencies "Initiation-", "Reverberation-"and "Average Frequency" are used to provide an estimate of the characteristic frequency before and after the peak-maximum, and of the complete signal. This should not be understood as exact frequency analysis since the number of threshold crossings is often very low and thus yields very inaccurate results. A similar measure can be obtained from the "Rise-"and "Decay Angle."

M.G.R. Sause, *In Situ Monitoring of Fiber-Reinforced Composites*,
Springer Series in Materials Science 242, DOI 10.1007/978-3-319-30954-5

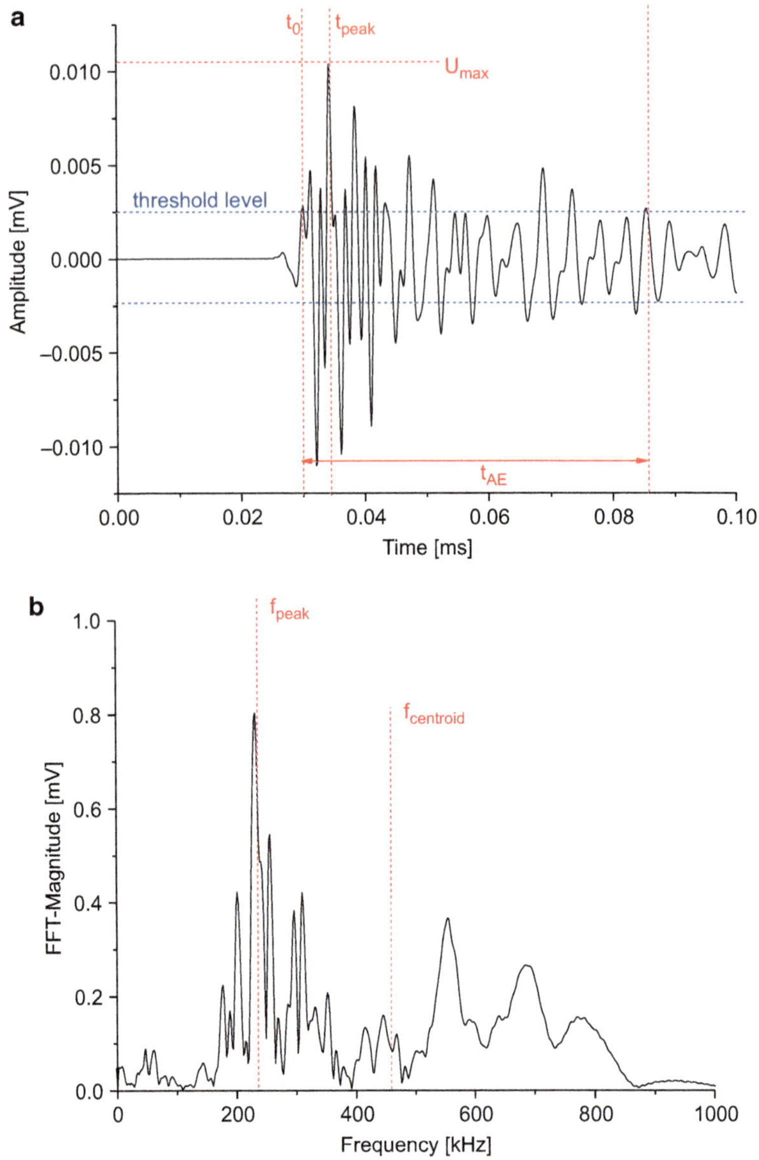

Fig. C.1 Examples of acoustic emission signal features in the time and frequency domain

These are used to describe the angle of the slope formed by the point of first threshold crossing and the signal maximum (rise) and the angle of the slope from signal maximum to the last threshold crossing (decay). The "Absolute Energy" is an absolute measure of the electrical energy measured for an acoustic emission signal and is given in Atto-Joule [10^{-18} J] due to their small energy. The "Frequency

Table C.1 Basic signal parameters used for feature definition

Feature	Definition	Unit
t_0	Time of first threshold crossing (arrival time)	[s]
N_{AE}	Number of threshold crossings	[#]
t_{AE}	Time between first and last threshold crossing of signal	[μs]
U_{max}	Maximum signal voltage	[mV]
t_{peak}	Time of maximum signal voltage	[μs]
f_{peak}	Frequency of maximum signal contribution	[Hz]
N_{peak}	Number of threshold crossings between t_0 and t_{peak}	[#]

Table C.2 Definition of acoustic emission signal features used within this book

Feature	Definition	Unit
Amplitude	$dB_{AE} = 20\log(U_{max}/1\mu V) - dB_{preamplifier}$	[dB]
Counts	N_{AE}	[#]
Duration	t_{AE}	[μs]
Rise time	$t_0 - t_{peak}$	[μs]
Counts to peak	N_{peak}	[#]
Root mean square (RMS)	$\mathrm{RMS} = \sqrt{\frac{1}{T_{RMS}} \int_{t_0}^{t_0+T} U^2(t)dt}$ with a characteristic time T_{RMS} for averaging ranging from 10 to 1000 ms	[mV]
Average signal level (ASL)	$\mathrm{ASL} = \sqrt{\frac{1}{T_{ASL}} \int_{t_0}^{t_0+T} (20\log(U(t)/1\mu V) - dB_{preamplifier})dt}$ with a characteristic time T_{ASL} for ranging from 10 to 1000 ms	[dB]
Average frequency	$\langle f \rangle = N_{AE}/t_{AE}$	[Hz]
Reverberation frequency	$f_{rev} = \frac{N_{AE} - N_{peak}}{t_{AE} - t_{peak}}$	[Hz]
Initiation frequency	$f_{init} = \frac{N_{peak}}{t_{peak}}$	[Hz]
Rise angle	$\phi_{rise} = \tan\left(\frac{U_{max}}{t_{peak}}\right)$	[rad]
Decay angle	$\phi_{decay} = \tan\left(\frac{U_{max}}{t_{AE} - t_{peak}}\right)$	[rad]
Absolute energy	$W_{AE} = \int_0^{t_{AE}} \frac{(U(t))^2}{10k\Omega} dt$ with 10kΩ input impedance of the recording equipment	[aJ]
Peak frequency	f_{peak}	[Hz]
Frequency centroid	$f_{centroid} = \frac{\int f \cdot \tilde{U}(f)df}{\int \tilde{U}(f)df}$	[Hz]
Weighted peak frequency	$\langle f_{peak} \rangle = \sqrt{f_{peak} \cdot f_{centroid}}$	[Hz]
Partial power	$\int_{f_1}^{f_2} \tilde{U}^2(f)df / \int_{f_{start}}^{f_{end}} \tilde{U}^2(f)df$ Frequency range of interest $[f_1;f_2]$ Frequency range of investigation $[f_{start};f_{end}]$	[%]

Centroid" is used to characterize the overall frequency content of an acoustic emission signal in a similar manner like the center of mass is used to describe the properties of geometric object with uniform density. Thus it can be understood as additional information on the characteristic average frequency of the signal and is in general not equal to the "Peak Frequency." This in turn is a highly efficient feature for discrimination of acoustic emission signals, since it is very discrete. To improve the representation of a characteristic frequency associated with an acoustic emission signal the "weighted Peak Frequency" was introduced to combine the discriminative efficiency of the "Peak Frequency" with the information on the average frequency content of the signal expressed by the "Frequency Centroid." A very similar approach is used by the different "Partial Power" levels. They measure the signals frequency contribution within a given interval and are thus used to capture the frequency composition of the signal in more than one characteristic value. For the "Partial Powers" used within this work, the frequency ranges of Table C.3 were selected as discretization of the full frequency range between $f_{start} = 0$ kHz and $f_{end} = 1200$ kHz.

Table C.3 Definition of Partial Power frequency ranges

Feature	Definition	Unit
Partial power 1	$f_{start} = f_1 = 0$ kHz; $f_2 = 150$ kHz	[%]
Partial power 2	$f_1 = 150$ kHz; $f_2 = 300$ kHz	[%]
Partial power 3	$f_1 = 300$ kHz; $f_2 = 450$ kHz	[%]
Partial power 4	$f_1 = 450$ kHz; $f_2 = 600$ kHz	[%]
Partial power 5	$f_1 = 600$ kHz; $f_2 = 900$ kHz	[%]
Partial power 6	$f_1 = 900$ kHz; $f_2 = f_{end} = 12000$ kHz	[%]

Appendix D: Definitions and Abbreviations

This Appendix is used to introduce all abbreviations and definitions used within this book. For better readability, symbols are grouped into different tables with reference to particular fields of science (Tables D.1, D.2, D.3, D.4, D.5, D.6, and D.7).

Table D.1 Symbols used for optics and geometry

Optics and geometry	
Small increment	Δ
Area	A
Volume	V
Angle (various uses)	θ, ψ
Distances (various uses)	a, b, h, s, d
Subset indices and counters (various uses)	i, j, k, l, m, n
Cartesian distance	$r = \sqrt{x^2 + y^2 + z^2}$
Cartesian vector, coordinates	$\mathbf{r}$, (x, y, z)
Vector normal to surface	$\boldsymbol{n}$
Number of entities	N
Intensity distribution reference subset	$\tilde{F}(\boldsymbol{r})$
Intensity distribution displaced subset	$\tilde{G}(\boldsymbol{r})$
Cost function	χ
Shape function	ξ
Scale and shift parameter	$\tilde{a}$, $\tilde{b}$
Point locations	$\overline{\boldsymbol{P}}, \bar{\boldsymbol{O}}$
Geometrical magnification	M
Pixel size	$\tilde{P}$
Focal spot size	$\tilde{F}$
Voxel size	$\tilde{V}$
Distortion matrix	$\boldsymbol{K}$
Parallel to fiber axis	$\parallel$
Perpendicular to fiber axis	$\perp$

M.G.R. Sause, *In Situ Monitoring of Fiber-Reinforced Composites*,
Springer Series in Materials Science 242, DOI 10.1007/978-3-319-30954-5

Table D.2 Symbols used for description of waves

Waves	
Scalar potential	X
Vector potential	$\boldsymbol{\psi}$
Sound velocity, longitudinal, transversal, Rayleigh, phase, group	c_i, c_L, c_T, c_R, c_P, c_G
Wave vector, number	$\boldsymbol{k}$, k
Wave length	λ
Time	t
Frequency, angular, complex angular	f, ω, ω^*
Acoustic impedance	Z
Coefficient reflection, transmission	Re, Tr
See (4.36)	$\tilde{A}$
See (4.37)	$\tilde{\mathrm{B}}$
Attenuation coefficient	α
Sampling interval, rate	ΔT, ΔT^{-1}

Table D.3 Symbols used for continuum mechanics

Continuum mechanics	
Load	L
Force	F
Displacement vector, magnitude	$\boldsymbol{u}$, u
Displacement, cartesian coordinates	u, v, w
Stress-tensor and vector	$\boldsymbol{\sigma}$
Normal stress	σ
Shear stress	τ
Deformation-tensor and vector	$\boldsymbol{\varepsilon}$
Normal strain	ε
Shear strain	γ
Elasticity-tensor, complex	$\boldsymbol{C}$, $\boldsymbol{C}^*$
Modulus	E
Shear modulus	G
Poisson number	υ
Lame's constants	κ, ς
Density	ρ
Cross-head displacement	$\tilde{\delta}$
Function input, output	$h_E(t)$, $h_A(t)$
Transfer function	$\Xi(t)$

Table D.4 Symbols used for fracture mechanics and failure theory

Fracture mechanics and failure theory	
Fracture toughness, dynamic fracture toughness	G_c, G_d
Stress-intensity factor, critical, dynamic	K_I, K_{Ic}, K_{Id}
Crack tip speed	c_{cr}
Fracture strength	R
Crack motion vector, normal	$\boldsymbol{b}, \boldsymbol{d}$
Moment tensor	$\tilde{\boldsymbol{M}}$
Green function	$\boldsymbol{\Gamma}$
Degradation function	$\breve{C}(\boldsymbol{r})$
Stiffness vector	$\tilde{\boldsymbol{k}}$
Unidirectional laminate strength, parallel, perpendicular	$R_{\parallel}^{\pm}$, $R_{\perp}^{\pm}$
Magnification factor	m
Inclination parameters	$p_{\perp\psi}^{\pm}$
Degradation parameter	$\tilde{\eta}$
Compliance calibration factor	m_c
Source emission time	t_e
Crack duration	t_{frac}

Table D.5 Symbols used for electrodynamics and piezoelectrics

Electrodynamics and piezoelectricity	
Electric field	$\boldsymbol{E}$
Magnetic field	$\boldsymbol{B}$
Electric displacement-field	$\boldsymbol{D}$
Polarization field	$\boldsymbol{P}$
Electric charge density	$\tilde{\rho}$
Electric current density	$\boldsymbol{J}$
Magnetizing field	$\boldsymbol{H}$
Magnetization field	$\boldsymbol{M}$
Electric charge	Q
Electric potential	U
Permittivity of free space, material, tensor	ξ_0, ξ_r, $\boldsymbol{\xi}$
Permeability of free space, material	μ_0, μ_r
Electric conductivity	$\tilde{\sigma}$
Electric impedance	η
Skin depth	δ
Charge relaxation time	$\tilde{\tau}$
Piezoelectric constants, inverse	$\boldsymbol{S}$, $\boldsymbol{S}'$

Table D.6 Pattern recognition and neural networks

Pattern recognition and neural networks	
Number of clusters	Ω
Number of features, minimum	K, $K_{\min}$
Number of objects in dataset	O
Number of feature combinations	N_{tot}
Separation measure between clusters	J
Feature vector, element	$\boldsymbol{q}$, q_i
Node weight vector, element	$\boldsymbol{w}$, w_i
Error surface function	$\tilde{L}$
Learning rate	ι
Rand index	RAND
Tou index	TOU
Davies-bouldin index	DB
Hubert's gamma statistics	GAMMA
Silhouette value	SIL

Table D.7 Common definitions for mathematical relations

Other	
Nabla-operator	$\nabla f = \left(\frac{\partial f}{\partial x} e_x + \frac{\partial f}{\partial y} e_y + \frac{\partial f}{\partial z} e_z\right)$
	$\nabla \cdot \mathbf{f} = \left(\frac{\partial f_x}{\partial x} + \frac{\partial f_y}{\partial y} + \frac{\partial f_z}{\partial z}\right)$
	$\nabla \times \mathbf{f} = \left(\frac{\partial f_z}{\partial y} - \frac{\partial f_y}{\partial z}\right) e_x + \left(\frac{\partial f_x}{\partial z} - \frac{\partial f_z}{\partial x}\right) e_y + \left(\frac{\partial f_y}{\partial x} - \frac{\partial f_x}{\partial y}\right) e_z$
Laplace-operator	$\boldsymbol{\Delta} = \nabla \cdot \nabla$
Mean value of O objects	$\langle x \rangle = \sum_{i=1}^{O} \frac{x_i}{O}$
Variance of O objects	$\mathrm{Var}(x) = \sum_{i=1}^{O} \frac{x_i - \langle x \rangle}{O}$
Standard deviation of O objects	$\mathrm{stdev}(x) = \sqrt{\frac{1}{O} \sum_{i=1}^{O} \left(x_i - \langle x \rangle\right)^2}$

Index

A
Acoustic emission, 44, 132, 134–146, 157, 190, 230, 231, 245–247, 250–281, 285–309, 323, 337
  detection, 131, 192, 226–249
  sensor aperture, 237–238
  sensor (sensor type)
    coupling medium (couplant), 245–247
    flat, 44, 230, 278
    multi-resonant, 230, 231, 278–280
    resonant, 230, 278, 279
  signal
    attenuation, 190
    classification, 250–281
    counting, 132
    feature extraction, 252, 255, 272, 306
    noise, 251–254, 337
    onset, 285–287
  source (mechanism, type)
    classical localization, 287–290
    Δt–mapping, 290
    identification, 305–309
    model, 138–146
    neural net based localization, 291–304
    radiation pattern, 157, 250
    rise time, 134–137
    strength, 135, 143, 190, 323
    uncertainty of localization (UoL), 132
  source-sensor distance, 135, 265–267
  test source (pencil lead break, piezoelectric pulser), 293–298
  waveguide, 239–245
Acousto-ultrasonics, 535, 596–599
Anomaly, 7, 10, 38, 69
  strain, 41, 42, 61, 69, 539
Artificial defect, 71, 74, 77–80, 82, 87–97, 103, 124
Auld, B.A., 202

B
Barrett, J.F., 464
Basan, R., 33
Baxter, M., 290
Bay, B.K., 493
Bleier, A., 33
Bunke, H., 257, 259
Bunsell, A.R., 24

C
Carrere, N., 21
Cervena, O., 134
Choi, H.-I., 553, 554
Classification, 6–18
  uncertainty of (UoC), 133, 260–263, 267–270, 278–280, 315, 342
Cluster (clustering)
  natural, 251, 256–260, 311, 322
  validity (measure, index), 252, 257–260, 262, 267
Cohesive zone (model, modeling, elements), 140, 141
Combination of methods (method combinations), 533–606
Cosslett, V.E., 457
Creep, 18, 28–29, 31, 32, 40, 111, 196
Cuntze, R., 20, 21, 33

M.G.R. Sause, *In Situ Monitoring of Fiber-Reinforced Composites*, Springer Series in Materials Science 242, DOI 10.1007/978-3-319-30954-5

D
Degradation function, 143, 146, 152, 156, 162, 167, 170, 174
Degrieck, J., 30
Delamination
  inter-ply (interlaminar), 13–16, 34, 74, 86, 89, 94, 95, 104, 149, 165, 167, 169, 170, 172, 189, 216, 224, 225, 306–309, 321, 324, 329, 387, 447, 450, 475, 476, 506, 562, 568–570, 572
  intra-ply (intralaminar), 13, 309
Detectability of failure mechanisms (detection sensitivity)
  in acoustic emission, 187–192
  in digital image correlation, 106, 124–126
  in electro magnetic emission, 538–540
  in X-ray computed tomography (detailvisibility), 493
Deuschle, H.M., 23, 27
Digital image correlation, 102
  artefact, 87, 100, 101
  resolution, 4, 67–69
  stereo vision, 62, 63, 67
  subset (neighborhood)
    intersection, influence of, 102
    size, influence of, 102
  system accuracy, 63–69
  3D, 58, 62–64, 67, 69, 71, 80–83, 101, 111, 113, 114, 116, 119, 124
Digital volume correlation, 110, 457, 493–498, 590, 605
Downs, K.S., 136

E
Eato, M.M.J., 290
Edge, E.C., 23
Elasticity tensor (stiffness tensor, elastic coefficients), 114, 134, 193, 205
  complex, 202, 205, 206
Electro magnetic emission (EME), 362, 363, 365–367, 371, 372, 374–376, 378, 380–383, 391, 401, 403–410, 452, 453
  sensor (detector)
    capacitive (capacitance plate), 362
    coil, 362, 401, 403–405, 407–408, 410
    wire, 403, 404, 406–407, 409, 410
  signal
    classification, 363
    counting, 132
  source
    models, 363, 365–367, 371, 372, 374–376, 378, 380, 453
    radiation pattern, 381–383, 391, 452, 453
  test source, 389–395
Electro magnetic shielding, 420–424, 426–427
  effectiveness, 401, 414, 415, 417–421, 424, 426, 451, 452
  reduction due to
    apertures, 420–422
    cable penetrations, 426–427
    seams, 425–426
    waveguides below cut-off, 423–424
Electro magnetic wave
  dispersion, 362, 396
  near-field, 397, 414
  propagation, 396

F
Failure, 19–21, 23, 26, 519, 520, 522, 581
  criterion
    anisotropic, 19
    differentiating, 19, 21
    global, 19, 21, 522
    Puck's, 20, 23, 26, 519, 520, 581
  mechanism, 1–3, 6–19, 21, 30, 35, 39, 43, 149, 361, 363, 386, 450, 475, 515, 533, 534, 538–542, 550, 552, 557, 562, 565, 567, 568, 575, 589, 590
  type, 1, 10, 11, 19, 33, 533–535, 538–540, 543, 552, 560, 567
Failure evolution (damage progress, damage progression), 3, 6, 24, 31, 35, 327–348
Fatigue, 18, 28–32, 40, 111
  R-ratio, 29
Felicity effect (Felicity ratio), 336–339, 605
Fiber
  breakage, 10, 11, 16, 30, 71, 73, 75–78, 82, 85, 87–89, 91–94, 99, 146, 149, 173, 184, 186–188, 190–192, 207, 209, 223, 279, 281, 302, 307–309, 316, 318–320, 325–327, 331, 334, 335, 339, 342, 387–389, 434, 453, 477, 538, 552, 553, 558, 560, 566, 570–573, 599
  bridging, 15, 149, 179–184, 189, 538–542

filament, 7–11, 15, 24, 33, 39, 43, 103, 131, 143, 145, 173, 174, 176, 186–189, 192, 216, 223, 316, 433–436, 477, 479, 495, 522, 526, 538, 539, 549, 572, 573, 590, 591
N-plets, 24
Fiber-matrix
debonding, 11–13, 39, 180, 538
pull-out, 11, 13, 552
Fiber reinforced (material, composite), 2, 3
fabric, 7, 8, 13, 17, 23
unidirectional, 7, 8, 14, 19, 38
Field of view
in digital image correlation, 69, 72, 100, 108, 116, 117, 125, 250, 314
In X-ray computed tomography, 493
Field testing, 346, 534, 543–547
Finkel, V.M., 365–367
First failure (failure onset), 2, 23, 34–35, 118, 309
Fracture, 158, 389
mechanics, 1, 5–7, 22, 27, 28, 36, 69, 92, 118, 119, 121, 140, 141, 309, 311, 317, 369, 372, 428, 603
surface, 13, 14, 16, 17, 368, 370–372, 375, 378, 381, 386–389, 400, 403, 405, 427, 431, 433, 439, 445–452, 477, 478, 503, 504, 507–513, 518–520, 526, 554, 555, 580, 581, 583, 585, 587, 588, 590, 594–596, 603
orientation, 158, 389
Freund, L.B., 369
Frid, V., 361, 367, 368
Full-field
information, 112, 557
measurement, 37, 38, 70, 71, 99, 107, 536

G
Gade, S.O., 371, 401
Gama, B.A., 32
Gary, J., 134
Geometrical unsharpness, 463
Geometric spreading, 192, 203, 205, 383, 583
Geometry extraction, 506–515
Goujon, L., 230
Griffith, A.A., 5
Guedes, R.M., 30
Guided waves
attenuation, 192, 204–206
dispersion, 192, 204
propagation, 196, 202, 212, 242, 250, 285, 543
testing, 32, 198, 199, 202, 596
Günter, S., 257, 259

H
Hamstad, M.A., 134, 135, 412
Hart-Smith, L., 23
Hashin, Z., 19
Haskell, N.A., 202
High-velocity (high strain rates), 18, 31–32
Hinton, M.J., 20
Hobrough, G.L., 57
Hora, P., 134

I
Image segmentation, 499, 506, 508
threshold, 499
Impact damage, 16, 17, 216, 479, 480
Impedance
acoustic, 195, 215, 230, 238, 239, 243
electric, 390
Indications, 44, 71, 79, 88, 110, 170, 288
In situ monitoring (in situ capabilities), 42, 43, 131, 536–537
Internal defects, 39, 43, 69, 71–97
Irvin, G., 5

J
Joe, H., 260

K
Kaddour, A.S., 20
Kaiser, J., 337
Kaiser effect, 337
Kastner, J., 501
Keat, N., 464
Kundu, T., 202

L
Lamb, H., 195
Lamb waves (plate waves), 135–137, 147, 197–201, 203, 204, 211, 212, 214, 218, 226, 237, 238, 277, 283, 288, 295
Laminate
cross-ply, 190, 271, 298, 299, 320–327
failure, 326

Laminate (*cont.*)
multi-axial, 18, 105, 167
quasi-isotropic, 298, 299, 301
unidirectional, 32, 73–77, 85–87, 123, 125, 173, 185, 186, 214, 298, 299, 317–320
Landau, L.D., 192
Lifschitz, E.M., 192
Lomov, S.V., 319
Lynnworth, L.C., 240

M
Magnification factor, 24, 68, 108, 458, 462–465, 552
Mal, A., 202
Matrix cracking, 180, 181, 192, 279–282, 309, 316, 318–320, 323–325, 327, 331, 334, 339, 340, 342–344, 538, 539, 541, 542, 552–558, 560–562, 568–572
Maxwell's equations, 364, 396, 401
Mesostructure, 21, 23
Microscopy
in situ, 4, 36–37
shadow, 458
Milligan, G.W., 260
Miner, M., 29
Miroshnichenko, M., 366
Misra, A., 365
Mohr, O., 19
Molotskii, M.I., 365
Moment tensor, 133, 250, 595

N
Nahas, M., 19
Nixon, W.C., 457

O
Obstacle
external, 214–226, 276
internal, 214–226, 276, 277, 557
Ohtsu, M., 134, 250
Ono, K., 134, 250
Optical extensometer, 111–117, 536
Orteu, J.J., 57

P
Pan, B., 495
Particle inclusion (foreign materials inclusion), 9, 71, 468
Parvizi, A., 22
Pattern recognition (method)
supervised, 251
unsupervised, 251, 257, 306, 309
Perfectly matched layer (PML), 144, 145
Peters, W.H., 57
Piezoelectric effect, 227, 228
Pinho, S., 21, 33
Poisson's ratio, 113–115, 143, 146, 153, 193, 245
Polikar, R., 257, 292
Pollock, A.A., 203
Porosity (voids), 7, 27, 71, 204, 479, 498–501, 504, 506, 519, 520
Pressure vessel, 29, 341, 344–347, 603
Probability of detection (PoD), 132, 192, 321, 346, 347, 363, 535
Prosser, W.H., 203
Puck, A., 20, 21, 23, 27, 317
Pullin, R., 290

Q
Qiu, W., 260

R
Radon, J., 461
Ramesh, K.T., 32
Ranson, W.F., 57
Rayleigh waves, 195, 196, 368, 369, 432
Reifsnider, K., 19
Representative volume element (RVE), 144–146, 148, 173, 174, 176, 178, 181, 185, 187
Resin rich area, 9, 10, 103, 316

S
Schreier, H., 57
Schürmann, H., 23, 28
Scott, A.E., 24
Scruby, C.B., 134, 369, 370
Seams, 0, 425, 426
Sendeckyj, G., 30
Shear-horizontal waves, 197
Shearography, 42, 70, 110, 535, 546
Shiwa, M., 412
Signal
attenuation, 362
onset, 591, 592
Signal-to-noise ratio
in acoustic emission, 281
in digital image correlation (detectability of failure mechanisms), 281–285

in electromagnetic emission, 44, 383, 451–452, 604
in X-ray computed tomography (quantummottle), 605
Skin depth (penetration depth), 416, 417
Soden, P.D., 20
Sound velocity
angular dependency, 383
group, 396, 575–577, 597
modal, 195, 196
phase, 396, 575–577
Speckle pattern, 37, 58, 64–69, 71, 73, 100–102, 112, 114–117, 119, 121, 494, 495, 512, 534, 536, 546, 602
Strain concentration (stress concentration), 69–110, 498, 555–557, 602
Stress rupture, 18, 28–29
Sutton, M.A., 57
Swolfs, Y., 522

T
Testing
bearing strength, 268, 269, 302–304
compressive, 113–116
double cantilever beam, 327–334
end-notched flexure, 121–123, 311–312
mechanical, 2, 32–44, 111, 224, 309, 314
short-beam shear, 34, 119–121, 310
single filament, 437
tapered double cantilever beam, 428
tensile, 35, 38, 40, 112–114, 334–339
transverse crack tension, 312–317
V-notched rail shear, 116–117
Thermoelastic dissipation (Akhieser dissipation), 203
Thermography, 40–42, 535, 537, 539, 541, 542, 558–563, 565
Thomson, W.T., 202
Transfer function, 228, 230, 233, 375, 391, 392, 401–403, 407–411, 434, 575, 577–579, 589, 604
Trappe, V., 27
Tsai, S.W., 19

U
Ultrasonics (active), 596–599
Uncertainty of classification (UoC), 363, 596
Undulation, 9, 10, 21, 538–543

V
Vibrometry (laser vibrometry), 32
nonlinear, 43

W
Wadley, H.N.G., 134, 250, 370
Wallner lines, 371, 513
Wave scattering, 215
Williams, W., 553, 554
Wires form, 429
Wöhler, A., 29
Wu, E.M., 19

X
X-ray, 459, 460, 462–464, 473, 476, 478, 479, 481–483, 492, 495, 502, 517, 518, 535, 571, 573, 605
attenuation, 486, 542
computed tomography
artifacts, 464
cone-beam, 459, 473
ex situ, 481, 502, 571
fan-shaped, 473
in situ, 482, 483, 492, 495, 535, 573, 605
voxel resolution, 483, 517
detector, 471, 482, 484
refraction, 415
source, 458–460, 482–484, 486
focal spot size, 459, 460, 462–464, 476, 478, 479, 483, 518, 605

Z
Zinoviev, P.A., 21

Printed by Printforce, the Netherlands